W9-AYD-968

Methods in Microbiology
Volume 32

Series Advisors

Gordon Dougan, Director of The Centre for Molecular Microbiology and Infection, Department of Biological Sciences, Imperial College of Science, Technology and Medicine, London SW7 2AZ

Graham J Boulnois, Schroder Ventures Life Sciences Advisers (UK) Limited, 71 Kingsway, London WC2B 6ST

Jim Prosser, Department of Molecular and Cell Biology, University of Aberdeen, Institute of Medical Sciences, Foresterhill, Aberdeen AB25 2ZD

Ian R Booth, Professor of Microbiology, Department of Molecular & Cell Biology, University of Aberdeen, Institute of Medical Sciences, Foresterhill, Aberdeen AB25 2ZD

David A Hodgson, Reader in Microbiology, Department of Biological Sciences, University of Warwick, Coventry CV4 7AL

David H Boxer, Department of Biochemistry, Medical Sciences Institute, Dundee DD1 4HN

Methods in Microbiology

Volume 32
Immunology of Infection
Second Edition

Edited by

Stefan HE Kaufmann

Max Planck Institute for Infection Biology
Berlin,
Germany

and

Dieter Kabelitz

Institute of Immunology
University of Kiel,
Germany

ACADEMIC PRESS

An imprint of Elsevier Science

Amsterdam   Boston   London   New York   Oxford   Paris
San Diego   San Francisco   Singapore   Sydney   Tokyo

This book is printed on acid-free paper.

Copyright 2002, Elsevier Science Ltd. All Rights Reserved.

No part of this publication may be reproduced or transmitted in any form or by any means, electronic or mechanical, including photocopying, recording, or any information storage and retrieval system, without permission in writing from the publisher.

Academic Press
An Imprint of Elsevier Science
84 Theobald's Road, London WC1X 8RR, UK
http://www.academicpress.com

Academic Press
An Imprint of Elsevier Science
525 B Street, Suite 1900, San Diego, California 92101-4495, USA
http://www.academicpress.com

ISBN 0–12–521532–0 (Hbk)
ISBN 0–12–402033–X (Comb)

A catalogue record for this book is available from the British Library

Cover picture: A colourised scanning electron micrograph of *Neisseria gonorrhoeae* on the surface of Hela (human cervix carcinoma) cells; courtesy of Dr Volker Brinkmann, Max-Planck-Institut für Infektionsbiologie, Mikroskopie, Schumannstr. 21/22, D-10117 Berlin, Germany.

Typeset by Phoenix Photosetting, Chatham, Kent
Printed and bound in Great Britain by Bookcraft, Bath

02   03   04   05   06   07   BC   9 8 7 6 5 4 3 2 1

Recent titles in the series

Contents

Contributors

Reinhard Andreesen Department of Haematology and Oncology, University of Regensburg, D-94042 Regensburg, Germany

Victor Appay Institute of Molecular Medicine, John Radcliffe Hospital, Windmill Road, Headington, Oxford OX3 9DU, UK

Mario Assenmacher Miltenyi Biotec, Friedrich Ebert Str. 68, D-51429 Bergisch Gladbach, Germany

Ingo B Autenrieth Institut fur Medizinische Microbiologie, Universitat Tubingen, Elfride-Aulhorn Strasse 6, D-72076, Germany

Jacques Banchereau Baylor Institute for Immunology Research, Suite 205, 3434 Live Oak, Dallas, TX 75204, USA

John T Belisle Mycobacterial Research Laboratories, Department of Microbiology, Immunology and Pathology, Colorado State University, B116 Microbiology Building, Fort Collins, CO 80523-1677, USA

Jay A Berzofsky Molecular Immunogenetics and Vaccine Research Section, Metabolism Branch, National Cancer Institute, NIH, Bldg. 10, Rm. 6B12, Bethesda, MD 20892, USA

James F Brown Dept. of Pathobiological Sciences, School of Veterinary Medicine, 2015 Linden Drive West, Madison, WI 53706, USA

Kevin Brunson University of Notre Dame, Mendoza College of Business, Notre Dame, IN 46556, USA

Jan Buer Gesellschaft für Biotechnologische Forschung (GBF), Mascheroder Weg 1, D-38124 Braunschweig, Germany

Dirk H Busch Institute of Medical Microbiology, Immunology and Hygiene, Technical University Munich, D-81675 München, Germany

Christophe Caux Schering-Plough, Laboratory for Immunological Research, 27 Chemin des Peupliers, BP 11, 69571 Dardilly, France

Emmanuel Claret Diaclone SA, 1Bd A Fleming, BP 1985, F-25020, Besancon Cedex, France

Andrea M Cooper Trudeau Institute, 100 Algonquin Ave, Saranac Lake, New York 12983, USA

MariPat Corr   Division of Rheumatology, Allergy and Immunology, and The Sam and Rose Stein Institute for Research on Aging, University of California, San Diego, 9500 Gilman Drive, La Jolla, CA 92093-0663, USA

Françoise Cottrez   INSERM U343, Hôpital de l'Archet, BP 79, Route de Saint Antoine de Ginestière, 06202 Nice Cedex 3, France

Nancy Craighead   NIDDK/Navy Transplantation & Autoimmunity Branch, NNMC/ AFRRI Building 46, Room 2417, 8901 Wisconsin Ave., Bethesda MD 20889-5607, USA

Charles J Czuprinsky   Dept. of Pathobiological Sciences, School of Veterinary Medicine, 2015 Linden Drive West, Madison, WI 53706, USA

Anne S De Groot   TB-HIV Laboratory, Box B-G 579 (Biomed Center, Rm 579), Brown University, 85 Brown Street, Providence, RI 02912, USA

Rene de Waal Malefyt   DNAX Research Institute for Molecular and Cellular Biology, 901, California Avenue, 94304-1104 Palo Alto, CA, USA

Colette Dezutter-Dambuyant   INSERM, U346, Hopital Edouard Herriot, 69374 Lyon, France

Stacey Drabic   Millennium Pharmaceuticals, 75 Sidney St, Cambridge MA 02139, USA

Rainer Duchmann   Gastroenterologie/Infektiologie/Rheumatologie, Universitätsklinikum Benjamin Franklin, Freie Univ., Hindenburgdamm 30, D-12200 Berlin, Germany

Stefan Ehlers   Division of Molecular Infection Biology, Research Center Borstel, Parkallee 22, D-23845 Borstel, Germany

Sabine Ehrt   Department of Microbiology & Immunology, Weill College of Cornell Medical School, Box 57 Room A-275A, 1300 York Ave, New York, NY 10021, USA

Elmar Endl   Molekulare Immunologie, Forschungszentrum Borstel, Parkallee 22, 23845 Borstel, Germany

Deborah Farlow   Millennium Pharmaceuticals, 75 Sidney St, Cambridge MA 02139, USA

Helmut Fickenscher   Abteilung Virologie, Hygiene Institut, Ruprecht-Karls-Universitat Heidelberg, Im Neuenheimer Feld 324, D-69120 Heidelberg, Germany

Bernhard Fleckenstein   Institut für Klinische und Molekulare Virologie, Universität Erlangen-Nürnberg, Loschgestraße 7, D-91054 Erlangen, Germany

Caroline Fossum   Faculty of Veterinary Medicine, Dept of Veterinary Microbiology, Swedish University of Agricultural Science, BMC, Box 585, S-751 23 Uppsala, Sweden

Kohtaro Fujihashi Immunobiology Vaccine Center and Departments of Oral Biology and Microbiology, The University of Alabama and Birmingham, USA

Johannes Gerdes Molekulare Infektionsbiologie, Forschungsinstitut Borstel, Parkallee 22, D-23845 Borstel, Germany

Geraldine M A Gillespie MRC Human Immunology Unit, Weatherall Institute of Molecular Medicine, The John Radcliffe, Headington, Oxford OX3 9DU, UK

Mercedes Gonzalez-Juarrero Mycobacteria Research Laboratories, Department of Microbiology, Immunology and Pathology, Colorado State Univ., Fort Collins, CO 80523-1677, USA

Siamon Gordon School of Pathology, University Oxford, South Parks Road, Oxford OX1 3RE, UK

Natascha K A Grzimek Institute for Virology, Johannes Gutenberg University, Obere Zahlbacher Str. 67 (Hochhaus am Augustusplatz), 55101 Mainz, Germany

Jose-Carlos Guiterrez-Ramos Millennium Pharmaceuticals, 75 Sidney St, Cambridge MA 02139, USA

Richard Haworth Pathology, GlaxoWellcome Research & Development, Park Road, Ware, Herts SG12 0DP, UK

Hans J Hedrich Institut für Versuchstierkunde, Medizinische Hochschule, Hannover, Germany

Thomas K Held Department of Hematology/Oncology, Charité/Campus Virchow-Klinikum, Humboldt University, Augustenburger Platz 1, 13353 Berlin, Germany

Takachika Hiroi Dept Mucosal Immunology, Research Institute for Microbial Diseases, Osaka University, Suita, Osaka 565-0871, Japan

Christine Hollmann AdnaGen AG, Ostpassage 7, 30853 Hannover, Germany

Rafaela Holtappels Institute for Virology, Johannes Gutenberg University, Obere Zahlbacher Str. 67 (Hochhaus am Augustusplatz), 55101 Mainz, Germany

Dieter Kabelitz Michaelisstr. 5, Inst. of Immunology, University of Kiel, D-24105 Kiel, Germany

Stefan H E Kaufmann Max Planck Institute for Infection Biology, Department of Immunology, Schumannstrasse 21/22, 10117 Berlin, Germany

Hiroshi Kiyono Dept Mucosal Immunology, Research Institute for Microbial Diseases, Osaka University, Suita, Osaka 565-0871, Japan

Stefan Krause   Department of Haematology and Oncology, University of Regensburg, D-94042 Regensburg, Germany

Pascale Kropf   Imperial College of Science, Tech. & Med., School of Medicine, Department of Immunology at St Mary's, Norfolk Place, London W2 1PG, UK

Michael Lahn   Department of Immunology, National Jewish Medical and Research Center, 1400 Jackson Street, Denver, Colorado, CO 80206, USA

Tamás Laskay   Institute for Medical Microbiology and Hygiene, Medical University of Lübeck, Ratzeburger Allee 160, D-23538 Lübeck, Germany

Jörg Lauber   Gesellschaft für Biotechnologische (GBF), Mascheroder Weg 1, D-38124 Braunschweig, Germany

Delphine J Lee   Division of Rheumatology, Allergy and Immunology, and The Sam and Rose Stein Institute for Research on Aging, University of California, San Diego, 9500 Gilman Drive, La Jolla, CA 92093-0663, USA

Jörg Lehmann   Institute of Immunology, College of Veterinary Medicine, An den Tierkliniken 11, D-04103, Leipzig, Germany

Yong-Jun Liu   Department of Immunobiology, DNAX Research Institute, 901 California Avenue, Palo Alto, CA 94304, USA

Karin Lövgren-Bengtsson   Faculty of Veterinary Medicine, Dept of Veterinary Microbiology, Swedish University of Agricultural Science, BMC, Box 585, S-751 23 Uppsala, Sweden

Jose Lora   Millennium Pharmaceuticals, 75 Sidney St, Cambridge MA 02139, USA

Erminia Mariani ˙ Instituto di Ricerca Codivilla Putti, Laboratorio di Immunologia e Genetica, Via di Barbiano 1/10, I-40136 Bologna, Italy

Elisabeth Märker-Hermann   Innere Medizin IV mit Scherpunkt Rheumatologie, Immunologie und Nephrologie, Dr.-Horst-Schmidt-Kliniken (HSK) Wiesbaden, Aukammallee 37, D-65191 Wiesbaden, Germany

Bill Martin   EpiVax, 365 Hope Street, Providence, RI 02906, USA

Jerry R McGhee   Immunobiology Vaccine Center and Departments of Oral Biology and Microbiology, The University of Alabama and Birmingham, USA

Andrew J McMichael   Institute of Molecular Medicine, John Radcliffe Hospital, Windmill Road, Headington, Oxford OX3 9DU, UK

Cornelius J M Melief   Department of Immunohematology and Blood Transfusion, Leiden University Medical Center, Albinusdreef 2, 2333 ZA Leiden, The Netherlands

Martin E A Mielke   Robert Koch Institut, Nordufer 20, D-13353 Berlin, Germany

Horst Mossman   Max-Planck-Institut für Immunbiologie, Stübeweg 51, D-79108 Freiburg, Germany

Ingrid Müller   Dept. of Immunology, Imperial College of Science, Technology & Medicine, St Marys, Norfolk Place, London W2 1PG, UK

Kerstin Müller   Institute for Medical Microbiology and Hygiene, Medical University of Lübeck, Ratzeburger Allee 160, D-23538 Lübeck, Germany

Werner Nicklas   Deutsches Krebsforschungszentrum, Heidelberg, Germany

Rienk Offringa   Department of Immunohematology and Blood Bank, Leiden University Medical Center, Albinusdreef 2, 2333 ZA Leiden, The Netherlands

Ian M Orme   Mycobacteria Research Laboratories, Department of Microbiology, Immunology and Pathology, Colorado State Univ., College of Veterinary Medicine, Fort Collins, CO 80523-1677, USA

A Karolina Palucka   Baylor Institute for Immunology Research, 3434 Live Oak, Suite 205, Dallas, Texas 75204, USA

Eric G Pamer   Laboratory of Antimicrobial Immunity, Memorial Sloan-Kettering Cancer Center, 1275 York Avenue, New York, NY 10021, USA

Graham Pawelec   Center for Medical Research, ZMF, University of Tübingen Medical School, Waldhörnlestr. 22, D-72072, Tübingen, Germany

Klaus Pechhold   NIDDK/Navy Transplantation & Autoimmunity Branch, NNMC/ AFRRI Building 46, Room 2417, 8901 Wisconsin Ave., Bethesda MD 20889-5607, USA

Leanne Peiser   School of Pathology, University of Oxford, South Parks Road, Oxford OX1 3RE, UK

Jürgen Podlech   Institute for Virology, Johannes Gutenberg University, Obere Zahlbacher Str. 67 (Hochhaus am Augustusplatz), 55101 Mainz, Germany

Andreas Radbruch   Deutsches Rheumaforschungszentrum, Schumannstrasse 21-22, 10117 Berlin, Germany

Eyal Raz   Division of Rheumatology, Allergy and Immunology, and The Sam and Rose Stein Institute for Research on Aging, University of California, San Diego, 9500 Gilman Drive, La Jolla, CA 92093-0663, USA

Christopher Reardon   Departments of Dermatology and Immunology, Univ. Colorado Health Sciences Center, Campus Box B153, 4200 E. 9th Ave, Denver, Colorado 80262, USA

Contributors

Matthias J Reddehase Institut für Virologie, Johannes-Gutenberg-Universität, Fachbereich Medizin, Obere Zahlbacher Str. 67, D-55101 Mainz, Germany

Michael Rehli Department of Haematology and Oncology, University of Regensburg, D-94042 Regensburg, Germany

Alan D Roberts Trudeau Institute, 100 Algonquin Ave, Saranac Lake, NY 12983, USA

Sarah L Rowland-Jones MRC Human Immunology Unit, Weatherall Institute of Molecular Medicine, The John Radcliffe Hospital, Headington, Oxford OX3 9DU, UK

Hakima Sbai TB/HIV Research Lab, Brown University, Providence, RI 02912, USA

Alex Scheffold Deutsches Rheuma-Forschungszentrum (DRFZ), Schumannstrasse 21-22, D-10117 Berlin, Germany

Dirk Schnappinger Department of Microbiology and Immunology, Weill Medical College of Cornell University, 1300 York Avenue, New York, NY 10021, USA

Gary K Schoolnik Department of Microbiology and Immunology, Stanford University, Beckman Center Room B241, 300 Pasteur Drive, Stanford, CA 94305, USA

Ulrike Seitzer Molekulare Infektionsbiologie, Forschungsinstitut Borstel, Parkallee 22, D-23845 Borstel, Germany

Raphael Solana Department of Immunology, Hospital Universitario Reina Sofia, Ave. Menedez Pidal s/n, ES-14004 Córdoba, Spain

Joanne Turner Mycobacteria Research Laboratories, Department of Microbiology, Immunology and Pathology, Colorado State Univ., College of Veterinary Medicine, Fort Collins, CO 80523-1677, USA

Martin I Voskuil Dept Microbiology & Immunology, Stanford University, Beckman Center Room B239, 300 Pasteur Drive, Stanford, CA 94305, USA

Daniela Wesch Institut für Immunologie, Universitätsklinikum Kiel, Michaelisstr. 5, 24105 Kiel, Germany

David Yowe Millennium Pharmaceuticals, Inflammation, 75 Sidney Street, Cambridge, MA 02139, USA

Hans Yssel INSERM U 454, Immunopathologie de l'asthme, Hopital Arnaud de Villeneuve, 371, Avenue Doyen Gaston Giraud, F-34295 Montpellier Cedex 05, France

Preface to the First Edition

It all began in a highly intertwined way. The golden age of medical micro-biology, which had its height just before the turn of the 19th century, not only witnessed the discovery of numerous medically important pathogens, but was also the cradle of immunological research. R. Koch discovered *Bacillus anthracis* and *Mycobacterium tuberculosis*, and was also the first to describe delayed type hypersensitivity reactions to bacterial products. The work of L. Pasteur was both instrumental in establishing the germ theory, and also laid the basis for rational vaccine development. E. Behring and S. Kitasato were not satisfied by their success in obtaining pure cultures of *Clostridium tetani*, and went on to develop vaccines against tetanus. Similarly, the seminal discoveries by E. Metchnikoff of phagocytosis and by J. Bordet of complement-mediated lysis, and the development of the side-chain theory by P. Ehrlich, to name but a few, were at the very interface between bacteriology and immunology. Thereafter, however, medical microbiology and immunology went their own ways.

After a period of quiescence, medical microbiology has been revital-ized by its amalgamation with molecular genetics and cell biology, to embark towards an understanding of the molecular cross-talk between the host cell and the microbial pathogen. Immunology dramatically broadened its scope from pure antimicrobial defence towards general topics concerned with its role in homeostasis and pathology within the mammalian organism. Today, the majority of immunological sub-disciplines have become virtually independent of their roots. Because of its broad scope, immunology has had to develop various methodologies of its own in addition to assimilating strategies from other fields. Immunology has been at the forefront in the development of several tech-niques of general importance. With the discovery of monoclonal antibody techniques, ELISA and ELISPOT assays for the detection of molecules and fluorescence-activated cell sorting (FACS) systems for the detection and separation of cells have gained general importance. Equally important is the assimilation of cell biology and molecular genetic techniques. Therefore, it is hardly possible for immunologists to master such a broad array of methodologies.

The need for comprehensive manuals on immunological methods has been fulfilled by several large volumes, including the *Immunology Methods Manual* (I. Lefkovits, Academic Press, London, 1997) and *Current Protocols in Immunology* (J. E. Coligan, A. Kruisbeek, D. H. Margulies, E. M. Shevach and W. Strober, Wiley, Chichester, 1995). These are complemented by more specific books for immunologists and for scientists working in related fields, such as *Antibodies – A Laboratory Manual* (E. Harlow and D. Lane, Cold Spring Harbor Laboratory Press, New York, 1988), the *Laboratory Manual on Manipulating the Mouse Embryo* (B. Hogen, R. Beddington, F. Constantini and E. Lacy, Cold Spring Harbor Laboratory Press, New York, 1994) and, last but not least, the famous standard work on *Molecular Cloning* (J. Sambrock, E. F. Fritsch and T. Maniatis, Cold Spring Harbor Laboratory Press, New York, 2nd edition, 1989). However, whilst being invaluable for immunologic laboratories, these manuals are often too broad for the microbiologist interested in using immunological methods to analyse host–pathogen relationships. It is these scientists for whom the present manual has been devised.

Accordingly, in preparing this volume, emphasis was placed on selection with the risk of bias. Sophisticated immunological techniques, as well as those mostly used in unrelated fields, such as tumour immunology or transplantation immunology, were omitted on purpose. Similarly, general techniques stemming from other fields, but now used widely in immunology, were omitted because excellent manuals are already available, and the reader is referred to those. These include general molecular cloning and cell biology techniques, as well as immunological techniques of broader application, such as the generation of monoclonal antibodies.

On which methodologies then did we focus? First, emphasis was placed on immune responses to bacteria, although many of the techniques described are equally applicable to antiviral and antiprotozoal immunity. Secondly, focus has been directed towards T cells, macrophages and cytokines. The T lymphocyte is the central regulator of the anti-infective immune response and the enormous increase in our knowledge about this cell has only been made possible by the recent development of appropriate techniques. T lymphocytes rarely combat infectious agents directly, but rather do so with the help of macrophages and the dialogue between these cells is mediated by cytokines. Equally large space, therefore, has been reserved for the characterization of cytokines and professional phagocytes. Although *in vitro* analyses are indispensible in investigating the immune response against infectious agents, *in vivo* studies remain of critical importance. In particular, evaluation of immunization stategies (such as vaccination) is impossible without appropriate *in vivo* models. With the advent of transgenic and gene deletion mouse mutants, novel strategies for characterizing the role of defined molecules and cells in the *in vivo* setting became feasible. Therefore we considered it essential to include chapters dealing with various aspects of experimental animal models.

We hope that this manual fills the gap between immunology and microbiology and helps to re-establish a closer relationship between the two disciplines. After all, infection is the outcome of the cross-talk between prokaryotic and eukaryotic cells, and in mammals the immune system has been given the task of being the major player.

We wish cordially to thank the care given by the editorial staff at Academic Press, particularly Tessa Picknett and Duncan Fatz, as well as our secretaries, Rita Mahmoudi and Constanze Taylor, for their great dedication. Last, but not least, we are grateful to all our colleagues who, by contributing to this manual, have generously let us share their extraordinary expertise.

Stefan H. E. Kaufmann
Max-Planck-Institute for Infection Biology, Berlin, Germany
Dieter Kabelitz
Paul-Ehrlich-Institute, Langen, Germany

Preface to the Second Edition

Four years have passed since the publication of the first edition of this volume. We are pleased that the success of the first edition, together with the rapid developments in the field, made a second edition of this book necessary. Progress in immunology continues to proceed rapidly, giving rise to new concepts. Hand in hand with this progress, novel technologies were developed that allow new approaches to, and new insights into, old as well as emerging scientific issues. It was therefore necessary not only to improve and update existing chapters of immunological methods, but to include new additions dealing with cutting-edge technologies. These include the genome-wide expression profiling of both the pathogen in the host and of the host response to infection, the cytometric analysis of cytokine secretion by immune cells and tetramer technology for the quantitative analysis of antigen-specific T-cell responses. These new additions as well as the existing chapters contain tried and tested cutting-edge protocols of experts in the field. We would like to thank the expert scientists who contributed to this manual, spared their valuable time and generously let us share their experience with scientists in the fields of microbiology, virology, mycology, parasitology and immunology.

We gratefully acknowledge the tremendous efforts of the editorial staff at Academic Press, notably Claire Minto and Tessa Picknett, in the preparation of this second edition. Last, but not least, we are very grateful to our editorial assistant Lucia Lom-Terborg and secretary Birgit Schlenga, without whom it would have been impossible to prepare this volume.

We hope that this new edition will directly benefit your research, and that it will be equally successful and beneficial as its predecessor. We also welcome your feedback which will help improve the succeeding edition of this book.

Stefan H. E. Kaufmann
Max-Planck-Institute for Infection Biology, Department of Immunology,
Berlin, Germany
Dieter Kabelitz
Institute of Immunology, University of Kiel, Germany

Introduction: The Immune Response to Infectious Agents

Stefan HE Kaufmann
Max-Planck-Institute for Infection Biology, Berlin, Germany

Dieter Kabelitz
Institute of Immunology, University of Kiel, Germany

◆◆◆

CONTENTS

◆◆◆◆◆◆ INTRODUCTION

It is the task of the immune system to protect the host against invading infectious agents and thereby to prevent infectious disease. A plethora of microbial pathogens exists (i.e. viruses, bacteria, fungi, parasites and helminths) that have exploited strategies to circumvent an attack by the immune system. Conversely, the immune system has evolved to provide appropriate defence mechanisms at various levels of 'unspecific' (innate) and 'specific' (adaptive) immune responses (Hughes, 2002). In many instances, an appropriate immune response to an infectious agent requires reciprocal interactions between components of the innate and the adaptive immune system.

The various micro-organisms have developed different strategies to invade their host. Viruses make use of the host cell's machinery for replication and are thus intracellular pathogens (Yewdell and Bennink, 2002). Helminths, the other extreme, are large organisms that cannot live within host cells but rather behave as extracellular pathogens (Pearce and Tarleton, 2002). In between are bacteria which, depending on the species, live within or outside host cells, fungi, and protozoa where the

Copyright © Elsevier Science Ltd
All rights of reproduction in any form reserved

extracellular or intracellular localization may depend on the stage of their life cycle (Kuhn *et al.*, 2002; Pearce and Tarleton, 2002; Romani, 2002).

The successful combat of an invading infectious agent largely depends on the host's capacity to mount an appropriate protective immune response. As a consequence, the analysis of such interactions between host and invading micro-organisms requires a broad spectrum of immunological methods. This book presents a collection of such methods, which are particularly useful for the *ex vivo* and *in vitro* analysis of murine and human immune responses towards infectious agents. Basic principles of the management of infected animals are described in the chapter by Mossmann and colleagues in Section II, and the current methods to establish bacterial infections in mice are discussed by Mielke *et al.* in Section II. The *ex vivo* analysis of the immune response to infection requires optimized protocols to detect immune responses *in situ*, some of which are described in the chapter by Seitzer *et al.* in Section III.

In this introductory chapter, a brief overview of the immune defence mechanism against micro-organisms is given to provide the reader with a guide to the subsequent chapters.

◆◆◆◆◆◆ THE INNATE IMMUNE SYSTEM

Several defence mechanisms exist that are ready to attack invading micro-organisms without prior activation or induction. They pre-exist in all individuals and do not involve antigen-specific immune responses. Hence they are referred to as components of the innate immune system. Among these components, granulocytes, macrophages and their relatives play an important role, especially during the early phases of the immune response. Surface epithelia constitute a natural barrier to infectious agents (Neutra and Kraehenbuhl, 2002). Apart from the mechanical barrier, surface epithelia are equipped with additional chemical features that help to restrain microbial invasion. Depending on the anatomical localization, such factors include fatty acids (skin), low pH (stomach), antibacterial peptides (defensins, intestine), and enzymes (e.g. lysozyme, saliva). Once the pathogen has crossed the protective epithelial barrier, cellular effector mechanisms are activated. Granulocytes and mononuclear phagocytes represent the most important effector cells of the anti-infective immune response.

Invading micro-organisms are sensed by surface receptors which recognize physico-chemical entities that are both unique for, and shared by, microbial pathogens. These entities have been termed pathogen-associated microbial patterns (PAMP). Amongst receptors for PAMP, the Toll-like receptors (TLR) are of particular importance. TLR not only react with the microbial pathogens but also transduce signals into the host cell causing the prompt activation of host defence. The group of granulocytes comprises neutrophils, eosinophils and basophils, which all possess high anti-infective activity. Neutrophils phagocytose microbes and can subsequently kill them. By means of Fc receptors for IgG and complement receptors, phagocytosis of microbes coated by antibodies or complement breakdown products is improved. Eosinophils and basophils primarily attack extracellular pathogens, in particular helminths, by releasing toxic effector molecules. Growth and differentiation of eosinophils are controlled by interleukin-5 (IL-5) and that of basophils by IL-4. These cell types express Fc receptors for IgE which provide a bridge between host effector cells and helminths.

The mononuclear phagocytes comprise the tissue macrophages and the blood monocytes. After activation by cytokines, particularly interferon-γ (IFN-γ), mononuclear phagocytes are capable of killing engulfed micro-organisms. However, in their resting stage, macrophages have a low anti-microbial potential and, thus, are often misused as habitat by many bacteria and protozoa. Killing and degradation of these intracellular pathogens by activated macrophages is achieved by a combination of different mechanisms. The most important ones are:

- Activated macrophages produce toxic effector molecules, in particular reactive oxygen intermediates (ROI) and reactive nitrogen intermediates (RNI), which often synergize in killing various intracellular bacteria and protozoa. Although RNI production is the most potent anti-microbial defence mechanism of murine macrophages, its production by human macrophages is still the subject of debate. However, an increasing amount of data supports RNI production by human macrophages during infectious diseases (MacMicking *et al.*, 1997; Chan *et al.*, 2001).
- Soon after engulfment of microbes, the phagosome becomes acidic and subsequently fuses with lysosomes. Lysosomal enzymes have an acidic pH optimum and, thus, express high activity within the phagolysosome. These lysosomal enzymes are primarily responsible for microbial degradation (Finlay and Cossart, 1997; Schaible *et al.*, 1999).
- Both the intracellular pathogen and the host cell require iron. Therefore, depletion of intraphagosomal iron reduces the chance of intracellular survival for various pathogens (Lieu *et al.*, 2001).
- Tryptophan is an essential amino acid for certain intracellular pathogens, such as *Toxoplasma gondii*. Accordingly, rapid degradation of this amino acid impairs intracellular replication of susceptible pathogens (Pfefferkorn, 1984).

Intracellular pathogens have developed various evasion mechanisms which prolong their survival inside macrophages (Schaible *et al.*, 1999). Some even persist within activated macrophages, though at a markedly reduced level. *Listeria monocytogenes* and *Trypanosoma cruzi* egress from the phagosome into the cytosol, thus escaping intraphagosomal attack. Several intracellular pathogens, such as *Mycobacterium tuberculosis*, remain in the phagosome. However, they prevent phagosome acidification and subsequent phagosome–lysosome fusion. To compete for the intracellular iron pool, some pathogens possess potent iron acquisition mechanisms, and to avoid killing by ROI or RNI, several microbes produce detoxifying enzymes. For example, catalase and superoxide dismutase directly inactivate ROI and indirectly impair RNI effects (Chan *et al.*, 2001).

In summary, living within macrophages provides a niche that protects intracellular pathogens from humoral attack. Yet, once activated, macrophages are capable of eradicating many intracellular pathogens and of restricting growth of more robust ones. Such pathogens may persist for long periods of time, thus causing chronic infection and disease (Munoz-Elias and McKinney, 2002).

In addition to their role in the non-specific anti-infective host response, macrophages contribute to the specific immune response against micro-organisms. Therefore, methods to evaluate the functional capacities of murine and human macrophages will be described in the chapters by Peiser *et al.* (Section II), and Krause *et al.* (Section III), respectively. Microbial degradation within macrophages delivers pathogen-derived antigenic fragments (peptides) which enter antigen-processing pathways leading to the cell-surface expression of 'foreign' microbial antigens in the context of appropriate major histocompatibility complex (MHC) molecules. Such antigenic peptides presented by MHC class I or class II molecules can then activate specific T lymphocytes. As discussed by De Groot and colleagues (Section I), the determination of MHC-binding peptide motifs of immunodominant antigens from infectious micro-organisms has important implications for vaccine development.

Among the humoral mechanisms of the innate immune system, the alternative pathway of complement activation is perhaps the most important one. While the classical pathway of complement activation requires the presence of specific antibodies (and hence is delayed upon microbial infection), the alternative pathway is initiated in the absence of antibodies (Carroll, 1998). A C3b homologue, generated through spontaneous cleavage of C3 present in the plasma, can bind to bacterial surfaces. Factor B of the alternative complement pathway binds non-covalently to C3b and is cleaved by a serum protease factor D, to yield a larger fragment, Bb, and a smaller fragment, Ba. The lytic pathway is triggered through binding of properdin (also called factor P) to C3bBb complexes. In contrast to host cells, bacteria lack complement controlling membrane proteins such as decay accelerating factor (DAF; CD55), homologous restriction factor (CD59), and membrane cofactor protein (CD46) (Liszewski *et al.*, 1996). Through the C3/C5 convertase activity of the C3bBb complex, further C3 molecules are cleaved, leading to the production of large amounts of C3b homologues which bind to bacterial surfaces and initiate the lytic pathway. Moreover, C3b deposition on microbial surfaces promotes the microbial uptake via complement receptors. Finally, the complement breakdown products C4a and C5a are chemoattractants for phagocytes, and are, therefore, termed anaphylatoxins. They induce phagocyte extravasation into foci of microbial implantation.

◆◆◆◆◆◆ THE ADAPTIVE IMMUNE SYSTEM

While the components of the innate immune system are appropriate as a first line of defence, the adaptive (or specific) immune system is activated if the invading micro-organism cannot be eliminated, or at least be neutralized, by the above-mentioned non-specific effector mechanisms. Two major features characterize the adaptive immune system. First, the immune response is antigen-specific; specificity is made possible through the usage of clonally distributed antigen receptors, i.e. surface immunoglobulin on antibody-producing B lymphocytes and T-cell receptors (TCR) on the surface of T lymphocytes. Secondly, the specific immune system develops memory (Ahmed *et al.*, 2002). This allows the rapid response of antigen-specific effector cells upon second encounter of the relevant antigen.

B cells recognize antigen in a fashion which is fundamentally different from that of T cells. The antibody expressed on the B-cell surface (and later, secreted by the plasma cell) directly binds to native, soluble antigen. By virtue of their antibody production, B cells contribute to the humoral immune defence against extracellular pathogens and neutralize virions before they enter the host cell. In contrast, T cells recognize antigen only if it is presented in the context of appropriate MHC molecules on the surface of antigen-presenting cells (APC). The surface-expressed TCR is non-covalently associated with the CD3 polypeptide complex which mediates signal transduction upon TCR triggering, leading to cytokine gene transcription and T-cell activation.

Two well-defined and some less well-defined subpopulations of T cells are involved in the specific cell-mediated immune response against infectious agents. The expression of relevant cell surface molecules allows their identification and phenotypic characterization by specific mono-clonal antibodies. Moreover, monoclonal antibodies can be used to separate subpopulations of cells on a fluorescence-activated cell sorter (FACS®) or by using magnetic beads. Accordingly, the chapter by Scheffold *et al.* (Section I) is devoted to a detailed description of these tech-niques. The dominant subsets of mature T cells are characterized by the reciprocal expression of CD4 and CD8 coreceptors. $CD4^+$ T cells recognize antigen in the context of MHC-II molecules; they produce various cytokines required for efficient activation of leukocytes – notably B cells and macrophages – and are therefore termed helper cells (Th cells). In contrast, $CD8^+$ T cells recognize antigen in the context of MHC-I molecules; one of their major tasks is to lyse virus-infected target cells. Hence, they are termed cytotoxic T lymphocytes (CTL). Recent tech-nological advances enable the quantitative assessment of antigen-specific T-cell responses. These studies revealed that the population of antigen specific T cells during primary and secondary immune responses are in the order of 1 and 10%, respectively. The use of tetramers composed of MHC-I heavy chain, β2 microglobulin, and the relevant antigenic peptide for the quantitative analysis of T-cell responses is described by Gillespie *et al.* (Section I)

The activation of T cells requires two signals. Signal one is mediated through the CD3/TCR molecular complex following antigen recognition. Signal two is a co-stimulatory signal that is delivered through receptor–ligand interactions. The major co-stimulatory molecules on antigen-presenting cells are CD80 (B7.1) and CD86 (B7.2). Both molecules can bind to CD28 on T cells, thereby exerting costimulation. In addition, CTLA4 (CD152) also binds to both ligands. The CD28/CD152 interaction with CD80/CD86 is complex. Recent evidence indicates that triggering of CD28 mediates co-stimulation, whereas binding to CD152 delivers an inhibitory signal (Salomon and Bluestone, 2001). An overview of some important cell surface molecules with relevance to anti-infective immunity is given in Table 1.

The vast majority of the CD4$^+$ and CD8$^+$ T cells express a TCR composed of α- and β-chain heterodimers. These T cells are termed conventional T cells. During the last few years, the existence of additional T-cell subsets has been appreciated. These include CD4$^-$CD8$^-$ 'double-negative' (DN) T cells that express either the αβ TCR or an alternative TCR composed of γ and δ chains. Furthermore, CD8$^+$ T cells expressing the αβ TCR restricted by MHC-I-like molecules, as well as CD4$^+$ T cells co-expressing the αβ TCR and natural killer (NK) receptors (NK1) have been identified. Increasing evidence suggests that there is a well-orchestrated interplay between all these subsets with a preponderance of any one of them, depending on the type of infection.

Table 1. Cell surface molecules with relevance to anti-infective immunity

Cell surface molecule	Function in anti-infective immunity
TCRαβ	MHC/peptide recognition by the major αβ T-cell population
TCRγδ	Ligand recognition by the minor γδ T-cell population
CD1	Presentation of lipids and glycolipids to αβ T cells
CD3	Marker of all T cells, signal transduction in T cells
CD4	Co-receptor with specificity for MHC class II, marker molecule of Th cell
CD8	Co-receptor with specificity for MHC class I, marker molecule of CTL
CD14	Pattern recognition receptor on macrophages which, for example, binds LPS from Gram-negative bacteria
CD40	Co-stimulatory molecule on B cells and antigen-presenting cells
CD154 (CD40L)	T-cell co-stimulation (ligand for CD40)
CD28	Co-stimulatory T-cell molecule (positive signal)
CD152 (CTLA-4)	Co-stimulatory T-cell molecule (negative signal)
CD80 (B7-1)	Ligand for CD28, CD152
CD86 (B7-2)	Ligand for CD28, CD152
CD95	Fas (Apo-1), a receptor which mediates an apoptotic signal

Conventional T lymphocytes

CD4$^+$ T cells expressing the αβ TCR recognize foreign peptides bound in the peptide-binding groove of MHC-II molecules. These peptides are generally derived from exogenous antigens (such as micro-organisms) that are taken up by phagocytosis or endocytosis. Immunogenic peptides of 13 or more amino acids in length are generated in endosomes, bound to MHC-II molecules, and transported to the cell surface. Presentation of antigen to CD4$^+$ T cells is restricted to APC that either constitutively express MHC-II antigens (monocytes/macrophages, dendritic cells, B cells), or can be induced to express MHC-II molecules (e.g. endothelial cells and activated human T cells).

There are two functionally distinct subpopulations of CD4$^+$ Th cells, i.e. Th1 and Th2 cells. These subsets are distinguished based on the characteristic spectrum of cytokines that they produce upon antigenic stimulation (Abbas *et al.*, 1996; Mosmann and Fowell, 2002). Th1 cells are characterized by their secretion of IFNγ and IL-2, whereas Th2 cells preferentially produce IL-4, IL-5 and IL-10. Differential cytokine expression also enables the separation of leukocyte populations. This most recent technological advance has been included in the chapter by Scheffold *et al.* in Section I. As these cytokines play characteristic but different roles in various types of immune responses, Th1 or Th2 CD4$^+$ T cells can dominate in a given situation, thus determining the outcome of infection. Th1 cells play an important role in the initiation of the cell-mediated immune response against intracellular pathogens, due to secretion of IFNγ (which activates macrophages) and IL-2 (which activates CTL) (Collins and Kaufmann, 2002; Scott and Grencis, 2002). In the mouse, IFNγ also stimulates the production of immunoglobulin subclasses (IgG2a, IgG3) that contribute to anti-microbial immunity by virtue of their complement fixing and opsonizing activities. On the other hand, Th2 cells produce cytokines (IL-4, IL-5) that control activation and differentiation of B cells into antibody-secreting cells. Th2 cells are thus important for the induction of humoral immune responses. IL-4 controls the immunoglobulin class switch to IgE and hence plays a central role in the immune defence against helminths and in the regulation of the allergic response. Moreover, IL-4 stimulates the production of IgG subclasses (IgG1 in the mouse) that neutralize but do not opsonize antigens. The Th2 cytokine IL-5, together with transforming growth factor β (TGF-β), induces B cells to switch to IgA, the major Ig subclass involved in local immune responses. In addition, IL-5 contributes to the control of helminth infection by activating eosinophils (Scott and Grencis, 2002). Figure 1 illustrates the major effector functions of Th1 and Th2 subsets and the cytokine-driven regulatory interactions.

Th1 and Th2 cells differentiate from an undetermined Th0 precursor cell. The differentiation into one or the other functional subset is driven by cytokines that are produced by cells of the non-specific immune system early after infection. In this regard, the rapid production of IL-12 and IL-18 by monocytes/macrophages, and of IFNγ by NK cells, drives the Th cell response into Th1 cells following bacterial infection. Conversely, the

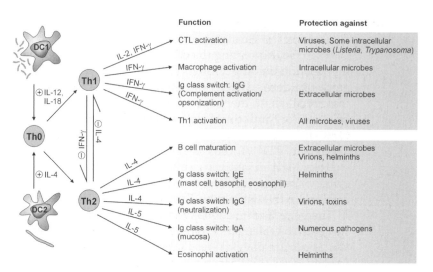

Function	Protection against
CTL activation	Viruses, Some intracellular microbes (*Listeria, Trypanosoma*)
Macrophage activation	Intracellular microbes
Ig class switch: IgG (Complement activation/ opsonization)	Extracellular microbes
Th1 activation	All microbes, viruses
B cell maturation	Extracellular microbes Virions, helminths
Ig class switch: IgE (mast cell, basophil, eosinophil)	Helminths
Ig class switch: IgG (neutralization)	Virions, toxins
Ig class switch: IgA (mucosa)	Numerous pathogens
Eosinophil activation	Helminths

Figure 1. Role of Th1 and Th2 cells in anti-infective immunity. The differentiation from Th0 precursor cells is driven by IL-12/IL-18 and IL-4, respectively. Th1 and Th2 cells are distinguished by a characteristic pattern of cytokine production. The key cytokines of Th1 and Th2 cells have differential roles in anti-infective immunity.

early production of IL-4 is a major force that drives Th cells along the Th2 differentiation pathway. The cellular source of the early IL-4 remains to be unequivocally defined.

In most instances, there is no absolute restriction in the activation of either Th1 or Th2 cells during the immune response to infectious agents. Nevertheless, in many situations, there is a clear dominance of one or the other Th-cell subset, and a (genetic) failure to activate the appropriate Th-cell subset may lead to a disastrous outcome after infection. A well-documented example for this is the infection of mice with the protozoan parasite *Leishmania major*. Hence, technical aspects of working with the leishmaniasis model are discussed by Kropf and colleagues (Section II). In resistant strains of mice, such as C57Bl/6, *L. major* causes a self-healing lesion, whereas in susceptible BALB/c mice, the infection is progressive and eventually fatal. It has been shown that, during infection, resistant mice produce high levels of IFNγ and little IL-4 (and thus display a Th1-type response), whereas the susceptible mice produce high amounts of IL-4 and little IFNγ (and thus display a Th2-type response). In view of the known role of IL-12 in driving Th0 cells into Th1 cells, attempts were made to prevent the fatal Th2 cell differentiation in *L. major*-infected BALB/c mice. In fact, the administration of leishmanial antigens together with IL-12 induced the appearance of *L. major*-specific Th1 cells in BALB/c mice. More importantly, these mice were protected from fatal infection when challenged with *L. major*.

Similarly, the protective T-cell response against mycobacteria is mediated by Th1 cells. Mycobacteria induce IL-12 and IL-18 in macrophages, and IFNγ secreted by Th1 cells is the major T-cell derived macrophage-activating mediator (Collins and Kaufmann, 2002). Tuberculosis is clearly dominated by a Th1 response, which, however, may be insufficiently protective. The spectrum of disease observed in leprosy patients can be partially explained on the basis of a Th1 or Th2 preponderance. Whereas malign lepromatous leprosy is frequently associated with the production of Th2 cytokines, the more benign tuberculoid form of leprosy is dominated by Th1 cytokine patterns. Murine models of tuberculosis are described by Roberts and co-workers (Section II).

The concept of functionally distinct CD4+ T-cell subsets being differentially involved in immune responses on the basis of their cytokine production, has greatly helped to delineate immune defence mechanisms in infection, and to devise therapeutic strategies. Attempts to identify surface markers that would allow unambiguous identification of Th1 and Th2 cells have not been too successful. A differential expression of the β2' subunit of the IL-12 receptor on Th1 versus Th2 cells was reported. It was found that IFNγ maintained the expression of IL-12R β2 and thereby the IL-12 responsiveness leading to Th1 differentiation, while IL-4 inhibited IL-12R β2 expression leading to the loss of IL-12 signalling, associated with a differentiation into Th2 cells (Rogge et al., 1997; Szabo et al., 1997). Reciprocally, a member of the IL-1 family termed T1/ST2 is characteristically expressed by Th2 cells (Löhning et al., 1998). Of similar importance is the differential expression of chemokine receptors on Th1 and Th2 cells, which provides the investigator with convenient means of distinguishing between Th1 and Th2 cells. More importantly, the differential chemokine receptor expression can direct Th1 and Th2 cells to sites of inflammation caused by microbes or helminths, respectively (Sallusto et al., 2000).

It is obvious that a precise analysis of cytokine production is of major importance for the understanding of the pathophysiology of infection, as well as for designing rational strategies for therapeutic intervention. Therefore, several methods to measure cytokines will be described in this book. Depending on the experimental system, cytokine detection methods may include ELISA, bioassays, (semi)quantitative RT-PCR, ribonuclease protection assay (RPA), or intracellular cytokine staining and FACS analysis (see the chapters by Assenmacher et al. (Section I), Ehlers et al. (Section II), Hiroi et al. (Section II) and Yssel et al. (Section III)).

The prime task of CD8+ CTL is the immune defence against intracellular pathogens with an emphasis on viruses. Experimental infection of mice with cytomegalovirus, one of the most elegant models for studying immune responses to viral pathogens, is described by Podlech et al. (Section II). Viruses are replicated by host cells. As a consequence, viral proteins are degraded in the cytosol of the cell. Virus-derived peptides are transported in association with the TAP molecules into the endoplasmatic reticulum, where they are introduced to MHC-I molecules (Szomolanyi-Tsuda et al., 2002). Upon transport to the cell surface, peptides of 8–10 amino acids in length are anchored in the peptide binding groove of MHC-I molecules, ready to be recognized by CD8+ CTL expressing the appropriate TCR. While the important role of CD8+ CTL in the elimination of virus-infected cells is well documented, there is clear evidence that MHC-I-restricted CD8$^+$ T cells also contribute to the immune defence against intracellular bacteria and protozoa (Collins and Kaufmann, 2002). Some intracellular microbes such as *Listeria monocytogenes* or *Trypanosoma cruzi* gain access to the cytosol, which causes their antigens to enter the MHC-I antigen processing pathway. In addition, it was found that bacteria-derived antigenic peptides can be introduced into the MHC-I pathway, despite the fact that the micro-organisms themselves remain in the phagosome. Taken together, there is a preponderance of CD4$^+$ T cells in the immune defence against phagosomal pathogens, and of CD8$^+$ T cells in the immune defence against cytosolic pathogens (Schaible *et al.*, 1999). In many instances, however, optimal protection against infectious agents requires the co-ordinated co-operation of CD4$^+$ and CD8$^+$ T cells.

CD8$^+$ CTL can eliminate infected cells, thereby limiting pathogen spread. Cytotoxicity is mediated through pore-forming proteins (perforins) and enzymes (granzymes) that are released by activated CTL upon cell contact-dependent recognition of relevant (e.g. virus-infected) target cells. Techniques to measure killer cell activity in murine and human systems are described in the chapters by Busch and Pamer (Section II), and Pawelec *et al.* (Section III), respectively. Recent experiments have led to the identification of a granulysin in CTL, which could directly contribute to antibacterial defence. Granulysin has been shown to kill directly a large variety of bacteria, fungi and protozoa (Kaufmann, 1999). Target cell perforation by perforin could enable granulysin to access microbes residing within target cells. This mechanism could therefore enable CD8 CTL to participate in direct combat of microbial infections.

CTL can also trigger programmed cell death (apoptosis) in target cells through receptor–ligand interactions. Upon activation, CTL are induced to express Fas-ligand (Fas-L), a member of the TNFα gene family. Fas-L interacts with the corresponding receptor Fas (CD95, APO-1) expressed on virus-infected target cells. The oligomerization of several Fas molecules triggers a rapid suicide programme which culminates in protease-dependent cell death, usually associated with fragmentation of genomic DNA into oligonucleosomal-sized fragments. Increasing evidence indicates, however, that the role of CD8$^+$ T cells in infection is not limited to their function as CTL. Like CD4$^+$ T cells, CD8$^+$ T cells are equipped with the capacity to produce cytokines. More specifically, the range of cytokine(s) secreted by CD8$^+$ T cells depends on the antigenic stimulation, in a manner comparable to the situation with CD4$^+$ T cells. This has led to the suggestion that CD8$^+$ T cells should also be divided into subsets (Tc1, Tc2), based on their cytokine secretion profile (Mosmann and Fowell, 2002).

The dominant role of T lymphocytes in the immune defence against infectious micro-organisms requires a broad spectrum of appropriate methods for analysis. Therefore, separate chapters are devoted to the isolation of lymphocytes from infected animals (Czuprynski and Brown, Section II), the accurate measurement of lymphocyte expansion (Pechhold *et al.*, Section I), and the establishment of T-cell lines and clones from murine (Reardon and Lahn, Section II) and human lymphocytes (Märker-

Herrmann and Duchmann, Section II). Methods to immortalize human T cells by herpesvirus saimiri are described by Fickenscher and Fleckenstein (Section III).

Unconventional T lymphocytes

Apart from the well-characterized CD4[+] and CD8[+] T-cell subsets that recognize antigenic peptides in an MHC-II or MHC-I restricted manner via the conventional αβ TCR, several additional T-cell populations can contribute to the immune defence against infectious micro-organisms. These additional T-cell subsets have been collectively termed 'unconventional' T cells (Kaufmann, 1996).

In mice, CD8[+] T cells have been described that express the conventional αβ TCR and recognize an unusual group of peptides in the context of MHC-I-like presenting molecules (Lenz and Bevan, 1996). The peptides carry the N-formyl-methionine (N-f-met) sequence that represents a characteristic signal sequence required for protein export in bacteria. N-f-met containing peptides are virtually absent from mammalian cells, with the exception of mitochondria. The MHC-I-like molecules that present N-f-met containing peptides to CD8[+] T cells are far less polymorphic than classical MHC antigens, and hence are broadly distributed. CD8[+] T cells with specificity for N-f-met containing peptides have been shown to mediate protection in experimental murine models of *L. monocytogenes* infection (Rolph and Kaufmann, 2000). The possible role of such cells during the immune response in humans is less clear, because homologues of the relevant murine MHC-I-like molecules have so far not been identified in man.

Additional unconventional T-cell subsets exist that recognize antigen in association with non-MHC molecules. Among those, T cells have been identified in humans that recognize non-peptide antigens in the context of CD1 gene products, which share some similarities with MHC-I molecules. These T cells express the αβ TCR. Interestingly, these T cells can recognize glycolipids derived from mycobacteria, including mycolic acid and lipoarabinomannan (Schaible and Kaufmann, 2000). There is only limited polymorphism of the presenting CD1 molecules. The group 1 CD1 molecules (CD1a–c) required for presentation of glycolipids to these T cells are expressed on the surface of human dendritic cells (DC), or can be induced to be expressed on APC. T cells with specificity for these glycolipids have not been described in mice possibly because the cognates of the human group 1 CD1 antigens are absent in this species. These T cells secrete IFN-γ and express cytolytic activity. Hence they could contribute to anti-microbial defence in a way similar to conventional Th cells.

Another subset of unconventional T cells is characterized by co-expression of αβ TCR and NK1, a characteristic marker of natural killer cells. These T cells recognize glycolipids in the context of group 2 CD1 molecules (CD1d) (Schaible and Kaufmann, 2000). Interestingly, the TCR repertoire of NK1 T cells is strikingly restricted. The TCR of these T cells is composed of an invariant α chain associated with a β chain that uses a limited set of variable (Vβ) elements, suggesting that NK1 T cells can recognize only a restricted array of antigens. It is likely that the NK T cells perform regulatory functions. Recently, β-galactosylceramide from a marine sponge, N-glycosyleramides from *Malaria plasmodia*, and phosphatidylinositol tetramannoside from mycobacteria have been identified as ligands for NK T cells presented by CD1d (Schaible and Kaufmann, 2000).

While the vast majority of CD3+ T cells express the 'conventional' TCR composed of an αβ chain heterodimer, a minor subset (1–10%) of CD3+ T cells expresses the alternative γδ TCR (Hayday, 2000). There are two major differences between αβ T cells and γδ T cells. First, the majority of γδ T cells lack the expression of co-receptor molecules CD4 or CD8, thus displaying a DN phenotype. Secondly, the number of germ-line gene elements that can be expressed to construct the variable regions of TCR chains is small for γ and δ when compared with α and β. Nevertheless, the available TCR repertoire of γδ T cells is as large as that of αβ T cells, because several non-germ-line encoded mechanisms such as N-region diversity, usage of alternative reading frames, etc., dramatically contribute to TCR diversity. Substantial evidence suggests that γδ T cells play a role in the immune defence against various infectious microorganisms (Kaufmann, 1996; Hayday, 2000). Human γδ T cells expressing the Vγ9/Vδ2 TCR are strongly activated by live or killed mycobacteria, as well as by several other intracellular or extracellular bacteria, or protozoa such as *Plasmodium falciparum* (Kabelitz et al., 2000). In several instances, a transient increase in circulating γδ T cells has been observed during acute infection (Kabelitz et al., 1999). γδ T cells express a functional repertoire similar to conventional αβ T cells. Thus, activated γδ T cells exert CTL activity and produce a range of cytokines, depending on the antigenic stimulation. A Th1 pattern of cytokines was produced by peritoneal γδ T cells when mice were infected with *L. monocytogenes*, whereas peritoneal γδ T cells produced Th2-type cytokines when mice were infected with *Nippostrongylus brasiliensis* (Ferrick et al., 1995). Similarly, human γδ T cells can be polarized into either Th1 or Th2 effector cells (Wesch et al., 2001). The microbial ligands recognized by human Vγ9/Vδ2 T cells have been recently characterized as non-proteinaceous, phosphate-containing low molecular weight compounds. The most active compounds are intermediates of the non-mevalonate ('Rohmer') pathway of isoprenoid biosynthesis which is restricted to certain micro-organisms (including *Mycobacterium tuberculosis*) and is not available in eukaryotic cells (Belmant et al., 1999; Altincicek et al., 2001). The recognition of these ligands by human γδ T cells is not restricted by classical MHC antigens or other presenting molecules (such as CD1) but requires some as yet ill-defined form of presentation. While the microbial phospholigands are potent activators of human γδ T cells, they do not appear to be recognized by their murine counterparts. Instead, there is evidence that some murine γδ T cells recognize heat shock proteins (hsp) derived from micro-organisms such as *M. tuberculosis*.

Despite their impressive *in vitro* reactivity towards certain ligands from infectious micro-organisms, the *in vivo* role of γδ T cells in infection is not precisely understood. In several experimental models of bacterial infection, a transient activation of γδ T cells during early phases of the immune response is observed. On the other hand, γδ T cells appear to contribute to protection against certain viral infections at later stages. In this context, it is interesting to note that characteristic changes in the expressed TCR repertoire of peripheral blood γδ T cells occur in HIV-infected individuals (Kabelitz and Wesch, 2001). A protective role of γδ T cells in experimental tuberculosis was revealed through the analysis of gene deletion mutant mice deficient in γδ T cells. It was found that these Cδ$^{-/-}$ knockout mice succumb to lethal infection with *M. tuberculosis*, but only if high inocula of *M. tuberculosis* are used (Ladel *et al.*, 1995; D'Souza *et al.*, 1997). Moreover, a protective role of human Vγ9Vδ2 γδ T cells in the anti-bacterial immune response was recently shown in an adoptive transfer model using SCID mice reconstituted with phosphoantigen-stimulated human PBL (Wang *et al.*, 2001). On the basis of the rapid response of γδ T cells (frequently preceding that of αβ T cells) and their limited germ-line TCR repertoire, it is assumed that γδ T cells provide a link between the innate and the adaptive immune system. In addition, a more general regulatory role of γδ T cells in inflammation appears likely.

Although unconventional T cells comprise only a minor fraction of peripheral T lymphocytes, they may play a decisive role in the immune defence against certain micro-organisms. Therefore, this aspect needs to be considered in the development of new subunit vaccines, e.g. against tuberculosis. In addition to well-defined peptides for protective αβ T cells, such subunit vaccines might include non-proteinaceous phospholigands stimulating γδ T cells as well as glycolipids for CD1-restricted T cells.

B lymphocytes

B cells express surface immunoglobulin (Ig) as their antigen-specific receptor molecules. Upon activation and differentiation into antibody-secreting cells, B cells produce and secrete large amounts of Ig with the same specificity as the membrane-bound Ig. T-cell-dependent B-cell activation requires cognate interaction between the two lymphocyte populations. In recent years, it has become obvious that the receptor–ligand interaction mediated between CD40 (expressed on B cells) and the corresponding receptor expressed on T cells (CD40-ligand; now termed CD154) is important for the initiation of humoral immune responses to T-cell-dependent antigens (Foy *et al.*, 1996). In addition, studies with gene deletion mutant mice lacking either CD40 or CD154 expression have shown that CD40/CD154 interactions are essential for secondary immune responses to T-cell-dependent antigens, as well as for the formation of germinal centres (Grewal and Flavell, 1996).

Proliferation and differentiation of B cells as well as the Ig isotype class switching are driven by cytokines (Stavnezer, 1996). In the mouse, IL-4 induces IgG1 and IgE secretion, while TGF-β and IL-5 trigger the IgA class switch. IFN-γ is known preferentially to induce IgG2a and IgG3 secretion. IgG3 (together with IgM) possesses complement-fixing activity. These Ig subclasses are thus involved in the initiation of the classical pathway of complement activation, leading to complement-mediated destruction of pathogens or infected cells. In addition, antibodies are required for antibody-dependent cellular cytotoxicity (ADCC) effector function. Lymphoid cells carrying receptors for the Fc portion of IgG (Fcγ receptor) such as large granular lymphocytes mediate ADCC of IgG-coated target cells.

The initial encounter of antigen-specific B cells with the appropriate Th cells occurs at the border of T and B cell areas in lymphoid tissues. Activated B cells migrate into a nearby lymphoid follicle where they form a germinal centre. In the germinal centres, somatic hypermutation occurs in rapidly proliferating B-cell blasts, thus giving rise to the selection of high-affinity antibodies (affinity maturation) (Rajewsky, 1996).

In addition to their unique role as antibody-producing plasma cells, B cells have the capacity to present antigen to T lymphocytes. Upon binding of soluble antigen to membrane-bound Ig with homologous specificity, antigen–antibody complexes are internalized and degraded in the endolysosomal compartment. Antigen-derived peptides are then introduced to the MHC-II-dependent processing pathway and can be presented to appropriate peptide-specific CD4$^+$ T cells (Watts, 1997).

◆◆◆◆◆◆ THE INTERPHASE BETWEEN INNATE AND ADAPTIVE IMMUNE SYSTEMS

During recent years it has become increasingly clear that the innate immune system, in addition to providing prompt host defence mechanisms, is also instrumental for the development of the adaptive immune response. At the early stage of encounter between host and pathogen the antigen-presenting cells, comprising dendritic cells (DC) and macrophages, are active. Macrophages are professional phagocytes which are also capable of presenting antigens to T cells. DC possess lower phagocytic activity but are most efficacious in presenting soluble antigens to T cells. Both macrophages and DC express non-clonally distributed pattern recognition receptors which react with unique microbial entities, notably TLR, the so-called pathogen associated microbial patterns (PAMP). Of critical importance also are Toll-like receptors (TLR). At least 10 TLR cognates exist, each responsive to different types of PAMP (Akira *et al.*, 2001). TLR-2 reacts with lipoproteins and lipoarabinomannans from mycobacteria. TLR-3 interacts with double-stranded RNA of various viruses, TLR-4 responds to LPS from Gram-negative bacteria, TLR-5 responds to bacterial flagellin and TLR-9 responds to oligodeoxynucleotides comprising unmethylated CpG nucleotide motifs. Some TLR form heterodimers to provide novel specificities. For example, TLR-2 and TLR-6 heterodimers can react with peptidoglycans from Gram-positive bacteria and zymosan from yeast. Importantly, interactions between PAMP and TLR initiate a signal transduction cascade which promptly mobilizes host defence mechanisms such as the production of RNI. In addition, it also induces the surface expression of co-stimulatory molecules such as CD40 and CD80/CD86, as well as the secretion of immunostimulatory cytokines such as IL-12. In this manner, sensing PAMP by means of TLR promotes the development of Th1 cells. Helminths lack such PAMP, but encompass other, so far ill-defined, molecular entities. In their presence, and in the absence of PAMP, Th2 cell development is induced.

DC express an enormous plasticity which allows them to develop distinct functional activities in response to different environmental stimuli (Reis e Sousa, 2001). In the presence of PAMP, TLR induce the maturation into DC that preferentially stimulate Th1 cells, and have therefore been termed DC1. In contrast, helminths cause the differentiation of DC2 through unknown mechanisms, and those, in turn, induce Th2 cell development. In this way DC are critical regulators of the adaptive immune response to various pathogens (see also Figure 1). The features of DC and methods to purify and culture these cells are described in the chapter by Palucka *et al.* (Section III). With regard to immunity against infectious agents, the lack of highly phagocytic and degradative activities of DC renders them dependent on co-operation with macrophages. It is likely that intracellular microbes are engulfed and degraded by macrophages which then produce vesicles containing antigenic cargo. This cargo can by taken up by bystander DC and presented in the most efficacious manner (Kaufmann, 2001). This co-operation is often termed cross-priming.

◆◆◆◆◆◆ CYTOKINES

Collectively, cytokines are soluble mediators that exert pleiotropic effects on cells of the immune system and transduce signals via specific surface receptors (see Table 2).

Cytokines primarily produced by cells of the immune system with known cDNA sequence are designated interleukins. As discussed above, T-helper cells are functionally differentiated into Th1 and Th2 subsets on the basis of their characteristic cytokine spectrum (Abbas *et al.*, 1996; Mosmann and Fowell, 2002). Upon appropriate activation, these Th subsets produce interleukins that are primarily required for immuno- logical control of intracellular pathogens (Th1) or the regulation of Ig class switching (Th2). In addition, cytokines produced by monocytes and macrophages (frequently termed monokines) have important roles in the immune defence against infectious agents. Cytokines produced by macrophages in response to stimulation with bacterial components include IL-1, IL-6, IL-12, IL-18 and TNFα.

IL-12 is a driving force for the differentiation of Th1 cells from un- determined Th0 precursor cells. Its action is supported by the more recently discovered IL-18 (Swain, 2001). IL-1, IL-6 and TNFα are pro- inflammatory and pleiotropic cytokines that induce a variety of effects on many different target cells. A large group of cytokines is collectively termed chemokines. These proteins recruit phagocytic cells and lympho-

Table 2. Cytokines with relevance for the anti-infective immune response

Cytokine	Major role in antimicrobial defence
Chemokines	Leukocyte attraction to site of microbial implantation
CXC chemokine	Granulocyte recruitment to site of microbial implantation
CC chemokine	Monocyte recruitment to site of microbial implantation
C chemokine	Lymphocyte recruitment to site of microbial implantation
IL-1	Proinflammatory, endogenous pyrogen
IL-6	Proinflammatory
TNF-α	Proinflammatory, macrophage co-stimulator, cachexia
IL-2	T-cell activation
IFN-γ	Macrophage activation, promotion of Th1 cells
IL-4	B-cell activation, switch to IgE, promotion of Th2 cells, activation of mast cells
IL-5	Switch to IgA, activation of eosinophils
IL-12	Promotion of Th1 cells
IL-18	Promotion of Th1 cells
IL-10	Anti-inflammatory
TGF-β	Anti-inflammatory

cytes to local sites of infection. Chemokines are characterized by four conserved cysteines forming two disulfide bridges. The position of the first two cysteines has been used to divide the chemokines into four families, the C-X-C, the C-C, the C, and the CX_3C chemokines (Baggiolini *et al.*, 1997; Zlotnik and Yoshie, 2000). IL-8 and NAP-2 are members of the C-X-C chemokine family that promote the migration of neutrophils. MIP-1β, MCP-1 and RANTES are members of the C-C family of chemokines that promote migration primarily of monocytes and T lymphocytes (Moser and Loetscher, 2001). The C and CX_3C chemokines comprise only one or a few members.

Chemokines can be produced by many different cell types in response to stimulation with bacterial antigens or viruses. As a consequence, local recruitment of phagocytic and effector cells due to the effect of chemokine release is a general feature of the immune response to infection. In addition, it has recently been discovered that chemokines and their receptors play important roles in the control of HIV infection of target cells (D'Souza and Harden, 1996).

Several cytokines possess anti-inflammatory activity. IL-10, produced by monocytes and Th2 cells, inhibits synthesis of IL-12 and Th1 cytokines IL-2 and IFNγ, thereby supporting the differentiation of Th2 cells, which is mostly promoted by IL-4. Similarly, TGF-β is a potent inhibitor of macrophage and lymphocyte activation. These cytokines contribute to the termination of ongoing inflammatory responses.

◆◆◆◆◆◆ CONCLUDING REMARKS

As briefly summarized in this chapter, the immune response to infectious agents involves a broad spectrum of mechanisms of the innate and the acquired immune system. Accordingly, the analysis of these mechanisms requires a similarly broad spectrum of sophisticated immunological technologies. A better understanding of the principal mechanisms underlying the protective immune response to infectious agents will also provide guidelines for the rational design of novel vaccination strategies. Technical aspects of vaccine development are covered in two chapters of this book. The use of naked DNA as novel vaccine candidates and the preparation and use of adjuvants for subunit vaccines are described by Corr *et al.* (Section II) and by Lövgren-Bengtsson (Section II), respectively.

Until recently, experimental approaches in immunology and microbiology have focused on changes in single or a few defined parameters. The elucidation of the genomes of numerous microbial pathogens and the completed elucidation of the human and murine genomes, has led to a paradigm shift. Using DNA chips we can now investigate the global changes in the transcriptomes of both the host and the pathogen. This approach broadens our interest from single molecules to the global signature responses that occur during infection. The methodological tools for these investigations are discussed by Yowe *et al.* (Section I) and by Ehrt *et al.* (Section I).

While some chapters of this book describe technologies that are generally appplicable to the analysis of murine and human immune responses, other chapters deal with methods that are particularly useful for the analysis of specific infection models such as tuberculosis, leishmaniasis or cytomegalovirus infection. Although the analysis of specific problems might require additional methods, we believe that this book provides a useful guide for the microbiologist wishing to investigate infection and infectious disease with immunological methods.

References

Abbas, A. K., Murphy, K. M. and Sher, A. (1996). Functional diversity of helper T lymphocytes. *Nature* **383**, 787–793.

Ahmed, R., Lanier, J. G. and Pamer, E. (2002). Immunological memory and infection. In: *Immunology of Infectious Diseases* (S.H.E. Kaufmann, A. Sher and R. Ahmed, Eds), pp. 175–190. ASM Press, Washington, DC.

Akira, S., Takeda, K. and Kaisho, T. (2001). Toll-like receptors: critical proteins linking innate and acquired immunity. *Nat. Immunol.* **2**, 675–680.

Altincicek, B., Moll, J., Campos, N., Foerster, G., Beck, E., Hoeffler, J. F., Grosdemange-Billiard, C., Rodriguez-Cencepcion, M., Rohmer, M., Boronat, A., Eberl, M. and Jomaa, H. (2001). Cutting edge: human γδ T cells are activated by intermediates of the 2-C-methyl-D-erythritol 4-phosphate pathway of isoprenoid biosynthesis. *J. Immunol.* **166**, 3655–3658.

Baggiolini, M., Dewald, B. and Moser, B. (1997). Human chemokines: An update. *Annu. Rev. Immunol.* **15**, 675–705.

Belmant, C., Espinosa, E., Poupot, R., Guiraud, M., Poquet, Y., Bonneville, M. and Fournie, J. J. (1999). 3-Formyl-1-butyl pyrophosphate: a novel mycobacterial metabolite activating human γδ T cells. *J. Biol. Chem.* **274**, 32079–32084.

Carroll, M. C. (1998) The role of complement and complement receptors in induction and regulation of immunity. *Annu. Rev. Immunol.* **16**, 545–568.

Chan, E. D., Chan, J. and Schluger, N. W. (2001). What is the role of nitric oxide in murine and human host defense against tuberculosis? Current knowledge. *Am. J. Respir. Cell Mol. Biol.* **25**, 606–612.

Collins, H. L. and Kaufmann, S. H. E. (2000). Acquired immunity against bacteria. In: *Immunology of Infectious Diseases* (S.H.E. Kaufmann, A. Sher and R. Ahmed, Eds), pp. 207–222. ASM Press, Washington, DC.

D'Souza, C. D., Cooper, A. M., Frank, A. A., Mazzaccaro, R. J., Bloom, B. R. and Orme, I. M. (1997). An anti-inflammatory role for γδ T lymphocytes in acquired immunity to *Mycobacterium tuberculosis*. *J. Immunol.* **158**, 1217–1221.

D'Souza, M. P. and Harden, V. A. (1996). Chemokines and HIV-1 second receptors. Confluence of two fields generates optimism in AIDS research. *Nature Med.* **2**, 1293–1300.

Ferrick, D. A., Schrenzel, M. D., Mulvania, T., Hsieh, B., Ferlin, W. G. and Lepper, H. (1995). Differential production of interferon-γ and interleukin-4 in response to Th1- and Th2-stimulating pathogens by γδ T cells *in vivo*. *Nature* **373**, 255–257.

Finlay, B. B. and Cossart, P. (1997). Exploitation of mammalian host cell functions by bacterial pathogens. *Science* **276**, 718–725.

Foy, T. M., Aruffo, A., Bajorath, J., Buhlmann, J. E. and Noelle, R. J. (1996). Immune regulation by CD40 and its ligand gp39. *Annu. Rev. Immunol.* **14**, 591–617.

Grewal, I. S., and Flavell, R. A. (1996). A central role of CD40 ligand in the regulation of CD4+ T-cell responses. *Immunol. Today* **17**, 410–414.

Hayday, A. C. (2000). γδ cells: a right time and a right place for a conserved third way of protection. *Annu. Rev. Immunol.* **18**, 975–1026.

Hughes, A. (2000). Evolution of the host defense system. In: *Immunology of Infectious Diseases* (S. H. E. Kaufmann, A. Sher and R. Ahmed, Eds), pp. 67–78. ASM Press, Washington.

Kabelitz, D. and Wesch, D. (2001) Role of γδ T-lymphocytes in HIV infection. *Eur. J. Med. Res.* **6**, 169–174.

Kabelitz, D., Wesch, D. and Hinz, T. (1999). γδ T cells, their TCR usage and role in human diseases. *Springer Semin. Immunopathol.* **21**, 55–75.

Kabelitz, D., Glatzel, A. and Wesch, D. (2000). Antigen recognition by human γδ T lymphocytes. *Int. Arch. Allergy Immunol.* **122**, 1–7.

Kaufmann, S. H. E. (1996). γ/δ and other unconventional T lymphocytes: What do they see and what do they do? *Proc. Natl. Acad. Sci. USA* **93**, 2272–2279.

Kaufmann, S. H. E. (1999). Killing versus suicide in antibacterial defence. *Trends Microbiol.* **7**, 59–61.

Kaufmann, S. H. E. (2001). How can immunology contribute to the control of tuberculosis? *Nat. Immunol. Rev.* **1**, 20–30.

Kuhn, M., Goebel, W., Philpott, D. J. and Sansonetti, P. J. (2002). Overview of the bacterial pathogens. In: *Immunology of Infectious Diseases* (S. H. E. Kaufmann, A. Sher and R. Ahmed, Eds), pp. 5–24. ASM Press, Washington, DC.

Ladel, C. H., Blum, C., Dreher, A., Reifenberg, K. and Kaufmann, S. H. E. (1995). Protective role of γ/δ T cells and α/β T cells in tuberculosis. *Eur. J. Immunol.* **25**, 2877–2881.

Lenz, L. L. and Bevan, M. J. (1996) H2-M3-restricted presentation of *Listeria monocytogenes* antigens. *Immunol. Rev.* **151**, 107–121.

Lieu, P. T., Heiskala, M., Peterson, P. A. and Yang, Y. (2001) The roles of iron in health and disease. *Mol. Aspects Med.* **22**, 1–87.

Liszewski, M. K., Farries, T. C., Lublin, D. M., Rooney, I. A. and Atkinson, J. P. (1996). Control of the complement system. *Adv. Immunol.* **61**, 201–283.

Löhning, M., Stroehmann, A., Coyle, A. J., Grogan, J. L., Lin, S., Gutierrez-Ramos, J. C., Levinson, D., Radbruch, A. and Kamradt, T. (1998). T1/ST2 is preferentially expressed on murine Th2 cells, independent of interleukin 4, interleukin 5, and interleukin 10, and important for Th2 effector function. *Proc. Natl. Acad. Sci. USA* **95**, 6930–6935.

MacMicking, J., Xie, Q. and Nathan, C. (1997). Nitric oxide and macrophage function. *Annu. Rev. Immunol.* **15**, 323–350.

Moser, B. and Loetscher, P. (2001). Lymphocyte traffic control by chemokines. *Nat. Immunol.* **2**, 123–128.

Mosmann, T. R. and Fowell, D. J. (2002). The Th1/Th2 paradigm in infections. In: *Immunology of Infectious Diseases* (S. H. E. Kaufmann, A. Sher and R. Ahmed, Eds), pp. 163–174. ASM Press, Washington, DC.

Munoz-Elias, E. J. and McKinney, J. D. (2002). Bacterial persistence: strategies for survival. In: *Immunology of Infectious Diseases* (S. H. E. Kaufmann, A. Sher and R. Ahmed, Eds), pp. 331–356. ASM Press, Washington, DC.

Neutra, M. R. and Kraehenbuhl, J. P. (2002). Regional immune response to microbial pathogens. In: *Immunology of Infectious Diseases* (S. H. E. Kaufmann, A. Sher and R. Ahmed, Eds), pp. 191–206. ASM Press, Washington, DC.

Pearce, E. J. and Tarleton, R. L. (2002). Overview of the parasitic pathogens. In: *Immunology of Infectious Diseases* (S. H. E. Kaufmann, A. Sher and R. Ahmed, Eds), pp. 39–52. ASM Press, Washington, DC.

Pfefferkorn, E. R. (1984). Interferon-γ blocks the growth of *Toxoplasma gondii* in human fibroblast by inducing the host cells to degrade tryptophan. *Proc. Nat. Acad. Sci. USA* **81**, 908–912.

Rajewsky, K. (1996). Clonal selection and learning in the antibody system. *Nature* **381**, 751–758.

Reis e Sousa, C. (2001). Dendritic cells as sensors of infection. *Immunity* **14**, 495–498.

Rogge, L., Barberis-Maino, L., Biffi, M., Passini, N., Presky, D. H., Gubler, U. and Sinigaglia, F. (1997). Selective expression of an interleukin-12 receptor component by human T helper 1 cells. *J. Exp. Med.* **185**, 825–831.

Rolph, M. S. and Kaufmann, S. H. E. (2000). Partially TAP-independent protection against *Listeria monocytogenes* by H2-M3-restricted CD8+ T cells. *J. Immunol.* **165**, 4575–4580.

Romani, L. (2002). Overview of the parasitic pathogens. In: *Immunology of Infectious Diseases* (S. H. E. Kaufmann, A. Sher and R. Ahmed, Eds), pp. 25–38. ASM Press, Washington, DC.

Sallusto, F., Mackay, C. R. and Lanzavecchia, A. (2000). The role of chemokine receptors in primary, effector, and memory immune responses. *Annu. Rev. Immunol.* **18**, 593–620.

Salomon, B. and Bluestone, J. A. (2001) Complexities of CD28/B7: CTLA-4 costimulatory pathways in autoimmunity and transplantation. *Annu. Rev. Immunol.* **19**, 225–252.

Schaible, U. E. and Kaufmann, S. H. E. (2000). CD1 and CD1-restricted T cells in infections with intracellular bacteria. *Trends Microbiol.* **8**, 419–425.

Schaible, U. E., Collins, H. L. and Kaufmann, S. H. E. (1999) Confrontation between intracellular bacteria and the immune system. *Adv. Immunol.* **71**, 267–377.

Scott, P. and Grencis, R. K. (2002). Adaptive immune effector mechanisms against intracellular protozoa and gut-dwelling nematodes. In: *Immunology of Infectious Diseases* (S. H. E. Kaufmann, A. Sher and R. Ahmed, Eds), pp. 235–246. ASM Press, Washington, DC.

Stavnezer, J. (1996). Antibody class switching. *Adv. Immunol.* **61**, 79–146.

Swain, S. L. (2001) Interleukin 18: tipping the balance towards T helper cell 1 response. *J. Exp. Med.* **194**, F11–F14.

Szabo, S. J., Dighe, A. S., Gubler, U. and Murphy, K. M. (1997). Regulation of the interleukin (IL)-12Rβ2 subunit expression in developing T helper 1 (Th1) and Th2 cells. *J. Exp. Med.* **185**, 817–824.

Szomolanyi-Tsuda, E., Brehm, M. A. and Welsh, R. M. (2002). Acquired immunity against fungi. In: *Immunology of Infectious Diseases* (S. H. E. Kaufmann, A. Sher and R. Ahmed, Eds), pp. 247–265. ASM Press, Washington, DC.

Wang, L., Kamath, A., Das, H., Li, L. and Bukowski, J. F. (2001). Antibacterial effect of human Vγ2Vδ2 T cells in vivo. *J. Clin. Invest.* **108**, 1349–1357.

Watts, C. (1997). Capture and processing of exogenous antigens for presentation on MHC molecules. *Annu. Rev. Immunol.* **15**, 821–850.

Wesch, D., Glatzel, A. and Kabelitz, D (2001). Differentiation of resting human peripheral blood γδ T cells toward Th1- or Th2-phenotype. *Cell. Immunol.* **212**, 110–117.

Yewdell, J. W. and Bennink, J. R. (2002). Overview of the viral pathogens. In: *Immunology of Infectious Diseases* (S. H. E. Kaufmann, A. Sher and R. Ahmed, Eds), pp. 53–64. ASM Press, Washington, DC.

Zlotnik, A. and Yoshie, O. (2000) Chemokines: a new classification system and their role in immunity. *Immunity* **12**, 121–127.

1 Phenotyping and Separation of Leukocyte Populations Based on Affinity Labelling

Alexander Scheffold and Andreas Radbruch
Deutsches Rheuma-Forschungtszentrum (DRFZ), Berlin, Germany

Mario Assenmacher
Miltenyi Biotec, Bergisch Gladbach, Germany

◆◆

CONTENTS

Introduction
Affinity-based fluorescent labelling
Cell sorting based on affinity labelling

◆◆◆◆◆◆ INTRODUCTION

The combination of monoclonal antibody technology with flow cytometry provides a powerful tool for detailed molecular phenotyping and isolation of individual cells according to the expression of specific proteins at resolutions down to single amino acid differences in protein sequence.

Specific fluorescent or magnetic labelling comprises not only antibodies but can be extended to all kinds of specific high-affinity ligand–receptor interactions and we will therefore replace 'immunolabelling' by the more general term 'affinity labelling'. We will only focus on techniques that allow quantitative labelling, i.e. labelling proportional to 'antigen' density. Quantitative labelling can be achieved with free ligands or ligands conjugated to colloidal magnetic particles (MACS-System (Miltenyi *et al.*, 1990)) or liposomes and will here be termed 'staining'. In contrast, larger particles in the micrometer range will allow only qualitative labelling ('all or nothing').

Cytometry, that is, affinity-based phenotyping in its traditional and more avantgardistic forms, which will be introduced here, allows specific quantification of nucleic acids, intracellular, surface and secreted proteins,

Copyright © Elsevier Science Ltd
All rights of reproduction in any form reserved

hormones and sugars. This is the analytical basis for the preparative approaches of fluorescence-activated cell sorting (FACS) and high-gradient magnetic cell separation (MACS), which allow isolation of sub-populations and individual cells of defined phenotype down to frequencies of less than one in 10^7 cells (Radbruch and Recktenwald, 1995).

◆◆◆◆◆◆ AFFINITY-BASED FLUORESCENT LABELLING

Basic considerations

Available parameters

Light scatter

Cytometry allows analysis of single cells according to light scattering and emission of fluorescent light. State-of-the-art flow-cytometers (such as, Becton-Dickinson's FACS-series Coulter´s ELITEs, Partec´s PAS and CyFlow or Cytomation´s CYAN) are detecting light scatter at an angle of 2–20° (forward scatter, FSC) and 90° (side scatter, SSC) relative to the axis of the illumination. Light scatter gives information about cell size (FSC) and granularity (SSC), allowing optical separation of leukocyte sub-populations, like monocytes, granulocytes, lymphocytes, cellular debris and cell aggregates.

Fluorescence

Apart from scattered light, flow cytometers detect fluorescent light, emitted from fluorochromes upon excitation by the illuminating light. The number of parameters (colours) is restricted by the number of available dyes which fulfil the following criteria: they must be excitable by the illuminating light source (usually one or more lasers with different emission wavelengths (see below)), they should provide high quantum yields (ratio of absorbed to emitted photons), their fluorescence emission has to be distinguishable from that of the other dyes used and it should be possible to conjugate them to proteins.

Research cytometers exist, which use up to three different light sources, e.g. krypton, argon, helium-neon, dye or diode lasers and mercury lamps, which allow simultaneous analysis of up to 11 fluorescent dyes (Baumgarth and Roederer, 2000). Here, we will focus on the standard combination of four different dyes using either an argon-laser (488 nm) alone or in combination with a second red laser (red diode or HeNe-laser, ~635 nm) as light sources.

Fluorescein (FL, excitation maximum (EX_{max}) 495 nm, maximum of emission (Em_{max}) 519 nm), a dye which can easily be conjugated to proteins, is commonly used together with the algal phycobiliprotein phycoerythrein (PE, EX_{max} 480 nm, 545 nm, 565 nm, EM_{max} 575 nm). Recently, alternative low-molecular weight dyes, e.g. Cy-2 or Alexa 488 with increased photostability and fluorescence intensity as compared with fluorescein have been introduced. PE displays brighter fluorescence than FL, due to its higher absorption and quantum yield. Another phyco-biliprotein, peridinin chlorophyll-a (PerCP, EX_{max} 470 nm, EM_{max} 680 nm), or tandem conjugates of PE with Cy5 (EX_{max} 650 nm, EM_{max} 666 nm) or related dyes can be used as third colour. Energy-transfer systems like PE/Cy5 absorb light via PE and the energy of the emitted light is directly 'transferred' to excite fluorescence emission of the second dye emitting, then fluorescence at longer wavelength. This results in wide separation of excitation and emission wavelength (Stokes´ shift). In addition, another phycobiliprotein, Allophycocyanin (APC, EX_{max} 650 nm, EM_{max} 660 nm), or the small molecular weight dyes Cy-5 (EX_{max} 644 nm, EM_{max} 665 nm) or Alexa 647 (EX_{max} 650 nm, EM_{max} 668 nm) which are excited by the red laser can be used as the fourth fluorescent label. An overview of standard dyes and their possible combinations for four colour measurement using a single argon-laser (Coulter 'ELITE') or an additional red laser (Becton-Dickinson's 'FACSCalibur' or 'LSR', Cytomation´s CyAn, Partec´s 'PAS'-series) is given in Table 1.

Alternatively, the third or fourth colour is often used to identify dead cells according to uptake of propidium-iodide (PI, EX_{max} 536 nm, Em_{max} 617 nm) for their exclusion from further analysis. PI exclusively enters dead cells via their damaged cell membranes and intercalates into DNA. It is excluded from viable cells with intact cell membranes. PI and other DNA stains can be used conveniently to correlate phenotype with stage of cell cycle, proliferation or apoptosis (Darzynkiewicz and Crissman, 1990; Ormerod, 1994).

The third colour can also be used to identify nucleated cells by staining with LDS (see protocol section), a vital DNA dye emitting in the far red, for example, for optical separation of leukocytes in the presence of an excess of non-nucleated cells or particles (erythrocytes, cellular debris). Other dyes are available for correlation of phenotype to biochemical parameters, like redox potential, pH or calcium influx (Rothe and Valet, 2000).

Benchtop cytometers using an additional UV laser have been recently introduced (BD 'LSR', Cytomation 'CYAN', Partec 'Cy-Flow') which allow combination of DNA measurement and immunofluorescence. However, conventional UV lasers (350 nm, 325 nm) are of limited use for immuno-fluorescence applications. Due to the high cellular autofluorescence (see 'Sensitivity' below) in that spectral region only strongly expressed antigens can be visualized using UV-excitable dyes, e.g. Alexa350.

Detection limit

Besides technical limitations, the detection limit of immunofluorescence is strongly influenced by cellular autofluorescence (see 'Sensitivity', below). For lymphocytes, several thousand surface molecules per cell are required for cytometric detection by conventional fluorochrome conju-gated antibodies (see below). For analysis of molecules expressed in lower frequencies, like cytokine receptors or surface cytokines enhancing tech-nologies have to be used (Assenmacher et al., 1996, Scheffold et al., 2000), e.g. magnetofluorescent liposomes (see later).

Table I.

Dye	Abbreviation, commercial names	EX$_{max}$ [nm]	EM$_{max}$ [nm]	Light source	Application
Alexa 350	Alexa 350	346	442	351 nm Ar-laser or 325 nm HeCd-laser	Additional colour for immuno-labelling using UV excitation
Fluorescein	FL	495	519	488 nm Ar-laser	Standard F1
Dialkylcarbocyanin-derivative (Cy-2)	Cy-2	489	505	488 nm Ar-laser	F1
Alexa-488	Alexa-488	491	515	488 nm Ar-laser	F1
Phycoerythrin	PE	480, 545, 565	575	488 nm Ar-laser	Standard F2
Dialkylcarbocyanin-derivative (Cy-5)	Cy-5	650	666	635 nm red diode laser or HeNe-laser	F4 for four-colour fluorescence (dual laser)
Alexa 647	Alexa 647	650	668	635 nm red diode laser or HeNe-laser	F4 for four-colour fluorescence (dual laser)
Phycoerythrin/Cy5	CyChrome, Tricolor, Red670	like PE + Cy5	666	488 nm Ar-laser + 635 nm red diode laser or HeNe-laser	F3 for three-colour fluorescence, F4 for four-colour fluorescence with single argon laser, *not* with red diode laser
Phycoerythrin/Sulforhodamine 101	PE / Texas Red, ECD	like PE + 596	625	488 nm Ar-laser	F3 for four-colour fluorescence single argonlaser
Peridinin-chlorophyll-a	PerCP	470	680	488 nm Ar-laser	F3 for four-colour fluorescence with Cy5 and APC or APC/Cy7
Peridinin-chlorophyll-a-Cy5.5	PerCP-Cy-5.5	470	695	488 nm Ar-laser	Like PerCP, enhanced photostability and fluorescence intensity
Allophycocyanin	APC	650	660	635 nm red diode laser or HeNe-laser	F4 for four-colour fluorescence (dual laser)
Allophycocyanin Cy7	APC – Cy7	650	767	635 nm red diode laser or HeNe-laser	F4 for four-colour fluorescence (dual laser) or in combination with APC for multicolour applications

Cell surface molecules

The most frequent application of affinity fluorescence is staining of surface molecules. Many of them are classified according to the CD (cluster of differentiation) nomenclature (Mason, 2002). Surface molecules, which include structural, transport and communication molecules, already provide detailed information about the functional and differentiation status of a cell. They can be stained on viable and on fixed cells, depending on the fixation protocol, and can be used to isolate live cells by affinity-label-based cell sorting with MACS or FACS.

Intracellular molecules

Since currently available affinity labels, like antibodies, cannot penetrate intact cell membranes, intracellular molecules are not accessible in live cells. Intracellular molecules of interest include molecules involved in signal transduction, gene regulation, physiology, and also molecules *en route* to secretion or surface display. For intracellular staining, cells have to be fixed, to preserve their structural integrity, and membranes have to be permeabilized, allowing the affinity labels to reach the intracellular space (see later). For analysis of molecules not expressed in all cells or expressed transiently, analysis by fluorescence and cytometry is an interesting technological option. Expression is quantified on the level of individual cells, the kinetics of expression can be determined, as well as the frequency of expressing cells, and expression can be correlated to expression of other markers of differentiation and function (Assenmacher, 2000).

Secreted molecules

Cells can be analysed and sorted according to expression of a defined secreted molecule, using the cytometric secretion assay, also termed cellular affinity matrix technology (Manz *et al.*, 1995; see below and Chapter I.2 by Assenmacher *et al.*). The concept is to create an artificial affinity matrix on the surface of all cells of a given population, by fixing an antibody or ligand specific for the secreted molecule on the cell membrane. This procedure does not affect cell viability. The labelled cells are then assayed for secretion *in vitro*. During secretion *in vitro*, secreted molecules bind to the affinity matrix on the surface of the secreting cells. Diffusion of secreted molecules to non-secreting cells is prevented by using a medium of high viscosity, plating the cells at low cell density, using short secretion times and adding soluble specific antibodies to the incubation medium. After washing, the secreted molecules can be stained like conventional surface antigens, using a second antibody or ligand, recognizing a different epitope than the affinity matrix antibody.

The cytometric secretion assay has been used for the identification of plasma cells (Manz *et al.*, 1997), the separation of Ig-producing hybridoma cells and for the isolation of cytokine producing T-cell subsets (Manz *et al.*, 1995). Recently, this technology has been adapted to the identification and isolation of viable, antigen-specific memory T cells according to cytokine secretion in reaction to re-stimulation with antigen (Brosterhus *et al.*, 1999). Because of the general interest of this application the details of the technique are described in the chapter by Assenmacher *et al.* in Section I.

Sensitivity

Sensitivity of affinity fluorescence depends on the physical constraints of the cytometric hardware, which is discussed in detail elsewhere (Shapiro, 1988; Melamed *et al.*, 1990). Fluorescence allows very sensitive detection, as the wavelength of the emission is increased compared with that of the exciting light, which allows background radiation to be eliminated by optical filters. In addition, fluorescence is measured perpendicular to the beam of exciting light, minimizing the amount of scattered light intensity. Further reduction of background light is achieved by setting a trigger threshold, usually FSC, according to forward scatter light, restricting the time of measurement.

In cytometry, light is detected by photomultiplier tubes (PMTs), which are able to detect single photons. The light signal is amplified in a linear or logarithmic scale.

While linear amplification is used in the case of signals which differ little in a defined range of intensity, like forward and side scatter or staining of DNA, signals with high dynamic range are recorded in logarithmic scales spanning 4–5 decades. This is usually the case for affinity labelling of cells, when up to 1000-fold differences among the cells analysed occur routinely and variables like volume and surface influence the staining.

The overall sensitivity of the cytometric instrument can be defined as the minimal number of dye molecules sufficient to separate two populations of standard beads (in general around 500 FL molecules). These particles contain defined numbers of FL equivalents and show minimal variation in signal intensity (coefficient of variation (CV)).

The sensitivity of measurement is also influenced by autofluorescence of the cells. Fluorescence is a general property of many organic compounds, mostly aromatic or polyunsaturated molecules, which are present to varying degrees in all cells, generating 'cellular auto-fluorescence'. Autofluorescence correlates with size and physiology of the cell, e.g. phagocytic activity. The intensity of affinity staining must reach at least the order of magnitude of autofluorescence to be detectable. The sensitivity of affinity labelling is strongly influenced by the number of fluorochromes conjugated to the affinity label. The number of dye molecules that can be attached to a single cellular epitope for the label increases in the order of direct reagents → indirect reagents → fluorescent particles, e.g. magnetofluorescent liposomes (see later).

Quantification

Ligand number

In flow cytometry, the fluorescent label of each cell is measured quantitatively over a large range, usually four logarithmic decades. Due to the limitation of sensitivity outlined above, the intensity values are relative, rather than absolute. Quantitative statements require standardization of the measurement. Standardization with unstained cells as reference for background intensity is a prerequisite for determination of relative fluorescence intensities. To determine the absolute numbers of stained target epitopes, the conjugation rate of fluorochromes to the affinity label has to be known, and the relative insight of fluorescence has to be calibrated to an absolute standard, i.e. calibration particles with known numbers of fluorochromes per bead. However, information on conjugation rates or affinity labels may not be readily available. Conjugation may also change the fluorescence of dyes and the affinity and avidity of binding to the target epitope of the label in an undefined way. To overcome these problems, calibration particles are nowadays used which have defined binding capacity for the affinity label in question. The fluorescence intensity of such beads, in comparison to the cells to be analysed, both stained under saturating conditions to eliminate differences in affinity and valency, can be directly translated into the corresponding number of cellular surface antigens. Quantification kits using this type of particles are available from various suppliers (e.g. 'Quantum Simply Cellular', Sigma, St Louis, MI, USA).

Cell frequency and cell count

The frequencies of stained subpopulations, as identified according to staining by affinity labels and light scatter, can easily be obtained by standard statistical analysis (see later). The large cell numbers that can be analysed by flow cytometry are indeed required to determine efficiently the frequencies of rare cells (Radbruch and Recktenwald, 1995).

Except for 'Partecs', 'CyFlow' and 'PAS' series, the sample volume is not an exactly defined parameter in flow-cytometric analysis. With other cytometers, absolute cell numbers can be determined by adding a defined number of easily identifiable calibration beads to a defined total volume of sample. The relative frequencies of beads versus cells can be used then to calculate the absolute cell number relative to the known number of beads in the sample (e.g. 'TRUCOUNT', Becton Dickinson, San Jose, CA, USA).

Staining reagents

Cells can be stained by affinity labels in two ways, either by using affinity ligands to which a fluorescent dye is covalently conjugated ('direct staining'), or by fluorescent reagents that stain specifically the primary affinity ligand ('indirect staining'). Such 'secondary' reagents can recognize 'haptens' like biotin, digoxigenin or nitrophenyl groups, which can easily be conjugated to proteins, and are detectable by avidin/streptavidin or specific antibodies. Frequently, antibodies are used which are directed against iso-, allo- or xenotype determinants of the heavy- or light-chain constant regions of the first, labelling antibody. Liposomal or colloidal magnetic particles conjugated to 'direct' or 'indirect' affinity ligands differ in various aspects from molecular staining reagents and are discussed separately.

Direct staining

Direct staining is preferable for routine purposes and commonly used antigens. It requires minimal manipulation of the cells, which is important to maintain cellular viability and to minimize loss of cells. It is easy to control, and it allows simultaneous labelling with different affinity labels. A multitude of antibodies conjugated to standard fluorochromes is commercially available from a large number of suppliers, or can easily be prepared from purified proteins (see later). Staining is controlled by antibodies of the same xeno- and isotype, but irrelevant specificity, conjugated to the same fluorochrome at preferably the same ratio (isotype control), by staining cells which do not express the respective antigen (cellular control) (see later) or by blocking the specific staining with an excess of the same unlabelled antibody or the soluble antigen (blocking control). It should be noted that the isotype control has to be handled with care. Since it is a different reagent, it may develop problems of its own, such as background staining due to aggregation or degradation.

Indirect staining

Compared with direct staining, indirect staining can provide higher sensitivity since several secondary antibodies can bind to one primary antibody, thereby increasing the number of fluorochromes per target molecule. In addition, second step reagents conjugated to exotic fluorescent dyes or protein dyes like phycobiliproteins (PE, PerCP, APC or Cychrome) are commercially available and provide variability and convenience of multiparameter staining.

Hapten systems are superior to allo-, iso- or xenotype-specific antibodies in terms of specificity and convenience, and they are easy to control. For the most common hapten–ligand system, biotin–streptavidin, a vast array of streptavidin–fluorochrome conjugates is commercially available.

Allo-, iso-, or xenotype-specific antibodies may be used as secondary reagents of primary antibodies which cannot or have not yet been purified from culture supernatant, ascites or sera. Although sensitive, secondary antibodies, especially polyclonal ones, have some drawbacks with respect to options for multiparameter staining, specificity and cross-reactivity. With respect to multiparameter staining, the various antigens have to be stained sequentially with the primary and secondary antibodies if the different primary antibodies happen to be of the same subclass. Free binding sites of each secondary antibody for one antigen have to be blocked, before the primary antibody for the next antigen is added. All staining and blocking steps have to be controlled individually, a tedious and complicated procedure.

Another problem of polyclonal and monoclonal iso-, allo-, or xenotype-specific-antibodies are low-affinity cross-reactivities with Ig of other types, e.g. anti-rat IgG cross-reacting with mouse IgG. Such cross-reactivities are the rule rather than the exception. For polyclonal antibodies, cross-reactive fractions of antibodies can be removed before staining by absorption on cells expressing the cross-reactive antigen or on affinity sorbents (see later). For cross-reactive monoclonal antibodies this option does not exist and such antibodies can therefore not be used.

Labelling with particles

Particles conjugated to affinity ligands show increased steric hindrance and reduced diffusion rates depending on their size, and this may require prolonged incubation times and result in non-quantitative labelling. Very small particles, like liposomes and colloidal magnetic particles, still allow quantitative, although not saturating labelling, with only a slight increase in staining time (see below).

Magnetic colloids and multiparameter sorting

Magnetic colloids, such as those used for high-gradient magnetic cell sorting, are superparamagnetic dextran-coated particles of irregular shape with a diameter of 50–100 nm, which is about 5–10 times the diameter of an IgG antibody. They bind to cells more slowly than protein ligands and are used under non-saturating labelling conditions, at staining times of 15 min. The weak magnetic label is sufficient to retain the cells in high-gradient magnetic fields. Remaining free epitopes allow fluorescent labelling for analysis of sorting. The bound particles do not interfere with cytometric evaluation or biological function.

An initial drawback of magnetic cell sorting, as compared with fluorescence-activated sorting, had been its restriction to one parameter. Today, however, magnetic beads are available which can be released enzymatically from the cell surface after sorting, allowing sequential high-gradient magnetic cell separation according to several parameters. An example of a magnetic two-parameter sort is given later.

Magnetofluorescent liposomes

Magnetofluorescent liposomes of defined size of 200–400 nm diameter, when conjugated to affinity ligands and filled with fluorochromes and magnetic particles, can enhance drastically the sensitivity of fluorescent labelling and at the same time allow magnetic separation of the cells. The magnetic label is also useful for sizing liposomes, i.e. depletion of small liposomes which could interfere negatively with efficient labelling by large liposomes (Scheffold *et al.*, 1995). Staining of cells with liposomes requires prolonged staining times of 30–60 min and gentle agitation during staining in order to maximize the probability for the liposomes hitting a cell.

Conjugation of fluorochromes and haptens to proteins

Basic principles

For convenient conjugation of proteins to small fluorochromes or haptens, activated forms of these molecules are required that react with available primary amino groups of the protein to form stable covalent bonds. Succinimidyl esters of fluorochromes and haptens are commercially available which react efficiently at pH 8.5, i.e. without damaging the protein. Whenever possible, haptens with an extra spacer arm of about six C-atoms should be used to reduce steric hindrance between protein and hapten-specific proteins. Sometimes isothiocyanate (ITC) derivatives of fluorochromes are used, such as fluorescein-ITC (FITC). This requires a pH greater than 9 which can damage some proteins. The protocols given below for conjugation of succinimidyl esters can also be used for isothiocyanates, using 0.1 M boric acid, 0.025 M sodium borate, 0.075 M sodium chloride, pH 9.5 (adjust by NaOH) as buffer for conjugation. Antibodies should be exposed to such high pH for as short a time period as possible.

Relevant parameters

Stability Succinimidyl esters are sensitive to hydrolysis and therefore should be stored dry in a desiccator at –20°C. Stored aliquots should be equilibrated to room temperature before opening. Alternatively, succinimidyl esters can be stored dissolved in *water-free* DMSO at high concentration (5 mg ml^{-1}), frozen in aliquots for single use. Aqueous solutions are prepared immediately before conjugation and instantly added to the protein solution. All buffers have to be free of strong nucleophils, like amines (e.g. Tris-buffer), azide or stabilizing proteins which would interfere with the coupling reaction.

Concentrations The conjugation ratio, i.e. number of haptens or fluorochromes conjugated to a single protein molecule, depends on (a) the absolute concentration of the reagents and (b) the molar ratio of dye or hapten to protein. Protein concentrations are optimal at about 1 mg ml^{-1} and dye concentration should be varied to determine the optimal ratio (for IgG, molar ratios of 10:1 to 50:1 usually give best results). Low conjugation ratios of 1–2 fluorochromes per protein molecule provide only low signal intensity, while ratios of more than 10 result in unspecific hydrophobic interaction with cells, inactivation of the binding site of the affinity reagent, or its precipitation. Precipitates can be removed by centrifugation (12 000*g*, 10 min).

> ### Conjugation of fluorochrome- and hapten-succinimidyl esters to proteins
>
> 1. Transfer protein into conjugation buffer (PBS or 0.1 M sodium carbonate pH 8.5) by dialysis or gel filtration on PD-10 (up to 2 ml) or NAP5 (up to 0.5 ml) columns (Pharmacia, Uppsala). Adjust concentration to about 1 mg ml^{-1} (at lower protein concentrations the conjugation will be suboptimal).
> 2. Dissolve succinimidyl esters in DMSO at 5 mg ml^{-1}. Water soluble 'sulfo'-reagents can be dissolved in buffer (precipitates may be removed by centrifugation, 1 min at 12 000g).
> 3. *Immediately* add as much of the dissolved succinimidyl ester to the protein solution as is required to obtain the appropriate molar ratio (e.g. for IgG start with a molar ratio of 20 : 1, see above) and mix.
> 4. After 1 h at room temperature, remove free fluorochromes or haptens by dialysis against PBS/0.05% sodium azide or by gel filtration.
> 5. The protein concentration and the average molar ratio of FL to protein (FL/P) in the conjugate can be determined according to the OD at 280 nm and the EX$_{max}$ of the fluorochrome (formula see below). Haptens like biotin or digoxigenin displaying no specific light absorption, require more tedious determination of the conjugation ratio. Alternatively such conjugates are tested only functionally, e.g. by titration (see Fig.1).
>
> Determination of conjugation ratio for IgG (MW 150 000):
>
> Fluorescein: FL/P = 2.9 × OD 495 / (OD 280 – 0.35 × OD 495)
> mg ml^{-1} = (OD 280 – 0.35 × OD 495) / 1.4
> Cy-5: FL/P = 0.68 × OD 650 / (OD 280 – 0.05 × OD 650)
> mg ml^{-1} = (OD 280 – 0.05 × OD 650) / 1.4
>
> 6. Test conjugates functionally by titration on test cells, as described below, and freeze stock solution in small aliquots at –70°C. Do not freeze/thaw repeatedly.
>
> ### Conjugation of phycobiliproteins to proteins
>
> The controlled cross-linking of proteins and purification of conjugates will not be discussed. Details can be found elsewhere (Hermanson, 1996). An easy protocol is provided in Mueller (2000). Detailed conjugation protocols are provided at http://www.drmr.com/abcon/index.html

Staining parameters

Variables of immunofluorescent labelling, such as cell type, viability, concentrations of cells and reagents, time, temperature, and buffer or labelling, have to be calibrated to obtain optimal and reproducible results. In general, those conditions should be selected that give the brightest

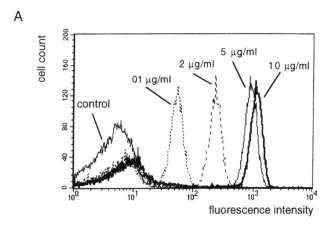

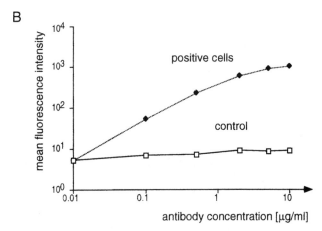

Figure 1. Titration of an anti CD4-PE conjugated antibody for staining human peripheral blood lymphocytes. The cells are stained with the concentrations indicated. (A) Plots of the fluorescence intensity versus the counted events; plots for all stainings are overlaid in one histogram. (B) Plots of the mean fluorescence intensities of the positive and negative populations versus the concentration of antibody; within the optimum range of concentrations (5–10 µg ml⁻¹) the staining is saturating, with minimum background.

positive signal possible at minimal background labelling. Cells should be treated gently with as few as possible steps of manipulation, to minimize stress for the cells, which could result in low viability, selective loss of particular cell types and artificial expression of antigen by the remaining cells.

Volume

The concentration of cells in the staining volume should be chosen such that cell viability is not impaired and that the staining reaction can be adapted to variable cell numbers by simply changing the volume of reaction. Neither the concentration of the label, nor the concentration of cells should be changed to provide reproducible standard conditions for the staining reaction. As a rule of thumb, up to 10^7 cells can be stained in 100 μl. 100 μl of a 5 μg/ml solution of antibody contain approximately 2×10^{12} antibody molecules, sufficient to label 10^7 cells for antigens expressed at a frequency of 10^5 per cell. In principle, the staining volume has to be adjusted to the number of positive cells and antigen density per cell. For rare positive cells, however, the volume of the majority of negative cells will limit the volume of the staining reaction. The practical lower limit of the cell numbers for cytometric analysis is about 10^5 cells per sample, if 10^4 cells are to be analysed, because there is always loss of cells due to washing and labelling. For routine analysis, 5×10^5 to 5×10^6 cells should be used or even more, if rare cells are cells of interest. The staining volume should be 50–100 μl, but not less than 30 μl, otherwise the staining reagents cannot be added in precise concentrations.

Concentration of reagents

Concentration of the staining reagents has a strong influence on signal intensity and background labelling. Reagents should be used at optimal concentration, i.e. those with the highest signal-to-noise ratio. Too low concentrations result in incomplete staining and low signal intensity, and thus yield suboptimal discrimination of positive and negative cells. Too high concentrations may result in unspecific staining, due to low affinity cross-reactions of the labelling reagent. The optimal concentration of each reagent has to be estimated by cytometric titration. Mixtures of positive and negative cells are stained under standard conditions, testing concentrations in the range of 0.1–10 μg ml^{-1} of labelling reagent. The stained cells are analysed by flow cytometry and the optimal concentration is determined by plotting the mean fluorescence intensity (see below) of positive and negative cells versus the concentration of labelling reagent (Figure 1).

Time

Specific binding of antibodies to cellular antigens occurs rapidly. After 1–5 min equilibrium is reached, and 10 min is sufficient for most applications. Longer staining periods may lead to increased background staining due to low-affinity cross-reactions, which have slower kinetics. Long staining periods may be required, however, if liposomes or magnetic beads are used as affinity labels (see below), and also for intracellular staining, when the antibody has to penetrate the cellular membranes and diffuse throughout the cell.

Buffer and temperature

In general, living cells are stained on ice. This reduces physiological reactions of the cells with the label, such as internalization, patching or capping and subsequent shedding of antigen and bound label. Apart from low temperature, sodium azide (0.02–0.05% in the staining buffer) can reversibly block such physiological reactions. The standard staining buffer is PBS, containing 0.5–1% bovine serum albumin (BSA) to saturate unspecific protein binding sites. To block unspecific binding, purified immunoglobulin (0.1–0.5 mg ml^{-1}) can be added to the buffer before or during the first step of staining. Specific blocking antibodies for Fc-receptors are also available and can be used instead.

Washing

During all staining steps, cells should be handled carefully. Damaged and dead cells absorb staining reagents and release 'sticky' DNA, trapping viable cells and clogging nozzles of flow cytometers and MACS columns. It should be avoided to let the cells stand pelleted after centrifugation or to blow air through cell suspensions. Cells should be kept on ice and cell pellets should be resuspended by gentle flicking of the tube, before adding new buffer, and not by pipetting the cells up and down. Washing steps should be minimized to avoid cell loss (10% per washing step) and mechanical stress. One washing step, i.e. addition of a 5- to 10-fold volume of washing buffer, followed by centrifugation at 300 g for 10 min, is sufficient to remove most of the unbound antibody. A second washing step will be necessary for indirect staining, where the primary antibody has to be removed completely, because otherwise it will react with the secondary reagent and reduce its available concentration.

Light

Most fluorescent dyes, especially phycobiliproteins, are sensitive to light. Absorption of light by the dye leads to the generation of reactive oxygen forms, which destroy dye molecules quickly by oxidation. To prevent this 'photobleaching', antibody conjugates and stained cells should be protected from light during the entire procedure whenever possible, and should be kept in the dark at 4°C.

Exclusion of dead cells

Since dead cells impair staining, analysis and sorting, they should be removed before the staining procedure, e.g. by centrifugation on a Ficoll density gradient. For analysis, they should be identified by addition of the DNA dye propidiumiodide (PI), and excluded. This highly fluorescent red dye can penetrate the damaged membranes of dead cells and stain DNA. Due to the fluorescence spectrum of PI, dead cells can be identified according to emission as recorded in F2- and F3-detectors (see below).

In the same way other DNA binding dyes which cannot penetrate viable cell membranes can be used, many of which are excited by UV light, e.g. DAPI (EX_{max} 345 nm, EM_{max} 455 nm) or by red light, e.g. TOTO-3, EX_{max} (642 nm, EM_{max} 661 nm).

Fixation

Fixation of cells before analysis or labelling is often used to limit the risk of infection and standardize the analysis. For intracellular affinity labellings it is necessary to stabilize the cellular structure by fixation and allow permeabilization of the cell membranes. Cell membranes are permeabilized for large molecular weight reagents by detergents. Standard fixation procedures use 0.5–2% formaldehyde if the cellular structure is to be preserved (for protein staining, details in protocols). For analysis of DNA or RNA, which are not well maintained in formaldehyde fixed cells, fixation in 70% methanol or ethanol/acetic acid (95/5 vol./vol.) is preferred, although the dehydration will change phenotype and light scatter. Most targets of labelling are not influenced by either type of fixation. Labelling should be performed after fixation and washing of fixed cells in PBS, to minimize effects of fixation on the label. In the case of high endogenous autofluorescence, fixation may not be advisable, since it increases this autofluorescence. Also, some antigenic determinants, e.g. those containing lysine residues, may be destroyed by reaction with aldehydes and will be no longer recognized by their affinity label. This has to be determined for each antigen. Cells that were dead prior to fixation can be identified cytometrically after fixation, either due to their light scatter or by staining with LDS 751 (Terstappen et al., 1988). However, LDS 751 staining is a tricky business and the technique is not suitable for routine use. After fixation apoptotic cells can also be recognized by PI staining, since their DNA staining is reduced, compared with viable $G_{0/1}$ cells, due to apoptic fragmentation of DNA.

Controls

Affinity labelling has to be controlled in terms of specificity. To this end, a labelling reagent as similar as possible to the reagent to be controlled, but with different specificity, should be used. For antibodies, isotype matched antibodies are routinely used (isotype control). However, this control is dangerous and may be misleading, since the differences in preparation of the proteins and conjugation to the fluorochrome may result in un-predictable differences in background labelling and thus misinterpreta-tion of the affinity labelling. An alternative is to stain cells which are similar to the cells of interest but which do not express the target molecule of the affinity label (cellular control). Obviously the quality of this control depends on the similarity of the two cell types. Genetic mutants are ideal, lacking the respective gene. In the case of transfectants, 'mock' transfected cells are the best control. A very informative specifity control is to block the staining of the affinity label with soluble target molecules, if available, or with unlabelled affinity reagent and control proteins. The latter control, however, can only be used with directly fluorochromated or haptenated affinity reagents.

Data acquisition and analysis

Acquisition

During cytometric data acquisition, the five or six available parameters (FSC, SSC and three or four colours (F1–F4)) are recorded for each particle which triggers measurement, i.e. which exceeds the preset threshold level for one of the parameters. Usually FSC is used to provide the trigger threshold.

Light scatter is analysed at linear amplification, while fluorescence is measured logarithmically, to allow display of up to 10 000-fold differences in intensity on one plot. Logarithmic amplification also emphasizes small differences in low numbers of antigen per cell, which are often of considerable biological relevance because they may determine the functional potential of a cell, reflecting expression or no expression, rather than gradual differences at high expression levels.

Apart from selection by trigger threshold, subsets of cells can be selected for analysis by setting upper and lower thresholds in any parameter. This 'live gating' may be necessary to obtain statistically significant numbers of rare cells for cytometric analysis. However, information about the excluded cells is invariably lost and unexpected or pathological situations may be misjudged. For routine analysis, live gating should be omitted. However, if it is used, an aliquot of the cells should be recorded without live gate, in parallel.

Compensation

Fluorescent light is emitted from a particular fluorochrome over a range of wavelengths characteristic for that dye. For dyes with overlapping emission spectra, like FL and PE, the fluorescence can not always be measured independently for each dye (Fig. 2). Spectral overlap can be corrected by compensation, i.e. subtraction of the relative contribution of the signal from the overlapping dye from the overall signal recorded in that parameter. Conventional flow cytometers have an electronic compensation circuit which can be corrected for spectral overlap, as illustrated in the following example.

The fluorescence intensity of FL, detected at 585 ± 21 nm (F2), is 10% of the fluorescence intensity detected at 530 ± 15 nm (F1). Therefore the correct fluorescence of phycoerythrin, as detected at 585 ± 21 nm (F2), is electronically obtained by subtracting 10% of the F1 signal intensity of the same cell (F2 (corrected) = F2 – 10% F1).

Compensation has to be established for each combination of dyes, by using a mixture of cells individually stained with the various dyes, at different concentrations, i.e. different degrees of brightness, determining the cross-talk of fluorescence emission in the various dyes, and adjusting the electronic compensation accordingly. Standardized calibration particles are also useful. Each dye has to be compensated against all others. Some instruments may not offer compensation for F1 versus F3 (>650 nm) and vice versa, since dyes are available without spectral over-

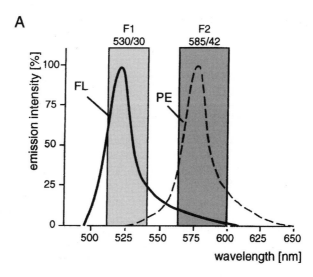

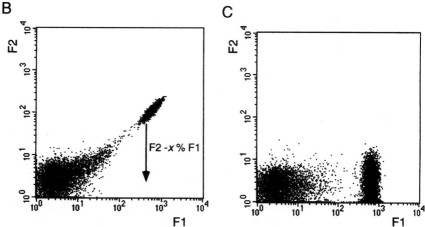

Figure 2. Spectral overlap and compensation. (A) The fluorescence emission spectra of fluorescein (FL) and phycoerythrin (PE), the bandwidths of the analysis filters (shaded bars). A certain percentage of the total fluorescence emission of fluorescein is detected in the F2 channel, and vice versa. (B, C) The effect of spectral overlap in a F1 versus F2 scatter plot for a single colour (F1) staining without (B) and with (C) electronic compensation of the F2 signal with $x\%$ of the measured F1 intensity subtracted from the F2 signal intensity.

lap at these wavelengths. Calibration of compensation has to be done at established PMT voltage. Changing the PMT voltage requires recalibration. Since the detection of fluorescence is not strictly linear over the entire range of four decades of logarithmic signal amplification, compensation is best established with mixtures of dimly and brightly stained cells.

Data evaluation

Plotting and presentation of data

Cytometric data are usually plotted either as a one-parameter histogram or as a two-parameter dot or contour plot. One-parameter histograms emphasize the quantitative aspects of flow cytometry (relative intensities, population size, mean fluorescence, coefficient of variation (CV), for a single parameter). Two-parameter plots provide information on the correlation of antigens expressed: in dot plots, every cell is indicated as a single dot, emphasizing rare cell populations, while large populations are poorly resolved. Contour plots provide good resolution in areas of high cell density, by delineating areas of equal cell density. Depending on the threshold limits, rare cells may not be depicted.

In any case, graphic data presentation, including documentation of live and analysis gates and statistical results, is preferable to exclusive presentation of statistical results, i.e. % positive cells in gates or mean fluorescence of cell populations. Gating and setting of statistical thresholds may cause significant differences in the final evaluation and may depend on the individual researcher. This is especially true in the case of weakly stained cells (see below).

Analysis gates

'Live' gates should be used with caution, and avoided whenever possible, since they preclude information from being recorded. 'Analysis' gates are a tool required for correct evaluation of the data recorded. They identify defined subpopulations of cells for statistical analysis. Most current state-of-the-art software still requires subjective decisions of the operator to define the gates. Gates can be verified either by using other cytometric parameters for control, or by cell sorting for microscopic or functional and molecular analysis. Analytical gating is described here for the analysis of white blood cells. In the example given, the first gate is set according to the light scatter of the cells. In the two-dimensional plot of forward versus side scatter, lymphocytes, monocytes and granulocytes appear as separate populations distinct from small cells and debris (Fig. 3 A–C). According to specific immune fluorescent labelling of lymphocytes, monocytes and granulocytes, gates can be defined for the forward versus side scatter, including all or most of the cells of these populations. These forward versus side scatter gates can now serve as analytical gates for future analyses. However, immunofluorescent staining controls are extremely important in the case of unexpected results, e.g. in pathological situations, when cells of interest from these populations change their scatter profile and escape the regular gates, and for analysis of rare cells, when confirmation is required. It should be kept in mind that activated leukocytes in general show increased forward light scatter. Scatter gating is also helpful for the exclusion of dead cells according to their reduced forward scatter, in addition to their exclusion by staining with PI (Fig. 3D).

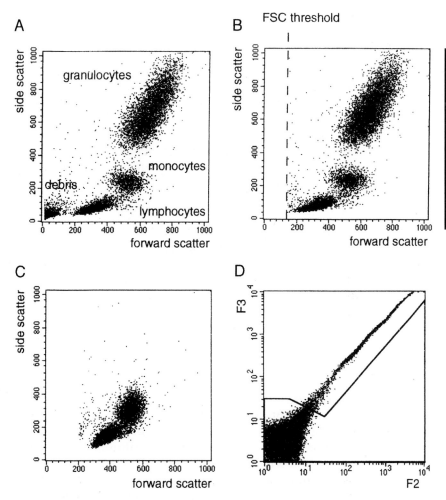

Figure 3. (A–C) FSC/SSC plots of peripheral blood leukocytes. Lymphocytes, monocytes and granulocytes can be identified by FSC/SSC properties. (A) After lysis of erythrocytes. (B) FSC threshold excludes debris from analysis. (C) Peripheral blood mononuclear cells; erythrocytes, granulocytes and dead cells are removed by centrifugation on a Ficoll gradient. (D) Exclusion of dead cells by PI staining. PI fluorescence is detected equally in the F2 and F3 channels, resulting in a diagonal of dead cells, which can easily be excluded from analysis by defining a gate in a F2/F3 scatter plot, as shown.

Statistical analysis

Statistical evaluation of cytometrically defined populations mostly concerns the frequency of cells in a given population, their scatter profile, mean fluorescence intensities (m.f.i.) and the respective coefficient of variation (CV). As a prerequisite the entire cell population has to be displayed 'on scale', since cells summed up in the first or last channel of the amplifyer cannot be analysed for mean fluorescence or CV. The statistical mean is defined as the arithmetic or geometric mean for linear and logarithmic scales, respectively. Variations within a population are the result of biological variation, e.g. cell size and antigen density, and of analytical variation, e.g. intensity and focus of the illuminating light, and orientation and movement of the cell during analysis in the focus. In logarithmic amplification a homogeneous population of cells results in a population symmetrically distributed around the mean value of fluorescence (Fig. 4). For such a distribution the coefficient of variation (CV), as a percentage, is half the width of the distribution at 0.6 times the maximum height.

$CV \cong 100 \times (10^{(a-b)/} * 256)$ with $s \cong$ channels per decade (usually 256), $a \cong$ upper and $b \cong$ lower channel with 0.6 maximum height.

The intensity of fluorescence can be converted to linear values for a better quantitative comparison of relative intensities between cell populations. The linear values L are calculated from channel values c as

$$L = 10^{(c/s)} \qquad s = channels/decade \ (usually \ 256)$$

Populations separated entirely from each other in one or more parameters are easy to compare statistically: the statistical threshold can be set anywhere between the two populations (see Fig. 5). If the two populations are overlapping, statistical evaluation becomes more complicated. One method commonly used is to set a statistical threshold, such that 99% of the negative control is included, and to consider all cells above this threshold as positive for that parameter. Although popular, this method cannot be recommended, since it will in any case falsify the results for entirely separated populations by at least 1%. For overlapping populations the results will be inaccurate by far more than 1% (see below). For shifted populations, i.e. an increase in mean fluorescence of all cells, compared with the control, the 99% threshold method does not reflect reality and will give entirely wrong results if the average fluorescence is weak as illustrated in Fig. 6. How can this be improved? For routine evaluation of overlapping populations, the statistical threshold can be set

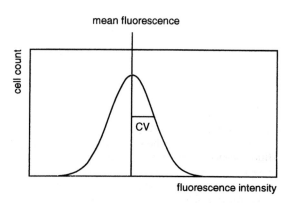

Figure 4. Fluorescence distribution of a homogeneous population of cells. Due to biological variations and variations in measurement, the cells are equally distributed around a mean value (mean fluorescence intensity (MFI)). The dispersion is described quantitatively by the coefficient of variation (CV). CV = SD/(MFI × 100), where SD is the standard deviation.

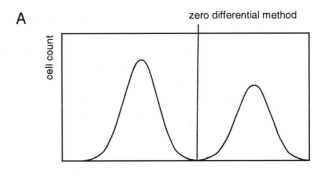

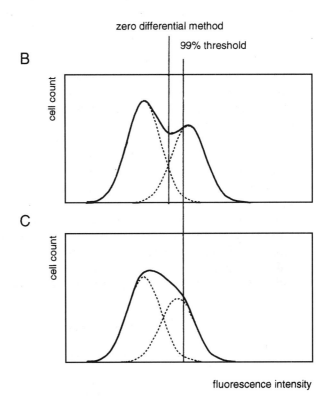

Figure 5. Statistical evaluation of separated (A) and bivariate (B) or asymmetrical
(C) populations. (A) For separated populations analysis can easily be done by
setting a statistical marker between the two populations (zero differential (valley)
method), which can then be analysed separately for relative cell number, mean
fluorescence intensity and coefficient of variation (CV). (B) Bivariate histograms
require the use of curve-fitting programs, but for routine analyses they can be
analysed using the valley method, assuming that the curve reflects two overlapping populations of similar size and CV so that false-positive and false-
negative cells, defined by the marker, would cancel out. (C) Asymmetrical
histograms, which may represent overlapping populations or kinetic transitions,
allow only very preliminary analysis by standard evaluation methods, but require
the use of curve-fitting programs. In all cases (A)–(C), the 99% threshold method
provides no advantages, but rather leads to misinterpretation of the data.

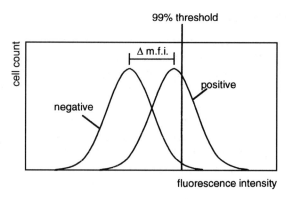

Figure 6. Shifted populations (histograms of stained and unstained cells are overlaid). All cells are stained without an increase in the coefficient of variation, and therefore setting statistical markers by any method would give false results. Such staining has to be described by the difference in the mean fluorescence intensity (Δ m.f.i.).

at the lowest point of the valley between the two peaks ('zero differential', see Fig. 5). In the case of overlapping populations of about equal size, 'false-positive' and 'false-negative' cells will then level out. If populations differ too much in size, e.g. for rare cell analysis, this method cannot be used. Ideally, statistical analysis of overlapping populations would require curve fitting programs such as Modfit (Verity) or Multicycle (Partec). However, even these require information about the number of populations to be expected and to what extent kinetic differences may be involved. In such cases it is preferable to improve staining and to add additional cytometric parameters for discrimination, to achieve better resolution of the populations. For rare cells, enrichment prior to analysis may help to identify the relevant population, using the physical sorting parameter as such an additional parameter (Radbruch and Recktenwald, 1995).

Standard Protocols: Staining and Instrumental Set up

Sample preparation

Flow cytometry *per se* provides single cell analysis and requires single cell suspensions, because otherwise immunofluorescence cannot be attributed to individual cells. Single cell suspensions are also required to prevent blocking of the flow channels. Cell aggregates can be removed by $1g$ sedimentation and/or filtration, and buffers containing protein (BSA) and EDTA will reduce aggregation of cells and unspecific staining. Details of how to obtain single cell suspensions from various tissues can be found in the chapter by Czuprynsky *et al.* (p. 233ff).

Set up of a flow cytometer

The protocol given below can be used for standard flow cytometers with stable alignment. It serves to determine the instrument settings of threshold, gain and compensation for three-colour analysis of FL (F1), phycoerythrin (F2) and CyChrome (F3) fluorescence. These basic principles can also be applied to multicolour approaches. Instead of cells stained for each colour separately, calibration particles as described above can be used. Since the optical alignment is stable and settings can be stored and recalled in state-of-the-art cytometers, calibration has to be performed in principle only once for a particular type of analysis and a particular instrument. However, the settings should be controlled before each acquisition to ensure the stability of the system. After changing the PMT voltage for any fluorescence parameter, the compensation has to be set up again. For further information about compensation see also: http://www.drmr.com/compensation/index.html

1. Select log-amplification for F1–F3 and linear amplification for FSC and SSC. Set all compensation levels to zero. Select FSC for threshold trigger, adjust trigger level approximately to channel 50.
2. Run unstained cells and adjust FSC/SSC until optimal separation of populations is obtained (with FSC around channel 300 for the smallest cells, e.g. lymphocytes). For leukocytes see Fig. 3. Increase forward scatter trigger level to exclude debris, with all cells still clearly being displayed.
3. Set a live gate around the population of the smallest cells (e.g. lymphocytes), with the lowest level of autofluorescence.
4. Adjust PMT voltage of F1–F3 to yield a mean relative fluorescence intensity (m.f.i.) of 3–4 (channel 120) for unstained cells. Some software allows to indicate live statistics simultaneously during acquisition, which is helpful. The entire population should be 'on scale' (see above).
5. Run a mixture of unstained and FL-stained cells, and display as F1/F2 plot. Reduce F2 signal of FL-stained cells by increasing F2 – (x % F1) compensation until m.f.i. of the FL-stained cells has reached the F2-m.f.i.-value of the unstained cells (m.f.i. 3–4).
6. Use mixtures of unstained and PE stained cells for adjusting F1 – (x % F2) and F3 – (x % F2) compensations, displaying them on a F2/F3 dot plot) and unstained/CyChrome stained cells for F2 – (x % F3) compensation by repeating the procedure as described for F2 versus F1.
7. Now run a mixture of all four populations. Cells should be 'positive' for just one colour, having a m.f.i. of around 3–4 for all other fluorescences. Record 10^4 cells as a master file, which serves to document the status of calibration and as source file for instrument settings, to be recalled for future acquisitions.

Conventional surface staining

Direct reagents

1. Wash cells once with PBS/0.5%BSA/0.02%NaN$_3$ (PBA), (10^6 cells ml^{-1}, 300 g).
2. Gently flick the cell pellet and add a minimum of 50–100 µl of PBA containing the fluorochrome-conjugated antibodies at concentrations determined beforehand by titration (up to three different colours, each has to be controlled separately).
3. Incubate for 10 min on ice, protect from light.
4. Wash with 1 ml PBA.
5. Take up cells in about 0.2–1 ml of PBA for immediate analysis or fix (optional) by adding 1 ml 0.5% paraformaldehyde in PBS to the flicked pellet and store at 4°C in the dark until analysis.

Indirect staining using haptenized primary labels

1. Wash cells once with PBA (10^6 cells ml^{-1}, 300 g).
2. Resuspend with 50–100 µl of haptenized antibodies diluted in PBA (concentration determined by titration) (10 min on ice).
3. Wash twice with 1 ml PBA.
4. Incubate with 50–100 µl of fluorochrome labelled anti-hapten reagents in PBA (concentration determined by titration) (10 min on ice).
5. Wash with 1 ml of PBA.
6. Take up cells in about 0.5–1 ml of PBA for immediate analysis or fix (optimal) as described above.

Indirect staining using isotype-specific reagents

1. Wash cells once with PBA (10^6 cells ml^{-1}, 300 g).
2. Stain with one primary antibody, e.g. murine IgG1, as described above and wash.
3. Stain with fluorochrome-labelled anti-isotype reagent, e.g. anti-mouse IgG1 at the concentration determined beforehand by titration and wash once.
4. Block free binding sites with unlabelled IgG1 (100 µg ml^{-1}, 10 min on ice). Control blocking efficiency by staining an aliquot of the cells with a fluorochrome labelled isotype control antibody, e.g. of the murine IgG1 subclass.
5. Add the second and third, haptenized or directly conjugated antibodies to obtain final concentrations, optimal for staining, as determined beforehand by titration. Stain and fix as described above.

Absorption of polyclonal reagents

To minimize unwanted cross-reactivities of polyclonal reagents against cell surface determinants of negative cells, they can be absorbed on unstained such cells prior to use.

(contd.)

46

1. Use stock solution of staining reagent at about 1 mg ml⁻¹.
2. Add 1–2 volumes of that solution to 1 volume of flicked cell pellet. Mix gently and incubate for 1–2 h on ice, flick from time to time.
3. Spin down the cells at 350 g and remove supernatant.
4. Spin down supernatant at 12 000 g for 20 min, to remove protein aggregates and debris. Titrate stock solution for staining and store aliquots of the stock solution at −70°C.

Exclusion of dead cells

If F2 and F3 are not both used for analysis of surface antigens, in a particular sample, or if the antigens stained for in F2 and F3 are expressed mutually exclusive on different subpopulations, propidiumiodide (PI), from an aqueous stock solution (0.1 mg ml⁻¹), at a final concentration of 1 μg ml⁻¹ can be added immediately (1 min) prior to analysis for exclusion of dead cells. PI is excluded by cell membranes of viable cells, but will enter dead cells and stain their DNA. It can not be used for fixed cells.

Exclusion of non-nucleated cells

LDS 751, a vital DNA-dye emitting light in the far red end of the spectrum (751 nm), can be used to label all nucleated cells (stock: 1 mg ml⁻¹ in methanol, dilute 1 : 100). Add it directly before analysis (Recktenwald, 1988). Its use has been proposed to identify cells in fixed samples, which were dead prior to fixation (Terstappen *et al.*, 1988).

Intracellular staining

Basic considerations

The basic principles of intracellular staining are discussed in the chapter by Yssel *et al.* (p. 707ff). A protocol for staining of murine cytokines and a list of appropriate reagents is given here. Specificity of staining can be controlled by staining cytokine gene transfectants and blocking with soluble cytokine (see earlier). In addition, the intracellular staining for most of the cytokines is concentrated in the Golgi compartment. Upon microscopic analysis, this localized staining provides further evidence for the specificity of staining. Intracellular staining of cytokines can also be controlled functionally, by using the 'cellular affinity matrix technology', which allows isolation of live cells secreting particular cytokines (see chapter by Assenmacher *et al.* (Section I)). Such cells should be specifically stained for the same cytokine intracellularly, and the specificity can be controlled further by analysis of the isolated cells *in vitro* or *in vivo*.

Standard protocol for intracellular staining of murine cytokines

Antibodies for intracellular staining of murine cytokines

Excellent fluorochrome-conjugated antibodies for intracellular staining of cytokines are now commercially available (see supplier list).

Fixation

1. Wash cells once with PBS and resuspend in PBS at 2×10^6 ml^{-1}.
2. Add 1 vol. 4% formaldehyde/PBS. Incubate for 20 min at RT.
3. Wash twice with PBA.
4. Resuspend in PBA at $1–2 \times 10^6$ cells ml^{-1} and store at $4°$C in the dark until staining.

Intracellular staining of one cytokine and one surface marker

1. Use about 1×10^6 fixed cells for each 1.5 ml test tube sample; spin down for 10 min at 300 g.
2. Incubate pellet with 50 µl DIG-conjugated anti-cytokine antibody and PE-labelled antibody against a surface antigen in PBA/0.5% saponin (from Quillaja bark, Sigma, St Louis, USA, S-2149) ('saponin-buffer') for 10–15 min at RT.
3. Wash twice with 1 ml saponin buffer.
4. Incubate pellet with 50 µl sheep anti-DIG-FL (Boehringer Mannheim), in saponin-buffer for 10–15 min at RT.
5. Wash with 1 ml saponin buffer and resuspend pellet in PBA.
6. Analyse by flow cytometry or fluorescence microscopy.

Detecting rare surface molecules by magnetofluorescent liposomes

Tips and tricks

In general labelling with liposomes follows the rules described above. For indirect labelling with liposomes, the primary labelling reagent has to be highly specific and staining conditions have to be optimized, to minimize background staining. Free primary or secondary reagents have to be removed carefully. Cells are washed twice after primary staining and liposomes are washed once prior to use. After staining, free liposomes have to be removed carefully by repeated washing (2–3 times). Single liposomes can be detected by flow cytometry, resulting in false-positive signals by coincidental measurement of cells with free liposomes. For the same reason, the speed of flow of the sample in the cytometer should be reduced to 100–500 events s^{-1}. On the other hand after washing cells labelled with liposomes have to be resuspended very gently, to avoid detachment of cell-bound liposomes.

> **Protocol for staining with liposomes**
>
> 1. To decrease unspecific binding, cells are preincubated for 10 min with a 100- to 1000-fold excess of unspecific 'blocking' IgG in PBA with 0.05% NaN_3, kept strictly on ice during the entire procedure.
> 2. Cells are labelled with haptenized antibody (usually 1–10 µg ml^{-1}) for 10 min on ice. Wash twice to remove any free antibody.
> 3. Liposome pretreatment: liposomes conjugated to hapten-specific antibodies are spun down at approximately 12 000 g, for 10 min, directly before use and resuspended in PBA. The optimal concentration has to be determined by titration beforehand.
> 4. The cells are resuspended in the liposome solution. Maximum concentration is about 10^7 cells/200 µl, with 200 µl minimal volume. Round-bottom tubes are used for staining and placed on a shaker for at least 30 min on ice. Make sure that the mixture is agitated. Cells are washed carefully at least twice.
> 5. Analysis by flow cytometry (flow rate 100–500 events s^{-1}) or isolation of labelled cells by MACS.

Surface cytokine detection and sorting

Some cytokines like IFN-γ (Assenmacher *et al.*, 1996) and IL-10 (Scheffold *et al.*, 2000) are expressed specifically on the surface of cells secreting those cytokines. They can be used to identify and isolate viable cells according to cytokine expression, e.g. T lymphocytes secreting IFN-γ. Surface cytokines are expressed transiently during the time of secretion by the cell, restricting their use to identification of cells activated *in vivo* or *in vitro* (for activation protocols see chapter by Yssel *et al.* (p. 707ff)). For staining, antibodies are used which do not recognize receptor-bound cytokines, i.e. antibodies blocking cytokine function, like AN 18-17.24 and GZ4 for murine and human surface IFN-γ respectively, and JES5-2A5 for murine surface IL-10. Cells are stained indirectly with digoxigenized cytokine-specific antibodies and anti-digoxigenin liposomes, as described above. Specificity of staining can be controlled functionally, by sorting and subsequent cell culture, followed by detection of secreted cytokines in culture supernatants, or fixation and intracellular staining of cytokines. Direct correlation of intracellular and liposome-surface staining is not possible, because liposomes are lysed by saponin buffer.

Analysis

Liposomes are 100- to 1000-fold more sensitive than conventional staining reagents, detecting as few as 50–100 molecules per cell. For analysis of cells stained with liposomes, some aspects have to be taken into account. The high sensitivity of the technology requires stringent controls, i.e. cells stained with liposomes alone and an isotype-matched digoxigenized control antibody. The latter provides the basis for statistical evaluation.

One bound liposome is sufficient to label a cell clearly for cytometric evaluation. For rare antigens of less than 300–400 molecules per cell, cells will be labelled with few liposomes. In that situation false-negative cells may falsify the analysis, i.e. although a subpopulation of positive cells is detectable, some of the unlabelled cells may be positive cells which did not even bind to a single liposome.

Flow cytometry of secretion

For the basic principles of the analysis of cells according to secreted molecules see earlier. Details of the technology are described in Chapter I.2 by Assenmacher *et al.* (p. 59ff).

Table 2.

A

Cytokine	Antibodies	Cytokine	Antibodies
IFN-γ	R4-6A2 or AN18.17.24	IL-4	11B11 or BVD4-1D11
IL-2	S4B6	IL-5	TRFK5
IL-3	MP2-43D11 + MP2-8F8	IL-10	JES5-2A5

B

Secreted molecule	Antibody pair (catching ab, detection ab)
Immunoglobulin	rat anti-mouse κ-light-chain, (e.g. clone R33-18.10) + rat anti-mouse Ig subclass
IL-2	JES6-1A12 + JES6-5H4
IFN-γ	R4-6A2 + AN18.17.24

◆◆◆◆◆◆ CELL SORTING BASED ON AFFINITY LABELLING

Introduction

Basic considerations

Affinity labelling, apart from its analytical potential, allows cells to be isolated according to this label for further analysis. Fluorescence activated cell sorting (FACS) and high-gradient magnetic cell sorting with super-paramagnetic colloids (MACS-system) are both based on quantitative labelling. Other, more qualitative methods based on physical (density centrifugation), biochemical (adherence, LME-lysis) or immunological differences (panning, macroscopic magnetic beads) may be useful for pre-enrichment. Some of these methods are described in the chapter by Czuprynski *et al.* in Section II.2, and by Esser (2000).

Sorting strategies

MACS or FACS

For fluorescence-activated cell sorting, closed systems with Piezo-controlled fluidic deflection systems (Partec, FACSort, FACSCalibur), which provide sort rates of below 10^6 per hour, and free flow in air systems with deflection of cells in charged micro-droplets (Coulter EPICS, Becton Dickinson FACS Vantage, or FACS DIVA, Cytomation MoFlo), with sort rates of more than 10^7 per hour, are available today. All standard systems allow sorting of cells according to light scatter and at least up to four different fluorochrome labels. The number of fluorochromes can be increased up to 12 in specialized research instruments. New instruments, e.g. Cytomation Moflo and Becton Dickinson FACS DIVA, can separate four different populations simultaneously from one sample. The basic principle of free flow in air sorters is the deflection of charged droplets containing single cells in an electric field, according to their fluorescence or light scatter. Higher sorting rates than 10^8 per hour cannot be used due to physical constraints. FACS is the technology of choice for isolation of cells according to complex multi-parameter labelling, intracellular staining and staining for biochemical parameters and DNA content. It provides the option to deflect individual cells or defined cell numbers and can separate cells according to subtle quantitative differences in labelling. High-gradient magnetic cell sorting, sorts cells in parallel rather than sequentially. Thus the sorting time is independent of the total cell number, i.e. with current MACS systems up to approximately 10^{11} cells can be sorted within minutes, with up to 10^9 magnetically labelled cells retained on a column. This makes MACS a useful tool for the isolation of rare cells. MACS can also be used for the separation of cells according to multiparameter labelling, by sequential labelling and separation (see later). Sorting for quantitative differences in magnetic labelling requires fine tuning of the labelling and separation conditions (see below). While MACS is an inexpensive, easy to use system, imposing little stress to the cells, free flow in air systems require experienced operators for alignment and sorting. Acceleration in the nozzle is stressful for the cells. Aerosol formation bears the risk of infection.

Positive or negative?

The first decision in developing sorting strategies is whether to sort for the wanted cells or against all other cells. Positive selection is usually more efficient in terms of purity. However, it requires a specific cell surface marker for the target cells, which, upon cross-linking by labelling, should not interfere with cell function.

'Negative' selection of distinct cells, by depletion of all other cells, will leave the enriched cells 'untouched', but the sorted population is usually less pure, because it is difficult to find labelling markers for all unwanted cells, especially in the case of rare positive cells.

Analysis

Sorting experiments are evaluated by cytometric analysis of the original, negative and positive fraction. The absolute numbers of (live) cells have to be determined in all fractions (e.g. by counting in a Neubauer chamber), to calculate recovery rates.

Sterility

MACS columns are provided sterile and the sorting procedure can easily be performed under sterile conditions. Sterile FACS sorting requires sterilization of the entire flow system.

FACS

The basic principles of free flow in air sorting are described elsewhere in detail (Melamed *et al.*, 1990; Radbruch, 2000). In any case, it is advisable to use this technology only in well-managed cytometric laboratories.

MACS

Principles of magnetic separation

The basic idea of magnetic separation in high gradient magnetic fields is to combine the advantages of labelling of cells with small, superparamagnetic particles with separation on a ferromagnetic matrix, magnetized by insertion into an external magnetic field. Cells labelled with superparamagnetic beads are attracted to the ferromagnetic matrix by the magnetic gradient generated. Unlabelled cells are eluted by washing. When the column is removed from the magnetic field, labelled cells can be eluted from the ferromagnetic matrix, since it is demagnetized and incapable of holding back superparamagnetic particles.

Parameters of MACS sorting

Factors that have to be considered for magnetic sorting include the quality of magnetic labelling and the choice of separation columns.

Magnetic label

The strength of magnetic labelling influences the efficiency of the separation. Like fluorescent staining, it is proportional to the density of antigen and, similarly, indirect reagents are more effective than direct labelling. Optimal concentrations have to be checked by titration.

Any background staining will lead to unspecific retention. Single-cell suspensions are essential to prevent clogging of the columns. Magnetic labelling should be performed in a refrigerator (6–12°C) for approximately 15 min (incubation on ice requires longer incubation times, i.e. about 30 min).

Separation columns

Columns are available with different capacities for processing different numbers of total cells and magnetically labelled cells). The size of the column has to be calculated to be able to retain the expected number of positive cells. For depletion, the column should have higher capacity to guarantee effective retention of all cells. Due to their geometry, ball-matrices (RS^+, VS^+, XS^+) have reduced unspecific binding sites, a smaller capacity will generate weaker magnetic gradients than wire matrices. They are ideally suited to isolate populations of high purity, especially if the total number of positive cells is small.

Apart from the choice of column, the speed of flow has a strong influence on the quality of sorting. High flow rates will not allow retention of weakly labelled cells, and thus give reduced background and high purity of the positive fraction ('enrichment'). However, not all positive cells will be retained. Slow flow-rates will result in quantitative retention of labelled cells (depletion). Two successive rounds of MACS, depletion and positive enrichment can be combined to obtain pure positive and negative populations.

Dead cells

Dead cells tend to take up magnetic beads and are therefore enriched unspecifically by MACS. They should be removed prior to MACS, especially for separation of rare cells.

Sensitivity

The sensitivity of magnetic separation is at least as high as that of optical separations obtained by immunofluorescent staining for the same parameter. Clearly stained cells can be isolated by MACS. Because the cells have no magnetic moment *per se*, magnetic sorting allows the separation of overlapping populations in fluorescent staining due to weakly expressed antigens, autofluorescence and natural CV (see above), if the magnetic label has been titrated carefully.

Multiparameter magnetic cell sorting

Magnetic sorting according to more than one parameter can be done by combining depletion for one and subsequent positive enrichment for a second parameter. Stringent depletion conditions are required to avoid enrichment of cells left over from the first labelling in the second round. For the enrichment, conditions should favour high purity, e.g. high flow rate. This approach has been efficiently used, e.g. for isolation of fetal cells from maternal blood (Buesch *et al.*, 1994). The positive enrichment of cells according to several surface markers can be done with MACS-Multisort reagents, which allow enzymatic cleavage of the magnetic beads after separation and subsequent positive or negative enrichment steps. One example for double positive enrichment, the purification of naive T cells from murine splenocytes, is given below.

Isolation of naive murine T cells

Detailed protocols for MACS are supplied together with the reagents. Here, we give a protocol for sorting of naive murine T helper cells according to the expression of CD4 and CD62L (L-selectin, (Gallatin *et al.*, 1983)) (protocol by M. Assenmacher and M. Loehning).

Reagents

Antibody conjugates:
- anti-murine CD4 mab: GK1.5 FITC (isomer-1, e.g. Sigma F-7250)
- anti-murine CD62L mab: MEL-14 biotin
- streptavidin phycoerythrin (PE)

Magnetic cell separation reagents (Miltenyi Biotec, Bergisch Gladbach, Germany):
- anti FITC (isomer-1) MultiSort Kit
- streptavidin MicroBeads

MACS columns for positive selection:
- MS$^+$/RS$^+$ column with MiniMACS (for up to 10^7 positive cells).
- LS column with MidiMACS, VarioMACS or SuperMACS (for up to 10^8 positive cells).

Protocol

First enrichment: separation of CD4$^+$ cells

1. Prepare single cell suspension of murine spleen cells, wash and stain with anti-CD4 FITC in PBS/BSA as described.
2. Wash once.
3. Incubate pellet with anti-FITC MultiSort MicroBeads, use 1:5 at about 2–5×10^8 cells ml^{-1} in PBS/BSA; 15 min, 10°C.
4. Wash cells in PBS/BSA.
5. Resuspend cells in PBS/BSA (MScolumns: 0.5–1 ml, LS columns: 1–2 ml) and apply cell suspension onto the appropriate MACS column (keep an aliquot: original fraction).
6. Rinse with PBS/BSA (MS: 3×0.5–1 ml, LS: 5×3 ml) (keep an aliquot: negative fraction).
7. Remove columns from magnetic separator and elute CD4$^+$ cells with PBS/BSA (MS: 1–2 ml, LS: 5 ml), (keep an aliquot: positive fraction I).

(contd.)

8. Check purity of positive cells by flow cytometry: separation may be repeated to increase purity to more than 98%.

Detachment of magnetic beads

1. Incubate CD4$^+$ cell fraction with MACS MultiSort Release Reagent, 20 µl ml^{-1} cell suspension; 10 min, 10°C.
2. Apply cell suspension onto a MS column, to remove any remaining magnetically labelled cells, rinse column and wash released fraction.
3. Incubate released CD4$^+$ cells with MACS MultiSort Stop Reagent, 30 µl ml^{-1} cell suspension, at 2×10^8 cells ml^{-1}, in PBS/BSA during the following labelling step.

Second enrichment: separation of CD4$^+$CD62L$^+$ cells

1. Stain CD4$^+$ fraction with Mel-14 biotin as described.
2. Wash cells in PBS/BSA.
3. Incubate pellet with Streptavidin MicroBeads, use 1 : 10 at about 2×10^8 cells ml^{-1} in PBS/BSA; 10 min, 10°C.
4. Add streptavidin PE; 5 min, 10°C.
5. Wash cells in PBS/BSA.
6. Resuspend cells in PBS/BSA, apply cell suspension onto the MACS column and proceed as described above.
7. Analyse by flow cytometry.

References

Assenmacher, M. (2000). Combined intracellular and surface staining. In: *Flow Cytometry and Cell Sorting*, 2nd edn (A. Radbruch, Ed.), pp. 53–58. Springer-Verlag, Heidelberg.

Assenmacher, M., Scheffold, A., Schmitz, J., Checa, J. A. S., Miltenyi, S. and Radbruch, A. (1996). Specific expression of surface interferon-gamma on interferon-gamma producing cells from mouse and man. *Eur. J. Immunol.* **26**, 263–267.

Baumgarth, N. and Roederer, M. (2000). A practical approach to multicolor flow cytometry for immunophenotyping. *J. Immunol. Methods* **243**(1–2), 77–97.

Brosterhus, H., Brings, S., Leyendeckers, H., Manz, R. A., Miltenyi, S., Radbruch, A., Assenmacher, M. and Schmitz, J. (1999). Enrichment and detection of live antigen-specific CD4(+) and CD8(+) T cells based on cytokine secretion. *Eur. J. Immunol.* **29**(12), 4053–4059.

Buesch, J., Huber, P., Pflueger, E., Miltenyi, S., Holtz, J. and Radbruch, A. (1994). Enrichment of fetal cells from maternal blood by high gradient magnetic cell sorting (double MACS) for PCR-based genetic analysis. *Prenatal Diagnosis* **14**, 1129–1140.

Darzynkiewicz, Z. and Crissman, H. A., (Eds). (1990). *Flow Cytometry. Methods in Cell Biology*. Academic Press, Ltd., London.

Esser, C. (2000). Powerful preselection. In: *Flow Cytometry and Cell Sorting*, 2nd edn (A. Radbruch, Ed.), pp. 133–140. Springer-Verlag, Heidelberg.

Gallatin, W. M., Weissman, I. L. and Butcher, E. C. (1983). A cell-surface molecule involved in organ specific homing. *Nature* **304**, 30.

Hermanson, G. T. (1996). *Bioconjugate techniques*. Academic Press, Inc., San Diego.

Manz, R., Assenmacher, M., Pflueger, E., Miltenyi, S. and Radbruch, A. (1995). Analysis and sorting of live cells according to secreted molecules, relocated to a cell-surface affinity matrix. *Proc. Natl. Acad. Sci.* **92**, 1921–1925.

Manz, R. A., Thiel, A. and Radbruch, A. (1997). Lifetime of plasma cells in the bone marrow. *Nature* **388**(6638), 133–134.

Mason, D. (Ed.) (2002). *Leucocyte Typing VII. White cell differentiation antigens.* Boston, Oxford University Press.

Melamed, M. R., Lindmo, T. and Mendelsohn, M. L., (Eds). (1990). *Flow cytometry and sorting*. Wiley Liss, Inc., New York.

Miltenyi, S., Mueller, W., Weichel, W. and Radbruch, A. (1990). High gradient magnetic cell sorting with MACS. *Cytometry* **11**, 231–238.

Mueller, W. (2000). Conjugation of fluorochromes, haptens and phycobiliproteins to antibodies. In: *Flow cytometry and cell sorting,* 2nd edn (A. Radbruch, Ed.), pp. 27–33. Springer-Verlag, Heidelberg.

Ormerod, M. G. (1994). Further application to cell biology. In: *Flow Cytometry* (M. G. Ormerod, Ed.), pp. 267–268. Oxford University Press, New York.

Radbruch, A. (2000). *Flow Cytometry and Cell Sorting,* 2nd edn (A. Radbruch, Ed.). Springer-Verlag, Heidelberg.

Radbruch, A. and Recktenwald, D. (1995). Analysis and sorting of rare cells. *Curr. Opin. Immunol.* **7**, 270–273.

Recktenwald, D. (1988). Method for analysis of subpopulations of blood cells. US Patent No. 4, 876,190.

Rothe, G. and Valet, G. (2000). Biochemical parameters of cell function. In: *Flow Cytometry and Cell Sorting,* 2nd edn (A. Radbruch, Ed.), pp. 100–120. Springer-Verlag, Heidelberg.

Scheffold, A., Miltenyi, S. and Radbruch, A. (1995). Magnetofluorescent liposomes for increased sensitivity of immunofluorescence. *Immunotechnology* **1**, 127–137.

Scheffold, A., Assenmacher, M., Reiners-Schramm, L., Lauster, R. and Radbruch, A. (2000). High sensitivity immunofluorescence for detection of the pro- and anti-inflammatory cytokines IFN-gamma and IL-10 on the surface of cytokine secreting cells. *Nat. Med.* **6**(1), 107–110.

Shapiro, H. M. (1988). *Practical Flow Cytometry*. Alan Liss, Inc., New York.

Terstappen, L. W. M. M., Shah, V. O., Conrad, M. P., Recktenwald, D. and Loken, M. R. (1988). Discrimination between damaged and intact cells in fixed flow cytometric samples. *Cytometry* **9**, 477–484.

Useful websites

http://www.cyto.purdue.edu/:
Purdue University homepage: provides all kinds of information around flow cytometry and related subjects. Provides access to the flow cytometry mailing list, a very active discussion forum.

http://www.drmr.com/compensation/index.html:
Tips and tricks for fluorescence compensation in flow cytometry.

http://www.drmr.com/abcon/index.html:
Protocols for the conjugation of fluorochromes to proteins.

http://www.probes.com:
The molecular probes homepage is a useful handbook of fluorescent molecules and provides very detailed information about available fluorochromes and applications.

http://www.pingu.salk.edu/flow/sitelink.html:
A collection of useful cytometry websites.

List of suppliers

Amersham International plc
http://www.apbiotech.com
Cyanine dyes, conjugates and labelling reagents

Bangs Laboratories Inc
http://www.bangslabs.com/
Fluorescent microspheres, calibration particles

BD Biosciences
http://www.bd.com/
Flow cytometers, equipment (e.g. calibration particles) and antibodies

BD Biosciences: Pharmingen
http://www.bdbiosciences.com/pharmingen/
Antibodies

Beckman-Coulter Corporation
http://www.beckmancoulter.com
Flow cytometers, equipment (e.g. calibration particles) and antibodies

Cyanotech Corporation
http://www.cyanotech.com
Antibodies and phycobiliproteins

Cytomation
http://www.cytomation.com
Flow cytometers

Flow Cytometry Standards Corporation
http://scooter.cyto.purdue.edu/pucl_cd/flow/vol2/6/fsc/index.htm
Calibration particles

Miltenyi Biotec GmbH
http://www.miltenyibiotec.com/
Magnetic cell separators (MACS), labelling reagents, columns and equipment, antibodies

Molecular Probes, Inc.
http://www.molecularprobes.com/
Biochemicals, fluorescent probes and labelling reagents. The catalogue is a useful handbook of fluorescence technology in biological science

Ortho Clinical Diagnostics
http://www.orthoclinical.com/ocdindex.asp
Flow cytometers and antibodies

Partec GmbH
http://www.partec.de/
Flow cytometers

Pierce
http://www.piercenet.com/
Protein purification, labelling and conjugation reagents

Polyscience, Inc.
http://www.polysciences.com
Calibration particles

R&D Systems
http://www.rndsystems.com/
Cytokines, anti-cytokine antibodies and immunoassays

Roche Diagnostics
http://www.roche.com/diagnostics/
Biochemicals and antibodies (e.g. digoxigenin labelling and detection reagents)

Sigma-Aldrich
http://www.sigmaaldrich.com
Biochemicals and labelling reagents

Sigma-Genosys
http://www.genosys.co.uk/
Haptens and hapten-specific antibodies

Southern Biotechnology Associates, Inc.
http://southernbiotech.com/
Antibodies

VERITY Software House, Inc.
http://scooter.cyto.purdue.edu/pucl_cd/flow/vol2/6/verity/home.htm
Software products for evaluation of flow cytometry data

2 Cytometric Cytokine Secretion Assay: Detection and Isolation of Cytokine-Secreting T Cells

Mario Assenmacher
Miltenyi Biotec GmbH, Bergisch-Gladbach, Germany

Alexander Scheffold and Andreas Radbruch
Deutsches Rheumaforschungszentrum Berlin, Berlin, Germany

◆◆

CONTENTS

◆◆◆◆◆◆ INTRODUCTION

The cell-surface affinity matrix technology, otherwise called Secretion Assay or capture assay, represents a new, innovative method for the analysis and enrichment of viable cells according to secreted molecules, such as antibodies or cytokines (Manz *et al.*, 1995). Recent modifications have led to an easy and fast procedure. In the original protocol the cytokine specific affinity matrix was created by cell surface biotinylation, followed by labelling with avidin-conjugated cytokine specific 'catch' antibody. In the actual protocol a cytokine-specific 'catch' antibody is directly attached to the cell surface of leukocytes as a conjugate with a CD45-specific monoclonal antibody. While originally medium of high viscosity, e.g. gelatine or alginate, was used in the secretion period, the actual protocol has been adapted to the use of normal medium.

Basically, the secreted product is retained on the cell surface of the secreting cell, making it accessible to powerful technologies for the detection of surface markers. The Cytokine Secretion Assay involves the following steps: (1) A cytokine specific Catch Reagent is attached to

Copyright © Elsevier Science Ltd
All rights of reproduction in any form reserved

the cell surface of all cells. (2) The cells are then incubated for 30–45 min at 37°C to allow cytokine secretion. The secreted cytokine binds to the cytokine-specific Catch Reagent on the secreting cells and (3) is subsequently labelled with a second cytokine-specific 'detection' antibody, which is usually conjugated to a fluorochrome such as phycoerythrin (PE) for sensitive analysis by flow cytometry. Optionally the captured cytokine is further magnetically labelled with specific antibody conjugated to super-paramagnetic particles for enrichment by MACS (Fig. 1).

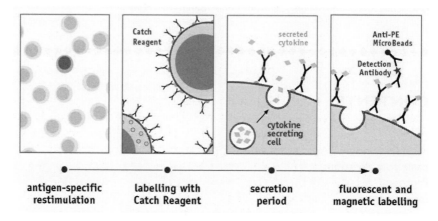

| antigen-specific restimulation | labelling with Catch Reagent | secretion period | fluorescent and magnetic labelling |

Figure 1. Principle of the Cytokine Secretion Assay for labelling of cytokine-secreting T cells after antigen-specific stimulation.

Analogous techniques for single-cell analysis but not for isolation of viable cytokine-secreting cells are ELISPOT (see chapter by Hiroi and Kiyono in Section II) or intracellular cytokine staining (ICS) (see chapter by Yssel *et al.* in Section III), which were developed in the late 1980s and early 1990s, respectively. Comparisons or direct correlations of the different techniques show comparable results with respect to the detected frequencies of cytokine-expressing cells (Oelke *et al.*, 2000b; Ouyang *et al.*, 2000; Pittet *et al.*, 2001). The Cytokine Secretion Assay combines advantageous features and overcomes several limitations of previously used methods.

- The Cytokine Secretion Assay allows the *isolation of viable cytokine-secreting cells* for cell culture, or other downstream experiments (Assenmacher *et al.*, 1998; Brosterhus *et al.*, 1999; Oelke *et al.*, 2000a,b; Ouyang *et al.*, 2000; Farrar *et al.*, 2001; Hu-Li *et al.*, 2001; McNeil *et al.*, 2001; Smits *et al.*, 2001) including adoptive transfer (Becker *et al.*, 2001).
- It allows the *sensitive multiparameter analysis of viable cytokine-secreting cells* down to frequencies of 10^{-6} due to the option of enrichment of cytokine-secreting cells by magnetic cell sorting (MACS) (Brosterhus *et al.*, 1999; Pittet *et al.*, 2001).

The Cytokine Secretion Assay is especially useful for the detection and isolation of viable antigen-specific T cells after a short restimulation with specific antigen *in vitro* to induce secretion of cytokines (Brosterhus *et al.*, 1999). T cells secrete cytokines only transiently upon stimulation, therefore normally only very few T cells actually secrete cytokines in peripheral blood or other tissues. However, memory/effector T cells rapidly restart to secrete cytokines after restimulation *in vivo* or *in vitro*.

With direct flow cytometric analysis the Cytokine Secretion Assay allows the very rapid detection of cytokine-secreting antigen-specific T cells down to frequencies of 0.01–0.1%. Combination with MACS enrichment greatly increases sensitivity of detection and cytokine-secreting antigen-specific T cells can be detected down to frequencies of 10^{-6} (0.0001%).

The Cytokine Secretion Assay combined with MACS enrichment also allows the isolation of viable cytokine-secreting antigen-specific T cells for expansion and functional characterization (Brosterhus *et al.*, 1999; Oelke *et al.*, 2000a, b; McNeil *et al.*, 2001).

The Cytokine Secretion Assay can also be used to analyse cytokine secretion by other cells, e.g. IFN-γ by NK cells or IL-10 by monocytes.

Cytokine-secreting cells can be analysed for co-production of two different cytokines by two-colour Cytokine Secretion Assays, i.e. using a cocktail of two different Cytokine Catch Reagents and Detection Reagents.

Cytokine Secretion Assays can directly be combined with peptide/MHC tetramer staining for functional characterization of antigen-specific T cells (see later).

Out of local areas of acute ongoing immune responses sometimes remarkable numbers of cells 'spontaneously' secreting cytokines can be detected and isolated with the Cytokine Secretion Assay, most likely after recent encounter of their relevant target antigens *in vivo*.

◆◆◆◆◆◆ BASIC CONSIDERATIONS

The Cytokine Secretion Assay is very useful for the detection and isolation of viable antigen-specific T cells after a short restimulation with specific antigen *in vitro* to induce secretion of cytokines.

Controls

A *negative control* sample, treated exactly the same as the antigen-stimulated sample but without addition of antigen (or with addition of control antigen), should always be included as a measure of spontaneous (probably *in-vivo* induced) cytokine secretion. (Optional) A *positive control* may be included in the experiment, e.g. using Staphylococcal Enterotoxin B (Sigma) at 1 µg ml^{-1} for 3–16 h. The addition of co-stimulatory agents such as anti-CD28 mAb or anti-CD49d mAb (usually 1 µg ml^{-1}) may

Detection and Isolation of Cytokine-Secreting T Cells

61

enhance the response to the antigen. If co-stimulatory agents are added to the antigen sample, they also have to be included in the negative control sample.

As a *high control* to verify that all cells are labelled with the Cytokine Catch Reagent and therefore are able to capture cytokine, immediately after labelling of the cells with the Cytokine Catch Reagent a small aliquot of the cells can be incubated in recombinant cytokine (typically at a final concentration of 200–1000 ng ml^{-1}) for 10 min on ice. After washing, the captured cytokine can be detected on all cells by staining with the fluorochrome-conjugated Cytokine Detection Antibody (Manz *et al.*, 1995; Assenmacher *et al.*, 1998; Ouyang *et al.*, 2000).

Kinetics and proposed time schedule

Upon stimulation with peptide, antigen-specific T cells can be analysed for IL-2, IL-4, IL-10 and IFN-γ secretion within approximately 3–6 h. Upon stimulation with protein, kinetics of cytokine expression are slightly slower (due to the time required for processing) and the cells can be analysed within approximately 6–16 h.

Human cells are typically collected and prepared for stimulation on the first day. Then antigen (protein) is either added in the late afternoon for overnight (≤16 h) stimulation (A) or cells are stored overnight (whole blood at RT; PBMC in culture medium at 37°C, 5–7% CO_2) and antigen (peptide) is directly added in the next morning for short (3 h) stimulation (B). Thus the Cytokine Secretion Assay procedure is started in the early (A) or late (B) morning of the second day.

Murine cells are typically collected and prepared for peptide stimulation early in the morning of the day. Then peptide is added for short (3 h) stimulation and the Cytokine Secretion Assay procedure is started in the early afternoon of this day. For protein stimulation typically in the afternoon of the first day, murine cells are collected and prepared for stimulation. Then protein is added in the late afternoon for overnight (≤16 h) stimulation and the Cytokine Secretion Assay procedure is started in the early morning of the second day.

If another type of *in vitro* stimulation or cell type, e.g. LPS stimulation of monocytes, is used, or if cells stimulated *in vivo* are directly analysed *ex vivo*, one can start the cytokine secretion assay with the stimulated cells at Labelling cells with Cytokine Catch Reagent stage of the protocol. Cytokines are usually only transiently secreted upon stimulation. Therefore the optimal time point after stimulation for analysis with the Secretion Assay has to be chosen in the individual experimental set up.

Counterstaining of cytokine-secreting T cells

To identify the cells of interest counterstaining with CD4 or CD8 is required. Upon activation of T cells TCR and associated molecules, like

CD3, are often downregulated and therefore are not useful markers. Exclusion of dead cells by staining with propidium iodide (PI) or 7-AAD will reduce non-specific background staining. For optimal sensitivity labelling of undesired non-T cells such as monocytes (human) or B cells (mouse) with antibodies conjugated to PerCP is recommended, e.g. using CD14.PerCP (human) or B220(CD45R).PerCP (mouse). These cells can then be excluded together with PI stained dead cells (see Figs 3 and 4).

On enrichment of rare antigen-specific T cells exclusion of dead cells is essential, because dead cells can severely disturb analysis of enriched cell fraction.

Critical parameters

The most important point within the Cytokine Secretion Assay is to prevent the capture of secreted cytokine by non-secreting cells during the secretion period. Not all of the secreted cytokine is caught by the affinity matrix of the secreting cells, i.e. some cytokine diffuses away from the secreting cell and accumulates in the culture medium. If the concentration of cytokine-secreting cells in the secretion phase is too high (approximately >5 × 10^4 cytokine-secreting cells ml^{-1}) and thereby the concentration of secreted and accumulated cytokine in the culture medium reaches a certain level (approximately >0.1–1 ng ml^{-1}), all cells start to catch a low amount of cytokine and subsequently are labelled with the Detection Antibody. The resulting fluorescent shift of all cells can be easily recognized in the flow cytometric analysis.

This problem (which similarly can occur in ELISPOT assays) can be prevented in two different ways. In the standard protocol the cell density in the secretion phase is adjusted according to the expected frequency of cytokine-secreting cells (among all cells), i.e. for <1–5% to 1–2 × 10^6 cells ml^{-1} and for >1–5% to 1–2 × 10^5 cells ml^{-1} (for >>20–50% to <10^5 cells ml^{-1}). In addition all washing and incubation steps (except the secretion phase) are done in the cold (ice-cold buffer, incubations on ice, refrigerated centrifuge) to prevent cells from secretion, when they are at high density, e.g. in the pellet or upon labelling with catch or detection antibody.

In the rapid protocol the assay is done without intervening washing steps on two samples with different cell densities (A: 10^6 cells in 200 µl and B: 10^5 cells in 200 µl). While sample A allows the detection of cytokine-secreting cells at frequencies from approximately 0.01 to 1% (and cells 'shift' at frequencies >>1%), sample B allows the detection of cytokine-secreting cells in the range of 1–10%. The second approach is similar to the serial dilution of cells in an ELISPOT assay.

These first two protocols describe the Cytokine Secretion Assay including magnetic enrichment starting from PBMC or equivalent murine/human leukocyte-containing populations or starting from whole blood. The third or fourth protocols are very rapid and simple procedures for the direct detection of cytokine-secreting antigen-specific T cells from small numbers of PBMC or equivalent cells or whole blood without magnetic enrichment.

♦♦♦♦♦♦♦ PROTOCOLS

Reagents

Buffer: Phosphate-buffered saline (PBS) pH 7.2, containing 0.5% (w/v) bovine serum albumin (BSA) and 2 mM EDTA (see recipe).

0.5 M EDTA stock solution: Dissolve 56 g sodium hydroxide (NaOH) in 900 ml distilled water. Add 146.2 g ethylenediamine-tetraacetic acid, adjust pH to 7.5, fill up to 1 l. Prepare buffer with, for example, 4 ml of 0.5 M EDTA stock solution per 1 l of buffer.

Culture medium: e.g. RPMI-1640 containing 10% AB or autologous serum for human cells or mouse serum for murine cells.
Additionally for protocols starting with whole blood (1.2 and 2.2):
Lysing solution (1×) (do NOT use FACS Lysing solution™ !)

10× stock solution: 41.4 g NH_4Cl (1.55 M), 5 g $KHCO_3$ (100 mM), 1 ml 0.5 M EDTA (1 mM), add 500 ml ddH_2O, adjust pH to 7.3. prepare 1× Lysing solution fresh from 10× Lysing stock solution.

Cytokine Catch Reagent: anti-cytokine monoclonal antibody conjugated to cell surface (CD45) specific monoclonal antibody (Miltenyi Biotec).
Cytokine Detection Antibody: anti-cytokine monoclonal antibody conjugated to PE or APC (Miltenyi Biotec).
Additional staining reagents, e.g. CD4-FITC or CD8-FITC and CD14-PerCP for human cells or CD45R(B220)-PerCP for murine cells (Becton Dickinson).
Propidium iodide (PI) or 7-AAD to exclude dead cells from the flow cytometric analysis.
Additionally for protocols with magnetic enrichment:
Anti-PE MicroBeads: colloidal super-paramagnetic MicroBeads conjugated to monoclonal mouse anti-PE antibody (Miltenyi Biotec).
MACS columns and separators (Miltenyi Biotec).
Tissue culture dishes or plates.
V-bottom tubes (depending on cell number: 1.5 ml to 15 ml) or deep well plates.
Refrigerated centrifuge, 37°C CO_2 incubator.

Cytokine Secretion Assay with enrichment and detection (PBMC or equivalent)

In vitro *stimulation of T cells with specific antigen*

1. Prepare human PBMC, murine spleen cells or other leukocyte-containing single cell preparations.

(contd.)

> *Do not use any non-human (non-murine) proteins (like BSA or FCS) in buffer or media upon cell preparation, stimulation or freezing to prevent non-specific stimulation.*

2. Wash cells by adding medium, centrifuge at $300\,g$ for 10 min, remove supernatant.

3. Resuspend cells in culture medium at 1×10^7 cells ml^{-1} and 5×10^6 cells cm^{-2} (e.g. 10^6 cells per 100 µl in a 96-well plate, 10^7 cells ml^{-1} in a 24-well plate or 10^8 cells per 10 ml in a petri dish).

   Add antigen immediately, or leave cells overnight at 37°C, 5–7% CO_2. High cell density is required for optimal antigenic stimulation.

4. Add antigen, e.g. peptide (often at 1–10 µg ml^{-1}) or protein (often at 1–100 µg ml^{-1}), mix well and incubate at 37°C, 5–7% CO_2 for either 3–6 h (peptide) or 6–16 h (protein).

5. Collect cells carefully, rinse dish/well with cold buffer.

Labelling cells with Cytokine Catch Reagent

1. Use 10^7 cells in a 15 ml closable test tube. For larger cell numbers, scale up all volumes accordingly. For less than 10^7 cells, use same volume.

2. Wash cells by adding 10 ml of cold buffer, centrifuge at $300\,g$ for 10 min at 4°C, remove supernatant completely. Optionally repeat washing step in ice-cold buffer.

3. Resuspend cells in 100 µl Cytokine Catch Reagent diluted in ice-cold medium (typically 10–20 µg ml^{-1}) per 10^7 cells, mix well and incubate for 5 min on ice.

   Keep the cells ice-cold to prevent secretion of cytokines at this step.

Cytokine secretion period

1. Add warm (37°C) medium to dilute the cells to 10^5–10^6 cells ml^{-1} depending on the expected frequency of cytokine-secreting cells (among all cells):

   <1–5%: 1–2×10^6 cells ml^{-1}, i.e. add 7 ml per 10^7 cells

   >1–5%: 1–2×10^5 cells ml^{-1}, i.e. add 70 ml per 10^7 cells

   (for >>20–50%: <10^5 cells ml^{-1}).

2. Incubate cells in (closed) preparation tube for 45 min at 37°C under slow continuous agitation/rotation or mix tube every 5 min to avoid sedimentation of the cells.

   It is important to prevent close contact of cells to avoid cross-contamination with cytokines.

Labelling cells with Cytokine Detection Antibody

1. Wash cells by adding a minimum of 1 volume of ice-cold buffer and place tube on ice. Centrifuge at $300\,g$ for 10 min at 4°C. Remove supernatant completely.

(contd.)

Keep the cells ice-cold to stop secretion of cytokines after secretion period.

2. Resuspend cells in 100 µl Cytokine Detection Antibody (and additional staining reagents) diluted in ice-cold buffer (typically 1–5 µg ml⁻¹) per 10^7 cells. Mix well and incubate for 10 min on ice.
3. Wash cells by adding 10 ml of cold buffer, centrifuge at 300 g for 10 min at 4°C, remove supernatant completely.

Magnetic labelling and separation (for PE conjugated Cytokine Detection Ab)

1. Resuspend cells in 80 µl of cold buffer per 10^7 cells, add 20 µl of anti-PE MicroBeads, mix well and incubate for 15 min at 8°C.
2. Wash cells by adding 10 ml of cold buffer, centrifuge at 300 g for 10 min at 4°C, remove supernatant completely.
3. Resuspend cells in 500 µl of cold buffer per 10^7 cells; for higher cell numbers use a dilution of 10^8 cells ml⁻¹. Take an aliquot for FACS analysis and cell count before enrichment.
4. Prepare two MS columns per sample by rinsing with 500 µl of cold buffer. Place the first column into the magnetic field of a suitable MACS Separator (e.g. MiniMACS).
5. Apply the magnetically labelled cells to the column, allow the cells to pass through and wash with 3×500 µl of cold buffer.
6. Remove the first column from the separator, place the second column into the separator, and put the first column on top of the second one. Pipette 1 ml of cold buffer on top of the first column, firmly flush out the retained cells using the plunger, directly onto the second one.
   Two consecutive column runs are required for optimal enrichment.
7. Wash with 3×500 µl of cold buffer.
8. Remove the column from the separator, place the column on a suitable collection tube. Pipette 500 µl of cold buffer on top of the column, firmly flush out the retained cells using the plunger.
   For subsequent cell culture, cells can also be eluted with medium. If part of the cells is analysed by flow cytometry, the medium should not contain phenol red.
9. Proceed to analysis and/or cell culture.

Cytokine Secretion Assay with enrichment and detection (whole blood)

1. Take 5 ml sodium heparinized whole blood in 50 ml conical polypropylene tube.
   Use sodium heparin as anticoagulant, __not__ EDTA or ACD. Calcium chelating anticoagulants prevent activation.
   Add antigen immediately, or leave blood overnight at RT.

(contd.)

2. Add antigen, e.g. peptide (often at 1–10 μg ml^{-1}) or protein (often at 1–100 μg ml^{-1}), mix well by vortexing and incubate at 37°C, 5–7% CO_2 for either 3–6 h (peptide) or 6–16 h (protein).
3. Add 45 ml of 1× Lysing solution. Mix gently and incubate for 10 min at RT. Rotate tube continously or turn tube several times during incubation.
4. Centrifuge cells at 300 g for 10 min at RT. Remove supernatant completely.
5. Resuspend cells in 15 ml of cold buffer and transfer into a 15 ml conical propylene tube.
6. Centrifuge cells at 300 g for 10 min at 4°C. Remove supernatant completely.
7. Resuspend cells in 200 μl Cytokine Catch Reagent diluted in ice-cold medium (typically 10–20 μg ml^{-1}), mix well and incubate for 5 min on ice.
   Keep the cells ice-cold to prevent secretion of cytokines at this step.
8. Add 7 ml warm (37°C) culture medium and incubate cells in (closed) preparation tube for 45 min at 37°C under slow continuous agitation/rotation or mix tube every 5 min to avoid sedimentation of the cells.
   It is important to prevent close contact of cells to avoid cross-contamination with cytokines.
9. Wash cells by adding 7 ml of ice-cold buffer and place tube on ice. Centrifuge at 300 g for 10 min at 4°C. Remove supernatant completely.
   Keep the cells ice-cold to stop secretion of cytokines after secretion period.
10. Resuspend cells in 200 μl Cytokine Detection Antibody (and additional staining reagents) diluted in ice-cold buffer (typically 1–5 μg ml^{-1}). Mix well and incubate for 10 min on ice.
11. Wash cells by adding 10ml of cold buffer, centrifuge at 300 g for 10 min at 4°C, remove supernatant completely.
12. Proceed to magnetic labelling and separation and then to analysis.

Rapid Cytokine Secretion Assay for detection only (PBMC or equivalent)

1. Prepare human PBMC, murine spleen cells or other leukocyte F1 containing single-cell preparations.
   Do not use any non-human (non-murine) proteins (such as BSA or FCS) in buffer or media upon cell preparation, stimulation or freezing to prevent nonspecific stimulation.
2. Wash cells by adding medium, centrifuge at 300 g for 10 min, remove supernatant.

(contd.)

3. Resuspend cells in culture medium at 1×10^7 cells ml^{-1}.
4. Place 10^6 cells in 100 µl culture medium in a >1.5 ml deep well plate or 1.5 ml tube labelled (A).

    Add antigen immediately, or leave cells overnight at 37°C, 5–7% CO_2.
5. Add antigen, e.g. peptide or protein (often at 1–10 µg ml^{-1}), mix well and incubate at 37°C, 5–7% CO_2 for either 3–6 h (peptide) or 6–16 h (protein).
6. Wash cells by adding 1 ml of medium, centrifuge at 300 g for 5 min at RT, remove supernatant.

    Remove supernatant from plates by simply inverting them <u>once without shaking</u> and let supernatant drain out.
7. Resuspend cells in 200 µl medium. Mix well and transfer 20 µl of cell suspension from well/tube (A) into a second well or tube labelled (B) containing 200 µl culture medium (B) to obtain a second sample with a higher dilution of the cells.

    Tube B is only required for the stimulated sample, but not for the control sample.
8. Add 20 µl Cytokine Catch Reagent (final concentration typically 5–10 µg ml^{-1}) to all wells or tubes (A+B), mix well and incubate cells for 30 min at 37°C under slow continuous agitation/rotation or mix every 5 min to avoid sedimentation of the cells.

    It is important to prevent close contact of cells to avoid cross-contamination with cytokines.
9. Add 20 µl Cytokine Detection Antibody (and additional staining reagents) (final concentration typically 1–5 µg ml^{-1}). Mix well and incubate for 10 min at RT.
10. Wash cells by adding 1 ml of cold buffer, centrifuge at 300 g for 5 min at RT, remove supernatant.
11. Resuspend cells in 500 µl of cold buffer and proceed with flow cytometric analysis.

Rapid Cytokine Secretion Assay for detection only (whole blood)

1. Place 300 µl sodium heparinized blood in a >1.5 ml deep well plate or 1.5 ml tube labelled (A).

    Use sodium heparin as anticoagulant, <u>not</u> EDTA or ACD. Calcium chelating anticoagulants prevent activation.

    Add antigen immediately, or leave blood overnight at RT.
2. Add antigen, e.g. peptide or protein (often at 1–10 µg ml^{-1}), mix well and incubate at 37°C, 5–7% CO_2 for either 3–6 h (peptide) or 6–16 h (protein).

    (contd.)

3. Mix well and transfer 30 µl of blood from well/tube A into a second well or tube labelled (B) containing 250 µl culture medium to obtain a second sample with a higher dilution of the cells.

    Tube B is only required for the stimulated sample, but not for the control sample.

4. Add 20 µl Cytokine Catch Reagent (final conc. typically 5–10 µg ml⁻¹) to all wells/tubes (A+B), mix well and incubate cells for 30 min at 37°C under slow continuous agitation or mix every 5 min to avoid sedimentation of the cells.

    It is important to prevent close contact of cells to avoid cross-contamination with cytokines.

5. Add 20 µl Cytokine Detection Antibody (and additional staining reagents) (final concentration typically 1–5 µg ml⁻¹). Mix well and incubate for 10 min at RT.

6. Add 1 ml Lysing solution (1×). Mix gently and incubate for 10 min at RT in the dark. Mix in between.

7. Centrifuge at 300 *g* for 5 min at RT. Remove supernatant.

    Remove supernatant from plates by simply inverting them <u>once without shaking</u> and let supernatant drain out.

8. Wash cells by adding 1 ml of cold buffer, centrifuge at 300 *g* for 5 min at RT, remove supernatant.

9. Resuspend cells in 500 µl of cold buffer and proceed with flow cytometric analysis.

Combination of Cytokine Secretion Assay with peptide/MHC tetramer staining

Cytokine Secretion Assays can directly be combined with peptide/MHC tetramer (or Dimer) staining for functional characterization of antigen-specific T cells (Pittet *et al.*, 2001) (see example later). Upon stimulation with antigen (or some mitogens like anti-CD3) the TCR on the specific T cells can be downregulated very rapidly. This dramatically reduces peptide-MHC tetramer staining. Therefore the peptide-MHC tetramer labelling should be done prior to the stimulation with peptide. After staining with peptide/MHC tetramer cells are stimulated with the specific peptide for <u>2–3 h</u> followed by the standard secretion assay procedures.

Depending on the type of peptide-MHC tetramer used, the peptide-MHC tetramer may (a) or may not (b) stimulate cytokine secretion. In the former case (a) the control sample is stained with peptide-MHC tetramer after the incubation (without peptide or with control peptide) and the cytokine secretion assay. Initially these two types of controls may be compared.

Flow cytometric analysis

1. Store samples at 4°C in the dark until analysis. Add PI (0.5 µg ml^{-1}) or 7-AAD immediately prior to FACS analysis.

   For enumeration of cytokine-secreting antigen-specific T cells (with or without enrichment):

2. Acquire 2×10^5 viable lymphocytes from the fraction before enrichment or from samples without enrichment.

   For optimal sensitivity, appropriate and same numbers of viable cells have to be acquired from the antigen-stimulated sample as well as from the control sample.

3. Acquire the complete positive fractions (from the antigen-stimulated sample as well as from the negative control sample) after enrichment.

   For isolation of cytokine-secreting antigen-specific T cells (e.g. for expansion of isolated cells):

4. Acquire up to 2×10^5 viable lymphocytes from the fraction before enrichment and acquire an aliquot of the positive fraction to determine the performance of the enrichment.

Enumeration of cytokine-secreting antigen-specific T cells (with or without enrichment)

(1) Without or before MACS enrichment frequencies of cytokine-secreting antigen-specific T cells were calculated based on the numbers of cytokine+ (CD8+ or CD4+) T cells and the numbers of (CD8+ or CD4+) T cells in the acquired fractions (see Fig. 2A):

% cytokine+ CD4+ T cells among CD4+ T cells =
(events in UR quadrant/events in UL + UR quadrants) × 100 =
0.257%

(2) After MACS enrichment frequencies of cytokine-secreting antigen-specific T cells were calculated as follows. First, the numbers of cytokine+ T cells in the magnetically enriched fractions were calculated on the basis of total number of cells and the frequency of cytokine+ T cells in the magnetically enriched cell fraction. Frequencies of cytokine+ T cells among total T cells were then calculated on the basis of the numbers of cytokine+ T cells in the magnetically enriched fractions and the numbers of T cells before enrichment of cytokine-secreting cells (see Fig. 2B)

(beside dot plot):
% cytokine+ CD4+ T cells among CD4+ T cells =
(total number of cytokine+ CD4+ T cells after enrichment/
total number of CD4+ T cells before enrichment) × 100 = 0.129%

(contd.)

While approach (2) gives highest sensitivity (down to 0.0001%), it is slightly less precise than approach (1), because not all of the cytokine+ T cells present in the sample before enrichment are recovered in the positive fraction after MACS enrichment. Usually the recovery is between 50–70%. The frequencies determined according to approach (2) are about half of the frequencies determined by approach (1). This is also illustrated in the enclosed example, where approach (1) results in 0.251% IFN-γ+ CMV-specific CD4+ cells among CD4+ cells and approach (2) gives 0.129% IFN-γ+ CMV-specific CD4+ cells among CD4+ cells.

Therefore, if the sensitivity of approach (1) (usually 0.1–0.01%) is sufficient to detect cytokine+ antigen-specific T cells it is preferred. If the sensitivity of approach (1) is not sufficient the extremely sensitive approach (2) should be used. The slight underestimation of the frequencies of cytokine+ antigen-specific T cells does not impair the analysis of antigen-specific T cell immunity, because comparisons between different donors or samples are not influenced and especially at low frequencies (0.01–0.0001%) a factor of 2 is probably biologically irrelevant.

Proposed criteria for a significant antigen-specific response are (1) minimal number of 10–20 cytokine+ T cells in the antigen-stimulated sample and (2) at least 3- to 5-fold higher frequency of cytokine+ T cells among total T cells in the antigen-stimulated sample than in the control sample.

◆◆◆◆◆◆ EXAMPLES

Detection and isolation of IFN-γ-secreting CMV-specific human Th cells

The first example shows the detection and isolation of IFN-γ-secreting CMV-specific CD4+ T cells using the Cytokine Secretion Assay (Fig. 2). PBMC from a CMV seropositive donor were incubated for 16 h with 5 μg ml⁻¹ CMV lysate (A+B) or without antigen (C+D) as described in the protocol above. The IFN-γ Secretion Assay was performed with 10^7 cells on both samples as described in the protocol B1.1 including enrichment of IFN-γ-secreting cells by a MACS separation over two sequential MiniMACS columns. Counterstaining of T cells was performed using CD4-FITC. Monocytes were stained with CD14-PerCP and dead cells were stained with PI. Flow cytometric analysis was performed on the original fractions and the enriched fractions. At least 200 000 events of the original fractions and the complete enriched fractions were acquired. A lymphocyte gate based on forward and side scatter properties (FSC/SSC) was set (like shown in Fig. 3A) and dead cells and monocytes were excluded according to PI and CD14-PerCP staining in a FL-2 versus FL-3 plot (as shown in Fig. 3B). For analysis PE (IFN-γ) versus FITC (CD4) staining of gated (viable) lymphocytes is displayed (Fig. 2A–D).

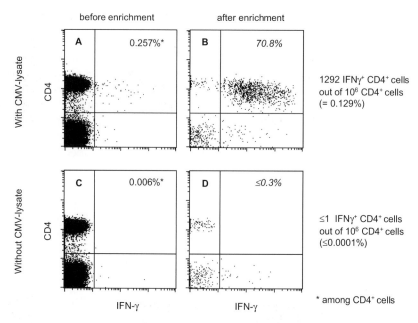

Figure 2. Detection and isolation of IFN-γ-secreting CMV-specific human Th cells.

Analysis of expression of IFN-γ and IL-2 by CMV-specific human CD8+ cells

The second example shows the analysis of CMV-specific CD8+ T cells for secretion of IFN-γ and IL-2 using a combination of Peptide/MHC Tetramer staining with Cytokine Secretion Assay (Fig. 3). PBMC from a CMV seropositive, HLA-A2+ donor were stained for 1 h at 4°C with a PE-labelled CMVpp65$_{495-503}$/HLA-A2 Tetramer (D, F and G, I) or incubated without Tetramer (E and H). Cells were then stimulated for an additional 2 h at 37°C with 5 µg ml^{-1} CMV peptide pp65$_{495-503}$ (E, F and H, I) or without peptide (D and G). The IFN-γ (D–F) or IL-2 (G–I) Secretion Assay was performed on the samples as described above, but without magnetic labelling/enrichment. Cells were stained with IFN-γ Detection Antibody (APC) (D–F) or IL-2 Detection Antibody (APC) (G–I), CD8.FITC, CD14.PerCP and PI. For flow cytometric analysis 200 000 events of each sample were acquired. A lymphocyte gate and an exclusion gate for dead cells and monocytes were set as described earlier (Fig. 3A+B). In addition CD8+ cells were gated based on FL-1 properties (Fig. 3C). For analysis APC (IFN-γ) versus PE (Tetramer) staining of gated (viable CD8+) lymphocytes is displayed (D–I).

While in this donor the majority (87%) of CMVpp65$_{495-503}$ specific CD8+ T cells secretes IFN-γ upon stimulation (Fig. 3F), only 16% secrete IL-2 (Fig. 3I).

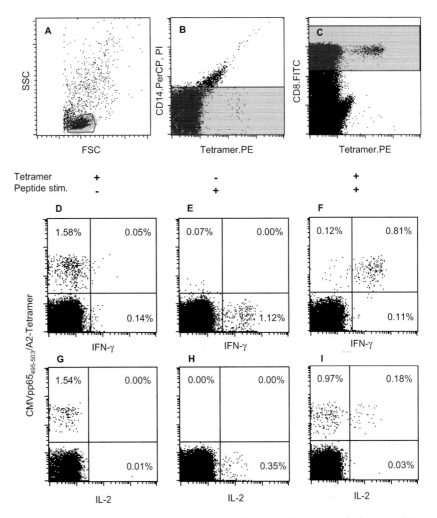

Figure 3. Analysis of expression of IFN-γ and IL-2 by CMV-specific human CD8+ cells.

Detection and isolation of IFN-γ-secreting HEL-specific murine Th cells

The last example shows the detection and isolation of IFN-γ-secreting HEL-specific murine CD4+ T cells using the Cytokine Secretion Assay (Fig. 4). BALB/c mice, female, 7–10 weeks old, were intraperitoneally (i.p.) immunized with 100 μg Henn eggwhite lysozyme (HEL) and 200 ng Pertussis Toxin (PT) in incomplete Freund's adjuvant (IFA) and an additional injection of 200 ng PT in PBS 24 h later. After 2–3 weeks spleen cells (SC) from immunized mice (C–F) or unimmunized control mice (G+H) were incubated *in vitro* for 16 h with 100 μg ml⁻¹ HEL (C+D and G+H) or without HEL (E+F) in RPMI 1640 medium supplemented with 5% murine (heat-inactivated) serum. IFN-γ Secretion Assays were performed with 10^7 cells on all samples as described earlier including enrichment of cytokine-secreting cells by MACS separation over two sequential MiniMACS columns. Counterstaining of T cells was performed using CD4.FITC. B lymphocytes were stained with CD45R(B220).PerCP and dead cells were stained with PI. Flow cytometric analysis was performed on the original fractions and the enriched fractions. Two hundred thousands events of the original fractions and the complete enriched fractions were acquired. A lymphocyte gate based on forward and side scatter properties (FSC/SSC) was set (Fig. 4A) and dead cells and B cells were excluded according to PI and CD45R(B220).PerCP staining in a FL-2 versus FL-3 plot (Fig. 4B). For analysis PE (IFN-γ) versus FITC (CD4) staining of gated (viable) lymphocytes is displayed.

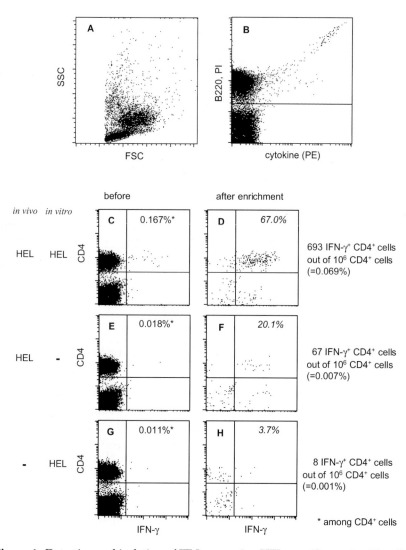

Figure 4. Detection and isolation of IFN-γ secreting HEL-specific murine Th cells.

References

Assenmacher, M., Löhning, M., Scheffold, A., Manz, R. A., Schmitz, J. and Radbruch, A. (1998). Sequential production of IL-2, IFN-γ and IL-10 by individual staphylococcal enterotoxin B-activated T helper lymphocytes. *Eur. J. Immunol.* **28**, 1534–1543.

Becker, C., Pohla, H., Frankenberger, B., Schüler, T., Assenmacher, M., Schendel, D. J. and Blankenstein, T. (2001). Adoptive tumor therapy with T lymphocytes enriched through an IFN-γ capture assay. *Nature Medicine* **7**, 1159–1162.

Brosterhus, H., Brings, S., Leyendeckers, H., Manz, R. A., Miltenyi, S., Radbruch, A., Assenmacher, M. and Schmitz, J. (1999). Enrichment and detection of live antigen-specific CD4+ and CD8+ T cells based on cytokine secretion. *Eur. J. Immunol.* **29**, 4053–4059.

Farrar, J. D., Ouyang, W., Löhning, M., Assenmacher, M., Radbruch, A., Kanagawa, O. and Murphy, K. M. (2001). An instructive component in T helper cell type 2 (Th2) development mediated by GATA-3. *J. Exp. Med.* **193**, 643–648.

Hu-Li, J., Pannetier, C., Guo, L., Löhning, M., Gu, H., Watson, C., Assenmacher, M., Radbruch, A. and Paul, W. E. (2001). Regulation of expression of IL-4 alleles: analysis using a chimeric GFP/IL-4 gene. *Immunity* **14**, 1–11.

Manz, R., Assenmacher, M., Pfluger, E., Miltenyi, S. and Radbruch, A. (1995). Analysis and sorting of live cells according to secreted molecules relocated to a cell-surface affinity matrix. *Proc. Natl. Acad. Sci. USA* **92**, 1921–1925.

McNeil, A. C., Shupert, W. L., Lyasere, C. A., Hallahan, C. W., Mican, J., Davey, R. T. Jr. and Connors, M. (2001). High-level HIV-1 viremia suppresses viral antigen-specific CD4+ T cell proliferation. *Proc. Natl. Acad. Sci. USA* **98**, 13878–13883.

Oelke, M., Moehrle, U., Chen, J. L., Behringer, D., Cerundolo, V., Lindemann, A. and Mackensen, A. (2000a). Generation and purification of CD8+ melan-a-specific cytotoxic T lymphocytes for adoptive transfer in tumor immunotherapy. *Clin. Cancer Res.* **6**, 1997–2005.

Oelke, M., Kurokawa, T., Hentrich, I., Behringer, D., Cerundolo, V., Lindemann, A. and Mackensen, A. (2000b). Functional characterization of CD8+ antigen-specific cytotoxic T lymphocytes after enrichment based on cytokine secretion: comparison with the MHC-tetramer technology. *Scand. J. Immunol.* **52**, 544–549.

Ouyang, W., Löhning, M., Gao, Z., Assenmacher, M., Ranganath, S., Radbruch, A. and Murphy, K. M. (2000). Stat6-independent GATA-3 autoactivation directs IL-4-independent Th2 development and commitment. *Immunity* **12**, 27–37.

Pittet, M. J., Zippelius, A., Speiser, D. E., Assenmacher, M., Guillaume, P., Valmori, D., Lienard, D., Lejeune, F., Cerottini, J. C. and Romero, P. (2001). Ex vivo IFN-gamma secretion by circulating CD8 T lymphocytes. Implications of a novel approach for T cell monitoring in infectious and malignant diseases. *J. Immunol.* **166**, 7634–40.

Smits, H. H., van Rietschoten, J. G., Hilkens, C. M., Sayilir, R., Stiekema, F., Kapsenberg, M. L. and Wierenga, E. A. (2001). IL-12-induced reversal of human Th2 cells is accompanied by full restoration of IL-12 responsiveness and loss of GATA-3 expression. *Eur. J. Immun.* **31**, 1056–1065.

3 Measurement of Cellular Proliferation

Klaus Pechhold and Nancy Craighead
Transplantation and Autoimmunity Branch, National Institute of Diabetes and Digestive, and Kidney Diseases, NIH

Daniela Wesch and Dieter Kabelitz
Institute of Immunology, University of Kiel, Germany

◆◆

CONTENTS

◆◆◆◆◆◆ INTRODUCTION

Activation and proliferation, along with differentiation into effector cells, are important responses of the cells of the immune system to a challenge by invading pathogens. This expansion is primarily meant to increase the number of effector cells capable of generating both the necessary inflammatory milieu and an increased frequency of pathogen-specific responder cells. Thus, proliferation is a critical response of lymphocytes to antigenic stimulation, accompanying or in many cases preceding their differentiation into effector cells and the formation of specific classes of immune responses.

Depending on the nature of the pathogen and the challenge to the host, the immune response may differ considerably in terms of the cellular components involved and the appropriate effector functions necessary to protect the host. Certain patterns of responses are characteristic or

Copyright © Elsevier Science Ltd
All rights of reproduction in any form reserved

prominent during the immune response to different pathogens. For instance, cytotoxic T lymphocytes (CTL) usually appear during viral infections, T helper 1 (Th1) responses are induced by intracellular bacteria such as mycobacteria, while Th2 patterns can be typically detected in association with helminth infections, and humoral (B lymphocyte) immune responses are frequently observed following immunization with protein. Yet co-operative action among the different components, which may change dynamically during the maturation of the immune response, is important and seems to be a hallmark of the function of the immune system. The regulation of the immune response is a major focus in bio-medical research.

The unfolding of immune responses *in vivo* is unique. Nevertheless, *in vitro* experiments using cells of the immune system remain an important supplement to the study of infectious diseases. Traditionally, pro-liferation assays are performed *in vitro* using isolated and highly purified lymphocyte subsets in order to dissect cellular complexities, and focus on the particular roles of the individual subsets. Unfortunately, studies on isolated cell subsets are always at risk of ignoring and underestimating potentially important regulatory influences.

The purpose of all *in vitro* measurements of cellular proliferation is to estimate the extent to which cells have entered the cell cycle, as evidenced by DNA synthesis, cell division, or an increase in overall cell numbers as a result of activation and specific culture conditions. Frequently these culture conditions contain antigen(s) or mitogens and are therefore stimu-latory, especially when using primary lymphocytes. Cell-based bioassays can be used to detect soluble mediators by means of characteristic growth factor dependency of defined indicator cell lines. In this case, cellular proliferation is used to interpret the nature and quantity of such soluble mediators, although these assays have largely been replaced by bio-chemistry-based methods such as the EIA (or ELISA). Additionally, the inhibition of proliferation by suppression, anergy and apoptosis has attracted increasing interest in recent years. Here, the decrease of growth rate and/or cellularity reflects the extent of growth inhibition.

The aim of this chapter is to provide an overview of various techniques used to assess the growth pattern of lymphocytes primarily *in vitro*. We will describe representatives of proliferation assays that focus on different stages during lymphocyte growth. For instance, incorporation of labelled nucleotide analogues such as [³H]-thymidine ([³H]TdR) or bromodeoxy-uridine (BrdU) can accurately determine the rate of *de novo* DNA synthesis of cells that have entered the cell cycle, whereas cleavage of the membrane-permeable tetrazolium salt MTT (or WST, one of its most recent successors) measures total metabolic activity. In contrast, FCM-based procedures, such as the tracking of cell divisions by CFSE fluorescence label decay or the enumeration of absolute cell numbers by standard cell dilution analysis (SCDA), permit the analysis of individual subsets of cells among a heterogeneous population. By pointing out strengths and limitations of the various methods we attempt to help the reader in choosing the most appropriate proliferation assay available for a given application.

◆◆◆◆◆◆ COLORIMETRIC ASSAY: ENZYMATIC CLEAVAGE OF MTT

Principle

MTT is a water-soluble and colourless tetrazolium salt, which has been widely used to measure redox potential of cells. The MTT cleavage based measurement of metabolic activity can be used to assess viability, proliferation, or cytotoxicity (Mosmann, 1983). Upon reduction, MTT, like other tetrazolium salts, forms a coloured formazan compound, which crystallizes, allowing the detection either as a precipitate in histochemical localization studies or photometrically upon dissolution of the crystals. In the photometric assay cells are incubated with the membrane-permeable MTT, allowing access to microsomal enzymes (dehydrogenases) of the respiratory chain, which cleave MTT to form the purple crystallization product. Following dissolution in organic solvents, quantification of the coloured cleavage product yields an estimate of cellular viability and metabolic activity.

Measurement of Cellular Proliferation

Specific equipment and reagents

- MTT (3-[4,5-dimethylthiazol-2-yl]-2,5-diphenyltetrazolium bromide) is solubilized at 5 mg ml^{-1} in PBS, and filter sterilized (0.2 μm). To avoid spontaneous precipitate formation, aliquots should be frozen if stored for an extended period.
- Acidic isopropanol (0.04–0.1 N HCl in isopropanol) or 0.01 N HCl in Tris/1% SDS detergent (HCL/SDS) to solubilize the formazan precipitate.
- Conventional scanning multiwell spectrophotometer (ELISA-reader) equipped with wavelength filters of 570 nm (specific readings) and 630–690 nm (background wavelength).

Assay

1. The MTT assay is performed in a flat-bottom 96 well microtitreplate. Depending on the cell type and assay conditions, 5×10^4–5×10^5 cells are cultured in 100–200 μl per well culture medium (e.g. RPMI-1640) supplemented with 10% FCS (CM-10) for the desired periods.
2. Culture supernatant in excess of 100 μl should be removed before adding 20 μl MTT solution (final concentration 0.5–1 mg ml^{-1}). After incubation for 4 h at 37°C a purple-blue coloured formazan precipitate should be visible in wells containing viable cells.

(contd.)

3. The reaction is stopped by addition of 100 µl acidic isopropanol or HCl/SDS to dissolve the dye crystals. Dissolution is complete within 30 min after addition of acidic isopropanol and repeatedly pipetting up and down, or will proceed more slowly when HCl/SDS is used (e.g. incubation overnight at 37°C). At this point, visual discrimination between positive (purple) and negative (yellow) wells is obvious (screening).
4. For quantification the O.D. can be read on an ELISA-reader using a 570 nm filter. A 630–690 nm filter should be used for background readings. The negative control wells (reference wells, blank) contain MTT in CM-10 but no viable cells, and do not show discolorization. O.D. readings of strongly proliferating cells can be as high as 0.5–2.0.

Because of considerable variations among cells types, it is recommended to test the desired responder cells or indicator cell lines for their capacity to reduce MTT in a 4-h assay.

Application

MTT has been extensively used in bulk cell viability and proliferation assays. Its principal application is a high throughput screening procedure for rapidly growing cell lines, especially immortalized cells, such as hybridomas or the indicator cell lines used in bioassays (e.g. cytokine detection bioassays) (Heeg et al., 1985). Advantages over alternative proliferation methods are: (a) no radioactivity is involved, (b) adherent cell lines can be used as readily as suspension cells, (c) fast and easy applicability and high-throughput analysis using a scanning multiwell spectrophotometer (ELISA-reader). Conversely, the nature of the MTT assay readily explains some of its disadvantages: (a) heterogeneous populations are difficult to analyse due to a considerable variability in the contribution of individual cell types in reducing MTT (variable size and metabolic activity), and (b) non-proliferating accessory cells even if irradiated and unable to synthesize DNA, may still metabolize MTT to some extent, which can obscure the specific analysis of the desired responder cells. Therefore, the applicability of the MTT assay may be limited due to culture condition and the cell populations investigated. Additionally, the MTT assay has been reported to give false results when cells are contaminated with mycoplasma (Denecke et al., 1999).

Although MTT remains the most common tetrazolium salt used in viability and proliferation assays, MTT analogues with water-soluble formazan cleavage products are now available: XTT: (2,3-bis[2-methoxy-4-nitro-5-sulfophenyl]-2H-tetrazolium-5-carboxanilide) (Roehm et al., 1991), MTS 3-(4,5-dimethylthiazol-2-yl)-5-(3-carboxymethoxyphenyl)-2-(4-sulfophenyl)-2-H-tetrazolium salt (Buttke et al., 1993), and WST-1 (4-[3-(4-iodophenyl)-2-(4-nitrophenyl)-2H-5-tetrazolio]-1,3-benzene disulfonate) (Wagner et al., 1997). While these newer analogues are more expensive than MTT, they have the distinct advantage of not requiring a solubilization step prior to analysis, thus reducing a potential source of error and decreasing sample processing time. The newer analogues are reported to have comparable or better sensitivity than MTT. Kits based on MTT, XTT, and WST-1 are available commercially (e.g. BioVision, Inc., PanVera Corporation, Promega Corporation, and Roche Molecular Biochemicals). The limitations of these assays are the same as those listed above for the MTT assay.

◆◆◆◆◆◆ INCORPORATION OF LABELLED NUCLEOTIDE-ANALOGUES DURING DNA SYNTHESIS

Principle

Many cells, especially lymphocytes, start vigorously to expand by repetitive mitosis cycles following appropriate stimulation. Each cycle of cell division requires cell cycle progression and DNA synthesis. Basically, each cell of a given population synthesizes a similar amount of DNA prior to its division. Thus a simple expansion of a cell population correlates quantitatively with its rate and extent of DNA synthesis. This basic relationship led to the wide application of assessing DNA synthesis as a measure of cellular proliferation.

◆◆◆◆◆◆ [^{3}H]-THYMIDINE

The labelled nucleotide [^{3}H]TdR at a sufficiently high concentration (i.e. 2–3 nM, 0.2 MBq ml^{-1}) competes with endogeneous thymidine for incorporation into newly synthesized DNA (Strong *et al.*, 1973). Convenient quantification is achieved by detecting radiolabelled, genomic DNA using glassfibre filter mats, which trap genomic DNA along with some insoluble cell fragments but spare oligonucleotides or single bases.

Specific equipment and reagents

- Tritiated methylthymidine ([^{3}H]TdR, specific activity 37 MBq ml^{-1} [1 mCi ml^{-1}]).
- Cell harvester unit to collect DNA from 96-well microtitre plate cultures in a single aspiration and washing cycle onto suitable glassfibre filter mats.
- β-emission counter, which is mostly based on liquid scintillation technology. They detect light generated in response to β-particle energy absorption by scintillator atoms. The light intensity measured correlates directly with the total DNA-bound [^{3}H]. Scintillation counters are available from several sources (e.g. PerkinElmer, Packard Bioscience).
- Appropriate containers to collect liquid (scintillation fluid and CM) and solid radioactive waste.

> **Assay**
>
> - [³H]TdR working solution (1.85 MBq ml⁻¹ [50 µCi ml⁻¹]) using CM without serum can be prepared in advance and stored for several weeks if kept sterile. Caution must be exerted when handling radioactivity.
> - Cells in 96-well microtitre plates are [³H]TdR-labelled by adding 20 µl working solution per 200 µl culture volume, and incubated for an additional 4–18 h at 37°C before harvesting. The optimal incubation period for individual applications should be established in pilot experiments, although generally there is little change of [³H]TdR uptake beyond 4–6 h of radioactive labelling.
> - Microtitre plates may be stored frozen at −20°C after labelling.
> - [³H]TdR-labelled microtitre cultures are harvested by automated aspiration and rinsing onto a glassfibre filter mat as recommended by the cell harvester's manufacturer. Current systems harvest an entire 96-well plate at one time.
> - Glassfibre filter mats must be completely dry before counting by any β-counter. They can be air-dried overnight at room temperature or by using a drying oven or microwave. Depending on the instrument used, filters are either soaked with scintillation fluid in a suitable container (vial or plastic bag) or directly counted according to the manufacturer's recommendations.
> - Radioactivity is measured as cpm and averaged to dpm. Dpm of experimental cultures ought to be compared with [³H]TdR-labelled control cultures devoid of any responder cells or containing un-stimulated cells (background).

Troubleshooting

Unexpectedly low counts are frequently due to bad timing of the [³H]TdR labelling. Too early or, more often, too late addition of [³H]TdR misses the maximal proliferative response even though the cultures may still look healthy microscopically. We strongly recommend determining the optimal response by repetitive testing of replicate cultures every day for a period of several days (e.g. days 3–6). Even once established, a 2–3 time-point kinetic assay is considered far more reliable when comparing very different stimulatory culture conditions (e.g. mitogens vs. antigens). False-negative measurements can also occur due to failure to collect all the cells from the culture well or from the generation of apoptotic DNA-fragments of newly synthesized [³H]-labelled DNA, which are not efficiently retained by the filter mats. Often, a sharply increased standard deviation among replicate cultures points to random heterogeneity in the cell collection process or to cell death due to overgrowth.

Rinsing of culture wells may be inadequate, especially if adherent cells are used. If microscopic examination of individual culture wells reveals residual plastic adherent cells after the first harvest, the authors recommend an additional trichloroacetic acid (TCA; 1% in PBS) or trypsin/ EDTA (0.25%/1 mM) detaching step followed by a second harvest cycle onto the same or on another filter mat, with dpm summation of respective culture well positions in the latter case. Obviously, cell collection failure may also reflect a dysfunction of the harvester, such as clogging of aspiration nozzles or tubing preventing effective aspiration and washing of the culture wells. In this case individual parts or the whole harvester circulation can be flushed with an appropriate cleansing reagent.

The glassfibre filter mats may not collect all DNA-bound radioactivity. This can be due to one of two reasons. First, insufficient thickness of the filter mat, which can be explored by using two filter mats together followed by separate counting. Secondly, stimulated cells may undergo apoptosis after a characteristic period of growth and DNA synthesis, resulting in oligonucleosomal fragmentation of their DNA. This may happen more frequently if the [³H]TdR labelling period is prolonged. Oligonucleotides may not be efficiently retained by the filtermats. In fact, an apoptosis/cytotoxicity assay has been designed to exploit exactly this feature. It measures the reduction of [³H]-activity collected by the filtermats from prelabelled cells exposed to an apoptotic stimulus (Matzinger, 1991). The presence of apoptotic cells may be confirmed using several FCM approaches for apoptosis (see below and elsewhere).

◆◆◆◆◆◆ BROMODEOXYURIDINE

A non-radioactive alternative to the [³H]TdR based cell proliferation assay uses the pyrimidine analogue 5-bromo-2'-deoxyuridine (BrdU). Both nucleotide analogues are believed to be similarly incorporated into newly synthesized DNA (thymidine analogue, Crissman and Steinkamp, 1987). While its usefulness in labelling proliferating cells *in vivo* followed by its detection with immunohistochemistry or flow cytometry (FCM) has been well established (Tough and Sprent, 1994), *in vitro* assays for cell proliferation have become available only recently (Maghni *et al.*, 1999). ELISA- or luminescence-based kits are available, which are marketed as a reliable substitute for conventional [³H]TdR assays.

Specific equipment and reagents

- Cell proliferation ELISA kit (Roche Molecular Biochemicals, AmershamPharmacia Biotech) or luminescence-based kits (Roche).
- Centrifuge with 96-well plate adapter.
- Spectrophotometer. The optical filter(s) required depend upon the substrate used.
- Microplate heater (60°C) to dry BrdU-labelled microcultures. A hairdryer can be used if a plate heater is not available.

Assay

1. Culture cells in 96-well flat bottom microtitre plates in a total volume of 200 μl. Due to differences to conventional [³H]TdR assays, an experimental optimization of the BrdU ELISA is highly recommended with respect to the time point of maximum proliferation and the number of responder cells per well. Additionally, a blank (no cells) and a background control (no BrdU label) are required to test for non-specific binding of the labelled anti-BrdU antibody.
2. Prepare BrdU labelling solution, stock and working solutions of peroxidase (POD)-conjugated anti-BrdU antibody, and ELISA wash buffer as directed by the manufacturer.
3. Add 20 μl of BrdU labelling solution to each well (10 μM BrdU final concentration). Return culture to the incubator for 2–4 h.
4. Centrifuge the plate at 300 g for 10 min. Remove the supernatant by flicking or suction. Heat the plate at 60°C for 1 h. (The dried plates can be stored for up to 1 week at 2–8°C.)
5. Add 200 μl per well of the fixing/denaturing solution. Incubate 30 min at room temperature. Remove the fixing/denaturing solution by flicking or suction.
6. Add 100 μl/well anti-BrdU POD working solution. Incubate 30–120 min at room temperature. Remove the anti-BrdU POD working solution by flicking or suction.
7. Wash wells three times with 200–300 μl ELISA wash buffer. Remove all of the washing solution by tapping the plate on a paper towel.
8. Add 100 μl substrate solution (provided by manufacturer) per well. Incubate at room temperature until colour development is sufficient to be detected photometrically (5–30 min).
9. Measure absorbence in scanning microwell spectrophotometer (ELISA reader) at the appropriate wavelengths. The cell proliferation ELISA Kit from Roche Molecular Biochemicals required 450 nm (specific) and 630 nm (reference) optical filters, respectively.

Application for both [³H]TdR and BrdU

Although [³H]TdR incorporation assays can be used almost universally to detect eukaryotic cell proliferation *in vitro*, there are some limitations to this method. First, and most importantly, [³H]TdR uptake does not allow measurement of the individual rates of DNA synthesis, which may occur among various subsets of heterogeneous cell populations. Rather, like MTT assays (see above), [³H]TdR uptake estimates the average DNA synthesis of all cells in the culture. Secondly, it has been convincingly demonstrated in recent years that proliferation as measured by [³H]TdR does not always reflect cellular expansion quantitatively. For instance, activation-induced cell death (AICD) has been shown to be a widely operative and important negative regulatory mechanism limiting cellular expansion especially of T cells (Kabelitz *et al.*, 1994). Activation followed by cell cycle progression and DNA synthesis may sensitize T cells to undergo AICD (Boehme and Lenardo, 1993). This in turn can result in an overall increase in [³H]TdR uptake while viable cell numbers decrease at the same time due to AICD (Kabelitz *et al.*, 1995).

BrdU cell proliferation assays are as convenient to perform as [³H]TdR assays, and unlike the latter involve no radioactivity. In the authors' hands, however, currently available BrdU-based cell proliferation kits (Cell Proliferation Elisa, Roche Molecular Biochemicals and Biotrak, AmershamPharmacia Biotech) can result in substantial differences when compared with standard [³H]TdR assays, especially when IL-2 driven T-cell responses are examined. One typical example is illustrated in Fig. 1. Measurement of antigen-specific proliferation of T-cell receptor (TCR) transgenic T cells at an early time point after activation (day 2 during a 4-day proliferation kinetic) yielded comparable results between [³H]TdR and BrdU. However, 24 h later the [³H]TdR results were substantially higher in the presence of recombinant IL-2 (up to two- to five-fold, Fig. 1, day 3, grey squares and circles). The reason is not clear, but may, in part, reflect assay-specific differences in the sensitivity of detection of labelled DNA under certain conditions, especially in the presence of ongoing apoptotic cell fragmentation. Thus, in view of the paucity of published reports on the usage of BrdU to determine cell proliferation *in vitro* (Maghni *et al.*, 1999), and without further thorough testing, the authors recommend that BrdU cell proliferation estimates be interpreted with caution.

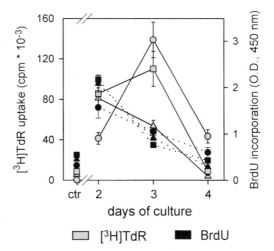

Figure 1. T-cell proliferation measured by DNA incorporation of nucleotide analogues: similarities and differences between [³H]TdR and BrdU incorporation. Purified naive LCMV-GP-specific TCR transgenic CD8+ T cells were stimulated with its cognate peptide (gp33) presented by dendritic cells (DC, circles) or B7-1-transfected fibroblast cell lines (FCL). FCL:T cell stimulation was further analysed without (triangles) or with a low dose of exogenous murine IL-2 (2 ng ml⁻¹, squares). Proliferation of triplicate cultures was determined on days 2, 3, and 4 by [³H]TdR-incorporation (grey symbols) or BrdU (cell proliferation kit, Roche, black symbols). Control wells were seeded identically except for an irrelevant peptide. Control cultures showed similar background proliferation on days 2 and 3, but decreased growth on day 4. Day 2 is shown (ctr).

◆◆◆◆◆◆ FLOW CYTOMETRY APPLICATIONS TO DETERMINE CELL NUMBERS AND CELL DIVISIONS

Tracking cell divisions by CFSE-labelling

Principle

Carboxyfluorescein diacetate succinimidyl ester (CFDA-SE or frequently just CFSE) is a fluorescein-based fluorochrome that meets standard excitation and detection settings used by conventional flow cytometers or fluorescence microscopes. CFSE can be used to label viable cells stably for weeks to months *in vivo* by virtue of its exquisite membrane permeability and its subsequent modification of fluorescein residues by intracellular enzymes, covalently coupling them to cellular proteins (Parish, 1999). There is little cytotoxicity at the moderate CFSE concentrations used for labelling. Importantly, cell proliferation results in distribution of CFSE fluorescence equally to daughter cells, while migratory patterns and proliferative responses seem not to be adversely affected, making it a powerful tool for both *in vitro* application and cell tracking studies *in vivo* (Hasbold *et al.*, 1999).

Here we will focus on the application of CFSE to measure proliferative T-cell responses (i.e. cell divisions) *in vivo* and *in vitro*.

Specific equipment and reagents

- CFDA-SE or, more conveniently, Vybrant CFDA-SE Cell Tracer Kit (Molecular Probes).
- Flow cytometer equipped with a standard argon laser (excitation at 488 nm) and a bandpass filter/detector capable of detecting fluorescein emission at ~530 nm, such as FACScan, FACSCalibur, Coulter XL. Most instruments have at least two additional filter/detector combinations available to measure PE (585 nm) and far red emissions (>650 nm, e.g. PI, PerCp, Cy5), which can be combined with CFSE for multicolour analysis.

Assay

1. CFSE stock solution is prepared at 5 mM CFSE in DMSO.
2. CFSE labelling according to manufacturer recommendation: centrifuged viable cells are carefully resuspended at $2–5 \times 10^6$ ml^{-1} in 37°C prewarmed PBS/CFSE (0.5–2 µM) for 15 min, centrifuged, resuspended in prewarmed (37°C) complete CM-10, and incubated at 37°C for an additional 30 min. Alternatively, equally effective labelling can be achieved by incubation for 5–10 min in prewarmed CM-5 containing CFSE, followed by two washing steps. However, a significantly larger concentration of CFSE is needed in the latter procedure (2–10 µM). Thus, the optimal labelling conditions should be established experimentally.

(contd.)

3. FCM analysis and staining, respectively, are performed according to standard procedures. Certainly, CFSE occupies the FL1 channel (515 nm band pass filter) such that FITC conjugated mAb cannot be used for multicolour staining. However, any other colour (e.g. PE, PerCP, Cy5, allophycocyanine) can be combined.

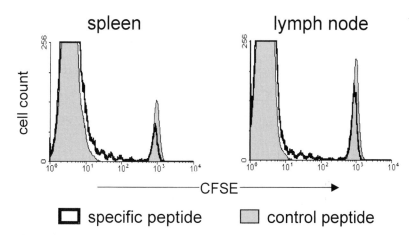

Figure 2. Detection of proliferation by CFSE cell tracking *in vivo*. T cells from LCMV-GP-specific TCR transgenic mice were purified, CFSE-labelled (15 min 37°C PBS containing 1 μM CFSE, washed once and incubated for 30 min in complete CM-10), and adoptively transferred i.v. (10^7 per mouse) into congeneic mice. 24 h later irradiated, B7-1+ FCL were injected i.p. (2×10^5 per mouse) that had been loaded with the cognate (gp33) or a control peptide (np396). Three days later, spleen (left) and lymph nodes (right) were harvested and CFSE fluorescence determined. Histograms show total CD8+ T cells (large unstained fraction are recipient T cells, whereas the smaller CFSE-positive population represents donor cells). T-cell proliferation was noted upon stimulation with cognate peptide-loaded FCL (thick black line), while control peptides did not show CFSE decay (shaded curves).

Application

Example of CFSE-labelled cells *in vivo*: Lymphocytic Choriomeningitis Virus (LCMV)-specific TCR-transgenic T cells recognize a peptide (gp33) derived from the LCMV-glycoprotein in the context of H-2D^b. We purified splenic T cells from TCR-transgenic naive mice, labelled them with CFSE (1 μM, 15 min PBS, 30 min CM-10), and adoptively transferred the labelled T cells i.v. into C57BL/6 mice (10×10^6/mouse). Twenty-four hours later a limited number of a syngenic fibroblast cell line (FCL, 0.2×10^6 per mouse) were either loaded with gp33 or control peptide and were injected i.p. Another 3 days later spleen and lymph node cells were analysed for CFSE decay. Figure 2 illustrates the strong proliferative response of a fraction of the CFSE labelled CD8+ T cells when immunized with fibroblasts presenting the cognate peptide (thick line) but not with control peptide (grey area). Thus antigen-specific proliferative responses in various tissues can be examined using CFSE cell-tracking technology.

Example of *in vitro* applications: purified human T cells were CFSE labelled (1 μM) and stimulated with irradiated autologous antigen presenting cells plus synthetic phosphoantigen isopentenyl pyrophosphate IPP (2 μg/ml) to selectively activate Vγ9-expressing γδ T cells, present at about 1–5% in the peripheral T-cell pool. For comparison, CFSE-labelled T

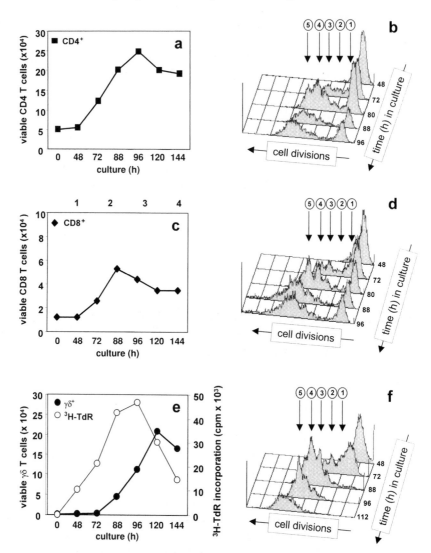

Figure 3. Comparison of SCDA and CFSE analyses *in vitro*. Purified human T cells were CFSE-labelled (right panels) or left unstained (left panels) and were stimulated with irradiated, autologous APC and SEA (a–d) or IPP plus IL-2 (10 U ml⁻¹) (e,f). Expansion, as determined by SCDA, and CFSE-fluorescence decay was closely monitored over a period of up to 7 days (168 h). In addition, γδ T-cell proliferation was determined in parallel cultures by [³H]TdR (e, ○). While both SCDA and CFSE analysis indicated that SEA-stimulated αβ T-cell expansion peaked around day 4 and ceased thereafter, γδ T cells growth occurred well beyond this point, showing expansion and division for as long as day 6. However, proliferation based on [³H]TdR incorporation peaked substantially earlier, around day 5, and sharply declined thereafter.

cells were activated by the superantigen staphylococcal enterotoxin A (SEA; 10 ng/ml) in separate cultures. Individual subsets were examined for their ability to be activated by the different stimuli and the kinetics of the resulting proliferative response. Figure 3b,d,f illustrates CFSE decay of CD4+ and CD8+ and γδ T cells.

While CFSE-labelling has been widely used in the recent past (Lyons, 2000), there are certainly limitations. First, although CFSE-labelling seems to be fairly well tolerated, it remains to be seen if certain cell functions are adversely affected, especially since strenuous cell separation procedures may be additionally applied to purify cell subsets for further labelling. Secondly, by halving the fluorescence with each division, background fluorescence levels are achieved with six or seven divisions depending on the initial labelling intensity. Thus, in many cases, analysis of proliferative response is possible for the first four to seven divisions, but as soon as the division-leading clones merge with background fluorescence levels, quantitative evaluations become impossible. Finally, while CFSE cell tracking is particularly useful for naive or resting cells, activated and proliferating cells or cell lines are not synchronized at the time of labelling and will, therefore, not demonstrate the typical discontinuous 'division peak' pattern. Rather, cycling cells more or less homogeneously lose their fluorescence intensity, allowing only for a rough estimate of the average proliferative response of the overall subset.

◆◆◆◆◆◆ DETERMINATION OF ABSOLUTE CELL NUMBERS: STANDARD CELL DILUTION ANALYSIS (SCDA)

Principle

Conventional FCM determines the relative frequency of cells using criteria defined by one or more parameters such as cell size (FSC vs. SSC), and fluorescence label (see also chapter by Scheffold et al. in Section I). Standard cell dilution analysis (SCDA) is a modified FCM approach, which enumerates phenotypically defined subsets of viable cells (Pechhold et al., 1994a). Basically, SCDA measures the frequency of the cell subset of interest relative to a constant number of reference cells (standard cells) added per sample. The principle is illustrated in Fig. 4. A similar principle underlies some methods of antigen quantification in radioimmunoassays (Feldman and Rodbard, 1971), and certain approaches for quantitative PCR (Siegling et al., 1994). In the former assay competitive binding takes place between monoclonal antibodies (mAb) and a mixture of radioactive labelled antigen of a known concentration and the test antigen. In the latter approach two slightly different PCR templates (i.e. the test template and the control template of a known abundance) compete for binding by the same primers, yielding two PCR products of distinguishable sizes. Similar to SCDA, the ratio of the results obtained from each pair of competitors directly correlates with their relative abundance.

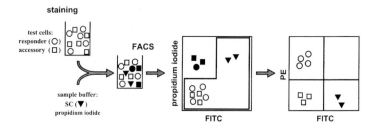

Figure 4. Principle of SCDA. Responder cells among a mixed population of T cells, APC, and accessory cells are stained by a PE-conjugated mAb (○). Following one washing step, the stained sample was mixed with the sample buffer containing SC (▼). PI (0.2 µg ml⁻¹) was added to the sample buffer to distinguish dead cells (●, ■). Finally, the samples are subjected to three-colour FCM. The two FACS displays of the SCDA sample and the electronic gates used are illustrated.

Specific equipment and reagents

SCDA requires the same equipment necessary for staining cells and FCM (see also chapter by Scheffold and Radbruch in Section I). Briefly, FACS staining is performed using appropriately diluted, fluorochrome-conjugated mAbs and staining buffer (PBS 1% BSA). Samples can be run on conventional cell sorters and cell analysers such as FACScan®, FACS Calibur® (Becton Dickinson) and Epics XL® (Beckman-Coulter). Fixation and storage after staining is not possible as it interferes with propidium iodide (PI)-facilitated live/dead cell discrimination.

● Standard cell (SC) preparation. The main criteria for useful SC are:
  (i) homogeneous FSC and SSC appearance comparable to sample cells
  (ii) SC must stain brightly for any single or a combination of several surface marker(s). The labelling is essential since it allows the discrimination of SC from both dead and unstained sample cells (see below). Figure 5a illustrates the FSC and SSC features of SC (black dots), while Figure 5c and 5d depicts the alignment of PI+ FITC+ SC among sample cells.

SC stocks can be prepared from freshly isolated human T cells or from various human or mouse cell lines. The authors have been using biotinylated anti-MHC class I mAb W6/32 plus FITC-avidin for the staining of human SC, although, more conveniently, CFSE-labelled cells (0.5-2 µM for 15 min in CM at 37°C) may be used as described above. Another possibility is the use of fluorescinating polysterene beads, commonly used for flow cytometer set up and calibration. These may be up to 8–10 µm in size but often require modifications in

(contd.)

the gate setting for deviations in FSC and SSC characteristics. Their passage through the instrument's nozzle may be slightly different from that of normal cells, and their application for SCDA needs to be tested and adapted. About 10^7–10^8 SC can be prepared and stored at 10^7 ml^{-1} in PBS 1% paraformaldehyde (PFA), protected from light, at 4°C for several months. Between 10^4 and 10^5 SC per sample are used depending on the staining procedures and sample sizes (see recommended staining procedures below). SC stocks may be counted using a conventional haemocytometer to control cell number stability during storage.

- Two 50–200 µl multichannel pipettes are a very useful aid in handling 96-well microcultures efficiently.

Assay

Sample buffer

The complete sample buffer consists of SC at a concentration of 1–10×10^5 ml^{-1} in PBS 1% BSA, and a low dose of PI (0.2 µg ml^{-1}). The SC concentration should be orientated at the expected number of responder cells being enumerated, which depends on the input cell number and the culture condition used. For statistical reasons, responder cells should not outnumber SC by more than 20–25-fold. Thus SC concentrations of 1–5×10^5 ml^{-1} are adequate for most assays using microcultures (96-well plates) or representative samples of larger cultures. PI concentrations to discriminate between live and dead cells should be as low as possible, because overlapping PI emission spectra can interfere with PE detection, rendering electronic fluorescence compensation difficult. The sample buffer may be prepared in advance and stored at 4°C. It is important to use the same complete sample buffer if titration experiments or kinetic SCDAs are carried out on different days.

Cell microcultures and stainings

These should be done in duplicate. The authors mostly apply SCDA to entire 96-microwell cultures enumerating cells per culture well rather than per aliquot volume. We recommend rinsing the culture wells and pipette tips with PBS after transfer of the cells to a staining vessel in order to minimize cell loss. Two multichannel pipettes greatly facilitate the microculture handling: one transfers the culture into 96V plates for staining in a maximal volume of 100 µl, and in a second step transfers from the same wells another 100 µl PBS added by the other pipette. It is important to note that only the cells contained in the final, stained sample can be enumerated by SCDA. In most cases one colour staining of the sample cells is performed using a fluorescence label distinguishable from both the SC and PI-stained dead cells (Pechhold et al., 1994a). A direct staining procedure using PE-conjugated primary mAbs is preferable over a two-step indirect staining. For two laser instruments additional colours may be possible.

FACS measurement

Samples are washed once after staining. Thoroughly resuspended sample buffer containing SC and PI is added at a constant volume (e.g. 100 μl) to each 96V well and transferred in 0.6 ml sample tubes (Greiner Bio-One, Inc.). After a short incubation of 10–20 min to allow uptake of PI by dead cells, samples have to be run on the flow cytometer.

Data analysis

SCDA data analysis is achieved using a four-parameter gate setting (Pechhold *et al.*, 1994a) as shown in Fig. 5b,c indicating the gate-excluded cells in grey. The first gate is applied to FSC and SSC and is placed more spaciously to ensure maximal sensitivity (no viable cells excluded). However, events with decreased FSC should be excluded by the gate, as they indicate cellular debris (Fig. 5a,b, left edge), while cells with moderately increased SSC and FSC should be included, resulting in the spacious right and upper gate edges. The superimposed second gate (Fig. 5c) is applied to the PI and FITC-signal. The main purpose here is to exclude PI+ cells (dead cells) but spare PI- and all SC (FITC+). Finally, the relative frequencies of the individual populations including SC, stained and unstained sample cells are obtained using the PE vs. FITC (Fl-2 vs. FL-1) display's quadrant statistic analysis (Fig. 5d).

The absolute number of viable cells $N_{[V]}$ of the stained subpopulation is calculated using the simple formula:

$$N_{[V]} = f_{[sample]}/f_{[SC]} \times C_{[SC]} \times V_{[SC]}$$

where $f_{[sample]}$ is the relative frequency of the test sample subpopulation, $f_{[SC]}$ is the relative frequency of the SC, $C_{[SC]}$ is the concentration of SC in the complete sample buffer (e.g. 10^5 ml^{-1}), and V[SC] is the sample buffer volume added to each sample (e.g. 0.1 ml).

The cell number calculation from the example shown in Fig. 5 is as follows: the number of the PE+ test cell population is

$$45.9\%/6.3\% \times 10^5 \text{ ml}^{-1} \times 0.1 \text{ ml} = 7.29 \times 10^4 \text{ per well.}$$

Both the PE-stained and unstained sample's absolute cell numbers can be calculated from the same quadrant analysis data set.

Troubleshooting

Controls

In addition to any applicable staining controls, two types of control are specifically recommended for SCDA. Always run plain complete sample

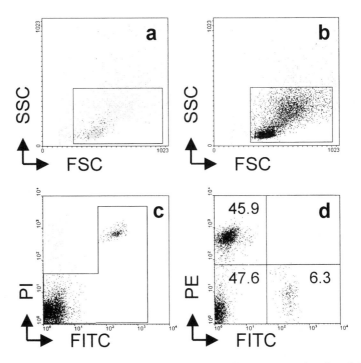

Figure 5. Example of SCDA. Purified human T cells were stimulated with T-cell depleted autologous APC and IPP in the presence of IL-2 (10 U ml⁻¹). On day 4 cells were stained with a PE-conjugated Vγ9-specific anti-TCR γδ mAb. After a washing step, cells were resuspended in sample buffer containing SC (10⁵ ml⁻¹, 0.1 ml per sample) and PI (0.2 µg ml⁻¹), and were immediately run on a FACSscan. (a) and (b) indicates the specific localization of the SC among all cells within the FSC/SSC gate, and of the overall population consisting of PE-stained γδ T cells, unstained PBL, and SC, respectively. Electronic exclusion of PI+ dead cells is illustrated in (c), resulting in the final quadrant analysis shown in (d).

buffer (without any sample cells). This should result in an absolute cell count of substantially less than 1% of the SC (ratio $f_{[sample]}/f_{[SC]} < 0.01$). Secondly, when stimulating cells for a certain period, it is helpful to run SCDA with the input cell population at the onset of culture. This will provide a useful base number for calculation of expansion characteristics. Finally, measurements in duplicates are sufficient, because the reproducibility of replicates is normally high.

Precautions

The precision of the analysis procedure can be monitored at several levels:

- Preparation of the SC stock solution is a critical factor. The fluorescence labelling of SC must be bright, homogeneous (Fig. 5d, lower right quadrant), and the concentration precise (counting three times independently using a haemocytometer). Counting should be repeated once in a while during storage periods. Note that PFA-fixed SCs have lost their membrane dye exclusion capability when using vital stains such as Eosin and Trypan Blue.

- Handling of complete sample buffer must be accurate. Thoroughly resuspend the sample buffer immediately before adding the same precise volume to each sample.
- Make sure to collect all cells from microculture wells by rinsing the wells and the pipette tips at least once. Any cell loss during handling will result in an underestimate of the real absolute cell numbers.
- Mix the sample cells and the SC in the sample buffer by gently vortexing immediately before running it on the flow cytometer.

Applications

Since application and effectiveness of every method grows with each usage designed to address new, different or modified questions, this paragraph is not meant to list all possible applications of SCDA. Instead, three major assays will be described and briefly compared with some other alternatively available techniques.

Growth characteristics of lymphocyte subsets

It has been difficult to study the antigen-stimulated responses of subsets of phenotypically composite cell populations such as Ficoll-Hypaque-purified mononuclear cells from peripheral blood. For instance, protein-aceous antigens derived from various bacteria stimulate proliferation of MHC class II restricted CD4+ T cells but not, or much less efficiently, MHC class I restricted CD8+ T cells, while γδ T cells but not NK cells are activated by non-proteinaceous antigens from mycobacteria. Thus total cell measurements such as [³H]TdR incorporation cannot selectively measure the extent of these responses (Pechhold et al., 1994b). Complementary measurements such as FACS analysis are required to dissect the subsets involved.

As an example, various proliferative T-cell responses were examined in vitro using SCDA (left panel) and CFSE-labelling (right panel) and are shown in Fig. 3. CFSE labelled or untreated peripheral human T cells were stimulated with SEA (Fig. 3a–d) or with IPP (Fig. 3e,f). Proliferation of CD4+ and CD8+ T cells was analysed by SCDA and compared with CFSE fluorescence decay and is detailed in Fig. 3a,b and c,d, respectively. Vγ9+ γδ T cells (Fig. 3e,f) were stimulated by IPP in the presence of IL-2 (10 U ml⁻¹). SEA-stimulation led to a limited expansion of a fraction of both CD4+ and CD8+ T cells, and exhibited approximately five divisions per responding cell corresponding to a four- to five-fold peak overall expansion occurring at day 4. In contrast, γδ T cell proliferation was significantly more sustained, showing at least six divisions with a peak well beyond 5 days. Since all Vγ9+ γδ T cells responded to IPP, their expansion was much more pronounced (>100-fold). Thus, both SCDA and CFSE labelling can contribute greatly to the repertoire of available methods to characterize proliferative T-cell responses in vitro.

Activation-induced cell death (AICD)

T cells can be induced to die upon antigenic or mitogenic stimuli. Thus, quantification of the subpopulations of T cells, which undergo AICD, is a prerequisite quantitatively to examine this important response to antigen stimulation (Kabelitz *et al.*, 1994). It is beyond the scope of this chapter to discuss cellular methods of measuring AICD or apoptosis in detail. However, apoptotic cells are often detected by FCM using annexin V staining, TUNEL assay (terminal deoxynucleotidyl transferase-mediated dUTP-biotin nick end-labelling), or eventually PI uptake. As an alternative, we have successfully applied SCDA to enumerate residual viable cells 12–24 h after activation. This approach has some advantages over other methods if the combined analysis of viable responder cell depletion due to AICD and the subsequent rebound of proliferative T-cell responses over the following 24–48 h are being studied. It is also very suitable if antigen-specific induction of apoptosis using APC is examined or if any accessory cells are present.

◆◆◆◆◆◆ CONCLUSION

We have discussed various principles to measure cellular proliferation and have introduced important representatives. There are advantages and disadvantages, strength and limitations to each assay. We have pointed out that the choice of any method to quantify proliferative responses depends on the experimental system. Therefore, we believe it is important to have an overview of available assays in order to make an educated choice of the most appropriate assay. Table 1 summarizes the authors' opinion about applicability of the various methods under certain common conditions.

Table I.

Assay	MTT	[³H]TdR	BrdU*	CFSE	SCDA
Adherent cell lines	+++	+	++	–	–
Suspension cell lines	+++	+++	+	–	++
T cells, high responder frequency (e.g. mitogen stimulation)	+	+++	+	++	+
T cells; responder cells low; accessory cells; debris; (e.g. Ag-specific cultures)	–	+	–	+++	+++
Growth inhibition; apoptosis	–	+	–†	+	+++

+++ recommended, ++ good, + applicable, – not recommended.
* depending on further validation of the approach in comparison with [3H]TdR.
† further evaluation of BrdU differential labelling during early stages of apoptosis may be worthwhile.

References

Boehme, S. A. and Lenardo, M. J. (1993). Propriocidal apoptosis of mature T lymphocytes occurs at S phase of the cell cycle. *Eur. J. Immunol.* **23**, 1552–1560.

Buttke, T. M., McCubrey, J. A. and Owen, T. C. (1993). Use of an aqueous soluble tetrazolium/formazan assay to measure viability and proliferation of lymphokine-dependent cell lines. *J. Immunol. Methods* **157**, 233–240.

Crissman, H. A. and Steinkamp, J. A. (1987). A new method for rapid and sensitive detection of bromodeoxyuridine in DNA-replicating cells. *Exp. Cell Res.* **173**, 256–261.

Denecke, J., Becker, K., Jurgens, H., Gross, R. and Wolff, J. E. (1999). Falsification of tetrazolium dye (MTT) based cytotoxicity assay results due to mycoplasma contamination of cell cultures. *Anticancer Res.* **19**, 1245–1248.

Feldman, H. and Rodbard, D. (1971). Principles of competitive protein-binding assays. In: *Mathematical Theory of Radioimmunoassay* (W. D. Odell and W. H. Doughaday, Eds), pp. 158–203. J.B. Lippincott Company, Philadelphia.

Hasbold, J., Gett, A. V., Rush, J. S., Deenick, E., Avery, D., Jun, J. and Hodgkin, P. D. (1999). Quantitative analysis of lymphocyte differentiation and proliferation in vitro using carboxyfluorescein diacetate succinimidyl ester. *Immunol. Cell Biol.* **77**, 516–522.

Heeg, K., Reimann, J., Kabelitz, D., Hardt, C. and Wagner, H. (1985). A rapid colorimetric assay for the determination of IL-2-producing helper T cell frequencies. *J. Immunol. Methods* **77**, 237–246.

Kabelitz, D., Oberg, H. H., Pohl, T. and Pechhold, K. (1994). Antigen-induced death of mature T lymphocytes: analysis by flow cytometry. *Immunol. Rev.* **142**, 157–174.

Kabelitz, D., Pohl, T, and Pechhold, K. (1995). T cell apoptosis triggered via the CD3/T cell receptor complex and alternative activation pathways. *Curr. Top. Microbiol. Immunol.* **200,** 1–14.

Lyons, A. B. (2000). Analysing cell division in vivo and in vitro using flow cytometric measurement of CFSE dye dilution. *J. Immunol. Methods* **243**, 147–154.

Maghni, K., Nicolescu, O. M. and Martin, J. G. (1999). Suitability of cell metabolic colorimetric assays for assessment of CD4+ T cell proliferation: comparison to 5-bromo-2-deoxyuridine (BrdU) ELISA. *J. Immunol. Methods* **223**, 185–194.

Matzinger, P. (1991). The JAM test. A simple assay for DNA fragmentation and cell death. *J. Immunol. Methods* **145**, 185–192.

Mosmann, T. (1983). Rapid colorimetric assay for cellular growth and survival: application to proliferation and cytotoxicity assays. *J. Immunol. Methods* **65**, 55–63.

Parish, C. R. (1999). Fluorescent dyes for lymphocyte migration and proliferation studies. *Immunol. Cell Biol.* **77**, 499–508.

Pechhold, K., Pohl, T. and Kabelitz, D. (1994a). Rapid quantification of lymphocyte subsets in heterogeneous cell populations by flow cytometry. *Cytometry* **16,** 152–159.

Pechhold, K., Wesch, D., Schondelmaier, S. and Kabelitz, D. (1994b). Primary activation of Vγ9-expressing γδ T cells by *Mycobacterium tuberculosis*. Requirement for Th1-type CD4 T cell help and inhibition by IL-10. *J. Immunol.* **152,** 4984–4992.

Roehm, N. W., Rodgers, G. H., Hatfield, S. M. and Glasebrook, A. L. (1991). An improved colorimetric assay for cell proliferation and viability utilizing the tetrazolium salt XTT. *J. Immunol. Methods* **142**, 257–265.

Siegling, A., Lehmann, M., Platzer, C., Emmrich, F. and Volk, H. D. (1994). A novel multispecific competitor fragment for quantitative PCR analysis of cytokine gene expression in rats. *J. Immunol. Methods* **177**, 23–28.

Strong, D. M., Ahmed, A. A., Thurman, G. B. and Sell, K. W. (1973). In vitro stimulation of murine spleen cells using a microculture system and a multiple automated sample harvester. *J. Immunol. Methods* **2**, 279–291.

Tough, D. F. and Sprent, J. (1994). Turnover of naive- and memory-phenotype T cells. *J. Exp. Med.* **179**, 1127–1135.

Wagner, S., Beil, W., Westermann, J., Logan, R. P., Bock, C. T., Trautwein, C., Bleck, J. S. and Manns, M. P. (1997). Regulation of gastric epithelial cell growth by *Helicobacter pylori*: offdence for a major role of apoptosis. *Gastroenterology* **113**, 1836–1847.

List of suppliers

Scanning Multiwell Spectrophotometer (Microplate Reader)

Bio-Rad Laboratories	*http://www.bio-rad.com*
Bio-Tek Instruments, Inc.	*http://www.biotek.com*
Dynex Technologies	*http://www.dynextechnologies.com*
Titertek Instruments, Inc.	*http://www. titertek.com*

Flow Cytometer

BD Immunocytometry Systems	*http://www.bdfacs.com*
Beckman Coulter, Inc.	*http://www.beckmancoulter.com*
Cytomation	*http://www.cytomation.com*

Liquid Scintillation Counter

Packard Bioscience Company*	*http://www.packardbioscience.com*
PerkinElmer, Inc.	*http://www.perkinelmer.com*
Laboratory Technologies, Inc.	*http://www.labtechinc.com*
Titertek Instruments, Inc.	*http://www.titertek.com*

Special Reagents and Supply

Molecular Probes, Inc.	*http://www.probes.com*
Roche Molecular Biochemicals	*http://www.biochem.roche.com*
Amersham Pharmacia Biotech, Inc.	*http://www.apbiotech.com*
Greiner Bio-One, Inc.	*http://www.greinerbiooneinc.com*

* Packard Bioscience Co. has been purchased by PerkinElmer, Inc. which may affect their websites and/or product palette in the future.

4 Use of Bioinformatics to Predict MHC Ligands and T-cell Epitopes: Application to Epitope-driven Vaccine Design

Anne S De Groot
TB/HIV Research Lab, Brown University, Providence, USA and Vaccine Design Unit, EpiVax, Inc., Providence, USA

Hakima Sbai
TB/HIV Research Lab, Brown University, Providence, USA

Bill Martin
Vaccine Design Unit, EpiVax, Inc., Providence, USA

Jay A Berzofsky
Metabolism Branch, National Cancer Institute, National Institutes of Health, Bethesda, USA

◆◆

CONTENTS

◆◆◆◆◆◆ INTRODUCTION

At the heart of the discipline of bioinformatics is the concept that biological entities can be described as being composed of patterns. These patterns can be discovered, described and interpreted using informatics. In the past decade, new information has emerged on the patterns of pathogen-derived peptides that bind to MHC (MHC binding motifs) and

Copyright © Elsevier Science Ltd
All rights of reproduction in any form reserved

interact with T-cell receptors to stimulate the immune system. These peptide 'epitopes' are pathogen-derived and, due to their intrinsic pattern of amino acids, MHC-specific (with rare exceptions). The presentation of these patterned, pathogen-derived T-cell epitopes in the context of MHC is pivotal to the development of cellular immune defence against infectious disease.

In recent years, a number of computer-driven algorithms have been devised to take advantage of the wealth of proteome information available in public and private databases to search for T-cell epitopes. These algorithms test each sub-sequence of a given protein for traits thought to be common to immunogenic peptides, thus locating regions with a greater likelihood of inducing a cellular immune response *in vitro*. The use of these algorithms not only significantly reduces the time and effort needed to identify putative T-cell epitopes, but also decreases the number of protein fragments requiring *in vitro* testing for immunogenicity, as well. Thus, the bulk of *in vitro* assays can be focused on evaluating selected candidate T-cell epitopes.

Informatics approaches to epitope mapping now make it possible to perform high-throughput screening of entire proteomes. For example, as described in this chapter, the use of epitope-mapping tools reduced the number of potential peptide epitopes to be screened in the entire proteome of *Mycobacterium tuberculosis* from more than 1.3 million to less than 500, in a first-pass analysis (De Groot *et al.*, 2001). Of those 500, 17 were selected for *in vitro* screening for potential immunogenicity.

T-cell epitopes derived using bioinformatics are now being used to construct and enhance vaccines. These vaccines are designed to imitate the natural development of cell-mediated immunity after initial exposure to a pathogen. Upon vaccination (or exposure to a pathogen) epitope-specific memory T-cell clones are generated, to respond more rapidly and efficiently upon subsequent exposure to the pathogen. The underlying paradigm is that an effective vaccine is able to elicit a number of epitope-specific memory T cells which drive the protective immune response upon re-exposure to the pathogen. A similar premise operates for B-cell epitope (antibody)-driven vaccines. Indeed, some vaccinologists would affirm that all vaccines are 'epitope-driven'.

This chapter will discuss bioinformatic tools and methods that are currently being used to mine proteomes for T-cell epitopes and being applied to the design of epitope-driven vaccines.

◆◆◆◆◆◆ ANTIGEN RECOGNITION BY T CELLS

T cells recognize peptides that are derived from endogenous and exogenous proteins and presented in the cleft of MHC class I or class II molecules at the surface of the antigen-presenting cell to the T-cell receptor. In general, MHC class I molecules present peptides 8–10 amino acids in length and are predominately recognized by CD8+ cytotoxic T lymphocytes (CTLs). Class I peptides usually contain an MHC I-allele specific motif sequence of two conserved anchor residues (Elliott *et al.*, 1991; Falk *et al.*, 1991; Rötzschke *et al.*, 1991). Peptides presented by class II molecules are longer, more variable in size, and have fewer defined anchor motifs than those presented by class I molecules (Unanue, 1992; Brown *et al.*, 1993; Chicz *et al.*, 1993). MHC class II molecules bind peptides consisting of 11–25 amino acids and are predominantly recognized by CD4+ T helper (Th) cells.

MHC class I molecules present peptides obtained from proteolytic digestion of endogenously synthesized proteins. Host- or pathogen-derived intracellular proteins are cleaved by a complex of proteases in the proteasome. Small peptide fragments are then transported by ATP-dependent transporters associated with antigen processing (TAPs) into the endoplasmic reticulum (ER), where they form complexes with nascent MHC class I heavy chains and beta-2-microglobulin. The peptide–MHC class I complexes are transported to the cell surface for presentation to the receptors of CD8+ T cells (Germain and Margulies, 1993).

MHC class II molecules generally bind peptides derived from the cell membrane or from extracellular proteins that have been internalized by antigen-presenting cells (APCs). The proteins are initially processed in an endosomal compartment called MIIC. Subsequently, the generated peptides are sorted by empty MHC class II molecules based on their affinity, in which HLA-DM catalyses peptide exchange (Appella *et al.*, 1995; Germain *et al.*, 1996). The class II molecules bound to peptide fragments are transported to the surface of APCs for presentation to CD4+ helper T cells. From these different antigen processing and presentation pathways, two different T-cell responses are generated: a CD4+ Th immune response and a CD8+ CTL immune response. Since these two T-cell responses are usually complementary, an efficacious epitope-driven vaccine should induce them both.

◆◆◆◆◆◆ **APPROACHES TO IDENTIFYING T-CELL EPITOPES FROM PROTEIN SEQUENCES**

The major pre-requisite for the development of an epitope-driven vaccine is the identification of allele-specific or promiscuous peptides that are associated with MHC class I or class II molecules and can be recognized by T cells. This section will present an overview of two different approaches to identifying T-cell epitopes: the standard overlapping approach and the bioinformatics approach.

Overlapping method

Identification of T-cell epitopes within protein antigens has traditionally been carried out through a variety of methods, including the commonly employed 'overlapping peptide' method, as well as the method of testing of whole and fragmented native or recombinant antigenic protein. The overlapping method involves the synthesis of overlapping peptides which span the entire sequence of a given protein antigen, followed by testing the capacity of the peptides to stimulate T-cell responses *in vitro*. This assay is the standard method for the identification of T-cell epitopes within protein antigens. Implementation of the overlapping peptide method is both costly and labour-intensive. For example, to perform an assay using 20 amino acid long peptides overlapping by 10 amino acids spanning a given antigen of length n (a small subset of all possible 20-mers spanning the protein), (n/10)-1 peptides would have to be constructed and tested. This method still does not ensure the identification of all possible T-cell epitopes considering that certain epitopes bridging the overlapping fragments could still be missed. The 'exhaustive' overlapping approach involves manufacturing and examining every possible overlapping peptide (often including 8-mer, 9-mer, and 10-mer versions of the same potential epitope), covering the entire protein sequence. While this approach is comprehensive, it is prohibitively expensive and labour-intensive for most laboratories, and unnecessary, given the precision of current epitope-mapping tools.

With the rapid accumulation of accurate sequence information for a wide array of protein antigens, the development of novel means of predicting antigenic sites from primary structure is particularly appealing. The decade spanning 1991 to 2001 witnessed the development of a large number of computer-driven algorithms that used alphabetic representation of protein sequence information to search for T-cell epitopes.

Bioinformatics approach to mapping epitopes

Prior to 1990, DeLisi and Berzofsky (1985), and Rothbard and Taylor (1988) were the first research teams to suggest searching for a conserved pattern of amino acids as a possible predictive tool for antigenicity, based on empirical observations of the periodicity of amino acid residues in T-cell epitopes. Predictions based on periodicity were a clear improvement over the brute force method of epitope mapping but still did not sufficiently narrow the options (Meister *et al.*, 1995).

Improved computer-driven algorithms were made possible by the discovery of MHC binding motifs (Rötzschke *et al.*, 1991). Our understanding of the nature of these binding motifs has been furthered by three major technological advances: crystallographic studies of MHC molecules in complex with peptides, the sequencing of naturally occurring MHC peptide ligands by Edman degradation, and tandem mass spectrometry (Falk *et al.*, 1991; Rötzschke *et al.*, 1991; Rammensee *et al.*, 1995).

The use of MHC binding motifs prospectively to identify T-cell epitopes and the first MHC-binding motif-based algorithms were described by Sette and co-workers in 1989, by Lipford *et al.* in 1993, by Falk *et al.* (1991) and Leighton *et al.* (1991), by Parker *et al.* in 1994, and by the TB/HIV Research Laboratory in 1995 (Meister *et al.*, 1995). A variation on predicting peptides using MHC binding motif-based algorithms known as 'extended MHC binding motifs' or 'peptide side chain scanning' was described by Sette *et al.* (1993) and Hammer *et al.* (1994). Other bioinformatics approaches to predicting T-cell epitopes include artificial neural networks (first described by Brusik *et al.*, 1994) and structural approaches (as explained by Rosenfeld *et al.*, 1995 and by Altuvia *et al.*, 1995). These methods allowed for the construction of a matrix taking into account all possible amino acid side-chain effects for a single MHC-binding motif (Davenport *et al*, 1995, Jesdale *et al.*, 1997). A similar approach using synthetic undecapeptides was developed by B. Fleckenstein *et al.*, and reported in 1996.

A 'matrix-based' approach to mapping T-cell epitopes was advanced by the contributions of Davenport and Ho Shon for one class II allele in 1995 and by Jesdale and De Groot for an array of class I and class II alleles in 1997. EpiMatrix, an algorithm developed by the TB/HIV Research Lab and licensed to EpiVax, ranks nine to ten amino acid long segments overlapping by eight to nine amino acids for a protein sequence based on the estimated probability of binding to a selected MHC molecule. This method for ranking prospective epitopes has been described in detail (De Groot *et al.*, 1997; Schafer *et al.*, 1998). The first trial of the EpiMatrix algorithm was to compare peptides it predicted to bind to selected MHC to 158 known ligands listed in Rammensee *et al.* (1995). An example of the comparison between EpiMatrix predictions for HLA B7 and published ligands is shown in Table 1.

Most recently, the teams of Hammer, Sturniolo *et al.* and Zhang, Anderson and DeLisi can be credited with another important advance in epitope mapping: the 'pocket profile' method. According to this method,

Table I. Prediction of published HLA B7 ligands utilizing EpiMatrix V 1.0 (1996)

HLA B7 Name of protein	Known ligand sequence	Protein length	EpiMatrix rank of ligand
Topoisomerase II	SPRYIFTML	1621	1
EBNA 3A	RPPIFIRRL	812	1
HLA-A2.1 signal sequence	APRTLVLLL	365	1
HLA-DP signal sequence	APRTVALTAL	258	1
Ribosomal S26 protein	APAPPPKPM	107	1
HLA-B7 signal sequence	LVMAPRTVL	255	2
HIV V3	RPNNNTRKRI	90	2
Histone H1	AASKERSGVS L	219	7
EBNA 3C	APIRPIPTRF	983	21

Known HLA B7 restricted ligands are shown for each of nine proteins. EpiMatrix was applied to the sequences of these proteins; 10 mers overlapping by 1 were scored for each of the original protein sequences. The rank of the peptide that corresponded to the known ligand is listed in the final column in this table. In the case of topoisomerase II for example, 1612 10 mers were evaluated and ranked by EpiMatrix score. The 10 mer that was scored the highest by EpiMatrix was identical in sequence to the published ligand for this protein. The published ligand corresponded to the highest ranked Epimatrix prediction for the same protein in four of the remaining eight cases. 4620 10 mers were scored and ranked by EpiMatrix for this analysis; in seven of nine cases the ligand would have been correctly identified had only the top two scoring EpiMatrix peptides been synthesized.
(Reprinted from Anne S. De Groot, Jesdale BM, Berzofsky JA, Prediction and determination of MHC ligands and T Cell Epitopes, Chapter 3, Immunology Methods, S.H.E. Kaufmann, Ed., *Methods in Microbiology*, Vol. 25, pp. 79–106, Academic Press, NY, 1998).

similarities in MHC binding constraints are reflected in commonalities in the composition of MHC binding pockets. This allows new motifs to be developed by mixing and matching binding pocket characteristics (Zhang *et al.*, 1998; Sturniolo *et al.*, 1999).

Crystallographic studies of peptide/HLA class II complexes have identified the presence of nine 'binding pockets' which determine a given allele's peptide specificity. Each binding pocket can be described in terms of the amino acids that are likely to bind, or not bind, in that pocket. This set of amino acids is termed the 'pocket profile' (by Sturniolo *et al.*, 1999). Although there are many different HLA alleles, the number of different binding pocket types (with a corresponding pocket profile) is much smaller. Different alleles appear to have evolved by a process of mixing and matching binding pockets. This observation led to the concept that the binding properties of an allele could be described by listing its constituent pocket types. Of the nine binding pockets, those located in positions 1, 2 and 3 appear to show very little difference between alleles. The greatest variation is observed in pockets 4 to 9. Due to the helical nature of MHC ligands, peptide side-chains are usually directed away from the binding groove in pockets 5 and 8. Consequently, these pockets play little part in peptide binding and the significant differences in binding affinity for most of the class II alleles appear to come from variation in pocket profiles for pockets 4, 6, 7 and 9. Sturniolo *et al.* published a set of pocket profiles describing pockets located in positions 4, 6, 7 and 9. By mixing and matching the pocket types, binding profiles were constructed for 35 class II alleles.

Careful review of the Sturniolo *et al.* data recently permitted EpiMatrix developers James Rayner and Bill Martin to re-generate these 35 binding profiles and to generate predictive matrices for 90 alleles, if subtypes are counted, or 76 unique class II alleles (James Rayner, unpublished data). These class II matrices are now included in the EpiMatrix repertoire.

The EpiMatrix algorithm has been made available on the Internet for the use of HIV researchers (De Groot *et al.*, 1997). EpiMatrix is just one of a number of tools available on the Internet that can be used for T-cell epitope searches. A list of these tools and their comparative features is provided in Table 2. Several epitope mapping tools are available to researchers for use via the Web, including the tool available at the SYF-PEITHI website (Rammensee *et al.*, 1999), and the HLA binding prediction tool available on the site authored by Ken Parker at the National Institutes of Health (BIMAS) (Parker *et al.*, 1994). None of these sites returns exactly the same predictions.

◆◆◆◆◆◆ APPLIED EPITOPE MAPPING: ANALYSING THE *MYCOBACTERIUM TUBERCULOSIS* PROTEOME

Epitope mapping tools can be applied to the analysis of single proteins or entire proteomes. One illustration of proteome mapping for vaccine

Table 2. Overview of epitope prediction tools as of 1 Jan, 2002

Name or group	EpiMatrix	Epimmune	BIMAS	SYFPEITHI	TEPITOPE	ProPred	EpiPredict	Predict
Chief Scientist	Anne S. De Groot	Alexander Sette	Ken Parker	Hans-Georg Rammensee	Jurgen Hammer	(none) G.P.S. Raghava	Buirkhard Fleckenstein	Vladimir Brusic
Corp. or Research Institution	EpiVax, Inc (US)	Epimmune, Inc (US)	NIH (US)	BMI-Heidelberg (Germany)	Vaccinome (US)	Institute of Microbial Technology (India)	EpiPredict (Germany)	Kent Ridge Digital Labs (Singapore)
Class I (number)	Yes (24)	Yes (?)	Yes (31)	Yes (12)	No (0)	No (0)	No (0)	Yes (?)
Class II (number)	Yes (74)	Yes (?)	No (0)	Yes (4)	Yes (51)	Yes (51)	Yes (6)	Yes (?)
Epitope Discovery Services Offered	Yes Also online algorithm (restricted demos for HIV and TB)	Yes No online algorithm	Online algorithm only	Online algorithm only	TEPITOPE software for sale. Free to academic researchers	Online algorithm only (identical to TEPITOPE, apparently unauthorized)	Online algorithm only	Yes Also online algorithm (restricted demo)
Website	www.EpiVax.com	www.Epimmune.com	bimas.dcrt.nih.gov/molbio/hla/bind/	syfpeithi.bmi-heidelberg.com	www.Vaccinome.com	www.intech.res.in/raghava/propred/	www.epipredict.de/	sdmc.krdl.org.sg:8080/predictdemo

design is provided in this section. As the complete sequence of the laboratory standard strain *of Mycobacterium tuberculosis* (Mtb) known as H37Rv has been sequenced by the Pasteur Institute, it can be directly obtained from the Sanger Centre website (http://www.sanger.ac.uk/Projects/M_tuberculosis/, accessed 1 Jan, 2002) (Cole *et al.*, 1998). The entire sequence of a clinical isolate of Mtb (CDC strain 1551) is available at the TIGR website (http://www.tigr.org/tigr-scripts/CMR2/GenomePage3.spl?database=gmt, accessed 1 Jan, 2002) (Delcher *et al.*, 1999). It is now possible to scan the entire Mtb proteome for putative epitopes. A pilot study using EpiMatrix to scan the 1551 and H37Rv proteomes for T-cell epitopes was recently performed by the TB/HIV Research Laboratory team in collaboration with EpiVax Inc. (De Groot *et al.*, 2001; De Groot *et al.*, manuscript in preparation). The approach is illustrated in the next few paragraphs and outlined in Fig. 1.

Rationale for Mtb genome scan

Cell-mediated immunity to *Mycobacterium tuberculosis* (Mtb) protein antigens appears to be the critical component of effective host defence against the development of TB disease. CD4+ T cells orchestrate the activation of Mtb-infected macrophages and marshal other components of cellular immune defence to the locus of Mtb infection; gamma-interferon, released by these T cells, activates Mtb-infected macrophages to activate cellular immune defences. It is generally believed that peptides or antigens (a) that are recognized by the CD4+ lymphocytes of Mtb-immune individuals, (b) that stimulate Th1-type responses including gamma-interferon secretion (Flynn *et al.*, 1993) and (c) that activate macrophages, would be ideal tuberculosis (TB) vaccine components.

The criteria for Mtb antigen selection for vaccine development have been vague at best. One correlate of protection has been T-cell response to antigens secreted by Mtb in culture filtrate (Boesen *et al.*, 1995). Until now, researchers have been proceeding with the evaluation of secreted Mtb antigens one fraction or protein at a time, using 2D gels to separate candidate antigens (Sonnenberg and Belisle, 1997). The current availability of the two whole Mtb proteomes has made it feasible to scan every one of more than 4000 putative genes for immunogenic regions.

Selection of secreted proteins from the proteome

The methods used to select a set of secreted proteins from the Mtb genome for epitope analysis are published and described on the web (http://tbsp.phri.nyu.edu/) (Gomez *et al.*, 2000), and reviewed briefly here. Proteins predicted from the genome sequence of *M. tuberculosis* strain H37Rv were analysed with bioinformatics tools that predict secretory signal peptides. These peptides were then aligned and compared with the CDC 1551 strain. A total of 73 putative secreted proteins conserved in both 1551 and H37Rv (with two exceptions that were only found in H37Rv) were identified in this manner.

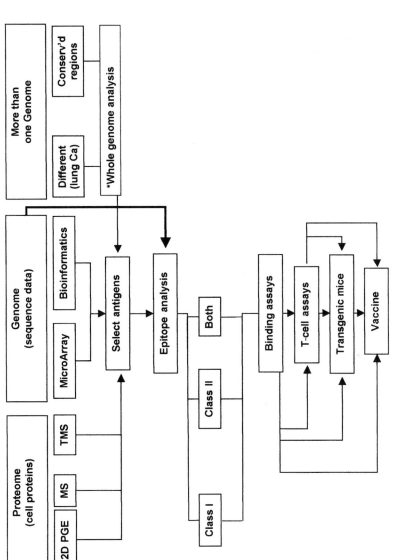

Application to Epitope-
driven Vaccine Design

Figure 1. Several genome-to-vaccine pathways are illustrated here. Departing from the whole proteome, researchers can use 2D SDS PAGE gels (2D PGE), Mass Spectrophotometry (MS) and/or Tandem MS (TMS) to select proteins of interest. Alternatively, departing from genome sequence data, microarrays or bioinformatics approaches (searching for secretion tags, for example) can be used to identify protein subsets of interest. Occasionally, a direct selection of potential epitopes can be performed on the entire genome, if the genome is not too large (viral genomes, for example). Where more than one genome is to be analysed, comparison of the proteomes to select regions of interest that may be different (as in normal lung tissue versus cancerous lung tissue) or conserved (as in quasispecies of HIV) may allow a narrowing of the potential proteins of interest. Selection of proteins is followed by epitope analysis (either Class I, Class II, or both) and *in vitro* assays (binding assays, T-cell assays) and studies of immune response to the potential components of the epitope-based vaccine construct in HLA transgenic mice.

Scoring the secreted proteins with EpiMatrix

All 73 putative secreted proteins were analysed with EpiMatrix using matrices for 74 class II alleles (developed by James Rayner at Epivax from pocket profiles, as described by Sturniolo *et al.*). Every match to a class II motif was counted and categorized by the class II allele match. These matches were then screened for promiscuity (single peptides matching successfully across a range of alleles) and clustering (groups of over-lapping, contiguous, or nearly contiguous peptides which collectively scored well across a range of alleles). A list of candidate epitopes that were ranked based on the number of motifs per length in amino acids (up to a maximum of 25 amino acids) was developed.

A total of 17 peptides were selected for synthesis based on their EpiMatrix rank. *In vitro* studies were performed using T cells collected from Mtb-immune subjects. Eleven (65%) of the 17 candidate Mtb epitopes stimulated gamma-interferon response from the T cells *in vitro*. One highly promiscuous epitope (MT2281-26-J) induced IFN-γ secretion in T cells collected from 15 of 27 Mtb-immune subjects (56%). This peptide, peptide J, is of particular relevance to Mtb vaccine development since it is both promiscuous (was recognized by the majority of individuals tested) and capable of stimulating gamma-interferon release *in vitro*.

A query of the TIGR and Sanger Center databases revealed that eight of the 17 proteins, scanned for epitopes, may have defined functions. The remaining nine proteins are classified as 'hypothetical proteins'. Thus, while some of these proteins have been ascribed a function, based on comparisons with other proteins in protein databases, most have not. In fact, the protein database assignments are approximate at best, and the true identities, structures and functions of the proteins from which these epitopes are derived *remain unknown*. In addition, some of the proteins selected for this first-pass evaluation may be completely silent in the human infection (despite their 'secretion tag'). For example, the proteins may be derived from archived and non-functional genes, or from genes that are not transcribed, and/or the proteins may be transcribed only under certain growth conditions or certain phases of the growth cycle. Recognition of 11 epitopes by T cells derived from Mtb-immune indi-viduals is a truly remarkable finding, considering the 1.3 million possible peptides from which the 17 candidates for this study were selected. It is clear from this pilot study that the pace of epitope discovery for vaccine design may dramatically accelerate when computer-driven epitope mapping tools are systematically applied to whole genomes and used in combination with *in vitro* methods for screening and confirming epitopes.

◆◆◆◆◆◆ EPITOPE-DRIVEN VACCINE DESIGN

The language analogy

Vaccinologists have begun to conceptualize vaccines built entirely out of epitopes linked together like a string of beads (Whitton *et al.*, 1993; Hanke et al., 1998; Thomson et al., 1998; Wang et al., 1998; Velders et al., 2001). Behind this vaccine approach is the following analogy: an *epitope* is to a *pathogen* as a *word* is to a *language*. Thus, even a single epitope may signal the presence of an infection to the immune system and stimulate a protective response, just as a single word (*bonjour*, for example) reminds the hearer of a certain language (Olsen et al., 2000). Due to genetic restriction, different words, or epitopes, are required to stimulate the immune system in the context of different MHC backgrounds. By extending the analogy, researchers have hypothesized that protective immune response to an entire pathogen might be generated by recognition of a repertoire of epitope 'words' derived from the proteins of the pathogen. This method of basing vaccine design on regions of pathogens that are presented by MHC molecules was recently termed 'reverse immunogenetics' by Hill and Davenport (1996). An alternative term suggested here would be 'epitope-driven vaccine design' (De Groot et al., 2001).

The epitope-driven vaccine is an attractive concept that is being successfully pursued in a number of laboratories (An and Whitton, 1997; Hanke et al., 1998). Complex vaccines containing T-helper and B-cell epitopes alongside CTL epitopes, all derived from a variety of pathogens (such as five viruses and one bacterium), have already been constructed and tested (Tine et al., 1996; An and Whitton, 1997).

A typical epitope-based vaccine construct contains a single start codon with epitopes inserted consecutively in the construct, with or without intervening spacer amino acids. *In vitro* studies of these constructs have confirmed that the epitopes are expressed, that they stimulate protective immune response, and that they do not interfere with one another (Whitton et al., 1993). Another epitope-driven vaccine approach is to mix several plasmids together, each of which contained genes for different proteins or different minigene epitopes. These vaccines have had no adverse effects, may have enhanced response, and may have shifted response toward the Th1 phenotype (Morris et al., 2000). These discoveries suggest that epitope-based vaccines are a reasonable and feasible approach to vaccine design.

Proof of the epitope-driven principle

Proof that epitope-driven minigene vaccination can stimulate protective immune responses has been obtained by researchers carrying out minigene or peptide vaccination studies in a range of animal models. For example, CTLs elicited by peptide immunization have been shown to afford protection against RSV challenge (Simmons et al., 2001) or against challenge by an HIV envelope recombinant vaccinia virus (Belyakov et al., 1998a,b). A similar epitope-based peptide vaccine delivered mucosally has led to a reduction in viral load and disease in macaques challenged intrarectally with a pathogenic chimeric HIV/SIV virus (Belyakov et al., 2001).

In a separate study, immunization of BALB/c and CBA mice with measles virus CTL epitopes resulted in *in vivo* induction of epitope-specific

CTL responses and conferred some protection against encephalitis following intracerebral challenge with a lethal dose of virus (Schadeck et al., 1999). In other studies, a conjugate peptide vaccine containing a CTL epitope from the HSV-1 immediate early protein elicited protection from intraperitoneal HSV challenge (Rosenthal et al., 1999). Vectored (Sindbis virus) minigene vaccines containing CTL epitopes have also induced epitope-specific CD8(+) T-cell responses and elicited a high degree of protection against infection with malaria or influenza A virus (Tsuji et al., 1998).

Similarly, immunization of BALB/c mice with three doses of a peptide construct containing an H-2(d)-restricted cytotoxic T-lymphocyte (CTL) epitope from a murine malaria parasite induced both T-cell proliferation and a peptide-specific CTL response mediating nitric-oxide-dependent elimination of malaria-infected hepatocytes *in vitro*, as well as partial protection of BALB/c mice against sporozoite challenge (Franke et al., 2000). Epitope-driven minigene vaccines have also been protective in the sheep model: vaccination with a minimal ovine CTL peptide epitope consistently induced epitope-specific CTLs; sheep whose CTLs were also capable of recognizing retrovirus-infected cells were fully protected when challenged with BLV (Hislop et al., 1998).

Even more important, these studies of epitope-based vaccines have demonstrated the importance of including T-helper epitopes with CTL epitopes (Shirai et al., 1994). An optimal class I-restricted CTL response required a class II-MHC-restricted helper response (Shirai et al., 1994) and could be improved by increasing the potency of the T-helper epitope (Ahlers et al., 1997, 2001). Further, some researchers have shown that specific and highly directed CTL induction is possible by unlinked minigene DNA immunization. Individually, CTL induction is not always sufficient enough to provide protection. However, it could be improved with the inclusion of Th epitopes (Fomsgaard et al., 1999). Mixed CTL and Th may be critical to protection. For example, immunization with a cocktail of peptides, consisting of a B-cell epitope, a T-helper epitope, and a CTL epitope, linked to a fusion peptide, resulted in a 190-fold reduction in RSV titre compared with the titre in unimmunized mice (Hsu et al., 1999).

Advantages over whole-protein vaccines

Epitope-driven vaccines offer several advantages over vaccines encoding whole-protein antigens. Not only are epitope-based vaccines capable of inducing more potent responses than whole-protein vaccines (Ishioka et al., 1999), they sidestep the propensity for the immune system to focus on a single immunodominant epitope by simultaneously targeting multiple dominant and subdominant epitopes (Oukka et al., 1996; Tourdot et al., 1997). This may be particularly important to the development of vaccines against pathogens like HIV and HCV, because the breadth of an immune response appears to be a critical determinant in the progression of the infection (Couillin et al., 1994; Missale et al., 1996; McMichael and Phillips, 1997; Cooper et al., 1999). The use of epitopes can also overcome any potential safety concerns associated with the vaccinating antigen (E6 and E7 in the case of HPV (Crook *et al.*, 1989; Hawley-Nelson *et al.*, 1989)).

Another aspect of vaccine development that needs to be considered, particularly where standard approaches have failed, is the probable heterogeneity of immunogenic epitopes. A large number of T-lymphocyte clones covering a wide range of specificities may have to be built up in order to afford good protection.

◆◆◆◆◆◆ OPTIMIZING EPITOPE-DRIVEN VACCINES

Spacers between epitopes

Enhancing the immunogenicity of multi-epitope vaccines has been approached in several ways. One approach has focused on how best to package multiple epitopes in a vaccine construct. One option is to present the epitopes as a 'string of beads', without any spacer sequences separating the individual epitopes. Indeed, some studies have indicated that flanking sequences have minimal effects on epitope presentation (An and Whitton, 1997). The actual role of flanking spacer sequences is currently under investigation, partly because our understanding of how multi-epitope constructs are processed remains incomplete.

In a 'string of beads' construct the individual epitopes are usually very closely apposed, without their natural flanking sequences. This has raised concern that their proteolytic processing may be compromised, and that peptides other than the specific peptides of interest may be generated as a result of the processing (Moudgil *et al.*, 1998; An and Whitton, 1999). However, there is some evidence that the introduction of spacer sequences to separate the individual epitopes may therefore help focus the immune response on the specific epitopes (Livingston *et al.*, 2001).

A study by Velders *et al.* compared the immunogenicity of two similar HPV epitope-string DNA constructs that differed only in the presence or absence of spacers between the epitopes. It was found that the addition of AAY spacers between the epitopes was crucial for epitope-induced tumour protection (Velders *et al.*, 2001).

And finally, some researchers have shown that the inclusion of flanking sequences around epitopes and of signal sequences dramatically increases the efficacy of epitope-string DNA vaccines against established tumours (Velders *et al.*, 2001). Extensive studies conducted in mice models have demonstrated that residues flanking an MHC class I epitope strongly influence its liberation efficiency by the proteasome system (Rodriguez and Whitton, 2000; Velders *et al.*, 2001). The specific role of flanking regions cannot be evaluated until data from human clinical trails become available.

Co-stimulatory molecules

Co-stimulatory molecules play a central role in the initiation of T-cell immune responses (Kuchroo *et al.*, 1995). CD28 and CTLA4 represent the co-stimulatory receptors on T cells, and B7 molecules represent their corresponding ligands on antigen-presenting cells. Several studies carried out in murine models suggest that the APC signal mediated via CD28 is required for TCR-mediated T-cell activation (Shahinian *et al.*, 1993). CTLA4, on the other hand, appears to play an antagonistic role in T-cell activation. Two primary members of the B7 family have been identified: B7-1 (CD80) and B7-2 (CD86). These two molecules show comparable affinity to CD28 molecules and may differentially activate Th1 or Th2 immune responses (Kuchroo *et al.*, 1995).

Lack of co-stimulation can lead to T-cell tolerance. Peptide epitopes presented by non-professional APCs may fail to activate T cells if signal 1 is delivered in the absence of signal 2, and instead may lead to anergy. Because of the role played by co-stimulatory molecules in the initiation of T-cell responses, they can be manipulated either to stimulate the immune system, in order to prevent infection, or to inhibit the immune system, for immunotherapy against allergies and autoimmune diseases. Schlom and co-workers have developed poxvirus vectors expressing three synergistic co-stimulatory molecules, B7-1, ICAM-1, and LFA-3 (TRICOM vectors), that have been used to markedly amplify T-cell responses (Hodge *et al.*, 1999; Zhu *et al.*, 2001).

Targeting peptides to MHC I or MHC II

One way to enhance CTL responses is to target the T-cell epitopes of interest to the proteasome of the host cell (Tobery and Siliciano, 1997). It is known that the ubiquitination of proteins acts as a tag to target proteins to the proteasome and to enhance the proteolytic degradation of the introduced epitopes within the cell. Similarly, Th responses can be enhanced by targeting the epitopes for either secretion or cell membrane expression. Past studies have exploited LAMPs (lysosome-associated membrane proteins), which, when fused to foreign antigens, can target antigens for lysosomal entry and destruction and enhanced class II presentation (Rodriguez and Whitton, 2000).

Both CD4+ and CD8+ T cells are critical for the generation of an effective immune response. The induction of Th responses will enable the enhancement of CTL responses and the secretion of cytokines that aid pathogen clearance. Therefore, an epitope-driven vaccine should provide T-cell epitopes presented by different MHC class I and class II alleles, as well as several epitopes for a single HLA allele.

Promiscuous epitopes

Multi-epitope T-cell vaccines should be designed to express not only highly conserved epitopes (to circumvent strain variation for a variety of pathogens) but also epitopes recognized by a broad spectrum of different HLA alleles. An ideal vaccine would be one that can be offered regardless of an individual's HLA phenotype.

Epitope search algorithms can be configured to find promiscuous T-cell epitopes, which can be presented in the context of more than one HLA molecule. For example, EpiMatrix incorporates the 'clustering' function from a previous epitope prediction algorithm (EpiMer, Meister *et al.*, 1995). EpiMatrix measures the MHC binding potential of each 10 amino acid long snapshot for a number of human HLA, and therefore can be used to identify regions of high density clusters of potential MHC binding sequences. Other laboratories have confirmed cross-presentation of peptides within HLA 'superfamilies' (such as the A3 superfamily: A11, A3, A31, A33 and A68) described by Walker, Sette, *et al.* (Threlkeld *et al.*, 1997; Sette and Sidney, 1998). Presumably, vaccines containing such 'clustered' or promiscuous epitopes will have an advantage over vaccines composed of epitopes that are not promiscuous.

Epitope enhancement

Because the affinity of natural viral or cancer epitopes for MHC molecules, or of the peptide–MHC complexes for T-cell receptors, is not necessarily optimal, it is possible to make improved vaccines by a process called epitope enhancement, in which the amino acid sequence of the epitope is modified to improve one or the other of these affinities (Berzofsky 1993; Berzofsky *et al.*, 2001). A number of cases in which viral or tumour antigen epitopes have been enhanced to improve binding to MHC molecules have been described (Pogue *et al.*, 1995; Parkhurst *et al.*, 1996; Ahlers *et al.*, 1997; Rosenberg *et al.*, 1998; Sarobe *et al.*, 1998; Berzofsky *et al.*, 1999; Irvine *et al.*, 1999; Ahlers *et al.*, 2001) and have even shown improved potency in human clinical trials (Rosenberg *et al.*, 1998). These enhanced sequences can be determined empirically (Boehncke *et al.*, 1993; Sarobe *et al.*, 1998) by using known information about primary and secondary anchor residues (Ruppert *et al.*, 1993). Combinatorial peptide libraries can also be used (La Rosa *et al.*, 2001).

In addition, EpiMatrix and similar bioinformatics approaches could be very useful for predicting sequence changes that would result in improved binding to MHC molecules. Recently, it was shown that such improved affinity of a helper epitope for a class II MHC molecule, when coupled to a CTL epitope in an epitope-based peptide vaccine, resulted not only in an increased CTL response and increased protection against virus, but also in a qualitatively different helper T-cell response (Ahlers *et al.*, 2001). The response was skewed more toward Th1 cytokine production, due to increased induction of CD40L on the helper T cells, which was shown to result in increased IL-12 production by the dendritic cells, making them more polarizing dendritic cells to polarize the T-helper cells toward the Th1 phenotype. Likewise, a number of examples have been published of modifications in the epitope sequence that result in higher affinity of the peptide–MHC complex for the T-cell receptor (Zaremba *et al.*, 1997; Fong *et al.*, 2001; Slansky *et al.*, 2001; Tangri *et al.*, 2001). In fact, a recent study from the group of Sette and co-workers defined specific positions in the peptide sequence at which conservative substitutions were more likely to result in higher affinity of the peptide–MHC complex for the T-cell receptor (Tangri *et al.*, 2001). Both of these types of epitope enhancement should allow the development of epitope-based vaccines that are more effective than the natural pathogen or tumour antigen proteins used in conventional vaccines.

Conserved epitopes

A number of pathogens have been shown to vary between individuals and also during the course of infection of a single individual. HIV and HCV are prime examples; both clades and subtypes (describing variation between infected individuals) and quasispecies (defining variation within a single individual) have been defined. The process of developing vaccines for variable pathogens is complicated by potential variation of key T-cell epitopes. However, the Conservatrix algorithm, another bioinformatics tool developed by the TB/HIV Research Lab, can define regions that are both conserved (across subtypes or quasispecies) and potentially immunogenic. Conservatrix accomplishes this by parsing every sequence in a given database into nine to ten amino acid long text strings. The algorithm then performs a simple string-of-text-based search, similar to the approach used by the 'find' function in word-processing programs. Each of these text strings is then ranked by the number of times it occurs in the set of text strings. Highly conserved peptide text strings are then input into EpiMatrix and ranked for immunogenicity by EBP. This tool has been applied to the analysis of HIV-1, hepatitis C, and human papilloma virus (De Groot *et al.*, 2001; and EpiVax, unpublished results).

◆◆◆◆◆◆ *IN VIVO* ASSESSMENT OF THE IMMUNOGENICITY OF A T-CELL VACCINE

In preparation for constructing an epitope-driven vaccine, selected peptide epitopes are synthesized, and MHC binding studies and T-cell responses to the peptides are evaluated *in vitro*. MHC binding can be evaluated using the T2 cell binding assay (Ljunggren *et al.*, 1990). T-cell responses to the peptides can be measured in standard T-cell assays (see chapter by Hiroi and Kiyono in Section II). Alternatively, newer techniques such as tetramers or intracellular cytokine staining can be performed. Once T-cell epitopes are confirmed *in vitro* (using T-cell assays as described) a vaccine is constructed from these epitopes. The most rapid approach is to clone the (DNA) coding sequences of the epitopes into a vector plasmid or viral vaccine vector in a tandem string.

Animal models for epitope-driven vaccines

After using *in vitro* T-cell assays to select for naturally processed T-cell epitopes, it is important to evaluate the ability of vaccines derived from these epitopes to induce an immune response *in vivo*. Transgenic mice expressing human MHC class I or class II molecules represent a suitable pre-clinical model for this purpose. The advantage of using HLA transgenic animals is that they can develop physiologically relevant HLA-restricted T-cell responses. Transgenic mouse strains that express either the entire HLA-A*0201 or DRB*0101 molecule have been developed. HLA-A2 transgenic mice have been used to assess the immunogenicity of peptides that bind to HLA-A2. A correlation has been found between CTL responses in infected individuals and CTL responses induced in immunized HLA transgenic mice (Newberg *et al.*, 1992; Sette *et al.*, 1994; Man *et al.*, 1995; Ressing *et al.*, 1995; Shirai *et al.*, 1995; Wentworth *et al.*, 1996; Diamond *et al.*, 1997; Sarobe *et al.*, 1998; Firat *et al.*, 1999).

Tetramers

Monitoring the specificity of the immune response following vaccination can also be accomplished using new epitope-specific reagents – MHC tetramers. These tools were first developed by Altman and colleagues (Altman *et al.*, 1996). These specialized constructs bear four MHC molecules complexed with beta-2-microglobulin and a specific pathogen-derived peptide ligand. Tetramers can bind directly to T cells that recognize the MHC–peptide complex. They can be used for direct *ex vivo* analysis of the frequency and phenotypes of epitope-specific T cells by flow cytometric technique. Tetramers permit the following types of experimental confirmations of epitope-specific T-cell responses *in vivo*: (a) direct quantitation of the number of epitope-specific T cells prior to and following vaccination; (b) phenotyping of responding T cells (examination for cell surface markers such as CD8, CD4, CD38 and additional activation markers); (c) monitoring of the immune response to specific epitopes following vaccination; and (d) direct evaluation of the effect of combinations of epitopes, epitope spacers or linkers and signal sequences on T-cell responses. These reagents will prove to be useful as epitope-driven vaccines move into clinical trials, as they provide a means of directly measuring and timing immune response to the vaccine.

◆◆◆◆◆◆ CONCLUSION

Bioinformatics is ushering in a new era of vaccine design. The ability to induce an immune response to a broad repertoire of epitopes that are universally recognized, across continents and across genetic backgrounds, is considered to be a critical characteristic of an effective vaccine. Opportunities for epitope discovery are expanding as the number of pathogens that are entirely sequenced approaches 100 and access to these data improves. Cancer therapy is another discipline that may benefit from the application of these tools to vaccine design.

The pace of vaccine design will accelerate as bioinformatics is systematically applied to whole genomes and used in combination with *in vitro* methods for screening and confirming epitopes. Given access to these bioinformatics, epitope-driven vaccines for a whole host of pathogens and cancers are now within reach.

◆◆◆◆◆◆ ACKNOWLEDGEMENTS AND DISCLOSURES

Two of the contributing authors, Bill Martin and Anne S. De Groot, are senior officers and majority shareholders at EpiVax, a privately owned vaccine design company located in Providence RI. These authors acknowledge that there is a potential conflict of interest related to their relationship with EpiVax and attest that the work contained in this review is free of any bias that might be associated with the commercial goals of the company.

Initial funding for the TB genome-to-vaccine analysis was provided by a subcontract to the TB/HIV Research Laboratory from the NIH (R01 AI 40125, Principal Investigator Robert Fleischmann). Funding for TB epitope analysis and T cell assays described in this manuscript was provided by the Sequella Global TB Foundation in the form of a core scientist award to Anne S. De Groot at EpiVax, Inc.

The authors wish to thank Laurence Vanleynseele (EpiVax) for her assistance with the manuscript and Julie McMurray (TB/HIV Research Lab) for providing assistance with the analysis and assays for the TB genome-to-vaccine project.

References

Ahlers, J. D., Takeshita, T., Pendleton, C. D. and Berzofsky, J. A. (1997). Enhanced immunogenicity of HIV-1 vaccine construct by modification of the native peptide sequence. *Proc. Natl. Acad. Sci. USA* **94**, 10856–10861.

Ahlers, J. D., Belyakov, I. M., Thomas, E. K. and Berzofsky, J. A. (2001). High affinity T-helper epitope induces complementary helper and APC polarization, increased CTL and protection against viral infection. *J. Clin. Invest.* **108**(11), 1677–1685.

Altman, J. D., Moss, P. A. H., Goulder, P. J. R., Barouch, D. H., McHeyzer-Williams, M. G., Bell, J. I., McMichael, A. J. and Davis, M. M. (1996). Phenotypic analysis of antigen, specific T lymphocytes. *Science* **27**(5284), 94–96.

Altuvia, Y., Schueler, O. and Margalit, H. (1995). Ranking potential binding peptides to MHC molecules by a computational threading approach. *J. Mol. Biol.* **249**(20), 244–250.

An, L. L. and Whitton, J. L. (1997). A multivalent minigene vaccine, containing B-cell, cytotoxic T-lymphocyte, and Th epitopes from several microbes, induces appropriate responses in vivo and confers protection against more than one pathogen. *J. Virol.* **71**(3), 2292–2302.

An, L. L. and Whitton, J. L. (1999). Multivalent minigene vaccines against infectious disease. *Curr. Opin. Mol. Ther.* **1**(1), 16–21.

Appella, E., Loftus, D.J., Sakaguchi, K., Sette, A. and Celis, E. (1995) Synthetic antigenic peptides as a new strategy for immunotherapy of cancer. *Biomed. Pept. Proteins Nucleic Acids* **1**(3), 177–184.

Belyakov, I. M., Derby, M. A., Ahlers, J. D., Kelsall, B. L., Earl, P., Moss, B., Strober, W. and Berzofsky, J. A. (1998a). Mucosal immunization with HIV-1 peptide vaccine induces mucosal and systemic cytotoxic T lymphocytes and protective immunity in mice against intrarectal recombinant HIV-vaccinia challenge Mucosal immunization with HIV-1 peptide vaccine induces mucosal and systemic cytotoxic T lymphocytes and protective immunity in mice against intrarectal recombinant HIV-vaccinia challenge. *Proc. Natl. Acad. Sci.* **95**(4), 1709–1714.

Belyakov, I. M., Ahlers, J. D., Brandwein, B. Y., Earl, P., Kelsall, B. L., Moss, B., Strober, W. and Berzofsky, J. A. (1998b). The importance of local mucosal HIV-specific CD8$^+$ cytotoxic lymphocytes for resistance to mucosal-viral transmission in mice and enhancement of resistance by local administration of IL-12. *J. Clin. Invest.* **102**(12), 2072–2081.

Belyakov, I. M., Hel, Z., Kelsall, B., Kuznetsov, V. A., Ahlers, J. D., Nacsa, J., Watkins, D. I., Allen, T. M., Sette, A., Altman, J., Woodward, R., Markham, P. D., Clements, J. D., Franchini, G., Strober, W. and Berzofsky, J. A. (2001). Mucosal AIDS vaccine reduces disease and viral load in gut reservoir and blood after mucosal infection of macaques. *Nature Medicine* **7**(12), 1320–1326.

Berzofsky, J. A. (1993). Epitope selection and design of synthetic vaccines: molecular approaches to enhancing immunogenicity and crossreactivity of engineered vaccines. *Annals NY Academy of Sciences* **690**, 256–264.

Berzofsky, J. A., Ahlers, J. D., Derby, M. A., Pendleton, C. D., Arichi, T. and Belyakov, I. M. (1999). Approaches to improve engineered vaccines for HIV and other viruses that cause chronic infections. *Immunological Rev.* **170**, 151–172.

Berzofsky, J. A., Ahlers, J. D. and Belyakov, I. M. (2001). Strategies for designing and optimizing new generation vaccines. *Nature Reviews Immunology* **1**, 209–219.

Boehncke, W. H., Takeshita, T., Pendleton, C. D., Sadegh-Nasseri, S., Racioppi, L., Houghten, R. A, Berzofsky, J. A. and Germain, R. N. (1993). The importance of dominant negative effects of amino acids side chain substitution in peptide-MHC molecule interactions and T cell recognition. *J. Immunol.* **150**(2), 331–341.

Boesen, H., Jensen, B. N., Wilcke, T. and Andersen, P. (1995). Human T-cell responses to secreted antigen fractions of *Mycobacterium tuberculosis*. *Infect Immun.* **63**(4), 1491–1497.

Brown, J. H., Jardetzky, T. S., Gorga, J. C., Stern, L. J., Urban, R. G., Strominger, J. L. and Wiley, D. C. (1993) Three-dimensional structure of the human class II histocompatibility antigen HLA-DR1. (1993). *Nature* **364** (6432), 33–39.

Brusik, V., Rudy, G. and Harrison, L. C. (1994). Prediction of MHC binding peptides using artificial neural networks, In: *Complex Systems, Mechanisms of Adaption.* (R. J. Stonier, X. S. Yu, Eds). Amsterdam, IOS Press, pp. 253–260.

Chicz, R. M., Urban, R. G., Gorga, J. C., Vignali, D. A. A., Lane, W. S. and Strominger, J. L. (1993). Specificity and promiscuity among naturally processed peptides bound to HLA-DR alleles. *J. Exp. Med.* **178**, 27–47.

Cole, S. T., Brosch, R., Parkhill, J., Garnier, T., Churcher, C., Harris, D., Gordon, S. V., Eiglmeier, K., Gas, S., Barry, C. E. 3rd, Tekaia, F., Badcock, K., Basham, D., Brown, D., Chillingworth, T., Connor, R., Davies, R., Devlin, K., Feltwell, T., Gentles, S., Hamlin, N., Holroyd, S., Hornsby, T., Jagels, K., Barrell, B. G., *et al.* (1998). Deciphering the biology of *Mycobacterium tuberculosis* from the complete genome sequence. *Nature* **393**(6685), 537–544.

Cooper, S., Erickson, A. L., Adams, E. J., Kansopon, J., Weiner, A. J., Chien, D. Y., Houghton, M., Parham, P. and Walker, C. M. (1999). Analysis of a successful immune response against hepatitis C virus. *Immunity* **10**(4), 439–449.

Couillin, I., Culmann-Penciolelli, B., Gomard, E., Choppin, J., Levy, J. P., Guillet, J. G. and Saragosti, S. (1994). Impaired cytotoxic T lymphocyte recognition due to genetic variations in the main immunogenic region of the human immuno-deficiency virus 1 NEF protein. *J. Exp. Med.* **180**(3), 1129–1134.

Crook, T., Morgenstern, J. P., Crawford, L. and Banks. L. (1989). Continued expression of HPV-16 E7 protein is required for maintenance of the transformed phenotype of cells co-transformed by HPV-16 plus EJ-*ras*. *EMBO J.* **8**(2), 513–519.

Davenport, M. P., Ho Shon, I. A. P. and Hill, A. V. S. (1995). An empirical method for the prediction of T-cell epitopes. *Immunogenetics* **42**(5), 392–397.

De Groot, A. S., Jesdale, B. M., Szu, E. and Schafer, J. R. (1997). An interactive web site providing MHC ligand predictions: application to HIV research. *AIDS Research and Human Retroviruses* **13**(7), 529–531.

De Groot, A. S., Bosma, A., Chinai, N., Frost, J., Jesdale, B. M., Gonzalez, M. A., Martin, W. and Saint-Aubin, C. (2001). From genome to vaccine: In silico predictions, ex vivo verification. *Vaccine* **19**(31), 4385–4395.

Delcher, A. L., Kasif, S., Fleischmann, R. D., Peterson, J., White, O. and Salzberg, S. L. (1999). Alignment of whole genomes. *Nucleic Acids Res.* **27**(11), 2369–2376.

DeLisi, C. and Berzofsky, J. A. (1985). T-cell antigenic sites tend to be amphipathic structures. *Proc. Natl. Acad. Sci. USA* **82**, 7048–7052.

Diamond, D. J., York, J., Sun, J. Y., Wright, C. L. and Forman, S. J. (1997). Development of a candidate HLA A*0201 restricted peptide-based vaccine against human cytomegalovirus infection. *Blood* **90**(5), 1751–1767.

Elliott, T., Cerundolo, V., Elvin, J., Townsend, A. (1991). Peptide-induced conformational change of the class I heavy chain. *Nature* **351**(6325), 402–406.

Falk, K., Rötzschke, O., Stevanovic, S., Jung, G. and Rammensee, H. G. (1991). Allele-specific motifs revealed by sequencing of self-peptides eluted from MHC molecules. *Nature* **351**(6324), 290–296.

Firat, H., Garcia-Pons, F., Tourdot, S., Pascolo, S., Scardino, A., Garcia, Z., Michel, M. L., Jack, R. W., Jung, G., Kosmatopoulos, K., Mateo, L., Suhrbier, A., Lemonnier, F. A. and Langlade-Demoyen, P. (1999). H-2 Class I knockout, HLA A-2. I-transgenic mice: a versatile animal model for preclinical evaluation of anititumor immunotherapeutic startegies. *Eur. J. Immunology* **29**(10), 3112–3121.

Flynn, J. L., Chan, J., Triebold, K. J., Dalton, D. K., Stewart, T. A. and Bloom, B. R. (1993). Role for interferon-gamma in resistance to Mtb infection. *J. Exp. Med.* **178**(6), 2249–2254.

Fomsgaard, A., Nielsen, H. V., Kirkby, N., Bryder, K., Corbet, S., Nielsen, C., Hinkula, J. and Buus, S. (1999). Induction of cytotoxic T-cell responses by gene gun DNA vaccination with minigenes encoding influenza A virus HA and NP CTL-epitopes. *Vaccine* **18**(7–8), 681–691.

Fong, L., Hou, Y., Rivas, A., Benike, C., Yuen, A., Fisher, G. A., Davis, M. M. and Engleman, E. G. (1998). Altered peptide ligand vaccination with Flt3 ligand expanded dendritic cells for tumor immunotherapy. *Proc. Natl. Acad. Sci. USA* **98**(15), 8809–8814.

Franke, E. D., Sette, A., Sacci, J. Jr, Southwood, S., Corradin, G. and Hoffman, S. L. (2000). A subdominant CD8(+) cytotoxic T lymphocyte (CTL) epitope from the *Plasmodium yoelii* circumsporozoite protein induces CTLs that eliminate infected hepatocytes from culture. *Infect. Immun.* **68**(6), 3403–3411.

Germain, R. N. and Margulies, H. (1993). The biochemistry and cell biology of antigen processing and presentation. *Ann. Rev. Immunology* **11**: 403–450.

Germain, R. N., Castolino, F., Han, R. C., Sousa, C. R., Romagnoli, P., SadeghNasseri, S. and Zhong, G. M. (1996). Processing presentation of endocytically acquired protein antigens by MHC class II and class I molecules. *Immunol. Rev.* **151**, 5–30.

Gomez, M., Johnson, S. and Gennaro, M. L. (2000). Identification of secreted proteins of *Mycobacterium tuberculosis* by a bioinformatic approach. *Infect. Immun.* **68**(4), 2323–2327.

Hammer, J., Bono, E., Gallazzi, E., Belunis, C., Nagy, Z. and Sinigaglia, F. (1994). Precise prediction of major histocompatibility complex class II-peptide interaction based on peptide side chain scanning. *J. Exp. Med.* **180**(6), 2353–2358.

Hanke, T., Schneider, J., Gilbert, S. C., Hill, A. V. S. and McMichael, A. (1998). DNA multi-CTL epitope vaccines for HIV and *Plasmodium falciparum*: immunogenicity in mice. *Vaccine* **16**(4), 426–435.

Hawley-Nelson, P., Vousden, K. H., Hubbert, N. L., Lowy, D. R. and Schiller, J. T. (1989). HPV16 E6 and E7 proteins cooperate to immortalize human foreskin keratinocytes. *EMBO J.* **8**(12), 3905–3910.

Hill, A. V. and Davenport, M. P. (1996). Reverse immunogenetics: from HLA-disease associations to vaccine candidates. *Molecular Medicine Today* **2**(1), 38–45.

Hislop, A. D., Good, M. F., Mateo, L., Gardner, J., Gatei, M. H., Daniel, R. C., Meyers, B. V., Lavin, M. F. and Suhrbier, A. (1998). Vaccine-induced cytotoxic T lymphocytes protect against retroviral challenge. *Nat Med.* **4**(10), 1193–1196.

Hodge, J. W., Sabzevari, H., Yafal, A. G., Gritz, L., Lorenz, M. G. and Schlom, J. (1999). A triad of costimulatory molecules synergize to amplify T-cell activation. *Cancer Res.* **59**, 5800–5807.

Hsu, S. C., Chargelegue, D., Obeid, O. E. and Steward, M. W. (1999). Synergistic effect of immunization with a peptide cocktail inducing antibody, helper and cytotoxic T-cell responses on protection against respiratory syncytial virus. *J. Gen. Virol.* **80**(6), 1401–1405.

Irvine, K. R., Parkhurst, M. R., Shulman, E. P., Tupesis, J. P., Custer, M., Touloukian, C. E., Robbins, P. F., Yafal, A. G., Greenhalgh, P., Sutmuller, R. P., Offringa, R., Rosenberg, S. A. and Restifo, N. P. (1999). Recombinant virus vaccination against 'self' antigens using anchor-fixed immunogens. *Cancer Res.* **59**(11), 2536–2540.

Ishioka, G. Y., Fikes, J., Hermanson, G., Livingston, B., Crimi, C., Qin, M. S., del Guercio, M. F., Oseroff, C., Dahlberg, C., Alexander, J. *et al.* (1999). Utilization of MHC class I transgenic mice for development of minigene DNA vaccines encoding multiple HLA-restricted CTL epitopes. *J. Immunol.* **162**(7), 3915–3925.

Jesdale, B. M., Deocampo, G., Meisell, J., Beall, J., Marinello, M. J., Chicz, R. M. and De Groot, A. S. (1997) Matrix-based prediction of MHC binding peptides: The EpiMatrix algorithm, reagent for HIV research, *Vaccines '97*, Cold Spring Harbor Press, Cold Spring Harbor, NY, 1997.

Kuchroo, V. K., Das, M. P., Brown, J. A., Ranger, A. M., Zamvil, S. S., Sobel, R. A., Weiner, H. L., Nabavi, N. and Glimcher, L. H. (1995). B7-1 and B7-2 costimulatory molecule activate differentially the TH1/Th2 developmental pathways: application to autoimmune disease therapy. *Cell* **80**(10), 707–718.

La Rosa, C., Krishnan, R., Markel, S., Schseck, J. P., *et al.* (2001). Enhanced immune activity of cytotoxic T-lymphocyte epitope analogs derived from positional scanning synthetic combinatorial libraries. *Blood* **97**(6), 1776–1786.

Leighton, J., Sette, A., Sidney, J., Appella, E., Ehrhardt C., Fuchs S. and Adorini, L. (1991). Comparison of structural requirements for interaction of the same peptide with I-Ek and I-Ed molecules in the activation of MHC Class II-restricted T cells. *J. Immunol.* **147**(1), 198–204.

Lipford, G. B., Hoffman, M., Wagner, H. and Heeg, K. (1993). Primary *in vivo* responses to ovalbumin. Probing the predictive value of the K^b binding motif. *J. Immunol.* **150**(4), 1212–1222.

Livingston, B. D., Newman, M., Crimi, C., McKinney, D., Chesnut, R. and Sette, A. (2001). Optimization of epitope processing enhances immunogenicity of multi-epitope DNA vaccines. *Vaccine* **19**(32), 4652–4660.

Ljunggren, H. G., Stam, N. J., Ohlen, C., Neefjes, J. J., Hoglund, P., Heemels, M. T., Bastin, J., Schumacher, T. N., Townsend, A., Karre, K., *et al.* (1990). Empty MHC class I molecules come out in the cold. *Nature* **346** (6283), 476–480.

Man, S., Newberg, M. H., Crotzer, V. L., Luckey, C. S., Williams, N. S., Chen, Y., Ridge, J. P., Huczko, E. L. and Engelhand, V. H. (1995). Definition of a human T cell epitope from influenza. A nonstructural protein 1 using HLA A2.1 transgenic mice. *Int. Immunol.* **7**, 597–605.

McMichael, A. J. and Phillips, R. E. (1997). Escape of human immunodeficiency virus from immune control. *Annu. Rev. Immunol.* **15**, 271–296.

Meister, G. E., Roberts, C. G. P., Berzofsky, J. A. and De Groot, A. S. (1995). Two novel T cell epitope prediction algorithms based on MHC-binding motifs; comparison of predicted and published epitopes from *Mycobacterium tuberculosis* and HIV protein sequences. *Vaccine* **13**(6), 581–591.

Missale, G., Bertoni, R., Lamonaca, V., Valli, A., Massari, M., Mori, C., Rumi, M. G., Houghton, M., Fiaccadori, F. and Ferrari, C. (1996). Different clinical behaviors of acute hepatitis C virus infection are associated with different vigor of the anti-viral cell-mediated immune response. *J. Clin. Invest.* **98**(3), 706–714.

Morris, S., Kelly, C., Howard, A., Li, X. and Collins, F. (2000). The immunogenicity of single and combination DNA vaccines against tuberculosis. *Vaccine* **18**(20), 2155–2163.

Moudgil, K. D., Sercarz, E. E. and Grewal, I. S. (1998). Modulation of the immunogenicity of antigenic determinant by their flanking residues. *Immunology Today* **19**(5), 217–220.

Newberg, M. H., Ridge, J. P., Vining, D. R., Salter, R. D. and Engelhard, V. H. (1992). Species specificity in the interaction of CD8 with the a3 domain of MHC class I molecules. *J. Immunol.* **149**, 136–142.

Olsen, A. W., Hansen, P. R., Holm, A. and Andersen, P. (2000). Efficient protection against *Mycobacterium tuberculosis* by vaccination with a single subdominant epitope from the ESAT-6 antigen. *Eur. J. Immunology* **30**(6), 1724–1732.

Oukka, M., Manuguerra, J. C., Livaditis, N., Tourdot, S., Riche, N., Vergnon, I., Cordopatis, P. and Kosmatopoulos, K. (1996). Protection against lethal viral infection by vaccination with nonimmunodominant peptides. *J. Immunol* **157**(7), 3039–3045.

Parker, K. C., Bednarek, M. A. and Coligan, J. E. (1994). Scheme for ranking potential HLA-A2 binding peptides based on independent binding of individual peptide side chains. *J. Immunol.* **1152**(1), 163–175.

Parkhurst, M. R., Salqaller, M. L., Southwood, S., *et al.* (1996). Improved induction of melanoma-reactive CTL with peptides from the melanoma antigen gp100 modified at HLA-A*0201-binding residues. *J. Immunol.* **157**(6), 2539–2548.

Pogue, R. R., Eron, J., Frelinger, J. A. and Matsui, M. (1995). Amino-terminal alteration of the HLA-A*0201-restricted human immunodeficiency virus pol peptide increases complex stability and in vitro immunogenicity. *Proc. Natl. Acad. Sci. USA* **92**(18), 8166–8170.

Rammensee, H. G., Friede, T. and Stevanoviic, S. (1995). MHC ligands and peptide motifs: first listing. *Immunogenetics* **41**(4), 178–228.

Rammensee, H. G., Bachmann J., Emmerich, N. N., Bachor, O. A. and Stevanovic, S. (1999). SYFPEITHI: database for MHC ligands and peptide motifs. *Immunogenetics* **50**(3–4), 213–219 (access via : http://www.uni-tuebingen.de/uni/kxi/),

Ressing, M. E., Sette, A., Brandt, R. M., Ruppert, J. *et al.* (1995). Human CTL epitopes encoded by human papillomavirus type 16E6 and E7 identified through in vivo and vitro immunogenicity studies of HLA-A* 0201-binding peptides. *J. Immunol.* **154**(11), 5934–5943.

Rodriguez, F. and Whitton, J. L. (2000). Enhancing DNA immunization. *Virology* **268**(2), 233–238.

Rosenberg, S. A., Yang, J. C., Schwartzentruber, D. J., Hwu, P., Marincola, F. M., Topalian, S. L., Restifo, N. P., Dudley, M. E., Schwarz, S. L., Spiess, P. J., Wunderlich, J. R., Parkhurst, M. R., Kawakami, Y., Seipp, C. A., Einhorn, J. H. and White, D. E. (1998). Immunologic and therapeutic evaluation of a synthetic peptide vaccine for the treatment of patients with metastatic melanoma. *Nature Medicine* **4**(3), 321–327.

Rosenfeld, R., Zheng, Q. and Delisi, C. (1995). Flexible docking of peptides to class I major-histocompatibility complex receptors. *Genet. Anal.* **12**(1), 1–21.

Rosenthal, K. S., Mao, H., Horne, W. I., Wright, C. and Zimmerman, D. (1999). Immunization with a LEAPS heteroconjugate containing a CTL epitope and a peptide from beta-2-microglobulin elicits a protective and DTH response to herpes simplex virus type 1. *Vaccine* **17**(6), 535–542.

Rothbard, J. B. and Taylor, W. R. (1988). A sequence pattern common to T cell epitopes. *EMBO J.* **7**, 93–100.

Rötzschke, O., Falk, K., Stevanovic, S., Jung, G., Walden, P. and Rammensee, H. G. (1991). Exact prediction of a natural T cell epitope. *Euro. J. Immunology* **21**(11), 2891–2894.

Ruppert, J., Sidney, J., Celis, E., Kubo, R. T., Grey, H. M. and Sette, A. (1993). Prominent role of secondary anchor residues in peptide binding to HLA-A2.1 molecules. *Cell* **74**(5), 929–937.

Sarobe, P., Pendleton, C. D., Akatsuka, T., Lau, D., Engelhard, V. H., Feinstone, S. M. and Berzofsky, J. A. (1998). Enhanced in vitro potency and in vivo immunogenicity of a CTL epitope from hepatitis C virus core protein following amino acid replacement at secondary HLA-A2.1 binding positions. *J. Clin. Invest.* **102**(6), 1239–1248.

Schadeck, E. B., Partidos, C. D., Fooks, A. R., Obeid, O. E., Wilkinson, G. W., Stephenson, J. R. and Steward, M. W. (1999). CTL epitopes identified with a defective recombinant adenovirus expressing measles virus nucleoprotein and evaluation of their protective capacity in mice. *Virus Res.* **65**(1), 75–86.

Schafer, J. A., Jesdale, B. M., George, J. A., Kouttab, N. M. and De Groot, A. S. (1998). Prediction of well-conserved HIV-1 ligands using a Matrix-based Algorithm, EpiMatrix. *Vaccine* **16**(19), 1880–1884.

Sette, A. and Sidney, J. (1998). HLA supertypes and supermotifs: a functional perspective on HLA polymorphism. *Curr. Opinion Immunol.* **10**(4), 478–482.

Sette, A., Buus, S., Appella, E., Smith, J. A., Chesnut, R., Miles, C., Colon, S. M. and Grey, H. M. (1989). Prediction of major histocompatibility complex binding regions of protein antigens by sequence pattern analysis. *Proc. Natl. Acad. Sci. USA* **86**(9), 3296–3300.

Sette, A., Sidney, J., Oseroff, C., del Guercio, M. F., Southwood, S., Arrhenious, T., Powell M. F., Colon, S. M., Gaeta, F. C. and Grey, H. M. (1993). HLA DR4w4-binding motifs illustrate the biochemical basis of degeneracy and specificity in peptide–DR interactions. *J. Immunol.* **151**(6), 3163–3170.

Sette, A., Vitiello, A., Reherman, B., Fowler, P. *et al.* (1994). The relationship between class I binding affinity and immunogenicity of potential cytotoxic T cell epitope. *J. Immunol.* **153**, 5586–5592.

Shahinian, A., Pfeffer, K., Lee, K. P., Kundig, T. M., Kishihara, K., Wakeham, A., Kawai, K., Ohashi, P. S, Thompson, C. B. and Mak, T. W. (1993). Differential T cell costimulatory requirements in CD28-deficient mice. *Science* **261**(5125), 609–612.

Shirai, M., Pendleton, C. D., Ahlers, J., Takeshita, T., Newman, M. and Berzofsky, J. A. (1994). Helper-CTL determinant linkage required for priming of anti-HIV CD8[+] CTL in vivo with peptide vaccine constructs. *J. Immunol.* **152**(2), 549–556.

Shirai, M., Arichi, T., Nishioka, M., Nomura, T., Ikeda, K., Kawanishi, K., Engelhard, V. H., Feinstone, S. M. and Berzofsky, J. A. (1995). Cytotoxic T lymphocyte (CTL) responses of HLA-A2.1-transgenic mice specific for hepatitis C viral peptides predict epitopes for CTL of humans carrying HLA-A2.1. *J. Immunol.* **154**, 2733–2742.

Slansky, J.E., Rattis, F. M., Boyd, L. F., *et al.* (2000) Enhanced antigen-specific anti-tumor immunity with altered peptide ligands that stabilize the MHC-peptide-TCR complex. *Immunity* **13**(4), 529–538.

Sonnenberg, M. G. and Belisle, J. T. (1997). Definition of *Mycobacterium tuberculosis* culture filtrate proteins by two-dimensional polyacrylamide gel electrophoresis, N-terminal aminoacid sequencing, and electrospray mass spectrometry. *Infect. Immun.* **65**(11), 4515–4524.

Sturniolo, T., Bono, E., Ding, J., Raddrizzani, L., Tuereci, O., Sahin, U., Braxenthaler, M., Gallazzi, F., Protti, M. P., Sinigaglia, F. and Hammer, J. (1999). Generation of tissue-specific and promiscuous HLA ligand databases using DNA microarrays and virtual HLA class II matrices. *Nature Biotech.* **17**(6), 555–561.

Tangri, S., Ishioka, G. Y., Huang, X., Sidney, J., Southwood, S., Fikes, J. and Sette, A. (2001) Structural features of peptide analogs of human histocompatibility leukocyte antigen class I epitopes that are more potent and immunogenic than wild-type peptide. *J. Exp. Med.* **194**(6), 833–846.

Thomson, S. A., Burrows, S. R., Misko, I. S., Moss, D. J., Coupar, B. E. and Khanna, R. (1998). Targeting a polyepitope protein incorporating multiple class II-restricted viral epitopes to the secretory/endocytic pathway facilitates immune recognition by CD4+ cytotoxic T lymphocytes: a novel approach to vaccine design. *J. Virol.* **72**(3), 2246–2252.

Threlkeld, S. C., Wentworth, P. A., Kalams, S. A., Wilkes, B. M., Ruhl, D. J., Keogh, E., Sidney, J., Southwood, S., Walker, B. D. and Sette, A. (1997). Degenerate and promiscuous recognition by CTL of peptides presented by the MHC class I A3-like superfamily. *J. Immunol.* **159**(4), 1648–1657.

Tine, J. A., Lanar, D. E., Smith, D. M., Wellde, B. T., Schultheiss, P., Ware, L. A., Kauffman, E. B., Wirtz, R. A., De-Taisne, C., Hui, G. S., Chang, S. P., Church, P., Hollingdale, M. R., Kaslow, D. C., Hoffman, S., Guito, K. P., Ballou, W. R., Sadoff, J. C. and Paoletti, E. (1996). NYVAC Pf7: a poxvirus-vectored, multiag, multistage vaccine candidate for *Plasmodium falciparum* malaria. *Infect-Immun.* **64**(9), 3833–3844.

Tobery, T. W. and Siliciano, R. F. (1997). Targeting of HIV-I antigens for rapid intracellular degradation enhances cytotoxic T lymphocyte (CTL) recognition and the induction of de novo CTL responses in vivo after immunization. *J. Exp. Med.* **185**(5), 909–920.

Tourdot, S., Oukka, M., Manuguerra, J. C., Magafa, V., Vergnon, I., Riche, N., Bruley-Rosset, M., Cordopatis, P. and Kosmatopoulos, K. (1997). Chimeric peptides: a new approach to enhancing the immunogenicity of peptides with low MHC class I affinity: application in antiviral vaccination. *J. Immunol.* **159**(5), 2391–2398.

Tsuji, M., Bergmann, C. C., Takita-Sonoda, Y., Murata, K., Rodrigues, E. G., Nussenzweig, R. S. and Zavala, F. (1998). Recombinant Sindbis viruses expressing a cytotoxic T-lymphocyte epitope of a malaria parasite or of influenza virus elicit protection against the corresponding pathogen in mice. *J. Virol.* **72**(8), 6907–6910.

Unanue, E. R. (1992). Cellular studies on antigen presentation by class II MHC molecules. *Curr. Opin. Immunol.* **4**(1), 63–69.

Velders, M. P., Weijzen, S., Eiben, G. L., Elmishad, A. G., Kloetzel, P. M., Higgins,

T., Ciccarelli, R. B., Evans, M., Man, S., Smith, L. and Kast, W. M. (2001). Defined flanking spacers and enhanced proteolysis is essential for eradication of established tumors by an epitope string DNA vaccine. *J. Immunol.* **166**(9), 5366–5373.

Wang, R., Doolan, D. L., Le, T. P., Hedstrom, R. C., Coonan, K. M., Charoenvit, Y., Jones, T. R., Hobart, P., Margalith, M., Ng, J., Weiss, W. R., Sedegah, M., de Taisne, C., Norman, J. A. and Hoffman, S. L. (1998). Induction of antigen-specific cytotoxic T lymphocytes in humans by a malaria DNA vaccine. *Science* **282**(5388), 476–480.

Wentworth, P. A., Vitiello. A., Sidney, J., Keogh, E., Chesnut, R. W., Grey, H., Sette, A. (1996). Differences and similarities in the A2. I-restricted cytotoxic T cell repertoire in humans and human leukocyte antigen-transgenic mice. *Eur. J. Immunol.* **26**(1), 97–101.

Whitton, J. L., Sheng, N., Oldstone, M. B. and McKee, T. A. (1993). A 'string-of-beads' vaccine, comprising linked minigenes, confers protection from lethal-dose virus challenge. *J. Virol.* **67**(1), 348–352.

Zaremba, S., Barzaga, E., Zhu, M., Soares, N., Tsang, K. Y. and Schlom, J. (1997). Identification of an enhancer agonist cytotoxic T lymphocyte peptide from human carcinoembryonic antigen. *Cancer Res.* **57**, 4570–4577.

Zhang, C., Anderson, A. and DeLisi, C. (1998). Structural principles that govern the peptide-binding motifs of class I MHC molecules. *J. Mol. Biol.* **281**(5), 929–947.

Zhu, M., Terasawa, H., Gulley, J., Panicali, D., Arlen, P., Schlom, J. and Tsang, K. Y. (2001). Enhanced activation of human T cells via avipox vector-mediated hyperexpression of a triad of costimulatory molecules in human dendritic cells. *Cancer Res.* **61**, 3725–3734.

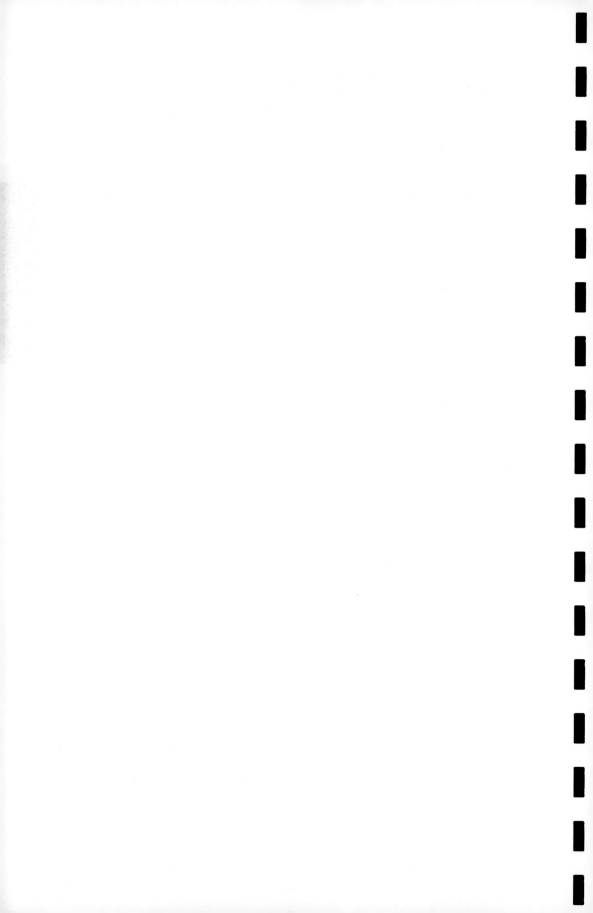

5 The Use of Tetramers in the Quantitative Analysis of T-cell Responses

Geraldine M A Gillespie, Victor Appay, Sarah L Rowland-Jones and Andrew J McMichael

MRC Human Immunology Unit, Weatherall Institute of Molecular Medicine, The John Radcliffe Hospital, Oxford, UK

◆◆

CONTENTS

The introduction of MHC class I tetramer technology undoubtedly represents a late twentieth-century milestone in the study of T-cell immunology. Prior to their development, cellular immunologists relied on methodology where antigen-specific T cell frequencies were calculated on the basis of specific functions. The generation of tetramers has revolutionized T-cell analysis by allowing enumeration by direct visualization, and has enhanced the functional analysis of specific T-lymphocyte populations. In this chapter we describe the principles of MHC class I tetramer generation and explore their applications in the quantitative analysis of CD8+ T cells and in the dissection of cellular functions.

METHODS IN MICROBIOLOGY, VOLUME 32
ISBN 0–12–521532–0

Copyright © Elsevier Science Ltd
All rights of reproduction in any form reserved

◆◆◆◆◆◆ **INTRODUCTION**

Antigen processing

The initiation of the cellular immune response relies on the ability of the αβT-cell receptor expressed on T lymphocytes to recognize its ligand, namely, an antigenic peptide epitope bound to a major histocompatibility complex (MHC) molecule on the surface of an antigen-presenting cell (APC). During the course of an infection, this occurs when virally infected cells process and present viral antigens, or exogenously derived antigens are cross-presented by professional APCs, to CD8+ T lymphocytes. The details of the processing pathway are complex and only a simplified account shall be presented here. In brief, newly synthesized viral proteins in the cytosol are targeted via the ubiquitin pathway of protein degradation to the immunoproteasome and cleaved into peptide epitopes ranging from 8–12 amino acids in length. From there, the peptide epitopes are actively transported into the endoplasmic recitulum (ER) via TAP transporter molecules. Through a complex series of events involving a number of chaperones and accessories molecules, a selection of transported peptides form part of a trimeric complex with the 12 kDa molecule β2M and MHC class I heavy chain molecule. Correctly assembled complexes are released from the ER and targeted via the Golgi stacks to the cell surface (Solheim, 1999; Yewdell *et al.*, 1999; York *et al.*, 1999).

MHC class I heavy chains are encoded by a gene complex that maps to chromosome 6. There are three loci, namely HLA-A, HLA-B and HLA-C. Despite the high overall structural homology, MHC class I molecules are highly polymorphic, particularly in regions associated with the binding of antigenic epitopes. It is this level of heterogeneity which ensures that a diverse pool of antigenic peptides is presented to CD8+ T lymphocytes. The three-dimensional structure of the MHC is such that peptide epitopes bind into a groove formed by the MHC class I alpha 1(α1) and alpha 2 (α2) helices. A proportion (usually two or three) of peptide amino acid side-chains act as 'anchor residues' and stabilize binding to MHC; a number of non-anchor residues point upwards and affect recognition by T lymphocytes (Ruppert *et al.*, 1994; Cerundolo *et al.*, 1995; Young *et al.*, 1995; Johansen *et al.*, 1997; Bouvier and Wiley, 1998; Wang *et al.*, 1998; Schueler-Furman *et al.*, 2000).

Recognition by T lymphocytes

Recognition of MHC class I–peptide complexes on the surface of the antigen-presenting cell (APC) relies on a heterodimeric T-cell receptor expressed on CD8+ T lymphocytes, comprising an alpha(A) and beta(B) glycoprotein chain component. During T-cell ontogeny, in a process similar to that described for antibodies, T-cell receptor diversity is generated through the splicing together of gene segments to generate regions of hypervariability, known as complementary-determining regions (CDRs), within both A and B chain components (Oettinger, 1992;

Thompson, 1995; Bogue and Roth, 1996; Willerford *et al.*, 1996; Gellert, 1997; Lewis and Wu, 1997; Schlissel and Stanhope-Baker, 1997; Nemazee, 2000). Thymocytes successfully expressing the re-arranged TCR are subjected to a complex selection process positively to select T cells with a low avidity for self-peptide–MHC complexes but to eliminate (negative selection) high avidity receptors that could potentially give rise to self-reactive T lymphocytes. If a TCR fails to interact with a self-peptide–MHC complex, the T cell will die by neglect. T cells expressing a TCR with a high affinity for self-antigen are either deleted, or continue to re-arrange the TCR A chain until a receptor displaying low avidity for self-peptide–MHC is expressed (Jameson and Bevan, 1998; McGargill and Hogquist, 2000; McGargill *et al.*, 2000).

The engagement of MHC–peptide complex by TCR–CD3 complex results in a signalling cascade that ultimately leads to T-cell activation and effector function. The T-cell co-receptor molecule CD8 is a transmembrane glycoprotein that binds the constant domain of MHC class I on the cell surface. CD8 also interacts with p56lck intracellularly (via the CD8α chain component) and plays an important role in the activation and localization of this kinase within cell membrane lipid rafts where CD3 phosphorylation is thought to occur. Once a sufficient number of TCR have engaged MHC–peptide complexes, activating kinases accumulate in lipid rafts with the concomitant exclusion of inhibitory phosphatases. Consequently, CD3 phosphorylation takes place, initiating the downstream signalling cascades, which ultimately results in T-cell activation and effector (e.g cytolytic activity, cytokine/chemokine production) function. Following engagement, MHC–TCR complexes are internalized in vesicles and destined for degradation (Qian and Weiss, 1997; Cantrell *et al.*, 1998; Alonso and Millan, 2001; Davis and van der Merwe, 2001; Harder, 2001; Heath and Carbone, 2001; Krauss *et al.*, 2001; van der Merwe, 2001).

Compared with antibody-antigen, the affinity of TCR for peptide-MHC is low (Davis *et al.*, 1998; Willcox *et al.*, 1999). Consequently, previous attempts to stain T cells using soluble monomeric forms of MHC proved unsuccessful. To compensate for this low affinity interaction, researchers focused on increasing the overall avidity by increasing the number of MHC class I molecules available to bind TCR. Consequently, tetrameric forms of MHC class I molecules were formed and this represented the breakthrough that allowed successful staining of antigen-specific T cells *in vitro* (Altman *et al.*, 1996).

◆◆◆◆◆◆ MHC CLASS I TETRAMER GENERATION

Method of MHC class I generation

The methodology for the generation of tetrameric MHC class I molecules is straightforward. The first step involves the generation of a bacterial expression construct incorporating a truncated form of the MHC class I heavy chain gene engineered to a Bir A biotinylation substrate motif

(BSM) (Schatz, 1993). Classical MHC class I constructs include the N terminus amino acid of the mature chain but are truncated beyond residue 275 to remove hydrophobic transmembrane residues (Garboczi *et al.*, 1992). They are further modified by the addition of glycine and serine residues to introduce a region of flexibility upstream of a 13 to 15 amino acid BSM which serves as a template for the bacterial-derived holo-enzyme synthetase Bir A. Bir A is an important enzyme in *Escherichia coli* where it plays an important role in biotin catabolism (Abbott and Beckett, 1993; Xu and Beckett, 1994; Xu *et al.*, 1995) and is responsible for catalysing the covalent ligation of biotin from an adenylate intermediate to a specific lysine residue located within the biotin carboxyl carrier protein of the enzyme acetyl-CoA carboxylase. The 15 amino acid substrate motif, LHHILDAQKMVWNHR, or the 13 amino acid LNDIFEAQKIEWH sequence contains sufficient consensus sequence to enable biotinylation *ex vivo* (Schatz, 1993).

Recombinant proteins are routinely expressed under the powerful T7 polymerase promoter and accumulate in the prokaryotic host as 'inclusion bodies'. These comprise primarily insoluble recombinant protein, either as misfolded protein aggregates or partially refolded structures, held together by non-covalent interactions. Fortunately, the majority of cloned MHC class I–BSM gene products are readily purified and solubilized and folded *ex vivo* using standard biochemical procedures. This involves the disruption of host cells either by chemical lysis or sonication and separation of the dense inclusion bodies from cellular debris by centrifugation. These are extensively washed in buffers containing a neutral detergent such as Triton-X-100 (0.5%) prior to solubilization in a chaotrophic reagent, either 8M urea or 6M guanidine. The renaturation conditions, although with slight variation between individual groups, generally comprise 100 mM Tris-HCL buffer (pH 8.0) containing a stabilizing agent such as 400 mM L-arginine, an oxido shuffling system, for example, reduced and oxidized glutathione (at a ratio of 10 : 1, with a starting concentration of 5 mM glutathione) to promote intrachain disulfide bond formation, and a cocktail of both aspartic acid and serine/cysteine protease inhibitors. Up to 30 mg of MHC class I heavy chain, 10 mg of β2M and 10 mg of peptide epitope are added to one litre of renaturation buffer; these concentrations are equivalent to approximately 10 μM of each of the individual protein constituents. β2M, peptide and MHC class I heavy chains are added to the renaturation buffer by dilution (i.e. the proteins are diluted straight from urea into the folding mix). To minimize protein aggregation and precipi-tation, we have modified this method by initially diluting the protein (in denaturant) into a small volume of renaturant buffer (one part of denaturant : five parts of renaturant) and this is subsequently added to the one litre renaturation buffer. This approach prevents a dramatic drop in denaturant concentration, and minimizes the precipitation of MHC class I heavy chain components which can occur when added by a single step dilution. The protein constituents are added sequentially; β2M is folded for up to 30 min prior to the addition of MHC class I heavy chain and peptide. Peptide epitopes are generally solubilized in urea prior to addition, but hydrophobic peptides are generally diluted in dimethyl sulfoxide (DMSO) (in less than 1/1000th the volume of the renaturation buffer) and then added to the folding buffer. The renaturation reaction is incubated for 24 to 48 h at 4°C.

Further to renaturation, MHC class I tetramers are biotinylated using commercially available Bir A enzyme and Biomix substrates containing biotin and ATP (purchased from Avidity) overnight at room temperature. Biotinylated MHC class I complexes are purified by size exclusion and ion-exchange chromatography. The class I monomers are then tetramer-ized and fluorescently labelled simultaneously, by the addition of avidin conjugated to a fluorochrome. Avidin possesses four high affinity-binding sites for biotin, and its addition, in the correct molar ratios, results in the formation of MHC class I tetrameric structures from the monomeric substrate. Phycoerythrin (PE) is generally the fluorochrome of choice due to its high fluorescent intensity, although a variety of other fluorescent labels have also been used.

Problems encountered during MHC class I generation and staining

MHC class I heavy chain expression

The prokaryotic T7 system is the method of choice for the expression of BSM-tagged MHC class I constructs and is incredibly efficient for the expression of MHC class I A allelic products. Many of the B alleles, how-ever, are often expressed at low levels or not at all. One possibility for a lack of eukaryotic gene expression in bacterial systems is the formation of stable hairpin structures in eukaryotic mRNA as a result of high G/C content. These structures are thought to interfere with ribosome binding and the translation of mRNA into protein product (de Smit and van Duin, 1990; Garboczi et al., 1992; de Smit and van Duin, 1994; Gao et al., 1998). This seems a likely explanation since the introduction of silent mutations to increase the A/T content in the immediate 5′ end of the MHC class I chain, and thus destabilize loop formation, has resulted in high titres of gene expression. In our experience, this was necessary for the expression of HLA-B*3501, HLA-B*5701, HLA-B*5703, HLA-B*5801 and HLA-Cw0402 heavy chain products.

MHC class I renaturation *in vitro*

Fortunately, the vast majority of class I molecules form native structures when manipulated *in vitro*. We, however, have encountered a number of HLA proteins which posed significant problems and are either impossible to fold, or yield low concentrations of protein product but large levels of protein aggregates (e.g HLA-B*5701 protein). Hydrophobic interactions and incorrect disulfide bond formation are thought to contribute to protein aggregation *in vitro*. Whereas the role of disulfide bond formation in protein aggregations remains controversial, it has been proposed that specific hydrophobic interactions lead to the formation of small protein aggregates and that these may act as a nucleation site onto which larger aggregates seed and are stabilized by disulfide bonds (Guise et al., 1996; de Bernardez Clark, 1998). Table 1 summarizes a list of common methods and reagents used to assist protein renaturation in vitro.

Recently we have attempted to optimize the folding of MHC class I B*5701 and B*5703 alleles. These molecules are of significant interest in our laboratory given their association with slow progression to AIDS in HIV-1 infection (Klein et al., 1998; Migueles et al., 2000). To improve their recovery in vitro we initially focused on the optimization of denaturant concentration, namely urea, present in the folding reaction to discover the final denaturant concentration sufficiently high so as to solubilize aggregates but low enough to ensure proper protein renaturation. This was performed over a broad range of denaturant concentrations (50 to 500 mM), but failed to increase the yield of folded B*57 complexes. We have also attempted to renature B*57 molecules by dialysis from high (1 Molar) to low concentrations of denaturant. This method allows the retrieval of B*57 complexes (E. Manting, personal communication) but fails significantly to increase the yields beyond that obtained by conventional methods of MHC class I renaturation. We have also, but to a lesser extinct, turned our attention to 'oxido shuffling' system which promotes oxidation and correct di-sulfide bond formation. Typically we use reduced and oxidized glutathione, at a ratio of 10:1 with a starting ratio of 5 mM reduced glutathione. Attempts to decrease the ratio, by decreasing the input of reduced glutathione, or the use of different 'oxido shuffling' reagents (e.g. cysteamine/cystamine) has not increased recovery of B*57 complexes.

Other methods routinely utilized to promote correct di-sulfide bond formation include the formation of mixed di-sulfide species or air oxidization; the latter appears to increase the recovery of certain MHC class II molecules (J. Wyer, personal communication), but has not increased the folding of B*57 in vitro. Occasionally, the presence of an unpaired cysteine residue can disrupt correct di-sulfide bond formation in vitro; this has been observed for HLA class I B*2705 (Allen et al., 1999). By mutating the unpaired cysteine at position 67 to serine, Allen and colleagues increased the recovery of folded B*2705 complexes. The renaturation of class I alleles can also be influenced by the affinity of peptide epitopes for MHC, and can be overcome, to some extent, by the addition of up to 10-fold excess peptide in the renaturation buffer (unpublished observation). The generation of β2M-peptide fusion proteins to compensate for low-affinity peptides is an alternative option. The β2M-peptide fusion system has been recently used to generate A*0201-Influenza A matrix complexes, and the resulting tetramers are functional and highly stable in vitro (Tafuro et al., 2001).

Influence of CD8 on binding of MHC class I tetramers to TCR

From a technical perspective, the influence of CD8 on the initial binding of tetrameric MHC class I to TCR ex vivo, is not fully understood. Certain

Table I. Factors that influence protein folding *in vitro*.

Factors	Comments	Limitations
A. Choice of renaturation strategy		
1. single-stage dilution	single-stage dilution from one buffer to another	
2. slow dialysis	proteins are dialysed against high and low concs. of denaturant	only suitable for proteins that do not aggregate at intermediate conc. of denaturants
B. Renaturation buffer additives		
1. Chaotrophic reagents, e.g. urea, guanidine	increase solubility of folding intermediates	chaotrophs may shift the competition between aggregation and renaturation to aggregation
2. Amino acids, e.g. L-arginine, glycine	as above	
3. Salts, e.g. ammonium sulphate	as above	
4. Sugars, e.g. glycerol, glucose		
5. Detergents/surfactants, e.g. Tween, Triton X-100, polyethylene glycol	prevent aggregation	detergents often interfere with downstream purification procedures, e.g. ion-exchange, ultrafiltration
C. Oxidation conditions		
1. air oxidation	promotes disulphide bond formation	slow, often inefficient with low protein recovery, unsuitable for proteins with unpaired cysteine residues
2. oxido-reshuffling system, e.g. reduced and oxidized glutathione, cysteamine/cystamine	promotes disulphide bond formation	need to determine the correct ratios of reduced to oxidized species emperically
3. mixed disulphide species, i.e. treat the reduced protein with oxidized species to generate mixed disulphides prior to renaturation and oxidization	increase protein solubility and prevents inter- and intradisulphide bond formation	
D. Chaperone-assisted protein refolding		
1. immobilized GroEL system	GroEL minichaperones are immobilized on a chromatographic resin to assist protein renaturation	only works with GroEL substrate
2. artificial chaperone-assisted folding	Proteins are refolded in the presence of non-ionic detergents to prevent aggregation; detergents are then stripped away, (using cyclodextrin) to promote renaturation	

For a detailed account of protein renaturation *in vitro* see Guise *et al.* (1996) and de Bernardez Clark (1998).

anti-CD8 monoclonal antibodies are known to interfere with the binding of tetramers to TCR, SK1 antibody for example, blocks tetramer binding to a T-cell clone if incubated with T cells prior to tetramer staining. If both SK1 and tetramer are added simultaneously, then tetramer binds but with a lower fluorescent intensity than if it was added first. The converse is true for the OKT8 antibody; if cells are incubated with OKT8 prior to tetramer staining they display enhanced tetramer binding when compared with samples incubated with both reagents simultaneously or with tetramer first (Campanelli et al., 2002). These, in addition to data from experiments on the decay kinetics of TCR-tetramer in the presence or absence of these antibodies, are interpreted to support a crucial role for CD8 in the initial binding and stabilization of the TCR–MHC interaction (Campanelli et al., 2002). However, HLA A*0201 tetrameric constructs mutated at residues crucial for interaction with CD8 (amino acid 245, 227 or 228 in MHC class I heavy chain $\alpha 3$ helix) bind T lymphocytes at similar intensities compared with their non-mutated counterparts (Bodinier et al., 2000; Purbhoo et al., 2001; Xu et al., 2001). Therefore the anti-CD8 antibody for certain tetramer preparations need to be selected with care, and possible artefacts introduced by the CD8-HLA class I protein interaction need to be excluded.

Binding of MHC class I tetramers to KIR and LILR receptors

MHC class I molecules are also ligands for receptors of the leukocyte-receptor complex (LCR) (Table 2). These comprise immunoglobulin-like receptor families of which the killer inhibitory receptors (KIRs) and leukocyte immunoglobulin-like receptors (LILRs), also known as ILT receptors, are well characterized. Whereas KIR expression is generally confined to NK and CD8+ T lymphocytes, LILRs are found on NK cells, B and T lymphocytes and myelomonocytic cells. Both KIRs and LILRs function as immunomodulatory receptors, and upon interaction with appropriate MHC class I ligands, deliver either inhibitory or activatory signals to NK cells and monocytes (Colonna et al., 1999; Moretta et al., 2000; Trowsdale, 2001; Young and Uhrberg, 2002). The expression of both KIRs and LILRs is mostly restricted to highly differentiated populations of memory T cells, and may play a role in the modulation of T-cell activation and function, control T-cell mediated immunopathology upon the resolution of infection, and influence T-cell survival (Tarazona et al., 2000; Ugolini et al., 2001; Young and Uhrberg, 2002).

Whereas LILRs are capable of binding to the $\alpha 3$ domain of a variety of classical and non-classical MHC molecules, binding to KIRs is limited to the classical MHC class I alleles, HLA-B and C. In addition, binding of the different KIR family members to different HLA-B alleles is dictated by the dimorphic amino acids 77–83, that separate the HLA-B into the Bw4 and Bw6 serology groups. Similarly, the dimorphisms in residues 77–80 of the $\alpha 1$ helix, that split HLA-C into groups 1 and 2, determine binding to different KIRs (Lanier, 1998; Colonna et al., 1999; Maenaka et al., 1999; Navarro et al., 1999; Long and Rajagopalan, 2000; Sawicki et al., 2001; Trowsdale, 2001; Martin et al., 2002; Young and Uhrberg, 2002).

Table 2. Summary of known LILR– and KIR–MHC ligand interactions.

Receptor	Official name	Ligand	Host cell
LILR family			
	LILRB1 (ILT2, LIR1)	HLA-A, HLA-B, HLA-G	monocytes/lymphocytes
	LILRB2 (ILT4, LIR2)	HLA-A, HLA-B, HLA-F, HLA-G	monocytes
	LILRA1 (LIR6)	HLA-B27	monocytes
KIR family			
	KIR2DL1	HLA-Cw2, -Cw4, -Cw5, -Cw6	NK cells
	KIR2DL2	HLA-Cw1, -Cw3, -Cw7, -Cw8	NK cells, T lymphocytes
	KIR2DL3	HLA-Cw1, -Cw3, -Cw7, -Cw8	NK cells, T lymphocytes
	KIR2DS4	HLA-Cw4	NK cells
	KIR2DL4	HLA-G	NK cells, T lymphocytes
	KIR3DL1	HLA-Bw4	NK cells, T lymphocytes
	KIR3DL2	HLA-A3, HLA-A11	NK cells
	KIR3DS1	HLA-Cw2, -Cw4, -Cw5, -Cw6	NK cells

Adapted from Martin *et al.*, 2002 (see also Allen *et al.*, 2001; Young and Uhrberg, 2002 and Moretta *et al.*, 2000).

Receptor nomenclature as defined by the Human Genome Organization (HUGO) – see www.gene.ucl.ac.uk for description of new LILR nomenclature.

LILR = leucocyte immunoglobulin-like receptors, (ILT = immunoglobulin-like transcripts), KIR = killer cell immunoglobulin-like receptors, NK = natural killer.

Tetramer-reactive cells are generally assessed as a function of CD8[bright] CD3+ cells, and binding to KIRs and LILRs present on NK cells and monocytes respectively, will not affect data interpretation. More importantly, however, we do not know if a proportion of tetramer binding on T lymphocytes reflects interactions with LILRs and KIRs. It is estimated that only a small percentage of peripheral blood CD8+ T cells (5%), express a given KIR family member (Mingari *et al.*, 1996; Speiser *et al.*, 1999). These issues may have been underestimated because of the disproportionate use of HLA-A2 tetramers; HLA-A2 does not bind to any KIR receptors. The problem may be greater with HLA-B and HLA-C tetramers. The proportion of T cells expressing LILRs may be higher, however, (Young *et al.*, 2001), and as these molecules bind HLA-A and HLA-B alleles it is worth investigating their interaction with LILRs. It is possible to block tetramer binding to LILRs, as has recently been demonstrated on LILR transfectants, by pre-incubating with anti-LILR specific antibodies (Allen *et al.*, 2001). Similar blocking experiments could be used to determine whether MHC class I tetramers bind to LILRs (and indeed KIRs) on CD8+ T lymphocytes.

◆◆◆◆◆◆ QUANTIFICATION OF CD8+ T LYMPHOCYTES

Other methods used to quantitate T lymphocytes

Prior to the development of MHC class I tetramer technology, methods to enumerate T-cell populations relied almost exclusively on functional activity, the most popular of these being the standard chromium-51 (^{51}CR) release cytotoxic T-cell assay (or CTL assay) and the limiting dilution analysis (LDA). The CTL assay involves testing the lytic capacity of T lymphocytes *in vitro*, when incubated with target cells presenting the nominal antigenic epitope. The target cells are pre-labelled with radioactive chromium, which is subsequently released during the assay if antigen-specific T cells lyse target cells. Using this assay to enumerate antigen-specific T cells is flawed for many reasons. It is not very sensitive and T-cell populations need to be enriched by cultivation *in vitro* prior to analysis. The expansion in culture of antigen-specific T cells is unreliable, making it impossible to quantify their frequencies (Brunner *et al.*, 1976). To some extent, the problem is addressed in the limiting dilution assay (LDA), where limiting dilutions of T lymphocytes are specifically expanded *in vitro* to obtain sufficient numbers to effect lytic activity in a standard ^{51}CR release assay (Owen *et al.*, 1982). Whilst this method allows the estimation of precursor frequencies it relies on the antigen-specific T cell to perform an effector function, and to retain the capacity to proliferate *in vitro*. We now know that many antigen-specific T lymphocytes display a limited ability to proliferate to antigen (Pan *et al.*, 1997; Burns *et al.*, 2000; Malaguarnera *et al.*, 2001; Speiser *et al.*, 2001). LDA analysis requires that a precursor cell should undergo approximately 10–15 cell divisions to be detected (McMichael and O'Callaghan, 1998).

Consequently, this technique often underestimates antigen-specific T-cell frequencies. It has recently been demonstrated that reducing the cell input number per well for LDA analyses improved its sensitivity, resulting in the estimation of frequencies comparable to those obtained by tetramer and ELISpot analysis (Goulder *et al.*, 2000a). The authors argue that in the normal LDA system high T-cell number input increases the proportion of non-specific T-cell input which, in turn, increases the risk of their over-growth in a given analysis well (non-specific T cells easily overgrow antigen-specific T cells under conditions of cytokine stimulation *in vitro*). Consequently, non-specific T-cell overgrowth would obscure antigen-specific cell proliferation and detection, resulting in an increase in the number of non-antigen specific wells per plate, and a decrease in the measure of antigen-specific T-cell frequencies. Despite the improved sensitivity in detection, this modified LDA is not practical given the very large numbers of replicate wells required per assay.

The Enzyme Linked Immunospot assay (ELISpot assay) (Czerkinsky *et al.*, 1991) has recently been modified to quantify antigen-reactive CD8+ T-cell frequencies (Lalvani *et al.*, 1997; Scheibenbogen *et al.*, 1997). In this system antigen-pulsed APCs are co-incubated with T lymphocytes and assess for the secretion of the anti-viral cytokine, generally interferon γ (IFNγ), which is captured in a sandwich-based ELISA system. Individual IFNγ-secreting cells give rise to a localized concentration of IFNγ, detected as a single 'spot' on the ELISpot, thus allowing the visual identification of specific T cell. This technique does not require long-term cultivation of cells *in vitro*, but it is limited by virtue of its reliance on a single cellular function, therefore making it likely to underestimate T-cell frequencies. The technique is occasionally hampered by high backgrounds of IFNγ production (occasionally observed in the setting of co-infections in HIV infection; unpublished observations) and there are often inconsistencies between individual laboratories in what represents acceptable criteria to define a positive result. In our laboratory we consider that the limits of detection are 20 antigen-specific T cells per million PBMCs, when the background is less than half this value (Kaul *et al.*, 2000), but other laboratories define positive criteria in other ways. Despite this, the assay has the advantage of sensitivity and specificity and is regularly used, in conjunction with tetramer analysis, to examine antigen-specific T cells.

Another cytokine production assay involves the intracellular staining of antigen-stimulated T cells (Jung *et al.*, 1993; Mascher *et al.*, 1999; Maecker *et al.*, 2001). Cells are incubated with peptide antigen, in the presence of a reagent that blocks protein egress from the ER (e.g Brefeldin A). Following 4 to 6 h of incubation, cells are permeabilized, stained intra-cellularly for cytokine production and analysed by flow cytometry. The method is similar in sensitivity to tetramer staining, although numbers may be lower because not all T lymphocytes make the cytokine, usually IFNγ. It has the advantage that tetramers do not need to be generated, and over-lapping peptide can be used (Maecker *et al.*, 2001).

Prior to the introduction of MHC class I tetramers, monoclonal antibodies directed against the variable (V) regions of TCR α and β chains were used to analyse T cells directly (Clarke *et al.*, 1994; Callan *et al.*, 1995,

1996; Moss *et al.*, 1996; Moss and Gillespie, 1997). This approach has allowed the identification of antigen-specific cells expressing certain V region segments, but it is difficult to estimate the proportion of cells expressing a given V region segment that represents the antigen-specific T-cell pool without sequencing the receptors. Using tetramers the antigen-reactive T-cell population can be specifically visualized by flow cytometric analysis. The technique has the advantage of being rapid; staining can be performed on small volumes (100–200 µl) of fresh blood, and standard protocols require that tetramer staining be performed at 37°C for 15 min only. The technique is sensitive with a cut-off value of approximately 0.02% of CD8+ T cells. The technique has some limitations, however, relating to the quality of individual tetramers and the skill of data interpretation. As a result of technical demands associated with their production *ex vivo*, and the requirement for precise epitope definitions, it is not possible to study all MHC class I–peptide complexes.

◆◆◆◆◆◆ FUNCTIONAL ANALYSIS OF TETRAMER-REACTIVE CELLS

Analysis of T-cell proliferation

The ability of a T cell to proliferate upon encounter with antigen marks the initiation of cellular activation. Traditionally, T-cell proliferation has been assessed using 3H-thymidine incorporation and involves the incubation of T lymphocytes with cognate antigen in the presence of 3H-thymidine. The degree of 3H-thymidine incorporation into newly synthesized DNA reflects the degree of cellular proliferation. This technique, however, only measured the proliferative capacity of the total T-cell pool. Recently, methods used to assess cellular proliferation by flow cytometry analysis have been combined with tetramer staining to dissect the activity of specific populations of T lymphocytes. For example, the thymidine analogue, bromodeoxyuridine (BrdU) is readily incorporated in proliferating cells and can really be detected following cell permeabilization and staining with anti-BrdU antibodies (Gratzner and Leif, 1981; Gratzner, 1982). The vital dye carboxyfluorescein diacetate succinimidyl ester (CSFE) is also a very useful reagent that can permeate cell membranes without cell damage. Upon cell proliferation, the dye is divided between daughter cells with incredible fidelity (Weston and Parish, 1990; Lyons, 2000) and the technique allows the tracking of many cycles of cell division (up to 10) before dye extinction. CFSE labelled cells remain viable, rendering it possible to include them in other functional analyses. Both BrdU and CFSE labelling have been combined with tetramers to tract T-cell turnover in mice and to monitor antigen-specific T-cell proliferation *ex vivo* (Doherty and Christensen, 2000; Turner *et al.*, 2001).

Other protocols assess the proliferative capacity of T cells directly *ex vivo*, and usually involve staining cells for intracellular 'markers' that correlate with cell proliferation (Gerdes *et al.*, 1984). The most common of these include the nuclear antigen, Ki67, the function of which remains unknown, but expression levels rapidly increase through the cell cycle. Ki67 staining has been used in conjunction with tetramers to analyse T cells (Tan *et al.*, 1999).

T-cell death/apoptosis

Following stimulation with antigen, the majority of activated T lymphocytes die by programmed cell death (apoptosis). There are two main apoptotic pathways. The first involves the engagement of cell surface death receptors by their respective ligands. For instance, the engagement of Fas ligand by CD95 expressed on T lymphocytes results in the delivery of a death signal to that cell (Green, 2000; Strasser *et al.*, 2000). The second involves the Bcl-2 superfamily of intracellular proteins. In this system the ratios of pro-apoptotic (e.g. Bad, Bax, Bid) to anti-apoptotic (Bcl-2 and Bcl-xL) molecules determine T cell fate (Dutton *et al.*, 1998; Grayson *et al.*, 2000). It is possible to assess the susceptibility of antigen-specific T cells to apoptosis. Tetramers have been successfully combined to investigate both pathways and are used routinely to co-stain for the cell surface expression of CD95 (Gillespie *et al.*, 2000b) or the intracellular levels of Bcl-2 (Grayson *et al.*, 2000). Both markers denote a cell's susceptibility to die, but do not actually indicate that a cell is in the process of apoptosis; Annexin V staining is generally used to confirm this. Annexin V binds an intracellular kinase (known as phosphatidylserine) that is cell surface exposed early on in the initiation of the apoptotic pathway (Bourez *et al.*, 1997; van den Eijnde *et al.*, 1997). Staining with this marker is readily combined with tetramer staining *in vitro* (Oxenius *et al.*, 2001).

Tetramer-reactive effector T-cell function

Cytokine and chemokine production

Tetramer staining analyses commonly estimate greater frequencies of antigen-specific T lymphocytes than those recorded by other techniques. These disparities suggest that not all tetramer-reactive T cells retain the capacity to proliferate and perform an effector function upon contact with antigen. To investigate this, methods combining tetramer staining with functional and proliferative analyses have been developed. For functional analysis, tetramer-reactive cells are assessed for their ability to produce important cytokines and chemokines that constitute mediators of the immune response. Cytokines considered important include IFNγ, TNFα, IL-2, in addition to the CC chemokines RANTES, MIP-1α and MIP-1β. Recently, protocols combining intracellular staining (ICS) and flow cytometric analysis have been developed to estimate cytokine secretion in response to antigen (Labalette-Houache *et al.*, 1991; Jung *et al.*, 1993; Mascher *et al.*, 1999; Maecker *et al.*, 2001) and following the introduction of tetramers, the proportion of antigen-specific cells expressing cytokines can be accurately assessed. Early attempts to combine both techniques, however, were hampered by the fact that activation of antigen-specific

T lymphocytes prior to incubation with tetramer leads to the down-regulation of functionally reactive TCRs, resulting in loss of TCR and lack of tetramer staining. Incubating cells with tetramer prior to cellular activation has resolved the problem (Appay *et al.*, 2000). In this instance tetramers bind a fraction of TCR on antigen-specific T lymphocytes, which are then internalized, and results in the fluorescent labelling of cells. The remaining unbound antigen-specific TCRs can subsequently be engaged and activated by peptide-pulsed APCs. This technique enables the successful identification and quantification of cytokine-secreting tetramer-reactive T cells and has been used to monitor IFNγ, TNFα, MIP-1b and IL-2 production by tetramer-reactive T cells (Appay *et al.*, 2000; Gillespie *et al.*, 2000b) (Fig. 1). A drawback to this technique is that it requires cell permeabilization, rendering the retrieval of live T cells impossible.

The combination of tetramer staining and the cell surface affinity matrix technique (so-called secretion/capture assay (Manz *et al.*, 1995)) allows the isolation of live antigen-specific T cells with a particular function (Pittet *et al.*, 2001). Cytokine-secreting cells are identified using a specific high-affinity capture matrix comprising a bi-specific antibody conjugate directed against the cytokine/chemokine under investigation (e.g. IFNγ) and the CD45 molecule which is ubiquitously expressed on all T lymphocytes (Merkenschlager *et al.*, 1988; Merkenschlager and Beverley, 1989). The bi-specific antibody allows binding to CD45 and 'positions' the second antibody to capture cytokine upon secretion. This method has been developed for the assessment of IFNγ production and the numbers of antigen-specific T cells detected by this method correlates closely to values obtained using both ELISpot and ICS (Pittet *et al.*, 2001). Importantly, the assay could represent a highly sensitive technique for the investigation of cytokine secretion in the absence of antigenic stimulation directly *ex vivo*.

Cytolytic T-cell function

The ability of CD8+ T lymphocytes to exert a cytolytic function is funda-mental to the cellular immune response. CD8+ T cells mediate cellular cytotoxicity primarily through a perforin-dependent pathway, or alterna-tively, and to a much lesser extinct, the Fas–FasL pathway (Kojima *et al.*, 1994; Takayama *et al.*, 1995). Perforin is found, in addition to granzymes, within lytic granules inside antigen-experienced T cells. Upon engage-ment of target cells, cytotoxic T cells release perforin, which forms pores in target cell membranes. This allows the subsequent entry of granzymes, which activate the apoptotic cascade resulting in death of the target cells (Liu *et al.*, 1995). In the Fas/FasL pathway, the engagement of FasL expressed on T cells, by Fas expressed on the target results in the delivery of a death signal to the target cell (Green, 2000; Strasser *et al.*, 2000). The ^{51}Cr release assay is an important method to assess actual cellular lysis but has limitations, namely, only the bulk population of T cells is assessed for lytic capacity and high frequencies (5% of the total CD8+) of cells are

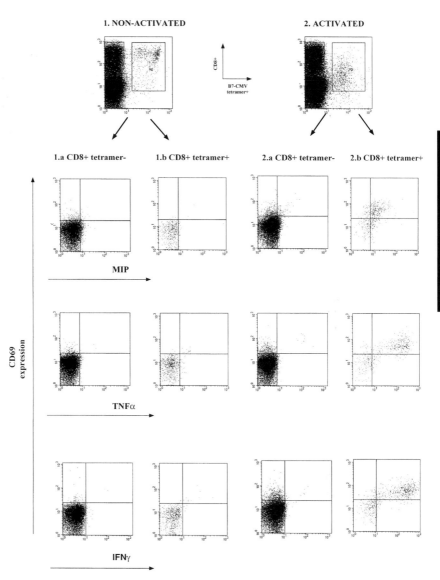

Figure 1. Intracellular cytokine staining in tetramer − and + CD8+ T lymphocytes. Peripheral blood-derived CD8+ T cells were stained with tetramer and anti-CD8 monoclonal antibody for 15 min at 37°C, washed and subsequently incubated for 6 h in the absence (1) or presence (2) of antigenic peptide. Cells were then permeabilized, and stained for extracellular expression of CD69 and intracellularly, using standard ICS protocols, for MIP, TNFα and IFNγ expression, and analysed by flow cytometry (Appay *et al.*, 2000). CD8+ tetramer- (1a and 2a) and CD8+ tetramer+ cells (1b and 2b) were gated in both the non-activated and activated T cell population, respectively. Cytokine expression is indicated along the *x*-axis and CD69 (a marker of T cell activation) on the *y*-axis. Abbreviations: MIP = macrophage inflammatory protein, TNFα = tumor necrosis factor α, IFNγ = interferon γ.

required to perform fresh killing assays directly *ex vivo*. The potential of T cells to exert a lytic function can also be assessed using flow cytometry, and antibodies specific for granzyme and perforin, and the protein granule membrane protein GMP-17 (also known as TIA-1), are readily combined with tetramer analysis (Appay *et al.*, 2000, 2002).

Phenotypic profiles of tetramer-reactive T cells

Antigen-specific CD8+ T lymphocytes are frequently analysed for the expression of cell surface or intracellular protein relating to 'activation' or 'maturation' status. It is of particular interest to determine whether T cells have encountered antigen, and are therefore 'antigen experienced', or bear the hallmarks of naïve T lymphocytes. Whilst precise cell surface protein that accurately delineates truly naïve and antigen experienced (effector/memory) T lymphocytes continues to be controversial, certain markers tend to associate with T-cell subsets (Hamann *et al.*, 1997; Campbell *et al.*, 2001; Champagne *et al.*, 2001; Appay *et al.*, 2002).

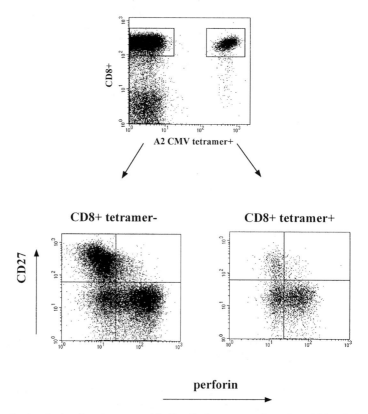

Figure 2. Analysis of antigen-specific T-cell phenotype. Peripheral blood-derived CD8+ T cells were stained with tetramer and anti-CD8 monoclonal antibody for 15 min at 37°C. Cells were then permeabilized, and stained for extracellular expression of CD27 and intracellularly, using standard ICS protocols, for perforin (Appay *et al.*, 2000). CD8+ tetramer- and CD8+ tetramer+ cells are gated. Perforin expression is indicated along the *x*-axis, and CD27 expression along the *y*-axis.

Originally, T-cell phenotyping could only be performed on the total CD8+ T cell pool or on cells divided according to TCR V region gene expression (Callan *et al.*, 1996; Moss *et al.*, 1996; Hamann *et al.*, 1997; Moss and Gillespie, 1997), but with the development of MHC class I tetramers the technique is routinely used to dissect the status of antigen-specific T lymphocytes (Fig. 2) (Callan *et al.*, 1998; Tan *et al.*, 1999).

◆◆◆◆◆◆ STAINING OF TISSUE SECTIONS USING MHC CLASS I TETRAMERS

Although routinely used to stain freshly isolated or frozen, MHC class I tetramers have also been used to stain fresh and lightly fixed tissue *in situ* (Skinner *et al.*, 2000). As demonstrated by Skinner and colleagues, MHC class I tetramers can be adapted to stain T cells in fresh tissue sections prepared from the spleens of transgenic mice. The technique involved embedding cut tissue sections in 4% low-melt agarose, which were then attached to vibratome blocks using adhesive. A vibratome 3000 was used to cut 200 µM sections, in a vibratome bath containing PBS. Staining was performed in 24 well tissue culture plates on free-floating tissue sections. Tetramer (conjugated to the FITC fluorochrome) and unconjugated anti-CD8 antibody staining was performed overnight at 4°C with rotation, then washed in PBS, and fixed in 2% paraformaldehyde for 30 min. Further to additional washes, sections were incubated with anti-FITC antibodies (to amplify the tetramer signal) and Cy3/Cy5 conjugated anti-CD8 antibodies in consecutive overnight incubation steps. Following extensive washing, sections were mounted to slides with glycerol gelatin containing n-propyl galate and analysed by confocal microscopy. The frequencies of tetramer-reactive cells estimated by this technique correlated with frequencies estimated by flow cytometry, and levels of non-specific (background) staining were low (0.4%). Tissue sections lightly fixed (in 2% paraformaldehyde or in 50% methanol/50% acetone) and stored overnight at 4°C, or samples frozen overnight were also successfully stained albeit with lower quality than using fresh tissue. This *in situ* staining technique could represent an important tool for the tracking of antigen-specific T cells in tissue.

◆◆◆◆◆◆ POSSIBLE CLINICAL APPLICATIONS OF TETRAMERS

MHC class I tetramer technology has clinical potential. In addition to diagnostic applications they can be used for the purification of T lymphocytes for propagation prior to T-cell-based therapy. The most common purification technique involves the sorting of tetramer-reactive T cell, generally using a fluorescence-activated cells sorter (FACS), and the resulting populations are propagated *ex vivo* (Dunbar *et al.*, 1998, 1999). Purification of tetramer positive cells can also be achieved using magnetically labelled beads, and is suitable for the isolation of T cell for the purposes of basic T-cell research (Gillespie, 2002).

Tetramers might also play an important role in the selection of high-quality clones for T-cell-based immunotherapy trials. A number of independent reports confirm that T-cell clones generated *ex vivo* bind tetramers with different avidities, and that different staining intensities correlate with functionally more efficient cells (Gillespie *et al.*, 2000a; Dutoit *et al.*, 2001; Palermo *et al.*, 2001). It is possible that tetramers could be used to control CD8+ T-cell mediated immunopathology in future. It was recently demonstrated that a tetrameric construct mutated to abrogate binding CD8, bound antigen-specific T cells and induced FasL expression comparable to normal tetramers, but differed by failing to induce significant CD3ζ chain signalling (Xu *et al.*, 2001). The authors proposed that this segregation of T-cell activation and FasL expression might represent an important mechanism to delete autoreactive T lymphocytes without inducing T-cell activation and immunopathology *in vivo*. In addition, MHC class I tetramers can modulate T-cell function when administered to mice, and depending on the quantities used can either enhance or abrogate antigen-specific T cell function (Maile *et al.*, 2001). Importantly too, MHC class I tetramers can be manipulated to target cells that down-regulate MHC class I for recognition by CD8+ T lymphocytes; this has been successfully achieved for MHC class I negative tumour cell lines *in vitro* (Ogg *et al.*, 2000; Robert *et al.*, 2000).

◆◆◆◆◆◆ IMPACT OF TETRAMER TECHNOLOGY ON OUR UNDERSTANDING OF VIRAL SPECIFIC T-CELL IMMUNITY

Frequency of virus-specific T lymphocytes in peripheral blood

The use of tetramers to re-evaluate T cells in a diverse range of viral infections has allowed immunologists to gain important information concerning T-cell frequencies (see Table 3 for a summary of common tetrameric MHC class I complexes used to quantify viral-specific T cells in human infections). The use of tetramer-based analyses has revealed that T-cell frequencies in diverse viral infection are much higher than estimated previously, particularly for persistent viruses. This was initially demonstrated in asymptomatic HIV-1+ patients (Altman *et al.*, 1996). Much lower frequencies of antigen-specific T cells are observed in patients receiving therapy and probably reflect the loss of antigen due to viral suppression (Gray *et al.*, 1999; Ogg *et al.*, 1999; Mollet *et al.*, 2000). More surprising, perhaps, are the frequencies of antigen-specific T lymphocytes detected in the setting of the latent herpesvirus infections. The β herpesvirus cytomegalovirus (CMV) and the γ herpesvirus Epstein-Barr virus (EBV) cause localized respiratory-based infection that is generally asymptomatic and establish latency in cells from the myeloid (Kondo *et al.*, 1994; Hahn *et al.*, 1998) and lymphoid lineages (Tierney *et al.*, 1994) respectively. We now know that, as estimated by tetramer analysis, frequencies of CMV and EBV specific T lymphocytes in healthy sero-positive individuals are higher than estimated by LDA (Borysiewicz *et al.*, 1988; Tan *et al.*, 1999) and often account for 5% of peripheral blood derived T lymphocytes in the latent phase of disease (Tan *et al.*, 1999, Gillespie, 2000b). These high frequencies of antigen-specific T cells are likely to reflect periodic viral reactivation from latency and support an important role for CD8+ T cells in 'latent' viral infections. Since the arrival of tetramers, the importance of CD8+ T cells in the acute phase of EBV infection has also been re-emphasized. The striking observation that a large percentage (up to 44%) of CD8+ cells in the periphery were directed against a single CTL epitope in acute EBV infection was not predicted by previous LDA analyses (Callan *et al.*, 1998). T-cell frequencies for viruses that cause limited infection of the host (e.g. influenza) are generally low (<0.2%) upon resolution of primary infection (Dunbar *et al.*, 1998). We have no data relating to viral-specific T-cell frequencies during acute disease.

On the basis of tetramer analyses, similar observations have been made in mice. Responses to persistent DNA viruses are generally of the greatest magnitude and levels of antigen-specific T lymphocytes are maintained at relatively high frequencies following resolution of primary infection (5–10% total CD8+ T cells), and T-cell responses to non-persistent infections are low or undetectable by tetramer following clearance (Doherty and Christensen, 2000).

Frequency of virus-specific T lymphocytes at the site of immunopathology

Tetramers are generally used to assess T-cell frequencies in the peripheral blood but also provide important data regarding virus-specific T numbers present at the site of pathology. During primary infection in mice, for example, antigen-specific T lymphocytes often comprise the majority (up to 80%) of CD8+ T lymphocytes present at the site of pathology with the majority accounting for antigen-specific activated T cells (Murali-Krishna *et al.*, 1998; Doherty and Christensen, 2000).

The large numbers of antigen-specific CD8+ T recruited to the liver of individuals with persistent hepatitis B virus (HBV) infection effectively control viral replication. Interestingly, in patients with high HBV and liver pathology the large cellular infiltrates might also comprise non-specific cells that contribute to the liver disease (Maini *et al.*, 2000). In certain infections, T cells specific for unrelated antigens are often recruited to the site of pathology and influence the outcome of disease (Chen *et al.*, 2001b).

T-cell phenotype and function in persistent infections

The combined analysis of viral specific T cells using tetramers, phenotypic profiling and functional analyses has enabled T-cell populations to be extensively assessed. We know, for example, that CMV-specific T

Table 3. Common MHC class I tetramers used to assess viral-specific CD8+ T lymphocytes in humans.

Virus	HLA restriction	Antigen source	Epitope sequence	Reference
HCV	A*0201	NS3 $_{1073-1081}$	CINGVCWTV	[Lechner et al., 2000]
	A*0201	NS3 $_{1406-1415}$	KLVALGINAV	[Lechner et al., 2000]
	A*0201	NS4B $_{1807-1816}$	LLFNILGGWV	[Lechner et al., 2000]
	A*0201	NS5 $_{2594-2602}$	ALYDWTKL	[Grabowska et al., 2001]
	B*0702	core	GPRLGVRAT	[Gruener et al., 2001]
	B*0702	core	DPRRRSRNL	[Gruener et al., 2001]
	B*0801	NS3 $_{1395-1403}$	HSKKKLDEL	[Lechner et al., 2000]
	B*0801	NS3 $_{1161-1618}$	LIRLKPTL	[Lechner et al., 2000]
	B*3501	NS3 $_{1359-1367}$	HPNIEEVAL	[Sobao et al., 2001]
HBV	A*0201	core $_{18-27}$	FLPSDFFPSV	[Maini et al., 1999]
	A*0201	polymerase $_{575-583}$	FLLSLGIHL	[Maini et al., 1999]
	A*0201	envelope $_{335-343}$	WLSLLVPFV	[Maini et al., 1999]
EBV	A*0201	BMLF1 (lytic) $_{280-288}$	GLCTLVAML	[Callan et al., 1998]
	A*0201	BMRF1 (lytic)	YVLDHLIVV	[Marshall et al., 2000]
	A*0201	LMP2 (latent) $_{426-434}$	CLGGLLTMV	[Marshall et al., 2000]
	A*0201	EBNA3C (latent) $_{284-293}$	LLDFVRFMGV	[Marshall et al., 2000]
	A*11	EBNA3B (latent) $_{416-424}$	IVTDFSVIK	[Tan et al., 1999]
	B*0702	EBNA3A (latent) $_{379-387}$	RPPIFIRRL	[Marshall et al., 2000]
	B*0702	EBNA3A (latent) $_{502-510}$	VPAPAGPIV	[Marshall et al., 2000]
	B*0801	BZLF1 (lytic) $_{189-196}$	RAKFKQLL	[Callan et al., 1998]
	B*0801	EBNA3A (latent) $_{325-333}$	FLRGRAYGL	[Callan et al., 1998]
	B*0801	EBNA3A (latent) $_{158-166}$	QAKWRLQTL	[Marshall et al., 2000]
	B*3501	EBNA1 (latent) $_{407-417}$	HPVGEADYFEY	[Blake et al., 2000]
	B*3501	EBNA3A (latent) $_{458-466}$	YPLHEQHGM	[Blake et al., 2000]
CMV	A*0201	pp65 $_{495-503}$	NLVPVATV	[Gillespie et al., 2000]
	B*0702	pp65 $_{417-423}$	TPRVTGGGAM	[Gillespie et al., 2000]
	B*0801	pp65 $_{516-524}$	DANDIYRIF	[Hassan-Walker et al., 2001]
	B*3501	pp65 $_{123-131}$	IPSINVHHY	[Hassan-Walker et al., 2001]

Table 3. (contd.)

Virus	HLA restriction	Antigen source	Epitope sequence	Reference
HIV-1	A*0201	Gag p17$_{77-85}$	SLYNTVATL	[Altman et al., 1996]
	A*0201	Pol$_{476-484}$	ILKEPVHGV	[Altman et al., 1996]
	A*11	nef$_{73-82}$	QVPLRPMTYK	[Appay et al., 2000]
	A*6802	pol$_{744-752}$	ETAYFILKL	[Appay et al., 2000]
	B*0702	env$_{843-852}$	IPRRIRQGL	[Appay et al., 2000]
	B*0801	nef$_{89-97}$	FLKEKGGL	[Appay et al., 2000]
	B*0801	Gag p24$_{128-136}$	DIYKRWII	[Appay et al., 2000]
	B*0801	Env$_{586-594}$	YLKDQQLL	[Appay et al., 2000]
	B*2705	Gag p24$_{131-139}$	KRWIIMGLNK	[Appay et al., 2000]
	B*3501	env$_{77-85}$	DPNPQEVVL	[Ogg et al., 1998]
	B*5301	Gag p24$_{48-56}$	TPQDLNMML	[Dorrell et al., 2001]
	B*5301	Gag p24$_{176-184}$	QASQEVKNW	unpublished
	B*5701	Gag p24$_{30-40}$	KAFSPEVIPMF	[Migueles et al., 2000]
	B*5703	Gag p24$_{30-40}$	KAFSPEVIPMF	[Gillespie et al., 2002]
	B*5801	Gag p24$_{108-117}$	TSTLQEQIGW	unpublished
	B*5801	Gag p24$_{176-184}$	QASQEVKNW	unpublished
	Cw0402	Gag p17$_{28-36}$	KYRLKHLVW	[Appay et al., 2000]
HIV-2	B*5301	Gag p24$_{48-56}$	TPYDINQML	[Dorrell et al., 2001]
	B*5801	Gag p24$_{108-117}$	TSTVEEQIQW	unpublished
HTLV-1	A*0201	tax$_{11-19}$	LLFGYPVYV	[Greten et al., 1998]
HPV	A*0201	E7$_{11-20}$	YMLDLQPETT	[Youde et al., 2000]
Influenza A	A*0201	matrix$_{58-66}$	GILGFVFTL	[Dunbar et al., 1998]
RSV	B*0702	nucleoprotein	NPKASLLSL	[Goulder et al., 2000b]

Abbreviations: HCV= Hepatitis C virus, HBV = Hepatitis B virus, EBV = Epstein-Barr virus, CMV = cytomegalovirus, HIV-1 = Human immunodeficiency virus type 1, HIV-2 = Human immunodeficiency virus type 2, HTLV-1 = human T cell lymphotrophic virus type 1, HPV = human papillomavirus virus.

lymphocytes from positive healthy persistently infected individuals are often heterogeneous in terms of phenotypic profiles (Vargas *et al.*, 2001) but mostly comprise mature 'effector-memory' like cells, display competent effector functions when stimulated *ex vivo* (Appay *et al.*, 2000; Gillespie *et al.*, 2000b). In terms of phenotypic and functional profiles, EBV-specific T cells express a predominant phenotype in acute infection, proliferate and exert direct cytolytic activity when assessed *ex vivo* (Callan *et al.*, 1998; Callan, 2000). Like CMV, the phenotype profiles of antigen-specific T cells are heterogeneous in healthy persistently infected individuals (Tan *et al.*, 1999; Hislop *et al.*, 2001). Functional analysis reveals that these EBV-specific T cells directed against either latent or lytic epitopes perform diverse effector functions irrespective of the expression patterns of the certain phenotypic markers (Hislop *et al.*, 2001).

Interestingly, combined tetramer, phenotypic and functional-based analyses have highlighted discrepancies between the quantity and quality of viral-specific T lymphocytes in certain infections. Most notably this has been demonstrated for virus-specific T cells in human infections such as HIV-1 (Appay *et al.*, 2000; Shankar *et al.*, 2000, Champagne *et al.*, 2001; Chen *et al.*, 2001a; Kostense *et al.*, 2001) and hepatitis B (Gruener *et al.*, 2001), and in SIV (Xiong *et al.*, 2001) and murine lymphocytic chorio-meningitis virus (LCMV) infections (Gallimore *et al.*, 1998). CD8+ T lymphocytes specific for HIV antigens, for example, often express lower levels of intracellular perforin and display a diminished cytolytic capacity *ex vivo* (Appay *et al.*, 2000). They occasionally fail to produce the anti-viral cytokine IFNγ (Shankar *et al.*, 2000, Kostense *et al.*, 2001), and there are also indications that the differentiation of CD8+ HIV-specific T cells is dys-functional or blocked, marked by the accumulation of 'immature'-like cells (Champagne *et al.*, 2001, Chen *et al.*, 2001a, Appay, 2002) (refer to Table 4 for a summary of T-cell phenotypes in persistent viral infections). A recent study demonstrates the susceptibility of these less differentiated HIV-specific T cells to apoptosis by the Fas/FasL pathway; and might partly explain the lack of highly differentiated HIV-specific T cells *in vivo* (Mueller *et al.*, 2001). In HBV infection, tetramer-reactive T lymphocytes often display limited functional ability early after acute infection and during the chronic phase of disease (Gruener *et al.*, 2001). The large

Table 4. Predominant phenotypes of peripheral blood-derived antigen-specific CD8+ T lymphocytes in persistent viral infections.

Virus	Phenotype (chronic infection)	Proposed stage of T-cell differentiation (antigen-experienced cells)
EBV	CD28+/CD27+/CD45RA−(+)/Perforin+^{L}/GrA+/CCR7+	EARLY
HIV	CD28−/CD27+/CD45RA−(+)Perforin+^{L}/GrA+/CCR7−	INTERMEDIATE
CMV	CD28−/CD27−/CD45RA+(−)/Perforin+$^{L/H}$GrA+/CCR7−	LATE

Adapted from Appay *et al.* (2002).

population of antigen-specific T lymphocytes induced in rhesus macaques following challenge with a pathogenic strain of SIV persist without apparent function, but activity can be restored upon incubation with IL-2 *in vitro* (Xiong *et al.*, 2001); indeed incubation with IL-2 is known to restore CD8+ T-cell function in HIV infection (Shankar *et al.*, 2000). The unresponsive or anergic state of CD8+ T cells in LCMV may be related to high antigenic dose (Gallimore *et al.*, 1998).

◆◆◆◆◆◆ CONCLUSION

The use of MHC class I tetramers has transformed our understanding of the immune responses to viruses. There is considerable potential for combinations of tetramer staining with functional and detailed phenotypic analyses.

References

Abbott, J. and Beckett, D. (1993). Cooperative binding of the *Escherichia coli* repressor of biotin biosynthesis to the biotin operator sequence. *Biochemistry* **32**, 9649–9656.

Allen, R. L., O'Callaghan, C. A., McMichael, A. J. and Bowness, P. (1999). Cutting edge: HLA-B27 can form a novel beta 2-microglobulin-free heavy chain homodimer structure. *J. Immunol.* **162**, 5045–5048.

Allen, R. L., Raine, T., Haude, A., Trowsdale, J. and Wilson, M. J. (2001). Leukocyte receptor complex-encoded immunomodulatory receptors show differing specificity for alternative HLA-B27 structures. *J. Immunol.* **167**, 5543–5547.

Alonso, M. A. and Millan, J. (2001). The role of lipid rafts in signalling and membrane trafficking in T lymphocytes. *J. Cell Sci.* **114**, 3957–3965.

Altman, J. D., Moss, P. A. H., Goulder, P. J. R., Barouch, D. H., McHeyzer-Williams, M. G., Bell, J. I., McMichael, A. J. and Davis, M. M. (1996). Phenotypic analysis of antigen-specific T lymphocytes [published erratum appears in *Science* 1998 Jun 19; **280**(5371): 1821]. *Science* **274**, 94–96.

Appay, V., Nixon, D. F., Donahoe, S. M., Gillespie, G. M., Dong, T., King, A., Ogg, G. S., Spiegel, H. M., Conlon, C., Spina, C. A., Havlir, D. V., Richman, D. D., Waters, A., Easterbrook, P., McMichael, A. J. and Rowland-Jones, S. L. (2000). HIV-specific CD8(+) T cells produce antiviral cytokines but are impaired in cytolytic function. *J. Exp. Med.* **192**, 63–75.

Appay, V., Dunbar, P. R., Callan, M., Klenerman, P., Gillespie, G. M. A., Papagno, L., Ogg, G., King, A., Lechner, F., Spina, C. A., Little, S., Havlir, D. V., Richman, D. D., Gruener, N. H., Pape, G. R., Waters, A., Easterbrook, P., Cerundolo, V., McMichael, A. and Rowland-Jones, S. (2002). Memory CD8+ T cells vary in differentiation phenotype in different persistent virus infections. *Nat. Med.* **8**(4), 379–385.

Bodinier, M., Peyrat, M. A., Tournay, C., Davodeau, F., Romagne, F., Bonneville, M. and Lang, F. (2000). Efficient detection and immunomagnetic sorting of specific T cells using multimers of MHC class I and peptide with reduced CD8 binding. *Nat. Med.* **6**, 707–710.

Bogue, M. and Roth, D. B. (1996). Mechanism of V(D)J recombination. *Curr. Opin. Immunol.* **8**, 175–180.

Borysiewicz, L. K., Graham, S., Hickling, J. K., Mason, P. D. and Sissons, J. G. (1988). Human cytomegalovirus-specific cytotoxic T cells: their precursor frequency and stage specificity. *Eur. J. Immunol.* **18**, 269–275.

Bourez, R. L., Mathijssen, I. M., Vaandrager, J. M. and Vermeij-Keers, C. (1997). Apoptotic cell death during normal embryogenesis of the coronal suture: early detection of apoptosis in mice using annexin V. *J. Craniofac. Surg.* **8**, 441–445.

Bouvier, M. and Wiley, D. C. (1998). Structural characterization of a soluble and partially folded class I major histocompatibility heavy chain/beta 2m heterodimer. *Nat. Struct. Biol.* **5**, 377–384.

Brunner, K. T., Engers, H. D. and Cerottini, J. C. (1976). The ^{51}CR release assay as used for the quantitative measurment of cell-mediated cytolysis in vitro. *In Vitro Methods in Cell-Mediated Tumor Immunity* **1**, 423–428.

Burns, J. B., Lobo, S. T. and Bartholomew, B. D. (2000). In vivo reduction of telomere length in human antigen-reactive memory T cells. *Eur. J. Immunol.* **30**, 1894–1901.

Callan, M. F., Fazon, C., Yang, H., *et al.* (2000). CD8(+) T-cell selection, function, and death in the primary immune response *in vivo*. *J. Clin. Invest.* **106**, 1251–1261.

Callan, M. F., Steven, N., Krausa, P., Wilson, J. D., Moss, P. A., Gillespie, G. M., Bell, J. I., Rickinson, A. B. and McMichael, A. J. (1996). Large clonal expansions of CD8+ T cells in acute infectious mononucleosis. *Nat. Med.* **2**, 906–911.

Callan, M. F., Tan, L., Annels, N., Ogg, G. S., Wilson, J. D., O'Callaghan, C. A., Steven, N., McMichael, A. J. and Rickinson, A. B. (1998). Direct visualization of antigen-specific CD8+ T cells during the primary immune response to Epstein-Barr virus in vivo. *J. Exp. Med.* **187**, 1395–1402.

Campanelli, R., Palermo, B., Garbelli, S., Mantovani, S., Lucchi, P., Necker, A., Lantelme, E. and Giachino, C. (2002). Human CD8 co-receptor is strictly involved in MHC-peptide tetramer-TCR binding and T cell activation. *Internat. Immunol.* **14**, 39–44.

Campbell, J. J., Murphy, K. E., Kunkel, E. J., Brightling, C. E., Soler, D., Shen, Z., Boisvert, J., Greenberg, H. B., Vierra, M. A., Goodman, S. B., Genovese, M. C., Wardlaw, A. J., Butcher, E. C. and Wu, L. (2001). CCR7 expression and memory T cell diversity in humans. *J. Immunol.* **166**, 877–884.

Cantrell, D., Bluestone, J., Vivier, E. and Tybulewicz, V. (1998). Signalling through the TCR. *Res. Immunol.* **149**, 866–867.

Cerundolo, V., Kelly, A., Elliott, T., Trowsdale, J. and Townsend, A. (1995). Genes encoded in the major histocompatibility complex affecting the generation of peptides for TAP transport. *Eur. J. Immunol.* **25**, 554–562.

Champagne, P., Ogg, G. S., King, A. S., Knabenhans, C., Ellefsen, K., Nobile, M., Appay, V., Rizzardi, G. P., Fleury, S., Lipp, M., Forster, R., Rowland-Jones, S., Sekaly, R. P., McMichael, A. J. and Pantaleo, G. (2001). Skewed maturation of memory HIV-specific CD8 T lymphocytes. *Nature* **410**, 106–111.

Chen, G., Shankar, P., Lange, C., Valdez, H., Skolnik, P. R., Wu, L., Manjunath, N. and Lieberman, J. (2001a). CD8 T cells specific for human immunodeficiency virus, Epstein-Barr virus, and cytomegalovirus lack molecules for homing to lymphoid sites of infection. *Blood* **98**, 156–164.

Chen, H. D., Fraire, A. E., Joris, I., Brehm, R. M., Welsh, R. M. and Selin, L. K. (2001b). Memory CD8+ T cells in heterologous antiviral immunity and immunopathology in the lung. *Nat. Med.* **2**, 991–993.

Clarke, G. R., Humphrey, C. A., Lancaster, F. C. and Boylston, A. W. (1994). The human T cell antigen receptor repertoire: skewed use of V beta gene families by CD8+ T cells. *Clin. Exp. Immunol.* **96**, 364–369.

Colonna, M., Nakajima, H., Navarro, F. and Lopez-Botet, M. (1999). A novel family of Ig-like receptors for HLA class I molecules that modulate function of lymphoid and myeloid cells. *J. Leukoc. Biol.* **66**, 375–381.

Czerkinsky, C., Andersson, G., Ferrua, B., Nordstrom, I., Quiding, M., Eriksson, K., Larsson, L., Hellstrand, K. and Ekre, H. P. (1991). Detection of human cytokine-secreting cells in distinct anatomical compartments. *Immunol. Rev.* **119**, 5–22.

Davis, M. M., Boniface, J. J., Reich, Z., Lyons, D., Hampl, J., Arden, B. and Chien, Y. (1998). Ligand recognition by alpha beta T cell receptors. *Annu. Rev. Immunol.* **16**, 523–544.

Davis, S. J. and van der Merwe, P. A. (2001). The immunological synapse: required for T cell receptor signalling or directing T cell effector function? *Curr. Biol.* **11**, R289–291.

de Bernardez Clark, E. (1998). Refolding of recombinant proteins. *Biochemical Engineering* **9**, 157–163.

de Smit, M. H. and van Duin, J. (1990). Secondary structure of the ribosome binding site determines translational efficiency: a quantitative analysis. *Proc. Natl. Acad. Sci. USA* **87**, 7668–7672.

de Smit, M. H. and van Duin, J. (1994). Control of translation by mRNA secondary structure in *Escherichia coli*. A quantitative analysis of literature data. *J. Mol. Biol.* **244**, 144–150.

Doherty, P. C. and Christensen, J. P. (2000). Accessing complexity: the dynamics of virus-specific T cell responses. *Annu. Rev. Immunol.* **18**, 561–592.

Dunbar, P. R., Ogg, G. S., Chen, J., Rust, N., van der Bruggen, P. and Cerundolo, V. (1998). Direct isolation, phenotyping and cloning of low-frequency antigen-specific cytotoxic T lymphocytes from peripheral blood. *Curr. Biol.* **8**, 413–416.

Dunbar, P. R., Chen, J. L., Chao, D., Rust, N., Teisserenc, H., Ogg, G. S., Romero, P., Weynants, P. and Cerundolo, V. (1999). Cutting edge: rapid cloning of tumor-specific CTL suitable for adoptive immunotherapy of melanoma. *J. Immunol.* **162**, 6959–6962.

Dutoit, V., Rubio-Godoy, V., Dietrich, P. Y., Quiqueres, A. L., Schnuriger, V., Rimoldi, D., Lienard, D., Speiser, D., Guillaume, P., Batard, P., Cerottini, J. C., Romero, P. and Valmori, D. (2001). Heterogeneous T-cell response to MAGE-A10(254–262): high avidity-specific cytolytic T lymphocytes show superior antitumor activity. *Cancer Res.* **61**, 5850–5856.

Dutton, R. W., Bradley, L. M. and Swain, S. L. (1998). T cell memory. *Ann. Rev. Immunol.* **16**, 201–223.

Gallimore, A., Glithero, A., Godkin, A., Tissot, A. C., Pluckthun, A., Elliott, T., Hengartner, H. and Zinkernagel, R. (1998). Induction and exhaustion of lymphocytic choriomeningitis virus-specific cytotoxic T lymphocytes visualized using soluble tetrameric major histocompatibility complex class I-peptide complexes. *J. Exp. Med.* **187**, 1383–1393.

Gao, G. F., Gerth, U. C., Wyer, J. R., Willcox, B. E., O'Callaghan, C. A., Zhang, Z., Jones, E. Y., Bell, J. I. and Jakobsen, B. K. (1998). Assembly and crystallization of the complex between the human T cell coreceptor CD8alpha homodimer and HLA-A2. *Protein Sci.* **7**, 1245–1249.

Garboczi, D. N., Hung, D. T. and Wiley, D. C. (1992). HLA-A2-peptide complexes: refolding and crystallization of molecules expressed in *Escherichia coli* and complexed with single antigenic peptides. *Proc. Natl. Acad. Sci. USA* **89**, 3429–3433.

Gellert, M. (1997). Recent advances in understanding V(D)J recombination. *Adv. Immunol.* **64**, 39–64.

Gerdes, J., Lemke, H., Baisch, H., Wacker, H. H., Schwab, U. and Stern, H. (1984). Cell cycle analysis of a cell proliferation-associated human nuclear antigen defined by the monoclonal antibody Ki-67. *J. Immunol.* **133**, 1710–1714.

Gillespie, G., Mutis, T., Schrama, E., Kamp, J., Esendam, B., Falkenburg, J. H., Goulmy, E. and Moss, P. (2000a). HLA class I histocompatibility antigen tetramers select cytotoxic T cells with high avidity to the natural ligand. *Hematol. J.* **1**, 403–410.

Gillespie, G. M., Wills, M. R., Appay, V., O'Callaghan, C., Murphy, M., Smith, N., Sissons, P., Rowland-Jones, S., Bell, J. I. and Moss, P. A. (2000b). Functional heterogeneity and high frequencies of cytomegalovirus-specific CD8(+) T lymphocytes in healthy seropositive donors. *J. Virol.* **74**, 8140–8150.

Gillespie, G. M. A., Kaul, R., Dong, T., Yang, H. B., Rostron, T., Bwayo, J., Kiama, P., Peto, T., Plummer, F. A., McMichael, A. J. and Rowland-Jones, S. (2002). Cross-reactive cytotoxic T lymphocytes against a HIV-1 p24 epitope in slow progressors with B*57. *AIDS*, **16**(7), 961–972.

Goulder, P. J., Tang, Y., Brander, C., Betts, M. R., Altfeld, M., Annamalai, K., Trocha, A., He, S., Rosenberg, E. S., Ogg, G., O'Callaghan, C. A., Kalams, S. A., McKinney, R. E., Jr., Mayer, K., Koup, R. A., Pelton, S. I., Burchett, S. K., McIntosh, K. and Walker, B. D. (2000). Functionally inert HIV-specific cytotoxic T lymphocytes do not play a major role in chronically infected adults and children. *J. Exp. Med.* **192**, 1819–1832.

Goulder, P. J., Lechner, F., Klenerman, P., McIntosh, K. and Walker, B. D. (2000b) Characterization of a novel respiratory syncytial virus-specific human cytotoxic T-lymphocyte epitope. *J. Virol.* **74**, 7694–7697.

Gratzner, H. G. (1982). Monoclonal antibody to 5-bromo- and 5-iododeoxy-uridine: A new reagent for detection of DNA replication. *Science* **218**, 474–475.

Gratzner, H. G. and Leif, R. C. (1981). An immunofluorescence method for monitoring DNA synthesis by flow cytometry. *Cytometry* **1**, 385–393.

Gray, C. M., Lawrence, J., Schapiro, J. M., Altman, J. D., Winters, M. A., Crompton, M., Loi, M., Kundu, S. K., Davis, M. M. and Merigan, T. C. (1999). Frequency of class I HLA-restricted anti-HIV CD8+ T cells in individuals receiving highly active antiretroviral therapy (HAART). *J. Immunol.* **162**, 1780–1788.

Grayson, J. M., Zajac, A. J., Altman, J. D. and Ahmed, R. (2000). Cutting edge: increased expression of Bcl-2 in antigen-specific memory CD8+ T cells. *J. Immunol.* **164**, 3950–3954.

Green, D. R. (2000). Apoptotic pathways: paper wraps stone blunts scissors. *Cell* **102**, 1–4.

Gruener, N. H., Lechner, F., Jung, M. C., Diepolder, H., Gerlach, T., Lauer, G., Walker, B., Sullivan, J., Phillips, R., Pape, G. R. and Klenerman, P. (2001). Sustained dysfunction of antiviral CD8+ T lymphocytes after infection with hepatitis C virus. *J. Virol.* **75**, 5550–5558.

Guise, A. D., West, S. M. and Chaudhuri, J. B. (1996). Protein folding in vivo and renaturation of recombinant proteins from inclusion bodies. *Molecular Biotechnology* **6**, 53–64.

Hahn, G., Jores, R. and Mocarski, E. S. (1998). Cytomegalovirus remains latent in a common precursor of dendritic and myeloid cells. *Proc. Natl. Acad. Sci. USA* **95**, 3937–3942.

Hamann, D., Baars, P. A., Rep, M. H., Hooibrink, B., Kerkhof-Garde, S. R., Klein, M. R. and van Lier, R. A. (1997). Phenotypic and functional separation of memory and effector human CD8+ T cells. *J. Exp. Med.* **186**, 1407–1418.

Harder, T. (2001). Raft membrane domains and immunoreceptor functions. *Adv. Immunol.* **77**, 45–92.

Heath, W. R. and Carbone, F. R. (2001). Cross-presentation, dendritic cells, tolerance and immunity. *Annu. Rev. Immunol.* **19**, 47–64.

Hislop, A. D., Gudgeon, N. H., Callan, M. F., Fazou, C., Hasegawa, H., Salmon, M. and Rickinson, A. B. (2001). EBV-specific CD8+ T cell memory: relationships between epitope specificity, cell phenotype, and immediate effector function. *J. Immunol.* **167**, 2019–2029.

Jameson, S. C. and Bevan, M. J. (1998). T-cell selection. *Curr. Opin. Immunol.* **10**, 214–219.

Johansen, T. E., McCullough, K., Catipovic, B., Su, X. M., Amzel, M. and Schneck, J. P. (1997). Peptide binding to MHC class I is determined by individual pockets in the binding groove. *Scand. J. Immunol.* **46**, 137–146.

Jung, T., Schauer, U., Heusser, C., Neumann, C. and Rieger, C. (1993). Detection of intracellular cytokines by flow cytometry. *J. Immunol. Methods* **159**, 197–207.

Kaul, R., Plummer, F. A., Kimani, J., Dong, T., Kiama, P., Rostron, T., Njagi, E., MacDonald, K. S., Bwayo, J. J., McMichael, A. J. and Rowland-Jones, S. L. (2000). HIV-1-specific mucosal CD8+ lymphocyte responses in the cervix of HIV-1-resistant prostitutes in Nairobi. *J. Immunol.* **164**, 1602–1611.

Klein, M. R., van der Burg, S. H., Hovenkamp, E., Holwerda, A. M., Drijfhout, J. W., Melief, C. J. and Miedema, F. (1998). Characterization of HLA-B57-restricted human immunodeficiency virus type 1. *J. Gen. Virol.* **79**, 2191–2201.

Kojima, H., Shinohara, N., Hanaoka, S., Someya-Shirota, Y., Takagaki, Y., Ohno, H., Saito, T., Katayama, T., Yagita, H., Okumura, K., *et al.* (1994). Two distinct pathways of specific killing revealed by perforin mutant cytotoxic T lymphocytes. *Immunity* **1**, 357–364.

Kondo, K., Kaneshima, H. and Mocarski, E. S. (1994). Human cytomegalovirus latent infection of granulocyte-macrophage progenitors. *Proc. Natl. Acad. Sci. USA* **91**, 11879–11883.

Kostense, S., Ogg, G. S., Manting, E. H., Gillespie, G., Joling, J., Vandenberghe, K., Veenhof, E. Z., van Baarle, D., Jurriaans, S., Klein, M. R. and Miedema, F. (2001). High viral burden in the presence of major HIV-specific CD8(+) T cell expansions: evidence for impaired CTL effector function. *Eur. J. Immunol.* **31**, 677–686.

Krauss, S., Brand, M. D. and Buttgereit, F. (2001). Signaling takes a breath – new quantitative perspectives on bioenergetics and signal transduction. *Immunity* **15**, 497–502.

Labalette-Houache, M., Torpier, G., Capron, A. and Dessaint, J. P. (1991). Improved permeabilization procedure for flow cytometric detection of internal antigens. Analysis of interleukin-2 production. *J. Immunol. Methods* **138**, 143–153.

Lalvani, A., Brookes, R., Hambleton, S., Britton, W. J., Hill, A. V. and McMichael, A. J. (1997). Rapid effector function in CD8+ memory T cells. *J. Exp. Med.* **186**, 859–865.

Lanier, L. L. (1998). NK cell receptors. *Annu. Rev. Immunol.* **16**, 359–393.

Lewis, S. M. and Wu, G. E. (1997). The origins of V(D)J recombination. *Cell* **88**, 159–162.

Liu, C. C., Walsh, C. M. and Young, J. D. (1995). Perforin: structure and function. *Immunology Today* **16**, 194–201.

Long, E. O. and Rajagopalan, S. (2000). HLA class I recognition by killer cell Ig-like receptors. *Semin. Immunol.* **12**, 101–108.

Lyons, A. B. (2000). Analysing cell division in vivo and in vitro using flow cytometric measurement of CFSE dye dilution. *J. Immunol. Methods* **243**, 147–154.

Maecker, H. T., Dunn, H. S., Suni, M. A., Khatamzas, E., Pitcher, C. J., Bunde, T., Persaud, N., Trigona, W., Fu, T. M., Sinclair, E., Bredt, B. M., McCune, J. M., Maino, V. C., Kern, F. and Picker, L. J. (2001). Use of overlapping peptide mixtures as antigens for cytokine flow cytometry. *J. Immunol. Methods* **255**, 27–40.

Maenaka, K., Juji, T., Nakayama, T., Wyer, J. R., Gao, G. F., Maenaka, T., Zaccai, N. R., Kikuchi, A., Yabe, T., Tokunaga, K., Tadokoro, K., Stuart, D. I., Jones, E. Y. and van der Merwe, P. A. (1999). Killer cell immunoglobulin receptors and T cell receptors bind peptide – major histocompatibility complex class I with distinct thermodynamic and kinetic properties. *J. Biol. Chem.* **274**, 28329–28334.

Maile, R., Wang, B., Schooler, W., Meyer, A., Collins, E. J. and Frelinger, J. A. (2001). Antigen-specific modulation of an immune response by in vivo administration of soluble MHC class I tetramers. *J. Immunol.* **167**, 3708–3714.

Maini, M. K., Boni, C., Lee, C. K., Larrubia, J. R., Reignat, S., Ogg, G. S., King, A. S., Herberg, J., Gilson, R., Alisa, A., Williams, R., Vergani, D., Naoumov, N. V., Ferrari, C. and Bertoletti, A. (2000). The role of virus-specific CD8(+) cells in liver damage and viral control during persistent hepatitis B virus infection. *J. Exp. Med.* **191**, 1269–1280.

Malaguarnera, L., Ferlito, L., Imbesi, R. M., Gulizia, G. S., Di Mauro, S., Maugeri, D., Malaguarnera, M. and Messina, A. (2001). Immunosenescence: a review. *Arch. Gerontol. Geriatr.* **32**, 1–14.

Manz, R., Assenmacher, M., Pfluger, E., Miltenyi, S. and Radbruch, A. (1995). Analysis and sorting of live cells according to secreted molecules, relocated to a cell-surface affinity matrix. *Proc. Natl. Acad. Sci. USA* **92**, 1921–1925.

Martin, A. M., Kulski, J. K., Witt, C., Pontarotti, P. and Christiansen, F. T. (2002). Leukocyte Ig-like receptor complex (LCR) in mice and men. *TRENDS in Immunology* **23**, 81–88.

Mascher, B., Schlenke, P. and Seyfarth, M. (1999). Expression and kinetics of cytokines determined by intracellular staining using flow cytometry [In Process Citation]. *J. Immunol. Methods* **223**, 115–121.

McGargill, M. A. and Hogquist, K. A. (2000). T cell receptor editing. *Immunol. Lett.* **75**, 27–31.

McGargill, M. A., Derbinski, J. M. and Hogquist, K. A. (2000). Receptor editing in developing T cells. *Nat. Immunol.* **1**, 336–341.

McMichael, A. J. and O'Callaghan, C. A. (1998). A new look at T cells. *J. Exp. Med.* **187**, 1367–1371.

Merkenschlager, M. and Beverley, P. C. (1989). Evidence for differential expression of CD45 isoforms by precursors for memory-dependent and independent cytotoxic responses: human CD8 memory CTLp selectively express CD45RO (UCHL1). *Int. Immunol.* **1**, 450–459.

Merkenschlager, M., Terry, L., Edwards, R. and Beverley, P. C. (1988). Limiting dilution analysis of proliferative responses in human lymphocyte populations defined by the monoclonal antibody UCHL1: implications for differential CD45 expression in T cell memory formation. *Eur. J. Immunol.* **18**, 1653–1661.

Migueles, S. A., Sabbaghian, M. S., Shupert, W. L., Bettinotti, M. P., Marincola, F. M., Martino, L., Hallahan, C. W., Selig, S. M., Schwartz, D., Sullivan, J. and Connors, M. (2000). HLA B*5701 is highly associated with restriction of virus replication in a subgroup of HIV-infected longterm nonprogressors. *Proc. Natl. Acad. Sci. USA* **97**, 2709–2714.

Mingari, M. C., Schiavetti, F., Ponte, M., Vitale, C., Maggi, E., Romagnani, S., Demarest, J., Pantaleo, G., Fauci, A. S. and Moretta, L. (1996). Human CD8+ T lymphocyte subsets that express HLA class I-specific inhibitory receptors represent oligoclonally or monoclonally expanded cell populations. *Proc. Natl. Acad. Sci. USA* **93**, 12433–12438.

Mollet, L., Li, T. S., Samri, A., Tournay, C., Tubiana, R., Calvez, V., Debre, P., Katlama, C. and Autran, B. (2000). Dynamics of HIV-specific CD8+ T lymphocytes with changes in viral load. The RESTIM and COMET Study Groups. *J. Immunol.* **165**, 1692–1704.

Moretta, L., Biassoni, R., Bottino, C., Mingari, M. C. and Moretta, A. (2000). Human NK-cell receptors. *Immunol. Today* **21**, 420–422.

Moss, P. A. and Gillespie, G. (1997). Clonal populations of T-cells in patients with B-cell malignancies. *Leuk. Lymphoma* **27**, 231–238.

Moss, P., Gillespie, G., Frodsham, P., Bell, J. and Reyburn, H. (1996). Clonal populations of CD4+ and CD8+ T cells in patients with multiple myeloma and paraproteinemia. *Blood* **87**, 3297–3306.

Mueller, Y. M., De Rosa, S. C., Hutton, J. A., Witek, J., Roederer, M., Altman, J. D. and Katsikis, P. D. (2001). Increased CD95/Fas-induced apoptosis of HIV-specific CD8(+) T cells. *Immunity* **15**, 871–882.

Murali-Krishna, K., Altman, J. D., Suresh, M., Sourdive, D. J., Zajac, A. J., Miller, J. D., Slansky, J. and Ahmed, R. (1998). Counting antigen-specific CD8 T cells: a reevaluation of bystander activation during viral infection. *Immunity* **8**, 177–187.

Navarro, F., Llano, M., Bellon, T., Colonna, M., Geraghty, D. E. and Lopez-Botet, M. (1999). The ILT2(LIR1) and CD94/NKG2A NK cell receptors respectively recognize HLA-G1 and HLA-E molecules co-expressed on target cells. *Eur. J. Immunol.* **29**, 277–283.

Nemazee, D. (2000). Receptor selection in B and T lymphocytes. *Annu. Rev. Immunol.* **18**, 19–51.

Oettinger, M. A. (1992). Activation of V(D)J recombination by RAG1 and RAG2. *Trends Genet.* **8**, 413–416.

Ogg, G. S., Jin, X., Bonhoeffer, S., Dunbar, P. R., Nowak, M. A., Monard, S., Segal, J. P., Cao, Y., Rowland-Jones, S. L., Cerundolo, V., Hurley, A., Markowitz, M., Ho, D. D., Nixon, D. F. and McMichael, A. J. (1998). Quantitation of HIV-1-specific cytotoxic T lymphocytes and plasma load of viral RNA. *Science* **279**, 2103–2106.

Ogg, G. S., Jin, X., Bonhoeffer, S., Moss, P., Nowak, M. A., Monard, S., Segal, J. P., Cao, Y., Rowland-Jones, S. L., Hurley, A., Markowitz, M., Ho, D. D., McMichael, A. J. and Nixon, D. F. (1999). Decay kinetics of human immunodeficiency virus-specific effector cytotoxic T lymphocytes after combination antiretroviral therapy. *J. Virol.* **73**, 797–800.

Ogg, G. S., Dunbar, P. R., Cerundolo, V., McMichael, A. J., Lemoine, N. R. and Savage, P. (2000). Sensitization of tumour cells to lysis by virus-specific CTL using antibody-targeted MHC class I/peptide complexes. *Br. J. Cancer* **82**, 1058–1062.

Owen, J. A., Allouche, M. and Doherty, P. C. (1982). Limiting dilution analysis of the specificity of influenza-immune cytotoxic T cells. *Cell Immunol.* **67**, 49–59.

Oxenius, A., Gunthard, H. F., Hirschel, B., Fidler, S., Weber, J. N., Easterbrook, P. J., Bell, J. I., Phillips, R. E. and Price, D. A. (2001). Direct ex vivo analysis reveals distinct phenotypic patterns of HIV-specific CD8(+) T lymphocyte activation in response to therapeutic manipulation of virus load. *Eur. J. Immunol.* **31**, 1115–1121.

Palermo, B., Campanelli, R., Mantovani, S., Lantelme, E., Manganoni, A. M., Carella, G., Da Prada, G., della Cuna, G. R., Romagne, F., Gauthier, L., Necker, A. and Giachino, C. (2001). Diverse expansion potential and heterogeneous avidity in tumor-associated antigen-specific T lymphocytes from primary melanoma patients. *Eur. J. Immunol.* **31**, 412–420.

Pan, C., Xue, B. H., Ellis, T. M., Peace, D. J. and Diaz, M. O. (1997). Changes in telomerase activity and telomere length during human T lymphocyte senescence. *Exp. Cell Res.* **231**, 346–353.

Pittet, M. J., Zippelius, A., Speiser, D. E., Assenmacher, M., Guillaume, P., Valmori, D., Lienard, D., Lejeune, F., Cerottini, J. C. and Romero, P. (2001). Ex vivo IFN-gamma secretion by circulating CD8 T lymphocytes: implications of a novel approach for T cell monitoring in infectious and malignant diseases. *J. Immunol.* **166**, 7634–7640.

Purbhoo, M. A., Boulter, J. M., Price, D. A., Vuidepot, A. L., Hourigan, C. S., Dunbar, P. R., Olson, K., Dawson, S. J., Phillips, R. E., Jakobsen, B. K., Bell, J. I. and Sewell, A. K. (2001). The human CD8 coreceptor effects cytotoxic T cell activation and antigen sensitivity primarily by mediating complete phosphorylation of the T cell receptor zeta chain. *J. Biol. Chem.* **276**, 32786–32792.

Qian, D. and Weiss, A. (1997). T cell antigen receptor signal transduction. *Curr. Opin. Cell Biol.* **9**, 205–212.

Robert, B., Guillaume, P., Luescher, I., Romero, P. and Mach, J. P. (2000). Antibody-conjugated MHC class I tetramers can target tumor cells for specific lysis by T lymphocytes. *Eur. J. Immunol.* **30**, 3165–3170.

Ruppert, J., Kubo, R. T., Sidney, J., Grey, H. M. and Sette, A. (1994). Class I MHC-peptide interaction: structural and functional aspects. *Behring Inst. Mitt.* **94**, 48–60.

Sawicki, M. W., Dimasi, N., Natarajan, K., Wang, J., Margulies, D. H. and Mariuzza, R. A. (2001). Structural basis of MHC class I recognition by natural killer cell receptors. *Immunol. Rev.* **181**, 53–77.

Schatz, P. J. (1993). Use of peptide libraries to map the substrate specificity of a peptide-modifying enzyme: a 13 residue consensus peptide specifies biotinylation in *Escherichia coli*. *Biotechnology (NY)* **11**, 1138–1143.

Scheibenbogen, C., Lee, K. H., Mayer, S., Stevanovic, S., Moebius, U., Herr, W., Rammensee, H. G. and Keilholz, U. (1997). A sensitive ELISPOT assay for detection of CD8+ T lymphocytes specific for HLA class I-binding peptide epitopes derived from influenza proteins in the blood of healthy donors and melanoma patients. *Clin. Cancer Res.* **3**, 221–226.

Schlissel, M. S. and Stanhope-Baker, P. (1997). Accessibility and the developmental regulation of V(D)J recombination. *Semin. Immunol.* **9**, 161–170.

Schueler-Furman, O., Altuvia, Y., Sette, A. and Margalit, H. (2000). Structure-based prediction of binding peptides to MHC class I molecules: application to a broad range of MHC alleles. *Protein Sci.* **9**, 1838–1846.

Shankar, P., Russo, M., Harnisch, B., Patterson, M., Skolnik, P. and Lieberman, J. (2000). Impaired function of circulating HIV-specific CD8(+) T cells in chronic human immunodeficiency virus infection. *Blood* **96**, 3094–3101.

Skinner, P. J., Daniels, M. A., Schmidt, C. S., Jameson, S. C. and Haase, A. T. (2000). Cutting edge: In situ tetramer staining of antigen-specific T cells in tissues. *J. Immunol.* **165**, 613–617.

Sobao, Y., Tomiyama, H., Nakamura, S., Sekihara, H., Tanaka, K. and Takiguchi, M. (2001). Visual demonstration of hepatitis C virus-specific memory CD8(+) T-cell expansion in patients with acute hepatitis C. *Hepatology* **33**, 287–294.

Solheim, J. C. (1999). Class I MHC molecules: assembly and antigen presentation. *Immunol. Rev.* **172**, 11–19.

Speiser, D. E., Valmori, D., Rimoldi, D., Pittet, M. J., Lienard, D., Cerundolo, V., MacDonald, H. R., Cerottini, J. C. and Romero, P. (1999). CD28-negative cytolytic effector T cells frequently express NK receptors and are present at variable proportions in circulating lymphocytes from healthy donors and melanoma patients. *Eur. J. Immunol.* **29,** 1990–1999.

Speiser, D. E., Migliaccio, M., Pittet, M. J., Valmori, D., Lienard, D., Lejeune, F., Reichenbach, P., Guillaume, P., Luscher, I., Cerottini, J. C. and Romero, P. (2001). Human CD8(+) T cells expressing HLA-DR and CD28 show telomerase activity and are distinct from cytolytic effector T cells. *Eur. J. Immunol.* **31**, 459–466.

Strasser, A., O'Connor, L. and Dixit, V. M. (2000). Apoptosis signaling. *Annu. Rev. Biochem.* **69**, 217–245.

Tafuro, S., Meier, U. C., Dunbar, P. R., Jones, E. Y., Layton, G. T., Hunter, M. G., Bell, J. I. and McMichael, A. J. (2001). Reconstitution of antigen presentation in

HLA class I-negative cancer cells with peptide-beta2m fusion molecules. *Eur. J. Immunol.* **31**, 440–449.

Takayama, H., Kojima, H. and Shinohara, N. (1995). Cytotoxic T lymphocytes: the newly identified Fas (CD95)-mediated killing mechanism and a novel aspect of their biological functions. *Adv. Immunol.* **60**, 289–321.

Tan, L. C., Gudgeon, N., Annels, N. E., Hansasuta, P., O'Callaghan, C. A., Rowland-Jones, S., McMichael, A. J., Rickinson, A. B. and Callan, M. F. (1999). A re-evaluation of the frequency of CD8+ T cells specific for EBV in healthy virus carriers. *J. Immunol.* **162**, 1827–1835.

Tarazona, R., DelaRosa, O., Alonso, C., Ostos, B., Espejo, J., Pena, J. and Solana, R. (2000). Increased expression of NK cell markers on T lymphocytes in aging and chronic activation of the immune system reflects the accumulation of effector/senescent T cells. *Mech. Ageing Dev.* **121**, 77–88.

Thompson, C. B. (1995). New insights into V(D)J recombination and its role in the evolution of the immune system. *Immunity* **3**, 531–539.

Tierney, R. J., Steven, N., Young, L. S. and Rickinson, A. B. (1994). Epstein-Barr virus latency in blood mononuclear cells: analysis of viral gene transcription during primary infection and in the carrier state. *J. Virol.* **68**, 7374–7385.

Trowsdale, J. (2001). Genetic and functional relationships between MHC and NK receptor genes. *Immunity* **15**, 363–374.

Turner, S. J., Cross, R., Xie, W. and Doherty, P. C. (2001). Concurrent naive and memory CD8(+) T cell responses to an influenza A virus. *J. Immunol.* **167**, 2753–2758.

Ugolini, S., Arpin, C., Anfossi, N., Walzer, T., Cambiaggi, A., Forster, R., Lipp, M., Toes, R. E., Melief, C. J., Marvel, J. and Vivier, E. (2001). Involvement of inhibitory NKRs in the survival of a subset of memory-phenotype CD8+ T cells. *Nat. Immunol.* **2**, 430–435.

van den Eijnde, S. M., Luijsterburg, A. J., Boshart, L., De Zeeuw, C. I., van Dierendonck, J. H., Reutelingsperger, C. P. and Vermeij-Keers, C. (1997). In situ detection of apoptosis during embryogenesis with annexin V: from whole mount to ultrastructure. *Cytometry* **29**, 313–320.

van der Merwe, P. A. (2001). The TCR triggering puzzle. *Immunity* **14**, 665–668.

Vargas, A. L., Lechner, F., Kantzanou, M., Phillips, R. E. and Klenerman, P. (2001). Ex vivo analysis of phenotype and TCR usage in relation to CD45 isoform expression on cytomegalovirus-specific CD8+ T lymphocytes. *Clin. Exp. Immunol.* **125**, 432–439.

Wang, F., Ono, T., Kalergis, A. M., Zhang, W., DiLorenzo, T. P., Lim, K. and Nathenson, S. G. (1998). On defining the rules for interactions between the T cell receptor and its ligand: a critical role for a specific amino acid residue of the T cell receptor beta chain. *Proc. Natl. Acad. Sci. USA* **95**, 5217–5222.

Weston, S. A. and Parish, C. R. (1990). New fluorescent dyes for lymphocyte migration studies. Analysis by flow cytometry and fluorescence microscopy. *J. Immunol. Methods* **133**, 87–97.

Willcox, B. E., Gao, G. F., Wyer, J. R., Ladbury, J. E., Bell, J. I., Jakobsen, B. K. and van der Merwe, P. A. (1999). TCR binding to peptide-MHC stabilizes a flexible recognition interface. *Immunity* **10**, 357–365.

Willerford, D. M., Swat, W. and Alt, F. W. (1996). Developmental regulation of V(D)J recombination and lymphocyte differentiation. *Curr. Opin. Genet. Dev.* **6**, 603–609.

Xiong, Y., Luscher, M. A., Altman, J. D., Hulsey, M., Robinson, H. L., Ostrowski, M., Barber, B. H. and MacDonald, K. S. (2001). Simian immunodeficiency virus (SIV) infection of a rhesus macaque induces SIV-specific CD8(+) T cells with a defect in effector function that is reversible on extended interleukin-2 incubation. *J. Virol.* **75**, 3028–3033.

Xu, X. N., Purbhoo, M. A., Chen, N., Mongkolsapaya, J., Cox, J. H., Meier, U. C., Tafuro, S., Dunbar, P. R., Sewell, A. K., Hourigan, C. S., Appay, V., Cerundolo, V., Burrows, S. R., McMichael, A. J. and Screaton, G. R. (2001). A novel approach to antigen-specific deletion of CTL with minimal cellular activation using alpha3 domain mutants of MHC class I/peptide complex. *Immunity* **14**, 591–602.

Xu, Y. and Beckett, D. (1994). Kinetics of biotinyl-5′-adenylate synthesis catalyzed by the *Escherichia coli* repressor of biotin biosynthesis and the stability of the enzyme-product complex. *Biochemistry* **33**, 7354–7360.

Xu, Y., Nenortas, E. and Beckett, D. (1995). Evidence for distinct ligand-bound conformational states of the multifunctional *Escherichia coli* repressor of biotin biosynthesis. *Biochemistry* **34**, 16624–16631.

Yewdell, J., Anton, L. C., Bacik, I., Schubert, U., Snyder, H. L. and Bennink, J. R. (1999). Generating MHC class I ligands from viral gene products. *Immunol. Rev.* **172**, 97–108.

York, I. A., Goldberg, A. L., Mo, X. Y. and Rock, K. L. (1999). Proteolysis and class I major histocompatibility complex antigen presentation. *Immunol. Rev.* **172**, 49–66.

Young, A. C., Nathenson, S. G. and Sacchettini, J. C. (1995). Structural studies of class I major histocompatibility complex proteins: insights into antigen presentation. *Faseb J.* **9**, 26–36.

Young, N. T. and Uhrberg, M. (2002). KIR expression shapes cytoxic repertoires: a development of survival. *Trends Immunol.* **23**, 71–75.

Young, N. T., Uhrberg, M., Phillips, J. H., Lanier, L. L. and Parham, P. (2001). Differential expression of leukocyte receptor complex-encoded Ig-like receptors correlates with the transition from effector to memory CTL. *J. Immunol.* **166**, 3933–3941.

6 Using Microarrays for Studying the Host Transcriptional Response to Microbial Infection

David Yowe, Deborah Farlow, Jose M Lora, Stacey Drabic and Jose-Carlos Gutierrez-Ramos
Millennium Pharmaceuticals, Cambridge, Massachusetts, USA

◆◆

CONTENTS

◆◆◆◆◆◆ **INTRODUCTION**

Transcriptional profiling provides a strategy for the global analysis of host gene expression in a parallel manner and allows small changes in expression to be detected. Several techniques (Schena *et al.*, 1995; Velculescu *et al.*, 1995; Staudt and Brown, 2000) including microarrays have been developed for the high throughput quantitation of RNA expression levels. Host–pathogen interactions can cause several changes to the host cell including RNA expression modulation, induction of target receptors, actin cytoskeletal rearrangements, and activation of signal transduction pathways (Galan and Bliska, 1996; Finlay and Cossart, 1997; Cummings and Relman, 2000). Post-transcriptional mechanisms are important in regulating many genes and their products, however, the majority of cell regulation occurs via RNA level modulation (Cohen *et al.*, 2000; Staudt and Brown, 2000). Both temporal and spatial changes occur to the host's RNA transcriptional levels upon interaction with a pathogen. The effect of pathogen infection on host cell RNA expression levels has been successfully investigated using microarrays and the reader is referred to several reviews (Finlay and Cossart, 1997; Cummings and Relman, 2000; Manger and Relman, 2000; Yowe *et*

Copyright © Elsevier Science Ltd
All rights of reproduction in any form reserved

al., 2001) for detailed information on this subject. These studies have determined the host's RNA expression levels before and after infection with the aim of identifying pathways potentially important in a host's response to infection.

Microarrays have been used to study the host's response to acute bacterial infection. In one study, RAW 264.7 cell gene expression was examined after *Salmonella typhimurium* infection or treatment with lipopolysaccharide, and of the 588 genes studied, 34 genes were detected which exhibited four-fold or greater induction (Rosenberger *et al.*, 2000). Cohen *et al.* (2000) has compared the transcriptional response of THP1 cells to *Listeria monocytogenes* infection using commercially available oligonucleotide chip technology versus two macroarray systems, with oligonucleotide chip technology being shown to detect three- to six-fold more genes than either of the macroarray systems. Using a microarray consisting of about 1.5% of the protein coding capacity of the human genome, 22 host genes have been identified that are induced upon *Pseudomonas aeruginosa* infection of A549 cells (Ichikawa *et al.*, 2000).

Several studies have used microarrays for studying a host cell's transcriptional response to viral infection. Zhu *et al.* (1998) have investigated the effect of human cytomegalovirus infection on fibroblasts and identified about 250 host gene transcripts that were modulated before the onset of viral DNA replication. An *in vitro* study on the global effects of HIV type 1 infection on the expression of $CD4^+$ T cell genes has shown that little change occurs in the host's gene expression 2 days after infection (Geiss *et al.*, 2000). However, at 3 days post-infection, 20 genes were identified as being differentially regulated (Geiss *et al.*, 2000). Taylor *et al.* (2000) have also used microarrays to investigate the *in vivo* transcriptional response of mouse mycocardium genes to mycocarditic coxsackievirus B3 infection, with 169 genes shown to exhibit a significant change in expression at one or more of the post-infection time-points.

The preceding paragraphs have shown that microarrays can provide extensive information regarding the spatial and temporal changes that occur in the host's transcriptional response to pathogen infection. Microarrays have identified host genes not previously implicated in host–pathogen interactions since experiments can be performed using arrays consisting of a large number of genes from diverse families. However, despite the wealth of transcriptional data that mircoarrays give us on a host's response to infection, the physiological relevance of these host genes for responding to infection needs to be determined. Insights into the pathogenesis of infection can be provided by the cluster analysis of host and pathogen (Cummings and Relman, 2000; Stingley *et al.*, 2000) genes that are modulated upon host–pathogen interaction. Furthermore, the recent advent of protein chip technology (Zhu *et al.*, 2000) can allow the high-throughput screening of protein biochemical activities. Such technology may provide an understanding at the protein level of the role of host genes that have been identified by microarrays.

This chapter focuses on using microarrays for analysing the host transcriptional response to microbial infection. An overview is given of the principles of microarray technology and the processes involved in its application. This is followed by a description of general protocols used in isolation and labelling of RNA for use as probes, nylon microarray construction/hybridization, bioinformatics and data interpretation.

◆◆◆◆◆◆ AN OVERVIEW OF THE MICROARRAY PROCESS

DNA microarray technology is based on the principle of massive parallel hybridization of complementary DNA (cDNA) populations to immobilized and orderly arrayed oligonucleotides or cDNAs (Fig. 1). RNA from a given cell type or tissue is isolated by standard methods and the first strand of cDNA prepared and labelled, commonly with fluorescent dyes (e.g. Cy3 or Cy5) or radioactive isotopes (e.g. ^{33}P). This process yields a labelled mixture of cDNAs that reflects the mRNA distribution for that particular starting material and which we will call hereafter 'probe'.

The labelled probe is then hybridized to a collection of cDNA fragments or oligonucleotides which are covalently attached to a solid support. This collection is distributed in an ordered array, creating a matrix of 'spots'. Each spot is therefore unequivocally identified by its

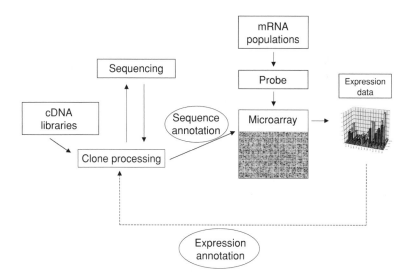

Figure 1. Schematic representation of the microarray process, from cDNA libraries to expression data. A central step in this process is clone processing, which involves the isolation and sequence characterization of clones from cDNA libraries. The cDNA clones are then arrayed onto a solid support to which a labelled probe is hybridized. The intensities of each of the spots on the microarray are quantitated and used to identify genes whose expression levels are modulated.

address in the matrix. Glass or nylon membrane solid supports are available, with each exhibiting comparable levels of sensitivity and array density. However, nylon microarrays offer several advantages over glass arrays. Nylon microarrays are the least expensive in terms of cost per array and the equipment needed to analyse the hybridized array and can be conveniently custom-made in-house to contain the desired genes of interest. Furthermore, nylon arrays only need 0.05 to 0.15 µg of mRNA, in contrast to glass arrays that require about 0.6 µg of mRNA.

After hybridization and washing, the microarray is scanned and the signal from each address is acquired and quantified. Before any meaningful biological information can be extracted, the raw data need to be filtered and normalized. The values obtained are then compared with a duplicate membrane hybridized in parallel with an identical probe. In general, every gene whose expression level varies by more than two-fold in duplicate microarrays is filtered out of the data set. Conversely, those genes with expression levels within the two-fold cut-off remain in the data set and the average value is subsequently used.

This brief description outlines the general process used to obtain the distribution and level of expression of each particular spot (gene) on the microarray in the starting mRNA population. In order to acquire relevant biological information the process is repeated simultaneously for a number of different starting cell types or tissues, and the absolute expression levels compared. In the simplest case, comparisons can be made between two distinct mRNA populations, such as two cell types or the same cell type under two different conditions. In this simple case, a pairwise comparison can be made to identify all the genes whose expression levels are higher in one sample compared with the other (commonly using two-fold as a cut-off). When a larger number of samples are compared, more powerful tools are needed including self-organizing maps (Tamayo *et al.*, 1999) and hierarchical clustering (Eisen *et al.*, 1998). Self-organizing maps are particularly useful to identify clusters of genes with biologically relevant expression patterns. Hierarchical clustering gives a snapshot of all expression patterns across the entire data set, and clusters not only the genes but also the probes used, giving a sense for the biological relevance of the conditions used within a specific experiment.

◆◆◆◆◆◆ MICROARRAY METHODOLOGY

This section describes methodology for designing and completing a transcriptional profiling experiment using microarray technology and the reader is referred to Fig. 1 for an overview of this process. The first part of this section describes criteria for the selection of RNA samples for use as probes and the preparation and handling of RNA. This is followed by the selection or production of microarray membranes, synthesis and hybridization of labelled cDNA, and the detection/quantitation of the hybridized signal. Lastly, criteria for the analysis, confirmation, and verification of the quality of microarray data is discussed.

Selection and preparation of the hybridization probe

Selection of RNA samples

When designing microarray experiments it is imperative to select the appropriate samples to isolate RNA for use as probes so the biological question(s) of interest can be addressed. In general, RNA from mock versus treated cells, tissues, or organisms are directly compared for transcriptional differences. RNA samples which can be used as microarray clone sets can be isolated from drug and mock treated cell populations; mock versus bacteria or virally infected cells; diseased and normal tissues; gene and mock transfected cells; as well as time courses of these situations. It is extremely important to include the appropriate positive and negative control RNA samples for any given microarray experiment to ensure the highest level of data output.

Total RNA isolation

It is imperative that all procedures involved in the isolation and handling of RNA are performed under conditions which eliminate exposure to ribonucleases. The use of DEPC-treated water and ribonuclease inhibitors, clean gloves, a 'clean' laboratory environment and equipment (e.g. ribonuclease-free tubes, other disposables and pipettes) is essential to prevent degradation of RNA which would lead to poor hybridization results. Precautions also need to be performed to minimize the degradation of RNA in the cell or tissue samples from which it is isolated. For instance, the cell or tissue samples collected for RNA preparation should be homogenized immediately upon collection in lysis buffer, snap-frozen in liquid nitrogen or a dry ice/ethanol bath, then stored at −80°C. The authors recommend the use of commercially available total RNA isolation kits since the reagents have been quality controlled for the absence of ribonucleases, are highly convenient, and have been optimized for the isolation of high-quality RNA from cell or tissue samples. For example, the Rneasy system (Qiagen, Valencia, CA, USA, catalogue # 75162) is recommended for total RNA isolation from primary cells or cell lines, while either the Trizol (Invitrogen Life Technologies, Carlsbad, CA, USA, catalogue # 10296028) or STAT systems (Tel Test, Inc., Friendswood, TX, catalogue # CS-112) are recommended for total RNA isolation from tissue samples. In either case, it is imperative that the manufacturer's protocols are followed exactly.

Preparation of radioactively labelled cDNA probes

Please note that the necessary precautions including the use of personal protective equipment and procedures must be used when working with radioactive materials. The following protocol is for the preparation of radioactively labelled cDNA probes from total RNA.

(contd.)

---

1. Resuspend 15 µg of total RNA in 10 µl of DEPC-treated water.
2. Add 3 µg of oligo-dT30 (Alphadna, Montreal, Canada).
3. Heat to 70°C in a heating block for 2 min.
4. Heat to 42°C in a heating block, then add

5X First Strand Buffer (Gibco BRL, Rockville, MD, catalogue # 11904-018) = 8 µl
0.1 mM DTT = 4 µl
Rnase Inhibitor (40 units/µl) (Boehringer Mannheim, Indianapolis, IN, USA, catalogue # 799025) = 1 µl
Superscript II Reverse Transcriptase (200 U/µl) (Gibco BRL, catalogue # 18604-022) = 2 µl
Solution containing 100 mM each of dATP, dGTP and dTTP = 2 µl
100 µCi [α^{33}P]-dCTP (Dupont, NEN, catalogue # NEG613H) = 10 µl.

Total Volume = 50 µl.

5. Mix by pipetting, then incubate for 60 min at 42°C.
6. Add 10 µl of Stop Solution (Gibco, BRL, catalogue # 11904-018) and heat at 70°C for 10 min.
7. Remove unincorporated nucleotides using commercially available columns (e.g. CHROMA SPIN +TE columns, Clontech, Palo Alto, CA, USA, catalogue # K1300-1) following the manufacturer's recommendations.
8. Determine the percent incorporation of radioactivity by measuring 1 µl of the final column eluate from Step 7 in 5 ml of scintillation fluid in a scintillation counter.

Counts per minute for a successful labelling reaction should be approximately 1×10^6 µl^{-1}.

Microarray selection and production

Due to the recent advent of commercially available and affordable microarrays (Cortese, 2000a,b) it has become convenient for individual investigators to purchase microarrays with predefined RNA samples or be custom-made. A wide variety of microarray formats are commercially available. The 'simplest' format is nylon membranes containing hundreds to a few thousand clones which can be analysed in individual laboratories using radioactively labelled cDNA. More sophisticated microarrays are silicon-based and contain tens of thousands of genes which are either spotted with photo-lithographically produced oligonucleotide sequences or PCR-generated cDNA inserts and are typically analysed by the company which supplied this array. It is up to the investigator to determine which array format is most suitable for their experiment.

Hybridization of labelled cDNA probes to microarrays

1. Pre-warm the Hybridization Solution ('Church buffer': 7% SDS; 250 mM Na phosphate pH 7.2; 1 mM EDTA; 0.5% filtered casein (ICN Biomedicals, Costa Mesa, CA, USA, catalogue # 101289) in DEPC-treated water) to 65°C.
2. Pre-anneal the ^{33}P-labelled cDNA probe at 65°C for 30 min in a tube containing

Human Cot1 DNA (Gibco BRL, Rockville, MD, USA, catalogue # 15279-011) = 10 µl
Poly dA (Pharmacia, Peapack, NJ, USA, catalogue # 27-7836-02) = 5 µl
Nuclease-free 20X SSC (Ambion, Austin, TX, USA, catalogue # 9762) = 17.5 µl
20% sodium dodecyl sulphate = 1.4 µl
^{33}P-labelled probe + DEPC-treated water (use the amount of probe for a final concentration in Step 5 to give approximately 2×10^6 cpm ml^{-1} hybridization solution) = 66 µl
Total Volume = 100 µl.

3. While the probe is pre-annealing, pre-wet the microarray membrane in 2X SSC then place the membrane in a hybridization oven bottle (Robbins Scientific, Sunnyvale, CA, USA) containing 100 ml of 2X SSC. Gently remove any air bubbles which may be present between the membrane and the bottle.
4. Discard the 2X SSC from the hybridization bottle and add 5 to 10 ml of pre-warmed Hybridization Solution. Make sure the membrane is completely covered with a 2 to 3 ml excess of solution.
5. Add the pre-annealed probe solution and hybridize at 65°C in a hybridization oven (Robbins Scientific, Sunnyvale, CA, USA, Model 400) for 18 h.
6. Discard the Hybridization Solution and wash the membrane using an excess amount of the following wash solutions at 65°C, discarding each wash solution after use

a) 2X SSC and 1% SDS for 15 minutes.
b) 2X SSC and 1% SDS for 15 minutes.
c) 0.2X SSC and 0.5% SDS for 30 minutes.
d) 0.2X SSC and 0.5% SDS for 30 minutes.
e) 2X SSC for 30 minutes.

7. Gently 'pat' dry the membrane with Whatman 3MM paper (VWR, catalogue # 21427-411), then carefully wrap in plastic film (e.g. Saran Wrap, Dow Brands LP, Indianapolis, IN). Immediately expose the membrane against a phosphorimager screen (Fuji, Tokyo, Japan, catalogue # BAS III) overnight. If the image is weak re-expose the membrane for a longer period.

Quantitation of the hybridized signal and quality control

The phosphorimager detection system (Fuji, Tokyo, Japan, catalogue # BAS 2500) is used for quantitation of the amount of radioactively labelled probe which has hybridized to each of the spotted samples on the microarray membrane. Precise gridding techniques are required to correlate the intensity detected at each address on the scan with the gene represented on the microarray membrane. The manufacturer's protocols and software should be utilized exactly as recommended to complete this part of the procedure successfully and to ensure quality control standards are met. It is highly recommended that each probe be hybridized to replicate membranes (Fig. 2) and that data are only considered valid when the coefficient of variation between the duplicate expression level values is acceptable. Scatterplot analysis of replicate membranes is typically used to identify experiments in which a high level of reproducibility or unacceptable variation exists between replicate membranes (Fig. 3).

Analysis of microarray data

It is essential to have the appropriate computing power and software packages to analyse effectively the extensive amount of data obtained from transcriptionally profiling hundreds to tens of thousands of genes in a number of different samples using microarrays. Appropriate software packages are supplied by companies which manufacture microarrays for commercial use and are available independently both on the Internet as well as through other commercial concerns. Two of the most widely applied methods of microarray data analysis are described conceptually as follows.

(A) 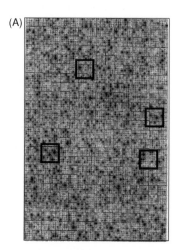 (B)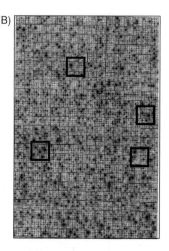

Figure 2. Representative images of replicate microarrays. The duplicate microarray membranes (A and B) contain more than 5500 individual cDNA clones and was hybridized with the same probe. Note the similarity of intensities of clones at corresponding locations on duplicate membranes. Square boxes have been placed over several locations of the membranes to highlight representative clones with similar intensities on the membrane replicates.

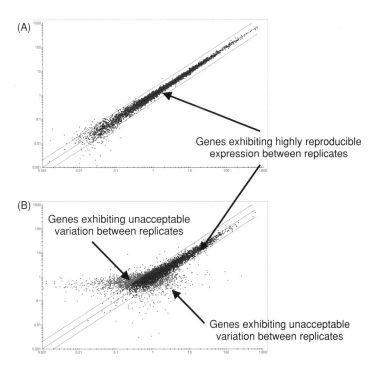

Figure 3. Scatterplot analysis of replicate microarray membranes. This figure shows a typical log–log phase scatterplot analysis of replicate hybridizations in which the same labelled RNA was used to probe two identical membranes. Each spot on the membrane represents an individual clone on the microarray. Diagonal ratio lines were set at 2. Genes whose expression level was reproducible from membrane to membrane (dark spots) and those genes whose expression levels appeared to be different on duplicate membranes (light spots) are indicated. (A) Scatterplot of replicate membranes which exhibit high reproducibility of data. (B) Scatterplot of unacceptable variation between replicate membranes.

Pairwise comparison analysis

The pairwise comparison (Fig. 4) of the differential levels of expression of specific genes in two relevant samples is the most straightforward and informative method of analysing a microarray experiment when the aim is directly to determine the effects of one type of treatment over another. Pairwise comparison analysis can be used to compare diseased versus normal tissue, as well as to determine the effect of a particular treatment on a cell population such as viral or bacterial infection, gene transfection, or exposure to a biochemical substance of interest.

Clustering analysis

It is often of interest to compare the differential levels of gene expression in more than two samples at any given time. There are many types of so-called 'clustering' algorithms including Self Organizing Maps (Fig. 5). This type of analysis is especially valuable for following differential gene

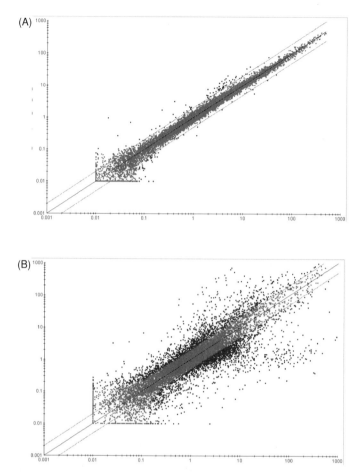

Figure 4. Pairwise comparison analysis. Log–log phase scatterplot analysis of the average values of pairwise comparisons in which different labelled RNA was used to probe replicate membranes. Each spot on the membrane represents the averaged values of each individual clone on the replicate membranes and the diagonal ratio lines were set at 2. Genes whose expression level was unchanged (light spots) or which differed significantly, i.e. greater than two-fold (dark spots) between the two RNA populations are shown. (A) Experimental conditions in which small changes in differential gene expression were observed. The majority of genes differed by less than two-fold in expression and are confined to the inner part of the diagonal ratio lines. (B) Experimental conditions in which large changes in differential gene expression were observed. A large number of genes differed by greater than two-fold in expression levels and are present outside the diagonal ratio lines.

expression over a series of time-points during *in vivo* or *in vitro* differentiation or development, or during a time-course of drug or other biochemical substance treatment.

Confirmation of microarray data

It is essential to independently verify the differential gene expression data obtained by microarray experiments because there exist limitations in the

166

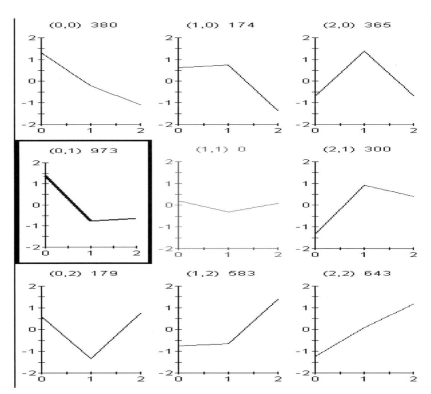

Figure 5. Typical clustering results using Self Organizing Map (SOM) analysis. SOM analyses representing nine distinct patterns of differential gene expression of three different RNA populations is presented. The numbers in parentheses above each SOM label each cluster, while the numbers to the direct right of each label describe the number of genes displaying the particular pattern shown. The x-axis (0, 1, and 2 at the bottom of each SOM) codes for each of the three experimental situations profiled. The y-axis (numbers to the left of each SOM) depicts the scaled expression levels.

sensitivity and accuracy of this type of data analysis. Several techniques and strategies are available including the use of previously published literature values, hybridization-based techniques (e.g. Northern blotting or RNase protection), or PCR-based analyses.

References

Cohen, P., Bouaboula, M., Bellis, M., Baron, V., Jbilo, O., Poinot-Chazel, C., Galiegue, S., Hadibi, E. H. and Casellas, P. (2000). Monitoring cellular responses to *Listeria monocytogenes* with oligonucleotide arrays. *J. Biol. Chem.* **15**, 11181–11190.

Cortese, J. D. (2000a). Array of options. *The Scientist* May 29, 26–29.

Cortese, J. D. (2000b). The array of today. *The Scientist* September 4, 25–28.

Cummings, C. A. and Relman, D. A. (2000). Using DNA microarrays to study host–microbe interactions. *Emerg. Infect. Dis.* **6**, 513–525.

Eisen, M. B., Spellman, P. T., Brown, P. O. and Botsein, D. (1998). Cluster analysis and display of genome-wide expression patterns. *Proc. Natl. Acad. Sci. USA* **95**, 14863–14868.

Finlay, B. B. and Cossart, P. (1997). Exploitation of mammalian host cell functions by bacterial pathogens. *Science* **276**, 718–725.

Galan, J. E. and Bliska, J. B. (1996). Cross-talk between bacterial pathogens and their host cells. *Ann. Rev. Cell Dev. Biol.* **12**, 221–255.

Geiss, G. K., Bumgarner, R. E., An, M. C., Agy, M. B., van't Wout, A. B., Hammersmark, E., Carter, V. S., Upchurch, D., Mullins, J. L. and Katze, M. G. (2000). Large-scale monitoring of host gene expression during HIV-1 infection using cDNA microarrays. *Virology* **266**, 8–16.

Ichikawa, J. K., Norris, A., Bangera, M. G., Geiss, G. K., van't Wout, A. B., Bumgarner, R. E. and Lory, S. (2000). Interaction of *Pseudomonas aeruginosa* with epithelial cells: identification of differentially regulated genes by expression microarray analysis of human cDNAs. *Proc. Natl. Acad. Sci. USA* **97**, 9659–9664.

Manger, I. D. and Relman, D. A. (2000). How the host 'sees' pathogens: global expression responses to infection. *Curr. Opin. Immunol.* **12**, 215–218.

Rosenberger, C. M., Scott, M. G., Gold, M. R., Hancock, R. E. and Finlay, B. B. (2000). *Salmonella typhimurium* infection and lipopolysaccharide stimulation induce similar changes in macrophage gene expression. *J. Immunol.* **164**, 5894–5904.

Schena, M., Shalon, D., Davis, R. W. and Brown, P. O. (1995). Quantitative monitoring of gene expression patterns with a complementary DNA microarray. *Science* **270**, 467–470.

Staudt, L. M. and Brown, P. O. (2000). Genomic views of the immune system. *Ann. Rev. Immunol.* **18**, 829–859.

Stingley, S. W., Garcia Ramirez, J. J., Aguilar, S. A., Simmen, K., Sandri-Goldin, R. M., Ghazal, P. and Wagner, E. K. (2000). Global analysis of herpes simplex virus type 1 transcription using an oligonucleotide-based DNA microarray. *J. Virol.* **74**, 9916–9927.

Tamayo, P., Slonim, D., Mesirov, J., Zhu, Q., Kitareewan, S., Dmitrovsky, E., Lander, E. S. and Golub, T. R. (1999). Interpreting patterns of gene expression with self-organizing maps: methods and application to hematopoietic differentiation. *Proc. Natl. Acad. Sci. USA* **96**, 2907–2912.

Taylor, L. A., Carthy, C. M., Yang, D., Saad, K., Wong, D., Schreiner, G., Stanton, L. W. and McManus, B. M. (2000). Host gene regulation during coxsackievirus B3 infection in mice. *Circul. Res.* **87**, 328–334.

Velculescu, V. E., Zhang, L., Vogelstein, B. and Kinzler, K. W. (1995). Serial analysis of gene expression. *Science* **270**, 484–487.

Yowe, D., Cook, W. J. and Gutierrez-Ramos, J-C. (2001). Microarrays for studying the host transcriptional response to microbial infection and for the identification of host drug targets. *Microbes Infect.* **3**, 813–821.

Zhu, H., Cong, J-P., Mamtora, G., Gingeras, T. and Shenk, T. (1998). Cellular gene expression altered by human cytomegalovirus: Global monitoring with oligonucleotide arrays. *Proc. Natl. Acad. Sci. USA* **95**, 14470–14475.

Zhu, H., Klemic, J. F., Chang, S., Bertone, P., Casamayor, A., Klemic, K. G., Smith, D., Gerstein, M., Reed, M. A. and Snyder, M. (2000). Analysis of yeast protein kinases using protein chips. *Nature Genet.* **26**, 283–289.

7 Genome-wide Expression Profiling of Intracellular Bacteria: The Interaction of *Mycobacterium tuberculosis* with Macrophages

Sabine Ehrt and Dirk Schnappinger
Department of Microbiology and Immunology, Weill Medical College of Cornell University, 1300 York Avenue, New York, NY 10021, USA

Martin I Voskuil and Gary K Schoolnik
Department of Microbiology and Immunology, Stanford University, Beckman Center, 300 Pasteur Drive, Stanford, CA 94305, USA

◆◆

CONTENTS

◆◆◆◆◆◆ INTRODUCTION

The number of fully sequenced bacterial genomes is large and increasing. The value of genome sequences is, however, limited by the fact that functional information is not available for 30% or more of the open reading frames (orfs) in any given genome. Genes are usually transcribed only during conditions under which their function is beneficial to the survival of a bacterium. The conditions that stimulate expression of a gene can therefore be used to characterize its function. In a typical microarray expression experiment two different RNAs (or their corresponding cDNAs), which have been prepared from a micro-organism under two different conditions of growth and were labelled with different fluorescent dyes, are hybridized to a single microarray. Binding to all hybridization probes on the array is quantified with a laser scanner that separately reports fluorescent intensities of the two dyes used for labelling. The fluorescence ratio of individual hybridization probes indicates the relative abundance of gene-specific RNAs and identifies genes, which are differentially expressed during the two conditions of growth. The capacity of microarrays readily accommodates the number of hybridization probes necessary to represent every individual orf of a bacterial genome. The relative expression status of every orf can therefore be determined in a single microarray expression experiment. In this chapter we focus on the application of microarrays for the analysis of *M. tuberculosis* inside host cells. More general methods used in microarray expression profiling have recently been described in detail elsewhere (Schoolnik *et al.*, 2001).

◆◆◆◆◆◆ THE DESIGN OF GENOME-WIDE EXPRESSION PROFILING EXPERIMENTS

The genome-wide expression profile of a cell is a complex description of its physiological status that includes most of the cell's metabolic activities and, therefore, gives access to an unusually holistic approach to the analysis of biological systems. In contrast, related but more reductionistic experiments such as reporter-gene assays only describe a narrow aspect of the cellular physiology. While genome-wide expression profiling contributes to a more complete understanding of cellular physiology it also demands a careful experimental design. All steps of an experimental protocol have to be investigated with respect to their impact on the expression profile because even routine procedures, such as the concentration of bacteria by centrifugation, can change the expression profile. Control samples are essential for the interpretation of an expression profile. Controls that are prepared at the same time and in the same manner as the experimental samples, except for the exposure to the stimulus or change of condition that is under investigation, will identify gene regulation occurring in response to the unavoidable handling of the bacteria during the experiment. In addition it may be necessary to perform control experiments to distinguish between direct and indirect changes of the expression profile. Antimicrobial drugs, for example, directly affect the genome-wide expression profile due to the inhibition of a particular biochemical pathway. The initial alteration of the expression profile can therefore be used to identify the pathway inhibited by a drug (Wilson *et al.*, 1998). After longer exposure to the drug, however, the expression profile changes more broadly, for example, in consequence of the cessation of growth. Time course experiments are helpful to separate direct from indirect changes of the expression profile. In this example control experiments in which the growth rate is changed by nutrient limitation may also identify indirectly regulated genes. Expression profiles are currently generated for a number of well-characterized growth conditions. The comparison of profiles generated in a pilot experiment with these reference profiles will identify stimuli that are applied in any given experiment and aid the selection of the most informative control experiments.

Spotted microarrays are generally used to compare two differently labelled mixtures of nucleic acids. In expression profiling experiments RNA that was prepared from the control or experimental sample is transcribed into cDNA and compared with a reference nucleic acid mixture. The most commonly used reference is derived from RNA that has been prepared from the bacterial culture before it was exposed to the experimental or control stimuli. This design allows the immediate identification of regulated genes from a single microarray experiment. Alternatively, chromosomal DNA can be used to prepare a reference that contains labelled fragments for every individual open reading frame of a genome. If chromosomal DNA is used the comparison of two microarray experiments is necessary to identify regulated genes. On the other hand, chromosomal DNA provides a constant reference sample whereas a reference prepared from RNA will vary depending on the exact culture conditions of individual experiments.

When microarrays were used for the first time, regulated genes were selected based on the magnitude of their change in expression. Recently, several laboratories demonstrated that this purely fold-change based approach has led to the incorrect identification of regulated genes. Moreover, other genes that failed to meet a particular fold-change criterion were misclassified as not having been significantly regulated (Ehrt *et al.*, 2001; Long *et al.*, 2001; Tusher *et al.*, 2001). Microarray experiments should, therefore, be designed to include statistical analysis. Several statistical procedures have been used to analyse microarray data. All statistical procedures depend on repeated measurements; however, given the cost of microarray experiments, an experimental design that allows statistical analysis may seem expensive. Fortunately, permutations of a limited number of repeated measurements as generated by the SAM (statistical analysis of microarrays) procedure minimize the number of experiments necessary for a statistical analysis. SAM has been demonstrated to reliably identify regulated genes in an expression profile, providing the data set contains four arrays per experimental condition representing two biological and two technical replicates (Tusher *et al.*, 2001). For the purpose of this discussion a biological replicate requires the use of RNA derived from a *de novo* experiment, whereas a technical replicate uses a second microarray to study the same RNA. Control samples that are generated at the same time as the experimental samples can further increase the sensitivity of most statistical procedures since they allow pairing of experimental and control samples. While the use of appropriate microarray-specific instruments will select a gene set that is significantly regulated in a statistical sense, it does not prove that such regulation is biologically significant. Proof that the regulation of a gene is biologically significant may require a combination of mutational, biochemical and animal model studies.

Expression profile libraries often contain millions of data points and much of the information they contain is likely to be overlooked.

Multivariate statistical analysis, such as clustering or linear decomposition (Eisen *et al.*, 1998; Alter *et al.*, 2000), can overcome this problem by associating genes of unknown function (annotated as genes coding for hypothetical proteins in the genome database) with functionally annotated genes that subserve a common cellular process or pathway. Clustering the expression profiles of cancer cells, for example, has been used for cancer typing and predicting the success of a particular cancer therapy (Bittner *et al.*, 2000; Scherf *et al.*, 2000). Multivariate statistical methods should be included in every analysis of microarray data. Therefore, it is important that microarray experimental design be compatible with the statistical tools the investigator plans to use during the data mining and analysis phase of the project. The bioinformatics aspect of microarray experimentation is a rapidly developing field and further discussion of it is beyond the scope of this article; additional information can be found elsewhere (Sherlock, 2000; Altman and Raychaudhuri, 2001).

◆◆◆◆◆◆ INFECTION OF MACROPHAGES WITH M. TUBERCULOSIS

Basic considerations

One of the most basic considerations when designing the infection experiment is that the amount of bacterial RNA extracted from infected tissue must be sufficient to perform a microarray experiment. Currently 0.5–1 μg of bacterial RNA is necessary for a single microarray experiment. Therefore, 2 μg should be available to allow analysis of the RNA for at least two independent hybridizations. At a multiplicity of infection of two to five bacilli, about 2×10^8 macrophages are necessary to allow the purification of a sufficient amount of *M. tuberculosis* RNA. This rather large number of cells imposes two restrictions on the experiment. First, primary human macrophages are often not available in the necessary quantity. Therefore, murine primary macrophages or human macrophage-like cell lines are typically used. Secondly, at very early time points, i.e. less than 4 h after infection, the numbers of intracellular bacteria are too low, even when 2×10^8 macrophages are used, to allow purification of sufficient amounts of bacterial RNA.

Another potential problem is that *M. tuberculosis* can form multicellular aggregates that, due to their size, interfere with their ingestion by macrophages. Therefore, in many infection experiments *M. tuberculosis* is exposed to rather harsh procedures, such as sonication, to reduce the size of the bacterial particles. These procedures should be avoided as they most likely change the expression profile of *M. tuberculosis*. Instead, growth conditions can be optimized to reduce the size of the bacterial particles. The particle size of *M. tuberculosis* is influenced by many factors including growth phase, culture density, medium components and strain characteristics. Many of the commonly used *M. tuberculosis* strains form sufficiently small particles in early log phase of growth if cultivated in the presence of detergent in a slowly moving (1.5 rpm) roller bottle.

Because genome-wide expression profiles are influenced by a large number of stimuli, nearly any difference in the experimental conditions in repeated measurements will lead to different expression profiles. To minimize these differences conditions of every individual experiment should be monitored as closely as possible. In the case of macrophage infection experiments the activation status of macrophages is one source of experimental variation. Accordingly, the activation status should be determined by measuring the generation of reactive oxygen and nitrogen intermediates and by monitoring the survival of intracellular *M. tuberculosis* (Ehrt *et al.*, 2001).

Macrophages used for infection with *M. tuberculosis* should form adherent monolayers of about 95% confluency. Murine bone marrow derived macrophages are readily obtained in large numbers, adhere well to tissue culture flasks even in large volumes and are not pre-activated during preparation. They are, therefore, well suited for microarray expression profiling of intracellular *M. tuberculosis*. A protocol for the isolation and maturation of murine bone marrow derived macrophages has been described previously in detail (Rhoades and Orme 1998; Roberts *et al.*, 1998). A brief summary of the method is given in Protocol 1.

Protocol I   Isolation of murine bone marrow macrophages

1. Isolate bone marrow cells from mouse femurs by flushing the femurs with DMEM (Invitrogen). Expect about 10^7 cells per femur. Harvest the cells by centrifugation at 1000 rpm for 10 min at 4°C.

2. To lyse the red blood cells, resuspend the cells in 5 ml cold 0.2% NaCl; after 30 s add 5 ml cold 1.6% NaCl and 10 ml PBS. Centrifuge cells as above.

3. Resuspend cells in complete cell culture medium containing 20% L929 fibroblast conditioned medium and count cell number with a haematocytometer.

4. Seed 5×10^7 cells per 175 cm^2 triple layer flask (Nunc) and incubate at 37°C in 5% CO_2. Induced by M-CSF from the L-cell conditioned medium, the cells will differentiate into macrophages over a period of 6–7 days. The cell number will approximately double over that time period and the monolayer should reach a confluency of more than 90%.

5. At day 4 after isolation feed the cells with one fourth volume complete cell culture medium.

6. At day 7 after isolation remove the medium and any non-adherent cells, wash the monolayers once with prewarmed PBS, then add complete cell culture medium containing 10% L-cell conditioned medium and no antibiotics. Add 100 U/ml recombinant murine interferon-γ (Genentech) if activation is desired. The macrophages are ready for infection 24 h or 48 h after the medium change.

Protocol 2   Preparation of *M. tuberculosis* culture and infection of macrophages

1. *M. tuberculosis* is grown in 7H9 medium with 10% ADNaCl and 0.05% Tween80 in 1l roller bottles rotated at 1.5 rpm. The bacteria are sub-cultured every 3–4 days (once they reach an OD_{580} between 0.6 and 0.8) at a 1 : 100 dilution into fresh medium. After about 2 months a new aliquot of *M. tuberculosis* is thawed and cultured.
2. For infection of macrophages, *M. tuberculosis* cultures are used at an OD_{580} of 0.3–0.4. Collect the appropriate amount of bacteria by centrifugation for 10 min at 3500 rpm.
3. Resuspend the bacteria in prewarmed complete cell culture medium and add to the macrophage monolayer at a multiplicity of infection of 5.[1] Determine the actual titre of the bacteria by plating serial dilutions onto 7H11 agar plates (BD Diagnostic Systems).

Protocol 3   Harvesting of *M. tuberculosis* grown in liquid culture for the preparation of RNA[2]

1. Transfer an appropriate volume of the liquid culture (for example 20–25 ml of a logarithmically growing culture of an OD_{580} of 0.2–0.4) to a 50 ml centrifugation tube containing the same volume of GTC solution.[3] Mix immediately and centrifuge for 10 min at 4000 rpm.
2. Pour off the supernatant; freeze the pellet on dry ice and store at −80°C.

◆◆◆◆◆◆ PREPARATION AND LABELLING OF RNA FROM INTRACELLULAR *M. TUBERCULOSIS*

The use of microarrays to analyse the expression profile of bacteria that are co-cultivated with host cells is obviously complicated by the eukaryotic RNA that is present in excess. Two approaches have been used to minimize the interference of eukaryotic RNA during hybridization. Talaat *et al.* (2000) designed a small number of short primers that can be used preferentially to label mycobacterial RNA that is present in a mixture with eukaryotic RNA. Butcher and colleagues instead took advantage of the very stable mycobacterial cell wall to develop a differential lysis procedure that allows the separation of mycobacteria from the eukaryotic tissue prior to purification of the mycobacterial RNA (Monahan *et al.*, 2001). The key step of this procedure is the incubation of the infected tissue with a guanidinium thiocyanate (GTC) solution that besides GTC contains detergent and β-mercaptoethanol as active ingredients. GTC is a highly denaturing salt, which has been used for the preparation of RNA especially from samples that contain high concentrations of RNase (Chirgwin *et al.*, 1979). Incubation of infected macrophages with the GTC solution immediately lyses the macrophages but does not lyse *M. tuberculosis*. Although it does not lyse *M. tuberculosis* the GTC solution kills the bacteria and stabilizes all transcripts that are present at the time the solution is added. After treatment with the GTC solution the bacteria can be safely exposed to procedures that would normally alter their expression profile. Because this procedure allows purification of bacterial particles prior to their lysis a selective labelling protocol is not needed.

Protocol 4   Harvesting intracellular *M. tuberculosis*

1. Inspect the macrophage culture for integrity of the monolayer and location of bacteria. Discard the medium and quickly wash with 50 ml of prewarmed (37°C) complete medium if a large number of obviously extracellular bacteria are present (relevant for early time points after infection). Remove medium before proceeding to step 2.
2. Pour 50 ml of GTC solution into the tissue culture flask and lyse the cells by shaking. The solution will become viscous due to the release of eukaryotic chromosomal DNA. Continue shaking till the viscosity is reduced to a point allowing transfer of the solution.
3. Transfer the GTC solution from the tissue culture flask into two 50 ml V-bottom plastic tubes and vortex for 1 min. At the end of the vortexing the viscosity of the solution should be about the same as fresh GTC solution.
4. Centrifuge for 15 min at 4000 rpm and room temperature to harvest the bacteria.
5. Decant the GTC solution and resuspend the cell pellet in 10 ml of fresh GTC solution.
6. Centrifuge as above and remove as much GTC solution as possible. The cell pellets can now be stored at −80°C or be used directly for the preparation of RNA.

Protocol 5   Preparation of RNA

1. Resuspend the bacterial pellets in 1 ml Trizol (Invitrogen) and transfer the suspension into a 2 ml screw cap tube containing 0.4 ml 0.1 mm silica beads (Biospec).
2. Shake the screw cap tube for 30 s in a bead beater (Biospec) at maximum speed and repeat this procedure twice. Cool the sample in between bead beating incubations if the temperature exceeds 37°C.
3. Centrifuge the sample for 20 s at 13 000 rpm at room temperature to separate the Trizol solution from the beads and bacterial debris.
4. Transfer the Trizol solution into a fresh 2 ml tube containing 350 μl chloroform: isoamylalcohol (24 : 1). Invert rapidly for 15 s and continue inverting periodically for 2 min.

(contd.)

5. Centrifuge for 5 min at maximum speed and room temperature. Remove the aqueous, upper layer and transfer it into a 1.5 ml tube containing 700 µl isopropanol and mix well.
6. Incubate for 2 h at –20°C and centrifuge for 20 min at 13 000 rpm and 4°C to precipitate nucleic acids. Remove isopropanol, add 1 ml of 75% ethanol, invert several times and centrifuge for 5 min at 13 000 rpm and 4°C. Remove ethanol and air dry for 5 min.
7. Resuspend the precipitated nucleic acids in water, determine their concentration and degrade DNA with DNase for 30 min at 37°C. Inactivate DNase for 15 min at 65°C or remove it by column purification using the RNeasy procedure (Qiagen). Store RNA at –80°C.

Protocol 6   Labelling of RNA

1. Bring 0.5–2 µg of RNA to a volume of 10 µl and add 2 µl (2 µg/µl) random hexamers (Roche).
2. Denature RNA for 10 min at 65°C. Chill on ice and centrifuge for 1 min.
3. Add 5.0 µl first-strand buffer (5X), 2.5 µl DTT (100 mM), 2.3 µl dNTP mix (5 mM dATP, dCTP, dGTP and 0.2 mM dTTP), 1.5 µl Cy3-dUTP or Cy5-dUTP (Amersham) and 1.0 µl reverse transcriptase (Invitrogen). Incubate for 5 min at room temperature followed by 90 min at 42°C. The labelled sample can be stored at –20°C; concentrate the sample using a microcon-10 (Ambion) or other concentration method before hybridization.

◆◆◆◆◆◆ CONCLUSIONS

A differential lysis procedure that stabilizes the bacterial RNA as soon as the lysis solution is added to the infected tissue is an ideal method to study the genome-wide expression profile of pathogens in contact with their host cells. Changes of the expression profile due to the collection of the bacteria are prevented and pathogen- and host-derived RNA are separated, significantly simplifying specific labelling of the pathogen's RNA. Unfortunately, the GTC solution used for the preparation of RNA from intracellular *M. tuberculosis* disrupts most other bacterial species. Thus, the GTC method described here cannot be used to study, for example, the interaction of Gram-negative bacteria with eukaryotic cells. Nonetheless, it may be possible to design alternative differential lysis solutions that can be used for the analysis of bacteria more fragile than *M. tuberculosis*. In contrast, a differential labelling procedure, using a small number of genome directed primers (Talaat *et al.*, 2000) or a large number of primers specifically designed for every orf in a genome, should be applicable for the analysis of the interaction of almost any pathogen with its host. It should be kept in mind though, even when this method of differential labelling is used, that the RNA of the pathogen must be stabilized as early as possible to avoid procedure-related changes of the expression profile.

Acknowledgements

Work by the authors has been supported by the Walter and Idun Berry Foundation (M.V.), National Institutes of Health grant AI44826 (G. K. S.), the Action TB Program of GlaxoWellcome (G. K. S.) and the Deutsche Forschungsgemeinschaft (D.S.). We thank Heran Darwin for her valuable comments on the manuscript.

Notes

1. We recommend determining the relationship of CFU and OD_{580} for the *M. tuberculosis* strain under study and spectrophotometer. In our hands an OD_{580} of 0.1 corresponds to 5×10^7 CFU.
2. This protocol should be used for the preparation of the reference samples if RNA is used as a reference in the microarray hybridization.
3. Use GTC at a concentration of 5 M instead of 4 M.

Reagents

Complete cell culture medium

DMEM (Invitrogen) with 4.5 g/l glucose, 0.584 g/l L-glutamine, 1 mM Na-pyruvate, 10% heat-inactivated FBS (HyClone), 10 mM Hepes buffer, 20% L929-fibroblast conditioned medium (contains M-CSF) and 100 U ml^{-1} penicillin G, 100 µg ml^{-1} streptomycin. Reduce L-cell conditioned medium to 10% and leave out the antibiotics, after differentiation of macrophages and for infection with bacteria.

L-929 conditioned medium

L-929 cells from American type culture collection (ATCC) are grown in DMEM (Invitrogen) with 4.5 g l^{-1} glucose, 0.584 g l^{-1} L-glutamine, 1 mM Na-pyruvate, 10% heat-inactivated FBS (HyClone) until confluent. Collect supernatant, filter through a 0.22-µm filter or centrifuge for 10 min at 1000 rpm and collect the supernatant. Freeze conditioned medium at –20°C in 50-ml aliquots.

GTC solution (4M)

Dissolve 50 g guanidine thiocyanate (Sigma) in 70 ml deionized water. Incubate the solution in a 50°C water bath until clear and add 0.5 g sodium N-lauryl sarcosine (0.5%), 0.75 g tri-sodium citrate (50 mM) and 0.7 ml 2-mercaptoethanol (0.1M). Adjust pH to 7.0 and volume to 100 ml. Keep the solution at room temperature and store it for no longer than 48 h.

References

Alter, O., Brown, P. O. and Botstein, D. (2000) Singular value decomposition for genome-wide expression data processing and modeling. *PNAS* **97**, 10101–10106.

Altman, R. B. and Raychaudhuri, S. (2001). Whole-genome expression analysis: challenges beyond clustering. *Curr. Opin. Struc. Biol.* **11**, 340–347.

Bittner, M., Meltzer, P., Chen, Y., Jiang, Y., Seftor, E., Hendrix, M., Radmacher, M., Simon, R., Yakhini, Z., Ben-Dor, A., Sampas, N., Dougherty, E., Wang, E., Marincola, F., Gooden, C., Lueders, J., Glatfelter, A., Pollock, P., Carpten, J., Gillanders, E., Leja, D., Dietrich, K., Beaudry, C., Berens, M., Alberts, D. and Sondak, V. (2000). Molecular classification of cutaneous malignant melanoma by gene expression profiling. *Nature* **406**, 536–540.

Chirgwin, J. M., Przybyla, A. E., MacDonald, R. J. and Rutter, W. J. (1979) Isolation of biologically active ribonucleic acid from sources enriched in ribonuclease. *Biochemistry* **18**, 5294–5299.

Ehrt, S., Schnappinger, D., Bekiranov, S., Drenkow, J., Shi, S., Gingeras, T. R., Gaasterland, T., Schoolnik, G. and Nathan, C. (2001). Reprogramming of the macrophage transcriptome in response to interferon-gamma and *Mycobacterium tuberculosis*. Signaling roles of nitric oxide synthase-2 and phagocyte oxidase. *J. Exp. Med.* **194**, 1123–1140.

Eisen, M. B., Spellman, P. T., Brown, P. T. and Botstein, D. (1998). Cluster analysis and display of genome-wide expression patterns. *PNAS* **95**, 14863–14868.

Long, A. D., Mangalam, H. J., Chan, B. Y., Tolleri, L., Hatfield, G. W. and Baldi, P. (2001). Improved statistical inference from DNA microarray data using analysis of variance and a Bayesian statistical framework. Analysis of global gene expression in *Escherichia coli* K12. *J. Biol. Chem.* **276**, 19937–19944.

Monahan, I. M., Mangan, J. A. and Bucher, P. D. (2001). Extraction of RNA from intracellular *Mycobacterium tuberculosis*: Methods, considerations, and applications. In: Mycobacterium tuberculosis *Protocols* (T. Parish and N.G. Stoker, Eds) pp. 31–42. Humana Press, Totowa.

Rhoades, E. R. and Orme, I. M. (1998). Similar responses by macrophages from young and old mice infected with *Mycobacterium tuberculosis*. *Mech. Ageing Dev.* **106**, 145–153.

Roberts, A. D., Belisle, J. T., Cooper, A. M. and Orme, I. M. (1998). Murine model of tuberculosis. In: *Immunology of Infection* (S.H.E. Kaufmann and D. Kabelitz, Eds) pp. 389–417. Academic Press, London.

Scherf, U., Ross, D. T., Waltham, M., Smith, L. H., Lee, J. K., Tanabe, L., Kohn, K. W., Reinhold, W. C., Myers, T. G., Andrews, D. T., Scudiero, D. A., Eisen, M. B., Sausville, E. A., Pommier, Y., Botstein, D., Brown, P. O. and Weinstein, J. N. (2000). A gene expression database for the molecular pharmacology of cancer. *Nat. Gen.* **24**, 236–244.

Schoolnik, G. K., Voskuil, M. I., Schnappinger, D., Yildiz, F. H., Meibom, K., Dolganov N. A., Wilson, M. A. and Chong, K. H. (2001). Whole genome DNA microarray expression analysis of biofilm development by *Vibrio cholerae* O1 E1 Tor. *Methods Enzymol.* **336**, 3–18.

Sherlock, G. (2000). Analysis of large-scale gene expression data. *Curr. Op. Immunol.* **12**, 201–205.

Talaat, A. M., Hunter, P. and Johnston, S. A. (2000). Genome-directed primers for selective labeling of bacterial transcripts for DNA microarray analysis. *Nat. Biotech.* **18**, 679–682.

Tusher, V. G., Tibshirani, R. and Chu, G. (2001). Significance analysis of microarrays applied to the ionizing radiation response. *PNAS* **98**, 5116–5121.

Wilson, M., DeRisi, J., Kristensen, H.-H., Imboden, P., Rane, S., Brown, P. O. and Schoolnik, G. K. (1998). Exploring drug-induced alterations in gene expression in *Mycobacterium tuberculosis* by microarray hybridization. *PNAS* **96**, 12833–12838.

List of Suppliers

Ambion, Inc.
2130 Woodward, Austin, TX 78744-1832, USA

Phone: +1 512-651-0200
Fax: +1 512-651-0201

Amersham Biosciences Corp.
800 Centennial Avenue, P.O. Box 1327, Piscataway, NJ 08855-1327, USA

Phone: +1 732-457-8000
Fax: +1 732-457-0557

Biospec Products, Inc.
P.O. Box 788, Bartlesville, Oklahoma 74005-0788, USA

Phone: (800) 617-3363
Fax: (918) 336-6060

BD Diagnostic Systems
1 Becton Drive, Franklin Lakes, NJ 07417, USA

Phone: 201-847-5772
Fax: 201-847-5757

Genentech, Inc.
1 DNA Way, South San Francisco, CA 94080, USA

Phone: 1-650-225-1000
Fax: 1-650-225-6000

HyClone Laboratories, Inc.
1725 S. HyClone Road, Logan, UT 84321, USA

Phone: 1-800-492-5663
Fax: 1-800-533-9450

Invitrogen Corporation
1600 Faraday Avenue, P.O. Box 6482, Carlsbad, CA 92008, USA

Phone: 1-800-828-6686
Fax: 1-800-331-2286

Nunc

75 Panorama Creek Drive, P.O. Box 20365, Rochester, NY 14602-0365, USA

Phone: (800) 625-4327
Fax: (716) 586-8987

QIAGEN Inc.

28159 Avenue Stanford, Valencia, CA 91355, USA

Phone: 800-426-8157
Fax: 800-718-2056

Roche Molecular Biochemicals

P.O. Box 50414, 9115 Hague Road, IN 46250-0414, USA

Phone: 800-428-5433
Fax: 800-428-2883

PART 11

Murine Models

◆◆

1 Management of Immunocompromised and Infected Animals

Horst Mossmann
Max-Planck-Institut für Immunbiologie, Freiburg, Germany

Werner Nicklas
Deutsches Krebsforschungszentrum, Heidelberg, Germany

Hans J Hedrich
Institut für Versuchstierkunde, Medizinische Hochschule, Hannover, Germany

◆◆

CONTENTS

◆◆◆◆◆◆ INTRODUCTION

Studies of the immune system have started with the vaccination of outbred animals, the examination of the immune response with respect to specific antibodies, immune cells and resistance to the corresponding infections. Inbred and congenic lines were developed, giving insight into the major histocompatibility complex (MHC) and allowing more detailed questions about the function of the immune system. In the course of inbreeding especially, many natural variants were found, including immunocompromised animals that served as models for human and animal disorders of the immune system. In the last decade a vast number of immunocompromised strains was created by genetic manipulation (Kühn and Schwenk, 1997; Croy *et al.*, 2001; Yu and Bradley, 2001) either directly or via the loxP-cre system allowing organ-specific deletions by the use of the respective promoters (Ray *et al.*, 2000; van Duyne, 2001). In addition, systems for artificially inducible deletions have been developed (Jaisser, 2000).

METHODS IN MICROBIOLOGY, VOLUME 32
ISBN 0–12–521532–0

Copyright © Elsevier Science Ltd
All rights of reproduction in any form reserved

In many areas of immunity, natural and induced immunodeficient variants have been used to study very different aspects of immunity such as T-cell function (Wang *et al.*, 1997; Croy *et al.*, 2000), autoimmunity (Shevach, 2000), cytokine effects (Kopf *et al.*, 1995; Assemann *et al.*, 1999; Kawachi *et al.*, 2000), cancer development (Bankert *et al.*, 2001; Berking and Herlyn, 2001; Recio and Everitt, 2001) and pathogenesis of infections (Mosier, 1996; Hess *et al.*, 2000; Schaible and Kaufmann, 2000). From these studies, the high complexity and considerable redundancy of the immune system became evident. A large number of the transgenic mutants used particularly in the field of immunology are immunodeficient, being more or less susceptible to infections. Particularly in these animals, known pathogens as well as ubiquitous micro-organisms may drastically reduce the breeding efficiency, endanger the maintenance and may counteract experimental manipulation. Reconstitution with immunocompetent cells and the use of antibiotics may only be a preliminary help. Finally, optimal hygienic standards are indispensable for these animals. The prerequisites are corresponding housing facilities, well-trained personnel, experience in re-derivation and the continuous monitoring of the stocks.

In addition to the microbiological status, the genetic background of the immunodeficient variant may strongly affect experimental results. Therefore, variants produced naturally or artificially have to be introduced into the desired background by a series of time-consuming backcrosses. Marker-assisted selection, also termed 'speed congenics', can accelerate this procedure.

Environmental conditions, food, water, handling, light cycle, etc. can alter experimental results. These factors have to be considered and standardized as far as possible. Furthermore, in some immunodeficient animal strains an ulcerative colitis (inflammatory bowel disease) may develop spontaneously (Bregenholt and Claesson, 1998).

We will try to point out here the special requirements for the management, breeding and housing of immunocompromised and infected animals, especially those for the mouse and rat.

◆◆◆◆◆◆ MICROBIOLOGICAL STANDARDIZATION

The quality of laboratory animals, mainly rodents, has improved during the last decade. First attempts at eliminating disease were made in the 1950s. At that time infectious agents were widespread in rodent colonies, and many experiments were interrupted by infections. It became obvious that classical veterinary approaches, such as improved husbandry, vaccination, antibiotics and chemotherapeutics, would not eliminate pathogens, and therefore gnotobiotic techniques were established, such as caesarean derivation and subsequent raising in isolation. This resulted in the elimination of various organisms, such as *Mycoplasma pulmonis*, which had previously been ineradicable. However, infections were still prevalent in many colonies. More sophisticated experimental procedures were increasingly sensitive to influence of viruses. Some viruses had been tolerated in the past as they have a low potential to induce clinical disease, but both scientists and breeders were aware of their presence. It was shown later that many of these agents, although clinically silent, can induce increased variation between individuals and can influence biochemical or immunological functions. Research complications occurred frequently, resulting in the need to eliminate also those agents that cause clinically silent infections, and to monitor colonies of rodents for the presence or absence of such organisms.

Today, it is generally accepted that good research requires animals that are free from micro-organisms that might influence the health of animals (or humans) or the results of animal experiments.

Influence of micro-organisms on research results

It is a well-known fact that research complications due to overt infectious diseases are significant and that clinically ill animals should not be used for scientific experiments. The effect of clinically silent infections, however, is frequently underestimated, but it may also be devastating because infections often remain undetected. Scientists in general are not well informed of such influences on their research. Only a small percentage of detected complications have been published. The literature is scattered across diverse scientific journals, and many articles are difficult to locate. To address the problem, conferences have been held on viral complications on research, and the knowledge available has been summarized in conference proceedings (Bhatt *et al.* 1986a; Hamm, 1986). The literature available has later repeatedly been reviewed (Lussier, 1988; National Research Council, 1991; Hansen, 1994; Baker, 1998; Nicklas *et al.*, 1999).

Research complications may occur in various ways. Although acute clinical signs may not be observed, infected animals may show altered behaviour, suppressed body weight, or reduced life expectancy, which may, for example, influence the tumour rate. Micro-organisms present in an animal may lead to contamination of samples and tissue specimens such as cells, tumours, sera and monoclonal antibodies. This may interfere with experiments performed with cells or isolated organs.

The experiment itself may be a stress factor and increase the sensitivity to an agent, and thus induce clinical disease or death. Environmental factors, such as increased temperature or relative humidity (for example in metabolic cages), may activate latent infections resulting in, for example, lung complications caused by *Pseudomonas aeruginosa, Klebsiella pneumoniae, Staphylococcus aureus* or *Pneumocystis carinii*, especially in immunodeficient animals. Naturally, various micro-organisms can interact and lead to clinical disease or research complications that are dependent on the combination of micro-organisms.

The disease rate is not only dependent on the host, but also on specific properties of the infectious agents. There are different strains of many viruses with different organotropism (e.g. hepatotropic, enterotropic and neurotropic strains of MHV). This influences the disease rate and the mortality, as well as the type and severity of pathological changes. As another example, the immunosuppressive variant of minute virus of mice

(MVMi) replicates in lymphocytes whereas the prototype strain (MVMp) replicates in fibroblasts, thus resulting in different effects on animals or experiments. Both variants usually do not induce clinical disease, but may affect various parameters such as wound healing, immunological parameters, tumour growth and development, embryonic development, and birth rate.

Various effects are possible on the function or the morphology of organs or cell systems. Histopathological changes that resemble adenomas have been observed in the trachea or bronchioles during the regenerative phase after a Sendai virus infection.

When pathogens infect laboratory animals, the immune system is activated regardless of the level of pathogenicity. Many micro-organisms have the potential to induce functional suppression or stimulation of the immune system. Sometimes, only T cells, or B cells, or specific sub-populations are influenced. Therefore, most virus infections and infections with bacteria or parasites are detrimental for immunological research and must be avoided.

Some micro-organisms have a specific effect on enzymatic or haemato-logical parameters. LDV can induce up to 100-fold increase in the activity of LDH and other enzymes in the plasma. Numerous reports exist in the literature about modulation of oncogenesis. Infectious agents may induce cancer, enhance chemical or viral carcinogenesis, or reduce the incidence of cancer. Some organisms even influence the growth rate of trans-plantable tumours.

Immunosuppressed and immunodeficient animals are usually more sensitive to infections than immunocompetent animals. Infections in immunodeficient animals frequently result in increased mortality due to a reduced or absent resistance to low pathogenic or even commensal micro-organisms.

It is important for various reasons that animals used for infectious studies are free from adventitious infections. An adventitious organism due to immunomodulation may influence the infection in question and, therefore, result in increased or reduced resistance to experimental infection. Micro-organisms resulting from a natural infection might contaminate viruses, bacteria, or parasites that are passaged in laboratory animals. Spontaneous infections may lead to false conclusions. For example, the first isolations of Sendai virus were made from mice that had been inoculated with diagnostic materials from humans and swine. In subsequent years, evidence accumulated to show that an indigenous virus of mice had been isolated (National Research Council, 1991).

Principles of health monitoring

The microbiological quality of laboratory animals is a direct result of colony management practices, and monitoring provides an after-the-fact assessment of the adequacy of those practices. Monitoring is, therefore, of greatest value in connection with maintenance of animals in isolation systems where vigorous microbiological control is applied.

Health monitoring procedures in animal populations differ from procedures used in human medicine. Especially in populations of small laboratory animals, such as mice and rats, a single animal has only a limited value. Health monitoring of laboratory rodents aims at detecting health problems or defining the pathogen status in a population rather than in an individual. Therefore, systematic laboratory investigations (health surveillance programmes) are necessary to determine the colony status and, most importantly, to prevent influences on experiments. Disease diagnosis differs from routine monitoring in that abnormalities are the subject of testing. This testing is not scheduled, and tests are directed towards identifying those pathogens most likely to cause the lesion.

Routine monitoring programmes will primarily focus on infectious agents. Most infections are subclinical, but can nevertheless modify research results. Therefore, detection of the presence of infectious agents, whether or not they cause clinical disease, is necessary. Monitoring must include animals in the colony and all relevant vectors by which micro-organisms may be introduced into a colony. It may therefore be necessary, particularly in experimental units, that monitoring is not restricted to animals, and that other materials that pose a risk (e.g. biological materials) should be monitored to prevent the introduction of agents into a facility.

Different methodologies are applicable to detect infections in a population. In general, absence of clinical signs has only limited diagnostic value because most infections in rodents are subclinical. While most parasites (ectoparasites, helminths, protozoans) are usually detected by microscopic methods, culture methods are still preferred for detecting most bacteria or fungi. Serology is the most commonly used routine method indirectly to detect viral and also bacterial infections (e.g. *Mycoplasma pulmonis*) in a population by demonstration of antibodies. Serological methods bear the risk of false-positive or false-negative reactions, and unexpected serological results should therefore be confirmed by a second (confirmatory) method, by a second laboratory, or by monitoring additional animals. Molecular methods have become increasingly important during the last few years (Compton and Riley, 2001). They are preferred for the detection of agents that are fastidious or cannot be cultured easily (e.g. *Pneumocytis carinii*, *Helicobacter* sp., viruses causing persisting infections or during the acute phase of an infection). Molecular methods are usually more sensitive than traditional methods. Meanwhile, a very broad spectrum of molecular tests is available for every pathogenic organism of interest (with the exception of prion diseases).

There is always a risk that infectious agents might be introduced into animal facilities, especially into experimental units. This risk has to be taken into consideration when the monitoring programme is designed. More frequent monitoring is reasonable if the risk of introducing unwanted organisms is high (e.g. if animals or biological materials are frequently introduced or if many persons need access to the animals). Simulation experiments have shown that small and frequent samples are more suitable to detect an infection than larger samples taken at less frequent intervals (Kunstyr, 1992).

Although efforts have been made since the 1960s, a universal testing strategy or reporting terminology for clear and consistent definition of pathogen status in rodent population does not exist. The need for health surveillance programmes is generally accepted, but there is a great diversity of opinions about their design. Each institution selects its own list of pathogens, test procedures, animal sampling strategy, frequency of sampling, and reporting terminology, and the terms used vary greatly in precision and meaning (National Research Council 1991; Lindsey, 1998; Jacoby and Homberger, 1999). Usually, an individual programme is tailored to the conditions it is to serve. Most importantly, although the programme is dependent on research objectives, numerous additional factors must be considered, such as physical conditions and layout of the animal house, husbandry methods and sources of animals. Number and quality of personnel as well as finances further influence the programme. It may even be necessary in a multipurpose unit to have a range of different programmes (e.g. one for isolator-housed and one for barrier-housed animals). Some aspects to consider when establishing a health surveillance programme have repeatedly been provided (Nicklas, 1996; Weisbroth et al., 1998). Detailed recommendations for monitoring of breeding or experimental colonies of rodents and rabbits were published by the Federation of European Laboratory Animal Science Associations (FELASA) (Nicklas et al. 2002).

The term most frequently used to describe the microbiological quality is 'specified pathogen free' (SPF), but this term requires explicit definition every time it is used. It means that the absence of individually listed micro-organisms has been demonstrated for a population by regular monitoring of a sufficient number of animals at appropriate ages by appropriate and accepted methods. SPF animals originate from germ-free animals. These are usually associated with a defined microflora and subsequently lose their gnotobiotic status by contact with environmental and human micro-organisms. Such animals are bred and housed under conditions that prevent the introduction of unwanted micro-organisms. SPF animals are morphologically and physiologically 'normal', and well suited for modelling the situation of a human population.

Animals

In general, the animals are the most crucial point in a monitoring programme. Their status has to be defined, and they are the most important source of infection. Proper sampling is therefore necessary to detect an infection in a given population as early as possible. Animals coming from outside have to be checked to assess or exclude the risk of introducing unwanted organisms, and animals already within the unit are monitored to define their status and to obtain information on the presence or absence of infectious agents in the colony. It is obvious that a sufficient number of animals has to be monitored. In general, the number of animals to be monitored is determined by the expected prevalence of an agent in a population. Based on a recommendation by the ILAR Committee on

Long-term Holding of Laboratory Rodents (1976), it has become common practice to monitor at least eight randomly sampled animals. This is (theoretically) sufficient to detect an infection with a 95% probability if at least 30% of a population is infected. Monitoring animals of different age is useful, because younger animals often have a greater parasitic or bacterial burden, whereas older animals (≥3 months) are more suitable to detect viral infections. Clinically ill animals are an import source of information and should be submitted to monitoring in addition to scheduled samples. Selwyn and Shek (1994) and Clifford (2001) have discussed strategies for sampling and calculation of sample sizes.

Sentinels/'control' animals

Random sampling for monitoring is not a serious problem in breeding colonies, but it is usually impossible in experimental units or not reasonable in the case of immunodeficient animals. These may not be able to produce sufficient amounts of antibodies so that their status can best be evaluated by the use of sentinels. It is therefore advisable to have sentinel animals in each experimental unit in order to evaluate the status of a population. Such animals should be kept in such a way that they receive maximum exposure to potential infections. If sentinels are not bred within the colony that is being monitored, they must be obtained from a breeding colony of known microbiological status, i.e. they must be negative for all agents to be monitored. For example when using sentinels to monitor immunodeficient animals, the sentinels must be initially free from *Pneumocystis carinii* or *Staphylococcus aureus*. The sentinel animals must be housed for a sufficiently long time in the population that is to be monitored to develop detectable antibody titres (for serology) or parasitic stages. It is common to house sentinels in a population at least for 4–6 weeks prior to testing, longer periods are even better. In most cases, outbred animals are used as sentinels, because they are cheaper and generally more resistant to clinical disease than are inbred animals. Inbred animals may in specific cases (e.g. for virus isolation) be more valuable as sentinels because they can be more sensitive to an agent and thus more likely to develop overt disease. In other cases, their extreme or even complete resistance to specific agents may be a reason to use specific strains with known characteristics. For example, C57BL/6 or DBA/2 mice are sensitive to clinical infections with mouse hepatitis virus (MHV) whereas A/J mice are resistant to this virus. On the other hand, C57BL/6 mice are resistant to ectromelia virus (Bhatt and Jacoby, 1987). This virus may cause high mortality with typical skin lesions in C3H, and high mortality but minimal skin lesions in CBA and DBA/2 mice. Use of immunodeficient animals, such as thymus-aplastic nude mice, as sentinels may increase the sensitivity, if specific bacterial pathogens such as *Pasteurella pneumotropica*, parasites (e.g. *Spironucleus muris*) or viruses are to be detected in a population. In the past, injection of cortisone or other immunosuppressive drugs (e.g. cyclophosphamide) to suppress the immune system was recommended. This results in overgrowth and easier direct detectability of bacterial pathogens. However, such tests have lost importance as direct demonstration of micro-organisms can now be performed more easily by means of molecular methods such as polymerase chain reaction (PCR).

A multitude of physiological characteristics can be influenced by introducing a transgene into the genome or by gene targeting. Changes of the immune status frequently arise, resulting in immune defects or immunosuppression. As a consequence, there may not only be altered sensitivity to pathogenic agents but also suppression or lack of antibody response. When monitoring an immunodeficient colony, to avoid false-negative results in serological tests, animals whose immune-responsiveness is well known (e.g. old vasectomized males, retired breeders) are to be used as sentinels in order to obtain reliable results. With respect to barrier protected facilities it is advisable to have sentinel animals in each animal room. The animals should be housed in various locations on the bottom shelves without filter tops. Each time the cages are changed, soiled bedding from different cages should be transferred to the sentinel cages.

During the last decade additional housing systems such as filter-top (microisolator) cages, individually ventilated cages or filter cabinets emerged. They offer the advantage of separating small populations from each other and are frequently used for housing immunodeficient, immunosuppressed, or infected animals. If handled properly, they prevent transmission of infectious agents very efficiently. Each isolator or microisolator cage must therefore be considered a self-contained micro-biological entity. Health monitoring under such housing conditions as well as monitoring of isolator-housed animals can be conducted properly only by the use of sentinels. Due to limited space, less than the recommended numbers of animals are available in many cases, which is acceptable if sentinels are properly housed. In the case of isolators, a realistic number of sentinel animals is housed in one or several cages (depending on the isolator size) on soiled bedding taken from as many cages as possible. In most cases, only three to five animals per isolator will be available for monitoring.

If animals are housed in micro-isolators or in individually ventilated cages, sentinels must be housed in filter top cages like other animals. When cages are changed in changing cabinets, soiled bedding from several cages is transferred into a specific cage that is then used to house sentinels. Weekly changes of donor cages will give a representative insight into the microbiological status of the whole population.

Frequency of monitoring

The frequency of monitoring will depend on various factors, but mainly on the importance of a pathogen, on the use of the population, and on the level of risk of infection for the population. Naturally, economic considerations do play a role as well. It is stated in the FELASA recommendation (Nicklas *et al.*, 2002) that monitoring be conducted at least quarterly. Most commercial breeders of laboratory rodents monitor more frequently (every 4–6 weeks). In most multipurpose units housing of immunodeficient or infected animals, a more frequent monitoring is also preferable as this will result in earlier detection of an infection. As a general rule, it is advisable to monitor a small number (e.g. three to five animals) from each unit every 4–6 weeks instead of 10 animals every 3 months. Under practical conditions, not every animal may be monitored for all micro-organisms. Depending on the factors already mentioned, the frequency of testing may be different for different agents. Monitoring for more frequently occurring organisms or for zoonotic or otherwise important agents will be performed more often (monthly), whereas testing for unusual organisms like K-virus or polyoma virus can be done less frequently (e.g. annually). Results obtained from monitoring sentinels are valid for all animals of the same species within a population, irrespective of the experiment or animal strain. Independent from animals that are scheduled for monitoring, all animals with clinical disease should be submitted for direct examination for micro-organisms (bacteria, parasites, viruses) and for histopathology.

Agents

A decision has to be made in each facility which organisms are acceptable or unacceptable. Lists of infectious agents to be monitored in routine programmes have been published by various organizations (Kunstyr, 1988a; National Research Council, 1991; Waggie et al., 1994; Nicklas et al., 2002) and can be used for guidance. Monitoring for all the agents mentioned (mycoplasma, bacteria, bartonellas, fungi, spirochaetes, protozoans, helminths, arthropods) on a routine base is neither realistic nor necessary. The most important micro-organisms are those that are indigenous and pose a threat to research or to the health of animals and humans and, in addition, those which can be eliminated. Therefore, oncogenic retroviruses are excluded as they integrate into the mammalian genome, and thus cannot be eradicated by the presently available methods. Other micro-organisms may be less important as they are unlikely to occur in good quality rodents due to repeated re-derivation procedures (e.g. *Brucella, Erysipelothrix, Leptospira, Yersinia*). Most cestodes are unlikely to be found, since they require an intermediate host. In the case of immunocompromised animals or in infectious experiments, however, monitoring for a comprehensive list of micro-organisms is reasonable. Various micro-organisms that usually do not cause clinical signs in immunocompetent animals (e.g. *Staphylococcus aureus, Pseudomonas aeruginosa, Pneumocystis carinii*) may cause serious problems in immunodeficient animals. Even agents such as *Burkholderia gladioli* may cause clinical signs in severely immunodeficient animals (Dagnaes-Hansen et al., 1991). It is thus necessary to monitor immunodeficient animals not only for strong or weakly pathogenic organisms, but also for opportunistic pathogens or commensals. Micro-organisms with low pathogenic potential can cause clinical signs of diseases if animals are infected with several agents (e.g. KRV and *Pasteurella pneumotropica* (Carthew and

191

Gannon, 1981)). In other cases, different micro-organisms of low clinical importance may interact and have a severe impact on research results, such as oncogenic viral expression (Riley, 1966).

Each institution should prepare a list of those organisms that are not acceptable in the colony or in parts of it. This list is easiest to establish for viruses. A large amount of information is available on their pathogenic potential and on their ability to compromise the object of research. Monitoring for viruses can be carried out selectively by serological methods. Only few exceptions exist, such as parvo-viruses, which cross-react in indirect immunofluorescence or ELISA tests (Jacoby et al., 1996) and at present cannot always be identified unequivocally. Assays using recombinant antigens that have an increased specificity are under development. For some viruses (e.g. K virus, polyoma virus) the only question is whether or not monitoring is necessary because they have been eradicated from the vast majority of rodent colonies many years ago. Only few new rodent viruses have been detected during the last few years, e.g. mouse parvovirus (MPV) and rat parvovirus (RPV) (Ball-Goodrich and Johnson, 1994; McKisic et al., 1995; Jacoby et al., 1996; Ball-Goodrich et al., 1998). One can expect that new rodent viruses will be isolated, although only occasionally.

Less is known about the ability of most parasites to influence research results. They are considered a hygiene problem and are therefore eradicated from rodent colonies. Some protozoans like trichomonads are occasionally detected in pathogen-free animals from commercial breeders. They are considered to be apathogenic, and nothing is known about their influence on the physiology of animals. They are, however, likely to be species-specific and thus might be an indicator of a leak in the system, or of the existence of direct or indirect contact to wild rodents. The most complex problems exist for bacteria. In contrast to viruses their importance for laboratory animals is usually estimated on the basis of their ability to cause pathological changes or clinical disease, since almost nothing is known about most rodent bacterial species with regard to their potential to cause other effects on their hosts and on experiments. Insufficient information exists on the taxonomy and proper identification for various rodent-specific bacterial species such as *Pasteurella pneumotropica* or other members of the Pasteurellaceae (e.g. *Haemophilus influenzaemurium, Actinobacillus muris*). Lack of detailed information on the characteristics of these organisms, together with the presently unclear taxonomic situation, often leads to misidentification, and the lack of knowledge about species-specificity impedes their elimination. The FELASA working group on animal health (Nicklas et al., 2002) therefore decided to recommend that rodents should be monitored for all Pasteurellaceae. There is, however, evidence that some growth factor-dependent Pasteurellaceae found in rodents are closely related to *Haemophilus parainfluenzae* and might therefore be transmitted by humans (Nicklas et al., 1993b). It is unclear if these bacteria can be eradicated permanently from barrier units, because exposure of barrier-produced animals to humans represents a permanent risk for re-infection. The same holds true for several members of the Enterobacteriaceae *(E. coli, Klebsiella,*

Proteus), *Staphylococcus aureus* and *Pseudomonas aeruginosa* for which humans serve as a reservoir. Another problem arises from the fact that many bacteria are presently being reclassified, resulting in changes of their names. For example, the mouse specific organism known as 'Citrobacter freundii 4280' has been reclassified as *Citrobacter rodentium* (Schauer *et al.*, 1995). Whole genera have been renamed, and additional bacterial species have been detected, such as *Helicobacter hepaticus*, *H. muridarum, H. bilis,* and *H. typhlonicus* (Lee *et al.*, 1992; Fox *et al.*, 1994, 1995; Franklin *et al.*, 1999). Some of these fastidious organisms are not detected or not properly identified by all monitoring laboratories. Adding such known pathogens to a list for which animals should be monitored may be unrealistic as long as proper methods for their detection and identification are not readily available in a monitoring laboratory.

A list of pathogens should contain all indigenous micro-organisms for which rodents are the infectious reservoirs and other micro-organisms that might be of importance for the research conducted with such animals. The list of these additional organisms may be long in the case of immunodeficient animals. The whole spectrum of micro-organisms as a concept is not a permanent list for all times; it rather represents a moving boundary in which old pathogens are eradicated and new pathogens are added. In practice, such lists of agents do not differ much between different facilities or commercial breeders. Monitoring for micro-organisms is usually done by commercial laboratories, and is thus determined by their capabilities (some of the larger research institutes and commercial breeders have dedicated diagnostic laboratories). It is important that all investigations should be performed in laboratories with sufficient expertise in microbiology or pathology of the relevant species. Serological tests also require technical competence to ensure sufficient standardization of tests (including controls) and accurate interpretation of results.

Testing of animals usually starts with necropsy and blood sampling for serology, followed by microscopic examination for parasites and sampling of organs for bacteriology, pathology, and, in rare cases, virological examinations. For financial reasons, bacterial culture is often restricted to very few organs. Monitoring more organs would, however, increase the probability to detect bacterial pathogens in an animal. Bacterial cultures should be made for the respiratory tract (nasal cavity, trachea, lungs), the intestinal tract (small and large intestine) and urogenital tract (vagina respectively prepuce, uterus, kidney). In the case of pathological changes, additional organs (liver, spleen, mammary gland, lymph nodes, conjunctiva, etc.) should be cultured.

Serology is easy and cheap to perform, and serum samples can be mailed easily. Whole body examinations including bacteriology and parasitology are more expensive, and living animals must be shipped to the monitoring laboratory. Therefore, many laboratories monitor only serologically. Although serological methods exist to detect some bacterial infections, these are not generally accepted and only few laboratories apply these methods. At present, the method of choice for the detection of most bacterial pathogens is bacterial culture, and thus should be part of each monitoring programme. During recent years PCR has been increasingly used for the detection of slowly growing or fastidious bacteria such as *Helicobacter* (Riley *et al.*, 1996; Mähler *et al.*, 1998) or fungi such as *Pneumocystis carinii* (Rabodonirina *et al.*, 1997; Weisbroth *et al.*, 1999). Meanwhile methods are available for almost all agents (Feldman, 2001).

Sources of infection

Keeping rodents free of pathogens in research facilities is a much more complex problem than in breeding colonies. Animals and various experimental materials need to be introduced into experimental facilities. In addition, more personnel must have access to animals due to the requirements of the experiments. This results in a higher risk of introducing pathogens.

The design of modern laboratory animal buildings is mainly based on microbiological concepts aimed at the prevention of infections. These measures are responsible for a high percentage of expenses arising from planning and constructing an animal house. Furthermore, high running costs are taken into account for energy, hygienic precautions, and personnel to avoid infections during operation.

In addition to constructive measures, an appropriate management system is necessary for the prevention of infections, as well as for their detection and control. It is a major task for the management of an animal facility to understand how micro-organisms might be introduced, or spread under the specific conditions given. Management of all animal facilities in an institution is best centralized. This warrants that all information dealing with the purchase of animals, use of experimental materials and equipment, as well as the performance of animal experiments, flows through one office. This reduces the opportunity for failures of communication. Centralized management can best establish comprehensive monitoring programmes to evaluate important risk factors such as animals and biological materials before they are introduced into a facility. Contamination of animals can happen in two ways. One has to distinguish between the introduction of micro-organisms coming from outside and transmission of micro-organisms within a colony. This can be influenced by the management and the housing system.

Animals

The greatest risk of contamination to any animal arises from another animal of the same species. Most facilities are multi-purpose and must therefore house a variety of strains coming from different breeding units. In addition, many specific strains or genetically modified animals are available only from research institutes. Animals are therefore the most important risk factor, even though their quality has constantly been improved during the last few decades. The importance of animals as sources of infections becomes obvious from a survey conducted in the USA in 1996. In this survey of 72 of the top 100 institutional recipients of NIH funds, MHV was 'on campus' in nearly 60% of the reporting institutions. Pinworms were reported to be present in >30% and parvo viruses in >25% of 'SPF' colonies (Jacoby and Lindsey, 1997).

As a general rule, all animals coming from sources of unknown microbiological status should be regarded as infected unless their status has been defined. This is especially important when genetically modified animals are introduced from other experimental colonies. These animals must be housed separately from others. The risk of introducing pathogens via animals from external sources is lower when animals are available from a few sources of well-known microbiological status, and if these animals have been protected from contamination during shipment. Direct transfer of such animals without quarantine into an experimental unit can be necessary; however, spot checks should be performed from time to time to redefine the status upon arrival. In many cases it is acceptable to introduce animals from microbiologically well-known (external) colonies into experimental units, but never into a breeding unit, especially if many different strains and/or transgenic lines are co-maintained. In the latter case new breeders should only be introduced via embryo transfer or caesarean section. Outbred mice are commonly used as surrogate and foster dams and can easily be bred in the transgenic unit, as is the case for the sterile males required to induce pseudocyesis in the surrogate dams.

It must be emphasized that a specific risk of transmitting microorganisms may arise from immunodeficient animals. Many virus infections (MHV, RCV/SDA, Sendai, PVM) are limited in immunocompetent animals and virus may be eliminated completely. Immunodeficient animals may, however, shed infectious virus for longer periods of time, or may be infected persistently (Barthold *et al.*, 1985; Weir *et al.*, 1990; Compton *et al.*, 1993; Gaertner *et al.*, 1995; Rehg *et al.*, 2001). Animals known to be infected must always be housed in isolation. This can best be done in flexible film isolators or, if proper handling is guaranteed, in microisolator cages or in individually ventilated cage racks.

The principles that are important for designing a quarantine programme have been thoroughly discussed by Rehg and Toth (1998).

Biological materials

Biological materials represent a high risk if they originate from or have been propagated in animals. In particular, tumours, viruses or parasites that are serially passaged in animals often pick up pathogens, and a high percentage of these thereby become contaminated. Such materials can be stored frozen without loss of infectivity and may be hazardous for humans or for laboratory animals even after decades. Immunodeficient animals (e.g. nude mice or rats, *Prkdc^scid* mice) are often used in xenotransplantation studies and are at risk to infections transmitted via transplanted tissue.

The problem of viral contamination in biological materials became obvious from studies by Collins and Parker (1972). They monitored 475 murine leukaemias and tumours and found viral contamination in 69% of the samples. The same percentage of contaminated mouse tumour samples after animal passages was found by Nicklas *et al.* (1993a). Many organisms disappear under *in vitro* conditions so that the contamination rate after these passages is lower. Among the contaminants, lymphocytic choriomeningitis virus (LCMV) (Bhatt *et al.*, 1986b) and hantaviruses (Yamanishi *et al.*, 1983) have repeatedly been found, and outbreaks in humans associated with infected animals or with contaminated tumour material have been reported (Kawamata *et al.*, 1987).

Pathogenic micro-organisms can be transmitted by other contaminated tissues or body fluids such as monoclonal antibodies (Nicklas *et al.*, 1988) or viruses (Smith *et al.*, 1983). Two recent outbreaks of ectromelia in the USA both resulted from use of contaminated serum samples (Dick *et al.*, 1996; Lipman *et al.*, 2000). In colonies of genetically modified animals, ES-cells, sperm and embryos should be considered as potential sources of infection. ES cells in particular are at increased risk of infection because they require growth factors that are usually supplemented by co-culture with primary mouse cells (Hogan *et al.*, 1994). Contamination of biological materials is not restricted to viruses. *Mycoplasma pulmonis* and other bacterial pathogens like *Pasteurella pneumotropica* or *Coxiella burnetti* (Criley *et al.*, 2001) have been found as contaminants. Additional pathogens (*Eperythrozoon* sp., *Haemobartonella* sp., *Encephalitozoon* sp.) can contaminate biological materials after animal-to-animal passages (Petri, 1966; National Research Council, 1991) and thus may be transmitted to recipient animals.

Biological materials have traditionally been tested for contaminating agents using the mouse or rat antibody test (MAP or RAP test). Meanwhile, PCR tests have been established to replace the MAP test as the preferred test for detecting viral contaminants in biological materials (Compton and Riley, 2001).

Humans

Humans can act as mechanical or biological carriers of micro-organisms. Humans are unlikely to be an appropriate host where murine pathogens can reside and replicate. However, the importance of humans as mechanical vectors should not be underestimated, and several human pathogens can cause infections in rodents, at least in immunodeficient animals. It has to be assumed that each micro-organism which is present in humans who have access to a barrier unit might colonize the animals sooner or later. Transmission certainly cannot be avoided in barrier-maintained colonies, even by wearing gloves and surgical masks and taking other precautions. It may only be avoided by establishing strict barriers as provided by isolator maintenance. Immunodeficient animals, at least animals used for breeding or in long-term experiments, that are known to have an increased sensitivity to infection with bacteria of human origin (*Staphylococcus aureus*, *Klebsiella pneumoniae*, *Escherichia coli*, etc.)

should preferably be housed in isolators or microisolators (respectively individually ventilated cages).

Little published information is available on the role of humans as mechanical vectors. There is no doubt that micro-organisms can be transmitted by handling (La Regina *et al.*, 1992). Micro-organisms can even be transported from pets to laboratory animals by human vectors (Tietjen, 1992). Such examples emphasize the need for proper hygienic measures and the importance of positive motivation of staff. It is an important task of the management of an animal facility to ensure that personnel coming into contact with animals have no contact with animals of lower microbiological quality.

Vermin

Vermin are another potential source of infections. Flying insects do not present a serious problem because they can easily be removed from the incoming air by filters or by insect-electrocuting devices. Crawling insects such as cockroaches are more difficult to control, and cannot definitely be excluded. The most serious problem arises from wild rodents, which are frequently carriers of infections. Wild as well as escaped rodents are attracted by animal diets, bedding and waste. Modern animal houses usually have devices, which normally prevent entry of vermin.

Possible routes of infection of laboratory animals have been discussed in more detail by Nicklas (1993).

Present status of laboratory rodents

Since serological testing was introduced in the 1970s, many laboratories have evaluated the viral status of murine colonies. Managing directors of animal facilities had to learn techniques for the prevention, control and eradication of infection, and the means of adapting the facilities for their own purposes. As a consequence, the diversity of viruses and the frequency with which they are detected has declined markedly. Virus infections have almost entirely been eradicated from most commercial breeding colonies. This gave animal care unit administrators and researchers the opportunity to procure and maintain virus-free stocks, and researchers to use better quality animals for research. However, this progress of eradication has not happened without periodic shut-downs at breeders' and users' facilities.

Reports on the prevalence of virus infections in rodents throughout the world have been published frequently. An overview given by the National Research Council (1991) indicates that the majority of colonies at that time was infected with three or four viruses. However, most facilities still house at least small numbers of animals that are infected or have an unknown status. Many small or decentralized facilities do not even monitor at all. In a retrospective study among French facilities the prevalence rate of some agents decreased, but some were still found to harbour viruses during the last years of the study (Zenner and Regnault,

2000). In a survey conducted in the USA among major biomedical research institutions MHV and mites were reported to be present in more than 10% of SPF mouse colonies, and more than 25% were reported to be positive for pinworms and parvoviruses. As expected, the prevalence of infections was even higher among non-SPF mice (Table 1). More than one-half of the replying institutions had endoparasites and MHV 'on the campus'. More than 10% were positive for at least seven other viral and bacterial infections. Surprisingly, serological evidence of ectromelia virus and lymphocytic meningitis virus (LCMV) were also reported (Jacoby and Lindsey, 1997, 1998). A similar profile to that of mice was revealed for rats (Table 2).

Table 1. Prevalence of infections in colonies of laboratory mice (Jacoby and Lindsey 1997, 1998)

Agent	Percentage of positive 'SPF' colonies	Percentage of positive 'non-SPF' colonies	Agent 'on campus' % colonies positive
Mycoplasma	3	17	12
Helicobacter	13	9	18
Mites	17	38	36
Pinworms	33	68	63
Adenovirus	2	8	8
Sendai	0	21	15
PVM	3	22	18
Reo	3	19	15
TMEV	4	35	29
Rota	6	28	24
MHV	12	74	57
Parvo	27	40	46

Table 2. Prevalence of infections in colonies of laboratory rats (Jacoby and Lindsey 1997, 1998)

Agent	Percentage of positive 'SPF' colonies	Percentage of positive 'non-SPF' colonies	Agent 'on campus' % colonies positive
CAR bacillus	6	22	19
Mycoplasma	0	36	26
Mites	4	12	12
Pinworms	33	68	53
Sendai	2	23	18
PVM	9	28	22
Reo	3	8	8
TMEV	8	18	15
Rota	2	7	4
Corona	9	38	32
Parvo	27	33	33

Various agents are still prevalent at a low level. They can emerge unexpectedly as seen a few years ago when a sudden outbreak of ectromelia was observed in the USA (Dick *et al.*, 1996). The situation is very similar for bacterial pathogens and parasites. Most of them were eradicated when principles of gnotobiology had been introduced into laboratory animal science. A few parasites (pinworms, mites, protozoans) are still endemic in various rodent colonies. Today, most of the primary bacterial pathogens (Salmonellae, *Corynebacterium kutscheri*, *Leptospira*, *Streptobacillus moniliformis*) are no longer detected in well-run facilities although they may re-emerge as shown recently (Wullenweber *et al.*, 1990; Koopman *et al.*, 1991). *Clostridium piliforme*, which is the causative agent of Tyzzer's disease, and *Mycoplasma pulmonis* are detected more often. Most experimental and some commercial breeders' colonies are positive for Pasteurellaceae like *Pasteurella pneumotropica* and *Actinobacillus muris*. The real prevalence of organisms belonging to this family is not definitely known due to difficulties in identification. The situation is also unclear for *Helicobacter* species because many colonies are not sufficiently monitored. It is to be expected that these micro-organisms are also widespread in laboratory rodents. It is, therefore, extremely important that germ-free, or gnotobiotic animals, rather than SPF animals, are used for hygienic rederivation to avoid this problem in future.

A number of additional disease agents, for example group B and G streptococci, *Staphylococcus aureus*, *Haemophilus parainfluenzae*, *Corynebacterium* spp. (inducing scaly skin disease), and others have been found in so-called pathogen-free rodents during the last few years. Rodents seem not to be the primary hosts for these organisms, which are more likely to be transmitted by humans. These infections have been named 'post-indigenous diseases' (Weisbroth, 1996). While agents such as '*Corynebacterium bovis*' might be tolerable in immunocompetent animals, clinical disease caused by these bacteria may be seen in immunodeficient animals (Clifford *et al.*, 1995; Scanziani *et al.*, 1997, 1998).

The presence of infectious agents, even if they are of low pathogenicity, may become a problem if animals from different sources are co-maintained. This occurs often, as transgenic animals are frequently exchanged between scientists from an almost unlimited number of sources. This is associated with a high risk of introducing different pathogens and thus of causing multiple infections. At present, infections that were common decades ago are re-emerging.

◆◆◆◆◆◆◆ IMMUNOCOMPROMISED ANIMALS

Natural variants (mutations, infectious agents)

Naturally occurring immunodeficient mouse strains express a variety of genetic defects in myeloid and/or lymphoid cell development. These strains have served as and are still valuable models for studying immune cell differentiation, mechanisms of transplant rejection, etc. Some of the most commonly used mutants are nude (*Foxn1^{nu}*), severe combined immunodeficiency (*Prkdc^{scid}*), beige (*Lyst^{bg}*) and X-linked immunodeficiency (*Btk^{xid}*). Sufficient information on the different variants produced by nature can be obtained from an ILAR Committee report (1989), Hedrich and Reetz (1990), Lyon *et al.* (1996), or more specifically by searching for defined mutations in databases, such as Mouse Genome Database (*http://www.informatics.jax.org*) and RATMAP (*http://ratmap.gen.gu.se*). Aside from their immunodeficient status, i.e. their inability to eliminate or neutralize foreign substances, some of the mutants also inherit a failure to discriminate between self and non-self.

In addition to the action of defined genes on the immune function there are several inbred strains or F1 hybrids harbouring genes that confer susceptibility or resistance to infectious or other immune system-related diseases. As an example, while C57BL/6 and related strains succumb to an infection with *Streptobacillus moniliformis*, AKR, BALB/c, DBA/2, and other mice survive, while BALB/c never show any sign of disturbance, nor even produce antibodies against this organism (Wullenweber *et al.*, 1990).

Induced immunodeficiencies

Aside from these immunodeficiencies there are other means to modulate the immune status of mice and rats. These induced deficiencies such as thymectomy, lethal or sublethal irradiation, depletion of various subpopulations of immune cells by antibodies require exactly the same management procedures as for any immunodeficient animal.

Genetic manipulation

The advent of transgenic rodent technology by transferring and over-expressing foreign genes under the control of specific vectors as well as directed mutagenesis by silencing specific genes has opened up new avenues for the study of innumerable factors affecting the immune system. One may search for these either by consulting literature databases, the Mouse Genome Database (MGD; *http://www.jax.org/resources*; check 'Induced Mutant Resources'), or the transgenic animal database (TBASE, *http://tbase.jax.org/*; mouse knock out and mutation database, *http://research.bmn.com/mkmd*). Again, as indicated above, identical phenotypes must not necessarily indicate identity of the genes. While in many respects phenotypically alike, the genetic factors controlling the expression of *Prkdc^{scid}* and *Rag1^{tm}*, respectively *Rag2^{tm}* deficient mice have been shown to be different. On the other hand, silencing of exon 3 of the *Whn*-gene has produced exactly the same phenotype as in *Foxn1^{nu}*-mice providing evidence that the fork head transcription factor is responsible for the nude and athymic phenotype (Nehls *et al.*, 1996).

PCR protocols by which mice carrying an induced mutation can be distinguished from normal wild type mice have been published in the respective descriptions. Some are available on the World Wide Web, e.g.

those maintained at the Jackson Laboratory (*http://www.jax.org/resources/documents/imr/protocols/index.html*; or through an e-mail inquiry to micetech@aretha.jax.org).

It should be noted that transgenic animals can only be maintained at or be supplied to laboratories, which comply with the national requirements of the respective host country for the use of genetically modified animals.

◆◆◆◆◆◆ MANAGEMENT OF COLONIES

Housing systems

The initial descriptions of housing systems for small rodents have not lost their principal validity (Spiegel, 1976; Otis and Foster, 1983; ILAR Committee, 1989; Heine, 1998) although many refinements have since then been introduced. On the basis of scientific demands, international standards and the risk for the own animal facility, the decision for the adequate hygienic status has to be made. In principle, this could be:

- *Germ-free*, designating a status in which no micro-organisms are present except those integrated into the genome.
- *Gnotobiotic*, in which the animals have a well-defined and specified flora, consisting mainly of anaerobic bacteria supporting the metabolism and possibly fertility. In addition, the gnotobiotic flora may induce some resistance to ubiquitous micro-organisms:
  SPF: specified pathogen-free, which describes by definition animals being free of pathogens to be specified.
- *Quarantine*, which was originally used for overcoming the potential latency period of infections, in this context, however, includes a different but not an acceptable SPF status, in particular for newly introduced strains from other institutions.
- *Infectious*, denominates the status of animals either infected naturally or artificially which could transmit pathogens not found in the animal's own colony or being a risk for humans or other species.

Isolators

Static micro-isolators are partially perforated boxes with a tight filter medium, covering the perforation and a cover, which has to be sealed (Kraft, 1958). They provide a micro-environment which is protected from adventitious contamination from outside. Micro-isolators are still in use (e.g. Han-Gnotocage[1]) for the transport of germ-free, gnotobiotic and SPF-founder animals. These can also be used for the short-term housing of germ-free fosters in a laminar flow cabinet. The disadvantage of these micro-isolators, however, is the impeded intracage ventilation. Due to an increase in humidity, ammonia and carbon dioxide concentrations and

1 HAN-Gnotocage, Firma EBECO, Hermannstrasse 2-8, D-44579 Castrop-Rauxel, Germany.

Management of Immuno-compromised Animals

with increasing animal density this intracage ventilation becomes intolerable.

Ventilated isolators in the *positive pressure* version are indispensable for breeding and maintaining germ-free and gnotobiotic animals. They consist of a closed construction with a HEPA-filter unit for air supply, a valve or a filter in the exhaust, long-arm gloves and a chemically steriliz-able lock for interconnection to the supply chamber (Trexler, 1983). For chemical sterilization of the isolator and the lock, freshly diluted peracetic acid[2], alkaline buffered peracetic acid[2] or hydrogen peroxide gazing should be used. In the case of pinworm contamination, an additional treatment cycle with an effective disinfectant (e.g. Chlorocresol[3]) should be carried out. Most of the required materials can be autoclaved in the supply chamber with control of the heating process by indicators (paper[4] and maxima thermometer) and in the retrograde by bioindicators (*Bac. stearothermophilus, Bac. subtilis* spore strips[5]). If dietary problems arise as a consequence of the food sterilization, gamma-irradiated diet (50 kGγ) can be used after chemical sterilization of the outside of the vacuum bags.

Germ-free animals are used for special experiments, for example to test the influence of the gut flora (Hirayama, 1999), the effect of lipopoly-saccharides on the immune system (Enss *et al.*, 1997) and as a back-up for foster mothers, which may be necessary to renew the gnotobiotic status. In this context, it has to be stressed that in germ-free mouse and rat strains very often a mega-caecum is found (Wostmann *et al.*, 1973), productivity can be decreased or lost and, when exposed to the outside environment, these animals may fall ill and die.

Gnotobiotic animals are normally derived from germ-free animals by oral application of a gnotobiotic flora[6] consisting mainly of anaerobic, well-defined micro-organisms (Dewhirst *et al.*, 1999). The reproductivity in this hygienic status may be restored and the resistance to outside environment may be improved in comparison to germ-free animals (Heidt *et al.*, 1990; van den Broek *et al.*, 1992). Therefore, this status can be recommended for immunocompromised strains – at least for their breed-ing stocks – and for foster mothers used for embryo transfer. A gnotobi-otic status can only be preserved in isolators, a fact that may restrict the expansion of colonies. This limitation may be circumvented by the use of individually ventilated cage systems, although contamination with other micro-organisms cannot be excluded totally.

The equipment of most commercially available isolators allows the alternative use in positive or negative pressure. Isolators in the *negative pressure version* protect in the first line the environment from infections inside the isolator by an HEPA-filter in the exhaust. In combination with

---

2 Peracetic acid: Kesla Pharma Wolfen GmbH, Thiuramstrasse 2, D-06803 Greppin, Germany.
3 Chlorocresol (Neopredisan ®): Menno-Chemie Vertrieb GmbH, D-22850 Norderstedt, Germany.
4 Indicator-paper: BAG Biologische Analysensysteme GmbH, D-35419 Lich, Germany.
5 Bio indicators: Werner, MBS, Untere Jasminstaffel 3, D-88069 Tettnang, Germany, Apex Laboratories, P.O. Box 794, NC 27502-0794, USA.
6 Gnotobiotic flora: Taconic: 273 Hover Avenue, Germantown, NY 12526, USA.

the use of a waste chamber, their use is obligatory for infectious experiments with high-risk pathogens and may be recommended for stocks of risk in quarantine if no appropriate barrier system is available.

Individually ventilated cages (IVCs)

By direct ventilation of individual cages with HEPA-filtered air, the presumptions for long-term bio-containment on the cage level can be accomplished. In addition, ventilation with *positive* pressure in the cages counteracts the leakiness of the system. *Negative* cage-pressure may prevent the escape of micro-organisms and allergens from the IVCs. The different aspects of ventilated cage systems are described in an overview by Lipman, 1999 and special topics are presented by Clough *et al.*, 1994, Perkins and Lipman, 1996, Hasegawa *et al.*, 1997, Tu *et al.*, 1997, Chaguri *et al.*, 2001, Gordon *et al.*, 2001, Höglund *et al.*, 2001, Reeb-Whitaker *et al.*, 2001 and Renström *et al.*, 2001.

Different versions are commercially available, in which the air is blown either directly into the cage or is passed through a wide mesh filter. In the latter, the intracage air velocity is lower but the desiccation of the bedding is reduced. In many systems the exhaust air passes a filter in the cage to retain dust from the exhaust pipes. In most of the different versions the intra-cage pressure can be adjusted to be *positive* or *negative*, respectively, allowing the use in different hygienic pretensions.

Handling of IVCs

This is the most critical and most underestimated procedure of running IVCs. Three different hygienic levels have to be considered:

1. The sterility level of the autoclaved material: *cage* with bedding, cover and lid as a whole, *diet* (or gamma-irradiated and outside sterilized) and *water* bottles, sterile transferred into the laminar flow changing station (in case of inside bottles).
2. The outside environment of the cage, e.g. the animal room.
3. The inside space of the cage containing the animals.

In a correct manipulation, these three levels have to be strictly discriminated. Several regimens for one (see Box) or two persons may be used.

The procedure of sterile handling of IVCs is labour intensive, however, it can be compensated at least in part by extending the cage change interval, due to the higher intracage ventilation. In addition, increasing the change interval reduces stress for the animals (Duke *et al.*, 2001; Reeb-Whitaker *et al.*, 2001).

IVCs can be used to breed and maintain animals within an SPF-unit to reduce the risk of contamination to the cage level at least theoretically and to improve the environmental conditions for the animals, which is of special interest in immunocompromised rodents. In addition, the personnel are fairly well protected from allergens and from smell if the exhaust of the IVCs is connected to the air outlet of the room. Problems may arise,

> ### Handling of IVCs
>
> 1. Laminar flow bench is running (30 min in advance).
> 2. Fast-acting sterilization compound (Recommendation: Clidox[1]) is freshly diluted for gloves and bench table and (in a separate vessel) for the forceps that can be used.
> 3. A filter top cage, fully equipped and autoclaved, is placed on the bench.
> 4. Sterile diet (kept in a filter-top cage) is filled in using a sterile ladle with a long handle.
> 5. Sterile water bottle is inserted using sterile pincers – in case of inside bottle location – or after disinfection of nipple and grummet – in case of outside bottle systems.
> 6. The cage to be changed is placed into the bench.
> 7. After removing the hoods of both cages, animals are transferred into the new cage, using sterile forceps.
> 8. Lids and filter tops are replaced and both cages removed.
> 9. The work-place and the gloves are disinfected after each working phase with Clidox.
> 10. In case of infectious animals, a biohazard, class II laminar flow bench has to be used and contaminated cages have to autoclaved with filter tops in place.

however, in health monitoring which has to be aligned to the cage level. IVCs are also particularly useful when the prerequisites for an *SPF-unit*, except an autoclave, are lacking and in *experimental areas*, where easy access to the animals by the scientists is indispensable. In addition, IVCs can be of help to preserve the hygienic status of the individual colonies. Barrier closed *quarantine* IVCs in *positive* pressure are ideal for containment of animals from different sources in their respective environment. For quarantine without additional barrier system and *infectious experiments* IVCs are run with *negative* pressure. In addition, *sealed* IVCs were developed, however, it should be kept in mind that sealing may not be absolutely tight and therefore hazardous experiments should be performed in isolators.

SPF-unit

By definition, animals are free of specified pathogens. However, no declaration on residual micro-organisms is given, implying the probability of extensive differences from one SPF-unit to another (Heine, 1980; O'Rourke *et al.*, 1988; Boot *et al.*, 1996). Therefore, it must be considered that when transfering animals from one SPF-unit into another, additional micro-organisms can be introduced which may disturb the microbiological equilibrium, particularly in immunocompromised animals (Ohsugi *et al.*, 1996).

1 Clidox: Outside Europe: Pharmacel Inc., Naugatuck, Connecticut 06770, USA
           Within Europe: Fi. Tecniplast, Gazzada 21020, Buguggiate (VA), Italy
           Dilution: 1 part basic component, 5 parts water and 1 part activator
           Ready for use after 15 minutes.

An SPF-unit is protected by a strict hygienic barrier system of air supply, materials, food, bedding and personnel (Otis and Foster, 1983; ILAR Committee, 1989; Heine, 1998). A conventional open caging system or IVCs may be used within the SPF-unit. After disinfection with formaldehyde or hydrogen peroxide (Krause *et al.*, 2001), gnotobiotic or SPF animals can be introduced via a chemical lock by external disinfection after covering the filter of the micro-isolator cage with a foil. *Standardized diet* is introduced by autoclaving whereby the diet has to be *fortified*, i.e. heat-sensitive vitamins are added in excess ensuring that sufficient amounts remain after heat treatment. The hardness after autoclaving must be controlled regularly. As an alternative, gamma-irradiated food (25 KGy) in vacuum bags can be introduced into the SPF-unit after external disinfection. The *drinking water* should be sterilized by heat, filtration or UV light and conserved by acidification (e.g. hydrochloric acid or acidic acid) to a pH of 3.0–2.5 or chlorination. For the latter, the pH should be adjusted to around 5 before adding stabilized hypochloride to reach 6–8 ppm of free chlorine (Leblanc, 2002). If problems arise from solubility of drugs to be added to the drinking water autoclaved tape water should be used. *Bedding* should be dust-free (<1% dust) and must be autoclaved with two or three vacuum cycles in advance. Pregnant females should be provided with nesting material such as autoclaved cellulose towels or nestlets (Van de Weerd *et al.*, 1997). The highest risk for the unit, however, is the *personnel* entering the barrier. Only a minimal number of well-trained caretakers (FELASA, 1995), having had no contact to external rodents for 4–7 days and being free of infections, should be allowed to enter the SPF-area.

The *microbiological status* of the SPF-area should be monitored regularly, sick animals removed from the unit and submitted to micro-biological examination/necropsy and sentinels should be checked at fixed intervals (see Frequency of monitoring, p. 190). The regular disinfection of floors, walls and racks is strongly recommended.

Nowadays, the vast majority of small rodents is raised in SPF-units. If properly managed (see Box), such systems may stay 'clean' for many years. However, it should be heeded that the outbreak of an infection is unlikely to be restricted to single cages when conventional cages are used. With proper handling this may be prevented by the use of IVCs.

Quarantine

As a consequence of the genetic manipulation, the exchange of breeding stocks between institutions has rapidly increased. Because of the presumably different hygienic constitutions, the single stocks should be preserved in their own microbiological status until re-derivation can be performed. This can be achieved by the use of IVCs in *positive* pressure within a separate barrier-unit in *negative* pressure. *Quarantine* precautions should also be established in testing unknown cellular material provided to be introduced into animals for contaminations, which could be of risk to the animal facility (Yoshimura *et al.*, 1997 and Biological materials,

205

┌───┐

Principles of proper colony management

1. During regular handling only one cage at a time should be managed. This will prevent accidental exchange of animals from different cages.
2. Animals that have escaped or dropped to the floor must never be returned to the suspected cage. Animals caught outside the cage should be killed or isolated, if identification is possible.
3. Cages and hoods should be in sufficient condition that no animal can escape or enter another cage, a problem more often encountered in mouse than in rat breeding units.
4. For ease of identification and in order to prevent an inadvertent mix-up, cage tags should have a strain-specific colour code and a strain-specific number (code).
5. Cage tags should always be filled out properly, including the strain name, strain number, parentage, date of birth and generation.
6. If a cage tag is lost, the cage should not be redefined except in the case of definite proof of identity through marked animals within the cage.
7. If at weaning the number of animals is larger than that recorded at birth the whole litter should be discarded or submitted to the genetic monitoring laboratory.
8. Any change in phenotype and/or increase in productivity should immediately be reported to the colony supervisor. The latter change should always be considered suspect for a possible genetic contamination.
9. Regular training programmes on basic Mendelian genetics, systems of mating and the reproductive physiology of the animals maintained should make animal technicians and caretakers conscious of the consequences any mistake will impose on the colonies. Further training should stress the importance of a search for deviants as potentially new models for biomedical research.

└───┘

p. 195). Animals, gamma-irradiated in an 'unclean' environment, should be submitted to quarantine as well.

Infections

Natural infections require a re-derivation in particular when rarely available stocks are concerned. IVCs in *negative* pressure should be used for containment until re-derivation is completed. In *experimental* infections the pathogenicity of the micro-organisms and the immune status of the animals determines the housing either in 'sealed' IVCs or in isolators.

Special considerations on immunocompromised animals

The consequences of gene manipulation on susceptibility to diseases cannot be predicted fully (Fernandez-Salguero et al., 1995). Therefore, the aim when creating new lines should be the highest possible level of hygiene, especially of the foster mother, the manipulated embryos and the management of the new colonies. Of course, this is of special importance when revitalizing immunocompromised strains. In practice, special staff should be available for these tasks. The risk of contaminating the clean side via the embryos is low if proper 'washing' of the embryos is carried out (see later). Adherence to a strict regimen offers the possibility of raising transgenic animals at a level of hygiene adequate for immunocompromised animals, thus avoiding time-consuming re-derivation.

While immunocompetent animals are able to overcome most infections and to eliminate the pathogen, immunocompromised animals are often unable to cope with the pathogen and may be a source of infection for their entire life. Furthermore, in immunocompromised animals bacteria from the gastrointestinal tract can pass through the epithelial mucosa into the organism (Ohsugi et al., 1996). The question arises as to whether the SPF standard is adequate for severely immunodeficient animals, or if a more stringent containment standard (germ-free, gnotobiotic), is advisable. It should, however, be taken into account that the immune system may depend on a general pre-stimulation which is lower in gnotobiotic and more so in germ-free animals. Therefore, experimental results should be interpreted with caution, when animals have been kept at different hygienic levels.

Mating systems

As mentioned earlier, the phenotype of a gene governing a state of immunodeficiency – either natural, induced or transgenic – may be seriously altered by its genetic background. While most of the established natural and induced mutants have been established in or transferred to an inbred background, many of the most recently developed transgenic and targeted mutants have a segregating mixed background, which should be back-crossed to more than one defined inbred strain in order to be able to make comparisons with the transgenic or targeted mutant and the modulating effects of different genetic backgrounds. There is sufficient information on the many mating systems for breeding rodents (Green, 1981; Silver, 1995).

Inbreeding

A unique advantage in working with mice and rats is the availability of standard inbred strains. By using this type of a strain, including an F1-hybrid, rather than an outbred stock or a strain with mixed genetic background, it is possible to eliminate genetic variability as a source of variation. By continuous brother by sister (BxS), or younger parent by offspring mating for a minimum of 20 generations this homogeneity can be obtained within a strain. After this period 98.02% of all loci within the genome of either animal of the particular strain should be homozygous. As of F12 the remaining heterozygosity within the (incipient) inbred strain will fall off by 19.1% per generation. The increase in homozygosity, respectively the loss of heterozygosity deviates from the expected value if there is any selective force (inadvertent or by purpose) towards a certain phenotype, or in case of mutations.

Congenic strains

In order to be able to identify effects of a particular locus the use of congenic strains is obligatory. Congenic animals represent attempts of genetic identity with the inbred partner strain except for the alleles at a single locus. The simplest approach is to produce an F1 hybrid from a cross between an animal carrying the allele of interest with the selected inbred partner. The resulting progeny is backcrossed to the inbred partner. This is repeated at least for a further nine back-cross generations. With this scheme one-half of the unwanted donor genome not linked to the differentiating locus is lost at every generation. With the availability of the many DNA-markers nowadays available (*http://waldo.wi.mit.edu/rat/ public/; http://www.informatics.jax.org/; http://ratmap.gen.gu.se/; http://www. otsuka.genome.ad.jp/ratmap/*) defining the locus of interest, or being tightly linked to it, other mating systems are in general no longer required. If a recessive allele in the homozygous state is lethal or induces sterility, a known heterozygote (as defined by genotyping) is back-crossed to the selected inbred (background) strain. Only when genotyping is difficult *in vivo* or in the aforementioned case cross–intercross matings have to be performed, whereby carriers are identified by the production of mutant offspring. Once identified, the heterozygote is crossed to the background strain and the resultant progeny again is intercrossed.

Speed congenics

By applying marker-assisted selection protocols, i.e. a genome-wide scan of genetic polymorphisms distinguishing donor and background strain, the production of genetically defined congenic strains is possible within a period of about 1.5 years (Wakeland et al., 1997). Apparently, with low density marker spacing of about 25 cM and screening only of male offspring of four litters at every generation a sufficient introgression is possible. This can be achieved after only five generations of back-crossing (Markel et al., 1997; Wakeland et al., 1997; Visscher, 1999). Moreover, the genome scan allows the identification of the chromosomal location of a transgene in N2 and may provide information on (unwanted) donor-derived regions. One has, however, to keep in mind that it is imperative that the marker set used for differentiation at the given interval does not exceed the upper limit (25 cM) and must be polymorphic unanonymously.

Propagation without inbreeding

Certain mutants cannot successfully be inbred or transferred to a specific inbred background in a fixed (homozygous) state. In these cases the mutation has to be maintained on a hybrid background such as an outbred stock, or descendants of an F1 hybrid. It is supposed that these animals with a heterogeneous background are hardier, more productive, faster growing and have a longer life expectancy. For example, it is extremely difficult to maintain the athymic-nude mutation of the rat (*Whn^{rnu}*, *Whn^{rnu-N}*) on DA and LEW backgrounds. These colonies have to be propagated by constant back-crossing since nude offspring quite often do not surpass weaning (Hedrich, unpublished).

Many of the targeted mutants are, therefore, maintained on the variable, mixed background composed of the ES-cell donor and recipient strain genome and sometimes another 'prolific' strain or stock genome. If a mutation affecting the immune system cannot successfully be inbred due to effects on viability and fertility there is no other means but to maintain it on a segregating background or by back-crossing the mutation onto two different standard inbred strains and by producing homozygous mutant F1 offspring by mating mutant bearing heterozygotes of either strain.

In all instances where research is to be carried out using animals from partially inbred or back-crossed strains or from non-inbred stocks one should be aware of the genetic variability of these experimental animals and therefore use as controls unaffected (heterozygous and +/+) littermates. If these littermates are not available F2 offspring derived from the two progenitor genomes provide the closest approximation in background genotype, while F1 hybrids will match least.

Genotype preservation

Cryopreservation of embryos, gametes and even ovaries is an important tool to secure, archive and distribute strains or stocks of laboratory animals. The techniques for the different types of germplasm to be preserved vary greatly and often depend on the skills and equipment available in the various laboratories. While most publications refer to the mouse, reports on other species are scarce. This is mainly due to the exponentially increasing number of induced mouse mutations, either by gene targeting or by chemical mutagenesis, that have been and are under development.

Embryo freezing

The freezing of preimplantation embryos is considered to be the proper means to cope with the multiplicity of strains of mice and rats presently available, to serve as a safeguard against loss, to allow for eradication of infections if the embryo transfer is performed under aseptic conditions onto barrier maintained surrogate dams, and to reduce the costs for valuable strains not currently used. Despite certain improvements, the freezing of murine embryos is a time-consuming and cost-effective task. While outbred stock and hybrids in general respond to superovulation by gonadotrophins with a high ovulation rate, inbred strains show a rather variable response. In addition revitalization results also vary substantially on a strain by strain basis and strongly depend on the skill of the personnel. Therefore, it has not been possible to preserve as many strains recently developed by molecular genetic methods as necessary.

The original technique of embryo freezing as described by Whittingham et al. (1972) and Wilmut (1972) requires a controlled slow freezing and slow thawing procedure with DMSO or glycerol as the cryoprotectant. Since this first description of successful freezing of eight cell mouse embryos various modifications in the use of cryoprotectants and freezing methods and freezing of other developmental stages have been reported (for an overview see Hedrich and Reetz, 1990).

Sperm freezing

Sperm freezing, although not well established, could assist in all cases where animal-holding space is limited. This primarily applies to, for example, ENU-mutagenesis programmes, or colonies of mice bearing mutations or transgenes. Although reports on sperm freezing associated with *in vitro* fertilization in mice claim that it is a successful means to alleviate the problems encountered with embryo freezing (Marschall and Hrabé de Angelis, 1999; Songsasen et al., 1997; Sztein et al., 2000), it is our experience that sperm freezing is reliable primarily in C3H mice, while results in other strains are rather poor (Sztein et al., 2001).

Ovary freezing

The transplantation of ovaries is a technique to maintain mouse strains with breeding problems established long ago (Russell and Hurst, 1995). Splitting the ovaries into halves further eases the surgical transfer (Stevens, 1957). This modification also increases the probability of success by using up to four recipients. Recently, Stein et al. (1999) reported on the successful orthotopic transplantation of frozen-thawed ovaries into syngeneic ovariectomized recipients. Homozygous *Prkdc^{scid}* mice will serve this purpose as well as syngeneic recipients (Hedrich, unpublished). This technique complements the techniques used in gamete banking.

Genetic monitoring

As well as differential fixation of alleles at early generations of inbreeding, mutations may alter the genetic constitution and thus the phenotype of an inbred strain. Many of the phenotypic differences detected between substrains have been shown to be due to these factors. Inadvertent outcrossing will alter a strain seriously, questioning its further use for research, since results are no longer comparable and repeatable. It is thus of utmost importance to separate strains that are not immediately to be distinguishable by their phenotypic appearance. If, however, due to shortage in shelf space and separate animal rooms several strains must be co-maintained in one room, regular screenings for strain discriminating markers as well as the differentiating locus (in case of congenic strains) are indispensable.

Proper colony management is the first step towards the provision of authentic laboratory animals (see Box). As repeated handling of animals during regular caretaking cannot be avoided, there is always the risk of mistakes. An animal might inadvertently be placed into a wrong cage, or a false entry put on the label. Assigning this type of work to well-trained and highly motivated animal technicians should be a matter of course. The colony set-up and structuring – nucleus colonies in a single (Festing, 1979) or parallel modified line system (Hedrich, 1990), pedigreed expansion colonies and multiplication colonies – should be self-evident, but strictly monitored. There are several publications dealing with the set up of colonies for maintenance and large-scale production (Green, 1966; Lane-Petter and Pearson, 1971; Hansen et al., 1973; Festing, 1979). In general, permanent monogamous mating is to be given preference, as this provides a constant colony output by minimal disturbance of the litters during the early postnatal period and by utilizing the chance that females are inseminated at the *post-partum* oestrus.

The measures required for genotyping a strain have to be adjusted to specific needs and may depend on the scientific purpose, the physical maintenance conditions and the laboratory equipment. Nevertheless, there are specific demands (although unfortunately not stringent rules) on how to authenticate a strain or to verify its integrity.

For any authentication it is necessary to determine a genetic profile that is to be compared with published data (as far as available), and which makes it possible to distinguish between (all) strains/stocks maintained in one unit. In general this profile is composed of monogenetic polymorphic markers, which may be further differentiated by the method of detection into immunological, biochemical, cytogenetical, morphological and DNA markers. Due to the recent rapid development of microsatellite markers (Simple Tandem Repeats, STRs) these have almost fully replaced the classical genetic markers in routine applications. A large number of primer pairs for mice and rats is available, for example through Research Genetics Inc., Huntsville, AL, USA (*http://www.resgen.com*). Other sources for primers are also available through the World Wide Web (see earlier). However, as with the classical markers it is indispensable to set up a genetic profile representing a random sample of the genome, which should be evenly spaced on the chromosomes, and which enables all strains maintained per separate housing unit to be identified. Unfortunately, this information is only partly available and not yet compiled in an accessible database. There are numerous publications and textbooks with protocols for PCR amplification and electrophoretic separation of the amplicons. Moreover, commercial suppliers of primers (e.g. Research Genetics) and of genetically modified animals (e.g. *http://informatics.jax.org*; check: Genes, markers and phenotypes, see

Polymorphism, or *http://www.jax.org/resources/documents/imr/protocols/index.html*) do provide PCR protocols. Nevertheless, it might be necessary to adjust temperature conditions as well as Mg^{2+} concentrations for each microsatellite marker. For routine screening separation on agarose gel and visualization by ethidium bromide will suffice. If separation of the amplicons is insufficient in agarose polyacrylamide gel electrophoresis should be performed. As radioactive labelling with ^{32}P uses a kinase reaction and since the half-life of isotopes is relatively short, a silver staining procedure is recommended. Information on RFLP polymorphisms as determined by a Southern blot (Sambrook, 1989) using a specific probe may also be found in the mouse genome database (MGD) maintained by The Jackson Laboratory.

Nevertheless, the classical markers are still relevant and may need to be verified, and sometimes allow for a faster and less expensive phenotyping. In this context immunological markers are of prime importance. This group is composed of cell surface markers, such as major histocompatibility antigens (*H2* in the mouse and *RT1* in the rat), lymphocyte differentiation antigens, red blood cell antigens, minor histocompatibility antigens, allotypes (immunoglobulin heavy chain variants) which can be determined by Trypan blue dye exclusion test (see the chapter by Czuprynski in Section II), flow cytometry (see the chapter by Scheffold *et al.* in Section I), immunodiffusion, ELISA (see the chapter by Yssel in Section III), immunohistochemistry ELISA (see the chapter by Ehlers *et al.* in Section II), using specific antibodies. The availability of antibodies depends on the specific marker and the species, with a broader spectrum available for mice. If it is too difficult to obtain or produce these antibodies certain markers might be demonstrated by applying published molecular biology techniques (see also: *http://www.informatics.jax.org/mgd.html*).

Further methods that can be applied easily and which depend on a specific phenotype may also be applied, as in the case of the lysosomal trafficking regulator (*Lyst^{bg}*, beige, expressing a pigmentation and platelet storage pool defect). The phenotype of homozygous beige mice can be determined by a prolonged bleeding time (20 min in homozygous *Lyst^{bg}* vs. 6 min in unaffected wild type or heterozygous controls), or a histochemical staining (checking for abnormal giant lysosomal granules detectable in all tissues with granule-containing cells; Novak *et al.*, 1985).

The determination of a profile is time-consuming and expensive, but strongly recommended as an initial check. In case of a variable segregating background genetic profiling is pointless as the typing results will only assist in determining the degree of heterogeneity. However, these results may provide hints on modifying genes, if the stock is being inbred and nearly homozygous.

Easy measures are still required to distinguish between those strains that are co-maintained and which clearly identify an outcrossing event. A critical subset of the markers (i.e. least amount of differentiating markers for a given strain panel) used to authenticate the strains maintained will provide reasonable information on the genetic quality of a strain. Unfortunately, with each strain added to a unit the number of markers in the critical subset increases. These critical subsets need to be verified at regular intervals (every 3 to 6 months). The intervals and the number of animals to be tested are incremented to the number of strains co-maintained and to the size of each colony.

Irrespective of these methods one of the most powerful aspects of an inbred strain lies in the demonstration of its isohistogeneity. This is best demonstrated by skin grafting. The technique is easy to perform. It is, however, time-consuming because of an observation period of about 100 days (for a description of the techniques see Hedrich, 1990). In certain immunodeficient mutants (e.g. $Foxn1^{nu}$, $Prkdc^{scid}$, $Rag1^{tm}$, $Rag2^{tm}$) a direct demonstration of isohistogeneity is impossible, as these animals are incapable of mounting an allorecognition response. Transferring grafts from these immunodeficient animals to their syngeneic background strains can circumvent this.

◆◆◆◆◆◆ MANAGEMENT OF INFECTED COLONIES

Quarantine and natural infections

Animals with an unknown microbiological status have to be kept in isolation. The degree of isolation should be the same as that for infected animals as already described. The need for re-derivation of both categories is obvious.

Re-derivation

Hysterectomy

As shown for most infections, the vertical transmission of viruses, bacteria and parasites can be avoided by this procedure. The most difficult part of this procedure is to achieve timed pregnancy, especially in poor breeding strains. This method (see Box) is recommended if embryo transfer cannot be performed due to lack of equipment and trained personnel, or if a donor strain is refractory to superovulation. Hysterectomy has the additional risk of intrauterine vertical transmission of infections, which is to be considered higher in immunodeficient than in immunocompetent animals.

Embryo transfer

Embryo transfer was shown to interrupt most vertically transmitted infections of viral, bacterial or parasitic origin with the exception of germ-line transmitted retroviral infections. The integrity of the zona pellucida is of decisive importance as shown for mouse hepatitis virus (MHV) infection (Reetz et al., 1988). The hygienic status of the foster mother should be of the highest level, especially when a new breeding unit is to be established. For routine procedures, the two-cell stage may be best suited because fertilization is no longer in question and a relatively high

> **Hysterectomy**
>
> 1. Mate foster mother (outbred or hybrid strain) in the clean area overnight; check for vaginal plug.
> 2. 24–48 h later, mate animals of the microbiologically contaminated strain; check for vaginal plug.
> 3. Install the dip tank filled with low-odour disinfectant before the expected date of birth of the foster mother.
> 4. Shortly before delivery, kill the pregnant dam of the strain to be rederived by cervical dislocation; carry out hysterectomy under aseptic conditions.
> 5. Transfer the uterus to the clean side through the disinfectant (38°C).
> 6. Wash the uterus intensively in physiological saline, and develop the pups.
> 7. As an extra safety precaution, the pups may be dipped again in disinfectant and washed again in physiological saline.
> 8. After gentle massage with a swab to induce spontaneous breathing and after warming up, transfer the pups to the nest of the foster mother after disposing her own offspring.
> 9. If coat-colour discrimination is possible, one or two of the foster mother's pups may be retained to assist in the induction of lactation.

number of embryos can be collected. The animals are timed mated without or after previous superovulation (for details see Reetz *et al.*, 1988; Hogan *et al.*, 1994; Schenkel, 1995). The latter method normally induces the production of higher numbers of embryos (other than by normal mating) especially if prepuberal females are used, and allows synchronized matings. Embryos are flushed from the oviducts of plug-positive mice on day 1.5. They are selected for integrity (intact zona pellucida), washed at least four times at different locations and in sufficiently large volumes of media (approx. 2 ml), before transfer to a clean area where the transfer into the oviducts of pseudocyetic surrogate dams (day 0.5) is performed by a different person. Pseudocyesis can be induced by mating the surrogate dam with either a vasectomized, or a genetically sterile male (Silver, 1985). It should be mentioned that there are strain-specific differences with respect to the optimal amount of injected hormones and the number of embryos. Problems with superovulation are also known for most inbred rat strains.

The embryo transfer offers certain advantages versus hysterectomy. It avoids the risk of intrauterine vertical transmission of infections, and allows easier timing especially by superovulation and cryopreservation of surplus embryos.

Furthermore, new lines shipped as cryopreserved embryos can be transferred to surrogate dams of the present SPF status, thus avoiding time-consuming quarantine and rederivation procedures.

Preventive treatment

The preventive treatment of immunocompromised breeders with immunocompetent cells can be of help in the propagation of highly immunocompromised strains (Wang *et al.*, 1997; Kawachi *et al.*, 2000). To avoid graft-versus-host reactions, immunocompetent cells of F1 hybrids of the strain to be reconstituted with an immunocompetent strain should be used. For reconstitution of *nude* mice, thymus homogenates can be injected intraperitoneally to overcome their defect. For homozygous SCID, RAG or Gamma-c mice, the injection of F1 spleen cells i.p. $(1–2 \times 10^7)$, or bone marrow cells $(2–5 \times 10^6)$ i.v. into juvenile animals improves their constitution and thus marks them as suitable breeders (Mossmann, unpublished).

Therapeutic treatment

In general, the administration of therapeutics influences the outcome of animal experiments and cannot be considered as a means to replace the improvement of hygienic standards. However, therapeutic treatment may be unavoidable after gamma-irradiation and in immunocompromised strains if the latter have to be maintained in 'dirty' conditions until rederivation is completed (Macy *et al.*, 2000). The success of treatment depends on several criteria: a correct diagnosis including antibiotic resistance (Hansen and Velschow, 2000); the consideration on species-specific toxicity; adverse reactions of the therapeutic; and an optimal dosage and regimen of application and accompanying hygienic procedures. Unfortunately, the dosage often refers to man or larger animals. For extrapolation to small rodents allometric parameters should be used, which increase the body–weight ratio by a factor of approx. 6 and 12 for rat and mouse respectively, in comparison to man (for review see Morris, 1995). By analogy, the half-life time of therapeutics is in general reduced in small rodents requiring more frequent application for maintaining an effective level of the therapeutic.

The treatment of parasitic invasions is in particular dependent on the accompanying hygienic procedures, e.g. use of gloves, chemical and/or physical disinfection of the animal rooms, cages, lids, bottles. In Table 3 some commonly used antiparasitics are summarized. For additional drug dosages, see Hawk and Leary (1995). In the case of parasitic eggs and oocysts, a chlorcresol[1] formulation has proven particularly valuable. It should be mentioned, however, that treatment may be associated with toxic effects (Scopets *et al.*, 1996; Toth *et al.*, 2000). Ivermectin induced long-lasting alterations, particularly in bone marrow derived macrophages (Mossmann and Modolell, unpublished).

Chemotherapeutic and antibiotic treatment of infections may induce resistance, especially when used on a large scale, on growth of other bacterial species (Hansen, 1995), adverse reactions by shifting the gut flora (for review: Morris, 1995), or derangements of physiological functions (el Ayadi and Errami, 1999). Commonly recommendable treatment procedures of infected animals are given in Table 4.

1 Chlorcresol (Neopredisan): Menno-Chemie Vertrieb GmbH, D-22850 Norderstedt.

Table 3. Treatment of common parasites (in combination with hygienic measures)

Generic name	Trade name	Application	Dose	Reference
Ectoparasites				
Ivermectin	Ivomec*	Topical spray	0.2–10 mg^{-1}	Hirsjärvi and Phyäiä (1995)
Endoparasites[+]				
Piperazine citrate	Piperazin*	Drinking water (for 12 weeks, every 2nd week)	0.2%	Maess and Kunstyr (1981)
Fenbendazole[‡]	Panacur* Coglazol*	Diet several months	150 ppm in diet or 8–12 mg/kg/day	Coghlan *et al.* (1993) Huerkamp *et al.* (2000) Wilkerson *et al.* (2001)
Ivermectin*	Ivomec*	Topical spray 2 ml/cage	1 mg ml^{-1}, 2 ml/cage, once weekly for 3 weeks	Le Blanc *et al.* (1993)
		Drinking water	2.9–4.0 mg kg^{-1} for 4 days, 3-day pause, 5 cycles	Klement *et al.* (1996)
Ivermectin–piperazine (combined)		Drinking water	7000 ppm, 2.1 mg ml^{-1} alternately every 2 weeks for several months	Lipman *et al.* (1994) Zenner (1998)

* Pharmazeutische Handelsgesellschaft, Siemensstr. 14, 30827 Garbsen.
[+] Especially *Syphacia obvelata* and *Aspiculuris tetraptera*.
[‡] Diet can be autoclaved without substantial loss of efficacy.

Table 4. Selected antibiotic therapies for small rodents

Disease/Species	Anti-infective	Application in drinking water	Dose	Reference
Pasteurellosis/mouse	Enrofloxacin (Baytril)	For at least 30 days	25.5–85 mg kg⁻¹	Goelz et al. (1996) Macy et al. (2000)
Mycoplasma/rat	Oxytetracycline	For at least 5 days*	3–5 mg ml⁻¹	Harkness and Wagner (1983)
	Tylosin	For 21 days	5 g l⁻¹	Carter et al. (1987)
Hepatitis-typhlitis/mouse (*Heliobacter* sp.[†])	Amoxillin[‡]	For 4 weeks – young mice	50 mg kg⁻¹	Russel et al. (1995)
	Amoxicillin[‡] Metronidozole Bismuth	For 2 weeks	200 mg l⁻¹ 138 mg l⁻¹ 37 mg l⁻¹	Foltz et al. (1996)
Pneumocystosis/mouse, rat	Sulfadoxine–trimethoprim (Borgal; Trimethosel, Cotrim K)	For 3 weeks	200 mg g⁻¹ l⁻¹	H.-J. Hedrich (unpublished)
Tyzzer's disease (*Clostridium piliforme*)/ mouse, rat, rabbit	No antibiotic therapy recommended; derivation Special disinfectant required[§]			

For additional drug dosages see Hawk and Leary (1995).

* Drinking water should not be acidified; addition of 1.35 g l⁻¹ potassium sorbat prevents growth of yeast.

[†] Especially immunodeficient mice.

[‡] Toxic for hamster and guinea-pigs.

[§] Chlorocresol (Neopredisan).

217

Infection experiments

General precautions

The safe operation of an animal laboratory is one of the main management responsibilities. Housing infected animals require precautions to prevent transmission of micro-organisms between animal populations and, in the case of zoonotic agents, to humans. The zoonotic risk arising from naturally infected rodents is low because most rodent pathogens do not infect man. Only few and seldom found agents like LCMV, Hantaviruses, or *Streptobacillus moniliformis* have the potential to cause severe infections in humans and might be prevalent in colonies of laboratory rodents. Severe disease outbreaks in humans associated with infected colonies of laboratory rodents have been reported (Bowen *et al.*, 1975; Kawamata *et al.*, 1987), and therefore safety programmes are necessary to prevent laboratory-associated infections and infections transmitted by laboratory animals.

Experimental infections are more likely to pose a risk for humans. A broad spectrum of infectious agents can be introduced accidentally with patient specimens, and many laboratory animals are still used for infectious experiments. In general, health precautions are very similar for clinical or research laboratories and for animal facilities. In many cases, however, an increased risk may arise from experimentally infected animals due to bite wound infections or when pathogens are transmissible by dust or by aerosols.

A number of recommendations exist from federal authorities for microbiological laboratories aiming at prevention of infections for laboratory personnel. Many programmes were developed in response to evaluations of laboratory accidents. Most laboratories have written control plans, which have been designed to minimize or eliminate risks for employees.

Reduction of the risk of disease transmission can be achieved by very general procedures, which are common practice in most well-run animal facilities housing animals behind barriers. Only major points can be discussed here; more details on general laboratory safety are given in many textbooks on clinical microbiology (Burkhardt, 1992; Strain and Gröschel, 1995) and in general recommendations for housing of laboratory animals (CCAC, 1980; Kunstyr, 1988b; Bruhin, 1989; BG Chemie, 1990; National Research Council, 1996, 1997; Smith, 1999).

Education is an important part of effective safety programmes. All safety instructions should be in written form and must be readily available at all times. The first point must be adherence to safety procedures and proper behaviour, like the use of personal protective clothes. Prohibition of eating, drinking, smoking, handling of contact lenses and the application of cosmetics in the laboratory are other basic rules, like the separation of food storage refrigerators from laboratory refrigerators. The most likely route of infection is direct contact with contaminated animals or materials. Micro-organisms do not usually penetrate intact skin. The risk of infection can therefore be reduced by repeated hand decontamination and by decontamination of surfaces or contaminated instruments.

Working with infectious agents should not be permitted in cases of burned, scratched or dermatitic skin. Needles and other sharp instruments should be used only when necessary, and handling of infected animals should be allowed only for experienced and skilled personnel to prevent bite wounds. Working in safety cabinets helps to avoid inhalation of infectious aerosols and airborne particles, which are easily generated in cages when animals scratch or play. Other procedures that might bring organisms directly on mucous membranes are mouth pipetting and hand–mucosa contact. Both must be strictly forbidden.

Microisolator cages are often used in animal facilities for transportation within the facility to avoid exposure of humans to allergens. Such cages, too, help to reduce the risk of spreading micro-organisms during transportation.

In most animal facilities containment equipment (microisolator cages, isolators) is used if immunosuppressed animals have to be protected from the environment or if infected animals might be a hazard for humans or other animals. Experiments with infectious agents will usually be conducted in separate areas which fulfil all safety requirements like ventilation (negative pressure in laboratories to prevent air flow into non-laboratory areas), or, better, in isolators which represent the most stringent containment system. For safety reasons, containment is generally necessary if animals are artificially infected with pathogenic micro-organisms. Various systems can be used depending on properties of the agents like pathogenicity, environmental stability, or spreading characteristics. In the case of low pathogenic organisms, microisolator cages might be sufficient. The risk of infection during handling is reduced if all work with open cages is conducted in changing cabinets or in laminar flow benches. Individually ventilated cages operating with a negative pressure are better suited than microisolators to prevent spreading of micro-organisms if they are properly handled. The highest level of safety can be achieved by using a negative pressure isolator. If handling through thick gloves is not possible, handling of animals can be performed in safety cabinets, which can be locked directly to the isolator.

An important part of safety programmes in laboratories, and especially in laboratory animal facilities, is waste management. In contrast to radioactive or chemical waste, infectious waste cannot be identified objectively. In many cases judgement as to whether or not waste from animals that are not experimentally infected is infectious is dependent on the person in charge. There is, however, no doubt if animals have been infected experimentally. In such cases the presence of a pathogen allows evaluation of the risk, which is dependent on the virulence and the expected concentration of an agent together with the resistance of a host and the dose that is necessary to cause an infection. The risk of pathogen transmission is increased by injuries with sharp items such as needles, scalpels, or broken contaminated glass. Segregation of such sharp items and storage in separate containers is necessary to reduce the infectious risk to a minimum.

Infectious waste from animal houses (bedding material, animal carcasses) can be submitted to chemical or thermal disinfection, but incineration and steam sterilization are the most common treatment methods. Incineration has the advantage of greatly reducing the volume of treated materials. The usually low content of plastic material in waste from animal housing and the high percentage of bedding material (e.g. wood shavings), resulting in a high energy yield, make incineration the method of choice.

Biosafety for housing laboratory animals

Biosafety criteria for housing vertebrates have been defined in the USA by CDC (1988) for biosafety levels 2 and 3 and later for all four biosafety levels (CDC/NIH 1993). Specific regulations for housing infected animals according to different safety levels also exist in other countries (e.g. for Germany see Gentechnik Sicherheitsverordnung Anhang V). Therefore, only general comments are given here.

Laboratory animal facilities may be organized in different ways. Sometimes, animal facilities are extensions of the laboratories and are managed under the responsibility of a research director. Large research institutions, companies or universities often have centralized laboratory animal facilities, which are managed by laboratory animal specialists. They are usually separated from laboratories or institutes. Such facilities usually fulfil more easily the legal requirements (animal welfare, safety) due to a more proficient management and specialized personnel, and their size. Centralized animal facilities are usually multipurpose, with a number of animal species or strains that are used for a variety of different experiments (short–long term) for different scientific disciplines (e.g. toxicology, immunology, biochemistry). Several housing systems (conventional units, barrier units, isolators) or microbiological quality standards (infected, pathogen-free, gnotobiotic) can be found in large facilities. Therefore, strict separation of animals used for different experiments (studies of infectious or non-infectious disease) or purposes (production and breeding, quarantine) is usually self-evident not only for safety reasons but in order to avoid research complications or influences between experiments. Traffic flow in centralized animal facilities is usually reduced to a minimum, thus minimizing the risk of cross-contamination. Such facilities are usually constructed in a way that facilitates proper cleaning and personal hygiene. Bedding material from animal cages is removed in a manner that avoids the formation of dust or aerosols and minimizes the risk of allergies, thus reducing the risk of airborne transmission of pathogens. Use of solid bottom cages helps to reduce dust formation and is absolutely necessary if experimentally infected animals are housed. The whole facility must be constructed in such a way that escape or theft of animals is impossible.

In general, biosafety levels recommended for working with infectious materials *in vitro* and *in vivo* are comparable. Some differences exist, because activities of the animals themselves can introduce new hazards by producing dust or aerosols, or they may traumatize humans by biting and scratching. Therefore, CDC/NIH (1993) established standards for activities involving infected animals which are designed 'animal biosafety levels' (ABSL) 1–4. These combinations describe animal facilities and practices applicable to work on animals infected with agents assigned to corresponding BL-1–4.

Housing animals of ABSL-1 is usually no problem if an animal facility, as well as operational practices and the quality of animal care, meet the standard regulations (CCAC, 1980; Bruhin, 1989; National Research Council, 1996, 1997). In contrast to experiments with non-infectious materials, additional hygienic procedures should be applied, such as decontamination of work surfaces after any spill of infectious material and decontamination of waste before disposal. Persons who may be at increased risk of acquiring infections should not be allowed to enter rooms in which infected animals are housed.

Additional practices are necessary for ABSL-2. Careful hand disinfection is necessary after handling live micro-organisms. All infectious waste must be properly disinfected (best by autoclaving), and infected animal carcasses should be incinerated. Cages and other contaminated equipment are disinfected before they are cleaned and washed. Whenever possible, infected animals will be housed in isolation to avoid the creation of aerosols. Physical containment devices are not explicitly required by the CDC/NIH (1993) for ABSL-2. Microisolator cages are not recommended because they do not reliably prevent aerosol formation and the transmission of micro-organisms. They should only exceptionally be used for housing and must be placed in ventilated enclosures (e.g., laminar flow cabinets). Therefore, the lowest level of biocontainment should be a ventilated cage with negative pressure. In many institutions negative-pressure isolators are considered the only suitable containment devices for housing animals infected with potential human pathogens. Special care is necessary to avoid infections during necropsy of infected animals. Necropsies as well as harvesting tissues or fluids from infected animals should therefore be carried out in safety cabinets.

As with BL-3 materials, access to an ABSL-3 facility is very much restricted. All laboratory personnel receive appropriate immunizations (e.g. hepatitis B vaccine). Physical containment devices are necessary for all procedures and manipulations. Animals must be housed in a containment caging system. Individually ventilated might be acceptable in specific cases, but negative pressure isolators or Class II biological safety cabinets offer a maximum of safety because supply and removal of infected materials is carried out in closed containers thus reliably avoiding a risk of transmission. Very few facilities house ABSL-3 animals. If this is really necessary, many more safety precautions will be taken than recommended by CDC/NIH (1993) (e.g. a one-piece positive-pressure suit that is ventilated with a life support system).

ABSL-4 is extremely uncommon and will be avoided whenever possible because transmission of extremely pathogenic organisms to humans can take place by scratching or biting. A maximum of access control and of hygienic measures are necessary.

References

Asseman, C., Mauze, S., Leach, M. W., Coffman, R. L. and Powrie, F. (1999) An essential role for interleukin 10 in the function of regulatory T cells that inhibit intestinal inflammation. *J. Exp. Med.* **190**, 995–1004.

Baker, D. G. (1998). Natural pathogens of laboratory mice, rats, and rabbits and their effect on research. *Clin. Microbiol. Rev.* **11**, 231–266.

Ball-Goodrich, L. J. and Johnson, E. (1994). Molecular characterization of a newly recognized mouse parvovirus. *J. Virol.* **68**, 6467–6486.

Ball-Goodrich, L. J., Leland, S. E., Johnson, E. A., Paturzo, F. X. and Jacoby, R. O. (1998). Rat parvovirus type 1: the prototype for a new rodent parvovirus serogroup. *J. Virol.* **72**, 3289–3299.

Bankert, R. B., Egilmez, N. K. and Hess, S. D. (2001). Human-SCID mouse chimeric models for the evaluation of anti-cancer therapies. *Trends Immunol.* **22**, 386–393.

Barthold, S. W., Smith, A. L. and Povar, M. L. (1985). Enterotropic mouse hepatitis infection in nude mice. *Lab. Anim. Sci.* **35**, 613–618.

Berking, C. and Herlyn, M. (2001). Human skin reconstruct models: a new application for studies of melanocyte and melanoma biology. *Histol. Histopathol.* **16**, 669–674.

BG Chemie (1990). *Tierlaboratorien*. Berufsgenossenschaft der chemischen Industrie, Merkblatt M007, 6/90.

Bhatt, P. N. and Jacoby, R. O. (1987). Mousepox in inbred mice innately resistant or susceptible to lethal infection with ectromelia virus. I. Clinical Responses. *Lab. Anim. Sci.* **37**, 11–15.

Bhatt, P. N., Jacoby, R. O., Morse, H. C. and New, A. (eds) (1986a). *Viral and Mycoplasma Infections of Laboratory Rodents: Effects on Biomedical Research.* Academic Press, New York.

Bhatt, P. N., Jacoby, R. O. and Barthold, S. W. (1986b). Contamination of transplantable murine tumors with lymphocytic choriomeningitis virus. *Lab. Anim. Sci.* **36**, 138–139.

Boot, R., van Herck, H. and van der Logt, J. (1996). Mutual viral and bacterial infections after housing rats of various breeders within an experimental unit. *Lab. Anim.* **30**, 42–45.

Bowen, G. S., Calisher, C. H., Winkler, W. G., Kraus, A. L., Fowler, E. H., Garmann, R. H., Fraser, D. W. and Hinman, A. R. (1975). Laboratory studies of a lymphocytic choriomeningitis virus outbreak in man and laboratory animals. *Am. J. Epidemiol.* **102**, 233–240.

Bregenholt, S. and Claesson, M. H. (1998). Increased intracellular Th1 cytokines in scid mice with inflammatory bowel disease. *Eur. J. Immunol.* **28**, 379–389.

Bruhin, H. (Ed.) (1989). *Planning and Structure of Animal Facilities for Institutes Performing Animal Experiments.* Society for Laboratory Animal Science Publ. No. 1, 2nd edition, Biberach.

Burkhardt, F. (1992). Sicherheit im Laboratorium. In: *Mikrobiologische Diagnostik.* (F. Burkhardt, Ed.). G. Thieme Verlag, Stuttgart 773–779.

Carter, K. K., Hietala, S. H., Brooks, D. L. and Baggot, J. D. (1987). Tylosin concentrations in rat serum and lung tissue after administration in drinking water. *Lab. Animal Sci.* **37**, 468–470.

Carthew, P. and Gannon, J. (1981). Secondary infection of rat lungs with *Pasteurella pneumotropica* after Kilham rat virus infection. *Lab. Anim.* **15**, 219–221.

CCAC (1980). *Guide to the Care and Use of Experimental Animals.* Vol. 1. Canadian Council of Animal Care, Ottawa, Ontario.

Centers for Disease Control (1988). Agent summary statement for human immunodeficiency virus and report on laboratory-acquired infection with human immunodeficiency virus. Addendum 1: Vertebrate animal biosafety criteria. *MMWR* 37 (Suppl. No. S-4), 11–15.

Centers for Disease Control/National Institute of Health (1993). Biosafety in microbiological and biomedical laboratories. 3rd edn, HHS Publication No. (CDC) 93-8395. U.S. Department of Health and Human Services.

Chaguri, L. C. A. G., Souza, N. L., Teixeira, M. A., Mori, C. M. C., Carissimi, A. S. and Merusse, J. L. B. (2001). Evaluation of reproductive indices in rats (*Rattus norvegicus*) housed under an intracage ventilation system. *Contemp. Topics* **40**, 25–30.

Clifford, C. B. (2001). Samples, sample selection, and statistics: living with uncertainty. *Lab. Animal.* **30** (10), 26–31.

Clifford, C. B., Walton, B. J., Reed, T. H., Coyle, M. B., White, W. J. and Amyx, H. L. (1995). Hyperkeratosis in nude mice is caused by a coryneform bacterium: microbiology, transmission, clinical signs, and pathology. *Lab. Anim. Sci.* **45**, 131–139.

Clough, G., Wallace, J., Gamble, M. R., Merryweather, E. R. and Bailey, E. (1994). A positive, individually ventilated caging system: a local barrier system to protect both animals and personnel. *Lab. Anim.* **29**, 139–151.

Coghlan, L. G., Lee, D. R., Psencik, B. and Weiss, D. (1993). Practical and effective eradication of pinworms (*Syphacia muris*) in rats by use of fenbendazole. *Lab. Animal Sci.* **43**, 481–487.

Collins, M. J. and Parker, J. C. (1972). Murine virus contamination of leukemia viruses and transplantable tumors. *J. Natl. Cancer Inst.* **49**, 1139–1143.

Compton, S. R. and Riley, L. K. (2001). Detection of infectious agents in laboratory rodents: traditional and molecular techniques. *Comp. Med.* **51**, 113–119.

Compton, S. R., Barthold, S. W. and Smith, A. L. (1993). The cellular and molecular pathogenesis of coronaviruses. *Lab. Anim. Sci.* **43**, 15–28.

Criley, J. M., Carty, A. J., Besch-Williford, C. L. and Franklin, C. L. (2001). *Coxiella burnetti* infection in C.B.-17 Scid-bg mice xenotransplanted with fetal bovine tissue. *Lab. Anim. Sci.* **51**, 357–360.

Croy, B. A., Di Santo, J. P., Greenwood, J. D., Chantakru, S. and Ashkar, A. A. (2000). Transplantation into genetically alymphoid mice as an approach to dissect the roles of uterine natural killer cells during pregnancy – a review. *Placenta* **21**, 77–80.

Croy, B. A., Linder, K. E. and Yager, J. A. (2001). Primer for non-immunologists on immune-deficient mice and their applications in research. *Comp. Med.* **51**, 300–313.

Dagnaes-Hansen, F., Pfister, R. and Bisgaard, M. (1991). Otitis media in scid mice due to infection with an atypical *Pseudomonas* bacterium. 7th International workshop on immunedeficient animals, A-10.

Dewhirst, F. E., Chien, C.-C., Paster, B. J., Ericson, R. L., Orcutt, R. P., Schauer, D. B. and Fox, J. G. (1999). Phylogeny of the defined murine microbiota: altered Schaedler flora. *Environm. Microbiol.* **65**, 3287–3292.

Dick, E. J., Kittell, C. L., Meyer, H., Farrar, P. L., Ropp, S. L., Esposito, J. J., Buller, R. M. L., Neubauer, H., Kang, Y. H. and McKee, A. E. (1996). Mousepox outbreak in a laboratory mouse colony. *Lab. Anim. Sci.* **46**, 602–611.

Duke, J. L., Zammit, T. G. and Lawson, D. M. (2001). The effects of routine cage-changing on cardiovascular and behavioral parameters in male Sprague-Dawley rats. *Contemp. Topics* **40**, 17–20.

el Ayadi, A. and Errami, M. (1999). Interactions between neomycin and cerebral dopaminergic and serotoninergic transmission in rats. *Therapie* **54**, 595–599.

Enss, M. L., Wagner, S., Liebler, E., Coenen, M. and Hedrich, H.-J. (1997). Response of germfree rat colonic mucosa. *J. Exp. Animal Sci.* **38**, 58–65.

FELASA (1995). Recommendations on the education of training of persons working with laboratory animals: categories A and C. *Lab. Anim.* **29**, 121–131.

Feldman, S. H. (2001). Diagnostic molecular microbiology in laboratory animal health monitoring and surveillance programs. *Lab. Animal* **30**(10), 34–43.

Fernandez-Salguero, P., Pinau, T., Hilbert, D. M., McPhail, R., Lee, S. T., Kimura, S., Nebert, D. W., Rudikoff, S., Ward, J. M. and Gonzalez, F. J. (1995). Immune system impairment and hepatic fibrosis in mice lacking the dioxin-binding Ah receptor. *Science* **268**, 722–726.

Festing, F. W. (1979). *Inbred Strains in Biomedical Research.* Macmillan Press, London.

Foltz, C. J., Fox, J. G., Yan, L. and Shames, B. (1996). Evaluation of various oral antimicrobial formulations for eradication of *Heliobacter hepaticus*. *Lab. Animal Sci.* **46**, 193–197.

Fox, J. G., Dewhirst, F. E., Tully, J. G., Paster, B. J., Yan, L., Taylor, N. S., Collins, M. J., Gorelick, P. L. and Ward, J. M. (1994). *Helicobacter hepaticus* sp. nov., a microaerophilic bacterium isolated from livers and intestinal mucosa scrapings from mice. *J. Clin. Microbiol.* **32**, 1238–1245.

Fox, J. G., Yan, L., Dewhirst, F. E., Paster, B. J., Shames, B., Murphy, J. C., Hayward, A., Belcher, J. C. and Mendes, E. N. (1995). *Helicobacter bilis* sp. nov., a novel *Helicobacter* species isolated from bile, livers and intestines of aged, inbred mice. *J. Clin. Microbiol.* **33**, 445–454.

Franklin, C. L., Riley, L. K., Livingston, R. S., Beckwith, C. S., Hook, R. R. Jr., Besch-Williford, C. L., Hunziker, R. and Gorelick, P. L. (1999). Enteric lesions in SCID mice infected with *Helicobacter typhlonicus*, a novel urease-negative *Helicobacter* species. *Lab. Animal Sci.* **49**, 496–505.

Gaertner, D. J., Jacoby, R. O., Johnson, E. A., Paturzo, F. X. and Smith, A. L. (1995). Persistent rat virus infection in juvenile athymic rats and its modulation by anti-serum. *Lab. Anim. Sci.* **45**, 249–253.

Goelz, M. F., Thigpen, J. E., Mahler, J., Rogers, W. P., Locklear, J., Weigler, B. J. and Forsythe, D. B. (1996). Efficacy of various therapeutic regimens in eliminating *Pasteurella pneumotropica* from the mouse. *Lab. Animal Sci.* **46**, 280–284.

Green, E. L. (1966). *Biology of the Laboratory Mouse.* McGraw-Hill, New York.

Green, E. L. (1981). *Genetics and Probability in Animal Breeding Experiments.* Macmillan, London.

Gordon, S., Fisher, S. W. and Raymond, R. H. (2001). Elimination of mouse allergens in the working environment: assessment of individually ventilated cage systems and ventilated cabinets in the containment of mouse allergens. *J. Allergy Clin. Immunol.* **108**, 288–294.

Hamm, T. E. (Ed.) (1986). *Complications of Viral and Mycoplasmal Infections in Rodents to Toxicology Research and Testing.* Hemisphere Publ. Co., Washington DC.

Hansen, A. K. (1994). Health status and the effects of microbial organisms on animal experiments. In: *Handbook of Laboratory Animal Science,* (P. Svendsen and J. Hau, Eds), Vol. 1, pp. 125–153, CRC Press Inc., Boca Raton.

Hansen, A. K. (1995). Antibiotic treatment of nude rats and its impact on the aerobic bacterial flora. *Lab. Anim.* **29**, 37–44.

Hansen, A. K. and Velschow, S. (2000). Antibiotic resistance in bacterial isolates from laboratory animal colonies naïve to antibiotic treatment. *Lab. Anim.* **34**, 413–422.

Hansen, C. T., Judge, F. J. and Whitney, R. A. (1973). *Catalogue of NIH Rodents,* DHEW Publication No. 74-606. US Department of Health, Education and Welfare, Washington, DC.

Harkness, J. E. and Wagner, J. E. (1983). *The Biology and Medicine of Rabbits and Rodents.* Lea & Fiebiger, Philadelphia.

Hasegawa, M., Kurabayashi, Y., Ishii, T., Yoshida, K., Uebayashi, N., Sato, N. and Kurosawa, T. (1997). Intra-cage air change rate on forced-air-ventilated micro-isolation system – environment within cages: carbon dioxide and oxygen concentration. *Exp. Anim.* **46**, 251–257.

Hawk, C. T. and Leary, S. L. (1995). *Formulary for Laboratory Animals.* Iowa State University Press, Ames, IA.

Hedrich, H. J. (1990). Testing for Isohistogeneity (Skin Grafting). In: *Genetic Monitoring of Inbred Strains of Rats* (H. J. Hedrich, Ed.) pp. 102–114. Gustav Fischer Verlag, Stuttgart.

Hedrich, H. J. and Reetz, I. (1990). Cryopreservation of rat embryos. In: *Genetic Monitoring of Inbred Strains of Rats* (H. J. Hedrich, Ed.), pp. 274–288. Gustav Fischer Verlag, Stuttgart.

Heidt, P. J., Koopman, J. P., Kennis, H. M., van den Logt, J. T., Hectors, M. P., Nagengast, F. M., Timmermans, C. P. and de Groot, C. W. (1990). The use of rat-derived microflora for providing colonization resistance in SPF rats. *Lab. Anim.* **24**, 375–379.

Heine, W. O. P. (1980). How to define SPF? *Z. Versuchstierk.* **22**, 262–266.

Heine, W. O. P. (1998). In: *Umweltmanagement in der Labortierhaltung. Technisch-hygienische Grundlagen, Methoden und Praxis* (Pabst Science Publishers) engerich.

Hess, J., Schaible, U., Raupach, B. and Kaufmann, S. H. (2000). Exploiting the immune system: toward new vaccines against intracellular bacteria. *Adv. Immunol.* **75**, 1–88.

Hirayama, K. (1999). Ex-germfree mice harboring intestinal microbiota derived from other animal species as an experimental model for ecology and metabolism of intestinal bacteria. *Exp. Anim.* **48**, 219–227.

Hirsjärvi, P. and Phyälä, L. (1995). Ivermectin treatment of a colony of guinea pigs infested with fur mite (*Chirodiscoides caviae*). *Lab. Anim.* **29**, 200–203.

Hogan, B., Beddington, R., Costantini, F. and Lacy, E. (1994). *Manipulating the Mouse Embryo.* Cold Spring Harbor Laboratory Press, New York.

Höglund, A. U. and Renström, A. (2001). Evaluation of individually ventilated cage systems for laboratory rodents: cage environment and animal health aspects. *Lab. Anim.* **35**, 51–57.

Huerkamp, M. J., Benjamin, K. A., Zitzow, L. A., Pullium, J. K., Lloyd, J. A., Thompson, W. D., Webb, S. K. and Lehner, N. D. M. (2000). Fenbendazole treatment without environmental decontamination eradicates *Syphacia muris* from all rats in a large, complex research institution. *Contemp. Topics* **39**, 9–12.

ILAR Committee on Long-term Holding of Laboratory Rodents (1976). *ILAR News* **XIX**, 4, L1–L25.

ILAR Committee on Immunologically Compromised Rodents (National Research Council) (1989). *Immunodeficient Rodents. A Guide to their Immunobiology, Husbandry, and Use.* National Academic Press, Washington, DC.

Jacoby, R. O. and Homberger, F. R. (1999). International standards for rodent quality. *Lab. Anim. Sci.* **49**, 230.

Jacoby, R. O. and Lindsey, J. R. (1997). Health care for research animals is essential and affordable. *FASEB J.* **11**, 609–614.

Jacoby, R. O. and Lindsey, J. R. (1998). Risks of infection among laboratory rats and mice at major biomedical research institutes. *ILAR J.* **39**, 266–271.

Jacoby, R. O., Ball-Goodrich, L. J., Besselsen, D. G., McKisic, M. D., Riley, L. K. and Smith, A. L. (1996). Rodent parvovirus infections. *Lab. Anim. Sci.* **46**, 370–380.

Jaisser, F. (2000). Inducible gene expression and gene modification in transgenic mice. *J. Am. Soc. Nephrol.* **11**, S95–S100.

Kawachi, S., Morise, Z., Jennings, S. R., Conner, E., Cockrell, A., Laroux, F. S., Chervenak, R. P., Wolcott, M., van der Heyde, H., Gray, L., Feng, L., Granger, D. N., Specian, R. A. and Grisham, M. B. (2000). Cytokine and adhesion molecule expression in SCID mice reconstituted with CD4[+] T cells. *Inflamm. Bowel Dis.* **6**, 171–180.

Kawamata, J., Yamanouchi, T., Dohmae, K., Miyamoto, H., Takahashi, M., Yamanishi, K., Kurata, T. and Lee, H. W. (1987). Control of laboratory acquired hemorrhagic fever with renal syndrome (HFRS) in Japan. *Lab. Anim. Sci.* **37**, 431–436.

Klement, P., Augustine, J. M., Delaney, K. H., Klement, G. and Weitz, J. I. (1996). An oral ivermectin regimen that eradicates pinworms (*Syphacia* spp.) in laboratory rats and mice. *Lab. Animal Sci.* **46**, 286–290.

Koopman, J. P., van den Brink, M. E., Vennix, P. P. C. A., Kuypers, W., Boot, R. and Bakker, R. H. (1991). Isolation of *Streptobacillus moniliformis* from the middle ear of rats. *Lab. Anim.* **25**, 35–39.

Kopf, M., Le Gros, G., Coyle, A. J., Kosco-Vilbois, M. and Brombacher, F. (1995). Immune responses of IL-4, IL-5, IL-6 deficient mice. *Immunol. Rev.* **148**, 45–49.

Kraft, L. M. (1958). Observation on the control and natural history of epidemic diarrhoea of infant mice (EDIM). *Yale J. Biol. Med.* **31**, 121–127.

Krause, J., McDonnell, G. and Riedesel, H. (2001). Biodecontamination of animal rooms and heat-sensitive equipment with vaporized hydrogen peroxide. *Contemp. Topics* **40**, 18–21.

Kühn, R. and Schwenk, F. (1997). Advances in gene targeting methods. *Curr. Opin. Immunol.* **9**, 183–188.

Kunstyr, I. (Ed.) (1988a). *List of Pathogens for Specification in SPF Laboratory Animals*. Society for Laboratory Animal Science Publ. No. 2, Biberach.

Kunstyr, I. (Ed.) (1988b). *Hygiene-Empfehlungen für Versuchstierbereiche*. Society for Laboratory Animal Science Publ. No. 5/6, Biberach.

Kunstyr, I. (Ed.) (1992). *Diagnostic Microbiology for Laboratory Animals*. GV-SOLAS Vol. 11. Gustav Fischer Verlag, Stuttgart.

La Regina, M., Woods, L., Klender, P., Gaertner, D. J. and Paturzo, F. X. (1992). Transmission of sialodacryoadenitis virus (SDAV) from infected rats to rats and mice through handling, close contact, and soiled bedding. *Lab. Anim. Sci.* **42**, 344–346.

Lane-Petter, W. and Pearson, A. E. G. (1971). *The Laboratory Animal – Principles and Practice*. Academic Press, London.

Le Blanc, R. Personal communication.

Le Blanc, S., Faith, R. E. and Montgomery, C. A. (1993). Use of topical ivermectin treatement for *Syphacia obvelata* in mice. *Lab. Animal Sci.* **43**, 526–528.

Lee, A., Phillips, M. W., O'Rourke, J. L., Paster, B. J., Dewhirst, F. E., Fraser, G. J., Fox, J. G., Sly, L. I., Romaniuk, P. J., Trust, T. J. and Kouprach, S. (1992). *Helicobacter muridarum* sp. nov., a microaerophilic helical bacterium with a novel ultrastructure isolated from the intestinal mucosa of rodents. *Int. J. Syst. Bacteriol.* **42**, 27–36.

Lindsey, J. R. (1998). Pathogen status in the 1990s: abused terminology and compromised principles. *Lab. Anim. Sci.* **48**, 557–558.

Lipman, N. S. (1999). Isolator rodent caging systems (state of the art): critical view. *Contemp. Topics* **38**, 9–17.

Lipman, N. S., Dalton, S. D., Stuart, A. R. and Arruda, K. (1994). Eradication of Pinworms (*Syphacia obvelata*) from a large mouse breeding colony by combination oral anthelmitic therapy. *Lab. Animal Sci.* **44**, 517–520.

Lipman, N. S., Perkins, S., Nguyen, H., Pfeffer, M. and Meyer, H. (2000). Mousepox resulting from use of ectromelia virus-contaminated, imported mouse serum. *Comp. Med.* **50**, 426–435.

Lussier, G. (1988). Potential detrimental effects of rodent viral infections on long-term experiments. *Vet. Res. Contrib.* **12**, 199–217.

Lyon, M. F., Rastan, S. and Brown, S. D. M. (eds) (1996). *Genetic Variants and Strains of the Laboratory Mouse*, 3rd edn. Oxford University Press, Oxford.

Macy, J. D. Jr., Weir, E. C., Compton, S. R., Shlomchik, M. J. and Brownstein, D. G. (2000). Dual infection with *Pneumocystis carinii* and *Pasteurella pneumotropica* in B cell-deficient mice: diagnosis and therapy. *Comp. Med.* **50**, 49–55.

Maess, J. and Kunstyr, I. (1981). Diagnose und Bekämpfung häufiger Parasiten bei kleinen Versuchstieren. *Tierärztl. Prax.* **9**, 259–264.

Mähler, M., Bedigian, H. G., Burgett, B. L., Bates, R. J., Hogan, M. E. and Sundberg, J. P. (1998). Comparison of four diagnostic methods for detection of *Helicobacter* species in laboratory mice. *Lab. Anim. Sci.* **48**, 85–91.

Markel, P., Shu, P., Ebeling, C., Carlson, G. A., Nagle, D. L., Smuko, J. S. and Moore, K. J. (1997). Theoretical and empirical issues for marker-assisted breeding of congenic mouse strains. *Nature Genet.* **17**, 280–284.

Marschall, S. and Hrabé de Angelis, M. (1999). Cryopreservation of mouse spermatozoa double your mouse space. *TIG* **15**, 128–131.

McKisic, M. D., Paturzo, F. X., Gaertner, D. J., Jacoby, R. O. and Smith, A. L. (1995). A nonlethal parvovirus infection suppresses rat T lymphocyte effector functions. *J. Immunol.* **155**, 3979–3986.

Morris, T. H. (1995). Antibiotic therapeutics in laboratory animals. *Lab. Anim.* **29**, 16–36.

Mosier, D. E. (1996). Small animal models for acquired immune deficiency syndrome (AIDS) research. *Lab. Animal Sci.* **46**, 257–265.

National Research Council, Committee on Infectious Diseases of Mice and Rats (1991). *Infectious Diseases of Mice and Rats.* National Academy Press, Washington, DC.

National Research Council (1996) *Guide for the Care and Use of Laboratory Animals.* National Academy Press, Washington, DC.

National Research Council (1997) *Occupational Health and Safety in the Use of Research Animals.* National Academy Press, Washington, DC.

Nehls, M., Kyewski, B., Messerle, M., Waldschutz, R., Schüddekopf, K., Smith, A. I. and Boehm, T. (1996). Two genetically separable steps in the differentiation of thymic epithelium. *Science* **272**, 886–889.

Nicklas, W. (1993). Possible routes of contamination of laboratory rodents kept in research facilities. *Scand. J. Lab. Anim. Sci.* **20**, 53–60.

Nicklas, W. (1996). Health monitoring of experimental rodent colonies; an overview. *Scand. J. Lab. Anim. Sci.* **23**, 69–75.

Nicklas, W., Giese, M., Zawatzky, R., Kirchner, H. and Eaton, P. (1988). Contamination of a monoclonal antibody with LDH-virus causes interferon induction. *Lab. Anim. Sci.* **38**, 152–154.

Nicklas, W., Kraft, V. and Meyer, B. (1993a). Contamination of transplantable tumors, cell lines, and monoclonal antibodies with rodent viruses. *Lab. Anim. Sci.* **43**, 296–300.

Nicklas, W., Staut, M. and Benner, A. (1993b). Prevalence and biochemical properties of V factor-dependent Pasteurellaceae from rodents. *Zentralbl. Bakt.* **279**, 114–124.

Nicklas, W., Homberger, F. R., Illgen-Wilcke, B., Jacobi, K., Kraft, V., Kunstyr, I., Mähler, M., Meyer, H. and Pohlmeyer-Esch, G. (1999). Implication of infectious agents on results of animal experiments. *Lab. Anim.* **33** (Suppl. 1, I) 39–87.

Nicklas, W., Baneux, P., Boot, R., Decelle, T., Deeny, A. A., Fumanelli, M. and Illgen-Wilcke, B. (2002). Recommendations for the health monitoring of rodent and rabbit colonies in breeding and experimental units. Recommendations of the Federation of European Laboratory Animal Science Associations (FELASA) Working Group on Health Monitoring of Rodent and Rabbit Colonies. *Lab. Anim.* **36**, 20–42.

Novak, E. K., McGarry, M. P. and Swank, R. T. (1985) Correction of symptoms of platelet storage pool deficiency in animal models for Chediak-Higashi syndrome and Hermansky-Pudlak syndrome. *Blood* **66**, 1196–1201.

Ohsugi, T., Kiuchi, Y., Shimoda, K., Oguri, S. and Maejima, K. (1996). Translocation of bacteria from the gastrointestinal tract in immunodeficient mice. *Lab. Anim.* **30**, 46–50.

O'Rourke, J., Lee, A. and McNeill, J. (1988). Differences in the gastrointestinal microbiota of specific pathogen-free mice: an often unknown variable in biomedical research. *Lab. Anim.* **22**, 297–303.

Otis, A. P. and Foster, H. L. (1983). Management and design of breeding facilities. In: *The Mouse in Biomedical Research*, Vol. 3 (H. L. Foster, J. D. Small and J. G. Fox, eds), pp. 18–35. Academic Press, Orlando, FL, USA.

Perkins, S. E. and Lipman, N. S. (1996). Evaluation of microenvironmental conditions and noise generation in three individually ventilated rodent caging systems and static isolator cages. *Contemp. Topics* **35**, 61–65.

Petri, M. (1966). The occurrence of *Nosema cuniculi* (*Encephalitozoon cuniculi*) in the cells of transplantable malignant ascites tumours and its effect upon tumour and host. *Acta Pathol. Microbiol. Scand.* **66**, 13–30.

Rabodonirina, M., Wilmotte, R., Dannaoui, E., Persat, F., Bayle, G. and Mojon, M. (1997). Detection of *Pneumocystis carinii* DNA by PCR amplification in various rat organs in experimental pneumocystosis. *J. Med. Microbiol.* **46**, 665–668.

Ray, M. K., Fagan, S. P. and Brunicardi, F. C. (2000). The Cre-loxP system: a versatile tool for targeting genes in a cell- and stage-specific manner. *Cell Transplant* **9**, 805–815.

Recio, L. and Everitt, J. (2001). Use of genetically modified mouse models for evaluation of carcinogenic risk: considerations for the laboratory animal scientist. *Comp. Med.* **51**, 399–405.

Reeb-Whitaker, C. K., Paigen, B., Beamer, W. G., Bronson, R. T., Churchill, G. A., Schweitzer, I. B. and Myers, D. D. (2001). The impact of reduced frequency of cage changes on the health of mice housed in ventilated cages. *Lab. Anim.* **35**, 58–73.

Reetz, I. C., Wullenweber-Schmidt, M., Kraft, V. and Hedrich, H.-J. (1988). Rederivation of inbred strains of mice by means of embryo transfer. *Lab. Animal Sci.* **38**, 696–701.

Rehg, J. E. and Toth, L. A. (1998). Rodent quarantine programs: purpose, principles, and practice. *Lab. Anim. Sci.* **48**, 438–447.

Rehg, J. E., Blackman, M. A. and Toth, L. A. (2001). Persistent transmission of mouse hepatitis virus by transgenic mice. *Comp. Med.* **51**, 369–374.

Renström, A., Björing, G. and Höglund, A. U. (2001). Evaluation of individually ventilated cage systems for laboratory rodents: occupational health aspects. *Lab. Anim.* **35**, 42–50.

Riley, V. (1966). Spontaneous mammary tumors: decrease of incidence in mice infected with an enzyme-elevated virus. *Science* **153**, 1657–1658.

Riley, L. K., Franklin, C. L., Hook, R. R. and Besch-Williford, C. (1996). Identification of murine Helicobacters by PCR and restriction enzyme analyses. *J. Clin. Microbiol.* **34**, 942–946.

Russel, R. J., Haines, D. C., Anver, M. R., Battles, J. K., Gorelick, P. L., Blumenauer, L. L., Gonda, M. A. and Ward, J. M. (1995). Use of antibiotics to prevent hepatitis and typhlitis in male SCID mice spontaneously infected with *Heliobacter hepaticus*. *Lab. Animal Sci.* **45**, 373–378.

Russell, W. and Hurst, J. (1995). Pure strain mice born to hybrid mothers following ovarian transplantation. *Proc. Natl. Acad. Sci.* **31**, 267–273.

Sambrook, J. Fritsch, E. F. and Marsiatis, T. (Eds.) (1989). *Molecular Cloning. A Laboratory Manual, 2nd edn.* Cold Spring Harbor Laboratory Press, New York.

Scanziani, E., Gobbi, A., Crippa, L., Giusti, A. M., Gavazzi, R., Cavaletti, E. and Luini, M. (1997). Outbreaks of hyperkeratotic dermatitis in athymic mice in Northern Italy. *Lab. Anim.* **31**, 206–211.

Scanziani, E., Gobbi, A., Crippa, L., Giusti, A. M., Pesenti, E., Cavaletti, E., and Luini, M. (1998). Hyperkeratosis-associated coryneform infection in severe combine immunodeficient mice. *Lab. Anim.* **32**, 330–336.

Schaible, U. E. and Kaufmann, S. H. (2000). CD1 molecules and CD1-dependent T cells in bacterial infections: a link from innate to acquired immunity? *Semin. Immunol.* **12**, 527–535.

Schauer, D. B., Zabel, B. A., Pedraza, I. F., O'Hara, C. M., Steigerwalt, A. G. and Brenner, D. J. (1995). Genetic and biochemical characterization of *Citrobacter rodentium* sp. nov. *J. Clin. Microbiol.* **33**, 2064–2068.

Schenkel, J. (1995). *Transgene Tiere*, pp. 61–108. Spektrum Akademischer Verlag, Heidelberg.

Scopets, B., Wilson, R. P., Griffith, J. W. and Lang, C. M. (1996). Ivermectin toxicity in young mice. *Lab. Animal Sci.* **46**, 111–112.

Selwyn, M. R. and Shek, W. R. (1994). Sample sizes and frequency of testing for health monitoring. *Contemp. Top. Lab. Anim. Sci.* **33**, 55–60.

Shevach, E. M. (2000). Regulatory T cells in autoimmunity. *Annu. Rev. Immunol.* **18**, 423–449.

Silver, L. M. (1985). Mouse t haplotypes. *Ann. Rev. Genet.* **19**, 179–208.

Silver, L. M. (1995). *Mouse Genetics. Concepts and Applications.* Oxford University Press, Oxford.

Smith, A. L., Casals, J. and Main, A. J. (1983). Antigenic characterization of tettnang virus: complications caused by passage of the virus in mice from a colony enzootically infected with mouse hepatitis virus. *Am J. Trop. Med. Hyg.* **32**, 1172–1176.

Smith, M. W. (1999). Safety and hygiene. In: *The Care and Management of Laboratory Animals*, (Poole, T. Ed.). 7th edn, Vol. 1, pp. 141–170. Blackwell Science Ltd., Oxford.

Songsasen, N., Betteridge, K. J. and Leibo, S. P. (1997). Birth of live mice resulting from oocytes fertilized in vitro with cryopreserved spermatozoa. *Biol. Reprod.* **56**, 143–152.

Spiegel, A. (1976). *Versuchstiere: Eine Einführung in die Grundlagen ihrer Zucht und Haltung.* Gustav Fischer Verlag, Stuttgart.

Stevens, L. C. (1957). A modification of Robertson's technique of homoitopic ovarian transplantation in mice. *Transplant. Bull.* **4**, 106–107.

Strain, B. A. and Gröschel, D. H. M. (1995). Laboratory safety and infectious waste management. In: *Manual of Clinical Microbiology*, (Murray, P. R. *et al.* Eds). 6th edn, pp. 75–85. American Society for Microbiology, Washington, DC.

Sztein, J. M., Sweet, H., Farley, J. S. and Mobraaten, L. E. (1999). Cryopreservation and orthotopic transplantation of mouse ovaries: New approach in gamet banking. *Biol. Reprod.* **58**, 1071–1074.

Sztein, J. M., Farley, J. S. and Mobraaten, L. E. (2000). In vitro fertilization with cryopreserved inbred mouse sperm. *Biol. Reprod.* **63**, 1774–1780.

Sztein, J. M., Noble, K., Farley, J. S. and Mobraaten, L. E. (2001). Comparison of permeating and nonpermeating cryoprotectants for mouse sperm cryo-preservation. *Cryobiol.* **42**, 28–39.

Tietjen, R. M. (1992). Transmission of minute virus of mice into a rodent colony by a research technician. *Lab. Anim. Sci.* **42**, 422.

Toth, L. A., Oberbeck, C., Straign, C. M., Frazier, S. and Rehg, J. E. (2000). Toxicity evaluation of prophylactic treatments for mites and pinworms in mice. *Contemp. Topics* **39**, 18–21.

Trexler, P. C. (1983). Gnotobiotics. In: *The Mouse in Biomedical Research* (H. L. Foster, J. D. Small and J. G. Fox, Eds), pp. 1–16. Academic Press, Orlando, FL.

Tu, H., Diberadinis, L. J. and Lipman, N. S. (1997). Determination of air distribution, exchange, velocity, and leakage in three individually ventilated rodent caging systems. *Contemp. Topics* **36**, 69–73.

Van den Broek, M. F., van Bruggen, M. C., Koopman, J. P., Hazenberg, M. P. and van den Berg, W. B. (1992). Gut flora induces and maintains resistance against streptococcal cell wall-induced arthritis in F344 rats. *Clin. Exp. Immunol.* **88**, 313–317.

Van de Weerd, H. A., Van Loo, P. L. P., Van Zutphen, L. F. M., Koolhaas, J. M. and Baumans, V. (1997). Preferences for nesting material as environmental enrichment for laboratory mice. *Lab. Anim.* **31**, 133–143.

Van Duyne, G. D. (2001). A structural view of cre-loxp site-specific recombination. *Annu. Rev. Biophys. Biomol. Struct.* **30**, 87–104.

Visscher, P. M. (1999). Speed congenics: accelerated genome recovery using genetic markers. *Genet. Res.* **74**, 81–85.

Waggie, K., Kagiyama, N., Allen, A. M. and Nomura, T. (eds) (1994). Manual of microbiologic monitoring of laboratory animals. 2nd edn. U.S. Department of Human Health and Human services, NIH Publication No. 94-2498.

Wakeland, E., Morel, L., Achey, K., Yui, M. and Longmate, J. (1997). Speed congenics: a classic technique in the fast lane (relatively speaking). *Immunol. Today* **18**, 472–477.

Wang, B., Simpson, S. J., Hollander, G. A. and Terhorst, C. (1997). Development and function of T lymphocytes and natural killer cells after bone marrow transplantation of severely immunodeficient mice. *Immunol. Rev.* **157**, 53–60.

Weir, E. C., Jacoby, R. O., Paturzo, F. X., Johnson, E. A. and Ardito, R. B. (1990). Persistence of sialodacryoadenitis virus in athymic rats. *Lab. Anim. Sci.* **40**, 138–143.

Weisbroth, S. H. (1996). Post-indigenous disease: changing concepts of disease in laboratory rodents. *Lab. Animal* **25**(9), 25–33.

Weisbroth, S. H., Peters, R., Riley, L. K. and Shek, W. (1998). Microbiological assessment of laboratory rats and mice. *IILAR J.* **39**, 272–290.

Weisbroth, S. H., Geistfeld, J., Weisbroth, S. P., Williams, B., Feldman, S. H., Linke, M. J., Orr, S. and Cushion, M. T. (1999). Latent *Pneumocystis carinii* infection in commercial rat colonies: Comparison of inductive immunosuppressants plus histopathology, PCR, and serology as detection methods. *J. Clin. Microbiol.* **37**, 1441–1446.

Whittingham, D. G., Leibo, S. P. and Mazur, P. (1972). Survival of mouse embryos frozen to –196°C and –269°C. *Science* **78**, 411–414.

Wilkerson, J. D., Brooks, D. L., Derby, M. and Griffey, S. M. (2001). Comparison of practical treatment methods to eradicate pinworm (*Dentostomella translucida*) infections from Mongolian gerbils (*Meriones unguiculatus*). *Contemp. Topics* **40**, 31–36.

Wilmut, I. (1972). The effect of cooling rate, warming rate, cryoprotective agent and stage of development on survival of mouse embryos during freezing and thawing. *Life Sci.* **11**, 1072–1079.

Wostmann, B. S., Reddy, B. S., Bruckner-Kardoss, E., Gordon, H. A. and Singh, B. (1973). Causes and possible consequences of cecal enlargement in germfree rats. In: *Germfree Research* (Heneghan, J.B., Ed.) pp. 261–270, Academic Press, New York.

Wullenweber, M., Kaspareit-Rittinghausen, J. and Farouq, M. (1990). *Streptobacillus moniliformis* epizootic in barrier-maintained C57BL/6 mice and susceptibility to infection of different strains of mice. *Lab. Anim. Sci.* **40**, 608–612.

Yamanishi, K., Dantas, J. R. F., Takahashi, M., Yamanouchi, T., Domae, K., Kawamata, J. and Kurata, T. (1983). Isolation of hemorrhagic fever with renal syndrome (HFRS) virus from a tumor specimen in a rat. *Biken J.* **26**, 155–160.

Yoshimura, M., Endo, S., Ishihara, K., Itoh, T., Takakura, A., Ueyama, Y. and Ohnishi, Y. (1997) Quarantine for contaminated pathogens in transplantable human tumors or infections in tumor bearing mice. *Exp. Animals* **46**, 161–164.

Yu, Y. and Bradley, A. (2001). Engineering chromosomal rearrangements in mice. *Nat. Rev. Genet.* **2**, 780–790.

Zenner, L. (1998). Effective eradication of pinworms (*Syphacia muris, Syphacia obvelata* and *Aspiculuris tetraptera*) from a rodent breeding colony by oral anthelmintic therapy. *Lab. Anim.* **32**, 337–342.

Zenner, L. and Regnault, J. P. (2000). Ten-year long monitoring of laboratory mouse and rat colonies in French facilities: a retrospective study. *Lab. Anim.* **34**, 76–83.

Management of Immuno-compromised Animals

2 *In Vitro* Analysis

2.1 Isolation and Preparation of Lymphocytes from infected animals for *in vitro* analysis

Charles J Czuprynski and James F Brown
Department of Pathobiological Sciences, School of Veterinary Medicine, University of Wisconsin-Madison, USA

◆◆

CONTENTS

◆◆◆◆◆◆ **INTRODUCTION**

Lymphocytes are the cells which provide specificity to host defence. Identifying the phenotype and antigen specificity of lymphocytes, that have been isolated from animals infected with microbial agents, is integral to understanding protective adaptive immunity. There are various methods to isolate, purify and characterize lymphocytes from the tissues of infected animals. Some of these are elegant and sophisticated procedures that rely on expensive instrumentation (i.e. flow cytometry) to yield highly purified and well-characterized cell populations. These procedures are not the principal subject of this chapter, which will focus largely on simple preparative techniques that can be used by nearly any microbiology laboratory. These techniques will yield populations of lymphocytes suitable for functional assessment *in vitro*, or adoptive transfer to recipient animals *in vivo*. For detailed description of more sophisticated techniques, the reader is referred to specific chapters in this text dealing with these subjects.

Copyright © Elsevier Science Ltd
All rights of reproduction in any form reserved

◆◆◆◆◆◆ SOURCES OF LYMPHOCYTES

Lymphocytes can be obtained from a variety of tissues of infected animals. The numbers of cells that can be recovered from each site vary depending on the type of tissue, the age and physiological status of the animals, the immune status of the donor, the virulence of the infectious agent, and the time during the infection at which the tissue is sampled. The investigator needs to decide what site is most relevant, convenient, and able to yield the required numbers of cells, before initiating a study. For systemic immunization or infection studies, the most common sources of lymphoid cells are the spleen, peritoneal cavity, or draining lymph nodes. If the investigator is interested in the mucosal immune response, then relevant sites include the mucosal associated lymphoid tissues (Peyer's patches, etc.), mucosal epithelium, or draining lymph nodes.

Thymus

The thymus is relatively large in young animals, and then involutes with age. It is easily removed and disrupted, yielding large numbers of a relatively pure population of T lymphocytes; few B lymphocytes are present. These T cells are largely immature cells (mostly CD4⁺CD8⁺), that have not yet been primed by exposure to antigen (Ritter and Boyd, 1993).

Spleen

Unlike the thymus, the spleen does not involute as the animal matures. The spleen can be readily disrupted, yielding mixed suspensions of T and B lymphocytes, macrophages, some neutrophils, dendritic cells and stromal cells (fibroblasts). The latter are usually a minority of the cells present. Splenomegaly is common during many types of microbial infection. This increases the numbers of cells that can be isolated, and changes the phenotypes of the cell populations present (Haak-Frendscho and Czuprynski, 1992).

Peritoneal exudate cells

The unperturbed peritoneal cavity contains a mixed population of mononuclear cells (both lymphocytes and macrophages). Neutrophils and eosinophils should be present in very low numbers. Mast cells are a minority population (generally 5% or less), but are prominent because of their large basophilic granules. After i.p. injection of microbes, microbial products, or other sterile irritants, there is a rapid influx of inflammatory cells into the peritoneal cavity. There is temporal sequence to the leuko-cyte infiltration (Czuprynski *et al.*, 1984). The earliest to arrive are neutro-phils, which predominate during the early phase of inflammation (4–12 h). These are later replaced by eosinophils and mononuclear cells (especially macrophages), although neutrophils continue to be present in significant numbers through 24 h after injection of sterile irritants such as thioglycollate or proteose peptone. The numbers and types of cells vary depending on the genetic phenotype and immune status of the recipient, and the nature of the agent injected. Injection of microbial antigens may elicit little cellular response in a non-immunized animal, whereas it can initiate a significant influx of inflammatory leukocytes in an immunized animal (Czuprynski *et al.*, 1985). However, it should be recognized that pattern recognition receptors on leukocytes can react to common motifs in certain microbial cell wall components (e.g. LPS, lipoprotein) (Medzhitov and Janeway, 2000) that might elicit substantial inflammatory response even in a non-immunized animal.

Lymph node cells

Lymph node cell suspensions contain B and T lymphocytes, as well as macrophages, dendritic cells and stromal cells. The numbers of lymphocytes recovered from lymph nodes depend on which nodes are collected, and whether there has been recent antigen stimulation. A single mouse lymph node will yield relatively few cells (generally not more than 1×10^6). Thus, recovery of adequate numbers of cells requires careful planning regarding the numbers of animals in each experimental group, and the number of lymph nodes that must be harvested to obtain the needed numbers of cells. Lymph nodes contain a tough capsule that must be mechanically disrupted with forceps, or a mesh screen, to release the lymph node cells from the network of stromal cells and extracellular matrix proteins.

Peyer's patch and other mucosal-associated lymphoid tissue

If one is interested in investigating the mucosal immune response, it is important to obtain lymphocytes from mucosal-associated lymphoid tissue (MALT) that has been exposed to the microbial agent or its antigens. The specific mucosal site sampled depends on the nature of the infectious agent being studied. Most investigations focus on the gut-associated lymphoid tissue (GALT), since at least some of its lymphoid aggregates (i.e. Peyer's patches) are visible to the naked eye and can be readily removed from the surrounding mucosal tissue. Similar principles would be followed if working with lymphocytes from other mucosal sites (e.g. respiratory or genitourinary tracts). These large lymphoid aggregates can be excised and mechanically dispersed as described for lymph node cells. In addition to the large lymphoid aggregates in the MALT, there are also intraepithelial lymphocytes (IEL) scattered throughout the mucosa. These can be released by simply incubating the tissue in medium, allowing the IEL to emigrate out of the tissue into the medium (Ishikawa *et al.*, 1993).

The following protocol has been reported for isolation of murine IEL (Ishikawa *et al.*, 1993).

1. Remove the small intestine and flush out the lumen contents. Invert the intestine with a piece of polyethylene tubing, and then cut it into three or four segments. Transfer up to 10 segments to a plastic box containing 250 ml RPMI-1640 with 2% FBS, 25 mM HEPES and penicillin-streptomycin (100 units/ml^{-1} and 100 µg ml^{-1}, respectively).
2. Place the box on an orbital shaker (150 rpm) in a 37°C incubator for 45 min.
3. Remove the non-adherent cell suspension and pass through a glass wool column to remove debris and adherent cells (i.e. macrophages and stromal cells).
4. Add the non-adherent cells to a discontinuous Percoll gradient (44% and 70%) and centrifuge for 30 min at 400 g. The IEL are removed from the gradient interface and washed twice in RPIM-1640 with 2% FBS.

◆◆◆◆◆◆ ISOLATION OF LYMPHOCYTES FROM TISSUES

Both mechanical and enzymatic techniques can be used to disrupt and release lymphocytes from normal tissue architecture. In most instances, the lymphocytes remain non-adherent and can be easily removed from the larger and denser parenchymal and stromal cells. The lymphocyte-enriched cell suspensions can then be further purified to obtain the cells needed. Specific procedures that can be used to release tissue lymphocytes are described below.

Mechanical disruption

Tissue can be mechanically disrupted using several methods. The simplest is to cut the tissue into convenient sized fragments, and then push these through a sterile nylon strainer using sterile forceps or the shaft of a plastic syringe. Some investigators prefer to tease the tissue apart using sterile forceps and scissors. With either procedure, the concept is the same: break open the tissue capsule, disrupt the stromal architecture and release the lymphocytes.

In the past, investigators frequently relied on the use of narrow mesh stainless steel screens for tissue disruption. These are effective, but present challenges regarding cleaning and sterilization. It is more convenient to use the disposable sterile nylon mesh screens that can be purchased from commercial suppliers (Falcon #2340, Becton Dickinson, Bedford, MA, USA). The lymphoid tissue (e.g. spleen or lymph node) is rubbed across and pushed through the screen into a small plastic dish that contains tissue culture media or balanced salts solution. The cell suspensions are then washed several times to remove debris and used as an unseparated cell suspension, or subjected to further purification as needed.

◆◆◆◆◆◆ CELL SEPARATION

Use of density gradients

The use of density gradients was initially described by Boyum (1968). Employing a mixture of Ficoll (Sigma #8016, St Louis, MO, USA) and sodium diatrizoate (Hypaque, a radio-contrast medium, Sigma #S-4506), one can obtain gradients of specified density and osmolarity suited to the separation of cell types based on buoyant density. The specific density of the gradient that is needed for successful cell separation depends on the species and cell type of interest. For example, human peripheral blood cells can be purified from diluted whole blood (using PBS as diluent, generally at two to three parts per volume of blood) using Ficoll-Hypaque of 1.077 density (290 mM osmolarity, Sigma #1077-1). This can also be used for mouse peritoneal or spleen cells, although some find that a Ficoll-Hypaque of 1.083 density (320 mM osmolarity, 'Lympholyte-M', #ACL-5030, Accurate Chemical, Westbury, NY, USA) is better suited for this purpose. Determining the optimum density for a particular cell from a given species is an empirical process. Starting with a 1.081 density medium, and centrifuging at $400\,g$ for 30 min at room temperature, is a good starting point. By altering the density, time and centrifugal force, a satisfactory separation can be achieved for many cell types.

The following procedure can be used to isolate murine lymphoid cells.

1. Suspend the cells in calcium- and magnesium-free PBS or HBSS, with 10 mM EDTA added to prevent cell clumping or fibrin deposition. The cells should preferably be placed in a round-bottom, rather than conical, disposable plastic polystyrene tube.
2. Using a long Pasteur pipette, or sterile needle with a syringe, underlay the diluted cell suspension with 0.5 ml Ficoll-Hypaque per ml of cell suspension. This should be added slowly, to avoid mixing.
3. The tubes should be centrifuged at $400\,g$ for 30 min at 22°C. The G-force needed will vary depending on the density of the Ficoll-Hypaque, and the type of cells being isolated. A nomograph for your centrifuge should be consulted to determine what speed is needed to obtain the desired G force. It is important that the centrifugation is carried out at 22–25°C, as the density of the gradient changes at refrigeration temperature (4–10°C).
4. The mononuclear cells (lymphocytes and mononuclear phagocytes) will accumulate as a hazy white band at the gradient interface, from which they can be readily aspirated.
5. Granulocytes (principally neutrophils) can be recovered from the cell pellet. The pellet can also contain erythrocytes, mast cells, eosinophils, immature leukocytes, and possibly stromal cells.

(contd.)

6. Contaminating red cells can be lysed with dilute ammonium chloride (8.29 g NH_4Cl, 0.37 g Na_2 EDTA, and 1 g $KHCO_3$ in 1 l distilled H_2O, pH 7.3). In some instances (i.e. bovine blood), red cells can be lysed by a brief hypotonic shock using dilute phosphate buffer (37 mM phosphate, pH = 7.20, no sodium chloride) for 45 s, before restoring isotonic conditions (by adding 0.1 volume 8.5% NaCl buffered with 37 mM phosphate, pH = 7.20).
7. Both the mononuclear and granulocyte populations should be washed two or three times in Ca^{2+} and Mg^{2+} containing HBSS, or tissue culture medium, to remove the contaminating Ficoll-Hypaque before the cells are used. Failure to do so could result in loss of viability or functional activity, as the Ficoll solution is somewhat toxic to cells.
8. Contamination with bacterial endotoxin can occur (Haslett et al., 1985), which is a concern in many cell culture systems. Each investigator should check their stocks of reagents, and the distilled water used to prepare them, with the *Limulus* assay (Biowhitakker, Walkersville, MD, USA) to estimate levels of endotoxin contamination.

An alternative density gradient procedure involves the use of Percoll, a polyvinylpyrrolidone coated colloidal silica in water. Percoll can be used to produce discontinuous or continuous density gradients, which can be used in a manner similar to Ficoll-Hypaque, to isolate various types of leukocytes (Harbeck et al., 1982).

Use of adherence to remove mononuclear phagocytes

The tendency of mononuclear phagocytes to adhere to glass or tissue culture plastic can be used to enrich for, or remove, macrophages from a mononuclear cell suspension. Mononuclear phagocytes obtained from different tissue sites are all adherent, but can differ in how strongly they adhere. In general, cells activated at sites of inflammation *in vivo* are more adherent than resident cells. The purity of the lymphocyte populations obtained after removal of adherent macrophages can vary, depending on the type of tissue culture grade plastic used, and the presence or absence of additional coating (e.g. serum proteins, synthetic extracellular matrix, poly-L-lysine) on the surface. However, commonly available tissue culture plates have all been designed to promote cell adherence, so that the differences among manufacturers should not pose significant problems in many applications. Adherence can be performed with or without the addition of serum or other proteins (e.g. bovine serum albumin). In general, mononuclear phagocyte adherence is higher in the absence of serum, but non-specific adherence of lymphocytes is also greater (Musson, 1979).

A general outline for depletion of mononuclear phagocytes, to yield lymphocyte-enriched cell populations is described below (Czuprynski *et al.*, 1985). To perform adherence, mononuclear cells are washed two or three times in HBSS or tissue culture medium.

1. The mononuclear cells are plated onto tissue culture flasks at a density not to exceed 2×10^7 cells per 25 cm^2 flask, or 2–4×10^6 cells per ml (or well), in a 24-well tissue culture cluster plate.
2. The cells are incubated at the appropriate temperature (usually 37 to 39°C) with 5% CO_2 for 1 to 2 h.
3. To recover a mononuclear phagocyte-depleted population of cells (i.e. largely lymphocytes), the non-adherent cells are gently remove by carefully decanting them into a new tissue culture flask. A second round of adherence can be performed, if needed, to try to remove additional mononuclear phagocytes.
4. The non-adherent cells are then centrifuged, counted with a haemocytometer and checked for viability (see section on viability and counting cells).
5. To confirm depletion of mononuclear phagocytes, a cytocentrifuge smear is prepared on a clean glass slide and a differential stain performed (Wright-Giemsa, or Diff-Quik). The esterase stain is a more specific staining technique, mononuclear phagocytes will usually stain darkly; whereas, lymphocytes will be negative or exhibit a discrete spot of staining (Koski *et al.*, 1980).
6. Mononuclear phagocyte depletion can be verified by staining with a fluorochrome-conjugated antibody directed against a monocyte/macrophage-specific cell surface antigen (i.e. F4380 in the mouse) (Springer, 1981), followed by flow cytometry or fluorescent microscopy. Assessing ingestion of latex beads, or opsonized yeast, is an additional functional assay, that can be performed to identify phagocytes.
7. To recover a mononuclear phagocyte-enriched population, the adherent cells can be physically removed with a plastic cell-scraper after two or three *gentle* washes of the adherent cells with 10–15 ml warm medium.
8. Alternatively, the adherent cells can be removed from the flask surface by incubation for 5–10 min at 37°C with 0.1% trypsin in Ca^{2+} and Mg^{2+} free PBS or HBSS with 10 mM EDTA. Because trypsin treatment also removes proteins from the cell surface, caution must be used when staining for surface antigens, or performing functional assays.

Differential staining (Wright-Giemsa or Diff-Quik) and estimation of viability (i.e. trypan blue exclusion) should always be performed. If there is a significant percentage of non-viable cells, these can be removed by pelleting the cell suspension through a Ficoll gradient, as described earlier. The viable mononuclear cells will remain at the gradient interface, whereas the dead cells will pellet at the bottom of the gradient.

Use of nylon wool to enrich for T lymphocytes

Nylon wool has long been used as a rapid preparative method to enrich T cells from complex mixtures of cells such as bone marrow, peripheral blood, or spleen (Julius *et al.*, 1973). The technique takes advantage of the relative 'stickiness' of mononuclear phagocytes, and B-cells, which tend to adhere when passed slowly through a column (usually a 10 ml syringe) of loosely packed nylon wool.

1. The nylon wool (Polysciences, Warrington, PA, USA) is pretreated, to remove toxic impurities, by boiling for 1 h in distilled water which has been made 0.2 N in HCl. The wool is then rinsed extensively with tissue culture-grade (18 mega-ohm) water by boiling for 20 min, in three successive changes of water.
2. The nylon wool is dried, packed loosely (to the 8 ml mark) in 10 ml polypropylene syringes, wrapped and autoclaved. A nylon wool column so prepared is sufficient to allow T-cell enrichment from a starting cell suspension of no more than 5×10^7 cells. The nylon wool can be removed from the column and reused by following the cleaning procedure outlined above.
3. To enrich for T lymphocytes, resuspend up to 5×10^7 mononuclear cells in 1.0 ml HBSS with 5% fetal bovine (HBSS-FCS) serum. The columns should be washed with at least 35 ml of warm medium before the cells are added in a volume of 1 to 2 ml. The cells are washed into the column with an additional 2 ml of warm medium, and the column incubated in a 37°C incubator for 45 min.
4. The T-lymphocyte enriched cell suspension is eluted by slowly washing the column with 20 ml of warm HBSS-FBS. The eluted cells are then centrifuged and resuspended at an appropriate concentration in the medium of choice (usually RPMI-1640 or DMEM).
5. The efficacy of macrophage depletion should be verified by microscopic examination of cytocentrifuge prepared smears that are differential stained with Wright-Giemsa or Diff-Quik. Monocyte contamination can be determined by esterase staining, and T-cell purity by fluorescent microscopy or flow cytometry, using FITC-Thy1 MAb.
6. Usually one can expect to obtain approximately 60–80% T cells (Julius *et al.*, 1973; Czuprynski *et al.*, 1985). Multiple rounds of nylon wool passage can improve this somewhat.

Use of complement-mediated lysis to remove cell populations

Complement-mediated lysis provides a powerful and convenient method for eliminating specific cell populations (Hathcock, 1991). The chief limitation is that an antibody must be used, specific for the surface marker of interest, of an isotype that can fix complement. Lymphocytes are suspended in RPMI (or HBSS) containing 5% FBS, at a cell concentration ranging from 1×10^7 to 1×10^8 cells ml^{-1}. The total volume should be kept small to conserve the amount of antibody used. The antibody used can be polyclonal or monoclonal, and either purified or unpurified (ascites, anti-serum, or culture supernatant). Appropriate specificity controls (an Ab of the same isotype that does not bind to the cell type of interest), must be included in all cases. If the antibody concentration, and complement fix-ing activity, are not known, these should be titrated beforehand, to deter-mine the minimum amount needed for effective lysis.

If the antibody for the marker of interest does not fix complement, then a complement fixing second antibody, which specifically recognizes the heavy or light chains of the first antibody, can be used. Concern about complement fixing ability is less of an issue when using polyclonal antibodies, since immunized rabbits or goats will generally produce some antibody clones which bind complement.

The complement source is important. Mouse serum is generally low in complement lytic activity (Ooi and Colten, 1979), hence a homologous system cannot be used. Baby rabbit serum (Low-Tox M, Accurate Chemical, Westbury, NY, USA) is an excellent source of high-titre complement, that generally works well with mouse and rabbit antibodies. Guinea-pig complement also usually works well. It may be necessary to absorb the complement source with agarose (see below) to remove non-specific toxicity before it is added to the antibody-coated cell suspension (Cohen and Schlesinger, 1970). Appropriate controls include use of an irrelevant antibody of the same isotype, plus the complement source, to demonstrate specificity of cell lysis.

To absorb rabbit or guinea-pig serum for use in complement depletion:
1. Thaw the serum on ice.
2. Add 1 g of molecular-biology grade agarose per 10 ml of serum.
3. Incubate on ice, with occasional mixing, for 15 min.
4. Centrifuge at $400\,g$ and carefully remove the supernatant. Store the absorbed serum in aliquots of –70°C. To prevent loss of lytic activity, avoid freeze-thawing, and keep on ice until used in an experiment.

The following protocol could be used to perform complement-mediated depletion of a T-cell subset.

1. Suspend mononuclear cells in RPMI containing 5% FBS at a cell concentration of $5–50 \times 10^6$ per ml.
2. Include a control antibody tube (same isotype that does not bind the cells of interest), and a complement-only control tube. Add the primary antibody at 1–2 µg per 10^6 cells (it is important to use a concentrated preparation of antibody to avoid excessive dilution of the cells).

(contd.)

3. Incubate on ice for 45 min, wash twice with media, and resuspend the cells in the original volume of medium.
4. If the primary antibody does not fix complement, add 1–2 μg of the second antibody (which recognizes the light or heavy chains of the first antibody), and incubate 30 min on ice.
5. Wash the cells twice with ice-cold RPMI-FBS, and resuspend in the original volume of the same.
6. Add an equal volume of the complement source (diluted to an appropriate concentration in medium) and incubate in a 37°C water bath for 30–45 min.
7. Wash the cell suspension twice with medium, and verify depletion by staining with FITC-labelled antibody for the marker of interest, followed by flow cytometry or fluorescent microscopy.

If cell depletion is not satisfactory, a second round of complement mediated lysis can be performed. If the cells exhibit clumping, strain through a 40 μm nylon screen, or gently pipette up and down through a 5 ml pipette, to disperse the cell clumps. This is particularly important if the cells are to be transferred to recipient animals, as i.v. injection of cell clumps can cause thrombosis and death.

Panning

Petri dishes or plastic flasks coated with an antibody against a particular cell surface marker can be used to immobilize that cell type on the plastic surface ('panning') (Wysocki and Sato, 1978). The technique works best for negative selection, but can be used for positive selection as well.

1. Use bacteriologic plastic petri dishes to reduce non-specific adherence of cells. Coat the dishes with 4 to 5 ml of a solution (150 mM NaCl with 50 mM Tris buffer, pH 9.5) containing Ab (10 μg per ml) specific for the surface marker of interest, and incubate overnight at 4°C.
2. Remove the fluid and wash the dish four times with PBS-1% FBS. Block non-specific binding sites by incubating the dish with PBS-1% FBS for 30 min at room temperature.
3. Wash the plate four times with HBSS-1% FBS. Add 20–200 × 10⁶ cells per dish (depending on cell type) in 5 ml cold HBSS-1% FBS.
4. Incubate the dish at 4°C for 1 h. Gently aspirate the medium and remove the unattached cells. Wash twice with 3 to 5 ml cold HBSS-1% FBS.
5. Centrifuge the cells at 4°C. Wash twice with the medium that will be used in subsequent experiments.

◆◆◆◆◆◆ QUALITY CONTROL

Once a cell suspension is obtained, it is important that it be assessed for the purity of the cell populations present and the viability of those cells. There are various ways this can be done. We will briefly discuss some of the simpler and more rapid methods below.

Viability

Trypan blue

Probably the most frequently used rapid method for assessing cell viability is the exclusion of trypan blue dye. Viable cells actively transport trypan blue out of the cell and remain refractile and colourless. In contrast, cells that are dead, or whose cell membrane is damaged, cannot eliminate trypan blue and will appear pale to dark blue (Caron-Leslie *et al.*, 1994). Although rapid, easy and inexpensive, the method suffers from several limitations. First, among these is the question of whether blue cells are truly dead, or exhibiting membrane damage that might be reversible. Secondly, if cells have been killed and completely lysed, then enumerating only the percentage of cells that take up trypan blue may grossly underestimate cell death, unless it is incorporated into a total cell count. A protocol for performing trypan blue exclusion is given below.

1. Make up 0.4% trypan blue in isotonic saline.
2. Add 1 drop per 0.2 ml of cell suspension.
3. Load into a haemocytometer and count at least 100 cells at 100× magnification. Score for both blue (dead) cells, and the total number of cells.
4. Calculate the percentage of dead cells, and the total number of cells present.

Propidium iodide staining

Propidium iodide is also excluded from viable cells. Dead cells become permeable to the dye, which then intercalates with their DNA. The resulting staining can be detected by flow cytometry or fluorescence microscopy (Shapiro, 1988). Using proper gating for the cell population of choice, this can provide a useful estimate of the proportion of dead cells in a cell suspension.

Vital dye staining

Some dyes (e.g. neutral red) are only taken up by viable cells. These dyes can be used to stain the cells, and estimate the viability of the cell population (monolayer or cell suspension) by the relative light absorbance (Kaufmann *et al.*, 1987). Inclusion of appropriate controls for background lysis, and maximal cell lysis, allows generation of standard curves from which one can extrapolate the number of viable cells.

Reduction of tetrazolium salts

Viable cells will take up various tetrazolium salts (MTT, DTT, XTT). These compounds are reduced to a formazan compound in the mitochondria of intact cells. This formazan product can be readily detected with a spectrophotometer, thus providing a means of measuring both cell viability and metabolic activity (Green *et al.*, 1984; Scudiero *et al.*, 1988). For cells that proliferate (lymphocyte blast transformation), MTT or XTT provides an attractive alternative to the use of radioisotopes like [^{3}H]-thymidine. For cells that are not proliferating, the absence of signal (i.e. decreased dye reduction) is an indication of cell death, or at least physiological inactivity. Inclusion of proper controls for background and maximal cell lysis, allows estimation of the percentage of viable cells.

---

1. Add cells (2×10^6 per ml) to wells of a 96-well microtitre tissue culture plate.
2. Prepare XTT (1 mg per ml) in warm medium without sera, and phenazine methosulfate (PMS) at 5 mM (1.53 mg ml^{-1}) in PBS.
3. Add PMS solution (1 : 20 volume) to XTT solution to make 0.25 mM PMS-XTT (both from Sigma, St Louis, MO, USA) solution.
4. Remove the medium from the cells and replace with PMS (0.25 mM)-XTT solution. Incubate for 4 h at 37°C. It might be necessary to incubate resting cells, that are metabolically less active, for a longer period.
5. Mix the medium, and read absorbance at 450 nm using a microELISA plate reader (e.g. Dynatech Model 600).

Differential staining

The types of cells present in a cell suspension can be estimated by staining the cells with one of several commercial stains. If the cells are adherent to a glass or plastic surface (i.e. coverslip), as for macrophages, this is a simple matter. If the cells are in suspension or non-adherent (as for lymphocytes), then it is first necessary mechanically to adhere the cells to a clean glass slide before they are stained. To do this, a small volume (usually 0.1 ml) of a cell suspension is added at an appropriate density (usually 0.5–2×10^6 ml^{-1}) to a chamber in a cytocentrifuge (Shandon-Lipshaw, Sewickley, PA, USA). The centrifugal force exerted during operation forces the cells onto the surface of the slide, where they form a circular ring that is readily visible. The slide is then fixed (by air drying or with ethanol) and stained with Wright-Giemsa or Diff-Quik stains. With a little training, these cells can then be examined microscopically at 400× magnification and scored for cell type (e.g. lymphocyte, neutrophil, macrophage). When so doing, at least 200 cells on each slide should be scored, and the results expressed as the percentage of each cell type in the total cell population. The absolute cell number should also be determined using a haemocytometer, or Coulter Counter, and the results expressed as the absolute number of each cell type.

References

Boyum, A. (1968). Isolation of mononuclear cells and granulocytes from human blood. *Scand. J. Clin. Lab. Invest. Suppl.* **21**, 77–89.

Caron-Leslie, L. M., Evans, R. B. and Cidlowski, J. A. (1994). Bcl-2 inhibits glucocorticoid-induced apoptosis but only partially blocks calcium ionophore or cycloheximide-regulated apoptosis in S49 cells. *FASEB J.* **8**, 639–645.

Cohen, A. and Schlesinger, M. (1970). Absorption of guinea pig serum with agar. *Transplantation* **10**, 130–132.

Czuprynski, C. J., Henson, P. M. and Campbell, P. A. (1984). Killing of *Listeria monocytogenes* by inflammatory neutrophils and mononuclear phagocytes from immune and nonimmune mice. *J. Leukoc. Biol.* **35**, 193–208.

Czuprynski, C. J., Henson, P. M. and Campbell, P. A. (1985). Enhanced accumulation of inflammatory neutrophils and macrophages mediated by transfer of T cells from mice immunized with *Listeria monocytogenes*. *J. Immunol.* **134**, 3449–3454.

Green, L. M., Reade, J. L. and Ware, C. F. (1984). Rapid colormetric assay for cell viability: application to the quantitation of cytotoxic and growth inhibitory lymphokines. *J. Immunol. Methods* **70**, 257–268.

Haak-Frendscho, M. and Czuprynski, C. J. (1992). Use of recombinant interleukin-2 to enhance adoptive transfer of resistance to *Listeria monocytogenes* infection. *Infect. Immun.* **60**, 1406–1414.

Harbeck, R. J., Hoffman, A. A., Redecker, S., Biundo, T. and Kurnick, J. (1982). The isolation and functional activity of polymorphonuclear leukocytes and lymphocytes separated from whole blood on a single percoll density gradient. *Clin. Immunol. Immunopathol.* **23**, 682–690.

Haslett, C., Guthrie, L. A., Kopaniak, M. M., Johnston, R. B., Jr. and Henson, P. M. (1985). Modulation of multiple neutrophil functions by preparative methods or trace concentrations of bacterial lipopolysaccharide. *Am. J. Pathol.* **119**, 101–110.

Hathcock, K. S. (1991). T cell depletion by cytotoxic elimination. In: *Current Protocols in Immunology* (J. Cooligan, A. Kruisbeek, D. Margulies, E. Shevach and W. Strober, Eds), pp. 3.4.1–3.4.3. Greene Publishing Associates and Wiley-Interscience, New York.

Ishikawa, H., Li, Y., Abeliovich, A., Yamamoto, S., Kaufmann, S. H. E. and Tonegawa, S. (1993). Cytotoxic and interferon γ-producing activities of γδ T cells in the mouse intestinal epithelium are strain dependent. *Proc. Natl. Acad. Sci. USA* **90**, 8204–8208.

Julius, M. H., Simpson, E. and Herzenberg, L. A. (1973). A rapid method for the isolation of functional thymus-derived murine lymphocytes. *Eur. J. Immunol.* **3**, 645–649.

Kaufmann, S. H. E., Hug, E., Vath, U. and De Libero, G. (1987). Specific lysis of *Listeria monocytogenes*-infected macrophages by class II-restricted L3T4+ T cells. *Eur. J. Immunol.* **17**, 237–246.

Koski, I. R., Poplack, D. G. and Blaese, R. M. (1980). A nonspecific esterase strain for the identification of monocytes and macrophages. In: *Methods for Studying Mononuclear Phagocytes* (D.O. Adams, P.J. Edelson and H. Koren, Eds), pp. 359–362. Academic Press, New York.

Leenen, P. J., de Bruijn, M. F., Voerman, J. S., Campbell, P. A. and van Ewijk, W. (1994). Markers of mouse macrophage development detected by monoclonal antibodies. *J. Immunol. Meth.* **174**, 5–19.

Medzhitov, R. and Janeway, C., Jr. (2000). Innate immune recognition: mechanisms and pathways. *Immunol. Rev.* **173**, 89–97.

Musson, R. A. and Henson, P. M. (1979). Humoral and formed elements of blood modulate the response of peripheral blood monocytes. Plasma and serum inhibit and platelets enhance monocyte adherence. *J. Immunol.* **122**, 2026–2031.

Ooi, Y. M. and Colten, H. R. (1979). Genetic defect in secretion of complement C5 in mice. *Nature* **282**, 207–208.

Ritter, M. M. and Boyd, R. L. (1993). Development in the thymus: it takes two to tango. *Immunol. Today* **14**, 462–469.

Scudiero, D. A., Shoemaker, R. H., Paull, K. D., Monks, A., Tierney, S., Nofziger, T. H., Currens, M. J., Seniff, D. and Boyd, M. R. (1988). Evaluation of a soluble tetrazolium/formazan assay for cell growth and drug sensitivity in culture using human and other tumor cell lines. *Cancer Res.* **48**, 4827–4833.

Shapiro, H. M. (1988). *Practical Flow Cytometry*. Wiley-Liss Interscience, New York, pp. 133–134.

Wysocki, L. J. and Sato, V. L. (1978). Panning for lymphocytes: a method for cell selection. *Proc. Natl. Acad. Sci. USA* **75**, 2844–2848.

Commercial Suppliers

Accurate Chemical
300 Shames Drive
Westbury, NY 11590, USA

Phone: 800-645-6264
Fax: 516-997-4948
Website: http://www.accuratechemical.com
E-Mail: info@accuratechemical.com

Lympholyte M separation medium
Low-Tox M serum

Becton-Dickinson
Two Oak Park
Bedford, MA 01730, USA

Phone: 800-343-2035
Fax: 617-275-0043
Website: www.bdbiosciences.com
E-Mail: labware@bd.com

Falcon cell strainers

BioWhittaker, Inc.
8830 Biggs Ford Road
Walkersville, MD 21793, USA

Phone: 800-638-8174
Fax: 301-845-8291
Website: www.cambrex.com

Limulus reagents and kits

Polysciences, Inc   Nylon wool
400 Valley Road
Warrington, PA 18976, USA

Phone: 800-523-2575
Fax: 800-343-3291
Website: www.polysciences.com
E-Mail: info@polysciences.com

Thermo-Shandon   Cytocentrifuge
117 Industry Drive
Pittsburgh, PA 15275, USA

Phone: 800-547-7429
Fax: 412-788-1137
Website: www.thermo.com
E-Mail: customerserviceusa@thermoshandon.com

Sigma   Ficoll-Hypaque
P.O. Box 14508
St. Louis, MO 63178, USA

Phone: 800-325-3010
Fax: 800-325-5052
Website: www.sigma-aldrich.com
E-Mail: Access via Website

Isolation and Preparation of Lymphocytes

2.2 Isolation and Development of Murine T-Lymphocyte Hybridomas and Clonal Cell Lines

Christopher Reardon
Departments of Dermatology and Immunology, University of Colorado Health Sciences Center, Denver, Colorado, USA

Michael Lahn
Department of Immunology, National Jewish Medical and Research Center, Denver, Colorado, USA

◆◆

CONTENTS

◆◆◆◆◆◆ INTRODUCTION

To study T lymphocytes that mediate many important functions of the immune system, analysing clonal cell populations has been necessary, because T cells and B cells express clonally distributed antigen receptors, reflecting their clonal specificities for antigens. As a means to study clonal antigen specificities of antibodies, the first cloning systems for B cells were developed to produce monoclonal antibodies (Köhler and Milstein, 1975a,b; Shulman *et al.*, 1978; Kearney *et al.*, 1979).

Cloning of T cells followed B-cell cloning, because the understanding of T-cell function has developed more slowly (Morgan *et al.*, 1976; Goldsby *et al.*, 1977; Kappler *et al.*, 1981; White *et al.*, 1989). However, it has been one of the most crucial methods for expediting the understanding of T-cell functions. Because T cells express their antigen receptors on their cell surface instead of secreting them, studies of T-cell specificity more often depend on the isolation of the reactive cells themselves. Cloning does not only represent an approach to studying individual T cells but for studying collections

Copyright © Elsevier Science Ltd
All rights of reproduction in any form reserved

Murine T-Lymphocyte
Hybridomas etc

of individual T-cell clones as well. The study of collections of T-cell clones has been used to understand mixed cell populations, T-cell receptor (TCR) repertoires and the development of immune responses. Furthermore, cloned T cells readily permit the establishment of correlations between separate properties, such as cytokine secretion and the expression of cell surface markers or the pairing of heterodimeric chains of TCRs.

However, approaches to cellular analysis of cloned T cells also suffer from inherent problems. Perhaps the most troublesome is the lack of control over selective forces *in vitro* which affect the population of clones that survive. Secondly, cloned cells, and especially hybridomas, are often unstable (Johnson *et al.*, 1982). Instability combined with non-physiological selection pressures can lead to misrepresentation of cells in a population because of changes in specificity and loss or occasionally even gain of cellular functions. Thirdly, cloning has not been universally successful. Some types of T cells have defied all attempts at cloning without clear evidence as to how they are different.

The goals of this chapter are to introduce methods for isolating T cells from different tissues and to describe the production of T-cell lines, T-cell clones, as well as T-cell hybridomas and transfectomas. T-cell proliferation assays, based on flow cytometry, are also described here. It is our hope that such methods will lead to studies that further our understanding of T lymphocytes and their functions.

◆◆◆◆◆◆ T-CELL ISOLATION METHODS

Background

Lymphocyte isolation with high purity is of major interest in studying the role of T cells in organs. While isolation of lymphocytes from spleen, thymus, peripheral blood and lymph nodes are mostly based on mechanical dissociation and physical separation, isolating lymphocytes out of organs is more complicated. In such cases, a cocktail of enzymes allows the isolation of organ lymphocytes. However, the challenge lies in finding the right balance between harsh enzymatic activity and the necessity to isolate lymphocytes from tightly organized tissue. This is a prerequisite for all the subsequent analysis with T cells isolated from organs.

Although there are a number of separation methods using density gradients (e.g. Percoll), we have observed that nylon wool columns not only lead to a higher purity of isolated T cells but also provide more viable T cells. This method is not associated with an activation of T cells that often occurs after positive and even negative sorting using flow cytometry. T cells rarely express CD25 or, even more importantly, CD69 after the nylon wool procedure while sorting causes an activation of T cells as determined by CD69 expression. Similar to cell sorting, T-cell purification using antibody-coated magnetic beads and passage over a magnet also may lead to cell activation. In sum, we feel that nylon wool isolation, although slightly more labour intensive, is the best way to obtain viable and non-activated T lymphocytes from different organs.

Isolation of T cells from the spleen and thymus

Materials

- Complete culture medium: Iscove's Dulbecco's modified Eagle's medium (Sigma I7633) plus sodium bicarbonate (2 g l^{-1}), sodium pyruvate (100×, Sigma S8636), modified Eagle's medium (MEM) non-essential amino acids (100×, Sigma M2025), MEM vitamins (100×, Sigma 6895), MEM essential amino acids (50×, Sigma M5550), L-glutamine (100×, G6392), 2-mercaptoethanol (5 × 10^{-5} m, Sigma M7522), 0.75% dextrose, penicillin/streptomycin (100×), gentamycin (10 µg ml^{-1}), 10% fetal calf serum.
  This medium is used in all assays below unless otherwise described.
- NH$_4$Cl RBC lysis buffer (Gey's solution, pre-made: Sigma G9779, or self-made: 0.155M NH$_4$Cl (Sigma A0171), 0.01M KHCO$_3$, 0.0005% phenol red (Sigma P0290)
- Hank's balanced salt solution (HBSS) (Sigma, H6136)
- Hank's balanced salt solution + 5% fetal calf serum (HBSS + FCS)
- 60- or 80-mesh (~2 µm pore size) screen, autoclavable (Cellector, E-C Apparatus Corp, EC587-60 or EC587-80)

Procedure

1. Isolate spleen or thymus from mouse.
2. Tease apart tissue on a cellector screen using a disposable 3–5 cc syringe barrel and resuspend mechanically dissociated cells in 5 ml HBSS + FCS.
3. Wash cell suspension with HBSS + FCS and centrufuge at 200g at room temperature for 5 min.
4. Discard the supernatant.
5. Add 1 ml HBSS alone to cell pellet from step 1, then add 2 ml Gey's solution to lyse red blood cells and resuspend gently the pellet.
   Note: Lyse red blood cells for spleen cell preparations which may not be necessary for thymic cell preparations.
6. Allow lysis to take place for 3 min.
7. Then add 10 ml HBSS + FCS to stop lysis and spin cell suspension at 200g at room temperature for 5 min.
8. Discard the supernatant.
9. Resuspend cells in 1 ml of HBSS for nylon wool purification or in medium if experiments with purified T cells is not desired.
10. For T-cell isolation, follow nylon wool column section below.

Isolation of T cells from non-lymphoid organs, e.g. lung

T cells are prepared from mouse lungs using a protocol modified by Jones-Carson *et al.* (1995).

Materials

- Complete culture medium as above
- Fetal Calf Serum (FCS)
- Hank's balanced salt solution (HBSS) (Sigma, H6136)
- Hank's balanced salt solution + 5% fetal calf serum (HBSS + FCS)
- Dispase II (Roche Diagnostics, Nr. 295 825), aliquoted and stored at −20°C
- Collagenase II (Sigma, C6885), 0.05 mg ml^{-1} in RPMI and store in 100 μl aliquots at −80°C until use
- Collagenase IV (Sigma, C5138), 0.05 mg ml^{-1} in RPMI and store in 100 μl aliquots at −80°C until use
- Gey's solution NH$_4$Cl RBC lysis buffer
- 60- or 80-mesh (~200 μm pore size) screen, autoclavable (Cellector, E-C Apparatus Corp, EC587-60 or EC587-80)
- Rotator or waterbath with movement

Procedure

1. Bleed anaesthetized mice by cutting aorta.
2. Remove heart.
3. Resect the lungs and place in HBSS + FCS.
4. Cut lungs into small pieces of about 0.5 cm in diameter using pair of scissors and place tissue in centifuge tubes.
5. Spin down tissue for 5 min at 200g.
6. Remove fluid without removing floating pieces of lung tissue.
7. Add digestion buffer:

   > 30 ml Iscove's medium with non-serum supplements
   > 5 ml FCS
   > 5 ml Dispase II
   > 80 μl of Collagenase II
   > 80 μl of Collagenase IV

8. Keep tissue in digestion buffer in a rotator at 37°C for 75 min.
9. Spin down tissue as before and remove carefully the supernatant fluid, collecting it in a 50 ml conical centrifuge tube.
10. Place tissue in a petri dish and with a syringe plunger mechanically dissociate tissue using screens, as above for spleen cells.
11. Place tissue in a 15 ml sterile centrifuge tube.
12. Spin tissue down again and remove and collect the supernatant fluid.

(contd.)

13. Discard the tissue and spin down the collected supernatant fluid.
14. Resuspend the cell pellet in HBSS + FCS and spin cells down for one wash.
15. Resuspend the cells in 1 ml of HBSS.
16. Add 2 ml Gey's solution for red cell lysis to the cells in 1 ml of HBSS and incubate at room temperature for 3 min.
17. Add 5 ml of HBSS + FCS and spin down as before and remove and discard fluid.
18. Resuspend cells in 2 ml HBSS + FCS and pass cells through a cellector screen. Place cell suspension on nylon-wool column or count cells and use as needed.

T-cell enrichment with nylon wool columns

Materials

- Complete culture medium
- HBSS
- HBSS + 5% Fetal Calf Serum (HBSS + FCS)
- Nylon-wool (Robbins Scientific Corporation, Scrubbed Combed Nylon Wool, 100 g Type 200L, Cat# 1078-02-0, 1250 Elko Drive, Sunnyvale, CA 94089, USA, Tel: 408-734-8500)
- Needles: Monoject 25 × 5/8 Gauge (0.5 mm × 16 mm)
- Rubber Stoppers
- Syringes: 10 cc Monoject syringes filled to 6 cc mark with 0.6 g nylon wool (0.1 g nylon wool per ml of syringe volume)

Procedure

Nylon wool column preparation:

1. Attach syringe needle to autoclaved syringe filled with nylon wool.
2. Place rubber stopper onto the end of the needle.
3. Fill the column completely with HBSS + FCS and top off to immerse nylon wool completely with this solution.
4. Place vacuum on the end of unstoppered needle and siphon the HBSS + FCS into the nylon wool, add more HBSS + FCS and repeat if necessary to completely immerse nylon wool with HBSS + FCS. Place rubber stopper onto the end of the needle.
5. Incubate column for at least 30 min at 37°C until needed.

(contd.)

Cell passage through column:

1. After incubation of nylon wool column for at least 30 min at 37°C, remove the stopper and siphon off the HBSS + FCS. It is important to remove all the HBSS + FCS and have the column dry. Place rubber stopper onto the end of the needle.
2. Resuspend the cells in 1 ml of HBSS (without FCS) and add drop-wise to the nylon wool column. Place rubber stopper onto the end of the needle.
3. Incubate for 20 min at 37°C.
4. Add now another 1 ml of HBSS at 37°C to the column.
5. Incubate for another 20 min at 37°C.
6. Remove rubber stopper and add HBSS at 37°C to the column to elute cells. If sterile cells are to be recovered, exchange the syringe needle with a new sterile one before adding the HBSS (but do not place the stopper back on the needle). Place a 15 ml tube under the column to collect the passaged cells.
7. If necessary, apply sterile suction tube briefly to the needle to start the elution.
8. Collect about 10–13 ml of cells in the tube.
9. Spin down cells and assess cell number. If necessary, determine cell composition by staining the cells for flow cytometry; e.g. anti-CD3ε mAb or anti-β-TCR mAb and/or other cell subpopulations (B220 for B cells, Mac-1 for macrophages, GR-1 for granulocytes/neutrophils). Expect enrichment of T cells from the spleen to be ≥70% T cells and about 20–30% B cells. Macrophages, as assessed by the Mac-1 stain, should not be more than 1%.

◆◆◆◆◆◆ T-CELL LINES AND CLONES

Background

Long-term *in-vitro* culture of normal lymphocytes became a possibility with the improvement of tissue culture conditions in the 1970s. Particularly important was the discovery that supernatants from stimulated short-term lymphocyte cultures contain growth factors for T cells (Morgan *et al.*, 1976). The first well-defined T-cell growth factor (TCGF) is now known as interleukin-2 (IL-2) (Aarden *et al.*, 1979). Further factors directly or indirectly influencing the growth of T cells *in vitro* are IL-1, IL-4, IL-7, IL-15, γ-interferon (IFN-γ), and various others (Coligan *et al.*, 1996; Okazaki *et al.*, 1989; Nishimura *et al.*, 1996).

Most commonly, instead of purified growth factors, mixtures of growth factors produced in other cell cultures are used to support long-term growth of T cells. Such conditioned media may be derived from mitogen- or alloantigen-stimulated cultures or from certain tumour cell cultures. Different cell cultures produce different mixtures of growth factors and consequently may support preferential growth of different

T-cell subsets. For example, secondary mixed lymphocyte cultures with mouse splenocytes abundantly produce IL-4 and IFN-γ but little IL-2, whereas concanavalin A stimulated spleen cells preferentially produce IL-2 (Gajewski et al., 1989a,b).

Although long-term lines and clones of T cells were derived originally by culture in the presence of growth factors without antigen stimulation (Haas et al., 1985), this approach is now used only if the antigen is not known, as is the case, for example, with most γδ T cells (Tsuji et al., 1996). T-cell culture in the presence of antigen and antigen-presenting cells, together with or alternating with growth factor stimulation, has proved far more reliable in generating long-term lines and clones with relatively normal features (Fathman and Hengartner, 1978; Glasebrook and Fitch, 1980; Ziegler and Unanue, 1981; Chestnut et al., 1982; Johnson et al., 1982; Kappler et al., 1982). Requirements are different for growing alloreactive and conventional antigen-reactive αβ T cells and for cytotoxic T lymphocytes and T-helper 1 and 2 cells. However, in all cases, antigen stimulation is required for long-term growth. Requirements for the culture of the far slower growing γδ T cells are not yet well established.

The description given below of culture conditions for the development and cloning of long-term αβ T-cell lines closely follows the more detailed protocols reported by Fitch and Gajewski (Coligan et al., 1996). With minor modifications, these methods seem applicable to the cloning of murine αβ T cells in general. By comparison, the culture conditions given for γδ T-cell clones are less well established and may not be generally applicable (Tigelaar et al., 1990; Tsuji et al., 1994).

Conditioned medium as a source of growth factors for T-lymphocyte lines

Conditioned medium from concanavalin A-activated spleen cells is a rich source of IL-2, but contains little IL-4 or IFN-γ. Conditioned medium from secondary mixed lymphocyte culture (MLC) reactions contains high levels of IFN-γ. EL-4 tumour cell conditioned medium contains IL-2 but not IFN-γ, although some lines produce IL-4 as well. EL-4 conditioned media can usually be used instead of concanavalin A-activated spleen cell conditioned medium.

Materials

- Concanavalin A (2.5 μg ml^{-1}, Sigma C5275)
- α-methylmannoside-conjugated agarose (0.2 g ml^{-1}) (Sigma M3139)
- Complete culture medium as above
- Phorbol myristate acetate (Sigma P148)

(contd.)

Procedure

Concanavalin A-activated spleen cell-conditioned medium

1. Culture 1.25×10^6 cells ml^{-1} rat or mouse spleen cells in a humidified 37°C, 5% CO_2 incubator, in complete medium in the presence of 2.5 µg ml^{-1} Con A for 24–48 h.
2. Collect culture supernatant and remove cells by centrifugation. Remove residual concanavalin A by absorption with a slurry of α-methylmannoside-conjugated agarose.
3. If necessary, assay concanavalin A-activated spleen cell conditioned medium for IL-2 content (e.g. with the HT-2 bioassay below or by ELISA). Store aliquots frozen at −70°C.

MLC-conditioned medium

1. Mix 2.5×10^5 C57BL/6-responding mouse spleen cells with an equal number of irradiated (2000 rad) DBA/2-stimulating mouse spleen cells in 20 ml complete culture medium. Culture the cells for 10–14 days at 37°C, 5% CO_2.
   Note: This particular combination of mouse strains is most effective in generating MLC-conditioned medium.
2. Collect cells from primary MLC in a 50-ml tube and wash twice with complete medium.
3. Prepare secondary MLC by mixing 6×10^6 primary MLC cells with 2.5×10^7 irradiated DBA/2-stimulating cells in 20 ml complete medium. Culture the cells for 36 h at 37°C, 5% CO_2.
4. Collect culture supernatant, pellet out cells and assess for cytokines.

EL-4-conditioned medium

This also can be purchased directly through Sigma (M8657)

1. Culture EL-4-IL-2 murine lymphoma cells (10^6 cells ml^{-1}) for 4 h in complete culture medium and 20 ng/ml phorbol myristate acetate (PMA).
2. Collect cells and wash three times. Incubate for 36 h at 37°C, 5% CO_2.
3. Collect culture supernatant, pellet out cells and assess for cytokines.

Production of alloreactive T-helper and cytotoxic T-lymphocyte clones

To obtain the highest frequencies of responding T cells, freshly isolated cells are first stimulated in primary allogeneic mixed lymphocyte bulk cultures (MLCs) and then cloned by limiting dilution.

Materials

- Complete culture medium as above
- 96-well flat-bottomed microtitre plates (Falcon)
- T-25 plastic culture flask (Falcon)
- Conditioned medium (see above)

Procedure

1. Prime alloreactive T cells *in vitro* (e.g. mix 2.5×10^7 responding mouse spleen cells with an equal number of irradiated (2000 rad) stimulating spleen cells in 20 ml culture medium. Culture in an upright T-25 plastic culture flask for 10–14 days at 37°C, 5% CO_2).
2. Prepare secondary MLC by mixing 6×10^6 washed cells from the primary culture with 2.5×10^6 irradiated splenic stimulator cells. Incubate for 36–48 h under the same conditions as for the primary culture.
3. Prepare cloning by plating 10^7 allogeneic spleen cells (2000–3000 rad) in 0.1 ml culture medium per well of 96-well flat-bottomed microtitre plates.
4. Resuspend T cells, derived from the secondary MLC, in conditioned medium (see above) at 1–10 cells ml^{-1}. Add 50 µl per well of this cell suspension to the prepared cloning plates. Prepare several plates with varying numbers of responding cells to account for possible differences in plating efficiencies. Incubate cloning plates for 4 days under the same conditions as for the primary MLC.
5. Add 50 µl conditioned medium and 50 µl culture medium to each well. Culture for 7–10 days until clusters of cells become evident.
6. To maintain the desired clones, transfer up to 10^5 cells in 100 µl fresh culture medium from the original microwell to a well in a 24-well flat-bottomed plate containing 6×10^6 irradiated allogeneic spleen cells in 0.9 ml culture medium. Add 0.5 ml conditioned medium (as described above), to reach a final volume of 1.5 ml. Passage cells at weekly intervals.

   Note: For the cloning of alloreactive αβ T cells, conditioned medium is best derived from allogeneic primary MLCs (as described above). The supernatant from the C57BL/6 anti-DBA/2 response is particularly rich in the required co-factors. The culture supernatant is used at a concentration of 20% (v/v) for cloning. Using responder cells from TCR-β 'knockout' mice, the same conditions may be applicable for the generation of alloreactive γδ T cell clones (A. Mukasa, personal communication).

Generation of T-helper cell clones reactive with soluble protein antigens

The isolation of such clones requires prior sensitization *in vivo*. Typically, mice are immunized with soluble protein antigens in complete Freund's adjuvant. *In vitro*, syngeneic spleen cells are used as antigen-presenting cells (APC). The following protocols briefly describe the generation of T-helper 1 and T-helper 2 clones reactive with chicken ovalbumin, as described in more detail by Fitch and Gajewsky (Coligan *et al.*, 1996).

Materials

- Antigen of choice (0.2–1.6 mg ml^{-1})
- Human rIL-2 (80 U ml^{-1}) (Sigma T3267)
- Mouse rIFN-γ (4000 U/ml) (Sigma I5517)
- HBSS
- HBSS + 5% Fetal Calf Serum (HBSS + FCS)
- Dulbecco's phosphate-buffered saline (PBS)
- Complete culture medium as above
- Concanavalin A-conditioned medium (see above)
- Freund's complete adjuvant (Sigma F5881)
- 24-well flat-bottomed culture plates (Falcon)
- 96-well flat-bottomed microtitre plates (Falcon)

Procedure

1. Immunize mice by subcutaneous injection of antigen (1:1 emulsion of antigen dissolved in Dulbecco's PBS or HBSS and Freund's complete adjuvant).
2. After 1 week, remove the draining lymph nodes and prepare a single-cell suspension in HBSS or complete culture medium.
3. Culture 2×10^6 lymph node cells for 6–8 days (37°C, 5% CO$_2$) in the presence of 6×10^6 irradiated syngeneic spleen cells plus antigen in 1.5 ml of complete culture medium in a 24-well flat-bottomed microtitre plate. For stimulation *in vitro*, protein antigens are typically used at concentrations of 50–400 µg ml^{-1}.

T-helper I clones

4a. To each well of a 96-well flat-bottomed microtitre plate add 50 µl irradiated syngeneic spleen cells (10^6 per well) in complete medium with 50 µl antigen (0.2–1.6 mg ml^{-1}) in medium plus 25 µl human rIL-2 (80 U ml^{-1}) and 25 µl mouse rIFN-γ (4000 U/ml).
5a. Resuspend T cells to be cloned at approximately 2000 cells ml^{-1} in complete medium. Add 50 µl to each prepared well from step (4a). Incubate for 1 week at 37°C, 5% CO$_2$. (Because of variable plating efficiencies, T cells should be titred so that plates with 100, 30, 10 and 3 cells per well are generated.)

(contd.)

6a. To each well, add 25 µl human rIL-2 (80 U ml^{-1}) and 25 µl mouse rIFN-γ (4000 U ml^{-1}). Incubate for one additional week or until bottoms of positive wells are almost covered with cells. Wells that have a single cluster of growing cells should be selected for expansion.

T-helper 2 clones

4b. To each well of a 96-well flat-bottomed microtitre plate add 50 µl irradiated syngeneic spleen cells (10^6 per well) in complete culture medium with 50 µl antigen (0.2–1.6 mg ml^{-1}) in medium plus 25 µl human rIL-2 (40 U ml^{-1}) or 50 µl concanavalin A-conditioned medium (40%). Note that some T-helper 2 cells require IL-1 for their growth.

5b. Resuspend T cells to be cloned at approximately 2000 cells ml^{-1} in complete culture medium. Add 50 µl to each prepared well from step (4b). Incubate for 1 week at 37°C, 5% CO$_2$. (Because of variable plating efficiencies, T cells should be titred so that plates with 100, 30, 10 and 3 cells per well are generated.)

6b. To each well, add 50 µl human rIL-2 (40 U ml^{-1}) or concanavalin A-conditioned medium (10% final). Incubate for one additional week or until the bottoms of the positive wells are almost covered with cells. Wells that have a single cluster of growing cells should be selected for expansion.

Maintenance of the T-helper cell clones

7. To each well of a 24-well flat-bottomed plate (final volume 1.5 ml) add 5 × 10^4 to 2 × 10^5 cloned cells in 100 µl complete culture medium, 6 × 10^6 irradiated syngeneic spleen cells in 0.9 ml medium and 50–400 µg ml^{-1} antigen in medium. For T-helper 1 clones, add human rIL-2 (25 U ml^{-1}) and mouse rIFN-γ (250 U ml^{-1}). For T-helper 2 clones, add human rIL-2 (25 U ml^{-1}). (For long-term cultures, use human rIL-2 at a final concentration of 10 U ml^{-1}.)

8. Culture cells at 37°C, 5% CO$_2$, in a humidified incubator. Passage every 7–10 days.
   Note: There are many variations to this basic protocol. For more detail, consult Coligan *et al.* (1996).

Production of αβ T-cell clones from mice challenged with a pathogen

The example used here is an immunization protocol against *Plasmodium berghei* sporozoites. Mice were immunized intravenously with *Plasmodium berghei* sporozoites (Tsuji *et al.*, 1990) or with peptide antigens representing epitopes of the circumsporozoite protein of *Plasmodium yoelii* (Takita-Sonoda *et al.*, 1996).

259

I

I

I

I

I

I

I

I

I

I

I

I

I

I

I

I

I

I

I

I

Materials

- NH$_4$Cl RBC lysis buffer (Gey's solution)
- Peptides derived from the circumsporozoite protein of *Plasmodium yoelii* or *Plasmodium berghei* sporozoites (quantities of antigens as above)
- Anti-mouse immunoglobulin (Jackson ImmunoResearch)
- Complete culture medium as above
- Hank's balanced salt solution (HBSS)
- Complete culture medium as above, but with 0.5% normal mouse serum instead of fetal calf serum
- Lympholyte-M (Cedarlane)

Procedure

1. Balb/c mice were immunized intravenously with *Plasmodium berghei* sporozoites or with peptide antigens representing epitopes of the circumsporozoite protein of *Plasmodium yoelii* (Takita-Sonoda et al., 1996).

2. Responder T cells from immunized mice were prepared from NH$_4$Cl-RBC lysis buffer-treated splenocytes (as described above) by removing B cells on anti-mouse immunoglobulin-coated petri dishes. Enriched T cells were adjusted to 3×10^6 cells ml^{-1} of culture medium and plated at 1 ml per well of a 24-well flat-bottomed tissue-culture plate.

3. Generation of antigen-presenting cells, NH$_4$Cl treated splenocytes of naive Balb/c mice were adjusted to 1×10^7, pulsed with antigen (1 h, 37°C), washed, irradiated (3300 rad) and plated at 5×10^6 cells per well.

4. Responder and stimulator cells were incubated for 5–6 days in culture medium containing 0.5% normal mouse serum.

5. The antigen-stimulated T cells were purified with lympholyte-M, resuspended at 3×10^6 per well (2 ml culture volume in 24-well flat-bottomed tissue-culture plates), and incubated for 2–3 days in complete culture medium, prior to re-stimulation with antigen-pulsed antigen-presenting cells.

6. From these bulk cultures, cell clones are produced by limiting dilution, as previously described. Such clones hopefully should recognize a plasmodial antigen in the context of the class II molecule and lead to a high degree of protection against sporozoite challenge in naive mice.

T-cell clones expressing γδ-TCR

Background

Preparing human γδ T-cell clones is not particularly problematic but generating their murine counterparts has proved to be more difficult. At present, only a few stable murine lines and clones are available. Murine γδ T-cell clones have been isolated and propagated under conditions avoiding possible competition with αβ T cells, presumably because αβ T cells adapt better to tissue culture and grow faster. The first reported murine clones were derived from congenitally thymus-deficient BALB/c nu/nu mice, after repeated immunization with Bl0.BR spleen cells (Matis *et al.*, 1987, 1989). The nearly complete absence of αβ T cells in the thymus-deficient mice allowed selective enrichment and cloning of the less thymus dependent γδ T cells. The clones isolated in this and a similar study were specific for major histocompatibility complex-encoded cell surface non-classical class I and class II molecules (Matis and Bluestone, 1991). They express Vγ4 in association with Vδ5, and TCR junctional differences appear to dictate ligand specificities (Rellahan *et al.*, 1991).

A natural tissue source of γδ T cells that lack contaminating αβ T cells is the murine epidermis (Nixon-Fulton *et al.*, 1986; Stingl *et al.*, 1987; Asarnow *et al.*, 1988). γδ T cells, also known as Thy-l⁺ dendritic epidermal T-cells (DETCs), are normally present in this tissue in rather large numbers, forming a loose network. In contrast, αβ T cells extravasate into the epidermal layers only under pathological conditions. In partially enriched preparation of epidermal cells, DETCs grow slowly and eventually form lines (Nixon-Fulton *et al.*, 1988). The epidermal γδ T cells all virtually express Vγ5 in association with Vδ1 without junctional variations (Asarnow *et al.*, 1988). The molecular nature of their ligands remains unknown. Murine γδ T-cell clones have also been derived from malaria-immunized mice (Tsuji *et al.*, 1994). Here, TCR-β 'knockout' mice were used as a source of γδ T cells, thus eliminating competition from αβ T cells. These spleen-derived γδ T-cell clones express either Vγl in association with Vδ5 or Vγ7 together with Vδ4 (Tsuji *et al.*, 1996). One clone protected mice from challenge with *P. yoehi* sporozoites, and all responded to restimulation with extracts of parasitized erythrocytes and spleen cells. Nevertheless, the specific trigger for the response has not been identified.

To our knowledge, no clones have yet been generated that express either Vγ2 or Vγ6 γδ TCRs. Only a few researchers have been successful in cloning murine γδ T cells, so generally applicable rules have not yet been developed. Thus, the procedures given below must be regarded as tentative. However, some of the peculiarities described, such as the extremely long incubation period prior to the appearance of γδ T-cell clones, should be taken into consideration when starting a cloning project involving murine γδ T cells. Detailed procedures for the isolation of DETC clones and for γδ T-cell clones derived from malaria-infected mice are given below.

Dendritic epidermal T-cell clones

Several epidermal γδ T-cell clones (Reardon *et al.*, 1995) have been produced using a method described by Nixon-Fulton *et al.* (1986) with some modifications.

Materials

- #22 scalpel (without handle) or curved forceps
- Depilatory (optional, Nair™)
- Trypsin (0.25%, Sigma T4799) + 0.1% glucose in phosphate buffered saline (sterile)
- 70% ethanol
- Complete culture medium as above
- Concanavalin A (2.5 µg ml^{-1}, Sigma)
- Recombinant mouse IL-2 (10 U ml^{-1}, R and D Systems or Sigma)
- HBSS + 5% fetal calf serum (HBSS + FCS)
- Dnase (0.1%) in trypsin as above
- Monoclonal anti-mouse antibodies: anti-γδ T-cell and anti-Vγ5 (Vγ3 by other nomenclature) antibody (Pharmingen)
- Nylon-wool columns as above
- 6-well tissue-culture plates (Falcon)
- 24-well flat-bottomed plates (Falcon)
- 96-well flat-bottomed plates (Falcon)

Procedure

1. Sacrificed mice are washed with a liquid antibacterial soap to remove hair oils and rinsed with water.
2. The wet intact trunk and ear skin is covered with an over-the-counter depilatory (without supplements, e.g. aloe vera), massaged into the hair with a gloved finger, and incubated for 15 min, after which the hair is removed gently with a #22 scalpel. Alternatively, the hair can be removed by shaving with the #22 scalpel alone or with electric clippers. The intact skin is washed again with the soap to remove the depilatory and excess hair and is rinsed with deionized water.
3. While holding the animal by the tail, the skin is rinsed with 70% ethanol, and the mouse placed on 70% ethanol-treated paper towels on a cork or styrofoam board.
4. After the limbs have been secured, the abdomen and chest skin is removed with sterilized scissors, cutting the skin over the supra-pubic areas, along the flanks and across the neck area. Care must be taken not to cut into the peritoneal cavity. The skin then is peeled off and placed into a sterile dry petri dish where it is rolled up with epidermis facing outwards to prevent desiccation of the dermal side. The animal is turned over, and the skin is cut across the caudal back and neck and again peeled from the fascia over-lying muscle. The skin is placed into the petri dish as before, while other animals are being prepared.
5. When all skin has been harvested, pieces of skin are turned over individually with the epidermal side down. Using scissors, the fat

(contd.)

and small vessels are teased and trimmed off the dermal side to reduce the possibility of contaminating the preparation with blood-derived T cells.

6. The trimmed skin is placed into new dry petri dishes with the dermal side facing down and left for about 15 min to adhere to the dish. Trypsin solution is gently placed into the dish until the epidermis is completely covered. Sterile metal screens or other like materials can be placed onto skin that has detached from the bottom to prevent it from floating. Alternatively, another approach is to cut the skin into 1 cm² squares and floating them epidermal-side down on the trypsin solution. The skin is left in trypsin overnight at 4°C.

7. The next day, the skin is removed from the petri dish and placed dermal side down into new dry petri dishes. Using a #22 scalpel or forceps, the epidermis is very gently scraped from the dermis.

8. This epidermal paste is placed in a 50-ml polystyrene tube, covered with 10 ml of the trypsin solution plus DNase, and incubated at 37°C for 15 min with gentle agitation of the tube to produce a single-cell suspension.

9. An equal volume of HBSS + FCS is added to this slurry of cells which is then passed quickly through a small plug of sterile nylon wool in a 10-ml syringe to remove the clumps of stratum corneum. The cells that are eluted from the nylon wool are washed with HBSS + FCS and placed in a medium- to large-sized flask in complete medium for overnight incubation to remove adherent keratinocytes and fibroblasts.

10. The next day, the non-adherent cells are passed over a nylon-wool column and incubated for 30 min at 37°C for the cells to adhere before elution to remove non-T cells.
    Note: The epidermal slurry can be incubated on the nylon-wool columns directly, bypassing the overnight incubation.

11. Eluted cells are cultured in 24-well flat-bottomed plates at 1×10^6 to 5×10^6 cells per well. They are stimulated with concanavalin A or placed in wells previously coated with 10 µg ml^{-1} pan-anti-γδ TCR monoclonal antibody, together with 10 U ml^{-1} recombinant IL-2.

12. Confluent wells are moved into six-well plates. Clonal T-cell lines are isolated by limiting dilution in 96-well flat-bottomed microtitre plates by diluting an aliquot of cells to 10 cells ml^{-1} and adding 0.1 ml of this to each well of one or more 96-well microtitre plates. Although antibody coating is no longer necessary, 10 units ml^{-1} IL-2 must be added to the medium.

13. Clones are expanded into 24-well plates and stained by flow cytometry to test for uniform TCR staining with pan-anti-γδ TCR monoclonal antibody. The cells also need to be stained with the anti-Vγ5 antibody to ensure that the cells are of the epidermal type, since other cells derived from contaminating peripheral blood may have been selected.

γδ T-cell clones from pathogen-challenged mice

The example displayed here is following immunization of mice with *P. yoeii* sporozoites. Immunization of αβ T-cell-deficient mice with *P. yoeii* sporozoites has been described in detail elsewhere (Tsuji *et al.*, 1994, 1996).

Materials

- Lympholyte-M (Cedarlane)
- EL4-conditioned medium added as a solution equivalent to 100 U of IL-2 per millilitre to all culture medium (Sigma M8657, as discussed above)
- HBSS + 5% fetal calf serum (HBSS + FCS)
- Monoclonal anti-mouse anti-γδ T-cell antibody (Pharmingen)
- NH_4Cl RBC lysis buffer (Gey's solution)
- Complete culture medium as above
- 12-well flat-bottomed tissue-culture plate (Falcon)
- 24-well flat-bottomed tissue-culture plate (Falcon)
- 96-well flat-bottomed tissue-culture plate (Falcon)

Procedure

1. Briefly, mice were immunized by bites of γ-irradiated, malaria-infected *Anopheles stephensi* mosquitoes daily for 2 weeks. Five days after the last exposure to the mosquitoes, spleens and livers were removed.
2. Responder cells from spleen and liver of immunized mice were cultured in the presence of feeder cells.
3. To prepare feeder cells, normal spleen cell suspensions were depleted of erythrocytes by treatment with NH_4Cl RBC lysis buffer, washed twice with HBSS + FCS, counted, and irradiated (3000 rad). Irradiated spleen cells (4×10^6) were added as feeder cells to each well of a 12-well flat-bottomed plate.
4. Spleen cells and non-parenchymal liver cells from immunized mice were prepared in the same fashion but omitting the irradiation. Responder cells (6×10^6) were added to each well, and the culture continued under standard conditions (37°C, 5% CO_2). Approximately half of the culture medium was exchanged every 2 days with fresh medium.
5. One week after beginning the culture, T-cell blasts were purified with lympholyte-M and placed into new cultures (6×10^6 cells per well of a new 12-well flat-bottomed plate). One week later, the cells were collected and counted.
6. Cloning was carried out by limiting dilution. Cells (10, 10^2 and 10^3) were placed in each well of a 96-well flat-bottomed plate (one plate for each dilution), together with 10^6 irradiated and erythrocyte-depleted normal spleen cells in each well as feeders.

(contd.)

7. Approximately half of the culture medium was replaced every 3 days. After 2 weeks, the feeder cells were replenished, continuing to replace half of the culture medium every 3 days.

8. After an additional 2 weeks, the feeder cells were replenished again, and the culture continued with medium changes as before. After an additional 1–2 weeks, colonies of γδ T cells appeared. These were expanded once or twice more in the presence of normal spleen feeder cells and then transferred into 24-well flat-bottomed plates at 10^5 cells per well.

9. Cell culture was continued from here on by changing approximately half of the culture medium every 3 days and re-stimulating the clones with splenic feeder cells every 2 weeks.

Note: Important details regarding this protocol include the following points. Erythrocytes must be lysed. The medium used for the initial bulk cultures should contain twice the concentration of EL4 conditioned medium (4%) compared with the medium (2%) used for expansion of αβ T-cell clones. Since γδ T cells grow much slower than αβ T cells, it usually takes 6–8 weeks or more for γδ T-cell colonies to become visible after cloning. After the number of cloned γδ T cells reaches more than 10^5 cells per well of a 24-well flat-bottomed plate, the purification of the γδ T cells one week after each stimulation may accelerate the growth of these cells. Unlike αβ T cells, the morphology of γδ T-cell clones is quite heterogeneous with large differences in cell sizes and shapes.

◆◆◆◆◆◆ T-CELL HYBRIDOMAS

Selectable fusion lines and problems with repertoire studies and functional competence

The basic technique of immortalizing lymphocytes by fusion to tumour cells was introduced about 20 years ago (Köhler and Milstein, 1975a; Taniguchi and Miller, 1978; Kappler *et al.*, 1981). The technique requires selectable fusion lines, efficient fusion agents and reliable selection and cloning protocols. Although practically all cells can be forced to fuse with other cells, only some are suitable for hybridoma generation (Shulman *et al.*, 1978; Galfré *et al.*, 1979; Kearney *et al.*, 1979). In fact, identifying a good fusion line is no small undertaking, as can be appreciated by the fact that there is still no efficient fusion line available for the generation of human T-cell hybridomas, despite considerable efforts to identify such cells.

One such fusion cell line, BW5147, is the most widely used T-cell fusion line for the generation of mouse T-cell hybridomas (White *et al.*, 1989). To eliminate the problem of endogenous TCR gene expression in BW5147, TCR gene loss variants have been generated, lacking either functional TCR-α rearrangements (BW/α⁻) or both TCR-α and functional TCR-β gene rearrangements (BW α⁻β⁻) (White *et al.*, 1989). BW α⁻β⁻ still contains non-functional TCR gene rearrangements for both α and β that give rise to partial gene transcripts, but they no longer contribute to surface-expressed TCR protein. BW $\alpha^-\beta^-$ has also been used for the generation of rat T-cell hybridomas, but it does not appear to be suitable for generating stable hybridomas with human T cells.

A γδ T-cell fusion cell line suitable for generating hybridomas with γδ T cells has not been identified, but BW α^- and BW $\alpha^-\beta^-$, in contrast to the original BW5147, can be used to generate hybridomas expressing γδ TCRs (Born et al., 1987). Although BW $\alpha^-\beta^-$ still contains its own endogenous TCR-γ gene rearrangements, they are non-functional and are not expressed on the cell surface. (δ genes are lost with the α genes.) The γδ T-cell hybridomas generated with BW $\alpha^-\beta^-$ which are comparable in stability to αβ T-cell hybridomas, were used to isolate γδ TCRs and to obtain partial protein sequences of surface-expressed TCR-δ (Born et al., 1987). Some of the first studies reporting γδT-cell responses to myco-bacterial antigens utilized γδ T-cell hybridomas (Happ et al., 1989; O'Brien et al., 1989). It should be noted, however, that hybridomas generated by fusions of γδ T cells with BW $\alpha^-\beta^-$ are heterohybrids, probably combining properties of αβ T cells with those of γδ T cells.

The primary purpose for generating hybridomas has been to immortalize and multiply individual B or T lymphocytes with properties of particular interest. However, cellular hybridization can also be used as a means of obtaining 'snapshots' of mixed cell populations, in developmental studies or when examining lymphocyte subset compositions during an immune response. For example, the sequence of TCR gene rearrangements during thymic maturation of T lymphocytes (both αβ and γδ) has been analysed using collections of thymocyte hybridomas representing subsequent stages of thymocyte development (Born et al., 1985, 1986; Haars et al., 1986). Incidentally, one such developmental study led to the discovery of the gene locus for TCR-δ and the peculiarly interspersed organization of TCR-α and TCR-β genes (Chien et al., 1987). Developmental patterns found in these hybridoma studies are reasonably well correlated with the findings of other studies based on antibody staining or the analysis of mRNA derived from bulk T-cell preparations (Raulet et al., 1985; Snodgrass et al., 1985). However, it cannot be assumed that this will always be the case. Hybridoma formation is clearly biased towards activated cells, so that within mixed cell populations such cells will be over-represented after hybridization. It also seems to be true that closely related cell types are more likely to form stable hybridomas, suggesting that the fusion line BW5147 probably favours certain types of T cells. Finally, some of the properties of the normal fusion partner, such as CD8 expression, may be suppressed after hybridization, leading to a distorted phenotype, even when the fusion was unbiased.

Although T-cell hybridomas have been used extensively to examine properties and functions of the T-cell receptor and other cell-surface molecules, as well as to study signalling pathways and consequences of signalling, such as T-cell anergy and apoptosis (Sloan-Lancaster et al., 1994; Brunner et al., 1995), there is little evidence that they retain the effector functions of the normal cells from which they are derived (Haas et al., 1985; Gu and Gottlieb, 1992; Gorczynski et al., 1996). With few exceptions, hybridomas cannot substitute for freshly isolated cells or clones in adoptive transfer experiments, since they typically lack cytolytic abilities and do not exhibit clear T-helper 1 and T-helper 2 phenotypes in terms of cytokine production.

Hybridoma production, cloning and maintenance

In modified form, a protocol for the hybridization of murine T cells with BW $\alpha^-\beta^-$ is suitable for use with any of the variants of this fusion line. For T-cell fusion experiments, typical numbers of approximately 2×10^7 fusion line cells are used, although the number of normal cells may vary. In some cases, successful fetal thymocyte fusions can be carried out with as few as 10^5 thymocytes with a yield of up to 20 hybridomas per fusion. Large numbers of hybridomas with a yield of up to several thousand are obtained with activated T cells in much larger numbers (purified T-cell blasts, up to 10^8 per fusion). Therefore, the number of hybridomas generated may vary greatly.

Materials

- Complete culture medium as above
- Sterile 50% polyethylene glycol 1540 (Sigma P7181) in 5 ml aliquots
- HAT solution (hypoxanthine, aminopterin, thymidine, 50×, Sigma H0262)
- HT solution (hypoxanthine, thymidine, 10×, Sigma H0137)
- HBSS
- 24-well flat-bottomed plates (Falcon)
- 96-well flat-bottomed microtitre plates (Falcon)

Procedure

1. Prior to each fusion procedure, PEG is placed in boiling water to avoid overheating for exactly 10 min. The melted PEG is kept at 37°C until use.
2. Thoroughly wash 2×10^7 BW $\alpha^-\beta^-$ cells (or its derivatives), as well as the normal fusion partner cells, using serum-free HBSS. Then combine the cells in a 50-ml conical polypropylene tube and wash once more. Suction off the supernatant, spin again for a short period of time (3 min) and remove all liquid supernatant.
3. Break up the wet cell pellet by gently flicking the tip of the tube, and then place the tube in a beaker containing dH_2O at 37°C in order to keep the pellet warm during the fusion.
4. Add to the pellet 1 ml of warm (37°C) 50% PEG solution, in a dropwise fashion over a period of 45 s. While adding the PEG, continuously turn the tube to ensure equal distribution of the fusion agent.

(contd.)

5. Incubate the tube for an additional 45 s in the 37°C water beaker. After a total of 90 s in 50% PEG, slowly and gradually dilute PEG with culture medium. We recommend the following method (adapted from John Kappler). Add 1 ml of culture medium (without supplements or serum) dropwise over 30 s, then 2 ml over the next 30 s, then 3 ml, then 4 ml, then, after a total time of 2 min, add 40 ml of culture medium, close the tube and incubate for an additional 5 min at 37°C in a waterbath. During the dilution procedure, it is necessary to gently turn or shake the tube to ensure that the 50% PEG solution is actually diluted further. The pellet should stay more or less intact to allow cell–cell contact in the presence of PEG for a prolonged period of time.

6. Centrifuge the tube at normal cell pelleting speed at 200g for 5 min at room temperature, remove the supernatant fluid, add 50 ml HBSS (if possible without dislodging the pellet) and wash twice more. This is necessary to remove as much as possible of the toxic fusion agent.

7. After the last wash, add 10 ml of culture medium containing all usual supplements and serum and gently break up the pellet using the same pipette. Add 20–40 ml of additional culture medium depending on the subsequent cloning step (see below).

8. The newly formed hybrids can be maintained in bulk culture or immediately cloned. For most purposes, an immediate cloning step is advisable and should be carried out before the actual selection begins. If immediate cloning is desirable, then continue the protocol from here on.

9. Cloning: With 96-well flat-bottomed tissue-culture microtitre plates, distribute the newly formed hybrids in 100-μl aliquots per well over several plates, depending on the anticipated number of selectable hybridomas. For example, for 100 hybridomas, four or five plates are considered adequate.

   Note: With less than 37% growth-positive wells, the Poisson distribution predicts that more than 95% are derived from single cells in each well, i.e. clonal hybridomas. Obviously, the choice of an appropriate number of plates requires some experience. In general, it is better to anticipate a larger number of hybrids because it becomes very difficult to deal with fusion experiments with a large predicted percentage of non-clonal hybridomas. Depending upon the purpose of the experiment, it might be possible to recover from such a problem by selecting a small number of potentially non-clonal cell lines and to subclone each one of them.

10. Selection: Hybridomas are selected using HAT. At 24 h after cloning, 50 μl HAT (diluted in culture medium to make a 3× solution) is added to each well of the 96-well flat-bottomed microtitre plates. Four days later, the medium must be changed,

(contd.)

using 1× HAT in culture medium. HAT-selectable hybridomas tend to begin to appear after 5–6 days of culture, although some hybrids seem to require as long as 20 days before they start to grow.

11. The growing hybrids, filling part of the well bottom (5×10^4 cells or more) can be picked and transferred into 24-well flat-bottomed plates. At this point, the culture medium used should no longer contain aminopterin but should be supplemented with HT medium at the same concentrations as before, for the subsequent two passages, regardless of how fast the cells grow. This is necessary because residual aminopterin can poison the new hybridomas if precursors of the alternative pathway of DNA synthesis are not provided in sufficient quantities. Only after at least two passages with HT can the new hybridomas be weaned into regular culture medium.

Note: If at this point many of the hybridomas die, weaning was probably begun too early. The cells should be transferred back into HT medium for several additional passages.

◆◆◆◆◆◆ T-CELL TRANSFECTOMAS

Background

Retroviral gene transfer is a means of introducing foreign genes into many cell types (Miller *et al.*, 1993). Recombinant retroviruses contain both selectable markers and strong promoters, such as the cytomegalovirus or long terminal repeat (LTR), which drive the transcription of the desired gene. Cells previously transfected with genes for the retroviral structural proteins *gag*, *pol* and *env* (packaging cell lines such as GP + E-86) can provide these in *trans* when a retroviral vector is introduced, so that they produce replication-defective virions continuously. Retroviral vectors commonly used have been described by Miller *et al.* (1993).

The murine stem cell virus vector, MSCV versions 2.1 and 2.2, which contains the murine stem cell virus LTR promoter/enhancer works well for infection of T-cell hybridomas (Hawley *et al.*, 1992). High viral titres ($>10^5$ cfu ml^{-1}) are made by first transfecting the packaging cell with the retroviral plasmid and then infecting naive packaging cells with viral supernatant harvested from the transfected cells. High viral titres are needed to transduce dividing recipient cells efficiently. Stable transfectants result when the viral sequences integrate into the target DNA. Precise integration of the retroviral vector is also possible, making this method useful for genetic studies (Miller *et al.*, 1993). One disadvantage of the technique is the size limitation of the gene that may be inserted in the retroviral vector of approximately 2 kb which limits transduction to small genes or cDNAs.

Gene transfection of retrovirus-packaging cells

The technique described by Miller *et al.* (1993) has been adapted for gene expression in T-cell hybridoma recipients. The most critical component of the calcium phosphate transfection method is the pH of the transfection buffer, a HEPES-buffered saline solution, in which the optimum pH is between 7.05 and 7.12. Most protocols recommend making this reagent fresh (Graham and van der Eb, 1973), although good results can be achieved using frozen aliquots stored at −80°C. The selection system utilizing the bacterial *neo* gene that confers resistance to G418 is convenient, but the sensitivity of cells to this drug varies greatly. The levels recommended here for T-cell hybridomas are higher than have been used for fibroblasts. A non-transfected control plate is always examined to ensure that positive selection is due to gene integration and expression, rather than cell resistance to the drug.

For infection and integration of the desired gene into T-cell hybridomas, the viral titre needs to be high. Although high titres may be obtained from the primary fibroblast transfectants, retroviral-producing fibroblasts generated by infection may produce higher titres. The use of the polycationic compound, hexadimethrine bromide, is used to increase the efficiency of infection of both the packaging cell and the gene-targeted cell, neutralizing the negative charges present on the cell surface so that virus and cells do not repel one another (Miller *et al.*, 1993). Tunicamycin, an inhibitor of glycosylation, is used to prevent viral envelope glyco-proteins produced by the packaging cell line from blocking the normal receptors for the retrovirus (Rein *et al.*, 1982). The need for tunicamycin can be overcome, if desired, by switching packaging lines.

Materials

- Complete culture medium as above
- Retrovirus packaging cells (GP + E-86 cells are recommended)
- Retroviral vector plasmid DNA (with *neo* resistance) and containing the gene of interest (CsCl-purified or equivalent)
- 5 M NH_4OAc/ETOH (1:5) for DNA precipitation
- Trypsin (0.25% solution in PBS)
- Geneticin (G418) (100 mg ml^{-1} in HBSS stock diluted to 1 mg/ml^{-1} G418 in medium final concentration)
- 2 M $CaCl_2$, sterile filtered
- N-[2-hydroxyethyl]piperazine-N'-[2-ethanesulfonic acid] (HEPES)-buffered saline (HBS): (For 50 ml quantity, use 5 ml 0.5 M HEPES (Sigma H0887) (pH 7.1), 6.25 ml 2 M NaCl, 0.5 ml 150 mM $NaPO_4$ (pH 7.0) and 38.25 ml sterile ddH$_2$O. Divide stock into 1 ml aliquots and store at −70°C or prepare fresh each time)
- Tunicamycin (100 µg ml^{-1} in sterile ddH$_2$O, Sigma T7765)
- Hexadimethrine bromide or Polybrene™ (400 µg ml^{-1} in HBSS, sterile filtered, Sigma H9268)
- 10-cm tissue culture dishes (Falcon)

Procedure

1. Harvest retrovirus packaging cells using trypsin solution to detach cells from the tissue culture plate by removing culture medium and covering cells with a small volume of trypsin solution for a few minutes at 37°C. Pipette cells repeatedly to obtain a single-cell suspension. Dilute 10-fold in medium, spin cells and discard the supernatant.

2. Plate the cells at 5×10^5 cells per 6-cm tissue culture dish in 4 ml culture medium per plate with an extra plate used for a mock transfection. Incubate overnight (37°C, 5% CO_2).

3. Precipitate 10 μg of DNA with five volumes of $NH_4OAc/ETOH$ to sterilize. Redissolve the DNA in 175 μl of sterile ddH_2O. Add 25 μl of sterile 2 M $CaCl_2$ to the DNA. Set up a control tube containing sterile water and $CaCl_2$ for the mock transfection.

4. Place 200 μl of freshly thawed HBS into a 1.7-ml Eppendorf tube. Add the DNA/$CaCl_2$ mixture dropwise to the HBS while gently vortexing the tube. Repeat with the control tube. Allow the tube to sit for 30 min at room temperature. A fine precipitate should form, since DNA and calcium phosphate co-precipitate.

5. Replace the medium on the retrovirus packaging cells with fresh medium. Add the DNA precipitate to the medium dropwise, while gently swirling the plates. Incubate cells for 16–20 h and then replace the medium. Incubate for a further 24 h.

6. Remove culture supernatant and place it into 15-ml conical tubes and spin at 200g for 5 min at 4°C to remove all cells and cell debris. This low titre viral transient transfection supernatant may be used to infect naive packaging cells, but this titre is generally too low to be useful. This material is stored at −70°C as a backup should the higher titre supernatant be lost.

7. Change G418-containing medium every 3–4 days. By day 6, a difference between the real and mock-transfected plates should be obvious. When a plate becomes confluent, replace the G418 medium with plain medium.

8. Culture for 2–3 days more, place the supernatant fluid from the plates in a 15-ml conical tube and spin at 200g to remove debris. This intermediate titre viral supernatant may be stored at −70°C or used immediately for viral infection of naive packaging cells. A good transfection will produce as many as several thousand colonies, but only one is needed to proceed.

9. Seed four 10-cm tissue culture dishes with 2×10^5 packaging cells on 10 ml culture medium. Incubate overnight.

10. Add 10 μl of tunicamycin to each dish and incubate overnight (37°C, 5% CO_2).

    Note: Tunicamycin is rather toxic, and a decrease in cell viability will probably be evident. It may be necessary to experiment with the dose of this drug if too many cells (>75 %) die.

(contd.)

Murine T-Lymphocyte Hybridomas etc

11. Replace medium and add 100 μl of 400 μg ml⁻¹ hexadimethrine bromide to each dish and swirl gently. Add 1 ml, 0.1 ml or 10 μl viral supernatant from the CaPO$_4$ transfection to each of three dishes. If the viral titre is expected to be high, add two more dilutions. The last dish will serve as the uninfected control. Incubate dishes for 16–24 h before replacing the media with 1 mg ml⁻¹ G418 medium. Medium should be changed every 2–4 days.
12. Colonies should be established in about 9 days and the non-infected cells will have died off. Colonies can be picked if desired, but, if not, when cells are confluent, replace the G418 medium with plain medium and culture for 2–3 days.
13. Place the supernatant in a 15-ml conical vial and spin at 200g for 5 min to remove all cell debris. The supernatant can be kept at −70°C until needed, preferably in aliquots to avoid repeated freezing and thawing.

Determining the retroviral titre in the supernatant material

Supernatants with titres of ≥10⁵ cfu ml⁻¹ are required for efficient infection of T-cell hybridomas. Determination of viral titres takes approximately 10 days. Repeated freezing and thawing of the viral supernatant can decrease the titre of the virus.

Materials

- NIH-3T3 or HeLa cells
- Supernatant from virally infected packaging cells
- G418 in medium as above
- Trypsin (0.25%)
- Tunicamycin as above
- Hexadimethrine bromide as above
- Coomassie Brilliant Blue stain/fix (CBB) (0.35 g Coomassie Brilliant Blue (Sigma B5133), 454 ml methanol, 92 ml glacial acetic acid, 454 ml dH$_2$O, sterile filtered)
- 6-cm tissue culture dishes (Falcon)

Procedure

1. To each six 6-cm tissue culture dishes add 1×10^5 NIH-3T3 cells in 4 ml medium. Incubate overnight. Five dishes will be used to dilute the viral titre, and the last dish as a non-infected control.

(contd.)

2. Add 40 µl of 400 µg ml⁻¹ hexadimethrine bromide to each dish. Remove 0.4 ml medium from the first dish and discard. Add 0.4 ml of viral supernatant to this dish, swirl and serially transfer 0.4 ml medium to the next dish. Repeat this for the next three dishes. Discard the last 0.4 ml medium. Incubate dishes for 16–20 h and then replace the medium with G418 medium.
3. Change the G418 medium every 3–4 days until the colonies are mature. This normally takes 9–10 days.
4. To calculate the number of cfu's per millilitre, visualize the colonies by washing with 2 ml HBSS and staining with Coomassie Blue. (Add 2 ml CBB and swirl the dishes until desired blueness is attained and then remove the stain.)

Viral infection of T-cell hybridomas with retrovirus-containing supernatants

Once a high-titre viral supernatant has been made, infection of the target cell can proceed. A number of amphotrophic viruses use the sodium-dependent phosphate symporters as cell surface retroviral receptors (Miller and Miller, 1994). For this reason, target cells are first incubated in phosphate free medium to upregulate the phosphate receptors. This is particularly important for T-cell hybridomas, which are difficult to infect compared with other cell types. T-cell receptor (TCR)-loss variants of hybridomas often are generated spontaneously in culture and can be isolated by cloning. This is useful if TCR genes are to be transfected.

Materials

- Exponentially growing T-cell hybridoma line
- High titre viral supernatant from packaging cells (>10⁵ cfu ml⁻¹)
- Culture medium as above
- G418 in medium as above
- 2×-concentrated G418 (2 mg ml⁻¹ G418 in medium final concentration from 100 mg ml⁻¹ in HBSS stock)
- Tunicamycin (100 µg ml⁻¹ in sterile ddH₂O)
- Hexadimethrine bromide (400 µg ml⁻¹ in HBSS, sterile filtered)
- HEPES buffered saline (HBS) as above
- Coomassie Brilliant Blue stain/fix, as above
- 24-well flat-bottomed tissue-culture plate (Falcon)
- 96-well flat-bottomed microtitre plate (Falcon)
- T-25 tissue-culture flask (Falcon)

Procedure

1. Culture 2×10^6 T-cell hybridoma recipient cells in 10 ml culture medium for 16–20 h.

2. Transfer 1×10^5 hybridoma recipient cells into a 15-ml conical vial for each infection. Prepare at least one extra tube as an uninfected control. Pellet cells as above, remove supernatant and resuspend cell pellet in 0.5 ml high-titre viral supernatant ($>10^5$ cfu ml^{-1}) or in plain medium for the negative control. Place in a single well of a 24-well flat-bottomed tissue-culture plate. Add 5 μl of 400 μg ml^{-1} hexadimethrine bromide. Incubate for 4–8 h. A longer incubation time increases the infection rate but decreases cell survival.

3. Transfer the infected cells to a 15-ml conical tube, rinse the well with 1 ml medium and add this to the tube. Spin down the cells and wash once with 1 ml medium.

4. Resuspend cell pellet in 7 ml medium containing 2×-concentrated G418. Dilute 6×10^3 cells (approximately 0.5 ml of cells) into 20 ml 2×-concentrated G418 medium. Add 100 μl of the cell suspension per well on two 96-well flat-bottomed plates. This will result in approximately 30 cells per well. Place the remaining cells into a T-25 tissue-culture flask. Repeat steps (3) to (5) for the uninfected control cells. Incubate all cells for 3–4 days.

5. After 4 days of incubation, all uninfected cells should be dead and can be discarded. Remove the medium from the 96-well flat-bottomed plates (can be flicked off into a sink in a clean environment) and replace it with standard G418 medium. For the cells in the flask, when the cells have settled, remove approximately 5 ml medium and replace with 5 ml standard G418-containing medium.

6. Check the flasks daily. By day 7, the bulk cultures should be mature if the transfection worked well. On day 8 or 9, clones in the 96-well flat-bottomed plate should appear that can be screened for transgene expression. For TCR transgenes, measuring IL-2 secretion following stimulation on anti-TCR antibody-coated microtitre plates is a convenient screening method, by using either the HT-2 assay (Kappler et al., 1981; Hansen et al., 1989) or by ELISA. To maximize gene expression, grow clones in G418-containing medium keeping the concentration of the cells low (2×10^5 cells ml^{-1} or less).

   Note: Although cells can be cloned from the bulk population flask, we have found that they usually do not express the desired gene at high levels. Also, different clones may not represent independent infection events. Instead, the bulk flask is used to monitor the overall success of the transfection before clones may be obvious. TCR-gene transfectants may produce only low levels of transfected genes but can often be rescued by fluorescent cell sorting, sometimes requiring several successive rounds of sorting.

◆◆◆◆◆◆ DETERMINATION OF LYMPHOCYTE DIVISION BY FLOW CYTOMETRY WITH CFSE LABELLING

This protocol is modified after that described by Lyons and Parish (1994). This method is useful for measuring proliferation of stimulated primary T cells or T-cell lines.

- 5- (and –6)-carboxyfluorescein diacetate succinimidyl ester (CFSE) 25 mg, (Molecular Probes C-1157). Make 1 mM CFSE dissolved in DMSO as a stock solution and store at −20°C
  Caution must be taken with this material as it is *light sensitive!*
- Concanavalin A (Sigma C5275)
- HBSS
- NaCl (0.9%), optional, see below
- Recombinant mouse interleukin 2 (rmIL-2) 10 U ml⁻¹, (R and D Systems or Sigma, I0523)
- Streptavidin-R-phycoerythrin (Jackson ImmunoResearch)
- 24-well flat-bottomed plates (Falcon)

Murine T-Lymphocyte Hybridomas etc

Procedure

For tissue culture:

1. Make a working solution of CFSE by warming stock solution at room temperature (use 10 µl of 1 mM stock solution into 10 ml HBSS to make a 1 µM solution). Keep solution in aluminium foil until used.
2. Wash nylon-wool purified T cells with sterile HBSS as above.
3. Resuspend cells in a 0.2 µM concentration of CFSE in HBSS at a concentration of 10^7 cells ml⁻¹ (resuspend cells in 1–2 µM CFSE when using them *in vivo* injections as described below).
4. Incubate cells with CFSE for 15 min at 37°C in water bath.
5. Spin cells down and wash three times with approximately 10 volumes of HBSS each time.
6. Resuspend cells at a concentration of 5×10^6 cells ml⁻¹ in tissue culture medium.
7. Add cells into 24-well flat-bottomed plate at 5×10^6 cells per well.
8. Add recombinant mouse or human IL-2.
9. Stimulate cells with concanavalin A (5 µg/ml).
10. Incubate cells for 48 h in complete media with IL-2.

(contd.)

275

11. Since CFSE fluoresces in the green spectrum, use a biotin-labelled primary or secondary antibody marker for γδ T cell subsets or αβ T-cell subsets followed by incubation with streptavidin-R-phycoerythrin.
12. Analyse cells by flow cytometry by gating first on live cells with subsequent gating on subset T cells.
13. Percentages of cells proliferating, i.e. dividing, were estimated by gating on populations with reduced CFSE fluorescence. The fluorescence peaks will follow the non-proliferating control peak, each representing a subsequent generation of proliferating cells.

*For **in vivo** intravenous injection:*

1. Use $5-10 \times 10^6$ cells ml^{-1} for αβ T cells (for smaller cell population, such as γδ T cells, more cells might be necessary) and resuspend in 0.9% NaCl or HBSS.
2. Wait at least three days for analysing dividing cells in the spleen. In other organs, e.g. lung, divisions may appear earlier than three days.
3. Use biotin-labelled primary or secondary antibody markers as above.
4. Analyse cells by flow cytometry as above.
5. Percentages of cells proliferating, i.e. dividing, were estimated by gating on populations with reduced CFSE fluorescence. The fluorescence peaks will follow the non-proliferating control peak, each representing a subsequent generation of proliferating cells.

Acknowledgement

We would like to thank Drs Willi Born and Rebecca O'Brien for their helpful suggestions in the preparation of this manuscript.

References

Aarden, L. A., Brunner, K. T., Cerottini, J.-C., Dayer, J.-M., deWeck, A. L., Dinarello, C. A., DiSabato, G., Farrar, J. J., Gery I., Willis, S., Handschumacher, R. E., Henney, C. S., Hoffman, M. K., Koopman, W. J., Krane, S. M., Lachman, L. B., Lefkovits, I., Mishell, R. I., Mizel, S. B., Oppenheim, J. J., Paetkau, V., Plate, J., Rollinghoff, M., Rosenstreich, D., Rosenthal, A. S., Rosenwasser, L. J., Schimpl, A., Shin, J. S., Simon, P. L., Smith, K. A., Wagner, H., Watson, J. D., Wecker, E. and Wood, D. D. (1979). Revised nomenclature for antigen non-specific T-cell proliferation and helper factors. *J. Immunol.* **123**, 2928–2929.

Asarnow, D. M., Kuziel, W. A., Bonyhadi, M., Tigelaar, R. E., Tucker, P. W. and Allison, J. P. (1988). Limited diversity of γδ antigen receptor genes of Thy-1$^+$ dendritic epidermal cells. *Cell* **55**, 837–847.

Born, W., Yagiie, J., Palmer, E., Kappler, J. and Marrack, P. (1985). Rearrangement of T-cell receptor β-chain genes during T-cell development. *Proc. Natl. Acad. Sci. USA* **82**, 2925–2929.

Born, W., Rathbun, G., Tucker, P., Marrack, P. and Kappler, J. (1986). Synchronized rearrangement of T-cell γ and δ chain genes in fetal thymocyte development. *Science* **234**, 479–482.

Born, W., Miles, C., White, J., O'Brien, R., Freed, J. H., Marrack, P., Kappler, J. and Kubo, R. T. (1987). Peptide sequences of T-cell receptor δ and γ chains are identical to predicted X and γ proteins. *Nature* **330**, 572–574.

Brunner, T., Mogil, R. J., LaFace, D., Yoo, N., Mahboubi, A., Echeverri, F., Martin, S. J., Force, W. R., Lynch, D. H., Ware, C. F. and Green, D. R. (1995). Cell-autonomous *Fas* (CD95)/*Fas*-ligand interaction mediates activation-induced apoptosis in T-cell hybridomas. *Nature* **373**, 441–444.

Chestnut, R. W., Colon, S. M. and Grey, H. M. (1982). Antigen presentation by normal B cells, B cell tumors, and macrophages: functional and biochemical comparison. *J. Immunol.* **128**, 1764–1768.

Chien, Y.-H., Iwashima, M., Wettstein, D. A., Kaplan, K. B., Elliott, J. F., Born, W. and Davis, M. M. (1987). T-cell receptor δ gene rearrangements in early thymocytes. *Nature* **330**, 722–727.

Coligan, J. E., Kruisbeek, A. M., Margolies, D. H., Shevach, E. M. and Strober, W. (eds) (1996). Production of T cell clones. In: *Current Protocols in Immunology*. John Wiley & Sons, Inc., New York, Vol. 1, Unit 3.13.

Fathman, C. G. and Hengartner, H. (1978). Clones of alloreactive T cells. *Nature (London)* **272**, 617–618.

Gajewski, T. F., Joyce, J. and Fitch, F. W. (1989a). Anti-proliferative effect of INF-γ in immune regulation. III. Differential selection of Th1 and Th2 murine helper T lymphocyte clones using rIL-2 and IFN-γ. *J. Immunol.* **143**, 15–22.

Gajewski, T. F., Schell, S. R., Nau, G. and Fitch, F. W. (1989b). Regulation of T cell activation: differences among T cell subsets. *Immunol. Rev.* **111**, 79–110.

Galfré, G., Milstein, C. and Wright, B. (1979). Rat × rat hybrid myelomas and a monoclonal anti-Fd portion of mouse IgG. *Nature* **277**, 131–133.

Glasebrook, A. L. and Fitch, F. W. (1980). Alloreactive cloned T cell lines. 1. Interactions between cloned amplifier and cytolytic T cell lines. *J. Exp. Med.* **151**, 876–895.

Goldsby, R. A., Osborne, B. A., Simpson, E. and Herzenberg, L. A. (1977). Hybrid cell lines with T-cell characteristics. *Nature* **267**, 707–708.

Gorczynski, R. M., Cohen, Z., Leung, Y. and Chen, Z. (1996). γδ TCR⁺ hybridomas derived from mice preimmunized via the portal vein adoptively transfer increased skin allograft survival *in vivo. J. Immunol.* **157**, 574–581.

Graham, F. L. and van der Eb, A. J. (1973). A new technique for the assay of infectivity of human adenovirus 5 DNA. *Virology* **52**, 456–462.

Gu, J. J. and Gottlieb, P. D. (1992). Inducible functions in hybrids of a Lyt-2⁺ BW5147 transfectant and the 2C CTL line. *Immunogenetics* **36**, 283–293.

Haars, R., Kronenberg, M., Gallatin, W. M., Weissman, I. L., Owen, F. L. and Hood, L. (1986). Rearrangement and expression of T cell antigen receptor and y genes during thymic development. *J. Exp. Med.* **164**, 1–24.

Haas, W., Mathur-Rochat, J., Kisielow, P. and Van Boehmer, H. (1985). Cytolytic T cell hybridomas. III. The antigen specificity and the restriction specificity of cytolytic T cells do not phenotypically mix. *Eur. J. Immunol.* **15**, 963–965.

Hansen, M. B., Nielsen, S. E. and Berg, K. (1989). Re-examination and further development of a precise and rapid dye method for measuring cell growth/cell kill. *J. Immunol. Methods* **119**, 203–210.

Happ, M. P., Kubo, R. T., Palmer, E., Born, W. K. and O'Brien, R. L. (1989). Limited receptor repertoire in a mycobacteria-reactive subset of γδ T lymphocytes. *Nature* **342**, 696–698.

Hawley, R. G., Fong, A. Z. C., Burns, B. F. and Hawley, T. S. (1992). Transplantable myeloproliferative diseases induced in mice by an interleukin 6 retrovirus. *J. Exp. Med.* **176**, 1149–1163.

Johnson, J. P., Cianfriglia, M., Glasebrook, A. L. and Nabholz, M. (1982). Karyotype evolution of cytolytic T cell lines. In: *Isolation, Characterization, and Utilization of T Lymphocyte Lines* (C. G. Fathman and F. W. Fitch, Eds), pp. 183–191. Academic Press, San Diego.

Jones-Carson, J., Vazquez-Torres, A., van der Heyde, H. C., Warner, T., Wagner, R. D. and Balish, E. (1995). Gamma delta T cell-induced nitric oxide production enhances resistance to mucosal candidiasis. *Nat. Med.* **1**, 552–557.

Kappler, J. W., Skidmore, B., White, J. and Marrack, P. (1981). Antigen-inducible H-2 restricted interleukin-2 producing T cell hybridomas. Lack of independent antigen and H-2 recognition. *J. Exp. Med.* **153**, 1198–1214.

Kappler, J., White, J., Wegmann, D., Mustain, E. and Marrack, P. (1982). Antigen presentation by Ia⁺ B cell hybridomas to H-2 restricted T cell hybridomas. *Proc. Natl. Acad. Sci. USA* **79**, 3604–3607.

Kearney, J. F., Radbruch, A., Liesegang, B. and Rajewsky, K. (1979). A new mouse myeloma cell line that has lost immunoglobulin expression but permits the construction of antibody-secreting hybrid cell lines. *J. Immunol.* **123**, 1548–1550.

Köhler, G. and Milstein, C. (1975a). Continuous cultures or fused cells secreting antibody of predefined specificity. *Nature (London)* **256**, 495–497.

Köhler, G. and Milstein, C. (1975b). Derivation of specific antibody-producing tissue culture and tumor lines by cell fusion. *Eur. J. Immunol.* **6**, 511–519.

Lyons, B. and Parish, C. R. (1994). Determination of lymphocyte division by flow cytometry. *J. Immunol. Methods* **171**, 131–137.

Matis, L. A. and Bluestone, J. A. (1991). Specificity of γδ receptor bearing T cells. *Sem. Immunol.* **3**, 75–80.

Matis, L. A., Cron, R. and Bluestone, J. A. (1987). Major histocompatibility complex-linked specificity of γδ receptor-bearing T lymphocytes. *Nature* **330**, 262–264.

Matis, L. A., Fry, A. M., Cron, R. Q., Cotterman, M. M., Dick, R. F. and Bluestone, J. A. (1989). Structure and specificity of a class II alloreactive γδ T cell receptor heterodimer. *Science* **245**, 746–749.

Miller, D. G. and Miller, A. D. (1994). A family of retroviruses that utilize related phosphate transporters for cell entry. *J. Virol.* **68**, 8270–8276.

Miller, A. D., Miller, D. G., Garcia, J. V. and Lynch, C. M. (1993). Use of retroviral vectors for gene transfer and expression. *Methods Enzymol.* **217**, 581–599.

Morgan, D. A., Ruscetti, F. W. and Gallo, R. (1976). Selective *in vitro* growth of T lymphocytes from normal human bone marrows. *Science* **193**, 1007–1008.

Nishimura, H., Hiromatsu, K., Kobayashi, N., Grabstein, K. H., Paxton, R., Sugamura, K., Bluestone, J. A. and Yoshikai, Y. (1996). IL-15 is a novel growth factor for murine γδ T cells induced by *Salmonella* infection. *J. Immunol.* **156**, 663–669.

Nixon-Fulton, J. L., Bergstresser, P. R. and Tigelaar, R. E. (1986). Thy-l⁺ epidermal cells proliferate in response to concanavalin A and interleukin 2. *J. Immunol.* **136**, 2776–2786.

Nixon-Fulton, J. L., Hackett Jr., J., Bergstresser, P. R., Kumar, V. and Tigelaar, R. E. (1988). Phenotypic heterogeneity and cytolytic activity of Con A and IL2-stimulated cultures of mouse Thy-l⁺ epidermal cells. *J. Invest. Dermatol.* **91**, 62–68.

O'Brien, R. L., Happ, M. P., Dallas, A., Palmer, E., Kubo, R. and Born, W. K. (1989). Stimulation of a major subset of lymphocytes expressing T cell receptor γδ by an antigen derived from *Mycobacterium tuberculosis*. *Cell* **57**, 667–674.

Okazaki, H., Ito, M., Sudo, T., Hattori, M., Kano, S., Katsura, Y. and Minato, N. (1989). IL-7 promotes thymocyte proliferation and maintains immunocompetent thymocytes bearing αβ or γδ T-cell receptors *in vitro:* synergism with IL-2. *J. Immunol.* **143**, 2917–2922.

Raulet, D. H., Garman, R. D., Saito, H. and Tonegawa, S. (1985). Developmental regulation of T-cell receptor gene expression. *Nature* **314**, 103–107.

Reardon, C. L., Heyborne, K., Tsuji, M., Zavala, F., Tigelaar, R. E., O'Brien, R. L. and Born, W. K. (1995). Murine epidermal Vγ5/Vγ1-TCR⁺ T cells respond to B cell lines and lipopolysaccharides. *J. Invest. Dermatol.* **105**, 58S–61S.

Rein, A., Schultz, A. M., Bader, J. P. and Bassin, R. H. (1982). Inhibitors of glycosylation reverse retroviral interference. *Virology* **119**, 185–192.

Rellahan, B. L., Bluestone, J. A., Houlden, B. A., Cotterman, M. M. and Matis, L. A. (1991). Junctional sequences influence the specificity of γ/δ T cell receptors. *J. Exp. Med.* **173**, 503–506.

Shulman, M., Wilde, C. D. and Köhler, G. (1978). A better cell line for making hybridomas secreting specific antibodies. *Nature* **276**, 269–270.

Sloan-Lancaster, J., Evavold, B. D. and Allen, P. M. (1994). Th2 cell clonal anergy as a consequence of partial activation. *J. Exp. Med.* **180**, 1195–1205.

Snodgrass, H. R., Dembic, Z., Steinmetz, M. and von Boehmer, H. (1985). Expression of T-cell antigen receptor genes during fetal development in the thymus. *Nature* **315**, 232–233.

Stingl, G., Gunter, K. C., Tschachler, E., Yamada, H., Lechler, R. L., Yokoyama, W. M., Steiner, G., Germain, R. N. and Shevach, E. M. (1987). Thy⁺ dendritic epidermal cells belong to the T-cell lineage. *Proc. Natl Acad. Sci. USA* **84**, 2430–2434.

Takita-Sonoda, Y., Tsuji, M., Kamboi, K., Nussenzweig, R. S., Clavijo, P. and Zavala, F. (1996). *Plasmodium yoelii:* peptide immunization induces protective CD4⁺ T cells against a previously unrecognized cryptic epitope of the circumsporozoite protein. *Exp. Parasitol.* **84**, 223–230.

Taniguchi, M. and Miller, J. F. A. P. (1978). Specific suppressive factors produced by hybridomas derived from the fusion of enriched suppressor T cell and a T lymphoma cell line. *J. Exp. Med.* **148**, 373–382.

Tigelaar, R. E., Lewis, J. M. and Bergstresser, P. R. (1990). TCR γδ⁺ dendritic epidermal T cells as constituents of skin-associated lymphoid tissue. *J. Invest. Dermatol.* **94**, 58S–63S.

Tsuji, M., Romero, P., Nussenzweig, R. S. and Zavala, F. (1990). CD4⁺ cytolytic T cell clone confers protection against murine malaria. *J. Exp. Med.* **172**, 1353–1357.

Tsuji, M., Mombaerts, P., Lefrancois, L., Nussenzweig, R. S., Zavala, F. and Tonegawa, S. (1994). γδ T cells contribute to immunity against the liver stages of malaria in αβ T-cell-deficient mice. *Proc. Natl Acad. Sci. USA* **91**, 345–349.

Tsuji, M., Eyster, C., O'Brien, R. L., Born, W. K., Bapna, M., Reichel, M., Nussenzweig, R. S. and Zavala, F. (1996). Phenotypic and functional properties of murine γδ T cell clones derived from malaria immunized αβ T cell-deficient mice. *Int. Immunol.* **8**, 359–366.

White, J., Blackman, M., Bill, J., Kappler, J., Marrack, P., Gold, D. and Born, W. (1989). Two better cell lines for making hybridomas expressing specific T cell receptors. *J. Immunol.* **143**, 1822–1825.

Ziegler, K. and Unanue, E. R. (1981). Identification of a macrophage antigen-processing event required for I-region-restricted antigen presentation to T lymphocytes. *J. Immunol.* **127**, 1869–1875.

2.3 Killer Cell Assays

Dirk H Busch
Institute of Medical Microbiology, Immunology, and Hygiene
Technical University Munich, Germany

Eric G Pamer
Laboratory of Antimicrobial Immunity
Memorial Sloan-Kettering Cancer Center, New York, USA

◆◆◆

CONTENTS

◆◆◆◆◆◆ INTRODUCTION

Cell-mediated cytotoxicity plays an important role in the host immune defence against pathogens localized within cells (Pamer, 1993). Intracellular pathogens escape antibody, complement and neutrophil-mediated defences. Infected cells must therefore be specifically identified and destroyed in order for the infection to be cleared, a role fulfilled by CD8$^+$ cytolytic T lymphocytes (CTLs). CTL recognize epitopes presented on the cell surface by MHC class I molecules (Zinkernagel and Doherty, 1974). Pathogen-derived proteins are first degraded into small peptides (8–10 residues) which are translocated via the transporter associated with antigen processing (TAP) into the endoplasmic reticulum (ER) (Pamer and Cresswell, 1998). Peptides with sufficient affinity for the MHC class I peptide-binding groove stabilize the newly synthesized molecules, forming MHC/peptide complexes that are transported to the cell surface. CD8$^+$ CTLs detect pathogen-derived peptides presented by MHC class I molecules with their specific T-cell receptor, resulting in activation of different effector mechanisms that induce death of the infected cell. Two major cytotoxic pathways utilized by CTL have been described: 1) release of cytotoxins from secretory granules that directly damage the host cell membrane (perforins), and 2) induction of programmed cell death

Copyright © Elsevier Science Ltd
All rights of reproduction in any form reserved

(apoptosis) of the target cell by secreted proteases (granzymes) or direct Fas/Fas-ligand binding (Berke, 1995).

We have used the murine model of *Listeria monocytogenes* infection to examine the CTL response to infection with an intracellular pathogen (Busch *et al.*, 1998, 1999). *L. monocytogenes* is a Gram-positive bacterium that survives and multiplies within the cytosol of infected cells. After phagocytosis by macrophages, *L. monocytogenes* lyses the phagolysosomal membrane by secreting listeriolysin (LLO) and enters the host cell cytosol (Bielecki *et al.*, 1990). Immunocompetent mice infected with a sublethal dose of *L. monocytogenes* clear the infection within a few days and develop long-lasting protective immunity. CD8+ cytotoxic lymphocytes play a major role in this rapid, extremely effective immune response (Kaufmann *et al.*, 1985).

In this chapter, we describe different strategies to characterize CTLs specific for *L. monocytogenes*, to identify pathogen-derived epitopes, and to study the efficiency of MHC class I antigen processing. These methods can also be easily modified for the detailed study of CTL responses to other intracellular infections, taking into account the specific character-istics (e.g. cell specificity, subcellular localization, persistence, and intra-cellular growth rate) of each pathogen.

◆◆◆◆◆◆ CHROMIUM RELEASE ASSAY

Although several assays have been established to detect and quantify CTL-mediated cell lysis (see below), the standard chromium release assay (CRA) remains one of the best and most convenient methods (Brunner *et al.*, 1968) because of its high sensitivity and specificity. The principle behind the CRA is to label target cells with radioactive ^{51}chromium. Most cell types take up Na_2CrO_4 when exposed to high concentrations and, as long as the cells remain viable, release of the chromium is very slow. The ^{51}Cr-labelled cells, which can be infected with a pathogen or coated with antigenic peptides, are then incubated together with CTLs. If the T cells, referred to as effector cells in these assays, detect their specific epitope on the surface of labelled target cells, they induce cell death and damage of the cell membrane, result-ing in the release of ^{51}Cr into the culture medium. The ratio of released to cell-associated ^{51}Cr is proportional to the degree of cell lysis.

Chromium release assay using infected target cells

As mentioned above, ^{51}Cr-labelled target cells can be infected with intra-cellular pathogens and tested for lysis by specific CTLs (Kaufmann *et al.*, 1986). These assays can determine if pathogen-specific CTLs are elicited by infection and, if epitope-specific T-cell clones or T-cell lines are avail-able, allow characterization of epitope presentation by actively infected cells. Although different cell types can be used as target cells in CRAs, the cell specificity of the pathogen may limit the range of available target cells. An additional restriction on the cell type that can be used in the CRA is that the majority of cells must be infected in order for the specific lysis to be interpretable. In the case of *L. monocytogenes* an infection efficiency of nearly 100% can be reproducibly achieved in both primary bone marrow macrophages and tumour cell lines. We have studied *Listeria*-specific CTL responses in laboratory mice with the H2^d haplotype (e.g. BALB/c) and have used the H2^d macrophage tumour cell line J774 (ATCC TIB 67) as a target cell (Pamer *et al.*, 1991). J774 cells are readily infected with *L. monocytogenes*, and bacteria enter the cytosol and multiply intracellularly. The J774 cell line is grown in conventional culture medium without the addition of supplementary growth factors, resulting in nearly unlimited quantities of target cells, the homogeneity of which makes this system highly reproducible.

The following section describes a typical protocol to test *Listeria*-specific CTL lines for specific lysis of J774 cells infected with live *L. monocytogenes*.

Reagents and equipment

- antibiotic-free culture medium (RP10$^-$): RPMI 1640 supplemented with L-glutamine plus 10%FCS
- ^{51}Cr sodium chromate (Dupont, MA, USA)
- gentamicin-sulfate (Gemini Bio-Products, CA, USA)
- 0.5% Triton X-100
- 96-well U-bottom microtitre plates
- J774 macrophage cell line (ATCC TIB 67), cultured in RP10$^-$
- bacterial culture: virulent *L. monocytogenes* (e.g. ATCC 43251) grown in trypticase soy broth (TSB)
- gamma-counter
- effector cells: *Listeria*-specific T-cell line

Labelling target cells with chromium 51

J774 cells are radioactively labelled by short incubation in the presence of high concentrations of ^{51}Cr. ^{51}Cr has a relatively short half-life (28 days) and is less hazardous to work with than most other isotopes. Nevertheless, work with ^{51}Cr must be performed following the appropriate radiation safety guidelines.

- pellet 1×10^6 J774 cells (500g, 5 min)
- resuspend cells in 100 µl RP10$^-$
- add 100 µCi [^{51}Cr] sodium chromate (usually equivalent to 100 µl fresh ^{51}Cr, calculate the actual activity considering a half-life of 28 days)
- incubate for 1 h at 37°C
- wash cells twice in 10 ml RP10$^-$
- resuspend cells in 10 ml RP10$^-$ (= 1×10^5 cells/ml)
- add 100 µl (= 1×10^4 cells) per well in a 96-well plate and allow macrophages to adhere for 30 min at 37°C

Infecting labelled cells with intracellular bacteria

^{51}Cr-labelled macrophages are infected by the direct addition of *L. monocytogenes* from a mid log-phase culture. After incubation, the medium is replaced by RP10$^-$ containing gentamicin, a membrane-impermeable antibiotic, to kill extracellular, but not intracellular bacteria.

- grow *L. monocytogenes* in TSB to early/mid log phase ($A_{600} = 0.1$); at this density, there are approximately 2×10^8 bacteria per ml.
- add 10 μl bacteria (= 2×10^6,) to wells with ^{51}Cr-labelled J774 cells designated as infected target cells
- add 10 μl TSB to ^{51}Cr-labelled J774 macrophages to be used as uninfected controls
- incubate for 25 min at 37°C
- carefully remove 80 μl of medium from each well, add 80 μl RP10$^-$ containing 10 μg ml^{-1} gentamicin

---

Assaying for specific lysis with CD8$^+$ T-cells using different effector to target ratios

To assay for specific lysis, CD8$^+$ T cells are incubated together with the prepared target cells (generation of antigen-specific CTLs is described in Chapter 2.2). The more antigen-specific T-cells/effector cells (E) are added to the assay, the greater the expected extent of specific target cell (T) lysis. However, high E : T ratios are often accompanied by high non-specific target cell lysis. Antigen-specific lysis is determined by comparison of the chromium release in parallel incubations of effector cells with infected and uninfected target cells. The optimal E : T ratio is difficult to predict, so it is advisable to test for specific lysis at several E : T ratios. Titration of the E : T ratio is usually achieved by changing the number of effector cells, keeping the number of target cells constant (here 1×10^4 cells) and an E : T ratio titration is achieved by adding different dilutions of effector cells.

A typical CTL assay contains the following controls and titrations.

Spontaneous release: Infected and uninfected labelled target cells are incubated in the absence of effector cells to control for the spontaneous release of chromium.

- add 100 μl RP10$^-$ to four wells each of uninfected and infected target cells

Maximum release: A detergent is added lysing all target cells in order to determine the maximum radioactivity that can be released.

- add 100 μl 0.5% Triton X-100 to four wells

(contd.)

> *CTLs plus infected or uninfected target cells:* CTLs added to uninfected and infected ⁵¹Cr labelled target cells at various E : T ratios to determine specific and background lysis.
>
> - add 100 µl of effector cells (in RP10⁻) to infected and uninfected target cells. Make several dilutions of effector cells to achieve E : T ratios of 30 : 1, 10 : 1, 3 : 1, 1 : 1 and 0.3 : 1
> - the final volume in all wells should be 200 µl
> - incubate the cells for 3 h at 37°C in a 5% CO_2 incubator
> - pellet the cells in the 96-well plate by gentle centrifugation at 400*g* for 5 min
> - carefully harvest 100 µl of supernatant from each well with a multichannel pipettor and count the released ⁵¹Cr with a gamma-counter.

Determining percentage specific lysis

Specific lysis is calculated by accounting for spontaneous chromium release (sr) and maximum release (mr) using the formula:

$$\% \text{ specific lysis} = 100 \times \frac{\text{cpm sample} - \text{cpm sr}}{\text{cpm mr} - \text{cpm sr}}$$

Lysis specifically related to the presence of the pathogen can be estimated by comparing lysis in response to infected vs. uninfected target cells at the same E : T ratios. A difference of greater than 10–20% between lysis against infected and uninfected cells can be interpreted as antigen-specific lysis.

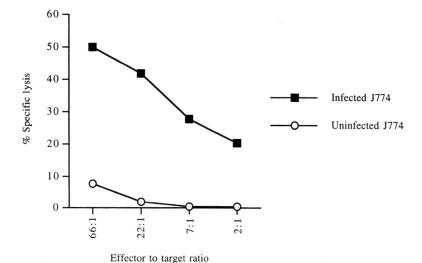

Figure 1. CTL derived from *L. monocytogenes* immunized mice specifically lyse infected J774 cells. Spleen cells from *L. monocytogenes* immunized BALB/c mice were restimulated *in vitro* with infected J774 cells. Five days following *in vitro* restimulation, CTL were assayed for specificity using *L. monocytogenes* infected or uninfected ⁵¹Cr labelled J774 cells, as described in the text. The % specific lysis was determined 3 h after addition of different CTL numbers to 10 000 target cells.

Limitations of direct CTL assays

Using target cells infected with intracellular pathogens as CTL targets has certain disadvantages that can limit the utility of these assays. For example, the pathogen will continue to multiply inside the infected target cells. Pathogens with a high intracellular growth rate might simply burst the infected cell within a relatively short time period, resulting in a rapid increase of spontaneous ^{51}Cr-release. Alternatively, some intracellular pathogens release lytic proteins that can cause high degrees of spontaneous lysis in the absence of CTL. If the spontaneous release of ^{51}Cr exceeds 30–40% of maximum release values, then the specific lysis values become very difficult to interpret.

Although the incubation time for CTL assays can be shortened to prevent exceedingly high spontaneous release values, there is a certain delay until specific target cell lysis is detectable; a minimum incubation time of 2–3 h is needed. Possible explanations for the delayed onset of specific lysis are that specific CTLs need time to find their target cells and to induce cell death, and that target cells may not immediately release all ^{51}Cr upon encounter with a specific CTL.

Chromium release assay using target cells coated with peptide extracts of infected cells

Several years ago, Rammensee and colleagues (Roetzschke *et al.*, 1990) exploited the ability to acid elute and HPLC purify MHC-associated peptides from infected cells, subsequently transferring the isolated peptides to uninfected cells in an assay for antigen-specific, CTL-mediated lysis. Although peptide–MHC class I complexes are relatively stable at a physiological pH, peptides rapidly dissociate at low pH. Pathogen-derived epitopes can be directly eluted from infected cells by lowering the pH to 2.0 using 0.1% trifluoroacetic acid (TFA). Eluted peptides are separated from high molecular weight material and fractionated by reverse-phase HPLC, and ^{51}Cr-labelled target cells are coated with HPLC-purified peptide fractions and tested for specific lysis in a standard CTL assay. This procedure is particularly useful when pathogens infect cells that cannot be used for CRAs or lyse cells within the time frame of a conventional chromium release assay (see also earlier). However, even when the performance of direct CTL assays with infected cells is feasible, the alternative of peptide extraction has several advantages. For example, the sensitivity of the assay is increased when epitopes can be extracted from large numbers of cells and added in relatively high concentrations to target cells; because peptides are protected from proteolysis by the MHC molecule, epitopes can even be extracted from cells lysed by the pathogen. When using peptides, it is also possible to choose target cells that optimize sensitivity (i.e. cells with high ^{51}Cr uptake, low spontaneous release, and a high capacity to bind exogenously added peptides) and give highly reproducible results. HPLC fractionation of antigenic epitopes allows one to estimate the complexity of the CTL response, assuming that most CTL epitopes will elute in different fractions. When fractionated peptides are loaded onto partially MHC-mismatched target cells, the MHC-restriction for individual epitopes can be determined. Finally, purified peptides can be sequenced to obtain more detailed information about the structure and origin of the epitopes.

Reagents and equipment (see also earlier)

- tissue culture plates (e.g. Falcon 3025, 150 mm)
- 0.1% trifluoric acid
- dounce homogenizer and sonicator
- Centricon 10 (Amicon, MA, USA)
- ultracentrifuge
- HPLC with C18-300A column (Delta Pak, 3.9 × 300 mm, 15 mm spherical beads)
- lyophilization unit
- a cell line that can be infected *in vitro* with the intracellular pathogen (e.g. macrophage cell lines J774, IC21, PU51R, RAW264.7)
- target cells: usually a tumour cell line expressing the appropriate MHC class I molecules (e.g. $H2^d$ positive mastocytoma cell line P815/ATCC TIB64)

Harvesting large numbers of cells infected with pathogenic organisms

For the extraction of *L. monocytogenes*-derived epitopes, we grow the bacteria in the macrophage cell line J774, which can be easily cultured and expanded *in vitro*. Large numbers of cells are grown to confluence in tissue culture plates and subsequently infected with the pathogen. The infected cells are harvested for peptide extraction after different time intervals of incubation; a kinetic analysis is necessary to determine the optimal time point for epitope extraction and also reveals valuable information about the dynamics of antigen presentation (see also later).

- grow J774 cells to confluence (approximately 10^8 cells per plate) in 27 ml RP10 in 150 mm tissue cultures plates⁻ (one plate for each time interval examined)
- add 3 ml of mid log-phase culture *L. monocytogenes* in TSB at A_{600} of 0.1 (= 6×10^8 bacteria) to each plate
- incubate for 30 min at 37°C
- replace medium with RP10⁻ containing 5 μg ml⁻¹ gentamicin
- incubate plates for different time intervals (e.g. 1, 3, 5, 7 hours)
- remove medium and harvest cells by scraping them into 5 ml PBS (10 μl of cells can be taken, diluted in PBS containing 0.1% Triton X-100, and plated out on TSB agar to estimate the number of bacteria per infected cell at each time point)
- pellet harvested cells by centrifugation, store the pellets at −80°C

Lysing infected cells and eluting peptides with TFA

Resuspension of cell pellets in 0.1% trifluoric acid (TFA) and homogenization results in peptide extraction from the infected cells. Insoluble material is pelleted by high speed ultracentrifugation, and the supernatants are depleted of high molecular weight molecules by passage through a Centricon-10 membrane.

> The following steps should be taken to keep the samples on ice
> - resuspend pellets in 10 ml of 0.1% TFA
> - homogenize cell suspension by dounce homogenization (roughly 20 strokes) and sonication (20 s cycles at an intermediate setting)
> - centrifuge for 30 min at 100 000g
> - freeze supernatant and lyophilize
> - resuspend lyophilized material in 2 ml of 0.1% TFA and pass the suspension through a Centricon-10 membrane (5000g)

HPLC fractionation of infected cell extracts

The material that passes through the Centricon-10 membrane is applied to a reverse phase HPLC C18 column and eluted with a 0–60% acetonitrile gradient in 0.1% TFA (0 to 5 min 0% acetonitrile, 5 to 45 min 0–60% acetonitrile). One millilitre fractions are collected at 1-min intervals and subsequently lyophilized. With this gradient, most MHC class I-associated peptides elute from the C18 column between fractions

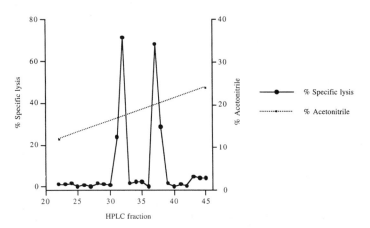

Figure 2. CTL epitopes can be TFA extracted from *L. monocytogenes* infected J774 cells. About 400 million J774 cells were infected with *L. monocytogenes* for 6 h and then harvested and TFA extracted as described in the text. Low molecular weight material was applied to a C18 reverse phase HPLC column and eluted with a shallow gradient of acetonitrile. HPLC fractions were lyophilized, resuspended in PBS and assayed for recognition by an *L. monocytogenes* specific, non-clonal T-cell line. P815 (H2^d) cells were labelled with ^{51}Cr and used as targets. The effector to target ratio in this experiment was 10:1. It can be seen that this CTL line recognizes two different peptides that elute from the C18 column with different acetonitrile concentrations.

(minutes) 24 and 40; if the gradient is changed, the elution time for the peptides will also be altered. It is possible to maximize the separation of different peptides by extending the acetonitrile/water gradient.

CTL assay

Peptide fractions are resuspended in PBS and used to coat target cells in a standard chromium release assay. Fractions obtained from 100 million extracted cells are typically resuspended in 200 µl PBS. Most tumour cell lines express well-characterized MHC class I molecules, and many of them are excellent candidates for chromium labelling. Thus, a wide range of possible target cells for different animal models and MHC haplotypes is available. In our system, we use the mouse mastocytoma cell line P815 (ATCC TIB 64) to test for H2-K^d-restricted epitope presentation.

- for detailed information, see above
- label P815 cells with ^{51}Cr and place 1×10^4 cells (in 50 µl RP10) in wells of a 96-well plate
- resuspend HPLC fractions in 200 µl of PBS and add 50 µl of each sample to a designated well of target cells
- incubate for 45–60 min at 30°C
- add CTLs at a constant E : T ratio in a volume of 100 µl RP10 medium per well (to determine an optimal E : T ratio, different effector cell dilutions should be tested in advance)
- incubate plates for 4–6 h at 37°C and harvest supernatants as described

---

Killer Cell Assays

Alternative protocol: Extraction of MHC class I-associated peptides from infected spleens

Essentially all studies of antigen processing have been performed *in vitro* in systems that are, at best, approximations of *in vivo* events; studies of *in vivo* antigen processing are difficult, because the quantity and concentration of antigen is very low. It is possible, however, to isolate MHC class I-associated peptides from infected spleens and to identify pathogen-derived epitopes (Pamer *et al.*, 1991). In the *L. monocytogenes* system, these experiments are generally only successful when mice are infected with a very high infectious dose and the peptides are extracted from spleens 48 h after infection. Because this method has great potential for direct correlation of *in vivo* antigen processing with T-cell responses, we briefly describe the acid-extraction of *Listeria*-derived peptides from infected spleens.

- infect BALB/c mice i.v. with 1×10^6 virulent *L. monocytogenes* (ATCC 43251)
- harvest 1–2 spleens 48 h post infection, homogenize the organs in 10 ml 0.1% TFA with a ground glass tissue grinder, followed by dounce homogenization and sonication as described above for infected J774 cells. The pH should be checked with pH paper; if the pH is greater than 2.0, it should be adjusted to pH 2.0 with 1% TFA
- centrifuge the homogenate for 30 min at $100\,000g$
- lyophilize the supernatant and then resuspend it in 2 ml of 0.1% TFA
- centrifuge the resuspended extract through a Centricon-10 membrane; because the extract can be quite viscous, passage through the membrane may require 3–4 h of centrifugation
- fractionate the Centricon-10 filtrate by HPLC as described above
- lyophilize individual fractions and resuspend in PBS for use in CTL assays as described above

QUANTITATION OF MHC CLASS I-ASSOCIATED EPITOPES IN INFECTED CELLS

Falk *et al.* (1991) were the first to determine the number of pathogen-derived epitopes in an infectious system, quantifying epitopes from influenza virus-infected cells following TFA extracting. This powerful method for quantifying the end result of the MHC class I antigen processing pathway has been extended to *L. monocytogenes*-infected cells (Villanueva *et al.*, 1994) and to cells expressing HIV proteins (Tsomides *et al.*, 1994). To quantify natural epitopes successfully in infected cells, it is necessary to: (a) obtain purified and precisely quantified synthetic peptides; (b) determine the efficiency of peptide extraction from cell pellets; (c) generate a standard curve that determines the % specific lysis by a particular CTL clone relative to the peptide concentration in the CTL assay; (d) determine the concentration of natural epitope in TFA-extracted HPLC fraction by comparison of the % specific lysis obtained with epitope-containing HPLC fractions to the synthetic peptide standard curve; the number of epitopes that were extracted per individual cell can be calculated from the starting number of infected cells that were extracted and correcting for the extraction efficiency for the CTL epitope. In the following sections we provide detailed information on each of these steps.

Precise quantitation of synthetic peptides

It is essential to use highly purified synthetic peptides for epitope quantitation. Since synthetic peptides frequently contain truncated versions of the peptide, HPLC purification is necessary following synthesis. This can be accomplished by subjecting approximately 100–200 µg of synthetic peptide to reverse-phase HPLC fractionation, using a gradient of water and acetonitrile with 0.1% TFA (as described in the previous section). The synthetic epitope, which should be readily identifiable as the predominant peak at an absorbance at 212 nm, is then collected and lyophilized.

It is important to remember that CTL can detect very small quantities of synthetic peptide. When HPLC-purifying large quantities of synthetic peptide, it is likely that the HPLC apparatus will become contaminated with the peptide. Subsequent runs on the same HPLC system will contain trace quantities of epitope that will interfere with the analysis of TFA extracts from infected cells. For this reason, we suggest the purification of synthetic epitopes is always performed on a HPLC system different from the one used for fractionating TFA extracts from infected cells.

HPLC-purified synthetic epitope should be assayed by mass spectrometry to determine the purity and mass of the peptide. The most accurate way to quantify the peptide is to subject it to hydrolysis and perform quantitative amino acid analysis. Although other methods to quantify peptides are available, they are highly dependent on the amino acid content of the peptide and are thus difficult to standardize.

Determining the extraction efficiency of epitopes from cellular pellets

Peptides are likely to be extracted from infected cell pellets with different efficiencies, depending on their hydrophobicity, stability and sensitivity to proteases. Furthermore, since pathogen-derived peptides elute in different HPLC fractions based on their size and sequence, the antigenic peptides compete with different endogenous peptides for MHC class I binding in subsequent CTL assays. It is therefore important to determine the overall efficiency of extraction and detection independently for each peptide. This is accomplished by spiking a pellet of 100–200 million cells with a small quantity of synthetic epitope (20, 50 and 100 μl of 10^{-10} M synthetic peptide), followed by TFA extraction of the pellet as described above. Following HPLC fractionation, the amount of peptide that is present in the epitope-containing fractions is titrated and compared with the synthetic peptide that was used to spike the pellet. We have found that the estimated extraction efficiency for different peptides generally falls in the range of 20–80%; the extraction efficiency for epitopes generally improves as the acetonitrile gradient is made shallower. It is important to re-determine the efficiency of peptide extraction and detection for individual epitopes if the HPLC gradient is changed, since changes in the gradient will change the family of endogenous peptide that the antigenic peptide travels with.

Preparing synthetic epitope standard curves

MHC class I-restricted CTL clones detect target cells in the presence of very low concentrations of synthetic peptide epitopes, but the sensitivity of different CTL clones for their antigenic peptides does vary. Depending on the length of time between the last stimulation and the assay, the peptide sensitivity and the degree of % specific lysis induced by an individual clone can also vary. Thus, when quantifying the amount of epitope in TFA extracts from infected cells, it is critical that the sensitivity of the CTL is determined concurrently using known concentrations of synthetic peptide. Our *L. monocytogenes*-specific CTL clones can detect their epitopes at concentrations as low as 10^{-12} M, and maximal lysis of target cells generally occurs at concentrations greater than $5–10 \times 10^{-11}$ M epitope. We have therefore made dilutions of the synthetic peptide in the range of 10^{-10} to 10^{-13} M, a concentration range where the % specific lysis correlates with the epitope concentration used to coat the target cells. In a typical assay, we generate a standard curve using epitope concentrations of 100, 80, 60, 40, 20, 10, 8, 6, 4, 2 and 1×10^{-12} M and assay each concentration in triplicate with a specific CTL clone.

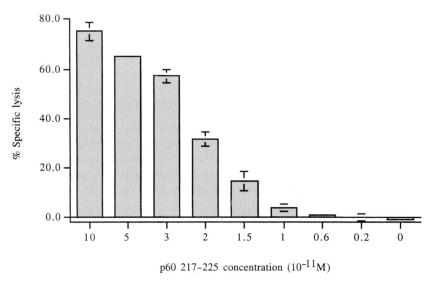

Figure 3. Percent specific lysis correlates with the peptide concentration present in the CTL assay. P815 target cells were labelled with ^{51}Cr and incubated in different concentrations of precisely quantified p60 217–225, an H2-K^d restricted CTL epitope. Target cells were assayed for recognition by CTL clone L9.6, which is specific for p60 217–225, at an effector to target ratio of 3 : 1. After 4 h the % specific lysis was determined and plotted against the peptide concentration.

Calculating the quantity of CTL epitopes in TFA extracts of infected cells and estimating the number of CTL epitopes per cell

In order to quantify epitopes from infected cells, it is necessary first to determine the epitope concentration in HPLC fractions of TFA extracts from infected cells. This quantification is performed by assaying, in triplicate, a volume of the HPLC fraction (generally 2 to 50 μl) that gives detectable but less than maximal lysis in a CTL assay. It is important that the % specific lysis obtained with HPLC fractions falls within the range of the standard curve determined with the synthetic peptide. By comparing the % specific lysis obtained with HPLC fractions to the standard curve, it is possible to determine the epitope concentration in the HPLC fraction.

By factoring in the volume of extract, the molar amount of epitope present in a given HPLC fraction can be calculated.

The average number of CTL epitopes per cell can be calculated with an equation that takes into account the starting number of infected cells, the extraction efficiency of the epitope from infected cells, the concentration of the epitope in the HPLC fraction, and the volume of the HPLC fraction. Using Avogadro's number, it is possible to convert the molar concentration of epitopes into the absolute number. The following formula takes these factors into account:

$$[(\text{epitope quantity}) \times (Y) \times (6.02 \times 10^{23})] / (\text{number of J774 cells}).$$

In this formula, Y is a factor to correct for the extraction efficiency of the epitope. If the extraction efficiency is 20%, $Y = 5$; alternatively, if the extraction efficiency is 50%, then $Y = 2$. Using this formula, it is possible to determine the absolute numbers of different epitopes that are present per cell at different times of infection or under varying circumstances (see below).

◆◆◆◆◆◆ KINETIC ANALYSES OF MHC CLASS I ANTIGEN PROCESSING IN INFECTED CELLS

The kinetics of antigen processing and presentation can be studied by quantifying epitope numbers at varying time intervals during the course of cellular infection. CTL epitope generation and presentation is a dynamic process that requires degradation of pathogen-derived proteins, transport of peptides into the ER, loading of empty MHC class I molecules, and translocation to the cell surface. Each step is characterized by distinct specificities, kinetics and efficiencies that may differ among epitopes. Furthermore, the overall quantity of a specific epitope present in an infected cell at a given timepoint also reflects peptide losses that may result from, for example, degradation by cytosolic or ER proteases prior to binding by MHC class I molecules or rapid dissociation from MHC class I molecules following binding. To study the kinetics of class I antigen presentation in more detail, the pathway can be blocked at distinct points to determine the effect on epitope generation. An advantage of studying antigen processing in cells infected with bacteria is that prokaryotic and eukaryotic protein synthesis can be independently inhibited. Thus, bacterial antigen synthesis can be turned off at defined timepoints, and the fate of the remaining antigen can be investigated. Furthermore, specific inhibitors of host cell proteolysis, protein synthesis, and membrane trafficking can be used to examine their impact on the MHC class I antigen processing pathway without affecting the production of antigen by the intracellular bacterium.

Isolation and quantitation of CTL epitopes from cells infected for varying time intervals

The number of epitopes per cell can be calculated by acid-elution and HPLC fractionation as described above. Following this approach, quantitative analyses can be performed at different timepoints during infection with the pathogen. We have used this approach to study antigen processing in J774 cells infected with *L. monocytogenes* for 1, 2, 3, 4 and 5 h (Villanueva *et al.*, 1994). Determination of epitope numbers and numbers of bacteria per cell as described (see earlier) revealed that there is a direct correlation between the secretion of antigenic proteins into the host cell cytosol and the generation of CTL epitopes.

Use of inhibitors of bacterial protein synthesis to examine MHC class I antigen processing

As mentioned above, prokaryotic protein synthesis can be specifically inhibited without affecting host cell metabolism. The antibiotic tetra-cycline (TCN) is one such inhibitor with specificity for a wide spectrum of Gram-positive and -negative bacteria. Unlike gentamicin, which is membrane-impermeable and can be used to kill extracellular bacteria, TCN is membrane-permeable and, therefore, affects intracellular bacteria. TCN rapidly inactivates bacterial ribosomes, terminating intracellular bacterial protein synthesis within several minutes of addition to the culture medium. To investigate the linkage between antigen synthesis and epitope generation in macrophages infected with bacteria, J774 cells were grown to confluence in 150 mm culture plates and infected with *L. monocytogenes* as described in the preceeding sections. After 3 h of infection 20 µg/ml TCN was added, control plates were left untreated. At different time intervals (e.g. 4, 5, 6 and 7 h postinfection), TCN-treated and -untreated cells were scraped into PBS, and epitopes were extracted and quantified as described (see above). Using this system, we found that the generation of epitopes is markedly and rapidly diminished in infected cells treated with TCN, indicating that epitopes are derived primarily from newly synthesized antigens (Villanueva *et al.*, 1994).

Using inhibitors of host cell proteolysis, protein synthesis, and protein transport to study pathogen interactions with the MHC class I antigen processing pathway

Antigen processing and epitope presentation by the host cell is a multi-step pathway that requires antigen degradation by cytosolic proteases, transport of peptide fragments into the ER, synthesis of MHC class I molecules, loading of epitopes onto the newly synthesized molecules, and transport of epitope/MHC class I complexes to the cell surface. Inhibitors that interfere with specific components in this pathway are available and provide an opportunity to study the contribution of each step to the efficiency of epitope generation. In the following sections, some of these approaches are summarized.

Inhibitors of host cell proteolysis

Cytosolic degradation of pathogen-derived antigens is mainly mediated by proteasomes, and distinct peptidase activities are associated with these multi-catalytic enzyme complexes (Orlowski *et al.*, 1993). Peptide aldehyde protease inhibitors, e.g. N-acetyl-Leu-Leu-norleucinal (LLnL) and (benzyloxycarbonyl)Leu-Leu phenylalaninal (Z-LLF) specifically inhibit the chymotrypsin-like activity of proteasomes (Rock *et al.*, 1994). These inhibitors are membrane-permeable and can be used to treat pathogen-infected cells; we have found that in cells infected with *L. monocytogenes*, these inhibitors do not impair host cell or bacterial protein synthesis. To determine the impact of proteasome inhibition on the generation of CTL epitopes in *L. monocytogenes*-infected J774, cells as were infected described in the preceding section and treated infected cells with either 250 µM LLnL or 10 µM Z-LLF. Infected cells were harvested at varying timepoints, and epitopes were extracted, HPLC purified, and quantified as described in the previous sections. Using this approach it was possible to determine that protein degradation of secreted bacterial proteins is tightly linked to epitope generation (Sijts *et al.*, 1996). Furthermore, the generation of different epitopes varied in sensitivity to these protease inhibitors, suggesting that more than one degradation pathway is used to generate different MHC class I-associated epitopes.

Inhibitors of host cell protein synthesis

As mentioned previously, MHC class I antigen processing involves multiple steps that occur in separate subcellular compartments. It is possible selectively to inhibit host cell protein synthesis with 50 µg ml^{-1} cycloheximide (CHX) and 30 µg ml^{-1} anisomycin (ANM) without affecting intracellular bacterial protein synthesis (Sijts and Palmer, 1997), an approach that allows analysis of the requirements of host cell protein synthesis for efficient CTL epitope generation. In the *L. monocytogenes* system, inhibition of host cell protein synthesis rapidly decreases CTL epitope present, likely due to the rapid depletion of available MHC class I molecules, and subsequent epitope degradation. Surprisingly, however, inhibiting host cell protein synthesis does not impair the degradation of bacterial proteins in the host cell cytosol, transport of peptides into the ER, or trafficking of MHC class I molecules to the cell surface.

Inhibitors of intracellular trafficking

After peptide is loaded onto MHC class I molecules in the ER, MHC/peptide complexes are rapidly transported via the Golgi complex to the cell surface. BrefeldinA (BfA) is a compound that disrupts transport via the ER and the Golgi complex, thereby preventing the translocation of ER contents to the cell surface. In the *L. monocytogenes* system, treatment of infected cells with 5 µg ml^{-1} of BfA does not affect intracellular bacterial growth, protein synthesis, or protein secretion (Sijts and Palmer, 1997). Thus, it is possible to differentiate between and compare the interaction of CTL epitopes with MHC class I molecules in the ER and on the cell surface.

◆◆◆◆◆◆ OTHER USEFUL ASSAYS FOR INVESTIGATING INTRACELLULAR PATHOGEN-SPECIFIC CTL

In order to circumvent some of the limitations of the conventional ^{51}Cr-release CTL assay, other methods have been established to quantify CTL-mediated lysis of infected target cells. In the following sections, some of these methods are briefly outlined and the reader is referred to appropriate references for detailed protocols.

Serine esterase release

Activated CTLs release cytotoxins and proteases from secretory granules in order to induce death of the target cells. One of these cytolytic proteases, the enzyme serine esterase, can be detected in the culture medium upon its release from CTLs (Taffs and Sikovsky, 1994). Enzyme activity in the supernatant correlates with the extent of CTL activation in the presence of epitope-presenting target cells. The detection of serine esterases is based on hydrolysis of N-α-benzyloxycarbonyl-L-lysine thiobenzyl ester, which is detected in a standard colorimetric assay using dithio-*bis*(2-nitrobenzoic acid). In this assay the amount of enzymatically active serine esterase released by antigen stimulation is compared with release in the absence of antigen and the maximal release obtained with a mild detergent. Thus, the serine esterase release is similar to conventional ^{51}Cr release assays, except that the condition of the target cell prior to interaction with CTL is less critical.

Sensitivity to membrane-impermeable antibiotics

When CTL attack a target cell, perforins damage the target cell membrane and create pores that allow otherwise impermeable substances to enter the cell (Berke, 1995). In the *Listeria monocytogenes* system, it is possible to take advantage of this mechanism to determine antigen-specific CTL activity. Specifically, target cells are infected with *L. monocytogenes*, and a CTL assay is performed in media containing the membrane-impermeable antibiotic gentamicin (Bouwer *et al.*, 1992). In the absence of CTL or if the CTL do not detect *L. monocytogenes* antigens, intracellular bacteria survive inside target cells and can be quantified by plating on TSB culture plates. If CTL lysis of infected cells occurs, gentamicin enters the lysed, infected cells and kills intracellular bacteria, and the number of bacteria that can subsequently be grown on TSB plates is diminished. Although this method is less precise than most other CTL assays, it provides an alternative to ^{51}Cr release assays when bacteria-induced ^{51}Cr release is a problem. One limitation of this method is that infected cells can begin to 'leak' after 4–5 h of infection, allowing the influx of antibiotics into target cells (see also earlier).

^{3}H thymidine-based CTL assay

Cytolytic T cells induce apoptosis in target cells, and the fragmentation of target cell DNA can be used to measure antigen-specific recognition

(Duke and Cohen, 1994). In the ^{3}H-thymidine-based CTL assay, DNA of target cells is labelled with ^{3}H-thymidine prior to interaction with CTLs, specific lysis is determined by measuring soluble ^{3}H in the supernatant (fragmented DNA) and cell-associated ^{3}H in the cell pellet (intact DNA) in the presence and absence of CTL. The strength of this assay is a relatively low spontaneous release that allows incubation times exceeding 10 h. Thus, lysis of cells via fas–fasL interactions, which require significantly more time then perforin/granzyme-mediated cytolytic mechanisms, can be detected.

References

Berke, G. (1995). The CTL's kiss of death. *Cell* **81**, 9–12.

Bielecki, J., Youngman, P., Connelly, P. and Portnoy, D. A. (1990). *Bacillus subtilis* expressing a haemolysin gene from *Listeria monocytogenes* can grow in mammalian cells. *Nature* **345**, 175–176.

Bouwer, H. G., Nelson, C. S., Gibbins, B. L., Portnoy, D. A. and Hinrichs, D. J. (1992). Listeriolysin O is target of the immune response to *Listeria monocytogenes*. *J. Exp. Med.* **175**, 1467–1471.

Brunner, K. T., Mauel, J., Cerottini, J.-C. and Chapuis, B. (1968). Quantitative assay of the lytic action of immune lymphoid cells on ^{51}Cr labeled allogeneic target cells in vitro: Inhibition by isoantibody and by drugs. *Immunology* **14**, 181–196.

Busch, D. H., Pilip, I. M., Vijh, S. and Pamer, E. G. (1998). Coordinate regulation of complex T cell populations responding to bacterial infection. *Immunity* **8**, 353–362.

Busch, D. H., Kerksiek, K. M. and Pamer, E. G. (1999). Processing of *Listeria monocytogenes* antigens and the *in vivo* T-cell response to bacterial infection. *Immunol. Rev.* **172**, 163–169.

Duke, R. C. and Cohen, J. J. (1994). Cell death and apoptosis. In: *Current Protocols in Immunology* (J. E. Coligan, A. M. Kruisbeek, D. H. Margulies, E. M. Shevach, W. Strober, Eds). John Wiley and Sons, Inc., unit 3.17, Chichester.

Falk, K., Roetzschke, O., Deres, K., Metzger, J., Jung, G. and Rammensee, H. G. (1991). Identification of naturally processed viral nonapeptides allows their quantification in infected cells and suggests an allele-specific T cell epitope forecast. *J. Exp. Med.* **174**, 425–434.

Kaufmann, S. H. E., Hug, E., Vaeth, U. and Mueller, I. (1985). Effective protection against *Listeria monocytogenes* and delayed-type hypersensitivity to listerial antigens depend on cooperation between specific L3T4+ and Lyt-2+ T-cells. *Infect. Immun.* **48**, 273–278.

Kaufmann, S. H. E., Hug, E. and De Libero, G. (1986). *Listeria monocytogenes*-reactive T lymphocyte clones with cytolytic activity against infected target cells. *J. Exp. Med.* **164**, 363–368.

Orlowski, M., Cardozo, C. and Michaud, C. (1993). Evidence for the presence of five distinct proteolytic components in the pituitary multicatalytic proteinase complex: properties of two components cleaving bonds on the carboxyl side of branched chain and small neutral amino acids. *Biochemistry* **32**, 1563–1569.

Pamer, E. G. (1993). Cellular immunity to intracellular bacteria. *Curr. Opin. Immunol.* **5**, 492–496.

Pamer, E. G. and Cresswell, P. (1998). Mechanisms of MHC class I-restricted antigen processing. *Annu. Rev. Immunol.* **16**, 323–358.

Pamer, E. G., Harty, J. T. and Bevan, M. (1991). Precise prediction of dominant class I MHC-restricted epitope of *Listeria monocytogenes*. *Nature* **353**, 853–855.

Rock, K. L., Gramm, C., Rothstein, L., Clark, K., Stein, R., Dick, L., Hwang, D. and Goldberg, A. L. (1994). Inhibitors of the proteasome block the degradation of most cell proteins and the generation of peptides presented on MHC class I molecules. *Cell* **78**, 761–771.

Roetzschke, O., Falk, K., Deres, K., Schild, H., Norda, M., Metzger, J., Jung, G. and Rammensee, H. G. (1990). Isolation and analysis of naturally processed viral peptides as recognized by cytotoxic T cells. *Nature* **348**, 252–254.

Sijts, A. J. A. M. and Pamer, E. G. (1997). Enhanced intracellular dissociation of MHC class I associated peptides: a mechanism for optimizing the spectrum of cell surface presented CTL epitopes. *J. Exp. Med.* **185**, 1403–1411.

Sijts, A. J. A. M., Villanueva, M. and Pamer E. G. (1996). CTL epitope generation is tightly linked to cellular proteolysis of a *Listeria monocytogenes* antigen. *J. Immunol.* **156**, 1497–1503.

Taffs, R. E. and Sitovsky, M. V. (1994). Granule enzyme exocytosis assay for CTL activation. In: *Current Protocols in Immunology* (J. E. Coligan, A. M. Kruisbeek, D. H. Margulies, E. M. Shevach, W. Strober, W. Eds). John Wiley and Sons, Inc., unit 3.16, Chichester.

Tsomides, T. J., Aldovini A., Johnson R. P., Walker B. D., Young R. A. and Eisen, H. N. (1994). Naturally processed viral peptides recognized by cytotoxic T lymphocytes on cells chronically infected by human immunodeficiency virus type 1. *J. Exp. Med.* **180**, 1283–1293.

Villanueva, M. S., Fischer, P., Feen, K. and Pamer, E. G. (1994). Efficiency of MHC class I antigen processing: a quantitative analysis. *Immunity* **1**, 476–489.

Zinkernagel, R. M. and Doherty, P. C. (1974). Restriction of in vitro T cell-mediated cytotoxicity in lymphocytic choriomeningitis within a syngeneic or semi-allogeneic system. *Nature (London)* **248**, 701–702.

Appendix: Company List

American Type Culture Collection (ATCC)
12301 Parklawn Drive
Rockville, MD 20852-1776, USA

Tel.: 011 800 638 6597
Fax.: 011 301 231 5826
cell lines and bacteria

Amicon, Inc.
72 Cherry Hill Dr.
Beverly, MA 01915, USA

Tel.: 001 800 343 1397
Fax.: 001 508 777 6204
Centricon-10

DuPont NEN Research Products
549 Albany Street
Boston, MA 02118, USA

Tel.: 001 800 551 2121
Fax.: 001 617 542 8468
radioactive ^{51}Chromium

Sigma Chemical Co., Sigma-Aldrich Techware
P.O. Box 14508
St Louis, MO 63178-9916, USA

Tel.: 001 314 771 5750
Fax.: 001 314 771 5757
protease inhibitors

Water Corporation
34 Maple Street
Milford, MA 1757, USA

Tel.: 001 800 252 4752
Fax.: 001 508 478 5839
HPLC, hardware

2.4 Quantitation of T-cell Cytokine Responses by ELISA, ELISPOT, Flow Cytometry and Reverse Transcriptase-PCR Methods

Takachika Hiroi and Hiroshi Kiyono
Department of Mucosal Immunology, Research Institute for Microbial Diseases, Osaka University, Japan

Kohtaro Fujihashi and Jerry R McGhee
Immunobiology Vaccine Center and Departments of Oral Biology and Microbiology, The University of Alabama at Birmingham, USA

◆◆◆

CONTENTS

◆◆◆◆◆◆ **INTRODUCTION**

Cytokines mainly produced by lymphocytes, and professional antigen-presenting cells including dendritic cells, B cells and macrophages play important roles in immunity and immunological tolerance. Thus, it has been reported that certain cytokines possess special assignments for the induction of immune responses to maintain immunological homeostasis. For example, cytokines such as IL-4, IL-5 and IL-6 produced by CD4-positive (CD4$^+$) T cells provide B cell help for immunoglobulin synthesis. IFN-γ production is closely related to cytotoxic activities. Recently, it was shown that CD4$^+$, CD25$^+$ T cells play an important role in down-regulation of inflammatory responses by the production of IL-10 and/or TGF-β. In this regard, it is essential to examine cytokine responses in order to characterize the nature of immune responses induced at different stages of host–parasite interactions.

In general, CD4$^+$ T helper (Th) cells are subdivided into at least two subsets, namely Th1 and Th2 cells, according to distinct cytokine profiles which account for two major functions, e.g. cell-mediated (CMI) and humoral- (antibody) mediated immunity in host immune responses,

METHODS IN MICROBIOLOGY, VOLUME 32
ISBN 0–12–521532–0
Copyright © Elsevier Science Ltd
All rights of reproduction in any form reserved

respectively (Mosmann and Coffman, 1989; Street and Mosmann, 1991). It is well established that Th1 cells secrete IL-2, IFN-γ, tumour necrosis factor-alpha (TNF-α) and TNF-β and function in CMI for protection against intra-cellular bacteria and viruses. In this regard, it has been shown that CD8[+] T cells, through their production of IFN-γ, are closely related to and play a central role in their cytotoxic functions. Further, Th1 cells also provide limited help for B-cell responses where IFN-γ supports μ → γ2a switches and IgG2a synthesis in mice. The Th2 cells preferentially secrete IL-4, IL-5, IL-6, IL-10 and IL-13 and provide effective help for B-cell responses, in particular for IgG1 (and IgG2b), IgE and IgA Ab synthesis (Murray et al., 1987; Bond et al., 1987; Coffman et al., 1987; Beagley et al., 1988, 1989; Harriman et al., 1988; Lebman and Coffman, 1988; Fujihashi et al., 1991). In the generation of these two subsets, several important cytokines influence the process of development of Th1 and Th2 cells. For example, IL-12 and IL-4 direct CD4[+] Th cell development down a Th1 or Th2 pathway, respectively, while later in development IFN-γ and IL-10 (together with IL-4) can reinforce Th1 or Th2 phenotype expansion (Seder and Paul, 1994). Along the same line, it was recently shown that T regulatory (Tr) cells and their derived cytokines, such as IL-10 and TGF-β play important roles in down-regulation of immune and inflammatory responses (Nakamura et al., 2001; Zhang et al., 2001). Although the exact phenotype of Tr cells is still controversial, these cells mainly express αβ TCR and are CD4[+], CD25[+].

It is now essential to measure the levels of Th1 and Th2 cytokines as well as cytokines produced by CD8[+] T cells in order to elucidate the precise mechanism for the induction and regulation of antigen-specific immune responses. The current technology allows the detection of T-cell cytokines at the level of the protein, the mRNA or the cell. For measure-ment of secreted cytokines, two distinct methods are currently available. The biological activity of cytokines can be measured by using certain cytokine-dependent cell lines (Helle et al., 1988; Sawamura et al., 1990; Slavin and Syrobe, 1978; Watson, 1979). Although this assay is the most sensitive in order to detect biologically active cytokines, the ELISA system is a simple, rapid and sensitive assay for the quantitative analysis of different cytokines using appropriate combinations of cytokine-specific monoclonal antibodies. Furthermore, using different substrates, the sensitivity of ELISA can be improved up to 10–100-fold. Thus, cytokine-specific ELISA is now widely used for the quantitation of cytokines in both in vivo and in vitro investigations.

In addition to cytokine-specific ELISA, the ELISPOT assay is also avail-able for the elucidation of Th1 and Th2 cytokine-producing cells. One advantage of this cytokine-specific ELISPOT assay is that this assay is able to determine the frequency of cytokine-producing cells in the single cell preparation (Taguchi et al., 1990a, b; Fujihashi et al., 1993a, 1999). Thus, both the frequency and total numbers of Th1 and Th2 cytokine-producing cells in different cell fractions (e.g. CD4[+] and CD8[+]) from various tissues (e.g. mucosal vs. systemic) can be determined. In addition, intracellular cytokine synthesis can also be detected by flow cytometry (Jung et al., 1993; Carter and Swain, 1997). By using this intracellular cytokine staining method, the frequency and numbers of cytokine producing cells can also be detected.

As discussed above for the cytokine-specific ELISA, and the ELISPOT and intracellular flow cytometry assays are used for the detection of cytokines at the protein and cellular levels, respectively. Further, Th1 and Th2 cytokines can be analysed at the molecular level by detecting cytokine-specific mRNA using northern blot, *in situ* hybridization, dot blot, mRNA protection and RT-PCR assays. Among these assays, RT-PCR is the most rapid and efficient method for the detection of Th1 and Th2 cell-specific mRNA, especially when dealing with small numbers of specific subsets of lymphocytes. Further, the recent development and wide use of quantitative RT-PCR has enhanced the capability of characterizing increases or decreases in specific cytokine mRNA (Wang *et al.*, 1989; Hiroi *et al.*, 1995; Marinaro *et al.*, 1995). This chapter will introduce the most updated cytokine-specific ELISA, ELISPOT, flow cytometry and RT-PCR methods for murine models which are routinely performed in our laboratories in order to detect cytokine synthesis at the protein, single cell or mRNA levels.

◆◆◆◆◆◆ MEASURING MURINE CYTOKINES

Background

Cytokines are important immunoregulatory proteins that mediate distinct functions in different immunocompetent cells including T and B cells, MØ, dendritic cells, NK cells and other cell types. The Th1 and Th2 cytokines are produced after stimulation of T cells via specific interactions with antigen-presenting cells (APC) through a two-step signalling pathway via the T-cell receptor (TCR) and MHC class II with processed peptide as well as through CD28 and B7-1/B7-2 molecule interactions. Techniques designed to detect cytokine expression have proven to be valuable for studies of immune responses directed against different infectious diseases and for development of effective vaccines. We will discuss three of these techniques to detect murine cytokines: cytokine-specific ELISA and ELISPOT and quantitative RT-PCR. Cytokine-specific ELISA and ELISPOT assays are used to measure cytokines at the protein and cellular levels, respectively. The former assay measures secreted cytokines, while the latter system is used to quantify the numbers of cells producing a particular cytokine. The frequencies of cytokine-producing cells are also quantified by detecting intracellular cytokine-positive cells by using flow cytometry. Finally, RT-PCR is used to measure the expression and quantity of cytokines at the mRNA level.

Mouse cytokine-specific ELISA

For analysis of various murine cytokines, a series of high-quality controlled ELISA kits are currently available from R&D, Amersham Pharmacia, Genzyme, BIOSource International and BD PharMingen (see Appendix). Using these kit systems, an array of murine cytokines including GM-CSF, IFN-γ, IL-1α, IL-1β, IL-2, IL-3, IL-4, IL-5, IL-6, IL-10, IL-12p40, IL-12p70, IL-13, IL-17 and TNF-α can be quantitated. Although this useful kit system is now available for daily experiments and is commercially available, we have summarized the basic protocol for this ELISA system used in the quantitation of secreted cytokines.

The ELISA protocol

The murine cytokine-specific ELISA assays are highly specific, simple and rapid procedures for the quantitative analysis of secreted cytokines. This assay quantitates cytokines produced by Th1- and Th2-type cells in culture supernatants and body fluids including serum and external secretions. Although this ELISA is a powerful technique to assess the precise levels of cytokines present in culture supernatants, results obtained using this method must be carefully interpreted. For example, it is possible that some cytokines produced by CD4+ Th cells may be consumed by neighbouring cells during the culture period. However, this is the most commonly used method for the characterization of Th1 and Th2 type responses, since two distinct profiles of cytokines remain as the most reliable markers for distinction of these subsets. Although an array of Th1 and Th2 cytokine-specific ELISA kits are now commercially avail-

Table I. Reagents used for cytokine ELISA and ELISPOT assays to detect murine cytokines.

| | | Monoclonal antibodies* (mAbs) | | | |
| | | Coating (rat anti-mouse) | | Detection (Biotin-rat anti-mouse) | |
Cytokine	Assay	Designation (clone)	Concentration (μg ml^{-1})	Designation (clone)	Concentration (μg ml^{-1})
IFN-γ	ELISA	RA-6A2	2.5	XMG1.2	0.3
	ELISPOT	RA-6A2	5.0	XMG1.2	0.1–2.5
IL-2	ELISA	JES6-1A12	2.5	JES6-5H4	0.4
	ELISPOT	JES6-1A12	5.0	JES6-5H4	0.2–2.5
IL-4	ELISA	BVD4-1D11	2.0	BVD6-24G2	0.2
	ELISPOT	BVD4-1D11	2.5	BVD6-24G2	0.1–2.5
IL-5	ELISA	TRFK-5	2.5	TRFK-4	4.0
	ELISPOT	TRFK-5	5.0	TRFK-4	0.2–2.5
IL-6	ELISA	MP5-20F3	2.0	MP5-32C11	0.5
	ELISPOT	MP5-20F3	5.0	MP5-32C11	0.5–2.5
IL-10	ELISA	JES5-2A5	2.0	JES5-16E3	0.3
	ELISPOT	JES5-2A5	5.0	JES5-16E3	0.2–2.5
IL-12	ELISA	C15.6	2.0	C17.8	0.5

* All mAbs were obtained from BD PharMingen.

able, cytokine-specific ELISAs to detect IL-2, IL-4, IL-5, IL-6, IL-10, IL-12 and IFN-γ can be performed with a standard set of specific monoclonal antibodies (mAbs) listed in Table 1 (VanCott *et al.*, 1996; Fujihashi *et al.*, 1999; Marinaro *et al.*, 1999). The following protocol is routinely used to detect cytokines in serum as well as in tissue culture supernatants.

1. Dilute the capture antibody (Table 1) in PBS and add 100 μl to the wells of a 96-well microtitre plate, Falcon Microtest III plates (Becton Dickinson, Oxnard, CA, USA) or NUNC MaxiSorp (Nalge Nunc International, Naperville, IL, USA). Incubate the plates overnight at 4°C.
2. Remove the solution from wells and block the remaining binding sites with PBS containing 1–3% BSA for 1 h at room temperature. Wash the plate three times by filling wells (200 μl per wash) with PBS and decanting the contents.
3. Generate standard curves using murine recombinant IFN-γ (rIFN-γ), rIL-2, rIL-4, rIL-5, rIL-6 and rIL-10 (R&D Systems Inc. Minneapolis, MN, USA); and rIL-12 (BD PharMingen, San Diego, CA, USA). Prepare two-fold serial dilutions of recombinant cytokine standards and unknown samples diluted in PBS containing 0.5% Tween 20 (PBS-T) and 1% BSA. Add 100 μl per well of IL-2 and IL-5 (diluted from 5–2000 pg ml^{-1}) and of IFN-γ, IL-4, IL-6 and IL-10 (diluted from 20–10 000 pg ml^{-1}). Prepare control wells for each cytokine standard by substituting a different standard as the only change. For example, for the IL-2 ELISA, use recombinant IL-4 as a background control. Incubate the plate overnight at 4°C.
4. Wash the plate six times with PBS-T and blot dry. Fully aspirate any remaining fluid from the wells by patting the bottom of the plate with dry absorbent paper.
5. Add 100 μl per well of appropriate biotinylated capture mAb diluted in PBS-T with 1% BSA (Table 1). Incubate overnight at 4°C.
6. Wash the plate six times with PBS-T, blot dry, and add 100 μl per well of peroxidase-labelled anti-biotin Ab (0.5 mg ml^{-1}; Vector Laboratories, Inc., Burlingame, CA, USA) for 1 h at room temperature.
7. Wash the plates six times with PBS, blot dry, and develop with the chromogenic substrate, ABTS [2,2'-azino-bis (3-ethylbenz-thiazoline-6-sulfonic acid)] (0.6 mg ml^{-1}) with 0.01% H$_2$O$_2$ (Moss, Inc., Pasadena, MD, USA) for 90 min at room temperature. Read the green colour absorbence at 414 nm.
8. The ELISA assays are capable of detecting 5 pg ml^{-1} of IL-2 and IL-5; 10–20 pg ml^{-1} of IFN-γ, IL-4 and IL-10; and 60 pg ml^{-1} of IL-6. If there are background problems associated with a particular biologic fluid such as serum, then add 2% rat serum to the dilution buffer in step 3 to block non-specific binding sites by the detection mAb.
9. Calculate the concentrations of unknown samples by reference to the linear portion of the standard curve.

An example for the detection of serum IL-12 by ELISA

Regulatory cytokines, i.e. IL-12 and IFN-γ vs IL-4, are effective for directing immunity through Th1 or Th2 pathways, respectively. IL-12 produced by APCs is a potent stimulator of IFN-γ production by NK and T cells (Seder et al., 1993; Trinchieri, 1995) and preferentially supports Th1-type responses. Use of IL-12 to direct Th1-type responses has important implications for the development of new therapy in the treatment of cancer and immunological diseases (Kobayashi et al., 1989; Robertson et al., 1992; Clerici et al., 1993). IL-12 has also been shown to induce T-cell precursors to develop into functional Th1-type cells (Hsieh et al., 1993; Seder et al., 1993). Further, it was interesting to note that nasally administered IL-12 can provide mucosal adjuvant activity for the induction of co-administered antigen-specific immune response (Boyaka et al., 1999; Marinaro et al., 1999). Thus, it is important to test the levels of IL-12 in serum when IL-12 is mucosally administered in order to define the role of IL-12 for the modulation of Th1-type responses. Therefore, following administration of IL-12 via nasal routes subsequent serum IL-12 levels were evaluated. Nasal administration of 10 ng of IL-12 did not result in detectable serum IL-12 levels (data not shown). Enhanced serum IL-12 levels were achieved when 100 ng of IL-12 was administered by the nasal route, and these effects were greater with 1000 ng of IL-12 per dose (Fig. 1). Further, serum IL-12 levels peaked at 2 h after the nasal treatment regardless of the amount of IL-12 given and declined within 24–48 h (Fig. 1).

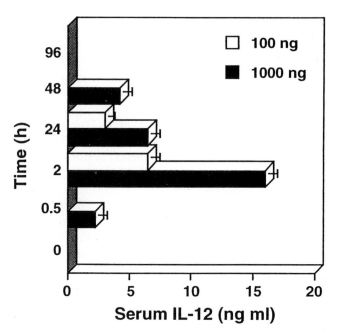

Figure 1. Kinetics of serum IL-12 following nasal administration. Groups of C57BL/6 mice were given 100 ng (□) or 1000 ng (■) of rmIL-12-liposomes on day 0 by the nasal route. Mice were bled both before rmIL-12 administration and 30 min, 2, 24, 48, and 96 h thereafter, and serum IL-12 levels were measured by ELISA.

Mouse cytokine-specific ELISPOT

For the elucidation of murine Th1 and Th2 cytokine-producing cells, a series of high-quality controlled ELISPOT kit systems are currently available from R & D (see Appendix). Using these kits, an array of murine cytokines including IFN-γ, IL-4 and TNF-α can be assessed. If it is necessary to examine other Th1 or Th2 cytokine-producing cells, the ELISPOT assay can be adapted by using a set of mAbs listed in Table 1. Although this useful ELISPOT kit is now available for daily experiments and is conveniently and commercially available, we have summarized a basic ELISPOT system for the quantitation of cytokine secreting T cells.

ELISPOT protocol

The cytokine-specific ELISPOT is used to detect and quantitate the frequency of individual cytokine-secreting cells. The ELISPOT is performed using the same concept as the ELISA with three key modifications: (a) the 96-well microtitre plate has a nitrocellulose base; (b) the unknown sample is a cell suspension rather than a biologic fluid or tissue culture supernatant; and (c) the concentration of detection mAbs is generally more concentrated. The following protocols were adapted to detect cytokine-secreting cells in single-cell suspensions from systemic sites, i.e. spleens, and mucosal sites (e.g. Peyer's patches, intestinal lamina propria, submandibular glands and epithelium) (Taguchi *et al.*, 1990b; Fujihashi *et al.*, 1993b, 1999; Xu-Amano *et al.*, 1993; Hiroi *et al.*, 1995; VanCott *et al.*, 1996).

1. Dilute the capture mAb in PBS and add 100 µl to the wells of a nitrocellulose-backed microtitre plate (Millititer-HA, Millipore Corp., Bedford, MA, USA; Table 1). Place the plate in a humidified chamber or carefully wrap the plate in saran wrap and incubate overnight at 4°C.
2. Remove fluids from the plate and block remaining binding sites with 100 µl per well of culture medium (RPMI 1640 containing 5% FCS or 10% rat serum, 25 IU ml^{-1} penicillin, 25 µg ml^{-1} streptomycin and 80 µg ml^{-1} gentamicin) for 1 h at 37°C in a 5% CO_2 incubator.
3. Rinse the plate three times with PBS without Tween 20 and shake off excess fluids from the plate.
4. Prepare five-fold dilutions of CD4$^+$ or CD8$^+$ T-cell suspensions prepared by flow cytometry sorting in culture medium starting at 10^6 to 10^7 cells ml^{-1}. Immediately add 100 µl per well of the cells to the capture mAb-coated microtitre plate. Incubate the plate with a plastic cover containing CD4$^+$ or CD8$^+$ T cells for 12 to 16 h at 37°C with 5% CO_2. The time required to purify CD4$^+$ or CD8$^+$ T cells from freshly isolated cell suspensions significantly reduces the number of detectable cytokine-secreting cells. Thus it is important to prepare single-cell suspensions in a prompt manner.

(contd.)

For the assessment of cytokine production by antigen-specific CD4$^+$ or CD8$^+$ T cells from mice, the cells are purified from different tissues and restimulated *in vitro* with the same antigen in the presence of irradiated feeder cells or virus-infected target cells (Welsh *et al.*, 1990; Kunisawa *et al.*, 2001). Between 1–6 days after antigen stimulation, CD4$^+$ or CD8$^+$ T cells are harvested and immediately added to capture antibody-coated plates as described above. In order to confirm the results obtained by ELISPOT assay, an aliquot of purified CD4$^+$ or CD8$^+$ T cells can be subjected to the cytokine-specific intracellular staining method and/or cytokine-specific mRNA by RT-PCR (see below).

5. Wash the plates three times with PBS followed by three washes with PBS-T in order to remove the CD4$^+$ or CD8$^+$ T cells from individual wells. Before the final aspiration, soak the plates in PBS-T for 10 min to allow any remaining cells to detach from the plate. Add biotinylated detection mAb diluted in PBS-T to each well (Table 1). Place a plastic cover over the plate and incubate it overnight at 4°C.

6. Wash the plate six times with PBS-T. Remove excess liquids by removing the plastic covering around the base of the plate and by vigorously hitting the plate over an absorbent towel. Place the plastic cover back on the plate and add peroxidase-labelled anti-biotin Ab (0.8 mg ml^{-1}) (Vector Laboratories, Inc.) diluted in PBS-T. Incubate at room temperature for 1 h.

7. Wash the plates six times with PBS without Tween as described above [see 6.]. Develop the spots with the substrate AEC (3-amino-9-ethylcarbazole; 0.3 mg ml^{-1}; Moss Inc.).

8. Count discrete red spots representing the respective cytokine producing CD4$^+$ or CD8$^+$ T cells using a dissecting microscope (SZH Zoom Stereo Microscope System, Olympus, Lake Success, NY, USA). The ability to determine true spots versus pseudo spots should be done by comparing experimental wells with control wells. Controls should include: (a) wells coated with PBS instead of an anti-cytokine Ab; (b) well coated with a different anti-cytokine Ab; (c) the substitution of a different biotinylated secondary Ab; (d) the addition of 50 μg ml^{-1} cycloheximide during the cell incubation period to inhibit *de novo* cytokine synthesis; and (e) the use of unstimulated and Con A-activated cells from unimmunized mice as negative and positive controls, respectively. In addition, the established and well characterized cytokine-specific cDNA transfected myeloma cell line, e.g. X63-Ag8-653X2 (IL-2), X63-Ag8-653X4 (IL-4) and X63-Ag8-653X6 (IL-6) can be used as positive controls (Fujihashi, 1993).

New developments in the elucidation of cytokine producing cells by ELISPOT

It is important to note that automated ELISPOT readers are now commercially available from Carl Zeiss (KS ELISPOT), Cellular Technology Ltd (distributed by BD PharMingen; The ImmunoSpot™ Series 1

Analyzer), Autoimmun Diagnostika GmbH (AID Elispot reader system) and AELVIS® (Automated ELISA-Spot Assay Video Analysis System®). These readers will help facilitate one to distinguish real spots from fake or antifactual spots and will help reduce background problems. When compared with visual observation of cytokine-specific spot forming cells, the advantage of these newly developed ELISPOT reader systems is an objective analysis of ELISPOT data at high resolution in a short period of time.

Analysis of antigen-specific Th1 and Th2 responses

The cytokine-specific ELISA and ELISPOT protocols were also used to distinguish and characterize recombinant (r) *Salmonella* and cholera toxin (CT) induced Th1 and Th2 cell responses (XuAmano *et al.*, 1993; Marinaro *et al.*, 1995; VanCott *et al.*, 1996). Mice were orally immunized with two different types of antigen delivery system. In the first, r*Salmonella* expressing fragment C (Tox C) of tetanus toxin was used and in the second, tetanus toxoid (TT) plus CT as mucosal adjuvant was employed. Four to six weeks later, CD4$^+$ T cells were isolated from the Peyer's patches of each group and cultured with an optimal amount of TT-coated beads (at a ratio of 1 : 20) and irradiated splenic adherent feeder cells ($1-2 \times 10^6$ ml^{-1}) from naive mice. After a 2–6-day incubation period, culture supernatants were harvested for the assessment of IFN-γ and IL-4 production as an indicator of Th1- and Th2-type responses, respectively. IFN-γ (Th1 cytokine) and IL-4 (Th2 cytokine)-specific ELISA revealed that orally administered r*Salmonella*-Tox C induced IFN-γ production, whereas CT plus TT elicited IL-4 production by Peyer's patch CD4+ T cells. In parallel, CD4$^+$ T cells were isolated from the culture for the enumeration of Th1 and Th2 cytokine producing, antigen-specific CD4$^+$ T cells by ELISPOT. Th1-type IFN-γ-producing cells (about 760/10^6 CD4$^+$ T cells) were dominant in mice orally immunized with r*Salmonella*-Tox C. In contrast, Th2-type IL-4-producing TT-specific CD4$^+$ cells (140/10^6 CD4$^+$ T cells) were induced in mice orally immunized with mucosal vaccine containing TT and CT (Fig. 2). Thus, the levels of secreted IFN-γ and IL-4 in culture supernatants correlated with the numbers of IFN-γ- and IL-4-producing TT-specific Th cells. Using both Th1 and Th2 cytokine-specific ELISA and ELISPOT assays, the type of Th1- and Th2-type responses induced by different forms of vaccine and antigen at the level of both protein and cell, respectively, can be elucidated.

Characterization of Th1 and Th2 type responses in oral tolerance

In addition to Th1 and Th2 responses in the antigen-specific immunity against infectious agent, it is also important to explore cytokine responses in the induction of oral tolerance. To this end, IL-10 and TGF-β produced by regulatory type T cells and immunological balance between Th1 and Th2 type responses can play important roles in the induction of oral tolerance as well as in general systemic unresponsiveness. In this regard, it is also important to examine antigen-specific Th1- and Th2-type

309

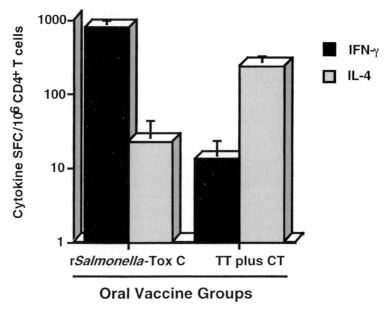

Figure 2. The frequency of Th1- and Th2-type cytokine spot-forming cells (SFC) in TT-stimulated Peyer's patch CD4[+] T cells from mice orally immunized with r*Salmonella* -Tox C or CT plus TT. CD4[+] T cells were harvested from *in vitro* cultures following 4 days of incubation. Numbers of IFN-γ- and IL-4- SFC were examined by cytokine-specific ELISPOT. To determine the numbers of TT-induced IFN-γ and IL-4 SFC, numbers of SFC in cultures without antigen (unstimulated) were subtracted from those in cultures with antigen.

responses in oral tolerance at the cellular level by using ELISPOT assay. For induction of oral tolerance, a standard single high dose protocol, 25 mg of OVA is commonly used. Seven days later, mice were immunized via the systemic route (intraperitoneal or subcutaneous) with 100 μg of OVA in 100 μl of CFA (OVA/CFA). Purified CD4[+] T cells from the spleens of mice previously fed OVA or PBS were incubated with or without 1 mg of OVA in the presence of autologous APCs. Four days later, the number of Th1 and Th2 cytokine-producing cells was determined by cytokine-specific ELISPOT assays (Fig. 3). Splenic CD4[+] T cells from mice systemically immunized with OVA in CFA included significant numbers of IFN-γ-producing Th1 type cells as well as IL-4-, IL-5-, IL-6- and IL-10-secreting Th2 type cells 4 days after *in vitro* Ag stimulation (Fig. 3) (Fujihashi *et al.*, 1999). Conversely, when mice received a tolerizing dose of OVA before parenteral challenge, the numbers of Ag-specific cytokine-producing cells were greatly reduced for both Th1- and Th2-type cytokines (Fig. 3) (Fujihashi *et al.*, 1999). These data indicate that Th1 and Th2 types of systemic responses are down-regulated in mice with oral tolerance induced by a large dose of protein Ag.

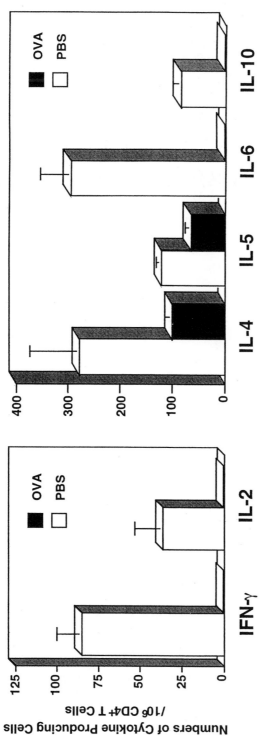

Figure 3. The frequencies of splenic CD4+ Th1 and Th2 cytokine producing cells from mice orally tolerized with OVA. Mice were given 25 mg of OVA (□) or PBS (■) orally and were challenged by the intraperitoneal route with 100 μg of OVA in 100 μl of CFA 7 days after feeding. Fourteen days after intraperitoneal challenge, splenic CD4+ T cells (2×10^6 ml^{-1}) from mice were cultured with 1 mg ml^{-1} of OVA in the presence of T cell-depleted and irradiated splenic feeder cells (4×10^6 per well). Non-adherent cells were harvested following 4 days of incubation and analysed by the respective cytokine-specific ELISPOT assays.

Mouse-specific intracellular cytokine analyses

Intracellular cytokine protocol

Using an intracellular cytokine staining method, the frequencies of cytokine-producing cells and their phenotype can be examined by flow cytometry analysis. We have adapted a general intracellular cytokine staining protocol to detect the frequencies of Ag-induced cytokine producing cells in both mucosal and systemic tissues. As an example, we have introduced analysis of IFN-γ production by antigen-specific CD4$^+$ T cells isolated from OVA-transgenic mice (Fig. 4).

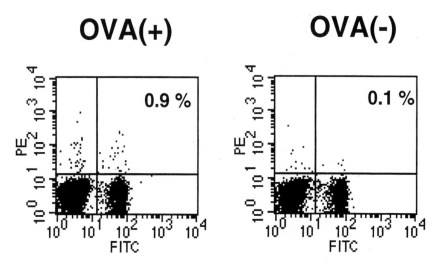

Figure 4. Intracellular staining of IFN-γ. The splenocytes from DO11.10 mice were cultured with or without 1 mg of OVA for 12 h. Non-adherent cells were harvested and cell surface staining was performed using the KJ1-26 mAb. Cells were further stained for detecting intracellular IFN-γ as described in the protocol.

1. Isolate mononuclear cells from DO11.10 (OVA transgenic) mice (Murphy *et al.*, 1990) and culture with 1 mg ml^{-1} of OVA for 6–48 h. During the last 4 h of incubation, add brefeldin A (10 µg ml^{-1}) in order to allow the accumulation of cytokines at the cell membrane.
2. Harvest mononuclear cells and suspend between $5 \times 10^5 - 1 \times 10^6$ cells in staining buffer (3% FCS and 0.1% NaN$_3$ in PBS) and wash with the same buffer.
3. Incubate with Fc Block mAb (CD16/32, BD PharMingen) for 15 min at 4°C.
4. Wash with staining buffer and stain with fluorescent anti-KJ1-26 mAb (CALTAG Laboratories, Burlingame, CA, USA) for detecting OVA-specific CD4$^+$ T cells for 15 min at 4°C.
5. Wash with 3 ml of staining buffer and fix the stained cells with 1000 µl of 2% paraformaldehyde for 10 min at room temperature (RT).

(contd.)

6. Wash with staining buffer (after washing, you may be able to stop at this step by resuspending the cells in PBS and storing them at 4°C for up to 2 days).
7. Permeabilize fixed cells with 500 μl of FACS Permeabilizing Solution (BD PharMingen; Cat. No. 340457) for 10 min at RT.
8. Centrifuge cells and discard FACS Permeabilizing Solution. Wash with 3 ml of staining buffer.
9. Add fluorescent anti-IFN-γ, -IL-4, -IL-6 or -IL-10 mAbs for intracellular cytokines or isotype controls corresponding to cytokine mAb isotype (BD PharMingen) for 30 min at RT.
10. Wash with staining buffer 2–3 times and resuspend the cells in PBS. The sample can be stored at 4°C. Flow cytometry analysis needs to be performed within 24 h.

Quantitative RT-PCR

Quantitative RT-PCR by use of a LightCycler

An adaptation of RT-PCR technology in the area of immunology has led to the detection of Th1/Th2 cytokine-specific mRNA in different subsets of T cells. Furthermore, the recent development of quantitative RT-PCR allowed us to elucidate alterations in the levels of Th1 and Th2 cytokine-specific mRNA. Thus, increased or decreased levels of specific cytokine expression can be examined. For example, it is now possible to assess the molecular amounts of Th1, Th2 or other type cytokine-specific mRNAs after *in vitro* stimulation of antigen-specific CD4$^+$ T cells from experimental animals mucosally and/or systemically immunized with different antigens or vaccines (Hiroi *et al.*, 1995; Marinaro *et al.*, 1995). On the other hand, several published papers have already dealt with the problems inherent in the quantitative RT-PCR (Wang *et al.*, 1989; Chelly *et al.*, 1990; Gilliland *et al.*, 1990; Katz and Haff, 1990; Lundeberg *et al.*, 1991). In the previous edition of this book, a basic protocol for the quantitative RT-PCR was described (Methods in Microbiology, 1st edition; Fujihashi *et al.*, 1998). At that time, a protocol for the quantitative RT-PCR method included the following crucial advantages: (a) specific detection of the amplified samples; (b) rapid, non-radioactive test system validated by an internal standard; and (c) quantitative evaluation of the assay, with calculation of the original amount in the sample (e.g. purified CD4$^+$ or CD8$^+$ T cells) obtained *in vivo* and/or *in vitro*. Since the methodology for quantitative RT-PCR has been improved over the last 3 years, our quantitative RT-PCR system which was described in the previous edition showed some new problems such as variations in the size of internal standards and target DNA. In addition, when amounts of synthesized cDNA were varied through some changes in conditions of the RT efficiency which are dependent on the purity of RNA in the sample, techniques and incubation conditions (Rappolee, 1990), mRNA gene expression cannot be reliably compared between different experimental samples. Thus, it has been difficult to accurately measure small amounts of synthesized cDNA. To circumvent this, it was important to adjust the amounts of template cDNA at the PCR step for quantitative RT-PCR in order to compare individual samples. Further, it was generally accepted that a quantitative RT-PCR system which uses internal standards was better than one employing external standards (Wang *et al.*, 1989), since different conditions of temperature control could influence the PCR reaction of the individual well in the thermal cycler. Further, a quantitative RT-PCR system that uses internal standards has additional disadvantages in the measurement of RNA content, since the system is limited to a narrow window of quantitation at a single time point. Finally, the previous quantitative RT-PCR system requires high amounts of total RNA in the sample. To overcome these disadvantages, the LightCycler™ (Roche Boehringer Mannheim) was introduced by adapting the glass capillary method. The real-time quantitative RT-PCR system improves the reproductivity of the same sample among different experiments. Thus, use of an external standard in the LightCycler™ system allows comparable quantitative results to be obtained which are achieved by using the internal standard system (Wittwer *et al.*, 1997).

Here we describe this new method which is both simple and sensitive for cDNA quantitation. This uses a fluorometric dye and quantitative PCR system in the LightCycler™ (Roche Boehringer Mannheim) with external controls constructed from murine cytokine genes (Wittwer *et al.*, 1997; Hiroi *et al.*, 2000). Detailed protocols for Th1 and Th2 cytokine-specific quantitative RT-PCR, which are routinely performed in our laboratories, are described below.

Materials

RNA isolation

- TRIzol® reagent (Life Technologies, Rockville, MD, USA)
- Chloroform
- Isopropanol
- 70% Ethanol
- Pellet Paint™ (Novagen, Darmstadt, Germany)

Reverse transcription of RNA

- 100 ng/µl Random primer (Life Technologies)
- 5× Reverse transcription buffer (250 mM Tris-HCl, pH 8.3, 375 mM KCl and 15 mM MgCl$_2$) (Life Technologies)
- 20 U µl^{-1} RNase Inhibitor (Amersham Pharmacia, Uppsala, Sweden)
- 10 mM dNTPs (Life Technologies)
- 0.1 M Dithiothreitol (DTT) (Life Technologies)
- SuperScript™ II (Life Technologies)
- 10 µg ml^{-1} RNase I (Promega, Madison, WI)
- 10 µg ml^{-1} RNase H (Life Technologies)

(contd.)

Quantification of synthesis of cDNA

- Oligreen® DNA quantitation Kit (Molecular Probes Inc., Eugene, OR, USA)
- Single-strand (ss) DNA (Oligonucleotide standard; Molecular Probes Inc.)
- 96 well-white microplate (Labsystems, Helsinki, Finland)
- Spectrofluorometer (Labsystem Fluoroskan II, Labsystems)

Construction of external standard DNA for murine Th1/Th2 cytokines

- PCR buffer [16.6 mM $(NH_4)_2SO_4$, 50 mM 2-mercaptoethanol, 6.8 mM EDTA, 67 mM Tris-HCl, pH 8.8, 0.1 mg ml^{-1} BSA] (Perkin-Elmer Cetus)
- 3 mM $MgCl_2$
- 0.2 mM of each dNTP (Life Technologies)
- 2.5 U Taq DNA polymerase (Perkin-Elmer Cetus)
- Wizard PCR Preps DNA Purification System (Promega)
- pGEM-T Vector (Promega)
- T4 DNA Ligase and 10× buffer (Promega)
- *Escherichia coli* competent cells (Promega)
- Wizard *Plus* Maxiprep (Promega)

Measurements of PCR products for murine Th1/Th2 cytokine gene

- FITC-labelled donor probe (0.2 µM), LC red 640-labelled acceptor probe (0.2 µM) (Roche Boehringer Mannheim)
- LC-FastStart DNA-Master-hybridization probes™ (Taq PCR buffer: 1×, Taq PCR polymerase: 0.02 U, nucleotides: 0.2 mM; Roche Diagnostics GmbH, Mannheim, Germany)
- LC FastStar -DNA-Master-SYBER Green I™ (Taq PCR buffer: 1×, Taq PCR polymerase: 0.02 U, nucleotides: 0.2 mM; Roche Diagnostics GmbH)
- A glass capillary tube and snaps sealed with a plastic cap (Roche Diagnostics GmbH)
- LightCycler™ (Roche Diagnostics GmbH)

Protocols

Isolation of total RNA from mononuclear cells

1. Total RNA is obtained using TRIzol® reagent (Life Technologies) from at least 1×10^4 cells from *in vitro* stimulated CD4⁺ T cells. For the precipitation step, it is better to use the co-precipitant reagent Pellet Paint™ (Novagen) for small numbers of cells.

(contd.)

2. Samples are diluted with 18 µl DEPC-H$_2$O and 2 µl of 20 U µl^{-1} RNase Inhibitor (Amersham Pharmacia) and 2 µl of 100 ng µl^{-1} Random primer (Life Technologies) are added at 70°C for 10 min and cooling on ice for 5 min (Fig. 5).
3. Samples are equally separated into two aliquots of RNA (11 µl sample: A and B) with reverse transcription primer in 1.5 ml micro-centritubes (Fig. 5).
4. Each sample is added to a reaction mixture containing 4 µl of 5× Fast Strand Buffer (Life Technologies), 1 µl of 10 mM dNTPs (Life Technologies), 2 µl of 0.1 M DTT (Life Technologies), 1 µl of 20 U µl^{-1} RNase Inhibitor (Amersham Pharmacia) (Fig. 5).
5. For the preparation of both synthesis cDNA and background samples, to one sample (A) is added 1 µl of SuperScript™ II (Life Technologies) and to the other sample (B) is added 1 µl of DEPC-H$_2$O instead of reverse transcription enzyme (Fig. 5). Reverse transcription is carried out by incubation at 42°C for 1 h, followed by heating at 90°C for 5 min to inactivate enzyme activity (Fig. 5).
6. Both samples (A and B) are treated with 1 µl of 10 µg ml^{-1} RNase I (Promega) and 1 µl of 10 µg ml^{-1} RNase H (Life Technologies) and incubated at 37 °C for 30 min (Fig. 5).

Quantification of cDNA synthesis

1. The synthesized cDNA solution in TE buffer is diluted to a final volume of 100 µl in 96 well-white microplate (Labsystems), and 100 µl of the aqueous working solution of OliGreen® reagent (1 : 200 dilution of the concentrated DMSO solution in TE buffer) is added to each sample.
2. The samples are then incubated for 2 to 5 min at room temperature, protected from light.
3. The samples are excited at 480 nm and the fluorescence emission intensity is measured at 520 nm using a Spectrofluorometer (Fluoroskan II, Labsystems).
4. Subtract the non-reverse transcription compartment from the absorbance measured at 520 nm and the difference is the component contributed by the specific cDNA synthesized in the sample. The measurement is performed by the analysis of the fluorescence of the sample using instrument parameters that correspond to those used when generating an oligonucleotide standard curve (synthetic 24-mer M13 sequencing primer, Molecular Probes, Inc.).

Construction of external standard DNA for murine Th1/Th2 cytokines

For the quantitation of cytokine-specific message, recombinant DNA (rDNA) is constructed as external standard (Hiroi et al., 1995). The external standard consists of 5′ cytokine-specific primer and 3′ cytokine-specific primer.

1. Synthesize one set of forward and reverse oligo DNA primers using a DNA synthesizer (Th1; IFN-γ and IL-2, Th2; IL-4, IL-5, IL-6 and IL-10) (Table 2).
2. Construct the first forward and reverse primers to the template gene by PCR. Mix 50 μl reaction mixture containing PCR buffer (16.6 mM $(NH_4)_2SO_4$, 50 mM 2-mercaptoethanol, 6.8 mM EDTA, 67 mM Tris-HCl, pH 8.8, 0.1 mg ml^{-1} BSA), 3 mM $MgCl_2$, 0.2 mM of each dNTP, 30 pmol of the forward and reverse primers, template gene and 2.5 U Taq DNA polymerase (Perkin-Elmer Cetus) and carry out 30 cycles of amplification: denature for 10 s at 94°C, anneal for 30 s at 59°C, extension for 45 s at 72°C; perform an additional extension for 5 min at 72°C after the last cycle.
3. Purify the PCR products with a Wizard PCR Preps DNA purification System (Promega).
4. Insert the PCR product (IFN-γ, IL-2, IL-4, IL-5, IL-6, IL-10 and β-actin) into pGEM-T Vector (Promega). Mix 1 μl T4 DNA Ligase 10× buffer (Promega), 1 μl pGEM-T vector (Promega), 1 μl 10× diluted PCR product, 1 μl T4 DNA Ligase (Promega) and 6 μl dH$_2$O, and incubate the mixture for 3 h at 15°C. Heat the reaction for 10 min at 70–72°C and allow to cool at 25°C.
5. Transform pGEM-T vector into *Escherichia coli* competent cells (Promega) for selection of a positive colony.
6. Amplify the plasmid containing internal standard by incubation in one litre of LB broth overnight at 37°C.
7. Purify the plasmid DNA (internal standard) by using Wizard *Plus* Maxiprep (Promega).

Measurements of PCR products for murine Th1/Th2 cytokine genes

We use rapid cycle PCR with continuous monitoring of Th1 (IFN-γ and IL-2) and Th2 (IL-4, IL-5, IL-6 and IL-10) cytokine gene PCR products synthesized from mRNA of experimental samples in order to compare them with standard DNA. Monitoring of fluorescence emission during the PCR is carried out with a LightCycler™ (Roche Diagnostics GmbH), an air thermal cycler with a built-in fluorometer (Wittwer *et al.*, 1997). Two different methods are available for fluorimetric online detection and subsequent evaluation of PCR reactions in glass capillaries. The PCR products formed may either be detected via:

1. a fluorophore that binds to double-stranded DNA molecules independently of the amplified sequence.
2. specific fluorophores coupled to sequence-specific oligonucleotides.

The first procedure determines the PCR products formed in a sequence-independent way. For this method, the use of the double-stranded DNA binding dye SYBER Green I [LC FastStar -DNA-Master-SYBER Green I™ (Roche Diagnostics GmbH)] has proven to be effective (Wittwer *et al.*, 1997). The latter method determines synthesized PCR products in a

Table 2. Th1 and Th2 cytokine-specific oligonucleotides for quantitative RT-PCR.

	Upper and lower primers*	Hybri-probes*	MgCl$_2$ final optimal concentration*
β-actin	5'TGGAATCCTGTGGCATCCATGAAAC-3' 5'TAAAACGCAGCTCAGTAACAGTCCG-3'	5'CTCCATCGTGACTCCTGCTTGCTGAT-FITC LCR-ACATCTGCTGGAAGGTGGACAGT-3'	4 mM
IFN-γ	5'-TGAACGCTACACACTGCATCTTGG-3' 5'-CGACTCCTTTTCCGCTTCCTGAG-3'	5'-AGACAATCAGGCCATCAGCAACA-FITC LCR-CATAAGGCGTCATTGAATCACACCTG-3'	4 mM
IL-2	5'-ATGTACAGCATGCAGCTCGCATC-3' 5'-GGCTTGTTGAGATGATGCTTTGACA-3'	5'-AAGGTGAGCATCCTGGGGAGTTT-FITC LCR-GGTTCCTGTAATTCTCCATCCTG-3'	3 mM
IL-4	5'-ATGGGTCTCAACCCCCAGCTAGT-3' 5'-GCTCTTTAGGCTTTCCAGGAAGT-3'	5'-CGTTTGGCACATCCATCTCCGT-FITC LCR-CATGGCGGTCCCTTCCTCCTGTG-3'	4 mM
IL-5	5'-GAAAGAGACCTTGACACAGCTG-3' 5'-GAACTCTTGCAGGTAATCCAGG-3'	5'-ACAGTACCCCCACGGACAGTT-FITC LCR-ATTCTTCAGTATGTCTAGCCCCCTG-3'	4 mM
IL-6	5'-GAACAACGATGATGCACTTGCAG-3' 5'-TTTTAGCCCACTCCTTCTGTGAC-3'	5'-TATCTGTTAGGAGAGCATTGG-FITC LCR-AATTGGGGTAGGAAGGACTAT-3'	4 mM
IL-10	5'-ATGCAGGACTTTAAGGGTTACTTGGGTT-3' 5'-ATTTCGGAGAGAGGTACAAACGAGGTTT-3'	5'-CTGCTCCACTGCCTTGCTCTTATT-FITC LCR-TCACAGGGGAGAAATCGATGACAG-3'	4 mM

* Upper and lower primers and one set hybri-probe and optimal MgCl$_2$ concentration are indicated.

sequence-dependent way using specific oligonucleotides acting as detection probes. For these experiments, the fluorophores LightCycler™-Red 640 and Fluorescein, have been found to be suitable (Wittwer et al., 1997).

The first detection system, SYBER Green I, has the disadvantage in measurement of the exact amounts of specific DNA genes. However, this system is convenient and useful in adjusting some experimental conditions (e.g., concentrations of primers, $MgCl_2$ and dNTPs) and temperature of thermal cycle because this dye can bind to primer dimers and non-specific amplification genes during thermal cycling. Among these different conditions, the system was preferentially affected by $MgCl_2$ concentration of the PCR reaction with LightCycler™ (Roche Diagnostics GmbH). Therefore, it is necessary to elucidate the optimal concentration of $MgCl_2$ for the method used by SYBER Green I. As a result, optimum concentrations of $MgCl_2$ were 3 mM for IFN-γ, IL-2 and IL-5 and 4 mM for IL-4, IL-6 and IL-10.

1. The composition of PCR reaction mixture by using hybri-probes is as follows: FITC-labelled donor probe (0.2 μM; Table 2), LC red 640-labelled acceptor probe (0.2 μM; Table 2), LC-DNA-Master-hybridization probes™ (Taq PCR buffer: 1×, Taq PCR polymerase: 0.02 U, nucleotides: 0.2 mM; Roche Diagnostics GmbH), water and sample cDNA or several concentrations of cytokine-specific standard template (1 fM~10 nM), and the amplification primers (0.2 μM each; Table 2) were contained in one glass capillary.
2. Each reaction sample is loaded in a glass capillary tube and snap sealed with a plastic cap. The PCR protocol is as follows: denaturation at 95°C for 2 min: 40 cycles of cDNA amplification, each cycle consisting of denaturation at 95°C for 1 s, annealing at 55°C for 15 s, and extension at 72°C for 20 s. The temperature transition rate is set at 20°C s^{-1}.

We generally use the software supplied with the LightCycler™ to analyse the acquired results. The fluorogenic hybridization probes are designed according to the technical guidelines recommended by the manufacturer (Idaho Technology Inc, ID, USA) (Fig. 5).

Evaluation of the cDNA measurement system

Initially, we test mRNA levels synthesized from the same material but with different amounts of sample. For example, a plot for the IFN-γ competitor or three different concentrations of IFN-γ targets are illustrated (Fig. 6A and B). Since three samples are isolated from the same original material (e.g. mononuclear cells isolated from a mouse spleen), the levels of mRNA expression are expected to have the same level per cDNA amount. mRNA expression of three samples resulted in almost equal levels after normalization against the same amount of cDNA (Fig. 6C). This result indicates that the cDNA measurement system is useful and reliable for mRNA quantification by real-time PCR (LightCycler™). The real-time PCR method includes the following crucial advantages: (a) specific detection of the amplified samples; (b) rapid, non-radioactive test system

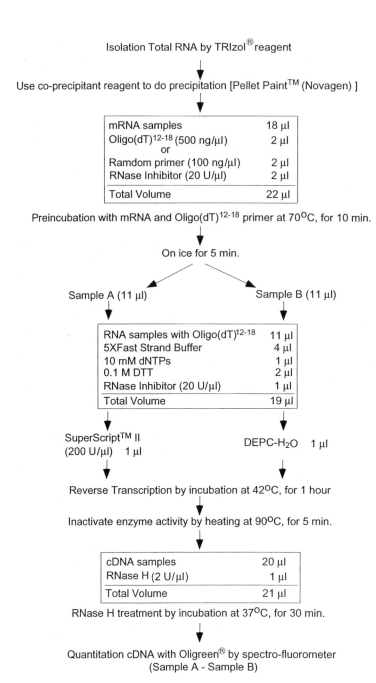

Figure 5. Schematic representation of the quantitative cDNA procedure.

validated by an external standard; and (c) quantitative evaluation of the assay, with calculation of the original amount in the sample from *in vitro* and *in vivo*. Furthermore, the LightCycler™ method, by using donor and acceptor probes, is a technique that can be used to detect low amounts of PCR products at levels in the attomolar range (Hiroi *et al.*, 2000).

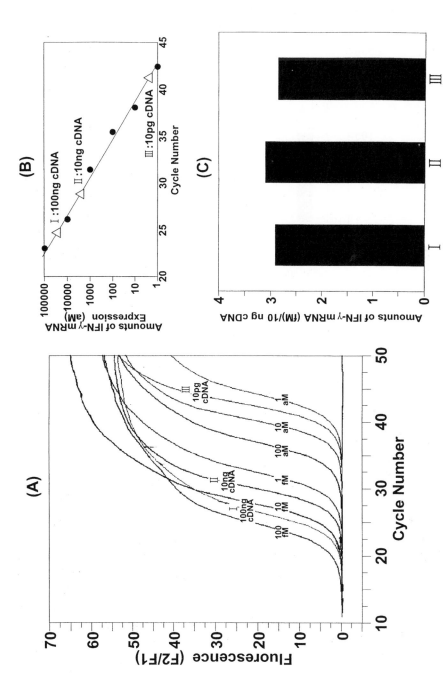

Figure 6. An example of quantitation of IFN-γ-specific mRNA. (A): Six different concentrations of IFN-γ standards (1 attoM, 10 attoM, 100 attoM, 1 fM, 10 fM and 100 fM) and three different amounts of synthesized cDNA isolated from one mouse spleen [10 pg (Ⅲ), 10 ng (Ⅱ) and 100 ng (Ⅰ)] are plotted. (B): Standard curves are constructed with plasmid dilutions ranging from 1 attoM to 100 fentoM. (C): Three tested samples of different amounts of cDNA [10 pg (Ⅲ), 10 ng (Ⅱ) and 100 ng (Ⅰ)] calculated by cDNA measurement system were applied to a LightCycler™. The result is expressed as mRNA concentration calculated after normalization against the same amount of cDNA.

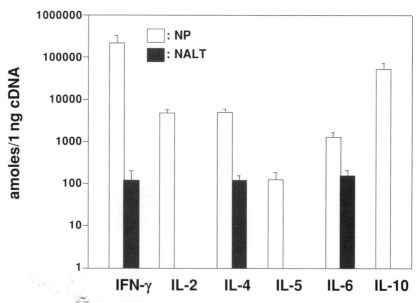

Figure 7. Analysis frequency of Th1- and Th2-type cytokine mRNA levels in NALT (nasopharyngeal-associated lymphoreticular tissues) and NP (nasal passage) isolated from normal mice. CD4$^+$ T cells were isolated from normal mice (C57BL/6, 8 weeks old) by flow cytometry. The levels of individual Th1 and Th2 cytokine specific mRNA were determined by quantitative RT-PCR using LightCycler™.

Analysis of Th1 and Th2 specific mRNA levels in the mucosal inductive and effector tissues

The nasal cavity is an important part of the mucosal immune system which is protected by secretory IgA (S-IgA). For the induction of mucosal S-IgA responses in the nasal cavity, the common mucosal immune system (CMIS), which consisted of inductive (NALT) and effector sites (e.g. nasal passage; NP), must be considered. This CMIS which interconnects between the NALT and NP is tightly regulated by Th1 and Th2 cytokines (Hiroi *et al.*, 1998). To elucidate the Th1 and Th2 cytokine profile at the mRNA level of murine NALT and NP, mRNA was purified from freshly isolated NALT and NP CD4$^+$ T cells by FACS sorting and subjected to the quantitative RT-PCR. When the levels of Th1 (IFN-γ and IL-2) and Th2 (IL-4, -5, -6 and -10) mRNA were examined and compared between NALT and NP, both Th1 and Th2 type cytokine mRNA levels were always lower in NALT when compared with NP (Fig. 7). On the other hand, high levels of both Th1- and Th2-type cytokine expression were detected in NP (Fig. 7). These results reflect two distinct immunological environments of NALT and NP as inductive and effector sites, since the later tissue is an active site for the generation of CTL and S-IgA responses, and high levels of Th1 and Th2 cytokine-specific mRNA expression were expected. Together with ELISA, ELISPOT and intracellular cytokine staining assay, the quantitation RT-PCR is a useful system for the elucidation of Th1 and Th2 responses.

◆◆◆◆◆◆ SUMMARY

In this chapter, we have described the protocols for Th1 and Th2 cytokine-specific ELISA, ELISPOT, intracellular cytokine and quantitative RT-PCR assays for mice. These four assay systems allow detection of different stages of cytokine production including secreted cytokine, cytokine-producing cells and cytokine-specific mRNA, respectively. Although each assay has unique advantages for the detection and quantitation of Th1 and Th2 cytokines at three distinct levels (e.g. protein, mRNA and cell) in the murine model, the use of individual assays in a separate manner may often not be sufficient for a thorough and accurate determination of the profile of Th1 or Th2 cytokine expression. For instance, when the levels of cytokine-specific mRNA are measured by quantitative RT-PCR only, the information concerning the biological activity of analysed cytokine is unknown. Further, examination of secreted cytokines in culture supernatants by ELISA may only detect residual cytokines produced by CD4+ T cells *in vitro*, since it is possible and even likely that secreted cytokines may be immediately taken up by neighbouring cells in the culture. To this end, it is best to perform at least two different cytokine assays to confirm the results describing the profile of Th1 and Th2 cytokine responses induced by mucosal and/or systemic immunization. An ideal situation would be to use these three different assays for the elucidation of Th1 and Th2 responses.

Quantitation of T-cell Cytokine Responses

Acknowledgements

We thank Drs Naoto Yoshino, Noriyuki Ohta and Naotoshi Kinoshita for helpful comments and discussion. We also thank Ms Noriko Kitagaki for helpful experimental assistance. We also thank Mr Kohichi Iwatani, Ms Sheila D. Turner and Pamela D. Thomas for preparation of the manuscript. This work is supported by US Public Health Service Grants AI 35932, DE 09837, DK 44240, DE 12242, AI 43197, AI 18958, AI 39816, AI 35544, AI 65298 and AI 65299 as well as grants from the Ministry of Education, Science Sports and Cultures, and the Ministry of Health and Welfare in Japan.

References

Beagley, K. W., Eldridge, J. H., Kiyono, H., Everson, M. P., Koopman, W. J., Honjo, T. and McGhee, J. R. (1988). Recombinant murine IL-5 induces high rate IgA synthesis in cycling IgA-positive Peyer's patch B cells. *J. Immunol.* **141**, 2035–2043.

Beagley, K. W., Eldridge, J. H., Lee, F., Kiyono, H., Everson, M. P., Koopman, W. J., Hirano, T., Kishimoto, T. and McGhee, J. R. (1989). Interleukins and IgA synthesis. Human and murine interleukin 6 induce high rate IgA secretion in IgA-committed B cells. *J. Exp. Med.* **169**, 2133–2148.

Bond, M. W., Shrader, B., Carty, J., Mosmann, T. R. and Coffman, R. L. (1987). A mouse T cell product that preferentially enhances IgA production. II. Physicochemical characterization. *J. Immunol.* **139**, 3691–3696.

Boyaka, P. N., Marinaro, M., Jackson, R. J., Menon, S., Kiyono, H., Jirillo, E. and McGhee, J. R. (1999). IL-12 is an effective adjuvant for induction of mucosal immunity. *J. Immunol.* **162**, 122–128.

Carter, L. and Swain, S. L. (1997). Single cell analyses of cytokine production. *Curr. Opin. Immunol.* **9**, 177–182.

Chelly, J., Montarras, D., Pinset, Ch., Berwald-Netter, Y., Kaplan, J. C. and Kahn, A. (1990). Quantitative estimation of minor mRNA by cDNA-polymerase chain reaction. Aplication to dystrophin mRNA in cultured mygenic and brain cells. *Eur. J. Biochem.* **187**, 691–698.

Clerici, M., Lucey, D. R., Berzofsky, J. A., Pinto, L. A., Wynn, T. A., Blatt, S. P., Dolan, M. J., Hendrix, C. W., Wolf, S. F. and Shearer, G. M. S. (1993). Restoration of HIV-specific cell-mediated immune responses by interleukin-12 in vitro. *Science* **262**, 1721–1724.

Coffman, R. L., Shrader, B., Carty, J., Mosmann, T. R. and Bond, M. W. (1987). A mouse T cell product that preferentially enhances IgA production. I. Biologic characterization. *J. Immunol.* **139**, 3685–3690.

Fujihashi, K., McGhee, J. R., Lue, C., Beagley, K. W., Taga, T., Hirano, T., Kishimoto, T., Mestecky, J. and Kiyono, H. (1991). Human appendix B cells naturally express receptors for and respond to interleukin 6 with selective IgA1 and IgA2 synthesis. *J. Clin. Invest.* **88**, 248–252.

Fujihashi, K., McGhee, J. R., Beagley, K. W., McPherson, D. T., McPherson, S. A., Hung, C-M. and Kiyono, H. (1993a). Cytokine-specific ELISPOT assay. Single cell analysis of IL-2, IL-4 and IL-6 producing cells. *J. Immunol. Methods* **160**, 181–189.

Fujihashi, K., Yamamoto, M., McGhee, J. R., Beagley, K. W. and Kiyono, H. (1993b). Function of $\alpha\beta$ TCR$^+$ intestinal intraepithelial lymphocytes: Th1- and Th2-type cytokine production by CD4$^+$ CD8$^-$ and CD4$^+$CD8$^+$ T cells for helper activity. *Int. Immunol.* **8**, 1473–1481.

Fujihashi, K., VanCott, J. L., McGhee, J. R., Hiroi, T. and Kiyono, H. (1998). Measuring cytokine responses by ELISA, ELISPOT and RT-PCR methods. In: *Methods and Microbiology*, vol. 25. Kaufmann, S. H. E. and Kabelitz, D. (Eds). pp. 257–286.

Fujihashi, K., Dohi, T., Kweon, M.-N., McGhee, J. R., Koga, T., Cooper, M. D., Tonegawa, S. and Kiyono, H. (1999). $\gamma\delta$ T cells regulate mucosally induced tolerance in a dose dependent fashion. *Int. Immunol.* **11**, 1907–1916.

Gilliland, G., Perrin, S. T., Blanchard, K. and Bunn, H. F. (1990). Analysis of cytokine mRNA and DNA: Detection and quantitation by competitive polymerase chain reaction. *Pro. Natl. Acad. Sci. USA* **87**, 2725–2729.

Harriman, G. R., Kunimoto, D. Y., Elliott, J. F., Paetkau, V. and Strober, W. (1988). The role of IL-5 in IgA B cell differentiation. *J. Immunol.* **140**, 3033–3039.

Helle, M., Boeije, L. and Aarden, L. A. (1988). Functional discrimination between interleukin 6 and interleukin 1. *Eur. J. Immunol.* **18**, 1535–1540.

Hiroi, T., Fujihashi, K. McGhee, J. R. and Kiyono, H. (1995). Polarized Th2 cytokine expression by mucosal $\gamma\delta$ and $\alpha\beta$ T cells. *Eur. J. Immunol.* **25**, 2743–2751.

Hiroi, T., Iwatani, K., Iijima, H., Kodama, S., Yanagita, M. and Kiyono, H. (1998). Nasal immune system: distinctive Th0 and Th1/Th2 type environments in murine nasal associated lymphoid tissues and nasal passage, respectively. *Eur. J. Immunol.* **28**, 3346–3353.

Hiroi, T., Yanagita, M., Ohta, N., Sakaue, G. and Kiyono, H. (2000). Interleukin 15 and IL-15R selectively regulate differentiation of common mucosal immune system-independent B-1 cells for IgA responses. *J. Immunol.* **165**, 4329–4337.

Hsieh, C. S., Macatonia, S. E., Tripp, C. S., Wolf, S. F., O'Garra, A. and Murphy, K. M. (1993). Development of Th1 CD4$^+$ T cells through IL-12 produced by *Listeria*-induced macrophages. *Science* **260**, 547–549.

Jung, T., Schauer, C., Heusser, C., Neuman C. and Rieger C. (1993). Detection of intracellular cytokines by flow cytometry. *J. Immunol. Meth.* **159**, 197–207.

Katz, E. and Haff, L. A. (1990). Rapid separation, quantitation and purification of products of polymerase chain reaction by liquid chromatography. *J. Chromatogr.* **512**, 433–444.

Kobayashi, M., Fitz, L., Ryan, M., Hewick, R. M., Clark, S. C., Chan, S., Loudon, R., Sherman, F., Perussia, B. and Trinchieri, G. (1989). Identification and purification of natural killer cell stimulatory factor (NKSF), a cytokine with multiple biologic effects on human lymphocytes. *J. Exp. Med.* **170**, 827–845.

Kunisawa, J., Nakanishi, T., Takahashi, I., Okudaira, A., Tsutsumi, W., Katayama, K., Nakagawa, S., Kiyono, H. and Mayumi, T. (2001). Sendai virus fusion protein mediates simultaneous induction of MHC class I/II-dependent mucosal and systemic immune responses via the nasopharyngeal-associated lymphoreticular tissue immune system. *J. Immunol.* **167**, 1406–1412.

Lebman, D. A. and Coffman, R. L. (1988). The effects of IL-4 and IL-5 on the IgA responses by murine Peyer's patch B cell subpopulations. *J. Immunol.* **141**, 2050–2056.

Lundeberg, J., Wahlberg, J. and Uhlen, M. (1991). Rapid colorimetric quantification of PCR-amplified DNA. *Biotechniques* **10**, 68–75.

Marinaro, M., Staats, H. F., Hiroi, T., Jackson, R. J., Coste, M., Boyaka, P. N., Okahashi, N., Yamamoto, M., Kiyono, H., Bluethmann, H., Fujihashi, K. and McGhee, J. R. (1995). Mucosal adjuvant effect of cholera toxin in mice results from induction of T helper 2 (Th2) cells and IL-4. *J. Immunol.* **155**, 4621–4629.

Marinaro, M., Boyaka, P. N., Jackson, R. J., Finkelman, F. D., Kiyono, H., Jirillo, E. and McGhee, J. R. (1999). Use of intranasal IL-12 to target predominantly Th1 responses to nasal and Th2 responses to oral vaccines given with cholera toxin. *J. Immunol.* **162**, 114–121.

Mosmann, T. R. and Coffman, R. L. (1989). Th1 and Th2 cells: different patterns of lymphokine secretion lead to different functional properties. *Annu. Rev. Immunol.* **7**, 145–173.

Murray, P. D., McKenzie, D. T., Swain, S. L. and Kagnoff, M. F. (1987). Interleukin 5 and interleukin 4 produced by Peyer's patch T cells selectively enhance immunoglobulin A expression. *J. Immunol.* **139**, 2669–2674.

Murphy, K. M., Heimberger, A. B. and Loh, D. Y. (1990). Induction by antigen of intrathymic apoptosis of CD4$^+$CD8$^+$ TCR^{low} thymocytes in vivo. *Science* **250**, 1720–1723.

Nakamura, K., Kitani, A. and Strober, W. (2001). Cell contact-dependent immunosuppression by CD4(+)CD25(+) regulatory T cells is mediated by cell surface-bound transforming growth factor beta. *J. Exp. Med.* **94**, 629–644.

Rappolee, D. A. and Werb, Z. (1990). mRNA phenotyping for studying gene expression in small numbers of cells: platelet-derived growth factor and other growth factors in wound-derived macrophages. *Am. J. Respir. Cell Mol. Biol.* **2**, 3–10.

Robertson, M. J., Soiffer, R. J., Wolf, S. F., Manley, T. J., Donahue, C., Young, D., Herrmann, S. H. and Ritz, J. (1992). Response of human natural killer (NK) cells to NK cell stimulatory factor (NKSF): cytolytic activity and proliferation of NK cells are differentially regulated by NKSF. *J. Exp. Med.* **175**, 779–788.

Sawamura, M., Everson, M. P., Shrestha, K., Ghanta, V. K., Miller, D. M. and Hiramoto, R. N. (1990). Characterization of 5B12.1, a monoclonal antibody specific for IL-6. *Growth Factors* **3**, 181–190.

Seder, R. A. and Paul, W. E. (1994). Acquisition of lymphokine-producing phenotypes by CD4$^+$ T cells. *Annu. Rev. Immunol.* **12**, 635–673.

Seder, R. A., Gazzinelli, R., Sher, A. and Paul, W. E. (1993). Interleukin 12 acts directly on CD4$^+$ T cells to enhance priming for interferon production and diminished interleukin 4 inhibition of such priming. *Proc. Natl. Acad. Sci. USA* **90**, 10188–10192.

Slavin, S. and Syrober, S. (1978). Spontaneous murine B cell leukemia. *Nature* **272**, 624–626.

Stordeur, P., Poulin, L. F., Cracium, L., Zhou, L., Schandene, L., deLavareille, A., Goriely, S. and Goldman, M. (2001). Cytokine mRNA quantification by real-time PCR. *J. Immunol. Methods* **259**, 55–64.

Street, N. E. and Mosmann, T. R. (1991). Functional diversity of T lymphocytes due to secretion of different cytokine patterns. *FASEB J.* **5**, 171–177.

Taguchi, T., McGhee, J. R., Coffman, R. L., Beagley, K. W., Eldridge, J. H., Takatsu, K. and Kiyono, H. (1990a). Detection of individual mouse splenic T cells producing IFN-γ and IL-5 using the enzyme-liked immunospot (ELISPOT) assay. *J. Immunol. Methods* **128**, 65–73.

Taguchi, T., McGhee, J. R., Coffman, R. L., Beagley, K. W., Eldridge, J. H., Takatsu, K. and Kiyono, H. (1990b). Analysis of Th1 and Th2 cells in murine gut-associated tissues: frequencies of CD4⁺ and CD8⁺ T cells that secrete IFN-γ and IL-5. *J. Immunol.* **145**, 68–77.

Trinchieri, G. (1995). Interleukin-12: a proinflammatory cytokine with immunoregulatory functions that bridge innate resistance and antigen-specific adaptive immunity. *Annu. Rev. Immunol.* **13**, 251–276.

VanCott, J. L., Staats, H. F., Pascual, D. W., Roberts, M., Chatfield, S. N., Yamamoto, M., Coste, M., Carter, P. B., Kiyono, H. and McGhee, J. R. (1996). Regulation of mucosal and systemic antibody responses by T helper cell subsets, macrophages, and derived cytokines following oral immunization with live recombinant *Salmonella*. *J. Immunol.* **156**, 1504–1514.

Wang, A. M., Doyle, M. V. and Mark, D. F. (1989). Quantitation of mRNA by the polymerase chain reaction. *Proc. Natl. Acad. Sci. USA* **86**, 9717–9721.

Watson, J. (1979). Continuous proliferation of murine antigen-specific helper T lymphocytes in culture. *J. Exp. Med.* **150**, 1510–1519.

Welsh, R. M., Nishioka, W. K., Antia, R. and Dundon, P. L. (1990). Mechanism of killing by virus-induced cytotoxic T lymphocytes elicited in vivo. *J. Virol.* **64**, 3726–3733.

Wittwer, C. T., Ririe, K. M., Andrew, R. V., David, D. A., Gundry, R. A. and Balis, U. J. (1997). The LightCycler: a microvolume multisample fluorimeter with rapid temperature control. *Biotechniques* **22**, 176–181.

Xu-Amano, J., Kiyono, H., Jackson, R. J., Staats, H. F., Fujihashi, K., Burrows, P. D., Elson, C. O., Pillai, S. and McGhee, J. R. (1993). Helper T cell subsets for immunoglobulin A responses: oral immunization with tetanus toxoid and cholera toxin as adjuvant selectively induces Th2 cells in mucosal associated tissues. *J. Exp. Med.* **178**, 1309–1320.

Zhang, X., Izikson, L., Liu, L. and Weiner, H. L. (2001). Activation of CD25(+)CD4(+) regulatory T cells by oral antigen administration. *J. Immunol.* **167**, 4245–4253.

Appendix

Amasham Pharmacia
Bjorkgatan 30 S-751 82
Uppsala, Sweden

Tel: 46(0) 18-16-50-00
Fax: 46(0) 18-14-38-20

Becton Dickinson Labware
Two Oak Park
Bedford, MA 01730-9902, USA

Tel: (800) 343-2035
Fax: (800) 743-6200

Biosource International
950 Flynn Road, Suite A
Camarillo, CA 93012, USA

Tel: (800) 242-0607, Outside USA: (805) 987-0086
Fax: (805) 987-3385

BD PharMingen
10975 Torreyana Rd.
San Diego, CA 92121, USA

Tel: (800) 848-6227
Fax: (619) 677-7749

Caltag Laboratories
1849 Old Bayshore Boulevard
#200 Burlingame, CA 94010, USA

Tel: (650) 652-0468
Fax: (650) 652-9030

ENDOGEN, Inc.
640 Memorial Drive
Cambridge, MA 02139, USA

Tel: (800) 487-4885, (617) 225-0055
Fax: (617) 225-2040

Genzyme Diagnostics
One Kendall Square
Cambridge, MA 02139, USA

Tel: (800) 788-1580
Fax: (617) 374-7300

Labsystems USA
8 East Forge Parkway
Frankin, MA 02038, USA

Tel: (800) 522-7633
Fax: (508) 520-2229

Life Technologies
8400 Helgerman Court
P.O. Box 6009
Gaithersburg, MD 20897, USA

Tel: (301) 840-8000

Milipore Corporation
80 Ashby Road
Bedford, MA 01730, USA

Tel: (800) 645-5476
Fax: (617) 533-8873

Molecular Probes Inc.
4849 Pitchford Ave.
Eugene, OR 97402-9165

Tel: (541) 465-8300
Fax: (541) 344-6504

Moss Inc.
P.O. Box 189
Pasadena, MA 21122, USA

Tel: (800) 932-6677
Fax: (410) 768-3971

Nalge Nunc International
2000 N Aurora Road
Naperville, IL 60563-1796, USA

Tel: (800) 288-6862
Fax: (708) 416-2556

Novagen
D-64271 Darmstadt
Germany

Tel: (0180) 570-2000
Fax: (0180) 570-2222

Olympus America
4 Nevada Drive
Lake Success, NY 11042-1179, USA

Tel: (800) 446-5260
Fax: (516) 488-3973

Perkin Elmer Cetus
761 Main Ave,
Norwalk, CT 06859-0012, USA

Tel: (203) 762-1000
Fax: (203) 762-6000

Promega
2800 Woods Hollow Road
Madison, WI 53711-5399, USA

Tel: (800) 356-1970
Fax: (608) 277-2516

R & D Systems
614 Mckinley Place N.E.
Minneapolis, MN 55413, USA

Tel: (800) 343-7475
Fax: (612) 637-0424

Roche Boehringer Mannheim
Sandhofer Strasse 116,
68298 Mannheim, Germany

Tel: (0621) 759-8545

Vector Laboratories, Inc.
30 Ingold Road,
Burlingame, CA 94010, USA

Tel: (415) 697-3600
Fax: (415) 697-0339

Quantitation of T-cell
Cytokine Responses

2.5 Isolation of and Measuring the Function of Professional Phagocytes: Murine Macrophages

Leanne Peiser and Siamon Gordon
Sir William Dunn School of Pathology, University of Oxford, Oxford, UK

Richard Haworth
Glaxo Wellcome Research & Development, Ware, Herts, UK

◆◆

CONTENTS

Abbreviations

BCG	Bacille Calmette-Guérin
BMMφ	Bone marrow-derived macrophages
BP	Bacteriologic plastic
BPMφ	Biogel-elicited peritoneal macrophages
c.f.u.	Colony forming units
CR3	Complement receptor type 3
DC	Dendritic cell
ECM	Extracellular matrix
EDTA	Ethylenediamine tetraacetic acid
FCS	Foetal calf serum
g	gravity
HBSS	Hank's buffered salt solution
IgG	Immunoglobulin gamma
IL	Interleukin
INFγ	Interferon-gamma
LCM	L929 cell conditioned medium
LPS	Lipopolysaccharide
LTA	Lipoteichoic acid

METHODS IN MICROBIOLOGY, VOLUME 32
ISBN 0–12–521532–0

Copyright © Elsevier Science Ltd
All rights of reproduction in any form reserved

Measuring the Function of Professional Phagocytes

Mφ	Macrophage(s)
mAb	Monoclonal antibody
M-CSF	Macrophage colony stimulating factor
MHC	Major histocompatibility complex
NO	Nitric oxide
PBMC	Peripheral blood mononuclear cells
PBS	Phosphate buffered saline
PMN	Polymorphonuclear neutrophils
RBMMφ	Resident bone marrow-derived macrophages
SR-A	Scavenger receptor class A
TBAC	Tris-buffered ammonium chloride
TCP	Tissue culture plastic
TPMφ	Thioglycollate broth-elicited peritoneal macrophages
Tris	Tri(hydroxymethyl)aminomethane

◆◆◆◆◆◆ INTRODUCTION

Professional phagocytes can be divided into macrophages (Mφ) and polymorphonuclear leukocytes (PMN) (Gordon, 2001). Monocytes and macrophages (Mφ) are detected in most tissues throughout the body. Mφ are sometimes referred to as the 'dustbins' of the body because of their efficiency at recognizing and removing foreign and host-derived debris. However, they are more than a vehicle for the elimination of waste and play a pivotal role in diverse processes including tissue homeostasis, inflammation and development. Mφ are also key cells in immunity as they are essential in innate protection against pathogens and can regulate the acquired immune system through interactions with B and T cells.

When investigating Mφ function, it is important to remember that as a population of cells they are extremely heterogeneous. Membrane receptor expression, biosynthesis and metabolic responses vary greatly between populations and during migration and maturation. Indeed, this variety of Mφ phenotype has provided obstacles to the use of Mφ as targets for drug delivery (Gordon and Rabinowitz, 1989). In this chapter, we will discuss *in vitro* methods for the measurement of classic Mφ functions such as phagocytosis and cytokine production. Those seeking information on PMNs are referred elsewhere (Leijh *et al.*, 1986), although protocols reported here are easily adapted for use on neutrophils. Also, a wide range of techniques is available for assessing the role of Mφ *in vivo*, and some of these are described elsewhere in this volume.

◆◆◆◆◆◆ ISOLATION OF MΦ

All animals should be treated and handled according to the guidelines dictated by the relevant home office.

Peritoneal Mφ

The mouse peritoneal cavity is a convenient source of primary Mφ. This site provides high yields of cells from which Mφ can be purified via adhesion. In addition, different phenotypes of Mφ can be isolated depending on the activation status and the stimulus used, if any, to recruit the cells. The populations that can be isolated can be divided phenotypically into resident, elicited or activated cells and are collected following peritoneal lavage.

Resident peritoneal Mφ

Isolation of resident peritoneal Mφ

1. The mice are sacrificed using carbon dioxide inhalation and then pinned down with their ventral surface uppermost.
2. Sterilize the skin with 70% ethanol in water. Using fine scissors, make a lateral cut in the skin over the abdomen. Do not break the body wall at this stage.
3. Pull back the skin from the incision to reveal the shiny surface of the body wall. It is important to keep this area sterile during the whole procedure.
4. Using a 19-gauge needle, inject approximately 10 ml of sterile saline into the peritoneal cavity. The needle should be injected bevel uppermost into the caudal half of the cavity and care should be exercised not to puncture any organs. As the needle is removed there may be a small leakage of fluid, but omental fat will usually block further leakage.
5. Agitate the filled cavity by rubbing the tube of a sterile 19-gauge needle over the external body wall a few times.
6. Remove the fluid using a 19-gauge needle attached to a 10 ml syringe. The needle should be inserted bevel downwards into the cranial half of the cavity to avoid fat blockage during aspiration. It is unlikely that the whole 10 ml of fluid will be recovered and if the cavity fills up with blood the animal should be discarded as it may contaminate the resident population.
7. Pellet the cells via centrifugation and culture in bacteriologic- (BP) or tissue culture (TCP) plastic vessels as appropriate.

Thioglycollate broth-elicited Mφ

Intraperitoneal injection of sterile inflammatory agents is a useful method for isolating large numbers of Mφ for *in vitro* assays. Thioglycollate-elicited peritoneal Mφ (TPMφ) are recruited to the peritoneal cavity following injection of 1 ml Brewer's complete thioglycollate broth (Difco Laboratories) (Johnson *et al.*, 1978). The cells are harvested, as described above for resident cells, 4–5 days following injection. TPMφ ingest large amounts of the inflammatory agent which contains agar, but retain active endocytic and phagocytic capabilities upon isolation. The use of protease peptone as a stimulant produces cells with a phenotype similar to that of TPMφ, but with fewer vacuoles and lower cell yields. An important consideration when analysing Mφ function using TPMφ is that the thioglycollate often contains small amounts of lipopolysaccharide (LPS) (0.5 ng ml^{-1}), which may alter Mφ responsiveness in the subsequent assays.

Biogel polyacrylamide beads-elicited Mφ

The need to obtain large numbers of elicited Mφ has resulted in the testing of other inflammatory stimuli. The first reported use of polyacrylamide beads for this purpose was by Fauve et al. (1983). They injected beads into subcutaneous pouches created in the dorsal skin of mice. In this model, 10^7 phagocytic cells (60% Mφ and 40% PMN) could be recovered from the resulting 'granuloma'. In our laboratory, Biogel beads have been successfully used to elicit a high yield (10^7 cells per animal) of peritoneal Mφ. The most suitable size of bead is P100 (hydrated size 45–90 nm), which Mφ are unable to ingest or digest extracellularly.

Preparation of BPMφ

1. Biogel polyacrylamide beads (Biogel P-100 (fine), Bio-Rad laboratories) are washed in phosphate buffered saline (PBS) by repeated centrifugation and autoclaved before use. Remember to use pyrogen-free laboratory equipment and endotoxin-free PBS to prevent LPS contamination.
2. Inject 1 ml of a 2% (v/v) suspension of Biogel, into the peritoneal cavity of the animal.
3. Isolate the Mφ from the peritoneal cavity as described above for resident cells 4–5 days following the injection of Biogel.
4. To purify the Mφ via adherence the cells can be routinely plated in medium at the appropriate density (Table 1) on TCP or BP.
5. After incubation at 37°C for 60–90 min when plated on TCP or 24 h on BP, the non-adherent cells can be removed by washing the culture dishes five times with PBS. Under these conditions, the adherent monolayers consist of >90% Mφ, and viability is usually >97% by phase microscopy and trypan blue exclusion.

Biogel-elicited peritoneal Mφ (BPMφ) have a number of features that distinguish them phenotypically from TPMφ. For example, following incubation overnight in serum-containing medium on TCP, 50% of BPMφ will have become completely non-adherent and the remainder will have rounded up. In contrast, TPMφ will remain flattened and tightly adherent to the substratum (M. Stein, unpublished observations). In addition, the culture medium used and the presence of serum has profound effects on

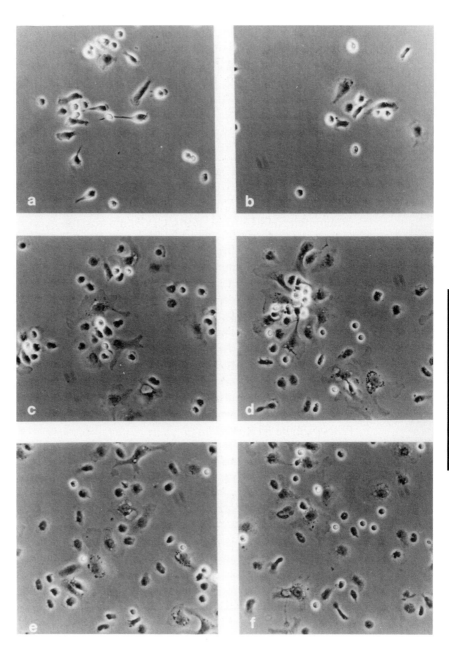

Figure 1. The effect of culture medium on BPMφ cultured on glass coverslips. Cells were harvested as described and allowed to adhere for 30 min in RPMI containing 10% FCS (a,b), in RPMI alone (c,d) or in Optimem (e,f). Cells were fixed in 2% paraformaldehyde and photographed under phase contrast. (a,c,e) Wild type cells; (b,d,f) cells from mice lacking SR-A. BPMφ spread more rapidly on this surface in media lacking serum (Optimem or RPMI). There is no discernible difference in the spreading morphology of wild-type cells and SR-A$^{-/-}$ Mφ. original mag. ×400.

the degree of spreading by BPMϕ (Fig. 1). Some differences in phenotype between these elicited and other Mϕ populations are listed in Table 1.

One of the most useful markers of murine Mϕ is defined by the monoclonal antibody (mAb) F4/80, which recognizes a 160 kDa glycoprotein on the surface of most mouse Mϕ populations (Austyn and Gordon, 1981; McKnight *et al.*, 1996). No function has yet been ascribed to F4/80, although it is a member of a growing family of EGF seven transmembrane molecules (Stacey *et al.*, 2000). Its expression is known to be down-regulated by INFγ and in response to Bacille Calmette-Guérin (BCG) infection (Ezekowitz *et al.*, 1981; Ezekowitz and Gordon, 1982) (Table 1).

The FA11 mAb recognizes macrosialin, the murine homologue of CD68, which is an endosomal marker for Mϕ and dendritic cells (DCs) (Rabinowitz and Gordon, 1991). CD68/macrosialin expression is a useful indicator of endocytic activity, and the data suggest that BPMϕ are less endocytic than TPMϕ. Therefore, BPMϕ may be useful in studies examining the entry and replication of facultative pathogens (e.g. *Mycobacterium tuberculosis*, *Mycobacterium leprae* and *Leishmania donovani*) within endosomal compartments. *In vivo*, elicited BPMϕ, largely unstimulated by lymphokines, are permissive host cells for the above-mentioned pathogens (Gordon, 1986). Therefore, BPMϕ, or other foreign body-elicited Mϕ may resemble cells recruited early to a focus of infection, may be appropriate populations for studies examining the regulation of Mϕ microbicidal activity. In addition, these cells respond well to cytokines *in vitro*, e.g. interleukin-4 (IL-4) and INFγ (Stein *et al.*, 1992). If sepharose, polystyrene or smaller polyacrylamide beads are used, then a higher percentage of PMNs will be recruited.

BCG-recruited Mϕ

BCG organisms (e.g. Pasteur strain) provide a suitable stimulus to recruit immunologically activated Mϕ to the peritoneal cavity (Ezekowitz *et al.*, 1981). BCG stocks are stored at −80°C and thawed immediately prior to use. The BCG organisms are resuspended in PBS and sonicated before

Table 1. The phenotype of different Mϕ isolated from the peritoneal cavity

	Resident	Thioglycollate	Biogel	BCG
Approx. total cellular yield per mouse (×10⁶)	7	21	17	10
% Mϕ	40	86	59	62
Adherence to TCP at 24 h	+	+	+/−	+
F4/80	++	+	+	+
Mannose receptor	++	++	++	+
Macrosialin	+	++	+	+
MHC II	−	+	+	+++
Respiratory burst	−	+	+	+
Constitutive NO production	−	−	−	+

use. Mice are inoculated with approximately 10^7 colony forming units (c.f.u.) in 0.2 ml PBS by intraperitoneal injection. Peritoneal Mφ are harvested by lavage, as described above, 4–6 days post-injection. Percoll gradients can be used at this stage in order to enrich the population for Mφ (Pertoft and Laurent, 1977).

These Mφ become activated *in vivo* under the influence of T-cell products, such as INFγ, and express high levels of cell surface major histocompatibility complex class II (MHC-II). In addition, they produce nitric oxide (NO) in serum-containing media in the absence of further stimulation, in contrast to the other elicited Mφ populations described above (Fig. 2). These cells are useful for investigating the activated Mφ response to bacterial cell products such as LPS and lipoteichoic acid (LTA), which can be added to the cells in culture.

Corynebacterium parvum-recruited cells

The use of inactivated *C. parvum* (also known as *Propionibacterium acnes*) provides a convenient, alternative method of recruiting activated Mφ. The use of this organism to recruit Mφ to the liver resulted in the cloning of IL-18, which induces the production of INFγ by T cells (Okamura *et al.*, 1995). The use of *C. parvum* avoids the need for using viable pathogenic organisms. Inactivated *C. parvum* whole cells can be purchased from RIBI Immunochemical Research Inc. The *C. parvum* is washed in non-pyrogenic

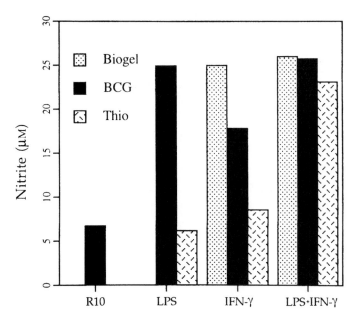

Figure 2. NO production by different peritoneal Mφ *in vitro*. TPMφ (thio), BPMφ (Biogel) and BCG-recruited cells were harvested as described in the text, and plated at 2×10^5 cells per well and treated with medium (R_{10}) alone, LPS (20 ng ml^{-1}) or IFNγ (100 units ml^{-1}) overnight. Nitrite levels were measured using the Griess reaction, as described in the text. All values shown are NMMA inhibitale.

saline twice, resuspended in PBS, and sonicated before use. Mice are inoculated with 500 µg in 0.2 ml PBS by intraperitoneal injection. Peritoneal Mφ are harvested by lavage 4–5 days post-injection.

Tissue Mφ

Resident Mφ can be isolated from a range of tissues and activated cells can be isolated from infected or inflamed organs using the enzymatic methods described below.

Spleen and thymus

Mφ can be isolated from the spleen and thymus as follows. The organs are removed from the mice and placed on ice in PBS until use. They are then digested in 0.05% collagenase (Roche-Boehringer Mannheim) and 0.002% DNase in RPMI 1640 at 37°C for 30 min, in the absence or serum. The organ fragments are then mechanically disrupted by vigorous pipetting and the suspension filtered through a cell strainer (Falcon, Becton Dickinson). The Mφ can then be purified via adhesion as described for BPMφ.

Note that in order to maintain Mφ cell integrity, it is important that sufficient digestion has taken place before mechanical disruption is used. It is worth noting that the Mφ in the spleen represent a heterogeneous population, and different approaches may be used to isolate the subpopulations.

Bone marrow

Both resident- (RBMMφ) and bone-marrow-derived (BMMφ) Mφ can be obtained from this tissue.

Resident bone marrow Mφ

Isolation of RBMMφ

1. Sacrifice the mice and sterilize the abdomen and hind legs with 70% ethanol in water.
2. The skin is dissected away from the abdomen and hind legs following a transverse cut through the skin of the abdomen.
3. Remove the muscles attaching the hind limb to the pelvis and those attaching the femur to the tibia using a pair of fine scissors.
4. Only when the femur is well exposed should the tibia be cut through just below the knee joint using strong scissors, and the femur freed from the mouse by cutting through the pelvis bone close to the hip joint.

(contd.)

---

5. The femurs can be stored in RPMI 1640 on ice until all the femurs are collected. Place the bones in a petri dish of 70% ethanol for 1 min to maintain sterility, before washing twice with PBS.
6. Next, each femur should be held firmly with forceps and, in a single motion, the expanded ends (epiphyses) cut off using strong scissors.
7. Using a 5-ml syringe attached to a 25-gauge needle, the bone marrow should be flushed out by forcing an RPMI solution containing 0.05% collagenase and 0.001% DNase down the central cavity. The bone will become white when all the marrow is expelled.
8. Resuspend the bone marrow plugs from two femurs in 10 ml of the same enzyme solution above and digest at 37°C with shaking for 1 h.
9. Add foetal calf serum (FCS) to a final concentration of 1% (v/v) to stop the digestion. At this stage the marrow plug fragments should no longer be visible, and a homogeneous suspension is obtained. Harvest and culture as appropriate.
10. Clusters of cells can be enriched by gravity sedimentation in RPMI containing 30% FCS or by use of a Ficoll-Hypaque cushion (Pharmacia) (Crocker and Gordon, 1985).
11. Wash purified cell clusters twice in RPMI by centrifugation at 100g for 10 min, suspended in RPMI containing 10% (v/v) FCS (R_{10}) and added to glass coverslips in TCP plates.
12. After 3 h incubation at 37°C, non-adherent cells are washed off with PBS, resulting in a population of adherent cells with the characteristic morphology of RBMMφ but contaminated with a varying population of monocytes and neutrophils.

Bone marrow-derived Mφ

The production of BMMφ is a simple method to obtain large numbers of primary non-activated Mφ. Typically one animal yields up to 2–6 × 10^7 Mφ. To obtain BMMφ, the femurs are flushed with PBS, with no enzymes added. The marrow plugs are disrupted mechanically by passage through a 19 gauge needle prior to centrifugation at 1000g for 5 min. The cells are then resuspended in RPMI-1640 containing 10 mM Hepes (N-[2-hydroxyethyl]piperazine-N^1-[2-ethanesulfonic acid]), 10% FCS and 15% (v/v) L-cell conditioned media (LCM) (Hume and Gordon, 1983) and plated into 15 cm BP dishes (two dishes per animal). Alternatively, recombinant Mφ colony stimulating factor (M-CSF) can replace the LCM. Fresh culture medium is added to the Mφ on day 3 and at day 6 all the medium is replaced with fresh R_{10} and 15% LCM. The cultures are routinely confluent after 7 days incubation, and cells can be harvested by incubating them with PBS containing 10 mM EDTA and 4 mg ml^{-1} Lidocaine-HCl for 10 min before removal of the loosened cells by pipetting. Although these cells represent mature Mφ they are proliferating, but can be used in a wide range of assays from phagocytosis to investigations of Mφ response.

Foetal liver

The foetal liver contains the richest source of Mɸ in the developing mouse. F4/80+ membrane processes of these cells interact extensively with developing haemopoietic cells, forming cell clusters *in vivo*. To investigate the interactions between erythroid cells and stromal Mɸ, isolation of haemopoietic cell clusters is recommended (Morris *et al.*, 1988).

Peripheral blood mononuclear cells

To obtain peripheral blood mononuclear cells (PBMCs), mice are sacrificed and bled by cardiac puncture into a heparinized syringe with a 25-gauge needle. The blood is diluted by adding an equal volume of 0.9% saline, and layered over a Nycoprep 1.077 Animal cushion (Nycomed Pharma AS). Cells are centrifuged at 1900 rpm (no brake) for 15 min. Mononuclear cells can then be collected from the interface between the plasma and the Nycoprep cushion. These cells are resuspended in tris-buffered ammonium chloride (TBAC) lysis buffer, which is made by mixing 0.15 M ammonium chloride and 0.17 M Tris at a ratio of 9:1 before adjusting the pH to 7.2 and filter sterilization. The red blood cells are lysed following 5 min incubation in TBAC buffer for 5 min at room temperature followed by three washes in RPMI 1640. The PBMCs are then resuspended in R_{10} before use.

Use of cell lines

A number of different cell lines can be used in assays for investigating the function of professional phagocytes (Ralph, 1986). Murine Mɸ-like cell lines include the widely available RAW 264 (Raschke *et al.*, 1978), J774 (Ralph *et al.*, 1975), and P388D1 (Koren *et al.*, 1975). These cells can be cultured in RPMI 1640 on either BP or TCP surfaces.

◆◆◆◆◆◆ CULTURE OF Mɸ

Following the isolation of primary Mɸ as outlined above, it is important to maintain the cells in culture under appropriate conditions, which will vary according to the functional tests required. For example, the substratum on which the cells are cultured may be important if adhesion studies are planned.

Substratum

Mɸ can be cultured in suspension using tissue culture vessels with a Teflon-coated surface (Nalgene). In contrast, Mɸ adhere firmly to TCP and BP. On BP, adherence is mediated via integrins, especially CR3 (complement receptor type 3), and can be readily detached using Lidocaine-HCl and EDTA as mentioned above for BMMɸ. However, on TCP, adherence

340

is mediated via integrins and SR-A (Mφ class A scavenger receptor), which is EDTA resistant, thus the cells are more difficult to detach and the viability of the Mφ should be examined after lifting. Incubation with Lidocaine-HCl and EDTA at 37°C for 10–30 min is effective in most cases (Rabinovitch and de Stefano, 1976), but in some instances Pronase may be used to remove the cells. Note that Lidocaine may be added directly to the culture medium as it is able to function in the presence of serum. Trypsin is ineffective at removing Mφ from TCP. All cells should be centrifuged and washed thoroughly with fresh medium before use.

Media and sera

Mφ are routinely cultured in a wide range of media in our laboratory, including RPMI 1640, modified Eagle's medium (MEM) and Dulbecco's modified Eagle's medium (DMEM) (Gibco). These media are supplemented with 2 mM glutamine, 50 IU ml^{-1} penicillin and 50 μg ml^{-1} streptomycin (Gibco). RPMI 1640 is also supplemented with 10 mM Hepes (pH 7.3). FCS (Sigma) is heat inactivated at 56°C for 30 min, filter sterilized through a 0.22 μM filter prior to use and used at 10% (v/v).

Primary Mφ can be cultured for a variable time under serum-free conditions. Specifically, in endocytic and phagocytic assays the use of OPTI-MEM (Gibco), which is a proprietary serum-free medium, has proven successful. However, it should be borne in mind that Mφ adhere to substrata under serum-free conditions by way of molecules that have not been fully identified, which may result in practical problems in harvesting the cells from the substrata prior to their use in assays.

◆◆◆◆◆◆ MEASURING Mφ FUNCTION

Adhesion phenotype

One of the key functions of a professional phagocyte is the ability to adhere both to other cells and to the extracellular matrix (ECM). In order to explore the cell surface molecules involved in this interaction, *in vitro* assays of adhesion of Mφ to artificial substrata (e.g. TCP) have been developed. These have provided a useful strategy for purifying Mφ from a mixed population (see above) and also for isolating reagents that interfere with this adhesion. For example, murine Mφ adhere to BP in the presence of serum in a divalent-cation-dependent fashion. Used as a screening strategy to develop novel mAbs, the ability of a hybridoma supernatant to inhibit this adhesion produced the 5C6 mAb (Rosen and Gordon, 1987). This antibody recognizes CR3, a leukocyte integrin. Subsequent studies have shown that this mAb has an *in vivo* role in adhesion, since 5C6 is able to block the adhesion of Mφ to inflamed endothelium and recruitment to immunologically non-specific stimuli (Rosen and Law, 1989; Rosen et al. 1989; Rosen, 1990).

In order to use this strategy to identify further molecules involved in adhesion, the investigator can vary (a) the phenotype of the Mφ added, (b) the presence or absence of chelators or other chemicals, or (c) the character of the substratum. For example, Mφ adhere to TCP in the absence of divalent cations (Fig. 3). The use of an adhesion assay, as outlined below, allowed the identification of SR-A (Fraser *et al.* 1993), which had no known adhesive function prior to these studies.

Adhesion assay

1. Plate Mφ at a density of 3×10^5 cells per well of a 96-well plate in the presence of various mAbs and chelators.
2. Incubate the plates at 4°C for 30 min, then at 37°C for 90 min before washing to remove non-adherent cells.
3. Fix the remaining adherent cells in methanol, and stain with 40% giemsa for 1 h.
4. Quantify the level of adhesion by solubilizing the dye in methanol and reading the optical density at 450 nm.

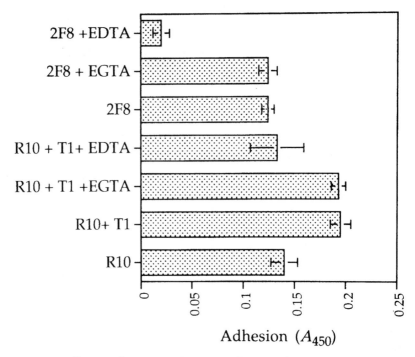

Adhesion (A_{450})

Figure 3. Adhesion phenotype of BPMφ. Adhesion of BPMφ to FCS coated TCP. Cells were plated at 3×10^5 Mφ per well of a 96-well plate in the presence of various mAb and/or chelators. Adhesion (mean ± SD) is represented as the absorbence at 450 nm (A_{450}), and is the result of quadruplicate wells. Significant adhesion occurs in the presence of an isotype matched control antibody (T1) in the presence of EDTA. In contrast, in the presence of 2F8 which blocks adhesion via SR-A. Significant inhibition of adhesion occurs. Note that EDTA needs to be present to observe this activity, since other mechanisms of adhesion (integrins) need to be inactivated.

Antigen expression

Changes in levels of expression of either cell surface or intracellular antigens (Table 2) can provide useful information regarding activation status or endocytic activity of Mφ. A highly sensitive method of analysis of individual cells is provided by immunostaining (Fig. 4) and flow cytometry. Immunohistochemistry is described elsewhere in this volume, and can be applied with success to defining resident and recruited Mφ populations *in vivo*.

Indirect immunofluorescent staining of cultured Mφ

1. Detach adherent cells from the culture dishes, if required, and fix the cells in 4% (w/v) Paraformaldehyde[a,b] in PBS buffered to pH 7.0 with 1M HEPES for 10 min on ice. Approximately $1-5 \times 10^6$ cells should be stained in analysing results by flow cytometry.
2. Harvest the cells by centrifugation and resuspend in a blocking solution containing 10% Normal serum[c,d] of the species of secondary antibody diluted in PBS. If staining an intracellular antigen, add permeabilization agents (0.25% (w/v) saponin or 0.1% (v/v) Triton in PBS) at this point to the blocking solution and keep present in all subsequent incubation steps.
3. After 30 min, resuspend the cells in blocking solution containing the correct dilution of primary Ab. Use antibodies at the manufacturer's recommended concentration or at 10 μg ml^{-1}. Incubate for 1 h.
4. Wash the cells three times with blocking solution before labelling with the secondary antibody diluted in the blocking solution.
5. After washing the cells three times with PBS, analyse on a flow cytometer or by microscopy if the cells were plated on a coverslip.

[a] Other fixatives, like acetone, may be used for microscopic analysis, but we recommend paraformaldehyde. Paraformaldehyde is best to use for flow cytometry.
[b] If staining unfixed cells then perform the staining at 4°C and begin the protocol from step 2.
[c] This is to block non-specific IgG binding sites.
[d] Use the normal serum of the primary antibody if doing direct immunofluorescent labelling.

Endocytosis

Endocytosis in Mφ can be mediated by ubiquitously expressed receptors, such as the transferrin receptor, or Mφ-restricted receptors including the MR and SR-A (Table 3; Fig. 4). The expression of many Mφ-specific receptors is regulated by the stage of differentiation of the Mφ and its activation state. Regulation may affect the levels of receptor expression, in addition to the rate of receptor trafficking and its processing in the endocytic path-

Table 2. Membrane antigens and the corresponding antibodies that can be used to define Mφ distribution and heterogeneity in murine tissues

Marker	Tissue distribution	Clone	Supplier
Macrosialin	Mφ and dendritic cell endosome membrane	FA-11	Rabinowitz and Gordon, 1991
CD11b (CR3)	PMNs, Mφ and NK cells	5C6	Serotec
CD14	PMN, monocytes, Mφ	rmC5-3	BD Pharmingen
MHC-II^d	Up-regulated on activated Mφ	TIB120	ATCC
FcR	Mφ	2.4G2	BD Pharmingen
F4/80	Mature Mφ	F4/80	Serotec
SR-A	Mature Mφ and hepatic endothelium	2F8	Serotec
Sialoadhesin	Stromal Mφ	3D6.112	Serotec

way. Ubiquitously expressed receptors, like the transferrin receptor, may give information on the basal rate of endocytosis. When testing the endocytic function of Mφ choose a well-documented receptor, like the transferrin- and LDL-receptor or MR. If testing the capabilities of a novel receptor, comparison with known receptors can give a wealth of information. Assays can be readily adapted to measure binding, internalization and degradation of ligand. Receptor-specific binding will be saturable, i.e. reaches a plateau, when background is subtracted.

When investigating a particular receptor the appropriate cognate ligand must be chosen. There are a large number of commercially available labelled ligands, but coupling of fluorochromes to proteins is quick and easy (see protocol below). Specificity can be shown by competition with saturating amounts of unlabelled ligand, which controls for alterations caused by the labelling procedure. A suitable ligand should not be degraded too quickly once internalized, especially when loading the late endosomes and lysosomes. Dextran is a good marker as its poly-(α-D-1-6-glucose) linkages make it resistant to degradation. See Table 3 for examples of commonly used ligands for analysis of the endocytic pathway. Suitable ligands for measuring pinocytosis, such as lucifer yellow, must not be recognized by any Mφ receptors. Horseradish peroxidase is commonly used as a pinocytic marker, however, it is not ideal for Mφ as it can have mannose residues which are recognized by the MR. Selected fluorochromes undergo pH-dependent shifts in their excitation and/or emission spectra so the acidification of endocytic compartments can be monitored.

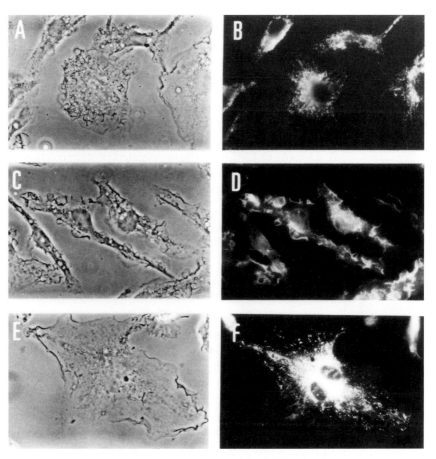

Figure 4. Immunofluorescent staining and DiI-AcLDL labelling of murine BMMφ. Murine BMMφ were plated in a 24-well plate containing 13 mm glass coverslips at a density of 2×10^5 cells per well.
(A–D) Cells were fixed in 4% paraformaldehyde, permeabilized and stained with anti-macrosialin (FA-11) (A+B) or anti F4/80 (F4/80) (C+D) followed by FITC-conjugated goat anti-rat IgG. Cells were viewed by fluorescence microscopy and photographs of the same field taken under phase contrast (A+C) or fluorescence (B+D). The staining highlights the predominantly intracellular localization of macrosialin compared to the cell surface expression of the F4/80 antigen. (E–F) Cells were labelled by incubation with DiI-AcLDL at a concentration of 5 µg ml⁻¹ for 3 h at 37°C, washed 4 times with PBS and subsequently fixed in 4% paraformaldehyde. Coverslips were viewed by fluorescence microscopy and photographs taken under phase contrast (E) and fluorescent (F) illumination.

Fluorescent labelling of proteins or particles

Coupling FITC^a to proteins

1. Mix 5.8 ml of 5.3% Na_2CO_3 with 10 ml 4.2% $NaHCO_3$.
2. Make the bicarbonate buffer by adding 1 volume of the mixture above to 9 volumes of 0.15M NaCl and adjust the pH to 9.5.

(contd.)

3. In separate tubes, dissolve the FITC and the protein[b,c] to be labelled in the bicarbonate buffer, at final concentrations of 1 mg ml^{-1} and 5 mg ml^{-1} respectively.
4. Mix the protein and FITC together at 0.3 ml FITC for each ml of protein.
5. Incubate in the dark for 2 h at room temperature.
6. Equilibrate a G-50 or G-25 sephadex column (Sigma) with PBS and run the FITC/protein mixture over the column.
7. Elute the FITC-conjugated protein with PBS.
8. Determine the OD of the conjugated protein fractions at 280 nm and 495 nm. The ratio of OD_{495}/OD_{280} should be approximately 1.
9. Determine the conjugated protein concentration using the following formula:

$$\text{Protein conc. (mg ml}^{-1}) = \frac{OD_{280} - (OD_{495} \times 0.35)}{1.4}$$

Coupling of Texas Red[a] to proteins

1. Prepare the bicarbonate buffer as above except adjust the pH to 9.
2. Dissolve the protein in the bicarbonate as above and add 1 mg Texas Red sulphonyl chloride (Molecular Probes) for every 10 mg protein in 1 drop of dimethylformamide.
3. Incubate for 1 h in the dark at room temperature.
4. Separate the conjugated protein on a sepharose column. Determine the optical density of the pooled fraction at 596 nm and 280 nm. The ratio of OD_{596}/OD_{280} should be approximately 0.8.

[a] There are many FITC and Texas Red derivatives available, use the most suitable for the individual requirements of the assay.
[b] Dialyse or exchange buffer by running a desalting column.
[c] Use zymosan at 10^9 particles per ml and live bacteria at 10^7 particles per ml.

Quantitation of Mφ endocytic function

Loading Mφ with a single tracer

1. Remove the culture medium from the cells and wash them twice in PBS.
2. Add OPTIMEM-I containing 100 µg ml^{-1} fluorescently labelled tracer to the wells.
3. Incubate the Mφ for 1 h at 37°C. Keep a control sample on ice.[b]
4. Stop the endocytic uptake by placing the Mφ on ice and wash the Mφ at least four times in ice-cold PBS.
5. If measuring the amount for uptake by flow cytometry, detach and fix the cells or if measuring the fluorescence on a fluorimetric plate reader, follow the steps below.

(contd.)

6. Lyse the cells with 1% Triton-X100 in 10 mM Tris buffer (pH 7.5). Incubate on ice for 30 min.
7. Scrape the Mφ from the bottom of the dish and transfer the lysed cells to an appropriate vessel to read on a plate reader.
8. Remove an aliquot of supernatant to determine the protein concentration and express the result as a function of the protein concentration or number of cells depending on analysis.

Dual endocytic tracer loading of Mφ

1. Perform Steps 1–3.
2. Remove the first tracer and wash the cells well in warmed RPMI (or the usual culture medium for the Mφ).
3. Chase the tracer into lysosomes, by incubating the Mφ overnight at 37°C.
4. Remove the culture medium and wash the Mφ twice in PBS.
5. Add culture medium containing 100 μg ml^{-1} of the second fluorescent-labelled tracer.[c] Place a control sample on ice.
6. Load the early endocytic compartments by incubating the Mφ at 37°C for 10–15 min.[d]
7. Remove the second tracer and cool the Mφ quickly by placing them on ice before washing with ice-cold PBS.
8. Follow steps 5–8 above remembering to take fluorometric readings for both the fluorochromes.

[a] This assay is easily adapted for analyses by flow cytometry or a plate reader.
[b] This controls for the non-specific sticking of the tracer to the extracellular surface of the cells and the tissue culture dishes. Subtract this value, after analysis, for a correct measurement of total endocytic uptake.
[c] This tracer should be labelled with a different fluorochrome than that used for the first tracer.
[d] Mφ are very endocytic cells so incubation for longer times will start to load later endocytic compartments as well as the earlier ones.

Phagocytosis

Commonly, phagocytic assays involve the addition of particles to Mφ followed by microscopic analysis of the number of particles bound and internalized by a cell. This type of analysis is time-consuming as large numbers of cells have to be counted manually to obtain statistically significant results. Therefore, we suggest adapting the assays so that the results may be analysed on a plate reader or flow cytometer which can collect information on large numbers of cells.

There are two types of basic assay: the first determines the number of particles associated with the Mφ, while the other monitors decreasing numbers of particles in the extracellular medium. Either assay is acceptable, but we will only discuss the former. Uptake assays can be adapted to measure cellular responses, such as the respiratory burst, by appropriate bulk or single cell methods (Baorto *et al.*, 1997) and to determine the survival or killing of ingested live organisms (see below). Appropriate safety precautions must be taken in handling living micro-organisms in all procedures.

The ligands expressed on a chosen particle will determine the receptors used for ingestion of that particle so must be appropriately chosen if investigating a particular receptor. Complex ligands, like bacteria, may be recognized by more than one receptor. Bacteria are easily fluoresceinated (see above) or some are available commercially (Table 3). If not investigating phagocytosis mediated by a particular receptor then complex ligands can be used. Latex beads, the receptors for which are unknown, are readily taken up and are suitable particles for phagocytosis. They are available in a wide range of sizes and can be coated by absorption or, in the case of carboxylated polystyrene latex beads, can be coupled directly to protein ligands to target them to specific receptors; some examples used previously are mannose BSA and lipoarabinomannan. However, even apparently single ligands may also be recognized by multiple receptors.

Besides latex beads, zymosan, derived from the cell wall of *Saccharomyces cerevisiae*, is a commonly used particle. It is highly

Table 3. Commonly used endocytic tracers and phagocytic particles

Probe	Receptor	Supplier
Texas Red Dextran (70000MW)	unknown	Molecular probes
FITC holo-transferrin	Transferrin receptor	Molecular Probes
Di-I LDL	LDL receptor	Biogenesis
Di-I acetylated LDL	SR-A, CD36, MARCO	Biogenesis
HRP	mannose receptor on Mφ or fluid phase in other cell types	Sigma
Lucifer Yellow	fluid phase	Sigma
Mannosylated BSA	mannose receptor	E-Y Lab
Latex Beads (polystyrene; with or without carboxylation)	unknown	Polysciences
Sheep erythrocytes	CR3 if coated with complement FcR if coated with IgG	Diamedix, Miami, FL
FITC-*E.coli* Bioparticles	multiple	Molecular Probes
FITC-*S.aureus* Bioparticles	multiple	Molecular Probes
Zymosan	β-glucan receptor	Sigma

mannosylated and is recognized by a number of receptors including β-glucan receptors (Brown and Gordon, 2001), CR3 and mannose receptors with or without opsonization. (It readily activates the alternative pathway of complement.) It is commercially available, though easy to prepare and label with fluorochromes. Zymosan should be boiled before use to destroy contaminating phospholipases. Erythrocytes coated with opsonins are widely used to analyse the function of opsonic receptors. Smaller particles may be taken up by macropinocytosis so when using latex beads ensure that the size used is larger than 1 μm in diameter and test the ability of phagocytic inhibitors on particle uptake. Inhibitors of ingestion, like cytochalasin B and D, and those that block ligand-binding should always be used as controls for phagocytosis.

Some ligands require opsonization by complement and antibodies. Mφ themselves may also produce opsonins like complement and fibronectin that could potentially influence uptake. The presence of serum can opsonize particles, so unless analysing general phagocytosis, use a serum-free protein-containing medium. If analysing specific opsonic receptors, coat the particles with the opsonin before the assay. Bacteria are easily opsonized by incubating them in an appropriate serum for 30 min at 37°C. Complement is only present in fresh serum and is destroyed by heat inactivation. Specific IgM and complement target CR3, but beware of IgG contamination of the IgM. IgG-coating targets the Fc receptors. Polyclonal Abs can be raised or where available, monoclonal Abs against erythrocyte antigens or hapten, for example, used with an appropriate isotype matched Ab control.

The differentiation of intracellular particles and those bound to the extracellular surface is crucial in any phagocytic assay and there are numerous modifications to existing methods available for this purpose. First, fluorescence on any extracellular particles can be quenched with appropriate agents such as ethidium bromide, crystal violet and trypan blue. It is not easy to control for total quenching of the extracellular fluorescence. In addition, the quenching agent must not be cell permeable. An alternative approach is to cleave or lyse the bacteria from the extracellular surface, for example, lysostaphin can lyse *S. aureus* and lysozyme can lyse *Micrococcus lysodeikticus*. Erythrocytes are easily lysed by brief osmotic shock (water or hypotonic solutions). Lastly, immunofluorescent techniques can be used to distinguish intra- and extracellular bacteria, with only external bacteria detected by antibodies. The distribution of the bacteria, with respect to the numbers found inside and bound to the cell, can be obtained by comparison between antibody staining of permeabilized and unpermeabilized cells.

Mφ are highly professional phagocytic cells and particle ingestion occurs rapidly. Generally, incubation times range between 10 min and 1 h. However, the kinetics of uptake should be determined by performing a time-course experiment before embarking on these assays. At time zero there should be no uptake and also at 4°C as the membrane is not fluid enough to mediate uptake. Following any uptake assay, the Mφ should be quickly cooled to 4°C to stop any further internalization and the rest of the

349

protocol should be performed in the cold. The optimal dose of particles should always be determined, especially when using live and virulent bacteria as too many bacteria may lyse or kill the Mφ. An initial particle to Mφ ratio of between 1:1 and 20:1 is recommended. The rate of ingestion should reach zero-order kinetics with increasing dose and is an important test of any assay method.

Particle contact with the Mφ may be enhanced by centrifuging them directly onto the cells, in special holders available commercially. If performing the assay on non-adherent cells, tumble the bacteria and Mφ together for optimal contact. The protocols detailed below are for adherent populations, although they are easily adapted for non-adherent assays.

Phagocytic uptake of dead bacteria or inert particles by Mφ

1. Remove the culture medium from the cells and wash twice in PBS to remove any non-adherent cells.
2. Add culture medium to the cells containing the phagocytic particle at the appropriate dose.
3. Centrifuge the particle onto the Mφ and incubate the cells at 37°C for 1 h.
4. Wash the cells well in ice-cold PBS to remove as many extracellular particles as possible before detaching and fixing the cells in 4% paraformaldehyde in PBS buffered to pH 7 using 1 m HEPES.
5. Stain the extracellular particles with the anti-particle antibody[a] and analyse the cells by microscopy or flow cytometry.

[a] Remember not to permeabilize the Mφ.

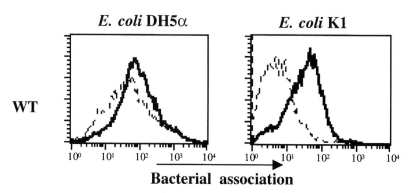

Bacterial association

Figure 5. The measurement of *E. coli* association with BMMφ by flow cytometry. BMMφ were incubated with paraformaldehyde-fixed fluorescently labelled *E. coli* DH5α or *E. coli* K1 (60 bacteria per Mφ) in the presence or absence of Poly I. —— depicts the fluorescence obtained by flow cytometry for the Mφ populations incubated with *E. coli* ; – – – – , Mφ populations incubated with *E. coli* in the presence of Poly I, a scavenger receptor inhibitor.

Ingestion of live bacteria

The association and ingestion of bacteria by Mφ is readily measured by colony assay. Important considerations are the 'stickiness' of the bacteria to the plastic surface, the cytotoxic effects of the organism on the Mφ, the ability of the cells to control the infection and the growth of the bacteria during the assay. All these factors must be controlled during the assay.

Measurement of bacterial association with Mφ using a colony assay

1. Seed Mφ onto 96-well plates at a density of 1×10^5 Mφ per well.
2. Wash the cells three times in Hank's buffered salt solution (HBSS) before use and add the appropriate culture medium with or without serum depending on particular assay conditions.
3. Harvest the bacteria from the agar plates or broth and resuspend in PBS. To remove large aggregates, bacteria can be centrifuged. Determine the concentration of bacteria and confirm by plating aliquots on solid media and counting the number of c.f.u.
4. Add the organisms to the cells and incubate at 37°C for a predetermined period. Remember to include the following controls: (a) for each assay condition, some wells containing bacteria and culture medium in the absence of Mφ are required to control for bacterial adhesion to the plastic; and (b) Mφ and bacteria incubated together for the length of the assay before lysing the total contents of the well with saponin (1% final concentration) to control for Mφ killing and bacterial growth.
5. Following the incubation wash all the wells, except for the bacterial growth controls, four times in HBSS to remove all organisms not associated with the cells.
6. Lyse the cells with saponin (1% final concentration). It is important to test that the bacteria are resistant to treatment with 1% saponin.
7. Estimate the total c.f.u. associated with the cells by making dilutions of the bacteria using PBS and count at least three dilutions for each well. The number of cell-associated bacteria can be determined after correction for attachment to exposed plastic by subtraction of the bacterial adhesion control wells from the experimental association wells as described previously (Virji *et al.*, 1991).

The number of internalized bacteria can be determined by including antibiotics to kill extracellular bacteria. Gentamicin is commonly used as it is considered to be impermeable, however, it has been suggested that it may be able to enter Mφ, and the antibiotic may not kill extracellular bacteria that are in close association with the cells. To measure internalized organisms only, incubate the Mφ as above with the bacteria and then the cells to remove most extracellular organisms. Culture medium containing 200 μg ml⁻¹ gentamicin is then added to each well for 0.5–1.5 h to eliminate the remaining extracellular bacteria. A control for gentamicin killing of bacteria must be included in each assay by incubating bacteria alone with the antibiotic. Following gentamicin treatment, the Mφ are washed in HBSS, lysed with saponin and the c.f.u. estimated as before. Adapting this assay to include a time course can give information on intracellular Mφ killing.

Secreted products

An important functional characteristic of Mφ is their conversion, under appropriate stimulation, from the resting to the activated state. Activated Mφ have increased numbers of lysosomal granules, more mitochondria and a greater capacity to phagocytose opsonized particles. In addition, the activated cell produces higher levels of certain cytokines (e.g. TNF-α) and has an increased capacity to generate superoxide anions. Assays for measuring superoxide, NO and cytokines *in vitro* are described below.

Superoxide

This assay provides an easy and convenient method for estimating microbicidal and cytocidal potential. Other assays of the respiratory burst include hydrogen peroxide release, chemiluminescence and fluorescence which are detailed elsewhere (Root *et al.*, 1975; Thrush *et al.*, 1978). The release of superoxide from murine Mφ is tightly regulated, and therefore freshly isolated cells produce negligible levels of superoxide in the absence of further stimulation and in our experience BMMφ do not release superoxide even after PMA stimulation.

Measurement of superoxide release

1. Plate Mφ in 24-well dishes. Suitable negative controls include a cell-free blank and wells containing superoxide dismutase at 30 µg ml^{-1}. Positive controls should include wells containing elicited cells stimulated with phorbol 12-myristate 13-acetate (PMA) (Sigma) at 10–100 ng ml^{-1} or zymosan at 100 µg ml^{-1}.
2. Wash the adherent cells with PBS and incubate with 450 µl of reaction mixture (HBSS, 80 µM ferricytochrome C (Sigma type IV), 2 mM sodium azide and 10 mM sodium phosphate buffer, pH 7.4) for 5 min at 37°C.
3. Add 50 µl HBSS containing the stimulant and incubate at 37°C for 1 h.
4. Remove 100 µl of the supernatant from each well and read the A$_{550}$ against a reaction mix cell blank.

Nitric oxide

Nitric oxide (NO) is a highly reactive molecule that mediates cytotoxic effects on micro-organisms and tumour cells (Saito and Nakano, 1996).

NO is an important player in innate immunity as a mediator of Mφ cytotoxicity against intracellular pathogens (Nathan and Shiloh, 2000). For example, the induction of NO production following BCG infection has been known for some time (Stuehr and Marletta, 1987). Because NO is rapidly converted to nitrite in the presence of oxygen, the secretory activity of cells can be estimated by determining nitrite concentrations by the colorimetric Griess reaction.

Colorimetric Griess reaction

1. Seed Mφ in 96-well plates at 1×10^5 cells per well and wash twice with PBS before use.
2. Add stimuli, e.g. LPS with or without INFγ at 50–100 U ml^{-1} (Serotec). For each stimulus set up a negative control with N-monomethyl arginine (NMMA) (Sigma) at the same time.
3. After incubation at 37°C, remove 50 μl of culture supernatant and add it to 50 μl Griess reagent (a 1:1 dilution of 1% (w/v) naphthylethylenediamine diHCl (Sigma) in distilled water and 1% (w/v) sulfanilamide in 5% (v/v) phosphoric acid (Sigma)). Set up sodium nitrite doubling dilutions for a standard curve starting with 1 mM nitrite.
4. Measure the absorbence measured at 550 nm and express the results as NMMA inhibitable accumulation of nitrite per 10^6 cells.

Cytokines

Murine Mφ secrete a wide range of cytokines *in vitro* and *in vivo*. It is possible to assay cytokine concentrations both from serum and from culture supernatant. Cytokines may be assayed by use of bioassay, enzyme-linked immunosorbent assays (ELISA) or by intracellular cytokine staining and flow cytometry. ELISA and intracellular cytokine assays are described elsewhere is this volume and in detail on the BD Pharmingen Website (www.bdbiosciences.com). A protocol for the measurement of bioactive TNF-α is given below as an example of a bioassay, where L929 cells are target cells susceptible to lysis when TNF-α reaches a critical concentration.

TNF-α bioassay

L929 cell culture

1. Grow L929 cells in Eagle's minimum essential medium (EMEM) or DMEM containing 24 mM HEPES 5% FCS at pH 7.4. It is important to split cultures frequently from a non-confluent status and change the medium every 3–4 days.

(contd.)

---

2. Use PBS to wash the cultures and incubate with 0.01% trypsin at 10 mM EDTA in PBS to detach the cells before harvesting using centrifugation.

L929 cytotoxic assay

3. Detach the L929 cells from a semi-confluent status and resuspend in assay medium at 4×10^4 cells in 100 μl medium in all wells of a 96-well plate except row 1, A–D (blanks). Incubate the cells for 18–20 h at 35°C in 5% CO_2.

4. Aspirate the culture medium and replace it in all wells with 100 μl assay medium, which contains culture medium supplemented with penicillin and streptomycin. Also include 1 μg ml⁻¹ actinomycin D.

5. Add in wells A2–A11, 50 μl of the samples to be tested and 50 μl of medium containing 2 μg ml⁻¹ actinomycin D. Add 50 μl of recombinant TNF-α 10 ng ml⁻¹ (Serotec) to well A12. Double dilute down the plate with an eight-channel pipette from row A to H and discard the last 100 μl.

◆◆◆◆◆◆ CONCLUSION

The assays described here will enable investigators to isolate various Mφ populations from the mouse and measure their functions in a range of assays. The choice of populations is an important part of the experimental design. Table 1 shows that the cellular phenotype can vary widely according to stimulus, even when the cells are isolated from the same site. Cell surface receptors expressed on different Mφ populations can be recognized using a variety of different surface markers (Table 2).

The mouse represents an important source of primary cells for use in studying the cellular response to micro-organisms. The development of transgenic and gene knockout mice has provided new tools for the study of the role of, for example, cell surface receptors or cytokines in the binding, uptake and killing of microbial pathogens. In addition, rapid progress is now being made in elucidating the molecular biology underlying many aspects of function. Further investigation using new tools will hopefully enhance our understanding and characterization of the host–pathogen relationship.

References

Austyn, J. M. and Gordon, S. (1981). F4/80, a monoclonal antibody directed specifically against mouse macrophages. *Eur. J. Immunol.* **11**, 805–815.

Baorto, D. M., Gao, Z., Malaviya, R., Dustin, M. L. *et al.* (1997). Survival of FimH-expressing enterobacteria in macrophages relies on glycolipid traffic. *Nature* **389**, 636–639.

Brown, G. and Gordon, S. (2001). A new receptor for β-glucans. *Nature* **413**, 36–37.

Crocker, P. R. and Gordon, S. (1985). Isolation and characterisation of resident stromal macrophages and hematopoietic cell clusters from mouse bone marrow. *J. Exp. Med.* **162**, 993–1014.

Ezekowitz, R. A. B. and Gordon, S. (1982). Down regulation of mannosy receptor-mediated endocytosis and antigen F4/80 in Bacillus-Calmette-Guérin-activated mouse macrophages. Role of T lymphocytes and cytokines. *J. Exp. Med.* **155**, 1623–1637.

Ezekowitz, R. A. B., Austyn, J., Stahl, P. and Gordon, S. (1981). Surface properties of Bacillus-Calmette-Guérin-activated mouse macrophages. Reduced expression of mannose-specific endocytosis, Fc receptors, and antigen F4/80 accompanies induction of Ia. *J. Exp. Med.* **154**, 60–75.

Fauve, R. M., Jusforgues, H. and Hevin, B. (1983). Maintenance of granuloma macrophages in serum-free medium. *J. Immunol. Methods.* **64**, 345–351.

Fraser, I., Hughes, D. and Gordon S. (1993). Divalent cation-independent macrophage adhesion inhibited by monoclonal antibody to murine scavenger receptor. *Nature* **364**, 343–346.

Gordon, S. (1986). Biology of macrophages. *J. Cell Sci. Suppl.* **4**, 267–286.

Gordon, S. (2001). Chapter 9: Mononuclear phagocytes in immune defence. In: *Immunology* (eds Roitt I., Brostoff J. and Male D.) pp. 1–13. Mosby, London.

Gordon, S. and Rabinowitz, S. (1989). Macrophages as targets for drug delivery. *Adv. Drug Delivery Rev.* **4**, 27–47.

Hume, D. A. and Gordon, S. (1983). Optimal conditions for proliferation of bone marrow-derived mouse macrophages in culture: the roles of CSF-1, serum, Ca2+, and adherence. *J. Cell Physiol.* **117**, 189–194.

Johnson, R. B. J., Godzik, C. A. and Cohn, Z. A. (1978). Increased superoxide anion production by immunologically activated and chemically elicited macrophages. *J. Exp. Med.* **148**, 115–127.

Koren, H. S., Handwerger, B. S. and Wunderlich, J. R. (1975). Identification of macrophage-like characteristics in a cultured murine tumor line. *J. Immunol.* **114**, 894–897.

Leijh, P. C. J., Van Furth, R. and Van Zwet, T. L. (1986). *In vitro* determination of phagocytosis and intracellular killing by polymorphonuclear and mononuclear phagocytes. In: *Handbook of Experimental Immunology* (D. M. Weir, Ed.) pp. 46.1–46.21. Blackwell Scientific, Oxford.

McKnight, A. J., MacFarlane, A. J., Dri, P., Turley, L. and Gordon, S. (1996). Molecular cloning of F4/80, a murine macrophage-restricted cell-surface glycoprotein with homology to the G-protein linked transmembrane 7 hormone receptor family. *J. Biol. Chem.* **271**, 486–489.

Morris, L., Crocker, P. R. and Gordon, S. (1988). Murine fetal liver macrophages bind developing erythroblasts by a divalent cation-dependent hemagglutinin. *J. Cell Biol.* **106**, 649–656.

Nathan, C. and Shiloh, C. U. (2000). Reactive oxygen and nitrogen intermediates in their relationship between mammalian hosts and microbial pathogens. *Proc. Natl. Acad. Sci. USA* **97**, 8841–8848.

Okamura, H., Tsutsui, H., Komatsu, T. and Kurimoto, M. (1995). Cloning of a new cytokine that induces INF-γ production by T-cells. *Nature* **378**, 88–91.

Pertoft, H. and Laurent, T. C. (1977). Isopycnic separation of cells and cell organelles by centrifugation in modified colloidal silica gradients. In: *Methods of Cell Separation* (N. Catsimpoolas, Ed.), p. 25. Plenum Press, New York.

Rabinovitch, M. and de Stefano, M. J. (1976). Cell shape changes induced by cationic anesthetics. *J. Exp. Med.* **143**, 290–304.

Rabinowitz, S. and Gordon, S. (1991). Macrosialin, a macrophage-restricted membrane sialoprotein differentially glycosylated in response to inflammatory stimuli. *J. Exp. Med.* **174**, 827–836.

Ralph, P. (1986). Macrophage cell lines. In: *Handbook of Experimental Immunology* (D.M. Weir, Ed.), pp 45.1–45.16. Blackwell Scientific, Oxford.

Ralph, P., Prichard, J. and Cohn, M. (1975). Reticulum cell sarcoma: an effector cell in antibody-dependent cell-mediated immunity. *J. Immunol.* **114**, 898–905.

Raschke, W. C., Baird, S., Ralph, P. and Nakoinz, I. (1978). Functional macrophage cell lines transformed by Abelson leukemia virus. *Cell* **15**, 261–267.

Root, R. K., Metcalf, J., Oshino, N. and Chance, B. (1975). Hydrogen peroxide release from human granulocytes during phagocytosis. (I) Documentation, quantitation, and some regulating factors. *J. Clin. Invest.* **55**, 945–955.

Rosen, H. (1990). Role of CR3 in induced myelomonocytic recruitment: insights from *in vivo* monoclonal antibody studies in the mouse. *J. Leuk. Biol.* **48**, 465–469.

Rosen H. and Gordon S. (1987). Monoclonal antibody to the murine type 3 complement receptor inhibits adhesion of myelomonocytic cells *in vitro* and inflammatory cell recruitment *in vivo*. *J. Exp. Med.* **166**, 1685–1701.

Rosen, H. and Law, S. K. A. (1989). The leukocyte cell surface receptor(s) for the iC3b product of complement. *Curr. Topics Microbiol. Immunol.* **153**, 99–122.

Rosen H., Milon G. and Gordon S. (1989). Antibody to the murine type 3 complement receptor inhibits T lymphocyte-dependent recruitment of myelomonocytic cells *in vivo*. *J Exp Med* **169**, 535–548.

Saito, S. and Nakano, M. (1996). Nitric oxide production by peritoneal macrophages of *Mycobacterium bovis* BCG infected or non-infected mice: regulatory roles of T lymphocytes and cytokines. *J. Leuk. Biol.* **59**, 908–915.

Stacey, M., Lin, H. H., Gordon, S. and McKnight, A. J. (2000). LNB-TM7, a group of seven-transmembrane proteins related to family-B G-protein-coupled receptors. *Trends Biochem. Sci.* **25**, 284–289.

Stein, M., Keshav, S., Harris, N. and Gordon, S. (1992). Interleukin 4 potently enhances murine macrophage mannose receptor activity: a marker of alternative immunologic macrophage activation. *J. Exp. Med.* **127**, 287–292.

Stuehr, D. J. and Marletta, M. A. (1987). Induction of nitrite/nitrate synthesis in murine macrophages by BCG infection, lymphokines or gamma interferon. *J. Immunol.* **139**, 518–525.

Thrush, M. A., Wilson, M. E. and van Dyke, K. (1978) The generation of chemiluminescence by phagocytic cells. *Methods Enzymol.* **57**, 462.

Virji, M., Kayhty, H., Ferguson, D. J., Alexandrescu, C. *et al.* (1991). Interactions of *Haemophilus influenzae* with cultured human endothelial cells. *Microb. Pathog.* **10**, 231–245.

List of suppliers

BD Biosciences (PharMingen)

Life Science Research, Europe, Tullastrasse 8-12, 69126, Heidelberg, Germany

Tel: +49-6221-305-521
Fax: +49-6221-305-531

Antibodies, ELISA kits and other immunochemical reagents

Biogenesis
7 New Fields Road, Stinsford Road, Poole, BH17 0NF, UK

Tel: +44-1202-660006

Modified lipoproteins

Bio-Rad Laboratories
2000 Alfred Nobel Drive, Hercules, CA 94547, USA

Tel: +1-510-741-1000
Fax: +1-510-741-5800

Biogel polyacrylamide beads

Difco Laboratories
PO Box 14B, Central Avenue, East Molesey, Surrey, KT8 0SE, UK

Tel: +44-1819799951

Thioglycollate broth, LPS and bacterial culture medium

E-Y Laboratories
Bradsure Biologicals, 15 Church Street, Market Harborough, Leicestershire, LE16 7AA, UK

Mannosylated ligands

ICN Pharmaceuticals Inc.
3300 Hyland Avenue, Costa Mesa, CA, 92626, USA

Tel: +1-714-854-0530
Fax: +1-714-641-7275

Tissue culture equipment

Jackson Immunoresearch Laboratories
872 West Baltimore Pike, PO Box 9, West Grove, PA 19390, USA

Tel: +1-215-367-5296
Fax: +1-215-869-0171

Secondary antibodies

Molecular Probes
4849 Pitchford Ave, Eugene, OR 97402-9165, USA

Fluorescent tracers

Nalgene
Box 20365, Rochester, NY 14602, USA

Tel: +1-716-264-3898
Fax: +1-714-264-3706

Tissue culture equipment

Nycomed Pharma
PO Box 5012, Majorstua, 0301 Oslo, Norway

Tel: +47-2-96-36-36
Fax: +47-2-96-37-13

Nycoprep for PBMC preparation

Ribi Immunochem Research
553 Old Corvalis Road, Hamilton, MT 59840, USA

Tel: +1-406-363-6214
Fax: +1-406-363-6129

Bacterial products (*C. parvum*)

Serotech Ltd
22 Bankside, Station Approach, Kidlington, Oxford, OX5 1JE, UK

Tel: +44-1865-852-700
Fax: +44-1865-373-899

Primary antibodies

Vector Laboratories
30 Ingold Road, Burlinghame, CA 94010, USA

Tel: +1-415-697-3600
Fax: +1-415-697-0339

Immunohistochemical reagents

3 *In Vivo* Analysis

3.1 The Immune Response in Mice Challenged with Bacterial Infections

Martin E A Mielke
Robert Koch-Institut, Germany

Ingo B Autenrieth
Institut für Medizinische Mikrobiologie, Universität Tübingen, Germany

Thomas K Held
Department of Hematology and Oncology, Charité, Humboldt University, Berlin, Germany

◆◆

CONTENTS

Introduction
Experimental infection and immunization of mice
Following the course of experimental infections
Identifying specific host responses *in vivo*
Manipulating the immune response *in vivo*
Methods for assessing immunological memory

I hope we never lose sight that all this was started by a mouse.
Walt Disney

◆◆◆◆◆◆ INTRODUCTION

Human interest in immunology dates back to ancient times when man first realized the phenomenon of acquired resistance against infections. Taking into consideration also the fact that severe infections are one of the major driving forces in the evolution of the immune system, since its beginnings the study of immunology has been closely related to studies of host defence mechanisms.

Due to extensive *in vitro* studies, much is known about the cells and molecules of the human immune system that are accessible in blood.

METHODS IN MICROBIOLOGY, VOLUME 32
ISBN 0–12–521532–0

Copyright © Elsevier Science Ltd
All rights of reproduction in any form reserved

Table 1. Well-established murine models of bacterial infections and intoxications in humans

Non-replicating agents

Systemically acting bacterial toxins:
- Tetanus toxin
- Botulinus toxin
- Toxic shock syndrome toxin-1 (TSST-1), *Staphylococcus* enterotoxin B (SEB) (*S. aureus*)

Locally acting bacterial toxins:
- Enterotoxins (*Escherichia coli, Clostridium difficile*)

Replicating agents

Peritonitis:
- *Escherichia coli*
- *Bacteroides fragilis*
- *Enterococcus faecalis*

Enterocolitis:
- *Salmonella enteritidis**
- *Yersinia enterocolitica**

Cystitis, pyelonephritis, cervicitis, pelvic inflammatory disease:
- *Chlamydia trachomatis* biovar mouse pneumonitis agent (MoPn)
- *Escherichia coli** (uropathogenic strains)
- *Proteus mirabilis**

Pneumonia:
- *Haemophilus influenzae* type B (bacteraemia, meningitis)
- *Klebsiella pneumoniae** (bacteraemia)
- *Pseudomonas aeruginosa** (bacteraemia)
- *Streptococcus pneumoniae** (bacteraemia, meningitis)
- *Coxiella burnetii*
- *Legionella pneumophila*
- *Mycobacterium tuberculosis*

Dermatitis, arthritis:
- *Staphylococcus aureus**
- *Pseudomonas aeruginosa*
- *Borrelia burgdorferi* (SCID mouse model)

Systemic infections

Extracellularly replicating agents:
- Gram positive
  - *Staphylococcus aureus* (endocarditis, arthritis)
- Gram negative
  - *Pseudomonas aeruginosa* (meningitis; immunocompromised (neutropenic) host)
  - *Haemophilus influenzae* type B (meningitis)

(contd.)

Intracellularly replicating agents:
- Gram positive
  - *Listeria monocytogenes** (meningitis/encephalitis)
  - *Rhodococcus equi* (pneumonia)
- Gram negative
  - *Bartonella bacilliformis*
  - *Brucella abortus*
  - *Francisella tularensis**
  - *Mycobacterium tuberculosis*
  - *Mycobacterium avium*
  - *Rickettsia akari, R. tsutsugamushi*
  - *Salmonella typhimurium*
  - *Yersinia enterocolitica**
  - *Yersinia pseudotuberculosis**

*Also described as spontaneous infections in mice.

However, relatively little is known about the way in which these cells and molecules behave or are controlled *in situ*, especially in the mucosa-associated lymphoid tissue, the skin, parenchymal (non-lymphatic) organs such as the liver and lung, or privileged sites such as the central nervous system, even though infections of concern in medicine are commonly those of the aforementioned organs. Important variations in local immunity may be demonstrated by comparing the host response to a particular agent in different organs as exemplified by the heightened susceptibility of the lung to *Mycobacterium tuberculosis*, the brain to *Cryptococcus neoformans*, or the placenta to *Brucella abortus*. On the other hand, studying the immune response in one organ system to various infectious agents will reveal typical patterns of response at this locality.

Although the era of modern immunology started with a controlled observation in humans (by Jenner), there are obvious reasons for using animals instead of humans to elucidate bacteria-induced mechanisms of protective immunity and inflammation (Table 1). The impact of animal models on the development of immunology can best be estimated by a historical overview, which demonstrates that most discoveries in the field of immunology have been made since the inauguration of experimental methods in animals.

The analysis of bacterial infections in animal models usually focuses on three topics:

- the mechanisms by which a bacterium induces the tissue lesions that result in the symptoms and signs of disease (pathogenesis)
- the mechanisms of host defence against the invasive organism or its toxins
- the evaluation of antimicrobial or immunomodulating agents such as antibiotics, adjuvants, cytokines, vaccines and antibody preparations.

While there is no doubt about the usefulness of animal models in the field of immunology, the question is which species is best suited to be used as an experimental subject.

Immune Response in Mice with Bacterial Infections

Murine models of medically important bacterial infections

From an evolutionary point of view, it is obvious that animals that have been exposed to a comparable selective pressure, that have developed an immune system of the same level, that live in close contact with man, and that have the same body temperature, for example rodents, are better suited as models than are birds, reptiles or amphibious animals. In fact, murine models of disease have contributed much to our understanding of infectious diseases and immunology. Nevertheless, mice are often chosen just because of habit, their inexpensiveness, their small size (easier handling and housing) and the relative lack of public sympathy for them. This, in turn, has at least led to a self-enhancing effect, resulting in the availability of genetically defined inbred strains, transgenic and gene knockout (GKO) mice, and a plethora of mouse-specific reagents that can characterize the immune response (e.g. monoclonal antibodies (mAbs) and recombinant cytokines). Consequently, a tremendous amount of knowledge about the immune system of the mouse has been accumulated.

However, before mice may be used as a model of human immune responses to bacteria, the following questions have to be answered:

- Does the experimental infection of mice mimic natural disease in humans, i.e. is the host response comparable?
- Is the degree of resistance against the challenging pathogen comparable to that in man?

Bacteria that are much less virulent, are unable to infect via a natural route, have a different organ tropism, or lead to different symptoms or histopathological changes in mice than in man should not be used to study pathogenesis and immune response.

Although there are a lot of similarities between the immune systems of man and mice, even at the molecular level (e.g. cross-activity of various cytokines, with the exception of interleukins-3, -4 and -12 (IL-3, IL-4, IL-12) as well as γ-interferon (IFN-γ)), there are some significant differences that are of importance in the interpretation of results obtained in murine models:

- *Blood.* There is an inverse relationship between the number of circulating polymorphonuclear cells (PMNs) and lymphocytes in man and mice (see Table 5). It is astonishing, however, that the inflammatory response in various tissues is often quite similar. This is an important example of the fact that the analysis of blood may be misleading in terms of characterizing host responses.
- *Spleen.* The spleen of the normal adult mouse routinely functions as a haematopoietic organ. One or more nodules of accessory splenic tissue are often embedded in the adjacent pancreas, which is of importance in the interpretation of results obtained after splenectomy. The vascular arrangement in the spleen is somewhat different in man and mouse, the latter demonstrating a higher proportion of blood circulation through tissue spaces rather than through sinuses.
- *Liver.* As is the case in spleen, the liver of mice is capable of haematopoiesis during adulthood. This may lead to confusion of, for example, megacaryocytes or clusters of erythoblasts with giant cells or inflammatory infiltrations. In contrast to human liver, trabeculae in the murine liver consist of monolayers

of hepatocytes that are not separated from blood by a basal membrane, which may result in different mechanisms of bacterial invasion.

- *Lung.* Bronchus-associated lymphoid tissue (BALT) is not present in normal mouse lung. There are no submucous glands in the lower respiratory tract of mice. There may, however, be sparsely distributed nodules of tightly packed lymphocytes within the lung parenchyma of normal mice, which should not be confused with inflammatory lesions.
- *Intestinal tract.* Likewise the gut-associated lymphoid tissue (GALT) differs in some respects from that of humans. The mouse lacks palatine and pharyngeal tonsils and has no vermiform appendix. Instead, the caecum of mice resembles the human appendix with regard to the lymphoid tissue. However, while the human intestine contains about 100 Peyer's patches within the ileum, each having 20–25 lymphoid follicles, the mouse intestine (throughout the whole small intestine) contains 8–12 Peyer's patches with 4–10 lymphoid follicles, depending on the mouse strain. The number of M cells within the follicle-associated epithelium of Peyer's patches is also species and even strain specific, and this may have profound effects on the invasion of bacteria from the gut.

In addition to these differences at the organ level, there are some systemic variations in the immune systems of man and mice:

- The cellular distribution of major histocompatibility complex (MHC) products is similar in man and mouse, but not identical. MHC class I gene products are expressed on virtually all nucleated cells in both species and on murine, but not on human, erythrocytes.
- The total amount of immunoglobulins in serum is lower in mice (2.3–6.6 mg ml^{-1}) than in humans (9.5–25.3 mg ml^{-1}).
- In mice, serotonin is more important than histamine. Thus, anaphylactic shock may not be as violent in mice as in other species.

These examples demonstrate that detailed knowledge about the anatomical and physiological differences between humans and mice may be critical for the correct planning and interpretation of animal experiments. Despite these differences, however, there are well-established murine models of medically important bacterial infections (Tables 1 and 2). These have been shown to be valid models of the host response to replicating antigens, and have contributed much to our knowledge about the immune system (Mielke *et al.*, 1997). Recently, transgenic mice proved to be suitable tools in order to generate 'humanized' mouse infection models (Lecuit *et al.*, 2001). In man and guinea-pig *Listeria monocytogenes* infection Listeria internalin protein binds to E cadherin of host cells thereby mediating enteroinvasion. The murine parenteral *Listeria* infection model was successfully applied for studying T-cell responses while intestinal infection and internalin function could not be studied. In fact, due to an animo acid exchange at position 16 of mouse (and rat) E cadherin the Listeria internalin does not bind to this receptor in mice and rats. In contrast in transgenic mice expressing human E cadherin internalin binds to this receptor and its interaction including *L. monocytogenes* enteroinvasion could be successfully studied.

Table 2. Infectious agents behaving differently in man and mice

Disease	Human pathogen	Murine equivalent
Gonorrhoea	*N. gonorrhoeae*	No equivalent
Syphilis	*T. pallidum*	No equivalent
Typhoid fever	*S. typhi*	*S. typhimurium*
Enteritis	*S. typhimurium*	*S. enteritidis*
	E. coli (enteropathogenic)	No equivalent
	C. jejuni	Not pathogenic if administered orally
	Shigella spp.	Not pathogenic if administered orally
Pneumonia	*M. pneumoniae*	*M. pulmonis*
	C. pneumoniae	*C. trachomatis* biovar MoPn
Cholera	*V. cholerae*	No equivalent (suckling mice may be used)
Erysipel	*S. pyogenes*	No equivalent
Reactive arthritis	*Y. enterocolitica*	No equivalent
Lepra	*M. leprae*	*M. lepraemurium*
Leptospirosis	*Leptospira interrogans*	Carrier status
Peptic ulcer, cancer	*Helicobacter pylori*	*H. hepatis?*

Here we focus on those basic methods that have a demonstrated value of providing important information about the immune system when mice are challenged with bacteria.

◆◆◆◆◆◆ EXPERIMENTAL INFECTION AND IMMUNIZATION OF MICE

Immunological experiments usually start with the immunization of animals. In contrast to several other antigens, viable bacteria are usually strong immunogens so that, in most cases, no additional adjuvants are needed for immunization. However, using viable agents for immunization, several parameters including the source, propagation, preparation and storage of the bacteria, have to be carefully controlled to ensure reproducibility and comparability of the data obtained in different laboratories or in the same laboratory at different times.

Sources and handling of infectious agents

Propagation, inactivation and storage of bacteria

Whenever working with a pathogen is being considered, the first thing to do is to look for guidelines about safety precautions and effective methods for the inactivation of the agent. As this information may change due to legal ramifications, we strongly advise the reader to look for the newest information available on this topic. It should be emphasized, however, that in any case the following bacteria have to be handled under the most stringent safety conditions because of the high incidence of accidental infection: *B. abortus*, *Coxiella burnetii*, *C. trachomatis*, *Francisella tularensis*, *Legionella pneumophila* and *M. tuberculosis*. Ideally, the bacterial species used to study experimentally induced host responses is a species that causes disease in humans (see Table 1). However, some species that cause human disease do not cause illness in mice (see Table 2). A possible solution to this problem may be to use a closely related bacterial species that produces symptoms in animals that resemble human disease. Another strategy is to administer the bacteria intravenously or intraperitoneally, thereby bypassing mechanisms of mucosal invasion and mucosal host defence. For studies on pathogenesis, immunodeficient animals such as suckling or infant mice or animals immunocompromised by corticosteroids or on a genetic basis (e.g. severe combined immunodeficiency (SCID) mutation) may be used, although their limitations should be kept in mind and the results should not be over-interpreted.

The choice of which strain of a particular pathogen to use also poses problems, since different strains of one species may vary considerably in their ability to cause disease. In fact, differences in virulence between the various strains used in different laboratories often explain major discrepancies in the role of certain defence mechanisms. Therefore, immunological studies using infectious agents are of no value without information about the route of infection and the LD$_{50}$ of the infective organism. Furthermore, pathogens may alter their phenotype considerably once they are moved from the human environment to artificial culture media. Therefore, it is advisable to preserve stock suspensions of the pathogen very soon after their primary isolation from humans. In any case, repeated passages *in vitro* must be avoided.

Sources of bacteria used in animal experiments are clinical isolates or bacteria obtained from culture collections such as American type culture collection (ATCC) (Rockville, MD) or its national equivalents. While clinical isolates often retain their virulence mechanisms, they are sometimes poorly defined and can give rise to results that cannot be reproduced in other laboratories working with another strain of the same species. On the other hand, bacteria obtained from culture collections may have lost their virulence due to serial subcultures in the absence of selective pressure. It may be that the only way out of this dilemma is to use and compare bacteria from both types of source.

While virulence in mice can often be increased and maintained by at least three initial cycles of *in vivo* passages, this method fails if virulence is encoded by genes on a plasmid that has been lost during *in vitro* culture. To obtain the initial culture, bacteria should be grown in rich media such as trypticase soy broth, brain–heart infusion or thioglycollate broth (e.g. Difco Laboratories, Detroit, MI) under optimal conditions. For *Borrelia* (BSK II medium + 12% rabbit serum, 32°C), *Legionella* (BCYE agar, 37°C) and mycobacteria special media are needed. *Chlamydia* (McCoy cell monolayer; overlay containing 1.5% agar) and *Rickettsia* have to be grown in tissue culture.

For experimental infections, bacteria from a log-phase culture are pelleted by centrifugation (e.g. 3000g, 15 min) and washed once in sterile phosphate buffered saline (PBS). Aliquots of this stock suspension can be maintained at −70°C, or in liquid nitrogen in PBS or fresh broth. In some cases it may be necessary to include 15% glycerol. For immunization or challenge, aliquots of frozen stocks are thawed and diluted to the appropriate concentration with PBS. Differences in culture conditions may have profound effects on virulence, as the growth temperature and certain deficiencies in the culture medium (e.g. a low iron content) may result in reduced or enhanced expression of virulence factors. In addition, one should be aware that even the washing procedure may influence some characteristics of the bacteria, including virulence factors. For example, the capsule of bacteria may be sheared off by harsh centrifugation. In such cases, a dilution of more than 1 : 50 of the broth culture in PBS may be used *in vivo*.

In vivo passage of bacterial strains is a prerequisite for reproducible *in vivo* analysis of host responses to bacteria. For this purpose, a sublethal dose of bacteria should be administered intraperitoneally or intravascularly. After a certain time period, depending on the growth characteristics of the bacterium used, the spleen is removed aseptically or a peritoneal lavage (5 ml sterile saline) is performed (see p. 389). The lavage fluid or organ homogenate is serially diluted and plated on agar, in addition to being inoculated into a series of broth cultures to obtain virulent bacteria. The bacteria should then be grown to log phase, again according to the requirements for the particular strain. Usually, a culture that shows only slight turbidity is chosen and the bacteria are collected by centrifugation, washed in PBS, frozen in aliquots and kept at −70°C or in liquid nitrogen. Prior to an experiment, the bacteria are thawed and grown to log phase again; alternatively, they may be injected directly. The viability and stability (in terms of virulence) of the frozen bacteria should be determined prior to the experiment (e.g. by determining the number of colony forming units (cfu) just before and 24 h after freezing, as well as after a longer period of frozen storage) because the percentage of non-viable cells present after storage may vary dramatically and may significantly influence the type of immunization or outcome of challenge studies due to non-specific stimulation of defence mechanisms by non-viable bacteria.

The actual number of experimentally administered bacteria should be determined by plating serial 10-fold dilutions of the inoculum and counting the number of cfu after appropriate incubation. It should be borne in mind that virulence factors determining adhesion or invasiveness may be lost if passages are done bypassing mucosal surfaces by parenteral infection. On the other hand, since most invasive bacteria finally reach the bloodstream, the response to such strains may be analysed successfully if they are administered intravenously or intraperitoneally.

Since the course of disease is critically determined by the relationship between the replication rate of the bacterium *in vivo* and the onset of effective defence mechanisms, non-reproducible results may often simply be due to differences in the growth phase and replication rate of the bacteria used. Therefore, great care should be taken to ensure that bacteria used for *in vivo* experiments are used in the midlogarithmic phase in a standardized culture medium under standardized (optimal) culture conditions.

Counting bacteria

The number of viable bacteria in suspensions can easily be determined by counting the number of colony forming units (cfu) on agar plates. However, to avoid falsely low bacterial numbers (e.g. due to clumping or adherence of bacteria), suspensions containing 0.1% Tergitol TMN 10 (Fluka, Buchs, Switzerland) and 0.1% bovine serum albumin (BSA) in PBS may be plated. A detergent in hypotonic PBS might also facilitate the release of intracellular bacteria from cells when numbers of bacteria in organ homogenates have to be determined. However, the effectiveness of this approach should be tested before use, as some bacterial species might be killed by the detergent.

Another method for determining the number of bacteria in a suspension just before infection is the nephelometric measurement of turbidity (e.g. at 660 nm in a Pharmacia LKB Ultraspec III). This has to be done in the midlogarithmic phase to ensure that turbidity is caused by viable organisms only. A retrospective confirmation by plating serial dilutions and determining the number of cfu is mandatory. Finally, bacteria may be counted in a Petroff–Hausser chamber by phase-contrast microscopy.

Preparation of non-replicating antigen from bacteria

Non-replicating preparations of bacteria may be used for the induction of an immune response, but more often they are used to elicit secondary responses such as delayed type hypersensitivity (DTH) reactions and granuloma formation in primed mice.

Killed bacteria (particulate antigen) can be prepared by incubating PBS-washed bacteria of a midlogarithmic phase culture for 1 h at 63°C in a water bath. The number of cfu has to be determined before the preparation, and may be adjusted to 1×10^{10} ml^{-1}. The killed preparations may be stored in aliquots at –70°C. Alternatively, bacteria may be inactivated by treatment with 0.1% formalin or glutaraldehyde in PBS for 1 h at 37°C and subsequent washings in PBS. Another method is γ-irradiation of the bacteria with 25 Gy. Sterility must be confirmed by subculturing aliquots both in liquid media and on agar plates.

The preparation of soluble antigens can be achieved by ultrasonication of bacteria in PBS (e.g. 1 g wet wt per 10 ml PBS) or by concentrating cell culture supernatants by ultrafiltration. Before use, soluble antigen preparations are filter sterilized using 0.2-μm filters. Progress in gene technology has enabled the generation of high amounts of pure proteins as recombinant antigens. These may be highly standardized tools for immunization and challenge. However, the lipopolysaccharide (LPS) content of

such a preparation should be tested before use. Proteins should be purified according to the instructions of the manufacturer of the vector system
used to produce the recombinant protein. Finally, the protein should, if
possible, be solubilized in PBS or saline in order to avoid toxicity from
compounds used during purification (e.g. urea). The amount of recombinant protein needed to generate humoral and/or cellular immune
responses can vary significantly, and depends on the biochemical properties of the protein, on its immunogenicity and on the adjuvant used (see
the chapter by Lövgren-Bengtsson, p. 551ff). For many proteins, 10 µg
administered parenterally or intranasally is sufficient to generate an
immune response. For orogastric immunization, a broad range of doses
should be tested (e.g. 10–100 µg per application).

Animals and animal facilities

Details about animal care and gene knockout as well as transgenic mice
are given in the chapter by Mossmann et al., p. 183ff. Here we focus on
those aspects that are particularly relevant to in vivo experiments involving viable bacteria.

Mice suited for immunological experiments can be purchased from the
Jackson Laboratory (Bar Harbor, MN, USA), or other well-reputed dis-
tributors such as Harlan Sprague–Dawley, Inc. (Charles River Wiga, IN,
USA), Harlan Olan Ltd, Bomhaltgard Breeding and Research Centre Ltd
(Ry, Denmark), etc., that provide a health certificate including data about
typical mouse pathogens such as mouse hepatitis or sendai virus. After
transport, the mice should be allowed to rest for a few days in order to
become acquainted with the new environment before the experiments are
started. Stress due to shipping and handling may lead to a disruption of
the normal circadian rhythm of corticosteroid secretion, which in turn
may profoundly influence the immune response. The same is true for the
day/night rhythm, which should be standardized in all experiments. In
addition, the water balance of the mice may be disturbed during travel,
and the mice should be adapted to the food used during the whole experiment. In order to prevent social stress, no more than 10 mice should be
housed in a 900-cm^2 cage.

For immunological studies, mice should at least have the standard of
monitored animals, i.e. the mice should be housed in a low-security barrier system (change of laboratory coat, disinfection of hands, use of gloves
and face mask, sterilization of cages) and should be demonstrated to be
free of known pathogens by sequential monitoring (see the chapter by
Mossmann et al., p. 183ff). Therefore, serum samples (see p. 388) should be
drawn on a regular basis and analysed at least for the presence of
antibodies against the most common pathogens such as mouse hepatitis
and sendai virus.

Unknown latent or persistent infections are the major cause of non-
reproducible experiments. Sources of infection that are often not realized
are hybridoma and tumour cells used in vivo, and infections transmitted
to animals from animal workers who have pets at home. The eradication

of a viral pathogen such as mouse hepatitis virus may stop scientific work for at least half a year. Therefore, no foreign animals should be obtained without being separated and controlled, and staff should be well educated.

Mice should be housed at 20–22°C (a temperature below 18°C or over 33°C will influence body temperature) at 60–70% humidity in air, and maintained on an alternating 12 h dark/light cycle. Mice may be fed heat-treated rodent chow and water *ad libitum*.

Experimentally infected mice should be housed in separate safety cabinets (Fig. 1). At the end of an experiment sacrificed animals should be collected in plastic bags and autoclaved (121°C, 20 min) before release from the laboratory.

Factors influencing susceptibility to infection

Susceptibility of mice to infections varies significantly with strain, age, body weight and sex. The immune response of very young mice differs from that of mature and of senescent mice. Adult levels of immune parameters are usually established after 9 weeks of age, and cell-mediated

Figure 1. Microisolation housing units.

immunity decreases after the age of 24 months. Therefore, mature mice aged 10 weeks to 12 months should be used for immunological studies. Since susceptibility to infection may differ significantly in male and female mice, it is prudent to use only one gender throughout the entire study.

Susceptibility to oral infection may be influenced by the composition of the intestinal flora. Therefore, for each animal facility the microbial status 'specific pathogen free' (SPF) should be characterized in order to achieve reproducibility of experiments. For example, the presence of *Citrobacter freundii* in the intestinal flora is of particular interest in studies of gut-associated lymphoid tissue as this bacterium may account for transmissible murine colonic hyperplasia. Treatment of mice with streptomycin (Wadolkowski *et al.*, 1990; Lindgren *et al.*, 1993) is a well-known method of modifying the composition of the intestinal flora by reducing the facultative anaerobic bacteria that normally colonize the mouse intestine and thus to increase susceptibility for mucosal infection. For example, streptomycin-treated mice become susceptible to infection with *E. coli* F-18 and K-12. Susceptibility of mice to mucosal infections may also be modified by a vitamin A deficient diet (Wiedermann *et al.*, 1993). Vitamin A is an important regulator of growth, differentiation and proliferation of epithelial cells. Vitamin A deficiency reduces mucosal immune responses. Furthermore, it may lead to breakdown of mucosal integrity, and thus may increase susceptibility to mucosal infections. Quality of rodent chow is therefore an important feature.

Germ-free mice may be used for intestinal infection experiments. However, the behaviour of an infective agent might significantly differ in these mice, as bacteria from the normal flora often provide important colonization factors in the intestinal microenvironment. Furthermore, the development and differentiation of the gut-associated lymphoid tissue, including Peyer's patches and intraepithelial lymphocytes, depends on the presence of normal gut flora. Germ-free mice have only small Peyer's patches, and the phenotypes of intraepithelial lymphocytes is changed (Umesaki *et al.*, 1993; Imakoa *et al.*, 1996). Likewise, the number of M cells within the follicle-associated epithelium (FAE) is reduced in these animals.

In some situations it might be helpful to initially immunosuppress animals by corticosteroids to get an infectious process started. This can be done by the subcutaneous injection of 125 mg kg^{-1} per animal once daily for 6 days.

Mouse inbred strains and spontaneous mutants

The discovery of spontaneous mouse mutants presenting with immunodeficiencies (e.g. nude or SCID mice) and the generation of inbred strains had a great impact on immunological research.

SCID mice lack the recombinase necessary for the rearrangement of T-cell receptor and immunoglobulin genes. Consequently, they lack functional lymphocytes and do not develop cellular or humoral immunity. Therefore, they have emerged as a model for the study of natural resistance (Bancroft and Stevens, 1996). However, some mice gradually develop a low level of T-cell function and some amount of serum immunoglobulin (Ig) ('leaky SCIDs') (see the chapter by Mossmann *et al.*, p. 183ff).

However, also in normal mice the susceptibility to infection and the type of immune response to various antigens are under genetic control. Therefore, great care has to be taken before an observation made or not made in one strain of mice is generalized to other strains. In immunological studies, at least C57BL/6, Balb/c, AKR and Swiss mice should be tested (Table 3). B10 mice may be of great value in evaluating peptide vaccines because of the availability of various H-2 haplotypes based on the same genetic background (Table 4).

For some purposes it may be useful to reduce the effects of LPS. Endotoxin-hyporesponsive C3H/HeJ mice fail to respond normally to endotoxin because their B cells and mononuclear phagocytes are hyporesponsive to the lipid A moiety of LPS.

Recently, the number of available transgenic or gene knockout (GKO) mice has been increasing dramatically, and these mice are most interesting tools in immunological research. It must be kept in mind, however,

Table 3. H-2 loci of commonly used mouse strains

Haplotype		K	Ab	Aa	Eb	Ea	D	Thy-1	CD8
A/J	a	k	k	k	k	k	d	2	2
C57BL/6J(B6)	b	b	b	b	b	–	b	2	2
DBA/2(J)	d	d	d	d	d	d	d	2	1
Balb/c(J)	d	d	d	d	d	d	d	2	2
C3H/He(J)	k	k	k	k	k	k	k	2	1
AKR/J	k	k	k	k	k	k	k	1	1
CBA/J	k	k	k	k	k	k	k	2	2
Swiss mice (SJL)	s	s	s	s	s	–	s	2	2

Table 4. H-2 loci of B10 mice

Haplotype		K	Ab	Aa	Eb	Ea	D	Thy-1	CD8
B10.BR	k	k	k	k	k	k	k	2	2
B10.D2	d	d	d	d	d	d	d	2	2
B10.Q	q	q	q	q	q	–	q	2	2
B10.A	a	k	k	k	k	k	d	2	2
B10.S	s	s	s	s	s	–	s	2	2
B10.A (1R)	h1	k	k	k	k	k	b	2	2
B10.A (2R)	h2	k	k	k	k	k	b	2	2
B10.A (3R)	i3	b	b	b	b/k	k	d	2	2
B10.A (4R)	ha	k	k	k	k/b	b	b	2	2
B10.A (5R)	i5	b	b	b	b/k	k	d	2	2
B10.T (6R)	y2	q	q	q	q	–	d	2	2
B10.S (7R)	t2	s	s	s	s	–	d	2	2
B10.S (8R)	as1	k	k	k	k/s	–	s	2	2
B10.S (9R)	t4	s	s	s	s/k	k	d	2	2

that certain genetic defects may result in profound changes in anatomy and physiology due to secondary disturbances in ontogeny. Therefore, detailed knowledge of these differences is important in order to prevent misinterpretation of data from experimental infections obtained using those mice (for details see the chapter by Mossmann et al., p. 183ff). In addition, using these mice studies on natural resistance are more easily performed than are studies on the immune effector phase or immunological memory.

Routes of infection and immunization

The course of infection, the host response and the effect of experimental immunomodulation may depend critically on the route of infection or immunization. In addition, as mentioned above, certain virulence factors may be critical only when bacteria are applied mucosally. The following routes of infection may be used (maximum volume applicable):

- Mucosal applications:
    - peroral   (300 µl)
    - intranasal   (50 µl)
    - intraurinary (50 µl)
    - intravaginal (30 µl)
    - rectal   (30 µl)

- Parenteral applications:
    - intradermal   (50 µl)
    - subcutaneous   (2 ml)
    - intramuscular (50 µl)
    - intraperitoneal (2 ml)
    - intravenous   (300 µl)
    - intracranial   (30 µl).

Application to mucosal surfaces

Orogastric application

Bacterial suspensions can be administered using a 0.86-mm polyethylene tube connected to a syringe fitted with a 20-gauge ½-inch needle (Fig. 2) or by a feeding needle. Successful oral administration requires thorough knowledge of the anatomical relationships of the oropharynx (Fig. 5), because the oesophageal orifice cannot be observed easily in the living mouse. The feeding needle (18 or 20 gauge) is introduced into the left diastema and gently directed caudally toward the right rami of the mandible. At this point, the mouse usually begins to swallow and the feeding needle can be inserted gently into the oesophagus.

In models of intestinal infection, there is usually much more variation than in models using parenteral infections. Besides reduced accuracy of application, this might be due to variations in the intestinal flora, which can interfere with the infective agent. Furthermore, individual and strain

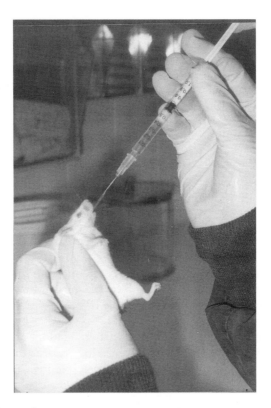

Figure 2. Oral inoculation.

variations of anatomical structures (e.g. the number of Peyer's patches varying between 6 and 12 per mouse) or functional parameters (e.g. stomach acid, amount of food in the stomach, enteral passage) may account for the diversity. The former might be critical for experiments with entero-invasive bacteria such as *Salmonella* and *Yersinia* species. As these bacteria enter the intestinal wall via M cells that occur only with the FAE overlying Peyer's patch tissue, the number of Peyer's patches correlates with the number of M cells and thus with the chance of invasion. Furthermore, it is very important to 'synchronize' mice prior to orogastric infection. For this purpose mice should be starved for 12–18 h to clear the bowel. To avoid coprophagy, which might result in reinfection by excreted bacteria, mice should be kept on grids (Fig. 3).

Ileal-loop model

The ileal-loop model allows the investigation of mechanisms operating in the very early phase of intestinal infections (Autenrieth and Firsching, 1996a). By focusing on a limited part of the intestine, sampling error is reduced. Mice should be starved for 18 h prior to the experiment. Anaesthesia can be accomplished by intraperitoneal injection of 50–70 mg kg^{-1} sodium pentobarbital (Abbott Laboratories, North Chicago, IL, USA). After a deep stage of anaesthesia has been reached (i.e. after

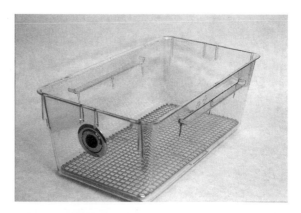

Figure 3. Cage with a grid preventing superinfection by coprophagy.

about 3–5 min, check by scratching the foot pad with a needle) a small midline incision of 1–1.5 cm in length is made down the abdomen in order to reveal the small intestine (Plate 2). Cutting into superficial abdominal or peritoneal vessels must be avoided as bleeding makes proper working impossible. To form a loop, two ligations are made in the ileum, leaving a 3- to 5-cm gap in between. This ensures that the loop contains a Peyer's patch if binding to a Peyer's patch or invasion of the Peyer's patch by bacteria is desired (e.g. *E. coli*, *Salmonella typhimurium*, *Shigella flexneri*, *Yersinia enterocolitica*, *Campylobacter jejuni*). The blood supply to the small intestine should always be carefully preserved. Vessels at the intestine and peritoneum are easily detectable. Bacterial suspension (0.1 ml) can be injected into the ileal loop via a 0.4-mm needle. Return the small intestine to the bowel. The incision is closed with a suture or two (e.g. prolene monofilament, Ethicon Ltd., Edinburgh, UK). During the whole procedure, mice should be warmed under a lamp, and the intestine should be kept wet with a towel soaked in prewarmed sterile saline 37°C. Disinfection is usually not required since a typical ileal loop experiment usually lasts only for a few hours.

Intranasal application

Infection of the respiratory tract can be achieved either surgically or non-invasively. Whenever possible, excess manipulations should be avoided, thus reducing stress and increasing reliability and reproducibility of the model.

Non-surgical methods. The murine trachea has a diameter of only 1–1.5 mm and the distance between nose and the lower parts of the lung is approximately 2.5–3 cm in adult mice. The inoculum size (50 µl) is thus a critical factor to the success of intranasal delivery. Inocula less than 10 µl are more likely to stick to the upper respiratory tract and thus may not reach the lungs. Another important parameter is the depth of anaesthesia. The mouse has to be anaesthetized deeply enough to suppress the swallowing reflex so that the inoculum actually reaches the lung and does not end up in the stomach or is expectorated by the mouse. On the other hand, delivering a bolus into the trachea induces a vagal response resulting in a reduced heart rate. If anaesthesia is too deep, cardiac arrest may result. Anaesthesia for this kind of manipulation should be performed by intraperitoneal administration of ketamine hydrochloride and xylacine (Rompun) together. The dose depends on the weight and strain of mice used. For example, CD-1 (ICR) mice may require doses almost 1.5 times higher than for C3H/HeN or C3H/HeJ mice. Usually, ketamine hydrochloride and xylacine are diluted in sterile, pyrogen-free PBS in a ketamine hydrochloride/xylacine ratio of 20:1, resulting in a dose of 120 mg kg^{-1} of ketamine hydrochloride and 6 mg kg^{-1} of xylacine. After reaching deep anaesthesia, which occurs usually within 5–10 min, mice are held upright and the inoculum is delivered by means of a pipette into one nostril (Fig. 4). This ensures that

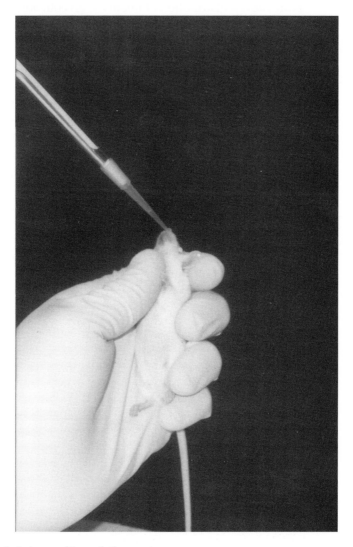

Figure 4. Intranasal inoculation.

the mouse can continue to breath through the contralateral nostril. The pipette tip should be placed about 1–2 mm apart from the nostril in order to obtain fine droplets of about 0.5–1 mm diameter which easily enter the nose and do not cloak major airways. The whole procedure should take no more than 1 min. Hyperventilation is a welcome side-effect, since it contributes to an even distribution of the inoculum throughout the lungs. The mouse should be held upright for another minute or two to ensure entry of the inoculum into the lungs. Then, the animal is placed back in its cage such that the head and thorax are slightly elevated. After 30–60 min, the animal usually awakes from anaesthesia. Using this approach, approximately 75–85% of the inoculum will end up in the lung. Plate 1 shows a lung after inoculation of 50 μl india ink to depict the distribution of the inoculum in the lungs. It must be stressed that the lungs cannot be inoculated separately by any method. As can also be seen in Plate 1, there is an uneven distribution of ink in the alveoli, despite homogeneous coating of the major bronchi. This means that the pattern of infection will be patchy, which is of consequence for the subsequent analysis of the tissue response.

Another way to administer inocula into the lungs of mice is by intubation using a small feeding tube (not more than 26 gauge) which is inserted slowly into the mouth (Fig. 5). However, this requires manipulation and bears the danger of injury. In addition, it takes more time to apply inocula by this technique. Thus, if large numbers of mice have to be processed, the intranasal route is preferred.

Surgical methods. After deep anaesthesia (see above) a small incision is made at the midline of the neck, directly above the trachea, not more than 1–1.5 cm in length. Only the skin should be cut, in order to avoid damage

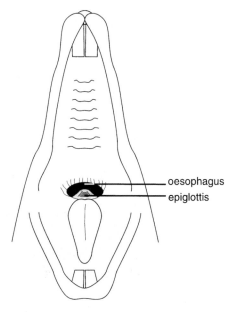

oesophagus
epiglottis

Figure 5. Murine oropharynx.

to the underlying thyroid gland. This organ as well as the muscles covering the trachea are prepared using atraumatic techniques and by carefully avoiding incision of blood vessels. The trachea is then exposed and a small cut (not more than 1–2 mm) is made between two cartilages of the trachea (Fig. 15). Either a small feeding tube (26 gauge or less) with a blunted end or a Teflon catheter (e.g. Abbocath) is inserted and the desired volume applied to the lungs. Again, even with this method, it is impossible to inoculate one lung selectively. The trachea should be closed with one suture. It is not necessary to close the muscles over the trachea; however, the skin should be closed with a suture.

Aerosols. A detailed description of infection by aerosols is given in the chapter by Roberts *et al.* (p. 433ff).

Urinary tract application

For urinary tract infection, mice are anaesthetized as described above and a polyethylene catheter (2.5 cm long, 0.61 mm outer diameter; Kebo Grave, Sweden) is inserted via the urethra into the bladder (Fig. 6). Prior to insertion of the catheter, the bladder should be emptied by gentle compression of the abdomen. The catheter should be fitted to a needle (0.4 × 0.22 mm gauge) on a 1-ml tuberculin syringe and up to 50 µl of a bacterial suspension can be administered.

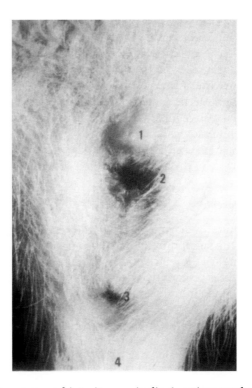

Figure 6. Rodent rectum and introitus vaginalis: 1, ostium urethrae externum; 2, ostium vaginae, 3, anus; 4, radix caudae (Olds and Olds, 1979).

Vaginal and rectal application

For vaginal and rectal administration of bacteria, mice should be anaesthetized. An inoculum of about 30 μl can then be introduced gently into the rectum or vagina using a 23-gauge needle with the end blunted with solder or a yellow tip (Fig. 6). After vaginal or rectal administration mice should be positioned with the vagina or rectum facing upwards for 30–60 min to reduce leakage of the inoculum. Alternatively, a swab soaked with the inoculum is left in the vagina for 20 min. For vaginal infections and immunizations, it is important to consider the oestrous status of the animals (Hopkins *et al.*, 1995). Mice at the late metestrus or diestrus exhibit stronger immune responses upon vaginal immunization. Therefore, female mice may receive 2.5 mg medroxyprogestrone (Depo Provera, Upjohn, Puws, Belgium) per dose subcutaneously in 100 μl PBS 7 days prior to intravaginal inoculation.

Routes of infection bypassing mucosal surfaces

If an infection cannot be achieved via the mucosal route, it may be helpful to bypass mucosal surfaces by parenteral injection.

Subcutaneous and intradermal injection

The subcutaneous route of administration may be utilized when prolonged release of a relatively large inoculum is required. Subcutaneous injections (1 ml) can be made into the loose skin over the flank using a 26- to 30-gauge, 0.5- to 1-inch needle. The needle should be inserted into the skin 0.25-inch caudal to the injection site, and then advanced through the subcutaneous tissues to the injection site in order to minimize leakage of the injected material.

Intradermal injection into the volar aspect of the hind foot pad is often used to investigate the immune response of the skin and to elicit secondary immunological responses such as the Arthus reaction or DTH (Fig. 7). A 30-gauge needle is inserted between pulvini and advanced just under the surface of the skin. Correct injection of up to 50 μl results in a pale bulla.

Intramuscular injection

Intramuscular injection should usually be avoided in the mouse because of the small muscle mass. The rate of absorption of aqueous solutions is similar following intramuscular and subcutaneous injections, and so the latter is preferred. If necessary, intramuscular injections of up to 50 μl may be made into the anterolateral thigh muscles (quadriceps femoris group) using a 22- to 26-gauge 0.25-inch needle. The needle should be directed away from the femur and sciatic nerve (Fig. 8).

Intraperitoneal injection

Intraperitoneal injection is the most convenient and simple technique to apply volumes up to 2 ml. To avoid puncture of the stomach, spleen or liver,

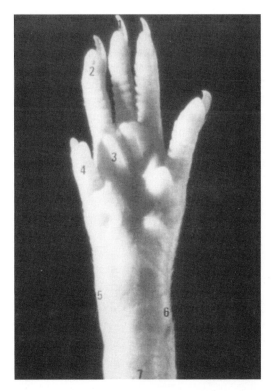

Figure 7. Rodent hind foot pad: 1, ungicula; 2, pulvini phalangici; 3, pulvini metatarsales; 4, hallux; 5, median side; 6, lateral side; 7, ankle (Olds and Olds, 1979).

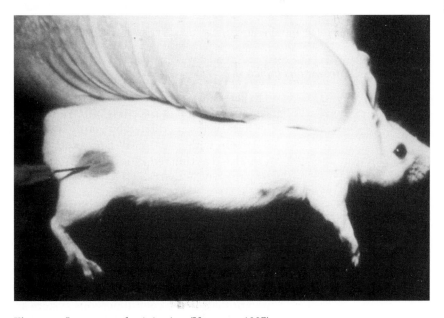

Figure 8. Intramuscular injection (Versteeg, 1985).

intraperitoneal injections are made into the lower right quadrant of the ventral abdomen (Fig. 9). The mouse is restrained, and the handler's wrist is rotated until the mouse's head and body are tilted in a downward direction, allowing the mouse's abdominal viscera to shift cranially. The needle (23- to 26-gauge, 0.25- to 0.5-inch) is then inserted through the skin, slightly medial to the flank and cranial to the inguinal canal, advanced cranially through subcutaneous tissue for 2–3 mm, and then inserted through the abdominal muscles. The needle and syringe should be held parallel to the mouse's vertebral column in order to avoid accidental retroperitoneal or intrarenal injection.

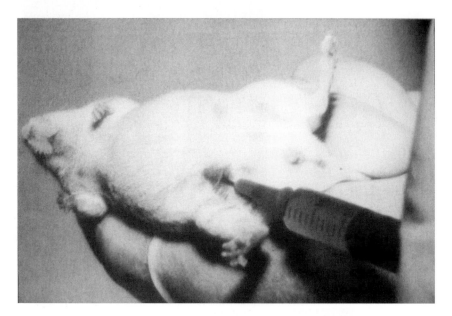

Figure 9. Intraperitoneal injection (Versteeg, 1985).

Intravenous injection

The lateral or dorsal tail veins are the usual sites for intravenous injection in mice (Fig. 10). Tail vein injection is easier if the veins are dilated by gently warming the mouse for 5–15 min with a 40–100 W light bulb. Lateral tail veins are visualized as thin red-blue lines. The needles used are 26- to 30-gauge, 0.25- to 0.5-inch.

Intracranial injection

Intracranial injection has been used to establish *in vivo* models of infections with neurotropic viruses, but may also be used for bacterial infections. This route may be helpful in investigating whether immunity induced in the periphery is expressed in the central nervous system as well. Mice are anaesthetized as described above, and the scalp is disin-

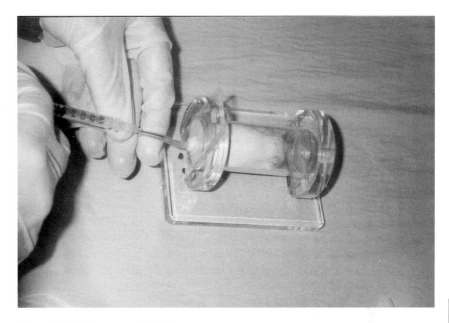

Figure 10. Intravenous injection.

fected with 70% ethanol. The needle (22- to 30-gauge) is inserted through the skin, over the midsection of the parietal bone slightly lateral to the central suture; this avoids puncture of the sagittal or transverse venous sinuses (Fig. 11). The needle is gently rotated until the bone is penetrated.

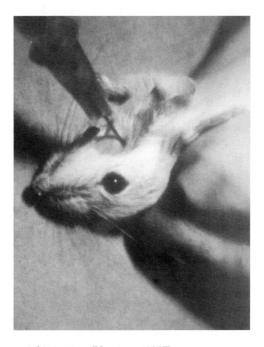

Figure 11. Intracranial injection (Versteeg, 1985).

The needle is then advanced to a depth of 1–4 mm, depending on the size of the mouse. Approximately 15–30 µl can be injected intracranially. Solutions injected intracranially should be at 37°C. After injection, the mice should be kept warm to reduce the possibility of shock.

◆◆◆◆◆◆ FOLLOWING THE COURSE OF EXPERIMENTAL INFECTIONS

Morbidity and mortality

Body weight

Usually, animals are infected and then observed over a certain period of time, the end-point being survival. Acquiring information in between these times may be difficult. Therefore, careful observation may provide important clues about the course of an infection and the best suited time points for more detailed investigation by invasive methods. After infection, mice may show non-specific signs of illness, such as inactivity, rough fur, conjunctivitis, dyspnoea or diarrhoea. Rolling, circling or paresis may be signs of neurological disease. In some cases mice may develop skin lesions or swelling. However, the onset of illness is seldom so obvious that time to illness would be a reliable parameter for measuring the kinetics of the infectious process. Moreover, quantification of severity is very difficult. One quantifiable parameter that can be observed during the course of infection is loss of body weight. Weighing the animals is a simple way to obtain this information. In bacterial infections in particular, animal weight may correlate not only with final outcome, but also with histopathology and bacterial organ load. Thus, measuring body weight daily or at even closer time points may be a reliable non-invasive method of acquiring the first information about the course of an infection.

LD$_{50}$ and time to death

Outcome and course of infection are critically influenced by the size of the inoculum. If a lethal challenge is used, the major parameters that can be studied are lethality and time to death. The estimation of the dose of a pathogen resulting in the death of a given percentage of animals is a global procedure – it can be used to compare both the virulence of different pathogens, and also a single pathogen administered by different routes or expressing different virulence factors in the case of isogenic mutants. The most commonly used end-point is the LD$_{50}$, which denotes the amount of pathogen necessary to kill 50% of the animals. To determine the LD$_{50}$, mice are infected with various doses of bacteria ranging from 10^1 to 10^9 cfu per animal. Mice are then observed over a period of time, ranging from days to several months depending on the infective agent. The LD$_{50}$ is calculated according to a method published in the late 1930s (Reed and Muench, 1938). Most often, 5–10 animals per dose of pathogen have to be used for calculation (see p. 399).

Methods requiring sacrifice of animals

Bacterial load in infected organs

Quantitative determination of the bacterial load in various organs allows a more detailed assessment of the infectious process. Furthermore, less animals are required for a definite result than for determination of LD_{50}.

The determination of numbers of bacteria per infected organ is the most direct parameter for following the course of an infection and defining protective host responses. It has to be kept in mind, however, that morbidity and lethality are the result of both bacterial replication and the inflammatory host response. Hence to obtain a complete picture, histological studies must be done in parallel.

A major disadvantage of the procedure is that mice have to be killed in order to obtain this information. However, recent work has demonstrated that bioluminescence may be applied to *in vivo* analysis of bacterial replication. Bacteria are transformed with a plasmid conferring constitutive expression of bacterial luciferase. Detection of photons transmitted through tissues of infected mice allows localization of bacteria by real time and non-invasive monitoring (Contag *et al.*, 1995).

It is worth mentioning that the use of high lethal doses of bacteria may result in the infection of unusual target organs (e.g. the brain) or host cells. Therefore, another advantage of the quantification of viable bacteria in infected organs is the ability to characterize sublethal infections. In a typical experiment mice are infected and a certain number of animals out of this group is killed by CO_2 and cervical dislocation at various time points after infection.

In order to obtain maximum information, organs of at least three mice may be cut into three equal pieces

- one for the determination of bacterial load
- one for histological analysis
- one for cellular and/or molecular analysis, such as the determination of soluble mediators or cell functions by cell culture studies.

However, whenever unequal distribution of the infection is expected, as in lung or gut after nasal or oral infection, respectively, the entire organ should be used for one single type of analysis (e.g. only cfu determination) in order to minimize sampling errors.

The basic method for determining bacterial load is easy to perform; the only difficulty being sterile preparation of certain organs such as Peyer's patches. Organs are removed aseptically (Plate 2) and parts or the whole organ are transferred, after weighing in a sterile Petri dish, into homogenizer tubes containing 5–10 ml sterile PBS (Fig. 12). Organs are dispersed for 1 min using a motorized homogenizer placed in a safety cabinet (Fig. 13). The homogenate is subsequently stored on ice until serial dilutions (1:10) in sterile PBS are plated on suitable and absorbent (predried) agar (e.g. TS or BHI agar). Bacterial colonies are counted after an appropriate incubation period, depending on the bacterium under investigation, and numbers may be normalized for organ or tissue weight.

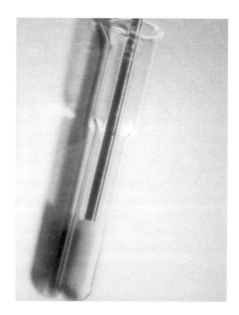

Figure 12. Glass tube with Teflon pistil (Versteeg, 1985).

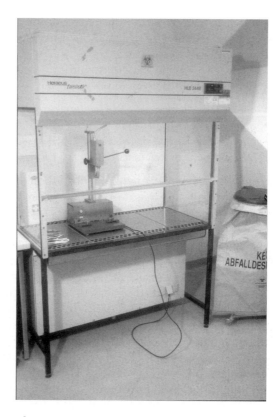

Figure 13. Potter homogenizer in a biosafety cabinet.

Microbial load in Peyer's patches

To remove Peyer's patches, the whole small intestine (Fig. 14) is excised and put into a Petri dish containing cold sterile PBS. The small intestine is then washed extensively with 10–20 ml cold buffer to remove intestinal contents and bacteria associated with the mucosal surface. Alternatively, gentamicin (10 µg ml^{-1}) may be added to kill bacteria located in the lumen of the intestine. Finally, Peyer's patches are excised by carefully cutting the gut longitudinally. As Peyer's patches are contaminated at their mucosal surface with bacteria from the intestinal flora, which cannot be removed completely by washing, the homogenates should be plated on selective agar media in order to suppress growth of contaminating bacteria (e.g. cefsulodin–irgasan–novobiocin agar for *Yersiniae* spp., SS agar for *Salmonella* spp. or *Shigella* spp).

Description and quantification of tissue lesions and inflammatory host responses

There is no simple way to quantify the degree of tissue destruction and inflammation *in vivo*. In infections afflicting the liver the determination of serum levels of hepatic enzymes may reveal some information, although we do not recommend their use.

Alanine aminotransferase (ALAT) is a leakage enzyme that is frequently used for the purpose of assessing hepatic injury. However, ALAT is not specific for liver. Therefore, its activity in serum should be considered together with other enzymatic data, such as aspartate aminotransferase (ASAT) and alkaline phosphatase (AP). In addition, it has to be kept in mind that the serum activities of leakage enzymes may decline in the presence of ongoing tissue destruction due to disturb-

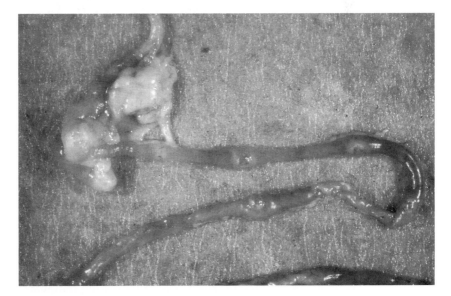

Figure 14. Peyer's patches in rodent gut.

ances in the synthesis of the enzymes as the injurious process intensifies. An inflammatory parameter correlating well with the severeness of infection is IL-6. This may easily be determined in serum and organ homogenates by ELISA or bioassay (see the chapter by Hiroi *et al.*, p. 301ff). The most reliable data, however, can be obtained by histological evaluation.

Another aspect that may be of interest in models of infection is to determine the host cell of intracellular bacteria *in vivo*. Careful histology is the only way to do this. However, only high numbers of bacteria can be detected by microscopy (usually $>10^6$ cfu per organ). In some cases only the suppression of host defence mechanisms (e.g. by γ-irradiation, corticosteroids, IL-10) will reveal the permissive host cell.

Histopathological examination

The detailed description of histopathological techniques is outside the scope of this chapter. For details, see Bancroft and Stevens (1996) and the chapter by Ehlers and Seitzer (p. 403ff). However, some aspects that may be helpful in planning an experiment are mentioned below.

Tissue fixation and embedding in plastic. The most powerful histological evaluation is possible by performing semi-thin sections from plastic embedded tissue and staining with haematoxylin and eosin or Toluidine Blue. Bacteria can be visualized by Gram stain, Ziehl–Neelsen stain, Dieterle stain or by Toluidine Blue, which is one of the advantages of the latter stain. Staining procedures and, more importantly, the interpretation of the slides, should be done in collaboration with an experienced pathologist. However, tissue preparation and fixation is decisive for the quality of histological slides and will therefore be described here. It is influenced by pH, temperature, osmolality, the concentration of the fixative and the duration of the process, as well as the size of the sample. Fixation for more than 24 h should be avoided.

Usually, tissues are removed, cut into small pieces (e.g. $5 \times 5 \times 5$ mm), put into cassettes and immediately immersed in 10% neutral formaldehyde containing 0.1 M sodium cacodylate, or 2% formaldehyde and 3% glutaraldehyde in cacodylate buffer, a fixative which has been developed for plastic embedding and ultra-thin sections. The tissue is allowed to fix for 12 h at room temperature before being washed for a total of 4 h with two changes of tap water. Dehydration is carried out for 1 h in 70% ethanol and for another hour in absolute ethanol using constant agitation. Subsequent embedding in acrylic resin and further processing should be done in collaboration with a pathologist.

The lung requires special attention, as it should be inflated with 1 ml of fixative via a 26-gauge needle inserted into the trachea prior to removal. The fluid should be retained in the lung by closing the trachea with a suture. Care has to be taken when the entire heart–lung package (Fig. 15) is removed, because the lungs are attached to the dorsal pleura at their cranial part via membranes. Lack of care during removal may thus result in tissue disruption and leakage of the fixative, and interpretation of lung

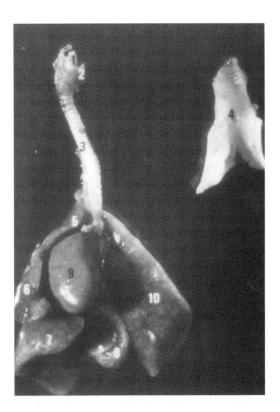

Figure 15. Rodent lung and heart: 1, epiglottis; 2, thyroid cartilage; 3, trachea; 4, thymus; 5–8, cranial lobe, medial lobe, caudal lobe, accessory lobe; 9, heart; 10, left lung (Olds and Olds, 1979).

lesions is complicated by atelectasis. Careful perfusion of the kidneys or liver via the renal or portal vein after incision of the vena cava inferior may, in some cases, increase the quality of histological examination. Tissue is cut into appropriately sized pieces, placed in plastic forms (Crymold, Miles Inc., Elkhart, IN), filled with OCT compound (Miles), and snap frozen in liquid nitrogen and stored at –70°C until further processing. For more details on this procedure, see the chapter by Ehlers and Seitzer (p. 403ff).

In order to obtain reproducible data, at least three slides from different levels of each organ should be examined by an investigator blinded to the experimental conditions. Lesions may be quantified on a number/area or area/area basis (e.g. 0.25 cm^2). Only slides prepared simultaneously and of identical thickness should be compared. Photography increases accuracy and should include calibration information (scale bars).

Investigating inflammatory responses in various compartments: collection of specimens

In addition to histological examination of infected tissues, inflammatory host responses may also be followed by detecting cellular and molecular changes in blood and serum, the peritoneal cavity or the alveolar space.

Since there are significant strain-, age- and method-dependent differences in the composition of inflammatory exudates of mice, normal values should always be established for each laboratory. The following data are presented only to give a rough idea of what can be expected. The values given herein represent data from adult mice.

Blood. For the collection of blood (maximum 1 ml per mouse), mice are sacrificed by CO_2 inhalation and bled within 1–2 min via the heart or brachial vessels. The heart is exposed via an incision in the ventral thoracic area and blood aspirated directly from the right ventricle using a 1- to 2-ml syringe and a 20- to 25-gauge 1-inch needle. Since murine blood clots very fast, it may be helpful to preinject 25 U of heparin in 0.5 ml PBS intraperitoneally 30 min before collection.

Retro-orbital bleeding. Retro-orbital bleeding is a method of obtaining serial blood samples from mice. It is performed with a Pasteur pipette or a capillary tube, and usually yields between 50 and 100 µl of blood. Anaesthesia (see above) should always be used. The Pasteur pipette or capillary tube is heparinized by dipping the pipette or tube into heparin (1000 IE ml^{-1}) and letting the heparin rise into the tube via capillary action. After blowing out the heparin, the pipette or tube is inserted between the nasal epicanthus and the eye and pushed slowly forward with rotation around the longitudinal axis. It is important not to use too much force since the pipette or tube may break. Any injury to the eye must be strictly avoided. Puncture of the retro-orbital venous plexus is achieved when blood starts to flow, sometimes with astonishing speed. After obtaining the blood sample by capillary action, the pipette or tube is removed. Usually, blood stops to flow on removal of the pipette. The animal is allowed to recover and can be bled again on the following day, but the contralateral eye should be used.

Blood can be analysed using a blood film slide spread. It is allowed to dry in air and subsequently stained with May–Grünwald–Giemsa or Diff Quick stain. It should be mentioned that the normal nuclear pattern of neutrophils in the mouse includes ring forms and that the cytoplasmic granulation is very fine. The nucleus of eosinophils is usually coiled and contains small acidophilic granules. The lymphocyte is the prevalent cell type in mouse blood. The nucleus of the monocyte is ameboid in shape, with stranded chromatin. Small vacuoles and a few acidophilic granules are usually present in the grey-blue cytoplasm (Table 5).

Serum. Serum is obtained after blood has been allowed to clot for 1 h at room temperature and subsequently has been centrifuged at 3000*g* to 4000*g*. For the determination of cytokines in serum, it is advisable to centrifuge heparinized blood immediately after collection at 4°C. Microcapillary tubes containing a separating gel (Microtainers, Becton Dickinson & Co., Rutherford, NY) may facilitate serum separation and increase the yield. For composition of serum see Table 6.

The inflammatory response to bacteria can be studied in exudates of the lung or the peritoneal cavity. However, there is significant strain dependency in the intensity of inflammation in response to the same

Table 5. Cellular composition of murine blood

Red cells	8×10^6 to $10 \times 10^6 \mu l^{-1}$
Leukocytes	3×10^3 to $22 \times 10^3 \mu l^{-1}$ (strain and source dependent)
Monocytes	1–14%
Neutrophils	7–28% (strain dependent)
Lymphocytes	65–80%! (strain dependent)
Eosinophils	1–4%
Basophils	< 1%
B cells	20% of lymphocytes
T cells	80% of lymphocytes
CD4$^+$ T cells	52% of lymphocytes
CD8$^+$ T cells	26% of lymphocytes
CD4$^-$CD8$^-$ T cells	0.5–2% of lymphocytes
NK cells	6–11% of mononuclear cells

NK, natural killer.

stimulus. Balb/c, A/J and DBA mice usually show low inflammatory responses, while C57BL/6 and C3H/HeJ are high responders.

Peritoneal cavity. Resident and elicited cells of the peritoneal cavity can be obtained by peritoneal lavage (Plate 3). After cleaning the abdomen with 70% ethanol, the skin is incised and the peritoneal cavity prepared by dissecting the skin cranially and caudally. Peritoneal cells, including resident macrophages, can then be harvested by lavaging the peritoneal cavity with 5–8 ml cold PBS supplemented with 10 U ml^{-1} heparin. A 19-gauge needle or a pipette is inserted in the umbilical region. The fluid should be withdrawn carefully in order to prevent aspiration of abdominal structures, which may result in obstruction, intestinal damage and bacterial contamination of the lavage fluid. Recovery of 70–90% of the injected fluid should be the goal. The cells are washed in PBS by centrifugation at 450g for 10 min (Table 7).

Table 6. Composition of murine serum

Total protein	4–5 g per 100 ml
Albumin	61–70%
α-Globulin	10–20%
β-Globulin	12–20%
γ-Globulin	5–10%
Total immunoglobulin	2.3–6.6 mg ml^{-1}
IgM	0.06–0.17 mg ml^{-1}
IgA	0.08–1.55 mg ml^{-1}
IgG$_1$	0.14–3.2 mg ml^{-1}
IgG$_{2a}$	0.89–3.7 mg ml^{-1}
IgG$_{2b}$	0.38–1.9 mg ml^{-1}
IgG$_3$	0.06–0.14 mg ml^{-1}

Table 7. Murine resident peritoneal cells as obtained by peritoneal lavage

Total cells	2×10^6 to 4×10^6
Macrophages	50–70%
Lymphocytes	25–50%
B cells	n.a.
T cells:	
CD4$^+$	74% of T cells
CD8$^+$	18% of T cells
CD4$^-$CD8$^-$	8% of T cells
Neutrophils	2%
Eosinophils	< 1%
Mast cells	< 1%
NK cells	< 0.5%

NK, natural killer.
n.a., not available.

Alveolar space. Lavage of the non-inflamed alveolar space is an excellent source for almost pure preparations of alveolar macrophages. Bronchoalveolar lavage obtained from mice suffering from pneumonia is helpful in characterizing the inflammatory response. This is easily performed using a 26-gauge Teflon catheter (e.g. Abbocath). After sacrifice, the mouse is placed on its back and fixed on the plate. After cleaning the skin with 70% ethanol, the skin is incised and removed from the mid-abdomen up to the chin. A laparotomy and thoracotomy is performed with sterile instruments. The trachea is prepared by carefully removing the thyroid gland and the muscles covering the trachea using atraumatic surgical techniques. Then, the scissors are inserted into the jugular fossa and the sternum is cut lengthwise towards the abdomen to expose the entire mediastinum. A small tunnel is prepared under the trachea by inserting closed scissors and pushing them forward using the forceps as a resistor. A suture is prepared by pulling an appropriate thread through the tunnel. Then, the Teflon catheter is inserted into the trachea and the hypodermic needle is removed. It is important to insert the catheter at the most cranial point possible so that, if the trachea is perforated accidentally, another, more caudal, insertion is still possible. The catheter is secured by fastening the previously laid suture. The lungs are then lavaged with a maximum of 1 ml sterile physiological saline containing 0.05 M ethylene diaminetetraacetic acid (EDTA). The lavage should be repeated up to 20 times, yielding a total of 20 ml lavage fluid. It is important not to apply too much pressure when injecting the fluid or too much suction when aspirating it, otherwise large blood vessels may rupture and the lavage becomes 'contaminated' with cells from the blood. A non-infected mouse yields about 1×10^5 to 5×10^5 alveolar macrophages in total. Plate 4 shows the typical cellular composition of a bronchoalveolar lavage as assessed by cytospin preparation and Diff Quick stain (Baxter Scientific, McGav Park, IL, USA). The presence of red blood cells indicates that some rupture of capillaries is unavoidable, but there should be no more than 5% of granulocytes in the lavage obtained from a healthy mouse.

◆◆◆◆◆◆◆ IDENTIFYING SPECIFIC HOST RESPONSES *IN VIVO*

B-cell response

The B-cell response to bacterial infections can be studied by determining the amount of specific antibodies in serum and/or mucosal secretions such as intestinal and bronchoalveolar fluid (see above) by the use of enzyme-linked immunosorbent assays (ELISAs) (see the chapter by Hiroi *et al.* (p. 301ff)). The specificity of antibody responses can easily be determined by Western blot analysis. Because in the mouse there is a close relationship between the IgG subclass and T-cell help by T helper 1 and 2 (Th1 and Th2) cells (IgG$_{2a}$/Th1 and IgG$_1$/Th2), determining subclass of specific antibodies may be the first hint of the dominant ongoing T-cell response. The frequency of antigen-specific B cells in blood, lymph nodes, spleen, lung, Peyer's patches or cells of the BALT can be determined using the ELISPOT assay (see the chapter by Hiroi *et al.*, p. 301ff).

Collecting faeces and intestinal fluid for detection of secretory IgA

Detection of secretory IgA in stool specimens can be performed as described by Haneberg *et al.* (1994). During the day, 1–10 stool specimens are collected. The samples are then dried using a speed-vac concentrator. After determination of their weight, the stool specimens are rehydrated and homogenized in PBS containing protease inhibitors such as bestatin (154 nM apotinin, 10 µM leupeptin, 200 µM 4-(2-aminoethyl)benzol-sulfonylfluoride-hydrochloride (AEBSF; molecular weight 239.5), 6 µM bestatin) and dry milk powder. The homogenate is centrifuged and the supernatants are collected. This extraction procedure should be repeated once, and the final extracts are then snap frozen in liquid nitrogen and stored at –70°C until analysis.

An alternative method involves the use of absorbant wicks (Polyfiltronics Inc., Rockland, MA, USA) to collect intestinal fluid (Haneberg *et al.*, 1994). After removing the wicks from the intestine, they are centrifuged for 5 min at 10 000*g* in 0.5-ml Eppendorf tubes, which are pierced at the bottom and placed in a large Eppendorf tube (1.5 ml) without a lid. After centrifugation the samples can be collected at the bottom of the large Eppendorf tube. Alternatively, protease inhibitors (see above) should be added in order to avoid degradation of IgA antibodies by proteases within the secretion fluid. Faecal samples can be dissolved in 5% powdered milk using a vortexer. The fluids should be aliquoted, snap frozen in liquid nitrogen and stored at –70°C, as secretory IgA antibody may form aggregates that make subsequent studies nearly impossible.

Gut washes

The small intestine is flushed with 2 ml PBS containing protease inhibitors (see above). The fluid is centrifuged for 5 min at 1000*g* to remove debris and bacteria. The remainder of the procedure is as described above.

T-cell-mediated immunity is expressed as specific cytotoxicity (see the chapter by Busch and Pamer, p. 281ff) or the release of (proinflammatory) cytokines (see the chapter by Hiroi *et al.*, p. 301ff). The characteristic *in vivo* manifestation of the latter is delayed-type hypersensitivity and granuloma formation. The reaction involves a specifically directed infiltration of mononuclear cells into an area where antigen is localized. The hypersensitivity is referred to as 'delayed' as it takes some 16–24 h to become fully apparent. An essential characteristic is that the reaction is independent of serum antibodies. Great care should be taken to differentiate the DTH reaction from the Arthus reaction. Arthus reactions, however, reach their maximum 4-6 h after application of the antigen, so that these time points should always be included in monitoring the skin response to intradermally injected antigen, especially when Gram-negative bacteria are used. Histologically, DTH is characterized by accumulation of mononuclear cells around small veins. Later, mononuclear cells may be seen throughout the area of the reaction, with extensive infiltration of the dermis. Polymorphonuclear cells constitute fewer than one-third of the cells at any time, and usually very few are present at 24 h or later unless the reaction is severe enough to cause necrosis, which is not typical in the mouse. Some of the swelling may be due to the packing of the tissues with these cells, although oedema and hyperaemia also play a role. The CD4[+] and CD8[+] subsets of T cells in the early perivascular areas are in the same ratio as in peripheral blood, but the cells in the late diffuse infiltrates in the dermis are predominantly CD4[+] T cells. In fact, the *in vivo* depletion of the latter (see p. 397) 24 h before challenge will ablate the reaction.

The tuberculin reaction, as discovered by Koch in 1891, is the prototype of these mechanisms. However, it has certain disadvantages, the main of which are that work with *M. tuberculosis* is hazardous and that the complex cell-wall composition of the tubercle bacillus complicates the study of mechanisms involved in sensitization.

A deeper insight into the mechanisms underlying DTH and cellular immunity developed after the classic work by Chase and Landsteiner. During the 1960s, the importance of cell-mediated immune responses in defence against bacterial infections was emphasized by Mackaness, mainly by use of the murine listeriosis model. Therefore, T-cell responses may well be studied *in vivo* by analysing protective mechanisms against *Listeria monocytogenes* and *Listeria*-induced T-cell-mediated inflammation (Mielke *et al.*, 1997).

Delayed hypersensitivity is a complex *in vivo* reaction, which cannot be duplicated *in vitro*, involving non-specific perivascular accumulation of primed T cells, activation of specific T cells by antigen presenting cells, production of cytokines, which subsequently induce the site-directed accumulation of monocytes mainly via cell derived chemokines and the induction of adhesion molecules on endothelial cells. The only *in vitro* correlate available reflecting part of the reaction is the production of IFN-γ by specific CD4[+] T cells (see the chapter by Hiroi *et al.*, p. 301ff). In the mouse, two types of DTH have to be differentiated: the tuberculin type and

Jones–Mote type. Tuberculin-type hypersensitivity, originally observed in tuberculous guinea-pigs, is the classical type of T cell-mediated inflammation, lasting 48 h even in mice. It is elicited in previously sensitized individuals by intradermal injection of protein antigens. Although it can also be elicited by heat-killed bacteria, the degree of non-specific swelling is generally higher when particulate antigen is used.

The term 'Jones–Mote reaction' originally referred to the delayed reappearance of a hypersensitivity reaction to serum proteins after the development and regression of an Arthus reaction, noted in humans by Jones and Mote in 1934. This term was extended to cover a transient form of delayed skin reaction to protein antigens occurring prior to antibody production in experimental animals. While the Jones–Mote type of lymphocyte-mediated inflammation may be suppressed by CD11b-specific mAb, the classical type cannot. It is not clear whether both types of DTH are mediated by different subsets of CD4$^+$ T cells (Ignatius et al., 1994).

The time course of the reaction is a very unreliable criterion to differentiate between these types of hypersensitivity reactions, so that histology should always be performed demonstrating pronounced PMN accumulation in the Jones–Mote type reaction.

Acquired resistance is not influenced by simultaneous measurement of DTH in the same animal, but challenge of primed animals with a large dose of bacteria or highly virulent micro-organisms may impair the expression of DTH. DTH reactions, although in most cases demonstrated in the skin, may be elicited in various organs when particulate antigen is injected intravenously. Under these conditions it will present as focal accumulations of monocytes and lymphocytes in tissue, i.e. granulomas. Therefore, challenging immunized mice with heat-killed bacteria and quantifying inflammatory tissue lesions will be an alternative method for investigating CD4$^+$ T-cell-mediated inflammation *in vivo*. While granulomatous lesions are difficult to determine in lymphatic organs such as lymph nodes and spleen, quantification is easy in a parenchymal non-lymphoid organ like the liver. Reliable quantification is possible if more than 10^6 particles are injected intravenously and the liver is removed 48–72 h after injection. However, no sequential determinations within the same animal are possible using this read-out. Antigen complexed to insoluble particles, such as latex beads, will also produce granulomas when injected into immunized animals, and therefore will enable the investigator to characterize the host response to defined antigens (Brocke et al., 1991). This may be important if non-degradable material is present in the bacterial preparation, as in mycobacteria, since under these conditions, T-cell-independent granulomas may occur. The great advantage of this method is easy histopathological analysis. The composition of granulomatous lesions may vary depending on the type of Th response (e.g. cercaria-induced granulomas (*Schistosoma mansoni*) may contain 50% eosinophils, 30% macrophages, 10–15% T cells, 5–10% other cells).

While CD4$^+$ T-cell-mediated responses such as DTH and granuloma formation can easily be detected, T-cell-mediated cytotoxicity is much more difficult to demonstrate. In fact, there is no model of bacterial infections in which the cytotoxic action of T cells has been shown directly using fresh, non-restimulated T cells *ex vivo*. Consequently, the effect of CD8+ T-cell-subset-depletion may be the only way to demonstrate their action *in vivo* (see chapter by Busch and Pamer).

Preparation of soluble antigen and antigen-coated latex beads

Soluble antigen from bacteria may be obtained by ultrasonication of heat-killed bacteria. Bacteria (1 g wet wt) is suspended in 10 ml PBS. This suspension is then sonicated 5–10 times for 30 s to 1 min each, on using, for example, a Branson Sonic Power sonifier placed in a safety cabinet. The sonicated suspension is centrifuged at 40 000g for 1 h and the soluble fraction is filter sterilized using a 0.22-µm pore filter. The protein concentration of this stock solution has to be determined and aliquots are stored frozen at –70°C.

An alternative method of obtaining soluble antigen, especially when secreted metabolic antigens are required, is to inoculate protein- and antibiotic-free RPMI 1640 cell culture medium with 1×10^8 bacteria. The culture is constantly agitated using Cellspin bottles (Bioscience, Inc., Tecnomara, Chino, CA, USA) and then incubated in an atmosphere of 5% CO_2 in air for time periods depending on the type of bacterium (16 h to 7 days). Subsequently, the suspensions are centrifuged and filter sterilized. The filter-sterilized supernatant may be concentrated by ultrafiltration (amicon, millipore).

DTH is usually elicited by intradermal injection (see above) with 3–5 µg of soluble antigen in PBS in a volume of 50 µl. The injection site should be inspected after 6, 24 and 48 h. The thickness of e.g. each hind foot is measured with a dial-gauge caliper (Fig. 16). Specific foot-pad swelling is reported as (Right minus left foot-pad thickness in primed mice) – (Right minus left foot-pad thickness in mice that received vehicle only).

Coating of latex beads (3 µm diameter, Sigma, Germany) with soluble antigen can be achieved by washing the native beads in PBS, collecting them by centrifugation at 1200g for 10 min, resuspending them at a

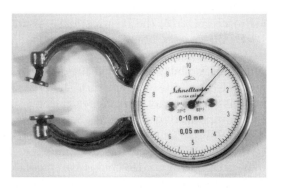

Figure 16. Dial-gauge caliper for measuring foot-pad thickness.

concentration of 10% vol./vol. in PBS containing the soluble bacterial antigen or an irrelevant protein such as ovalbumin at a concentration of 30 µg ml^{-1}. The beads are then co-incubated with the antigen for 1 h at room temperature, washed twice in PBS, and subsequently injected intravenously at a dose of 200 µl of a 10% vol./vol. suspension per mouse (Brocke *et al.*, 1991).

An alternative method to elicit T-cell-mediated monocyte accumulation in the liver is to use heat-killed bacteria injected intravenously at a dose of 2×10^8 per animal. This dose will result in a quantifiable density of granulomas per liver section (Mielke *et al.*, 1992).

Cytokine response

Bacterial infections usually induce strong cytokine responses that can be determined both systemically (in serum) and locally at both the organ and the cellular level *in situ*. The cytokine response to bacteria starts almost immediately after infection, as macrophages begin to transcribe message from a number of early response genes and cytokines such as IL-1, IL-6, IL-10, IL-12, macrophage chemotactic protein (MCP)-1, macrophage inflammatory protein (MIP)-1α, MIP-1β, granulocyte, macrophage and granulocyte–macrophage colony-stimulating factors (G-CSF, M-CSF, GM-CSF) and tumour necrosis factor α (TNFα), which in turn induce the production of IFN-γ by natural killer (NK) cells.

Global cytokine responses may be monitored by measuring cytokine levels in serum (see the chapter by Hiroi *et al.*, p. 301ff). Therefore, at the organ level, cytokines may best be determined by semiquantitative polymerase chain reaction (PCR) or RNase protection assays (see the chapter by Ehlers and Seitzer, p. 403ff). Another relatively reliable method of obtaining information about infection-induced cytokines is to measure spontaneous and antigen-induced cytokine production in spleen cell cultures (see the chapter by Hiroi *et al.*, p. 301ff) obtained at various time points after infection. The production of cytokines at the cellular level may be demonstrated by immunohistology or by *in situ* hybridization (see the chapter by Ehlers and Seitzer, p. 403ff).

For the determination of cytokines in organ homogenates, tissues should be weighed and suspended in 10 times their weight of cold RPMI 1640 containing 0.2% BSA, 0.02% sodium azide and 1% wt/vol. (cholamidopropyl)-dimethylammonio)-1-propane sulfate (CHAPS; Wako Pure Chemical Co., Osaka 541, Japan). In this solution, tissue is homogenized on ice and the homogenate is centrifuged at 1000g for 10 min at 4°C. The supernatant may be filter sterilized using low protein binding filters and stored in aliquots at –70°C until determination of cytokine levels using specific ELISAs (see the chapter by Hiroi *et al.*, p. 301ff).

For mRNA analysis, organs should be excised, weighed and immediately homogenized in guanidine isothiocyanate (4 M guanidium isothiocyanate, 0.5% N-lauroylsarcosine, 25 mM sodium citrate, 100 mM 2-mercaptoethanol) and frozen at –70°C for future use in preparing mRNA (see the chapter by Hiroi *et al.*, p. 301ff).

◆◆◆◆◆◆ MANIPULATING THE IMMUNE RESPONSE
IN VIVO

Up to now descriptive methods have been discussed. However, pheno-typical analysis of the various aspects of the host response does not tell us all about the functional role in inflammation and defence. In fact, temporal correlations have been shown to be unreliable in terms of elucidating a causal relationship. Only the loss of a certain function after depletion of one component of the immune system allows some causal inference. In this respect, both the depletion of cells and the neutralization of soluble mediators can be achieved by means of monoclonal antibodies or specific heterologous sera. This approach competes with the use of gene knock-out (GKO) mice (see the chapter by Mossmann *et al.*, p. 183ff), but has certain advantages, especially in models in which the period of observation is short (up to 4 weeks). In contrast to GKO mice, the mAb-mediated depletion of certain cells or the neutralization of cytokines has no long-lasting effects that may favour compensatory mechanisms such as the increase in NK-cell activity in T-cell-deficient mice, or may even result in anatomical differences such as a lack of Peyer's patches (PP) in GKO mice deficient in TNF-RI. Therefore, whenever a transient, temporally defined immunomodulation is of interest, studies using antibodies for depletion or neutralization may be of advantage. In particular, in situations in which the role of a cell population either in the induction or the expression of immunity is to be explored, monoclonal antibodies are ideal tools. On the other hand, GKO mice are well suited to the demonstration of the compensatory potential of the still active cells and mechanisms.

Depleting PMNs, macrophages and NK cells

The SCID mouse is a well-established model for studying mechanisms of innate resistance, due to its inability to generate specific antigen receptors. In this context it may be interesting to abolish certain aspects of non-specific resistance by depleting PMNs, monocytes, resident macrophages or NK cells.

While the depletion of PMNs and NK cells can be achieved relatively easily by *in vivo* application of certain mAbs, the selective depletion of monocytes and/or resident macrophages is more difficult to secure. To deplete mice of neutrophils, mAb RB6-8C5 (R. Coffman, DNAX Research Institute, Palo Alto, CA, USA) is administered intraperitoneally in a dose of 0.25 mg, 1 day prior to initiating infection. It has been shown that treating mice in this way renders them severely neutropenic for at least 3 days. Control mice should receive normal rat IgG (Sigma Chemical Co., St Louis, MO, USA).

The depletion of NK cells can be achieved by intraperitoneal application of 400 μg rabbit anti-asialo-GM1 antibody (Wako Pure Chemical Co.) in 0.5 ml pyrogen-free PBS 2 days before challenge with antigen. Normal rabbit globulin should be used as a control. A single injection results in a marked reduction in splenic NK activity for at least 7 days.

Recently, a method for selectively depleting animals of resident macrophages by liposome preparations has been described. Multilamellar liposomes containing either dichloromethylene diphosphonate (Cl$_2$MDP; Boehringer Mannheim GmbH, Mannheim, Germany) or PBS may be prepared as described by van Rooijen and Sanders (1994). Briefly, 75 mg phosphatidylcholine (Lipoid GmbH, Ludwigshafen, Germany) and 11 mg cholesterol (Sigma Chemical Co., St Louis, MO, USA) are dissolved in chloroform, evaporated by rotation under vacuum at 37°C, dispersed by mixing with 10 ml PBS containing 2.5 g Cl$_2$MDP for 10 min, left at room temperature for 2 h, sonicated for 3 min at room temperature in a water bath sonicator, and left at room temperature for 2 h. The resulting liposomes containing Cl$_2$MDP (L-Cl$_2$MDP) are washed twice in PBS by centrifugation at 100 000g for 30 min to remove non-encapsulated Cl$_2$MDP. PBS-containing liposomes (L-PBS) are prepared by the same procedure, except that the phosphatidylcholine–cholesterol mixture is dispersed in 10 ml PBS without Cl$_2$MDP. The L-Cl$_2$MDP and L-PBS are each suspended in 4 ml PBS, and 0.2 ml of either preparation is injected intravenously per mouse.

Depleting T-cell subsets

An efficient method of elucidating the role of T-cell subsets *in vivo* is to utilize their selective depletion (Waldmann, 1989). However, antibodies induced against the injected monoclonal antibodies may limit their repeated use in long-term experiments. The only exception is the depletion of CD4$^+$ T cells by mAb GK 1.5 (ATCC TIB 207), for which it has been shown that no such antibodies will be produced. CD8$^+$ T cells may be depleted by IgG$_{2b}$ antibodies from the rat hybridoma 2.43 (ATCC TIB 210), and for Thy 1.2$^+$ cells rat IgG$_{2b}$ antibodies from clone 30H12 (ATCC TIB 107) can be used. Intraperitoneal injection of 400 µg of the mAbs per mouse will result in a profound depletion lasting about 7 days.

In order to prevent non-specific effects on the macrophage system by cell destruction, and to guarantee efficacy, mAbs should be injected 2 days before the start of the experiment. The efficacy of depletion has to be verified by fluorescence-activated cell sorting (FACS) analysis of splenocytes or Peyer's patches (see the chapter by Scheffold and Radbruch, p. 23ff) (Tables 8 and 9), at least at the end of the experiment. In order to prevent

Table 8. Cellular composition of murine spleen

Total number of cells	0.5×10^8 to 2×10^8 per spleen
B cells	65% of lymphocytes
T cells	35% of lymphocytes
CD4$^+$ T cells	20% (65% of T cells)
CD8$^+$ T cells	15% (32% of T cells)
CD4$^-$CD8$^-$ T cells	0.5–2% of lymphocytes (3% of T cells)
Monocytes	10% of total cells
NK cells	5% of total cells

NK, natural killer.

Table 9. Cellular composition of murine Peyer's patches

T cells:	20%
CD4+	12%
CD8+	8%
CD4-CD8-	50% of lymphocytes in intestine
B cells	80%
IgG+ B cells	70%
Phagocytes	3%

failure of cell detection by blockage of surface markers due to the antibodies given *in vivo*, a sandwich method should be used to label the cells.

Neutralization of cytokines

Cytokines can be neutralized *in vivo* by intraperitoneal administration of mAbs. The neutralizing capacity of the antibodies should be determined in order to verify the neutralizing effect, since the amount of antibody required varies significantly, ranging from 125 µg to 4 mg per mouse. In variation of the procedure used for cell depletion, neutralizing antibodies may be administered up to 4 h before infection or immunization. Equal amounts of normal rat or rabbit immunoglobulin of the same isotype must be injected as a control.

The effects of administration of cytokines *in vivo* is even more dependent on dose and route of application. Some cytokines are very potent immunomodulators and act in minute concentrations, while other cytokines need to be present in high concentrations to promote immunomodulatory effects *in vivo*. Furthermore, co-administration of an infective agent may dramatically increase the effects of a cytokine. Cytokines may be administered intravenously or intraperitoneally, but usually act in a more protracted manner after subcutaneous injection. Compared to the value of *in vivo* neutralization, the usefulness of recombinant cytokines is much lower due to problems of pharmacokinetics, distribution and secondarily induced cytokines.

All reagents used *in vivo* should be tested for endotoxin contamination. Best suited for this purpose is the Limulus amoebocyte lysate assay (Associates of Cape Cod, Woods Hole, MA, USA; Bio Whittaker, Walkersville, MD, USA; Sigma), which detects amounts as low as 10 pg ml^{-1}. The goal is to inject no more than 10 pg LPS per dose.

◆◆◆◆◆◆ METHODS FOR ASSESSING IMMUNOLOGICAL MEMORY

Infectious processes may be divided into (a) an early phase, in which mechanisms of innate resistance are expressed and immunity is induced, and (b) an immune effector phase, which is rapidly switched off. Finally, a state of long-term immunological memory develops. If memory reflects the extent of previous lymphocyte proliferation that is maintained, the optimal method of measuring this may be to determine the frequency of specific T and B cells (see the chapter by Hiroi *et al.*, p. 301ff). If memory represents a change in the physiological status of, for example, IL-producing or cytotoxic T cells, rather than just an increase in numbers, differences in cytokine patterns and cytotoxic activity should be studied (see the chapter by Busch and Pamer, p. 281ff).

Protection against secondary challenge is chosen as the most important measure of memory *in vivo*. Kinetics of a challenge infection should be monitored in naive versus primed or adoptively immunized animals. However, because of the complexities of the latter parameters, these *in vivo* read-outs are sometimes difficult to interpret. For example, it may be particularly troublesome to evaluate the roles of B cell, Th or cytotoxic T-cell memory. Under these conditions, depletion studies (see above) are most helpful. In any case, to get a complete picture, both high- and low-dose challenges should be analysed in animals immunized with both high and low doses of bacteria.

In contrast to cytokine production by T cells, there is at present no *ex vivo* assay available for measuring directly the cytolytic activity of *in vivo* primed cytotoxic memory T cells with specificity for bacteria, making acquired immunity to secondary challenge the only reliable read-out.

Adoptive transfer of immunity

While studies based on the depletion of a certain immunologically important component will reveal whether it is *necessary*, the adoptive transfer of a phenomenon by serum or cells will reveal whether they are *sufficient*. It has to be kept in mind, however, that adoptive transfer studies correlate more with mechanisms expressed during secondary challenge than with those active during the eradication of a primary infection. The ability to transfer protection adoptively by means of 200–400 µl hyperimmune serum points to a role of antibodies in immunity to a secondary challenge.

Adoptive transfer of specific T cells may demonstrate their functional potency in mediating immunity when given before challenge. However, the interval after infection in which adoptive transfer of cell-mediated immunity is possible may be quite short, since it is often restricted to the immune effector phase, i.e. just before most activated cells undergo apoptosis and immunological memory develops. Cells able to transfer immunity may be enriched in the peritoneal cavity 3 days after the intraperitoneal injection of a sterile inflammatory irritant such as 10% proteose pepton in PBS. Peritoneal exudate T cells (PETLs) may then be obtained by peritoneal lavage (see above) and enriched by using methods (see the chapter by Czuprynski and Brown, p. 233) based on the adherence of macrophages and B cells to plastic surfaces. Usually 5×10^6 PETLs suffice to transfer a certain degree of immunity. Another method of transferring cellular immunity is based on spleen equivalents. The easiest way to do this is to disperse the cells of a spleen of a donor mouse (see the chapter by Czuprynski and Brown, p. 233) and to inject up to 1×10^8 cells intravenously in 0.3 ml PBS into the recipient. Adoptively transferred activated macrophages are usually unable to confer protection. However, in infections with intracellular bacteria, viable bacteria may be transferred with the cell suspension if macrophages are not depleted.

In order to track lymphocytes in tissues after adoptive transfer, cells can be labelled with fluorescent dyes such as PKH2-GL (Dianova, Hamburg, Germany). After incubation of cells with this compound the cells should be washed and thereafter injected intravenously into the animal. PKH2-GL allows tracing of cells in organs by FACScan or fluorescence microscopy for up to 4 days. Alternatively, cells can be labelled with ^{51}Cr and traced *in vivo* with a γ-counter (Hamann *et al.*, 1994).

References

Autenrieth, I. B. and Firsching, R. (1996a). Penetration of M cells and destruction of Peyer's patches by *Yersinia enterocolitica*: an ultrastructural and histological study. *J. Med. Microbiol.* **44**, 285–294.

Autenrieth, I. B., Kempf, V., Sprinz, T., Preger, S. and Schnell, A. (1996b). Defense mechanisms in Peyer's patches and mesenteric lymph nodes against *Yersinia enterocolitica* involve integrins and cytokines. *Infect. Immun.* **64**, 1357–1368.

Bancroft, J. D. and Stevens, A. (Eds) (1996). *Theory and Practice of Histological Techniques*, 4th edn. Churchill Livingstone, Edinburgh.

Brocke, S., Chakraborty, T., Lombardi, O., Hahn, H. and Mielke, M. (1991). Listeriolysin negative mutants of *Listeria monocytogenes* specifically stimulate T-lymphocytes mediating protection and granulomatous inflammation. *Z. Bakt.* **359**, 758–761.

Contag, C. H., Contag, P. R., Mullins, J. I., Spilman, S. D., Stevenson, D. K. and Benaron, D. A. (1995). Photonic detection of bacterial pathogens in living hosts. *Mol. Microbiol.* **18**, 593–603.

Hamann, A., Andrew, D. P., Jablonski-Westrich, D., Holzmann, B. and Butcher E. C. (1994). Role of α 4 integrins in lymphocytes homing to mucosal tissues *in vivo*. *J. Immunol.* **152**, 3282–3293.

Haneberg, B., Kendall, D., Amerorigen, H. M., Apter, F. M., Kraehenbuhl, J.-P. and Neutra, M. R. (1994). Induction of specific immunoglobulin A in the small intestine colon–rectum, and vagina measured by a new method for collection of secretions from local mucosal surfaces. *Infect. Immun.* **62**, 15–23.

Hopkins, S., Kraehenbuhl, J. P., Schödel, F., Potts, A., Peterson, D., De Grandi, P. and Nardelli-Haeflinger, D. (1995). A recombinant *Salmonella typhimurium* vaccine induces local immunity by four different routes of immunization. *Infect. Immun.* **63**, 3279–3286.

Ignatius, R., Mielke, M. and Hahn, H. (1994). *Mycobacterium bovis* (BCG)-induced immunomodulation of the DTH to SRBC results in CD11b-independent, CD4[+] T-cell mediated myelomonocytic cell recruitment. *Cell. Immunol.* **156**, 262–266.

Imakoa, A., Matsumoto, S., Okada, Y. and Umesaki, Y. (1996). Proliferative recruitment of intestinal intraepithelial lymphocytes after microbial colonization of germ-free mice. *Eur. J. Immunol.* **26**, 945–948.

Lecuit M., Vandormael-Pournin, S., Lefort, J., Huerre, M., Gounon, P., Dupuy, C., Babinet, C. and Cossart, P. (2001). A transgenic model for listeriosis: role of internbalin in crossing intestinal barrier. *Science* **292**, 1722–1725.

Lindgren, S. W., Melton, A. R. and O'Brian, A. D. (1993). Virulence of enterohemorrhagic *Escherichia coli* O91:H21 clinical isolates in an orally infected mouse model. *Infect. Immunol.* **61**, 3832–3842.

Mielke, M. E. A., Rosen, H., Brocke, S., Peters, C. and Hahn, H. (1992). Protective immunity and granuloma formation are mediated by two distinct TNF-α and IFN-γ dependent T cell–phagocyte interactions in murine listeriosis: dissociation on the basis of phagocyte adhesion mechanisms. *Infect. Immun.* **60**, 1875–1882.

Mielke, M. E. A., Peters, C. and Hahn, H. (1997). Cytokines in the induction and expression of T cell-mediated granuloma formation and protection in the murine model of listeriosis. *Immunol. Rev.* **158**, 79–93.

Olds, R. and Olds, J. (1979). *A Colour Atlas of the Rat – Dissection Guide*. Wolfe Medical, London.

Reed, L. J. and Muench, H. (1938). A simple method of estimating fifty per cent endpoints. *Am. J. Hyg.* **27**, 493–497.

Umesaki, Y., Okada, Y., Matsumoto, S. and Setoyama, H. (1993). Expansion of alpha-beta T cell receptor bearing intestinal intraepithelial lymphocytes after microbial colonization in germ-free mice and its independence from the thymus. *Immunology* **79**, 32–37.

Van Rooijen, N. and Sanders, A. (1994). Liposome mediated depletion of macrophages: mechanism of action, preparation of liposomes and applications. *J. Immunol. Methods* **174**, 83–93.

Versteeg, J. (1985). *A Colour Atlas of Virology*. Wolfe Medical, London.

Wadolkowski, E. A., Sung, L. M., Burris, J. A., Samuel, J. E. and O'Brian, A. D. (1990). Acute renal tubular necrosis and death of mice orally infected with *Escherichia coli* strains that produce Shiga-like toxin type II. *Infect. Immun.* **58**, 3959–3965.

Waldmann, H. (1989). Manipulation of T-cell responses with monoclonal antibodies. *Ann. Rev. Immunol.* **7**, 407–444.

Wiedermann, U., Hanson, L. A., Holmgren, J., Kahn, H. and Dahlgren, U. I. (1993). Impaired mucosal antibody response to cholera toxin in vitamin A deficient rats immunized with oral cholera vaccine. *Infect. Immun.* **61**, 3952–3957.

3.2 Measuring Immune Responses *in vivo*

Stefan Ehlers
Division of Molecular Infection Biology, Research Center Borstel, Germany

Jörg Lehmann
Institute of Immunology, College of Veterinary Medicine, Leipzig, Germany

Kerstin Müller and Tamás Laskay
Institute for Medical Microbiology and Hygiene, Medical University of Lübeck, Germany

Jan Buer and Jürgen Lauber
Gesellschaft für Biotechnologische Forschung (GBF), Braunschweig, Germany

◆◆

CONTENTS

Measuring Immune Responses in vivo

◆◆◆◆◆◆ INTRODUCTION

When analysing the immune response during infection in human patients or in experimentally infected mice, there are two major goals: (a) to detect differences in the kinetics and magnitude of, for instance, cytokine or chemokine expression in individual tissues, and (b) to define the cellular localization of this response.

The standard procedure to allow detection of molecules associated with individual cells within structurally intact tissue is immunohisto-chemistry. More often than not, detection at the protein level is, however, not feasible because the abundance of the target structure *in situ* is too small to be accessible to immunohistochemical or ELISA-based detection. *In vitro* restimulation of *ex vivo* isolated cell populations may then be used to boost protein expression which can subsequently be visualized by intracellular fluorescence staining and flow cytometric analysis.

An alternative strategy is to use highly sensitive radioactive probes in a format that allows simultaneous detection of multiple mRNAs expressed in infected tissues, e.g. ribonuclease protection assays. Even more sensitive

METHODS IN MICROBIOLOGY, VOLUME 32
ISBN 0–12–521532–0

Copyright © Elsevier Science Ltd
All rights of reproduction in any form reserved

procedures use amplification strategies of reverse-transcribed mRNAs for cytokines and chemokines (qualitative or semi-quantitative RT-PCR). These methods have the distinct disadvantage that the cells expressing the mRNA can no longer be directly identified. However, a combination of the approaches outlined in this chapter will often lead to a complete picture of the nature of the infection-induced immune response.

◆◆◆◆◆◆ IMMUNOHISTOCHEMISTRY

Tissue preparation

Frozen tissues

1. Remove tissues and place into a flat-bottomed cylindrical sealable plastic tube filled with sterile PBS or normal saline to fully cover the tissue sample. Antigenicity is best preserved under these conditions, although some researchers prefer commercial embedding media (e.g. Tissue freezing medium, Leica Instruments).
2. With a long forceps, immediately place the sealed container in an upright position into liquid nitrogen until frozen. It is important to snap-freeze fresh specimens to prevent the formation of ice crystals in the tissue. These will appear as slits or fracture lines and give rise to artefactual staining. Containers may be stored at –70°C indefinitely. Tissues should always be maintained in the frozen state. If they are allowed to thaw, refreezing will result in extensive damage to the tissue as well as loss of antigenicity.
3. Prepare cryostat sections of 4–5 μm thickness.
4. Air dry sections for 4 –24 h. Fix for 15–30 min in acetone followed by 15–30 min in chloroform. (If the cryostat sections are not meant to be stained the following day, it is advised to also air dry them for 4–24 h, fix them for 10 min in acetone and store them at –70°C. When needed, thaw the sections covered with a paper towel in order to prevent water condensation on the frozen slides, and fix in acetone and chloroform as detailed above.)
5. For some staining procedures (particularly to detect cytokines), fixing of specimens with paraformaldehyde/saponin is preferred. For this purpose, fix the air-dried slides for 15 min in 4% paraformaldehyde in phosphate buffered saline without NaCl (pH 7.4–7.6), followed by a 15-min treatment in 0.1% saponin/PBS to remove cholesterol from the membranes. Wash thoroughly with PBS before immunostaining.

Paraffin-embedded tissues

1. Prepare sections from paraffin blocks onto pre-cleaned, grease-free slides.

(contd.)

---

2. Dewax routinely processed paraffin sections by submerging them for 10 min in xylene followed by 10 min in acetone and 10 min in a 1 : 1 mixture of acetone and Tris-buffered saline (TBS: 50 mM Tris, 150 mM NaCl, pH 7.5). Keep the sections in TBS until staining.
3. Deparaffinate the sections for 5 min in xylene and rehydrate them in a series of acetones at 5-min intervals (100%, 70%, 40%). Rinse under running tap water.

Microwave pre-treatment

1. Transfer the slides to a plastic staining jar filled with 10 mM sodium citrate buffer (pH 6).
2. Place the plastic staining jars into a microwave oven and heat for 5 min at 720 W. After 5 min it is essential that the staining jar be refilled with buffer to prevent the specimens from drying out. The frequency of the heating steps depends on the embedding procedure performed previously and must be optimized for each laboratory individually. Thus, the time of formalin fixation and the quality of the paraffin may influence antigen retrieval. In most instances, heating the sections three to six times for 5 min will be sufficient to give good results.
3. After microwaving, let the slides cool down in the staining jar for approximately 20 min at room temperature and rinse briefly in TBS before proceeding with immunostaining.

Pressure cooker pre-treatment

1. Fill a normal household pressure cooker with enough 10 mM sodium citrate buffer to cover the slides (approx. 2 l). Bring the buffer to a boil before submerging the slides, close the lid and heat until the top pressure is reached. Boil the sections at this pressure for 1 to 10 min. As with microwaving, the boiling time must be optimized depending on the embedding procedure used, because variables such as time of formalin fixation, quality of paraffin and section thickness may influence antigen retrieval.
2. Cool the pressure cooker under running cold water. (Take extreme care when performing this step and absolutely follow the manufacturer´s instructions for opening the cooker.)
3. Transfer the sections into TBS and wash in TBS three times before proceeding with immunostaining.

Staining procedure: Peroxidase method

For a more detailed outline of various immunoenzymatic methods, please refer to the chapter by Seitzer, Endl, Hollmann and Gerdes in Section III. When working with mouse tissue, the primary antibody is most likely a rat monoclonal antibody or a rabbit polyclonal antiserum. Below is a description of primary and secondary reagents necessary for the horse-radish peroxidase (HRP) method which should be a useful guideline when adapting other protocols for use in animal tissues.

All incubations with antibodies are performed in a level humid chamber at room temperature, using approximately 100 µl of antibody solution. Excess humidity should be avoided since water condensation on the slides will interfere with the staining reaction. On the other hand, insufficient humidity will dry out the antibody on the sections, resulting in false-positive staining. Drying is usually most apparent at the edge of the sections (rim effect). Between incubations, wash the slides twice in TBS in staining jars and drain off excess fluid by capillary action using a paper towel.

---

1. Always include negative controls without primary antibody and with an irrelevant primary antibody of the same immunoglobulin class (or with pre-immune serum). The specificity of the primary antibody can be analysed by neutralization experiments in which blocking of antibody binding results in negative staining. For this purpose, pre-incubate the antibody with recombinant or purified antigen for 30 min at 37°C before applying it to the slides.
2. Before incubating the slides with primary antibody, block endogenous peroxidase activity in the tissue by pre-incubating the slides for 20 min in a light-protected staining jar with 1% H_2O_2 TBS.
3. Add the rabbit primary antiserum or rat monoclonal antibody in the appropriate dilution in TBS containing 10% FCS. (To find the appropriate concentration of primary antibody, perform three-fold serial dilutions to optimize the ratio of background and specific staining.) Incubate for 30 min.
4. Add the first secondary HRP-conjugated goat anti-rabbit-IgG antibody (Dianova; diluted 1 : 30 in TBS/10% FCS and 20% mouse serum). When a rat monoclonal antibody is used as the primary step reagent, add a HRP-conjugated rabbit-anti-rat-IgG antibody (Dianova, diluted 1 : 30 in TBS/10% FCS and 20% mouse serum). The optimal concentration of secondary reagents is usually indicated by the manufacturers.
5. Incubate for 30 min.
6. Add the second secondary antibody: HRP-conjugated rabbit-anti-goat-IgG, or HRP-conjugated goat-anti-rabbit-IgG (Dianova; diluted 1 : 50 in TBS/10% FCS) depending on the primary and first secondary reagents used. Incubate for another 30 min.
7. Prepare the developing buffer by dissolving 6 mg 3,3´-diamino-benzidine-tetrahydrochloride (Sigma) in 10 ml TBS. Add 100 µl H_2O_2 directly before use. Mix well.
8. Incubate the slides with developing buffer in the dark in the humid chamber for 3 to 15 min. The degree of development can be checked microscopically and is terminated at the desired point.

(*contd.*)

---

9. Counterstain with haematoxylin for 90–105 s. Haematoxylin stock solution is prepared from 1 g haematoxylin (Merck 4305), 0.2 g NaJO₃ (Merck 6525), 50 g aluminium potassium sulfate dodecahydrate (KAl(SO₄)₂ x 12 H₂O (Merck 507 A 95744) in 1 l distilled water. After these reagents are dissolved, add 50 g chloralhydrate (Merck 2425) and 1 g of citric acid (Merck 244). Before staining, pass the solution through a sterile filter. Haematoxylin solution may be reused a few times.
10. Rinse the slides in running tap water and leave in tap water for 5 min before mounting with pre-warmed (56°C) Kaiser's glycerol-gelatine.

Typical results using this method are shown in Plate 5.

◆◆◆◆◆◆ FLOW CYTOMETRIC APPROACHES

In situ immunohistological techniques are often not suitable for the simultaneous staining of an intracellular pathogen (and/or protein) and additionally one or more specific surface markers to identify the infected (or protein-expressing) cell type. In contrast, flow cytometry allows for the simultaneous analysis of several parameters, including size (forward scatter) and granularity (side scatter), as well as the expression of surface and intracellular molecules (via fluorochrome-labelled antibodies) on a single cell level.

Several sample materials inherently represent single-cell suspensions and can therefore be stained without time-consuming preparation procedures (i.e. blood, lymphoid fluid, bone marrow, cerebrospinal fluid, or broncho-alveolar lavage). Cells from a range of other tissues or organs can also be easily prepared for intracellular flow cytometric analysis. For example, spleen or lymph node cells can be isolated simply by mechanical manipulation (passage through a sieve of defined mesh size) without enzymatic disintegration (Lehmann *et al.*, 2001). Epithelial cells, endothelial cells or cells from the connective tissue (i.e. synovial fibroblasts, macrophages, T cells) may be obtained by enzymatic disintegration (Zimmermann *et al.*, 2001) or in certain cases also by 'growing-out' methods (Lehmann *et al.*, 2000). Interestingly, a broad range of surface marker antigens have been shown to be inert against treatment with trypsin or collagenase, both of which are often used for tissue disintegration.

Intracellular immunofluorescent staining of bacterial or viral antigen in murine phagocytes and dendritic cells *ex vivo*

1. Gently fix 10⁷ target cells in 500 µl cold phosphate-buffered 4% paraformaldehyde (PFA; Serva) for exactly 10 min. Permeabilize with the detergent saponin (0.1%; Sigma).

(contd.)

Measuring Immune Responses *in vivo*

2. Following fixation, wash cells twice in 10 ml cold phosphate-buffered saline (PBS) containing 1–5% fetal calf serum (FCS; Seromed®, Biochrom). Pellet by centrifugation for 10 min at 350g, 4°C. Depending on the antigen, the cells can be stored as a pellet with about 100 μl supernatant under sterile conditions for up to one month. This may be of interest for the storage of ready-prepared and fixed standard cells (e.g. a permanently infected cell line) for use as a positive control. However, the stability of any individual antigen has to be confirmed for each specific application.

3. Immediately before staining, wash cells once again. After removing the supernatant, resuspend pellet in 10 ml cold permeabilization buffer (PBS containing 0.1% saponin, 1% FCS and 0.01 M HEPES) and centrifuge for 10 min at 350g, 4°C. Repeat this step twice.

4. Adjust cells to $(5 \times 10^6 \text{ ml}^{-1})$ in the required volume. Centrifuge again.

5. In order to block non-specific binding of fluorochrome-conjugated antibodies to FcγRII/III, resuspend the cell pellet in 100 μl permeabilization buffer supplemented with purified 2.4G2 antibody (anti-CD16/CD32; 1 μg per 10^6 cells; Fc Block™, BD-Pharmingen) and preincubated for 15 min at room temperature. Adjust cell suspension to the required volume in permeabilization buffer.

6. To 5–10 μl adequate antibody combination (e.g. FITC-conjugated anti-*Salmonella* + PE-conjugated anti-CD11b, usually at 1–5 μg ml^{-1}; if necessary, dilute in permeabilization buffer) in 4 ml polystyrene tubes (BD-Falcon®), add 100 μl cell suspension. Stain cells in permeabilization buffer for 30 min at 4°C. Wash three times in permeabilization buffer to remove unbound antibody.

7. Resuspend stained cells in 250 μl fixation buffer containing 1% formaldehyde (CellFix™, BD) for storage at 4°C until flow cytometric analysis.

A typical example of stained *Salmonella enteritidis* in murine splenic phagocytes is shown in Plate 6A. In an analogous procedure, some intracellular viral antigens, such as Borna Disease Virus and Pestivirus, may also be successfully stained (Qvist *et al.*, 1990; Bode *et al.*, 1995).

Intracellular immunofluorescent staining of cytokines in murine splenocytes *ex vivo*

To stain for T-cell derived cytokines (e.g. IFN-γ, TNF-α, IL-2, IL-4) induced during infection *in vivo*, murine splenocytes have to be re-stimulated with phorbol myristate acetate (PMA) [10 ng ml^{-1}] + ionomycin [1 μM] (Sigma) at a cell density of [5×10^6 ml^{-1}] at 37°C and 5% CO$_2$ for 5 h. Alternatively, the cells may be stimulated with plate-bound anti-mouse CD3 [25 μg ml^{-1}] + anti-mouse CD28 [2 μg ml^{-1}]. For measurement of intracellular cytokines in monocytes/macrophages (e.g. IL-1, IL-6, TNF-α, IL-10, IL-12, IL-18), the cells have to be stimulated with LPS [1 μg ml^{-1}], or

LPS [100 ng ml^{-1}] + IFN-γ [100 U ml^{-1}], or heat-killed salmonellae [10^8–10^9 cfu ml^{-1}] for 5 h. During stimulation of T cells or monocytes, cytokine secretion must be inhibited by monensin [2 μM] (GolgiStop™, BD-Pharmingen).

Afterwards, the cells are harvested, washed twice in PBS (1–5% FCS) and fixed for 10 min in cold phosphate-buffered 4% PFA. In contrast to microbial or viral antigens, cytokines are much more sensitive to protein degradation and should therefore be stored no longer than 24 h at 4°C at this stage. In our experience, only IFN-γ may remain stable up to 48 h. To avoid loss of signal, it is strongly recommended to stain the cells within 24 h after fixation.

Discrimination of murine Th1/Tc1 and Th2/Tc2 cells by using intracellular immunofluorescent staining of IFN-γ and IL-4 in murine splenocytes *ex vivo*

Cell stimulation

1. Sacrifice the mouse and isolate the whole spleen aseptically.
2. Cut the spleen tissue into small pieces and isolate splenocytes by mincing and passing the tissue through a sieve of 100 μm mesh size (Cellstrainer™, BD-Falcon®).
3. Red cells can be depleted by density gradient or simply by hypotonic lysis using ammonium chloride buffer (0.15M NH$_4$Cl; 0.1 mM EDTA disodium salt dihydrate; 10 mM NaHCO$_3$, or a commercial product such as Lysing Reagent Ortho-mune®, Ortho).
4. Pipette 1 ml cell suspension in culture medium (RPMI 1640, 10% FCS) at [5 × 10^6/ml] in each of two wells of a 24-well tissue culture plate (BD-Falcon®).
5. To one of these wells add:

   - 10 μl PMA [1 μg ml^{-1}] ⇒ final concentration: [10 ng ml^{-1}]
   - 10 μl Ionomycin [100 μM] ⇒ final concentration: [1 μM]
   - 10 μl Monensin [200 μM] ⇒ final concentration: [2 μM]

   (Always prepare stimulators and Golgi transport inhibitor fresh from stock solution kept in absolute ethanol at –20°C.)
6. Vortex briefly to mix.
7. The other well serves as unstimulated control and is left untreated.
8. Incubate the cells for 5 h in 5% CO$_2$ at 37°C.

Fixation and permeabilization

1. Resuspend and transfer cells into 4 ml Falcon® tubes.
2. Wash cells 3× in washing buffer (PBS, 1–5% FCS, w/o Ca^{2+}/Mg^{2+}, pH 7.2). Use 3 ml/wash.
3. Pellet by centrifugation at 350g at 4°C for 5 min.

   (contd.)

4. Thoroughly resuspend the sedimented cells in 250 μl cold fixation buffer. This is 4% phosphate-buffered paraformaldehyde in PBS (1% FCS, w/o Ca^{2+}/Mg^{2+}), pH 7.4 (stable for 8 weeks at 4°C), prepared from a stock solution of 40% (w/v) paraformaldehyde (Serva) in H$_2$O bidest. by 1 : 10 dilution in phosphate buffer. Paraformaldehyde is toxic and of low solubility, therefore it should be dissolved by heating at 70°C in a covered Erlenmeyer flask for several hours within a fume hood. Finally, a few drops of 2 M NaOH are added until the solution becomes clear, followed by addition of 0.54 g glucose. Phosphate buffer is prepared by mixing 83% (v/v) solution A (22.6 g NaH$_2$PO$_4$×H$_2$O in 1 l H$_2$O bidest.) + 17% (v/v) solution B (25.2 g NaOH in 1 l H$_2$O bidest.)
5. Incubate for 10 min at 4°C.
6. Add 4 ml cold washing buffer immediately after incubation. Spin 5 min at 350g.
7. Wash cells 2× (3 ml per wash, 350g, 4°C, 5 min). Aspirate supernatant.
8. At this stage, cells can be stored up to 24 h at 4°C.

Multicolour staining for intracellular cytokines and cell surface markers

1. Resuspend cells and wash 2× in permeabilization buffer (2 ml per wash). This is 0.1% saponin in PBS (1% FCS, 10 mM HEPES), freshly prepared from a sterile stock solution of 10% (w/v) saponin in 1× PBS (1 M HEPES) by 1 : 100 dilution in 1× PBS (1% FCS).
2. Spin 4 min at 350g. Aspirate supernatant.
3. Resuspend the cell pellet in 200 μl permeabilization buffer.
4. Add 10 μl anti-mouse CD16/CD32 (*Fc Block*™, BD-Pharmingen) to the cell suspension. Mix well.
5. Incubate at RT for 15 min.
6. Meanwhile, pipette the following antibody combinations into fresh 4-ml Falcon® tubes:

Sample-No.	cells	Staining (cytokine antibody/surface marker antibody)
1	unstimulated	IgG1-FITC / -PE / -Biotin / IgG2a-FITC / -PE / -Biotin
2		anti-IFN-γ-FITC / anti-IL-4-PE / anti-CD3-Biotin
3		anti-IFN-γ-FITC / anti-CD3-PE / anti-CD4-Biotin
4		anti-IFN-γ-FITC / anti-CD3-PE / anti-CD8a-Biotin
5		anti-IL-4-PE / anti-CD3-FITC / anti-CD4-Biotin
6		anti-IL-4-PE / anti-CD3-FITC / anti-CD8a-Biotin
7		
|	stimulated	analogous to 1–6
12		

7. If there is more than one cell preparation (e.g. cells from different animals or animal groups), sample no. 1–5 has to be prepared for any cell preparation separately, thus the number of samples is multiplied by the number of cell preparations.

(*contd.*)

8. Pipette 2.5–5 μl of the individual antibody per tube (the antibody optimum has to be determined for any special application and any individual antibody).
9. Fill the cell suspension up to 1 ml with permeabilization buffer, resuspend. Add 100 μl of cell suspension to every tube with antibodies ⇒ 1 : 40 to 1 : 20 dilution of antibody in permeabilization buffer.
10. Vortex well.
11. Incubate at 4°C for 30 min in the dark.
12. Wash 3× with cold permeabilization buffer (3 ml per wash). Aspirate supernatant.
13. Resuspend cell pellet in 100 μl permeabilization buffer by vortex.
14. Add 3 μl Streptavidin-RPE-Cy5 (DAKO) to every tube. Vortex to mix.
15. Incubate at 4°C for 30 min in the dark.
16. Wash 3× with cold permeabilization buffer (3 ml per wash). Aspirate supernatant.
17. Resuspend cells in 250 μl 1% formaldehyde in PBS (1% FCS) or *CellFix*™ (BD). The fixed cells should be stored at 4°C in the dark until analysis.

Flow cytometric analysis (see Plate 6B)

1. Perform acquisition and analysis on a flow cytometer (e.g. *FACScan*™ or *FACSCalibur*™, BD) using the software *CellQuest*™ (BD).
2. Analysis of Samples 2 and 8 allow the discrimination between Th1/Tc1 (IFN-γ^+/IL-4$^-$), Th2/Tc2 (IFN-γ^-/IL-4$^+$), and Th0/Tc0 cells (IFN-γ^+/IL-4$^+$), while gating onto CD3$^+$ cells.
3. If the analysed cells definitely belong to one of both subtypes Th1/Tc1 or Th2/Tc2, the analysis of samples 3–6 and 9–12 further allow to identify the cytokine-producing T cells as CD4$^+$ (i.e. CD4$^+$/IFN-γ^+ ⇒ Th1; CD4$^+$/IL-4$^+$ ⇒ Th2) or CD8$^+$ cells (i.e. CD8$^+$/IFN-γ^+ ⇒ Tc1; CD8$^+$/IL-4$^+$ ⇒ Tc2).

◆◆◆◆◆◆ *EX VIVO* mRNA ANALYSIS

General precautions

When working with RNA, every attempt should be made to keep reagents and utensils RNAse-free. This involves making buffers and stock reagents with diethylpyrocarbonate (DEPC)-treated water, double-autoclaving all utensils, and using disposable articles wherever possible. Gloves must be worn at all times and should be frequently changed. As long as you work with RNA, keep all solutions, tubes etc. on ice. It is useful to clean bench surfaces routinely with bleach solutions or commercial RNAse destroying agents such as RNAse off (Natutec).

Harvesting tissues

1. Remove tissues to be analysed under aseptic conditions. In kinetic studies, it is useful to always sample the same part of the organ, e.g. upper right liver lobe, lower left lung lobe. The size of the removed tissue should be roughly the same throughout the course of the experiment.
2. Place samples immediately into chilled lysis buffer (approx. 1 ml per 10–100 mg of tissue):
   4 M guanidine-isothiocyanate (Merck)
   25 mM Na-citrate (pH 7)
   0.5% N-lauroylsarcosine (Sigma)
   100 mM 2-Mercapto-Ethanol.

Alternatively, commercially available lysis buffers may be used, e.g. TRIzol or TriFast Fl (Gibco BRL; peqlab).

3. Samples can now be snap-frozen on liquid nitrogen and stored at –70°C until homogenization.
4. Alternatively, samples can now be homogenized using either an ultra-turrax at full speed or teflon pestles and tight-fitting glass tubes. Use appropriate biosafety hoods when homogenizing infectious tissue. If you must reuse homogenizing tools, rinse at full speed (a) for 10 s in a beaker containing a large volume of distilled water, (b) for 10 s in a beaker containing a large volume of 95% ethanol, (c) in lysis buffer, before additional tissue is homogenized.
5. After homogenization, store 500-µl aliquots in Eppendorf tubes for multiple work-ups. Mark tubes with indelible ink writers. Homogenized tissues in lysis buffer can be stored at –70°C indefinitely.

Preparation of RNA

1. To the tissue homogenates in lysis buffer (500 µl), add:
   50 µl of 2 M Na-acetate (pH 4)
   500 µl of acid phenol (water-saturated phenol)
   100 µl of a chloroform-isoamylalcohol mix (49 : 1).
2. Vortex thoroughly. Allow to sit on ice for 15 min.
3. Centrifuge at 14 000g for 15 min at 4°C.
4. In steps of 100 µl each, remove aequous supernatant (approx. 400 µl) to another labelled tube, take care not to disturb the interface.
5. Add 400 µl chilled isopropanol, vortex, and let precipitate for at least 2 h at –20°C. For some tissues like liver, it is advisable to perform two sequential phenol-chloroform extractions. Precipitates can be stored at –20°C indefinitely.

Alternatively, commercially available RNA isolation kits (Qiagen, Eppendorf) may be used.

Some researchers prefer to re-suspend and re-precipitate the RNA before performing reverse transcription. For this purpose, centrifuge samples for 15 min at 14 000g and remove the supernatant entirely. The RNA pellet should be white-translucent. Let the pellet dry (an evacuated desiccator works fine) for 5–10 min, and resuspend in DEPC-treated water. Add approx. 100 µl of water per 50 mg of tissue. Then add 3 volumes of absolute ethanol and 1/10 volume of 3 M Na-acetate (pH 5.2). Reprecipitate at –20°C.

---

1. To determine concentration of nucleic acid in your sample, centrifuge precipitate at 14 000g for 15 min at 4°C.
2. Remove supernatant, and add 500 µl chilled 70% ethanol (in DEPC-treated water).
3. Flick pellet briefly, and centrifuge at 14 000g for 15 min at 4°C.
4. Remove the supernatant and let the RNA pellet dry in an evacuated desiccator for 5–10 min.
5. Resuspend the RNA pellet in 100 µl DEPC-treated water and measure O.D. 260/280 in a 1 : 100 dilution. 1 mg RNA has an O.D. reading of 1.0 at 260 nm. (To calculate the concentration of RNA in µg/ml: O.D. reading x dilution factor x 40.) You may wish to check the integrity of your RNA on a formaldehyde gel at this point.

◆◆◆◆◆◆◆ *EX VIVO* RNase PROTECTION ASSAY

The RNase Protection Assay (RPA) is a highly sensitive and specific method that allows the simultaneous detection and quantitation of several mRNA species in a single sample of total RNA.

RPA is based on the hybridization of radiolabelled anti-sense RNA probes to the given mRNA species in the experimental sample. As first step, single-stranded anti-sense RNA probes are synthesized on cDNA templates using DNA-dependent RNA polymerases from bacteriophages T7 or SP6. The cDNA templates (linearized plasmids containing the given sequences) are assembled into multi-probe template sets where the individual template sequences differ in their lengths. Following synthesis of radiolabelled anti-sense RNA probes, the probes are hybridized to total cellular RNA. In the subsequent step, non-hybridized single-stranded RNA is digested by RNase treatment. Double-stranded RNA is resistant to RNase. Consequently, radiolabelled probes that hybridize to mRNA are protected from being digested by RNase. These protected anti-sense RNA hybrids are separated on denaturing polyacrylamide gels based on the differences in their lengths. The quantitation of radiation intensity of RNA bands is performed by autoradiography or phosphorimaging. The RPA offers the possibility to compare different mRNA species within samples. In addition, individual mRNA species can be compared between samples by incorporating probes for house-keeping genes such as L32 and GAPDH.

Recently, we developed two RPA template sets that allow the mRNA analysis of the murine chemokines CXCL12/SDF-1, XCL1/lymphotactin, CCL20/exodus-1, CCL25/TECK, CX3CL1/fractalkine, CXCL1/KC, CCL22/MDC, CXCL9/MIG, CCL9/10/MIP-1γ, CXCL13/BLC, CCL12/MCP-5, and CCL19/ELC (Müller *et al.*, 2001). Other RPA template sets for the detection of murine cytokine gene expression are commercially available from BD PharMingen (San Diego, CA, USA). It is recommended to perform the RPA using the Riboquant® *In vitro* Transcription kit and the RPA kit (BD PharmMingen) according to the standard protocol as described below. In order to reduce the radiation hazard to the personnel handling the radiolabelled probes, the high-energy nuclide [α-^{32}P]UTP may be replaced by [α-^{33}P]UTP which has a significantly lower energy.

Synthesis of radiolabelled anti-sense RNA probes

1. Bring [α-^{32}P]UTP or [α-^{33}P]UTP, nucleotide pool, DTT, 5× transcription buffer, and RPA template set to room temperature. For each probe synthesis, add the following reagents to one tube:

   1 μl RNasin (40 U)
   1 μl GACU nucleotide pool (275 μM each)
   2 μl DTT (10 mM)
   4 μl 5 × transcription buffer
   1 μl RPA template set (15 ng of each plasmid)
   10 μl [α-^{32}P]UTP or [α-^{33}P]UTP (3000 Ci/mmol, 10 mCi/ml)
   1 μl T7 RNA polymerase (20 U)
   Always add T7 RNA polymerase last, and return the enzyme stock to the freezer immediately.

   Incubate the reaction mixture at 37°C for 30 min.
2. Add 2 μl DNase (2 U) to terminate the reaction and incubate at 37°C for 30 min.
3. To perform a phenol-chloroform extraction of radiolabelled RNA probes, add the following reagents to the tube:

   26 μl 20 mM EDTA
   25 μl Tris-saturated phenol, pH 8.0
   25 μl chloroform-isoamyl alcohol (50 : 1)
   2 μl yeast tRNA

   Vortex into an emulsion and spin in a microfuge at 14 000g for 5 min at room temperature.
4. Transfer the upper aqueous phase to a new tube and add 50 μl of chloroform-isoamyl alcohol (50 : 1). Vortex into an emulsion and spin in a microfuge at 14 000g for 5 min at room temperature.
5. Transfer the upper aqueous phase to a new tube. Precipitate the RNA probes by adding 50 μl 4 M ammonium acetate and 250 μl of ice-cold 100% ethanol. Mix by inverting the tube several times and incubate for 2 h at −70°C. Spin in a microfuge at 14 000g for 30 min at 4°C.

   (contd.)

---

6. Carefully remove the supernatant. Wash the RNA pellet by adding 100 µl of ice-cold 90% ethanol. Spin in a microfuge at 14 000g for 10 min at 4°C.
7. Carefully remove the supernatant completely. Air dry the pellet (do not use a vacuum evaporator centrifuge) and add 50 µl of hybridization buffer (Riboquant® RPA Kit). Solubilize the pellet by gentle flicking and quick spin in a microfuge.
8. Quantitate duplicate 1 µl samples in the scintillation counter. Add scintillation fluid to the aliquot prior to quantitation when [α-^{33}P]UTP is used for probe synthesis. Store the probe at –20°C until use.

Hybridization and RNase treatment

1. Add the appropriate amounts of target RNA (5–20 µg depending on the cellular source of RNA) to tubes. A tube containing yeast tRNA should be included as a negative control. In addition, it is recommended to use positive-control RNA included in the commercially available RPA template sets.
2. Dry RNA completely in a vacuum evaporator centrifuge without heat.
3. Add 8 µl of hybridization buffer to each sample. Solubilize the RNA by gentle flicking and quick spin in a microfuge.
4. Add 2 µl of radiolabelled probe (1.5–3 x 10^5 cpm µl^{-1}) to each RNA sample and mix by gentle flicking. Overlay the samples with a drop of mineral oil and quick spin in a microfuge.
5. Place the samples in a heat block or a water bath pre-warmed to 90°C and turn the temperature to 56°C. Incubate the samples for 12–16 h. Set the temperature to 30°C allowing it to ramp down slowly.
6. Prepare the RNase cocktail per 24 samples:

   2.5 ml RNase buffer
   6 µl RNase A (200 ng/ml) + RNase T1 (600 U/ml) mix

   Pipette 100 µl of the RNase cocktail to each sample. Mix by gentle flicking, quick spin in a microfuge and incubate for 45 min at 30°C.
7. Prepare the Proteinase K cocktail per 24 samples:

   390 µl Proteinase K buffer
   30 µl Proteinase K (1.5 mg/ml)
   30 µl yeast tRNA (20 µg/ml)

   Add 18 µl of Proteinase K cocktail to the tubes, vortex, quick spin in the microfuge and incubate for 15 min at 37°C.
8. Add 65 µl Tris-saturated phenol (pH 8.0) and 65 µl chloroform-isoamyl alcohol (50 : 1). Vortex into an emulsion and spin in the microfuge at 14 000g for 5 min at room temperature.

(contd.)

Measuring Immune Responses *in vivo*

9. Transfer the upper aqueous phase to a new tube. Add 120 µl of 4 M ammonium acetate and 650 µl of ice-cold 100% ethanol. Invert the tubes several times to mix and incubate for 2 h or overnight at –70°C. Spin in the microfuge at 14 000*g* for 30 min at 4°C.
10. Carefully remove the supernatant and add 100 µl of ice-cold 90% ethanol. Spin in the microfuge for 10 min at 4°C.
11. Carefully remove the supernatant and air dry the pellet completely. Add 5 µl of 1× loading buffer, solubilize by flicking and quick spin in a microfuge.

Separation of protected probes on denaturing 5% polyacrylamide gels

1. Clean a set of sequencing gel plates (>40 cm in length) thoroughly with water followed by 96% ethanol and acetone. Siliconize the short plate and clean again. Assemble the plates and the spacers (0.4 mm).
2. Combine the following to give a final concentration of 5% acrylamide: 74.5 ml acrylamide solution (final 19 : 1 acrylamide/bis acrylamide):

> 8.85 ml of 40% acrylamide
> 9.31 ml 2% bis acrylamide
> 35.82 g urea
> 7.45 ml 10 × TBE
> bring to a volume of 74.5 ml with DEPC-treated water
> 450 µl of 10% ammonium persulfate (APS)
> 60 µl N,N,N⁻,N⁻ -tetramethyl-ethylene-diamine (TEMED)

> Dissolve the urea in 30 ml DEPC-treated water and TBE by heating up to 50°C. Add acrylamide/bis-acrylamide and bring to a volume of 74.5 ml with DEPC-treated water. The mixture should be at room temperature prior to adding TEMED and APS. Gentle mix the solution avoiding air bubbles and immediately pour the gel. Insert an appropriate comb (e.g. 5-mm well width).

3. After polymerization (1–2 h), remove the comb and place the gel in the gel electrophoresis chamber. Flush the wells thoroughly with 0.5 × TBE and pre-run the gel at 40 W constant power for ~1 h with 0.5 × TBE as running buffer.
4. Dilute the anti-sense RNA probe in loading buffer to a concentration of 1000–2000 cpm/lane to serve as a size marker.
5. Prior to loading the samples and the size marker on the gel, heat them for 3 min at 90°C and place them immediately in an ice bath.
6. Load the samples and run the gel at 50 W constant power until the front dye of the loading buffer reaches ~30 cm.
7. Disassemble the gel mould, remove the siliconized short plate, and adsorb the gel to filter paper. Cover the gel with Saran wrap and place it between two additional pieces of filter paper. Dry the gel under vacuum for 1 h at 80°C.

(contd.)

8. For autoradiography, expose dried gels to Kodak X-AR film either at room temperature when using [α-³³P]-labelled probes or with intensifying screens at −70°C when using [α-³²P]-labelled probes. For quantitation, radiation intensity of RNA bands can be analysed by phosphorimaging.

9. Use the undigested probes as marker to plot a standard curve of migration distance versus log nucleotide length on semi-log graph paper. With this curve establish the identity of the RNase protected bands in the samples.

A typical example of results obtained with this RPA protocol is shown in Fig. 1.

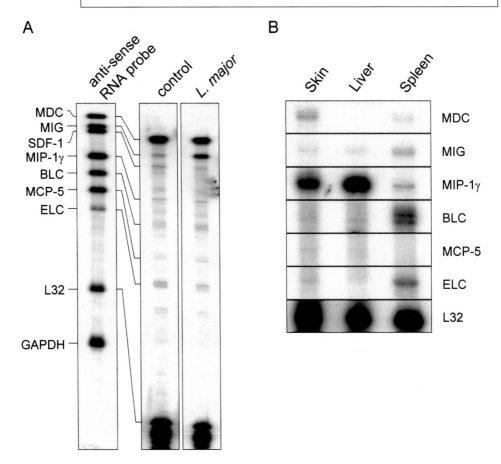

Figure 1. Detection of murine chemokine gene expression using RPA

A. Chemokine expression was analysed in the popliteal lymph nodes (LN) of CD-1 mice during the early phase of infection with the protozoan parasite *Leishmania major* and in popliteal LN from non-infected mice by RPA followed by autoradiography. Autoradiogram of a representative RPA gel is shown. The anti-sense probe bands have a higher molecular weight than the hybridized probe fragments since the probe itself contains portions of the plasmid multiple cloning site that do not hybridize to the target sense RNA and are thus not protected during RNase digestion.

B. Constitutive chemokine expression in skin, liver and spleen of C57BL/6 mice were assayed by RPA. Autoradiogram of the RPA gel is shown. The mRNA expression of the corresponding housekeeping gene L32 is included to normalize for gel loadin. (Adapted from Müller *et al.*, 2001, © Elsevier).

417

Troubleshooting

Poor probe labelling

1. Decreased probe recovery and increased lane background may occur if [α-^{32}P] and [α-^{33}P] have decayed beyond one half-life.
2. Always store DTT stock solution in small aliquots at −20°C to avoid repeated freeze-thawing.
3. Remove ethanol carefully and completely from the precipitated probe.
4. Store probe set in small aliquots at −70°C to keep quality.

High levels of RNA degradation in the gel

1. Degradation of mRNA is natural in cells so that some protected probe fragmentation is normal. Excessive breakdown products may occur if probe set has lost its quality.
2. The RNase A can recognize certain single base pair mismatches in RNA–RNA hybrids and may cleave the molecule at these positions leading to multiple bands of smaller size.
3. Residual RNase present in the organic interphase may lead to RNA degradation. Thus, the aqueous phase should be carefully removed from the phenol-chloroform extraction.

◆◆◆◆◆◆ *EX VIVO* REVERSE TRANSCRIPTION OF mRNA AND POLYMERASE CHAIN REACTION OF cDNA USING CYTOKINE-SPECIFIC PRIMERS (RT-PCR)

General precautions

Cross-contamination is the biggest problem in all PCR procedures, and great care should be taken to avoid contamination of test samples with extraneous sources of DNA. Therefore, only use aerosol-resistant pipette tips, keep separate pipetors for reagents and stock solutions, and use separate pipetors for PCR and RNA work. It helps to aliquot source reagents in a room far away from where RNA extraction and PCR is performed, and RNA work-up and PCR pipetting should be performed in separate rooms as well. Most importantly, pipetting amplicons (i.e. for electrophoresis) should be carried out with a set of altogether different pipetors (preferably old and used ones, so that nobody else uses them) and in a completely different room from the rest. If contamination has occurred, it helps to clean surfaces and pipetors with a solution of 0.1 N HCL.

Reverse transcription (RT)

1. Add 1–5 µg of RNA to an Eppendorf tube, add 2 µg of oligo-dT (12–18mer, GIBCO), and fill up the volume to 12.5 µl with RNAse-free-water. Be careful not to use too much RNA in the RT reaction.
2. Denature the mixture at 70°C for 10 min, then place on ice.
3. Prepare RT-mix as follows (per reaction):

5× first strand buffer	4 µl
100 mM DTT	2 µl
10 mM dNTPs	1 µl
RNAsin (40 000 U ml⁻¹; Promega)	0.5 µl

4. Add the RT-mix to the RNA oligo-dT mixture and incubate at 42°C for 2 min. Add 1 µl of reverse transcriptase (e.g. Superscript II RT, 200 U µl⁻¹; GIBCO) and continue the reaction for 60 min.
5. Stop the reaction by adding 40 µl TE (10 mM Tris/HCl, 1 mM EDTA, pH 7.5) buffer and incubate at 95°C for 5 min. cDNA may be stored at –20°C indefinitely or for up to two weeks at 4°C.

Qualitative PCR

For PCR primer selection, remember: primers should be complementary to sequences in separate exons, or should span exon–exon boundaries, to be mRNA/cDNA specific. 5′primers should contain sequences not more than 1000 bp removed from the poly A-tail of the mRNA to avoid misrepresentation due to inefficient reverse transcription. Primers should be checked for unique complementarity using a computer-assisted search (Laser Gene, DNAStar); for most purposes, annealing temperatures around or higher than 60°C will guarantee specificity.

1. For each PCR, add to one tube:

dNTPS (10 mM)	0.5 µl
MgCl$_2$ (50 mM)	1.25 µl
water	10.75 µl
10× buffer	2.5 µl
TAQ polymerase (5 U µl⁻¹; Perkin-Elmer; preferably HotStart)	0.1 µl
primers (mix of sense and antisense), 1 µM	5 µl
cDNA	5 µl

2. It is best to first put 5 µl of each primer mix to the bottom of the tube, then make a mix of the other PCR reagents for all reactions and vortex thoroughly. Add 20 µl of this reaction mix to the side of each tube.

If you wish to perform a PCR for different cDNA samples at the same time, first put 5 µl of primer mix to the PCR tubes, then make PCR mix leaving out the cDNA, add 15 µl of the reaction mix to one side of the tube and 5 µl of the cDNA to the other side. After closing and numbering the tubes, spin briefly to make sure all reagents mix at the bottom of the tube. If you use a thermal cycler without a heated lid, add one drop of mineral oil on top of the reagent mix to all tubes (you need not spin first; spinning will allow the aqeuous reagents to pass through the oil phase). Prepare a chart to identify tube numbers with primer and cDNA contents.

A typical protocol consists of 25 to 30 cycles

- 95°C for 45 s (to denature dsDNA)
- 60°C for 45 s (to let cDNA and primer sets anneal)
- 72°C for 30 s (to allow for elongation of strands by the Taq polymerase).

Electrophoresis and detection of PCR products

1. Prepare 2% agarose gel in 0.5x TBE buffer. (You may add 0.5 µg ml^{-1} ethidium bromide at this point to visualize amplicons on a UV tray during electrophoresis, but a more even staining of the gel is obtained when soaking the gel after electrophoresis in 0.5x TBE containing 0.5 µg ml^{-1} ethidium bromide.)
2. Add 5 µl of 5x loading buffer to each well of a microtiter plate.
3. Remove 15–20 µl of PCR solution, mix with loading buffer in the wells, and add to individual slots in the gel. (You may use less PCR solution if you want to perform additional studies on your amplicon. A suitable control for the specificity of your amplicon involves a restriction enzyme digest of the PCR product resulting in the predicted pattern of bands after electrophoresis.)
4. Add molecular weight markers to first and last rows. If you have many samples, it is useful to double-comb the gels.
5. Run at 120 V for approx. 2–3 h, until dye front reaches bottom end of the gel.
6. Use any form of photodocumentation to visualize bands on a UV transilluminator.

A typical example of results obtained with this type of protocol is shown in Fig. 2. Primer sequences used for this particular experiment are given by Ehlers *et al.* (1992).

Semi-quantitative PCR

The purpose of semi-quantitative RT-PCR is to compare RNA samples from different sources for their relative content of specific mRNAs. Since no absolute quantitation is involved, no 'absolute' calibration using *in vitro* synthesized RNA is necessary. However, only cDNAs normalized for content of house-keeping genes (i.e. β2-microglobulin or Ribosomal Protein S9) may be directly compared with one another.

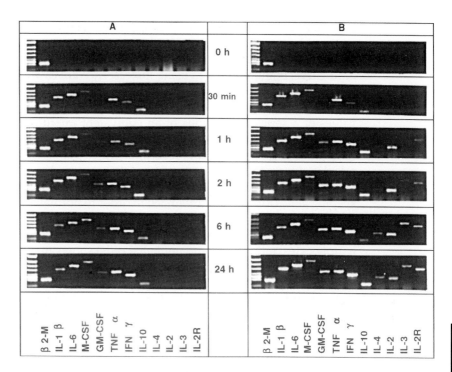

Figure 2. Qualitative RT-PCR of the livers of naïve (A) and *Listeria*-immune (B) mice

C57BL/6 mice were challenged intravenously with 1×10^7 cfu *L. monocytogenes*, and liver samples were processed for RT-PCR at various time points after infection according to the protocol. (From Ehlers *et al.* (1992). Copyright © 1992 The American Association of Immunologists.)

Semi-quantitative PCR using 'Sybr-Green Real Time' RT-PCR

There are a number of instruments on the market capable of performing real-time PCR. These machines monitor the fluorescence emitted during the reaction as an indicator of amplicon production during each PCR cycle as opposed to the endpoint detection in conventional quantitative PCR protocols, such as by gel electrophoresis. The real-time RT-PCR does not detect the size of the amplicon and thus does not allow the differentiation between DNA (e.g. primer–dimer) and cDNA amplification. Real-time PCR quantification eliminates post-PCR processing of PCR products which helps to increase throughput and reduce the potential for 'carry-over contamination'. In comparison to conventional RT-PCR, real-time PCR offers a much wider dynamic range of up to 10^7-fold compared with 1000-fold in conventional RT-PCR.

The real-time PCR is based on the detection and quantification of a fluorescent reporter. The signal increases in direct proportion to the amount of PCR product in a reaction. By recording the amount of fluorescence emission (by means of a CCD camera or a photomultiplier) at the end of each cycle, it is possible to monitor the PCR reaction online during the exponential phase. Quantification is based on the threshold cycle (C_t), defined as the cycle number at which the fluorescence (and thus the PCR product) becomes first detectable above background noise. The C_t value can then be translated into a quantitative result by constructing a standard curve. Standard curves using fluorescence are easily generated due to the linear response over a wide dynamic range. The standard curve for a gene 'A' (e.g. a cytokine gene) is generated by plotting individual C_t values of a series of dilutions of an equivalent mixture of the different samples (i.e. X and Y) against the logarithm of the corresponding volumes. The ratio of expression between the two samples X and Y is then calculated from the linear regression of that standard curve (as shown in Fig. 3). The same calculations as for 'A' must be performed for a housekeeping gene (Ribosomal Protein S9) in order to normalize for differences in the amount of total RNA starting material. Our own experience studying gene expression in both murine and human immune cells (B and T lymphocytes, monocytes, and epithelial cells) shows that the use of RPS9 as housekeeping gene is particularly suitable to normalize for RNA content in these cell types.

There are two general methods for the quantitative detection of the amplicon: fluorescence probes (TaqMan probes, molecular beacons) or DNA-binding agents (SYBR Green). TaqMan probes make use of labelled oligonucleotide probes and a 5′ nuclease PCR assay to generate a fluorescent signal during PCR. In this approach, a specific oligo probe is used with a reporter and a quencher dye attached. During the PCR reaction, the probe is cleaved by the 5′ nuclease activity of the Taq DNA Polymerase, separating the reporter dye from the quencher dye. This generates a sequence-specific fluorescent signal that increases with each cycle.

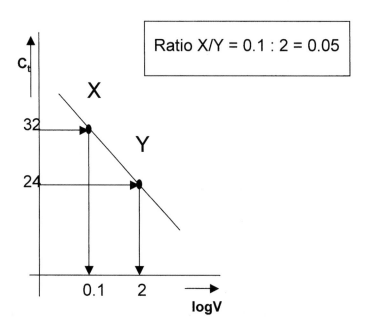

Figure 3. Calculation of the ratio of gene expression for a particular gene in samples X and Y.

Molecular beacons also contain fluorescent and quenching dyes at either end but they are designed to adopt a hairpin structure while free in solution to bring the fluorescent dye and the quencher into close proximity. The close proximity of the reporter and the quencher in this hairpin configuration suppresses reporter fluorescence. When the beacon (specific oligo probe) hybridizes to the target during the annealing step, the reporter dye is separated from the quencher and the reporter fluoresces. Molecular beacons remain intact during PCR and must rebind to their target every cycle for fluorescence emission. Both TaqMan probes and molecular beacons allow detection of multiple DNAs ('multiplex PCR') by use of different reporter dyes on different probes/beacons.

DNA dye-binding assays are the cheapest option for performing real-time RT-PCR assays, as they do not require an amplicon-specific probe (which can make gene expression profiling experiments in immunology very costly). SYBR Green is a minor groove binding dye which does not bind single-stranded DNA. The major problem with SYBR Green is that non-specific amplicons cannot be distinguished from specific ones. The primers shown in Tables 1 and 2 have therefore been extensively tested for non-specific amplification and primer–dimer complex formation and are suitable for a broad range of applications in immunology of infection (Hess *et al.*, 2001). Further primer sequences can be obtained upon request at jla@gbf.de.

Primers for SYBR green real-time RT-PCR

When searching primers on the computer (several software packages are commercially available which provide reasonable results), the following parameters are recommended:

- Primer length: 17–24 bases
- Primer melting temperature: 52° to 65°C
- 'Optimal' annealing temperature: 54°, 58° or 60°C

Product length

The optimal amplicon length is less than 100 bp. It is recommended to generate amplicons no longer than 90 to 150 bp, but a product length of up to 250 bp may give good results. Shorter amplicons amplify more efficiently than longer ones and are more robust towards reaction conditions.

Annealing temperature

In order to optimize throughput, primers should be designed that give optimal results with one out of three annealing temperatures that match most of the primers used. In case sufficient amounts of product at the annealing temperature suggested by the software cannot be generated, a gradient PCR to find the most suitable temperature should be performed.

Table I. Real-time PCR primers for studies with mouse cDNA

Name	Sequence	Concentration (mM)	Product length (bp)	Ann. temp. (°C)
FAS 5´	CTACTGCGATTCTCCTGGCTGTG	900	126	58
FAS 3´	CATAGGCGATTTCTGGGACTTTGT	50		
FASL 5´	CAACAGGTCAGCTATCCTTCATTT TC	300	94	58
FASL 3´	CCTCAGTCACAGACCTCCTTCCATC	900		
IFNγ 5´	AGGAACTGGCAAAAGGATGGTGA	300	106	58
IFNγ 3´	TGTTGCTGATGGCCTGATTGTCTT	50		
IL-2 5´	TGAGTGCCAATTCGATGATGAGT	300	96	55
IL-2 3´	TATTGAGGGCTTGTTGTTGAGATGATGC	50		
IL-4 5´	GGACGCCATGCACGGAGAT	900	94	58
IL-4 3´	AGCACCTTGGAAGCCCTACAGAC	900		
IL-5 5´	CTCTTCGTTGCATCAGGGTCTCA	300	127	55
IL-5 3´	AAAGGATGCTAAGGTTGGGTATGTG	900		
IL-10 5´	CTGGACAACATACTGCTAACCGACTC	900	134	58
IL-10 3´	ATTTCTGGGCCATGCTTCTCTGC	50		
IL-12 p35 5´	GGTGCCCTCCATCGCTTCT	50	106	58
IL-12 p35 3´	TCTGCTAACACATTGAGGGGAGGAC	50		
IL-13 5´	GGGGGCTCCAGCATTGAAG	300	148	58
IL-13 3´	GCCGTGGCAGACAGGAGTG	300		
IL-16 5´	AGGCGGCCAATGCTAAGACAA	900	122	58
IL-16 3´	CAGCAGCTACCAGAGGCACACC	900		
IL-18 5´	GGTGGGGGTTCTCTGGGTTC	300	117	55
IL-18 3´	GTTTGAGGCGGCTTTCTTTGTC	300		
RPS9 5´	CTGGACGAGGGCAAGATGAAGC	900	143	54–61
RPS9 3´	TGACGTTGGCGGATGAGCACA	50		
TGFβ1 5´	ACCTGGGTTGGAAGTGGAT	50	139	58
TGFβ1 3´	GAAGCGCCCGGGTTGTGTTGGTT	900		
TNFα 5´	CAATGCACAGCCTTCCTCACAG	300	117	58
TNFα 3´	CCCGGCCTTCCAAATAAATACAT	50		
TNF-R1 5´	TCCCGCGATAAAGCCACACC	900	102	58
TNF-R1 3´	CTTTGCCCACTTTTCACCCACAG	900		

FASL = FAS ligand; IFNγ = Interferon gamma; IL- = Interleukin; RPS9 = ribosomal protein S9; TNFα = tumour necrosis factor alpha; TNF-R = tumour necrosis factor receptor.

424

Table 2. Real-time PCR primers for studies with human cDNA

Name	Sequence	Concentration (mM)	Product length (bp)	Ann. temp. (°C)
FAS 5′	TGGCATGCTAAGTACCCAAATAGG	900	141	58
FAS 3′	AGTGGGGTTAGCCTGTGGATAGAC	50		
FASL 5′	CCTCCAGGGGGACTGTCTTTCAGA	300	119	58
FASL 3′	AGTGGGGCATTGGCCTCTTTC	900		
IFNγ 5′	TAGCAACAAAAAGAAACGAGATGACT	300	145	58
IFNγ 3′	CC TTTTTCGCTTCCCTGTTTTAG	50		
IL–1α 5′	TGCTGAAGGAGATGCCTGAGATACC	900	113	58
IL–1α 3′	TGGATGGGCAACTGATGTGAAATAG	900		
IL–1β 5′	CAGGGACAGGATATGGAGCAACAA	50	133	58
IL–1β 3′	CATCTTTCAACACGCAGGACAGGT	300		
IL–4 5′	TCTGTGCACCGAGTTGACC	300	136	58
IL–4 3′	ACCCAGGCAGCGAGTGT	900		
IL–5 5′	GAAGAAAGACGGGAGAGTAAACCAA	900	136	51
IL–5 3′	AAATGTCCTTTCTCCTCCAAAATCT	900		
IL–10 5′	CCCTAACCTCATTCCCCAACCAC	50	135	58
IL–10 3′	CCGCCTCAGCCTCCCAAAGT	50		
IL–10R 5′	GACCCCCACCCCTCTGCCAAAGTA	900	146	58
IL–10R 3′	TGCTGCCTCTCCAATAAATAAATG	300		
IL–12p40 5′	ATGCCGTTCACAAGCTCAAGTATG	300	179	58
IL–12p40 3′	TGTCAGGGAGAAGTAGAAATGTGG	50		
IL–13 5′	GCCTGTGCTGCCCGTCTTCA	300	141	58
IL–13 3′	TCTAGGGCAGGGGAGGGGTGTA	50		
IL–15 5′	GAGTCCGGAGATGCAAGTATTCA	900	125	58
IL–15 3′	TTTTCCTCCAGTTCCTCACATTCT	300		
IL–16 5′	CCATGCCCGACCTCAACTCCT	900	130	58
IL–16 3′	CCCAGCCCTGCCGACATC T	900		
RPS9 5′	CTGCTGCGGCGGCTGGTC	50	111	58
RPS9 3′	CTGGGTCTGCAGGCGTCTCT	50		
TNFα 5′	AGGCCAAGCCCTGGTATGAGC	300	145	58
TNFα 3′	CACAGGGCAATGATCCCAAAGTAG	50		
TNF–R1 5′	GGGGCCCCTGGTTCGTC	300	134	58
TNF–R1 3′	CAGGCA GAGGGCACAGGAGT	300		
TNF–R2 5′	GTAGCCTTGCCCGGATTCTGG	300	118	58
TNF–R2 3′	ACCCTGCCCCTGCTCTGCTA	300		

FASL = FAS ligand; IFNγ = Interferon gamma; IL– = Interleukin; RPS9 = ribosomal protein S9; TNFα = tumour necrosis factor alpha; TNF-R = tumour necrosis factor receptor.

5′–3′ relation

For the SYBR Green method it is crucial not to generate non-specific PCR products (see above). To avoid these non-specific products every primer pair has to be optimized regarding its concentration. For this purpose, a so-called 'primer-matrix' is performed in which three different concentrations (900, 300 and 50 mM) for every 5′-primer with the corresponding concentrations from the 3′-primer with and without template are tested. The appearance of non-specific amplification products in the presence and absence of template is monitored by plotting the first derivative of the melting curve. By doing this crucial optimization step, one can raise the efficiency of the reaction, which increases the difference in threshold (C_t) between the specific and non-specific amplicons.

Exon–intron

One should aim to identify primers that bind to separate exons to avoid false-positive results arising from amplification of contaminating genomic DNA. Unfortunately, the genomic sequence is not known for all mRNAs. Moreover, sometimes targeting of an intron-less gene is desired. In this case the RNA may be treated with RNAse-free DNAse.

Position of the PCR product in the template

If possible we try to place the amplicon as far to the 3′ end of the template as possible. By doing so one can assure that the reverse transcriptase is able to process the transcription up to a length which is sufficient for the PCR. To verify the origin of the PCR products they should be digested with appropriate restriction enzymes or sequenced!

Real-time PCR

The methodology described here has been developed using the ABI GeneAmp 5700, a high-throughput system that utilizes a 96-well format. It has a typical run time of 2 h. The formation of PCR products is monitored fluorimetrically via the incorporation of SYBR Green.

Typical protocol

> **Comparison of cytokine gene expression in two samples X and Y**
>
> 1. For a standard reaction, mix the following to make up the master-mix (ABI SYBR Green Kit):
>
> *(contd.)*

10x buffer	5 µl
MgCl$_2$ (25 mM)	6 µl
dNTPs (2.5 mM each)	4 µl
Taq (Hotstart)	0.25 µl
AmpErase	0.5 µl
5′ Primer	a µl (see Tables 1 and 2)
3′ Primer	b µl (see Tables 1 and 2)
Template	0.01 to 1 µl (see below, point 5)
dH$_2$O	up to 50 µl

Reaction volumes may be scaled down to as low as 10 µl for use in a Light Cycler (Roche). The amounts of template (see below) and primers depend on the abundance of the gene of interest in the sample and the primer matrix. Home-made mixes work less well, as they are often not as sensitive and robust.

2. For temperature cycling, use a two-step protocol (no extension step) with primer-specific annealing temperatures and an initial activating step (94°C, 10 min) for the Hotstart-Enzyme, e.g.

> 94°C, 1 min (10 s when using a Light Cycler)
> 58°C, 1 min (10 s when using a Light Cycler)
> for 40 cycles.

3. Produce a 'standard' by mixing equivalent amounts of cDNA from samples X and Y.
4. Run a real-time PCR for the cytokine gene and for the house-keeping gene RPS9 with serial dilutions from the 'standard', corresponding to e.g. 1.0 to 0.01 µl of the mix. Generate a standard curve by plotting C_t values obtained against the logarithm of the corresponding volumes.
5. Perform a normal PCR reaction under the same conditions as in the real-time PCR to determine the approximate amount of template to be used in the real-time PCR and estimate the amount of template by comparing PCR products on an agarose gel. If you see a broad band on the agarose gel, take 0.01 µl from the cDNA for the real-time PCR reaction, if you see a weak or no band take 1 µl or more.
6. Run a real-time PCR with primers for the cytokine gene and a second real-time PCR for RPS9 for samples X and Y, always performing duplicates (better triplicates).
7. The obtained C_t values from samples X and Y for the cytokine gene and for the house-keeping gene are used to calculate a 'volume' based on the corresponding standard curve by linear regression (Fig. 3). Normalize by dividing the 'volume' for the cytokine gene of sample X by the 'volume' of the house-keeping gene of X. Do the same for sample Y.
8. The ratio of cytokine gene expression between samples X and Y is calculated by dividing the normalized values from X and Y.

> **Points to remember**
>
> - be careful to isolate sufficient amounts of good quality RNA
> - when pipetting try to hold the reaction at 4°C (setting-up a 96-well plate may easily take more than 2 h)
> - take a Hot-Start enzyme
> - use a stepper for the mastermix
> - make sure that the C_t values of the samples are in the linear range of the corresponding standard
> - always check for the presence of non-specific products (perform melting curve analysis after the PCR is complete)

Limitations

Any RT-PCR measures steady state mRNA levels, i.e. quantitative differences may only be found when there is induced/repressed transcription of mRNA or increased/repressed turnover of mRNA. Whenever the amount of cytokine secreted is highly regulated by translational modifications (as is for example the case with TNFα and TGFβ) it may be difficult to measure mRNA differences although at the protein level major differences are observed. It is, however, true that where there is no mRNA there usually is no protein; thus RT-PCR always is a good indicator what cytokine *not* to look for by ELISA in tissue homogenates or by immunohistochemical analyses.

Remember that the kinetics of induction for mRNA species vary tremendously and always precede peaks of protein production; therefore, a detailed kinetic study is often warranted to define where to best look for differences when comparing experimental settings.

Always be aware that these procedures allow for the relative quantitation of mRNA for any given cytokine but do not allow for comparison of mRNA levels between different cytokines.

References, further reading and typical examples for the described procedures

Immunohistochemistry

Autenrieth, I. B., Kempf, V., Sprinz, T., Preger, S. and Schnell, A. (1996). Defense mechanisms in Peyer´s patches and mesenteric lymph nodes against *Yersinia enterocolitica* involve integrins and cytokines. *Infect. Immun.* **64**, 1357–1368.

Benini, J., Ehlers, E. M. and Ehlers, S. (1999). Different types of pulmonary granuloma necrosis in immunocompetent vs. TNFRp55-gene-deficient mice aerogenically infected with highly virulent *Mycobacterium avium. J. Pathol.* **189**, 127–137.

Cattoretti, G., Piteri, S., Parraricini, C., Becker, M. A. G., Poggi, S., Bifulco, C., Key, G., D´Amato, L., Sabattini, E., Fendale, E., Reynolds, F., Gerdes, J. and Rilke, F. (1993.) Antigen unmasking on formalin-fixed, paraffin-embedded tissue sections. *J. Pathol.* **171**, 83–98.

Ehlers, S., Kutsch, S., Benini, J., Cooper, A., Hahn, C., Gerdes, J., Orme, I., Martin, C. and Rietschel, E. T. (1999). NOS2-derived nitric oxide regulates the size, quantity and quality of granuloma formation in *Mycobacterium avium*-infected mice without affecting bacterial loads. *Immunology* **98**, 313–323.

Flow cytometric approaches

Bode, L., Zimmermann, W., Ferszt, R., Steinbach, F. and Ludwig, H. (1995). Borna disease virus genome transcribed and expressed in psychiatric patients. *Nat. Med.* **1**, 232–236.

Jung, T., Schauer, U., Heusser, C., Neumann, C. and Rieger, C. (1993). Detection of intracellular cytokines by flow cytometry. *J. Immunol. Methods* **159**, 197–207.

Lehmann, J., Jüngel, A., Lehmann, I., Busse, F., Biskop, M., Saalbach, A. *et al.* (2000). Grafting of fibroblasts isolated from the synovial membrane of rheumatoid arthritis (RA) patients induces chronic arthritis in SCID mice – a novel model for studying the arthritogenic role of RA fibroblasts *in vivo*. *J. Autoimmun.* **15**, 301–313.

Lehmann, J., Bellmann, S., Werner, C., Schröder, R., Schütze, N. and Alber, G. (2001). IL-12p40-dependent agonistic effects on the development of protective innate and adaptive immunity against *Salmonella* Enteritidis. *J. Immunol.* **167**, 5304–5315.

Picker, L. J., Singh, M. K., Zdraveski, Z., Treer, J. R., Waldrop, S. L., Bergstresser, P. R. and Maino, V. C. (1995). Direct demonstration of cytokine synthesis heterogeneity among human memory/effector T cells by flow cytometry. *Blood* **86**, 1408–1419.

Qvist, P., Aasted, B., Bloch, B., Meyling, A., Rønsholt, L. and Houe, H. (1990). Flow cytometric detection of bovine viral diarrhea virus in peripheral blood leukocytes of persistently infected cattle. *Can. J. Vet. Res.* **54**, 469–472.

Sander, B., Andersson, J. and Andersson, U. (1991). Assessment of cytokines by immunofluorescence and the paraformaldehyde-saponin procedure. *Immunol. Rev.* **119**, 65–93.

Sander, B., Hoiden, I., Andersson, U., Möller, E. and Abrams, J. S. (1993). Similar frequencies and kinetics of cytokine producing cells in murine peripheral blood and spleen. Cytokine detection by immunoassay and intracellular immunostaining. *J. Immunol. Methods* **166**, 201–214.

Zimmermann, T., Kunisch, E., Pfeiffer, R., Hirth, A., Stahl, H. D., Sack, U., Laube, A., Liesaus, E., Roth, A., Palombo-Kinne, E., Emmrich, F. and Kinne, R. W. (2001). Isolation and characterization of rheumatoid arthritis synovial fibroblasts from primary culture – primary culture cells markedly differ from fourth-passage cells. *Arthritis Res.* **3**, 72–76.

Ribonuclease Protection Assays

Gilman, M. (1993). Ribonuclease protection assay. In *Current Protocols in Molecular Biology*, Vol. 1 (F. M. Ausubel, R. Brent, R. E. Kingston, D. D. Moore, J. G. Seidman, J. A. Smith and K. Stuhl, Eds), pp. 4.7.1–4.7.8, John Wiley and Sons, Inc., New York.

Kim, Y., Sung, S. J., Kuziel, W. A., Feldman, S., Fu, S. M. and Rose, C. E. Jr. (2001). Enhanced airway Th2 response after allergen challenge in mice deficient in CC chemokine receptor-2 (CCR2). *J. Immunol.* **166**, 5183–5192.

Müller, K., Ehlers, S., Solbach, W. and Laskay, T. (2001). Novel multi-probe RNase Protection Assay (RPA) sets for the detection of murine chemokine gene expression. *J. Immunol. Methods* **249**, 155–165.

Sambrook, J., Fritsch, E. F. and Maniatis, T. (1989). Synthesis of RNA probes by *in vitro* transcription of double-stranded DNA templates by bacteriophage DNA-dependent RNA polymerases. In *Molecular Cloning: A Laboratory Manual*, 2nd edn (N. Ford, C. Nolan and M. Ferguson, Eds). Cold Spring Harbor Laboratory Press, New York, pp. 10.27–10.37.

Vester, B., Müller, K., Solbach, W. and Laskay, T. (1999). Early gene expression of NK cell-activating chemokines in mice resistant to *Leishmania major*. *Infect. Immun.* **67**, 3155–3159.

Yang, L., Cohn, L., Zhang, D., Homer, R., Ray, A. and Ray, P. (1998). Essential role of nuclear factor κB in the induction of eosinophilia in allergic airway inflammation. *J. Exp. Med.* **188**, 1739–1750.

Semi-quantitative RT-PCR, including primers

Bouaboula, M., Legoux, P., Pessegue, B., Delpech, B., Dumont, X., Piechaczyk, M., Casellas, P. and Shire, D. (1992). Standardization of mRNA titration using a polymerase chain reaction method involving co-amplification with a multi-specific internal control. *J. Biol. Chem.* **267**, 21830–21838.

Bustin, S. A. (2000). Absolute quantification of mRNA using real-time reverse transcription polymerase chain reaction assays. *J. Mol. Endocrinol.* **25**, 169–193.

Ehlers, S., Mielke, M. E., Blankenstein, T. and Hahn, H. (1992). Kinetic analysis of cytokine gene expression in the livers of naive and immune mice infected with *Listeria monocytogenes*. The immediate early phase in innate resistance and acquired immunity. *J. Immunol.* **149**, 3016–3022.

Gilliland, G., Perrin, S., Blanchard, K. and Bunn, H. F. (1990). Analysis of cytokine mRNA and DNA: Detection and quantitation by competitive polymerase chain reaction. *Proc. Natl. Acad. Sci. USA* **87**, 2725–2729.

Hänsch, H. C. R., Smith, D. A., Mielke, M. E. A., Hahn, H., Bancroft, G. J. and Ehlers, S. (1996). Mechanisms of granuloma formation in murine *Mycobacterium avium* infection: the contribution of CD4+ T cells. *Int. Immunol.* **8**, 1299–1310.

Hess, S., Rheinheimer, C., Tidow, F., Bartling, G., Kaps, C., Lauber, J., Buer, J. and Klos, A. (2001). The reprogrammed host: *Chlamydia trachomatis* induced up-regulation of glycoprotein 130 cytokines, transcription factors, and antiapoptotic genes. *Arthritis Rheum.* **44**, 2392–2401.

Rhoades, E. R., Cooper, A. M. and Orme, I. M. (1995). The chemokine response in mice infected with *Mycobacterium tuberculosis*. *Infect. Immun.* **63**, 3871–3877.

Svetic, A., Finkelman, F. D., Jian, Y. C., Dieffenbach, C. W., Scott, D. E., McCarthy, K. F., Steinberg, A. D. and Gause, W. C. (1991). Cytokine gene expression after *in vivo* primary immunization with goat antibody to mouse IgD antibody. *J. Immunol.* **147**, 2391–2397.

Wang, A. M., Doyle, M. V. and Mark, D. F. (1989). Quantitation of mRNA by the polymerase chain reaction. *Proc. Natl. Acad. Sci. USA* **86**, 9717–9721.

Wynn, T. A., Eltoum, I., Cheever, A. W., Lewis, F. A., Gause, W. C. and Sher, A. (1993). Analysis of cytokine mRNA expression during primary granuloma formation induced by eggs of *Schistosoma mansoni*. *J. Immunol.* **151**, 1430–1440.

Suppliers of unique reagents

BD PharMingen	www.pharmingen.com
	www.bdfacs.com
Biochrom	www.biochrom.de
Dianova	www.dianova.de
Eppendorf, Hamburg	www.eppendorf.de
Gibco; Serva	www.lifetech.com
Jackson ImmunoResearch	www.jacksonimmuno.com
Leica Instruments	www.leica.com
Merck Co.	www.merck.com
Natutec, Frankfurt	www.natutec.de
Peqlab, Erlangen	www.peqlab.de
Perkin-Elmer	www.perkin-elmer.com
Promega	www.promega.com
Sigma Chemical Co.	www.sigma-aldrich.com

4 Specific Models

4.1 Murine Model of Tuberculosis

Alan D Roberts and Andrea M Cooper
Trudeau Institute, Saranac Lake, New York, USA

John T Belisle, Joanne Turner, Mercedes Gonzalez-Juarerro and Ian M Orme
Mycobacteria Research Laboratories, Department of Microbiology, Immunology and Pathology, Colorado State University, USA

◆◆

CONTENTS

Handling and cultivation of mycobacteria
Propagation methods for the isolation of immunologically reactive protein fractions
Infection protocols
Culturing and infecting macrophages
Identifying key protein recognized by immune T cells
Following the course of infection
DTH and non-specific resistance
Measuring the cellular response *in vivo*

Murine Model of Tuberculosis

◆◆◆◆◆◆ HANDLING AND CULTIVATION OF MYCOBACTERIA

Introduction

The isolation and cultivation of *M. tuberculosis* in both the clinical and research laboratory setting has been well described in numerous publications and books (Vestal, 1975; Youmans, 1979; Kubica and Wayne, 1984; Bloom, 1994; Rom and Garay, 1995). The methods described in many of these references have been in use for many years and have changed very little. For example, many laboratories routinely cultivate *M. tuberculosis* cultures in Proskauer and Beck liquid medium which is a modified medium first developed by Youmans and Karlson (Youmans and Karlson, 1947). However, this chapter will not attempt to outline all of the well-established methodologies which have been used by various investigators,

Copyright © Elsevier Science Ltd
All rights of reproduction in any form reserved

but instead will concentrate on those utilized in our own laboratory for the cultivation of *M. tuberculosis* for research purposes.

In this laboratory we routinely cultivate strains of *M. tuberculosis* for use in the following activities: (a) the *in-vivo* determination of disease host response and pathogenesis, primarily using a murine airborne infection model; (b) the *in vivo* determination of bactericidal effectiveness of novel immunotherapy, vaccination, or chemotherapeutic agents using both the murine and guinea-pig airborne infection models; and (c) the *in vitro* testing of potential chemotherapeutic agents using a murine bone marrow derived macrophage infection model. Each of these procedures requires the cultivation of both 'laboratory strains' (such as Erdman, H37Rv) and newly acquired clinical isolates of *M. tuberculosis*.

While very small inocula of *M. tuberculosis* can be handled safely in appropriate 'Class II' biosafety cabinets, any larger operations require stringent BSL-3 biosafety conditions. Although everyone seems to have their own definitions of this level, our recommended 'minimum requirements' are a facility with shower in/shower out (which also avoids transmission of mouse viruses), a safety cabinet, a glove box, air filtered both in and out through HEPA filters, and a graduated air-flow handling system (a second, back-up system is good but not essential). Animals should be kept in separate rooms away from equipment because of animal bedding dust. Personnel should wear protective 'over-clothing' or surgical scrubs (we prefer the latter; over-clothes heat you up quickly and can also contribute to a feeling of claustrophobia), and should be skin tested about every 6 months. BCG vaccinated personnel should consider having a chest X-ray every 1–2 years.

Industrial rather than surgical masks should be worn at all times. Infection procedures, tissue homogenization, and plating can all potentially create significant aerosols, which, other than needle sticks, are the primary source of danger.

Each BSL-3 should have an Exposure Control Plan in place before starting operation, and all personnel should be familiar with this. In our experience accidents are very rare, but when they do occur often as not it is due to some mental preoccupation instead of some technical lapse. In other words, only work under BSL-3 conditions when your mind can be kept completely on the task at hand.

Wherever possible BSL-3 laboratories should be stand alone or part of a larger animal facility. If there are regular open laboratories on floors above the BSL-3 then the air-handling systems should be very carefully checked on regular occasions (however, we do not recommend this type of architectural arrangement). If the facility includes any form of aerosol generation device, this should be kept separately in a room under the lowest air pressure relative to the rest of the BSL-3 facility.

Receiving new cultures

The cultivation of *M. tuberculosis* in our laboratory begins with the receipt of clinical isolates from other research institutions/hospitals. When shipments arrive the shipping container should be examined to make sure that it has not been damaged while in transit. All aetiologic agents such as tuberculosis need to be shipped (within the USA) in accordance with the interstate quarantine regulations (Federal Register, Title 42, Chapter 1, part 72, revised July 30, 1972). The shipping container should consist of an outer heavy cardboard shipping box with adequate packing to absorb any crushing. Inside this box should be a leak-proof canister (such as heavy polypropylene biocontainment canisters with rubber seals) with absorbent packing material inside. Within this canister there should reside an additional canister with more absorbent material and the tubes containing the tuberculosis isolates on Lowenstein-Jensen slants. The outside of the shipping container should have all appropriate biohazardous materials markings as well as addresses/phone numbers for the shipper and recipient. The shipping box should also contain appropriate documentation describing the contents as per strains included and possible drug resistance data.

Upon receipt and initial inspection of the package it should be taken to an appropriate biosafety level 3 (BL-3) containment room or facility. The technician wearing appropriate facility specific clothing and protective respirators should open the package within a biological safety cabinet. The package is carefully opened and examined for any damage or spillage onto the absorbent materials. If leakage has occurred the materials are thoroughly soaked with disinfectant (5% Lysol or Amphyl) and allowed to let stand for 30 min before proceeding. If the interior tubes are damaged it is best carefully to bag all materials in autoclave bags and autoclave immediately. Wipe down all work surfaces with disinfectant and notify sender of the condition of the shipment. If the contents of the package are intact, inventory what was received versus the documentation that came with the package.

Cultivation

It is essential that stock cultures of M. tuberculosis used in research experiments should be standardized in terms of the culture medium as well as the cultural conditions (initial inocula, temperature, aeration, agitation, subculture times, and so on). This is important because it assures that each individual isolate has been cultivated under similar physiological conditions during growth, which is critical if you plan to compare data between strains and experiments.

Each laboratory has its own favourite broth mediums; we prefer Proskauer and Beck (Youmans and Karlson, 1947), or GAS (glycine alanine salts) broth. Both of these liquid mediums are simple minimal salt solutions with glycerol as a carbon source. Our approach limits the amount of extraneous additives in our cultures found in other broth mediums, such as 7H9 broth, which requires enrichment with Oleic Acid Albumin Dextrose Complex (OADC). This is particularly important when trying to isolate secreted proteins or metabolites from a mycobacterial culture so as to avoid contaminating proteins from the culture medium.

Mycobacteria in general, and *M. tuberculosis* in particular, have the tendency to clump or raft when cultivated in liquid broth media. This is a problem which is overcome by the addition of the detergent Tween 80 to the culture which disperses the bacilli and provides a smooth even suspension. When cultures are grown in Tween-containing media there can be a more accurate determination of the true number of viable bacilli by optical absorbence or serial dilution plating on solid media such as 7H11 agar.

There has been some discussion as to the merits of adding detergents to the culture medium as it may affect the pathogenicity or viability of the bacilli (Davis and Dubos, 1948; Dubos and Middlebrook, 1948; Dubos, 1950; Collins *et al.*, 1974). We recognize this concern but feel that in order to achieve dispersed cultures free from large clumps, Tween is a necessary evil. For example, the necessity for smooth and evenly dispersed cultures is critical for the accurate delivery of bacilli for *in vivo* studies when the infection is given by aerosol. The aerosol device utilizes a glass venturi system to form tiny droplets containing the bacilli which are then delivered to a chamber containing the test animals over a specified period of time. If inocula from different isolates with different degrees of clumping were to be used, the data we would generate would be meaningless because of the wide variation in uptake, as well as the variation in particle size (i.e. size of rafts). On the other hand, with smooth evenly dispersed cultures the uptake parameters are highly reproducible.

Another critical parameter is that all cultures used in biological experiments should be harvested at the same phase of growth. We like to harvest broth cultures when they are still in the log growth phase but before the culture enters into the late-log growth/stationary phase. At this point the culture is at its optimum concentration and viability. Cultures harvested at this time routinely have colony forming unit (cfu) counts at 5×10^7 up to a maximum of 10^9 colonies per millilitre of broth. Higher concentrations are not recommended; when the culture gets in the range of 10^{10} cfu per millilitre there is more clumping of the bacilli and many bacilli are probably dead. If necessary, the presence of non-viable bacilli in cultures can be detected by the presence of autolytic enzymes such as isocitrate dehydrogenase.

As described above, all manipulations with *M. tuberculosis* should be done in a biological safety cabinet (Class II) within a biosafety level 3 facility. The establishment of a primary seed culture collection or repository is essential in maintaining the integrity of the individual clinical isolates as they are received. Proper documentation as to the source, date received, accession number, strain name, history of isolate, drug susceptibility-resistance, virulence data, should be maintained in some form of database. When a clinical isolate is received and unpacked it first needs to be evaluated as to whether there is sufficient growth for immediate subculture. In most instances clinical isolates of *M. tuberculosis* arrive in the research laboratory on Lowenstein-Jensen (LJ) slants (Jensen, 1932), but there are occasions when samples arrive in sealed serum vials containing broth cultures (1.0–2.0 ml).

If cultures arrive on LJ slants they have usually been incubated prior to shipment and have sufficient colony growth to allow immediate subculture into broth. On occasion there is very little visible growth and these slants should be further incubated at 37°C for 1–2 weeks to establish visible colonies. With samples which have been shipped as broth cultures, they are usually at a high CFU density and can be subcultured directly into broth media. It is generally more desirable to ship LJ slants because of the reduced risk of leakage while the package is in transit.

The first step in subculture of clinical isolates is the preparation of the work area in the biological safety cabinet. The cabinet is turned on and allowed to run for at least 20 min to allow the blower motors to reach optimum operating speed and allow the air balance to be achieved. The interior work surfaces are then disinfected prior to use with 5%Lysol™ or Amphyl™ solution followed by 70% ethanol. All cleaning materials are deposited into a small autoclave bag which is then taped to the inside wall of the cabinet. This bag is used as a waste bag for soiled gloves, Kimwipes, etc., while working in the cabinet and is closed and removed when work is complete. The work surface within the cabinet is covered with bench blotters (i.e. plastic backed paper toweling) which have been sprayed with disinfectant solution (Lysol™/Amphyl™). This reduces the potential hazard resulting from unnoticed spatters which may occur during manipulations. A metal pipette boat with 2–3 cm of 5% Lysol™ is placed in a convenient working location for the technician (usually in the centre toward the back, but not too close or over the air returns of the cabinet). All materials which will be needed for manipulations, such as sterile pipettes, inoculation loops, sterile tubes, flasks, extra gloves, etc., should be located close to the cabinet on a bench or laboratory cart. When all materials are assembled work can then proceed.

To establish broth seed cultures we use sterile 150 × 25 mm screw-capped culture tubes which have been prefilled with 20 ml of either Proskauer and Beck or GAS liquid medium supplemented with 0.05% Tween 80. Each tube contains a very small plastic-coated magnetic stir bar for culture aeration and agitation. Using a sterile disposable inoculating loop, a sample from the LJ slant is inoculated into the broth within the tube. It is important only to inoculate one loop; if you seed the culture with an overly heavy amount of bacilli the clumps will not dissipate and you have the potential of having a larger amount of dead bacilli in your seed stock upon harvest. The inoculation loop is then discarded carefully into the pipette boat containing the Lysol™. The reason for using plastic disposable loops is that you minimize the potential aerosol hazard which may occur with flaming a metal inoculation loop. Metal loops can be used, but should be first speared into a 70% ethanol sand bath to remove any remaining material before being flamed in an electric loop incinerator.

With broth samples, the top of the vial is first swabbed with 70% ethanol and then flamed with a Bunsen burner to sterilize. A 1 cc tuberculin syringe fitted with a 26G needle is then used carefully to remove 1.0–2.0 ml of the inoculum from the vial. The inoculum is then carefully injected down the side of the culture tube, taking care not to create an aerosol. The inoculated tubes are each labelled with the following information: Species (*M. tuberculosis*), Strain, Media, Date, Technician, Drug Resistance (Yes/No).

The seeded culture tubes are carefully placed into a plastic test-tube rack which has been secured on top of a magnetic stir plate within a 37°C incubator. The original LJ slants are deposited in a 4°C refrigerator within the biosafety laboratory and retained. All instruments during subculture are wiped down with disinfectant prior to removal from the cabinet. All waste materials are bagged in autoclave bags and removed immediately for sterilization in a steam autoclave (121°C at 15 lbs/in^2 for 40 min). The interior surfaces of the biological safety cabinet are thoroughly disinfected with 5% Lysol followed by 70% ethanol.

The seeded cultures are checked twice a week for growth and possible contamination. Usually the cultures reach a density of 5×10^7 to 5×10^8 within 1–2 weeks after initial seeding. With experience a trained technician can visually estimate the approximate CFU growth within a tube of broth. Generally a tube less than 10^6 CFU ml^{-1} is clear and growth cannot be visually detected. At 10^7 CFU ml^{-1} there is a discernible cloudiness which can be visually seen. When the culture reaches 10^8 CFU ml^{-1} there is a very hazy growth and printed text can be detected when placed behind the tube but individual letters cannot be visualized. A culture which has reached 10^9 CFU ml^{-1} or greater is very turbid/milky and some clumping or 'stringy' growth may be detected (resulting from the Tween being metabolized by the bacilli).

An optical density measurement at A580 can also be made to determine the relative concentration. When the cultures reach mid–late log growth, approximately 10^8 CFU ml^{-1}, they are then subcultured into 150 ml of fresh broth in disposable 250 ml polycarbonate Erhlenmyer flasks with screw type closures (Corning 25600). The flasks are labelled as above and incubated at 37°C on orbital shakers (slow speed with gentle agitation), or stationary with gentle swirling twice weekly for an additional 1–2 weeks. When the density has once again reached approximately $1–5 \times 10^8$ CFU ml^{-1} the culture is removed from the incubator to be aliquoted into serum vials for laboratory seed cultures. The mycobacterial culture is aliquoted in volumes of 1.5 ml into 2 ml sterile serum vials which are fitted with sterile butyl rubber stoppers, which are then crimped into place with sterile aluminium seals (Wheaton Glass). The vials are disinfected by immersion in 70% ethanol and then allowed to air dry. Each vial is then labelled with small Avery™ labels printed with information as to the species, strain, date (lot), media, and technician. The vials are then placed in labelled storage boxes which are then placed in a −70°C freezer for long-term storage.

After the vials have been frozen, a random sampling of vials from the storage box is taken out the following day for the determination of the viability of the lot. The vials are thawed, vortexed vigorously, and the butyl rubber septum disinfected with 70% ethanol followed by flaming with the Bunsen burner. A sterile 24-well tissue culture plate is set up with 0.9 ml of sterile tween-saline (0.05% Tween 80 in 0.85% NaCl solution) in each well. With a 1 ml sterile tuberculin syringe, the suspension of bacilli is removed from the serum vial via the butyl stopper. A 0.5 ml sample of this suspension is plated onto a Trypticase Soy or Blood Agar plate to check sterility. Into the first column of the 24-well plate, 100 μl of sample is deposited. With a P-200 Gilson™ Pipetman set at a 100 µl volume, and fitted with a barrier type pipette tip, the samples are serially diluted through a series of 10-fold dilutions starting at column 1 and ending at column 6. Column 1 corresponds to a 1 : 10 dilution of the original vial as column 6 in the series corresponds to a 1 : 1 000 000 dilution. A volume of 100 µl is removed from each dilution well and plated onto quadrant-type petri plates containing Middlebrook 7H11 agar. The plates are bagged in polypropylene zip lock sandwich bags (four plates per bag) and incubated in the dark at 37°C for 3 weeks. Colonies are then counted and the CFU of the mycobacterial frozen seed lot determined.

Once the CFU has been established for each seed culture lot, these data are entered into the culture collection log, notebook, or database. For each particular strain approximately 25–30 vials are retained as 'primary' seeds and not used except for subculture to establish more working stock. The repetitive or continuous subculture of any strain of *M. tuberculosis* should be discouraged as this may cause physiological changes in the strain over time.

If a strain needs to be subcultured or expanded it is advisable to start from one of the 'primary' seeds in the culture collection or if available from the original LJ slant (Note: LJ slants can be maintained at 4°C and subcultured onto fresh LJ for maintenance of cultures). The additional 70 vials or so from each initial subculture are then used as your working stocks.

Liquid media

Proskauer and Beck (PB) Liquid Medium

Into 1 l of distilled water add each of the following ingredients in the order listed, making certain that each salt is dissolved before the next is added: KH_2PO_4 5.0 g; asparigine 5.0 g; $MgSO_4 \bullet 7H_2O$ 0.6 g; magnesium citrate 2.5 g; glycerol 20 ml.; Tween 80 0.5 ml (optional for dispersed cultures, 0.05% final concentration). The pH is then adjusted to 7.8 by the addition of 3–5 ml 40% NaOH. The medium is then autoclaved at 121°C for 15 min on slow exhaust, after which the pH should be 7.4. On occasion there is a precipitate which forms after autoclaving; if so, the media should be allowed to cool and the precipitate removed by filtration. The media is then autoclaved once again at 121°C for 15 min (or can be sterile-filtered through a 0.2 µm filter unit/cartridge (Gelman)).

GAS (Glycerol-Alanine-Salts)

To prepare GAS, each of the following ingredients is dissolved in the following order into 990 ml of distilled water: Bacto Casitone (Pancreatic Digest of Casein (Difco)) 0.3 g; ferric ammonium citrate 0.05 g; K_2HPO_4 4.0 g; citric acid (anhydrous) 2.0 g; L-Alanine 1.0 g; $MgCl_2 \bullet 6H_2O$ 1.2 g; K_2SO_4 0.6 g; NH_4Cl 2.0 g; Glycerol 10 ml. The pH is adjusted to 6.6 by the addition of 1.8 ml 40% NaOH. The GAS media is then autoclaved at 121°C for 15 min on slow exhaust.

Solid media

Middlebrook 7H10 and 7H11 Agar (Difco)

Both 7H10 and 7H11 agar media are routinely used for the cultivation and enumeration of M. tuberculosis by limiting dilution plating. The 7H11 agar contains a pancreatic digest in the base, whereas 7H10 does not; this additive seems to enhance the growth of many strains of mycobacteria and hence is the preferred medium in our laboratory. Both formulations, however, produce acceptable colony growth over a 3-week incubation at 37°C.

To prepare 1 l of 7H11 or 7H10, place the following ingredients into a 2-l Erhlenmyer flask: 7H11 (or 7H10) agar base 21 g; asparagine 1.0 g; glycerol 5.0 ml; distilled water 900 ml. All of the ingredients are thoroughly dissolved and the flask is capped with heavy aluminium foil which is then secured into place with autoclave tape. The agar is then autoclaved for 15 min at 15 p.s.i. and 121°C on slow exhaust cycle. Remove the flasks from the autoclave and check the colour of the agar. The agar should be an emerald green, if, however, it appears to be olive drab or brown, discard this batch of agar as it may have been 'overcooked' during autoclaving and the malachite green has degraded. (The breakdown of malachite green produces compounds which inhibit the growth of mycobacteria.) After checking the colour of the agar, place the flask into a 57°C water bath and allow the agar to equilibrate for at least 45 min. While the agar is cooling, set up the following materials in a sterile media prep bench:

- One sterile 10 ml Cornwall Syringe set to dispense 5.0 ml aliquots (if dispensing into X-plates; if using non-divided plates, disregard the Cornwall Syringe).
- 60 #1009 X-plate petri dishes for each litre of agar.
- One wire basket to support the 2-l flask.
- Sharpies to code plates as to possible drug additives.
- 100 ml of OADC enrichment media per litre of agar (described below).

When the agar has equilibrated to 57°C, remove the flask from the water bath and place it in the wire basket in the bench. Carefully remove the foil and aseptically pour 100 ml of the OADC enrichment media into the litre of agar. Gently swirl the flask to mix the OADC throughout the agar. At this time, drugs such as TCH or INH can be added by filter sterilization. Carefully unwrap the Cornwall syringe from its foil envelope and aseptically place the tubing into the flask. Prime the syringe and proceed to dispense the agar into the petri dishes with strict aseptic technique. Dispense 5.0 ml of agar per quadrant per plate, or pour approximately 20 ml of agar into a non-divided plate. After dispensing the media, allow the plates to stand at room temperature for at least 2 h. This is sufficient time to allow the agar to solidify. Cover the plates with a dark plastic bag or aluminium foil to protect the media from the light.

When the agar has solidified put the plates into plastic storage trays and place these trays in a dry 37°C incubator for 24–48 h. This is a 'curing' step in which the excess moisture is removed from the agar and allows an opportunity for any contaminants to become apparent. After the incubation of the plates is accomplished they are placed into zip lock polypropylene bags (five plates/bag) and labelled as to type of agar, drugs, date and name. If the plates are not to be used immediately, they are stored at 4°C until use. Plates can be stored for up to 1 month; use after this time is not recommended.

Middlebrook Oleic Acid Dextrose Complex (OADC)

To prepare 4.0 l of OADC the following ingredients need to be dissolved in 3.8 l of distilled water in the following order: NaCl 32.4 g; Bovine Albumin Fraction V 200 g; dextrose (D-glucose) 80 g. The pH of the mixture is then adjusted to 7.0 with 4% NaOH when everything is in solution.

To this mixture a sodium oleate solution is then added. This solution is prepared with the following ingredients: distilled water 120 ml; 6 M NaOH 2.4 ml; oleic acid 2.4 ml. Once the sodium oleate solution is prepared it is warmed in a 56°C water bath until the solution is clear. The complete OADC enrichment media is then filter sterilized through 1.0 μM and then 0.2 μM cartridge filters (Gelman) using a peristaltic pump system. The media are dispensed in a laminar flow tissue culture hood into sterile 100 ml bottles which are capped, heated in a 56°C water bath for 1 h, and then incubated at 37°C overnight. The heating and incubation is then repeated the following day. After inspecting the OADC media for contamination, the bottles are stored at 4°C for up to 2 months.

Adding drugs to media

Acriflavine

- Working Concentration: 100 μg ml^{-1} = 100 mg l^{-1}
- Preparation: Dissolve 100 mg of acriflavine in 10 ml of sterile distilled water and filter sterilize. Add all 10 ml to 1 l of agar.

Isoniazid

- Working Concentration: 0.2 μg ml^{-1} = 200 μg l^{-1} = 0.2 mg l^{-1}
- Preparation: Dissolve 20 mg of INH in 10 ml of sterile distilled water. Dilute this 1 : 10 in sterile distilled water to give a final concentration of 2000 μg ml^{-1} and filter sterilize. Add 1 ml of the 2000 μg ml^{-1} stock to 1 l of agar.

2-Thiophenecarboxylic acid hydrazide

- Working Concentration: 1 μg ml^{-1} = 1 mg l^{-1}
- Preparation: Dissolve 10 mg of TCH in 10 ml of sterile distilled water and filter sterilize. Add 1 ml of this stock to 1 l of agar.

◆◆◆◆◆◆ PROPAGATION METHODS FOR THE ISOLATION OF IMMUNOLOGICALLY REACTIVE PROTEIN FRACTIONS

Equipment and reagents

The growth of *M. tuberculosis* described in this section is for 500–1000 ml cultures, i.e. much higher volumes than for infecting inocula. Because of this some growth conditions are different from above.

The equipment required includes a roller-bottle apparatus or platform shaker for gentle agitation of cultures and a class II biological safety cabinet. *M. tuberculosis* is a class III pathogen and its propagation must be performed in a certified BSL-3 laboratory (Centers for Disease Control, 1993). Several types of broth media are available for the growth of *M. tuberculosis*; however, media containing supplemental proteins such as BSA or yeast extract should be avoided. We generally use glycerol alanine salts (GAS) broth (see above). This media allows for ample growth of the tubercle bacilli and is easily prepared.

Growth of *M. tuberculosis*

For initial growth of *M. tuberculosis* a 1-ml aliquot from frozen stock is inoculated on a Middlebrook 7H11 agar plate and incubated at 37°C for 2–3 weeks. Once colonies are visible a liberal inoculum of cells is scraped from the plate and placed in GAS broth (20 ml). This is allowed to incubate at 37°C with gentle shaking for 2 weeks. An aliquot (10 ml) of this seed culture is transferred to 100 ml of GAS broth. Following 2 weeks of incubation, the entire 100 ml culture is used to inoculate 1 l or is split for 500 ml GAS broth cultures. These cultures are incubated in roller bottles or fernbach flasks with gentle agitation for 2 weeks. It is our experience that 2 weeks of incubation provides cultures that are in a mid- to late-logarithmic phase of growth (Sonnenberg and Belisle, 1997).

Harvesting culture filtrate proteins (CFPs) and preparation of subcellular fractions

Equipment and reagents

The separation of the cells and the culture supernatant is performed in a class II biological safety cabinet in a BSL-3 laboratory. The supernatant is harvested using 0.2 µm ZapCap filters (Schleicher and Schuell, Keene, NH, USA). Vacuum for this procedure is provided by a vacuum pump (maximum vacuum 26 in mmHg) with an in-line 0.2 µm filter (Gelman, Ann Arbor, MI, USA, catalogue #4251) placed between the ZapCap and the vacuum pump. Concentration of the filtrate is accomplished with an Amicon apparatus (Beverly, MA, USA). Harvesting of the cells requires a table top centrifuge such as a Sorvall RT-6000 with sealed buckets for containment of potential aerosols. Lysis of the cells requires a French press or probe sonicator.

Breaking buffer

- 10 mM Tris-Cl, pH 7.4
- 150 mM NaCl
- 10 mM EDTA
- 100 mg ml^{-1} DNase
- 100 mg ml^{-1} RNase
- proteinase inhibitors.

Harvesting of CFPs

The bacilli in 500 or 1000 ml cultures are allowed to settle and the supernatant is filtered through a 0.2 μm ZapCap filter into 4-l bottles. To minimize plugging of the filter it is important to use ZapCap with a prefilter and decant a minimal amount of cells along with the supernatant. Sodium azide is added to the filtrate to a final concentration of 0.04% w/v and this material is stored at 4°C. After filtration, the culture supernatant is considered sterile. Nevertheless, before further use, a 1-ml aliquot of this filtrate is plated on Middlebrook 7H11 agar and incubated for 3 weeks to ensure the absence of viable bacilli. The filtrate is concentrated to approximately 2% of its original volume using an Amicon apparatus with a low protein binding, 10 kDa molecular weight cut off membrane. This concentrate is dialysed extensively against 10 mM ammonium bicarbonate, and the protein concentration estimated by the BCA protein assay (Pierce, Rockford, IL, USA). The final CFP preparation is aliquoted and stored at −70°C. Typically, 4–5 mg of CFPs are obtained from 1 l of a 2-week culture of *M. tuberculosis*.

Isolation of subcellular fractions of *M. tuberculosis*

The cells from which the culture supernatant has been decanted are collected by centrifugation at 3000g. The cell pellet is washed with sterile H$_2$O and inactivated. To preserve the integrity of proteins γ-irradiation is recommended for the killing of *M. tuberculosis*. A radiation dose of 2.4 megaRad renders all exposed bacilli non-viable whilst not affecting the large majority of the enzymatic functions of the cells (Hutchinson and Pollard, 1961). Alternatively, *M. tuberculosis* cells can be killed by heating at 80°C for 1 h. This is best done in an autoclave with the capacity to perform low temperature isothermal cycles. However, heat killing is much more damaging to the integrity of the *M. tuberculosis* proteins. As with the CFP preparations the lack of viable bacilli should be checked before cells are removed from containment facilities.

Several techniques for the generation of *M. tuberculosis* subcellular fractions have been reported (Hirschfield *et al.*, 1990; Lee *et al.*, 1992; Trias *et al.*, 1992; Wheeler *et al.*, 1993). Although some of these protocols may result in cleaner separation of subcellular fractions, we have found that the methods used by Hirschfield *et al.* (1990) and Lee *et al.* (1992) provide a rapid and simple means by which to obtain crude fractions of cell wall, membrane and cytosol. Lysis of *M. tuberculosis* cells is accomplished by suspending the cell pellet at a concentration of 2 g of cells per ml of breaking buffer and passing this suspension through a French press at 1500 psi, five to seven times. The lysate is diluted with 1 volume of the breaking buffer and unbroken cells are removed by centrifugation at 3000*g* for 15 min. Cell wall material is harvested from the supernatant of the low speed spin by centrifugation at 27 000*g* for 30 min. The cell wall pellet is washed with 10 mM ammonium bicarbonate, dialysed against the same, aliquoted and stored at −70°C. Supernatant from the 27 000*g* spin is further separated into cytosolic and membrane fractions by centrifugation at 100 000*g* for 1 h. The membrane pellet is washed, suspended in 10 mM ammonium bicarbonate and stored at −70°C. The cytosolic fraction is dialysed against 10 mM ammonium bicarbonate, aliquoted and stored. Extraction of proteins from the cell wall or membrane fraction may be performed using a number of detergent or chaotropic agents that are easily removed prior to use of the proteins in immunological assays (Hjelemeland, 1990; Thomas and McNamee, 1990; Scopes, 1994).

◆◆◆◆◆◆ INFECTION PROTOCOLS

Intravenous infection of mice

The proper handling and restraining techniques for mice should be mastered prior to any attempt at intravenous inoculation with virulent strains of *M. tuberculosis*. The technician should seek proper training which is provided by most institutional animal facilities in order to assure the safety of both the technician and the laboratory animal. Proper humane methods for handling laboratory mice should be learned by the technician and a certain degree of skill and confidence achieved prior to performing any work involving the use of *M. tuberculosis* in an animal model. The technician needs to be fully aware of the risks of accidental infection via a bite wound, scratch, accidental needle stick, or aerosol exposure when restraining and inoculating mice with *M. tuberculosis*. Because of this, it is extremely important that the technician develop a methodical approach when performing such techniques.

All infections of laboratory animals with *M. tuberculosis* should be done in an appropriate Animal Biosafety Level-3 laboratory (ABL-3) with the technician wearing appropriate facility clothing and personal protective equipment. Double disposable surgical gloves should be worn when handling mice to prevent accidental exposure due to tearing of the outer glove (mice can bite through the outer glove, but rarely through both).

To infect mice intravenously the technician will require a preparation of *M. tuberculosis* from a stock of known CFU ml^{-1}. Working under BSL-3 conditions, the bacilli are thawed and carefully removed from the vial via a 1-ml tuberculin syringe fitted with a 26G needle. The bacteria should then be diluted in normal saline or PBS to achieve the required CFU concentration per ml for intravenous challenge. When the mycobacterial suspension is dispensed from the tuberculin syringe into the dilution tube it is done in a manner in which the suspension is slowly dispensed down the inside wall of the tube. This is done in order to minimize the possibility of the creation of an aerosol from the needle. After dispensing the mycobacterial suspension, the syringe is carefully placed in an appropriate 'sharps' container for disposal.

The mycobacterial suspension is serially diluted and a sample plated on 7H11 agar to verify the concentration. Pipetting operations during the serial dilution process are carried out with extreme care in order to prevent aerosol generation. Any small spills or drops during the pipetting process should be immediately disinfected with 5% Lysol™. All pipettes and other materials used serially to dilute the mycobacteria are placed in a pipette boat containing 5% Lysol™ solution. The biological containment hood is disinfected after use. The culture tubes, stock vials and pipette boat are double bagged in autoclave bags and taped with heat-sensitive autoclave tape. These materials are then placed in a large autoclave pan located near the ABL-3 autoclave for subsequent sterilization.

In our laboratory, the standard intravenous challenge dose is 10^5 CFU per mouse. This dose was chosen because it is sub-lethal and allows the mouse to make a very strong cellular response. This is delivered intravenously in a volume of 0.2 ml from a working concentration of 5×10^5 CFU ml^{-1} suspended in normal saline or PBS. The injectate is loaded into a 1-ml tuberculin syringe fitted with a 26G needle with the needle bevel at the same orientation as the syringe graduations.

Prior to injecting mice a full-face shield should be put on to prevent accidental exposure from injectate inadvertently spraying back during injection (a common, and scary event, to the novice). Although a simple shield is probably sufficient, our common handling of drug-resistant isolates has prompted us to use a RACAL AC-3 (Racal Health and Safety) respirator which consists of a belt with a blower unit fitted with HEPA filters which delivers sterile air to a helmet face-shield. This unit provides clear visibility and does not fog up, unlike traditional face respirators.

Once sufficient syringes have been prepared, a cage of mice is placed under a heat lamp to warm the mice gently and increase their venous circulation. It is essential that the worker pay close attention that the mice are warmed but are not heat stressed while under this lamp. While the mice are warming, a mouse restrainer device is set up. We use a metal cone type of restrainer which bolts to the edge of the laboratory bench. When the mice are sufficiently warm, one mouse is removed from the cage by gently grasping the tail. It is then placed into the restrainer with the tail sliding through the slot in the top of the restrainer. While holding the tail, swab the area to be injected with a 70% ethanol soaked gauze sponge or swab. Allow the ethanol to air dry. Identify a lateral tail vein and keeping the needle bevel up gently insert the needle parallel to the vein 2–4 mm into the lumen.

Once you are sure you are in the vein, inject the *M. tuberculosis* suspension slowly. No bleb should be visible if the needle was properly inserted into the lumen of the vein; if a bleb appears, indicating failure to locate the vein, additional attempts may be made proximally. It is desirable to make the first attempt at injection as close to the tip of the tail as possible.

NEVER force the syringe if you miss the vein as spray back will occur, or the needle hub may dislodge from the syringe or tail and cause a subsequent aerosol of *M. tuberculosis*. When finished, withdraw the needle slowly and apply pressure at the injection site with gauze if necessary to achieve haemostasis.

Aerosol infection of mice

Because the natural route of tuberculosis infection is the lung, the most precise animal models try to mimick this route. Intranasal or intratracheal inoculation of mice can give rise to pulmonary infection, but the most reproducible technique is to generate an aerosol that is inhaled by the animal. Aerosol devices are available in two main types; one is the sealed cabinet system pioneered by Middlebrook, the other a newer device that fits over the nose of the animal. Whatever your choice, the device must be kept under stringent ABL-3 conditions due to the very high aerosol danger.

Our preference is the Middlebrook Airborne Infection apparatus which has been used for this specific purpose for the past 30 years. This instrument is currently manufactured by Glas-Col, Inc. of Terre Haute, Indiana and has changed very little in configuration since the original design. The instrument consists of a large circular tank (aerosol chamber) which contains a circular basket/cage with five pie-shaped compartments in which animals are placed. Each of the compartments can accommodate as many as 25 mice, two 500-g guinea-pigs, or one rabbit. The aerosol chamber has a heavy acrylic lid with four locking handles that lock tightly against a heavy-duty rubber gasket. The lid also has two ultraviolet lamps on its underside and these lamps are turned on during the decontamination cycle of instrument operation.

The front of the instrument consists of a control panel, which allows the machine to be programmed; two air-flow meters; two air-control knobs; and a series of switches which control power. Also, on the front of the instrument are three tygon hoses with hose clamps for attaching the glass venturi unit (nebulizer) which is fitted in a holding bracket and filled with the bacterial suspension. When the instrument is in operation, compressed air flows through the nebulizer and produces a very fine mist of the bacterial suspension, which is then carried by a larger volume of air flowing into the aerosol chamber. The airflow then exits the chamber through two HEPA filters and a super-heated exhaust stack wherein it is incinerated.

The programme, controlled via the control panel, is used to determine the duration of the various cycles involved in the 'aerosol' process. The first parameter to be entered is the length of the 'pre-heat' cycle, this cycle allows the incinerator to attain suitable temperature prior to nebulization. The second parameter is the length of time required for the nebulization cycle, this cycle does not engage until the pre-heat cycle is complete. During nebulization the compressed air comes on and is routed through the venturi of the nebulizer to create the fine bacterial mist. The third parameter to be entered is the time required for the 'cloud decay' cycle and it is during this cycle that the aerosol chamber is purged with fresh air and the bacterial mist dissipated. The final parameter is the length of the 'ultraviolet cycle', in which the UV lamps are switched on and decontaminate the top surfaces of the basket. The length of time suggested for the incinerator warm up and the UV decontamination cycles is 15 min each. The time required for the cloud decay cycle is 40 min. The time required for the nebulizing cycle should be determined empirically for each animal type.

Once all cycles are complete the instrument is carefully examined to make sure hoses and gaskets are still in place. The technician, wearing appropriate safety equipment (a RACAL safety helmet with HEPA filters is recommended (Orme and Collins, 1994)), then opens the chamber lid and removes the animals. The basket is carefully removed, wrapped in autoclave bags, and autoclaved to sterilize. The interior of the chamber is then disinfected with 5% Lysol™ followed by 70% ethanol. The glass venturi nebulizer is carefully removed and placed in a stainless steel pan containing 5% Lysol, covered with a lid, wrapped in an autoclave bag, and sterilized via autoclaving. All disinfecting materials are likewise bagged and autoclaved.

◆◆◆◆◆◆ CULTURING AND INFECTING MACROPHAGES

Establishing macrophage cultures *in vitro*

The establishment of primary cell cultures of murine bone marrow derived macrophages is an important *in vitro* technique, which is used extensively in our laboratory. These macrophages can be used for numerous types of studies ranging from macrophage responses to infection (nitric oxide or chemokine production for example), to evaluating the ability of drugs to prevent intracellular replication, to identifying protein targets of IFN-secreting T-cell subsets.

Bone marrow macrophage cultures are established by first euthanizing mice in accordance with the methods approved by your institutional animal care and use committee. We recommend placing mice in a large glass jar which has some pellets of dry ice on the bottom covered with gauze for carbon dioxide asphyxiation. Once the animals have been euthanized they are placed on a clean dissection board and saturated with 70% ethanol. The skin at the mid back is clipped with sterile scissors and then pealed back to expose the lower part of the body including the hind legs. The knee is then dislocated by holding the thigh with one hand and the shin with the other. Dislocation is accomplished by firmly pulling apart the knee joint. When this is done bend the knee back, holding the leg on the shin. Clip the muscle under the knee and bend the knee in the direction it normally would not bend. Pull down on the shin, and the femur will emerge through the muscle. Clip the femur from the hip, removing the excess muscle and place it in a tube containing ice-cold, sterile media (see below). Clip the foot off, and place the sharp edge of the scissors between the tibia and fibula, and slide the scissors up the shin, cutting the muscle. Peel the tibia out of the muscle, and clip it from the knee, removing the excess muscle. The tibia is also placed in the tube containing ice-cold media.

The tube containing the bones is then taken to the tissue culture area of the laboratory and placed in a laminar flow biological containment cabinet for the extraction of the bone marrow. The tube containing the bones and media is carefully emptied into a large sterile petri dish. Sterile forceps and scissors are used to pick up the bones, remove any excess muscle tissue, and to clip the ends to expose the marrow. Using a 10-ml syringe filled with media and fitted with a 26G needle, rinse the bone marrow out of both ends of the bones into a 50-ml conical tube which contains 5–10 ml of media. Rinse with about 4 ml of media per bone to ensure complete removal of all bone marrow cells.

A 10-ml pipette is then used gently to mix the marrow suspension up and down until all the clumps have been broken up. The bone marrow cells are then pelleted by centrifugation ($150g/7$ min/4°C). The supernatant is decanted and the cells resuspended (use 2-ml media per animal harvested). Pass the cells through 100-μm sterile nylon mesh to break up cell clumps. Count the cells in 3% acetic acid/PBS (to lyse red cells) and then adjust the cell suspension to a concentration of 2×10^6 nucleated cells per millilitre of medium. Each mouse should provide $1–2 \times 10^7$ bone marrow cells. These are then plated onto petri dishes, flasks, or well-type tissue culture plates as shown in Table 1.

The bone marrow cell cultures are then placed in a 37°C incubator which is supplemented 95% humidified air with 5–7% CO_2. The media in the cultures are changed at 48 h post-seeding and then again changed at day 4–5. The macrophages should have differentiated and formed a confluent monolayer by day 7 to 9. At this point the macrophages are mature and ready for use in experiments. If these bone marrow macrophages are to be used for infection with M. tuberculosis the media needs to be changed to antibiotic-free media 48 h prior to the experiment.

Table I. Bone marrow macrophage culture

Tissue culture system	Seed density (nucleated cells ml⁻¹)	Volume required (ml)
75-cm TC-flasks	5.00×10^5	30.00
150-mm TC dishes	1.00×10^6	20.00
100-mm TC dishes	1.00×10^6	12.00
12-well TC plates	1.00×10^6	2.00
24-well TC plates	1.00×10^6	1.00
96-well TC plates	1.00×10^6	0.20

TC-flasks, tissue culture flasks that have a 75 cm² culture area; TC, tissue culture.

Bone marrow medium

- Complete DMEM (Sigma #D5530) supplemented with 10% Heat-inactivated FCS (Intergen) and 10% L-929 conditioned medium
- + HEPES buffer (Sigma #H0887)
- + L-glutamine (Sigma #G7513) MEM Non-essential amino acids (Sigma #M7145)
- + Penicillin/streptomycin (Sigma #P-0781)
- (omit for Antibiotic-Free Bone Marrow Media).

L-929 conditioned medium

- L-929 cells from ATCC are grown up at 4.7×10^5 cells total in a 75 cm^2 T-flask with 55 ml of DMEM (Sigma)
- 10% FCS (Intergen)
  1. Allow cells to grow for 7 days or until confluent.
  2. On day 7 collect the supernatant, filter through a 0.45 µm Nalgene filter and freeze at –20°C in 40 ml aliquots.

N.B. The use of fetal calf serum in these cultures requires that each new batch be assessed prior to purchase. The characteristics of the macrophages can differ depending on the constituents of the FCS.

Infecting macrophage cultures

The *in vitro* infection of murine bone marrow derived macrophages is routinely used in our laboratory to evaluate the macrophage response to different clinical isolates, and also to determine the efficacy of various novel antibiotic compounds on the intracellular growth of *M. tuberculosis*.

In such studies, the macrophage monolayers are washed once with sterile phosphate buffered saline (PBS) and then supplemented with fresh antibiotic-free bone marrow macrophage medium as described above. The cultures are then allowed to incubate for 48 h prior to infection with *M. tuberculosis*. If the protocol involves cytokine treatment of the cells, these are added at this time.

The resulting macrophage monolayer contains approximately 1.0×10^7 cells after 8–10 days of incubation (24-well plate method). On the day of infection the media are removed from the macrophage monolayer and immediately replaced with 200 µl of antibiotic-free media containing 1.0×10^6 CFU (5.0×10^6 CFU ml^{-1}) of *M. tuberculosis*. The plates are returned to the 37°C humidified air–CO$_2$ incubator for 4 h. After this incubation period, the monolayers are washed gently four times with 1 ml sterile PBS to remove any bacilli that were not ingested. (Great care should be taken not to generate aerosols during this procedure.)

The cells are now ready for further use. Production of oxygen or nitrogen radicals can be measured by colorimetric assays, and supernatants can be collected to measure the secretion of chemokines or cytokines. The cells can be lysed in Ultraspec (Biotexc) and processed for the presence of mRNA for molecules of interest. Alternatively, growth of the bacteria can be followed by plating lysates (distilled water plus 0.05% Tween 80, then diluted in PBS) and plotting CFUs versus time.

It has been claimed that such cultures can be taken out for as long as a month. Certainly the cultures can look good under phase-contrast microscopy for up to 2 weeks in our hands, but even then some detached macrophages are seen ('floaters') which can make the bacterial numbers appear lower than they really are. For this reason, we do not go further than 8 days in these assays.

There is a similar caveat with regard to using radioactive uracil as opposed to counting CFU in macrophage cultures. In our hands reduction in uracil uptake usually only reflects stasis, not cidal activity, so any claims that a reduction in uracil counts equals killing should be backed up by showing a reduction in CFU counts (Rhoades and Orme, 1997).

◆◆◆◆◆◆ IDENTIFYING KEY PROTEINS RECOGNIZED BY IMMUNE T CELLS

The bone marrow-derived macrophage system can also be utilized in immunological assays involving antigen processing and presentation to T cells. In our laboratory we have had quite variable results based on cell proliferation assays, and so, given our demonstration of protective T cells as cells that secrete IFN (Orme et al., 1992; Cooper et al., 1993), we developed the following assay to detect the presence of these cells, and to obtain a picture of the proteins they are recognizing. Details of these results are published elsewhere (Roberts et al., 1995).

The system uses bone marrow-derived macrophages to present antigen to T cells (or subsets thereof) harvested from syngeneic mice which have been intravenously infected with 10^5 CFU of M. tuberculosis 15 days earlier.

This is the timing of the peak response; other times can be chosen if one is interested in the kinetics of response to a given antigen (Orme and Collins, 1994). T cells can also be harvested from the lung, as described below, and used in the T-cell overlay method.

Tissue culture plates (96 well) are seeded with 2×10^5 bone marrow cells and left for 7 days. Samples of individual proteins or protein fractions are dissolved in sterile pyrogen free water to a concentration of 1 mg ml^{-1}. The medium in the bone marrow macrophage cultures is carefully removed and replaced with fresh medium. For pulsing macrophage cultures the stock protein samples must be diluted to a working concentration of 10 µg ml^{-1}. Ovalbumin is used as an irrelevant negative control (there is some background IFN secretion, usually in the picogram range). The macrophage cultures are then incubated overnight at 37°C and 6% CO_2 to allow processing and presentation of the test antigens.

The following day infected mice are euthanized and spleen (or lung, see below) cell suspensions prepared. These are then incubated at 37°C in 5% CO_2 for 1 h in plastic flasks to allow macrophages to adhere. After incubation, the plate is removed from the incubator and gently rocked/swirled to resuspend the non-adherent spleen cell population. The non-adherent cells are gently pipetted off into a sterile 50-ml conical tube and centrifuged at 200g for 7 min. The supernatant is gently decanted and the cell pellet is resuspended with 5 ml of ACK lysing buffer (see below) per spleen in order to remove red blood cells from the non-adherent spleen cell population. The cells are incubated for 5 min at room temperature with occasional shaking. (Note: ACK lysing buffer needs to be used at room temperature in order to work properly.) After the ACK incubation is complete, the tube is filled to 45 ml with tissue culture medium and again centrifuged for 7 min. This wash process is then repeated one additional time. After the last wash the supernatant is decanted and the cell pellet is resuspended with monoclonal antibody supernatant preparations from clones TIB-210 (clone 2.43 anti-CD8.2) and TIB-183 (clone J11d.2 anti-B-cell, granulocyte, immature T-cell supernatant).

The exact working dilutions need to be determined for each lot of antibody supernatants. After incubation at 37°C for 30 min Cederlane™ Low-Tox-M Rabbit Complement (1 : 16 Final Concentration) is added to the cell suspension and incubated for a further 1 h. The cells are then centrifuged at 200g for 7 min, and repeated twice. These cells should be an enriched population containing predominantly CD4 T cells, which can be checked by flow cytometric analysis.

Alternatively, CD4 or CD8 T cells can be purified using MACS cell separation beads (Miltenyi Biotec) and by following the manufacturer's protocol. Briefly, cells are incubated with MACS beads specific for CD4 or CD8 for 15 min at 4°C, washed, and passed over a cell separation column (Miltenyi Biotec). Eluted cells are then collected, counted and resuspended 1×10^6 cells ml^{-1} in media containing 20 units of IL-2 per ml to maintain T-cell viability. The T cells are then added in volumes of 0.1 ml to the wells containing the macrophages and incubated for 72 h at 37°C in 5%CO_2. Supernatants are then removed and assayed for IFN-γ by ELISA.

ACK lysing buffer

1. NH$_4$Cl 8.29 g (0.15 M)
2. KHCO$_3$ 1.0 g (1.0 mM); Na$_2$EDTA 37.2 mg (0.1 mM)
3. Add 800 ml H$_2$O and adjust pH to 7.2–7.4 with 1N HCl
4. Add H$_2$O to 1.0 l
5. Filter sterilize through 0.2 µm filter and store at room temperature.

Complete tissue culture medium

- Complete DMEM (Sigma #D5530) supplemented with 10% Heat-inactivated FCS (Intergen)
- + HEPES buffer (Sigma #H0887)
- + L-glutamine (Sigma #G7513) MEM Non-essential amino acids (Sigma #M7145)
- + Penicillin/streptomycin (Sigma #P-0781).

◆◆◆◆◆◆ FOLLOWING THE COURSE OF INFECTION

Following the course of the tuberculosis infection by plating tissues and counting bacterial colonies can provide valuable information about the kinetics of expression of the host response (Orme, 1995). You should also be aware, however, that the assay is rather crude, with a counting error of about 20%. Because of this variance one cannot read too much into small changes in numbers; as an example, late stages of the infection often appear to be 'chronic', i.e. flat-line. The infection may truly be chronic, but could also be increasing and waning over a small range; there is simply no way to tell.

Bacterial counts can also be used to see the effects of a treatment, such as chemotherapy, vaccination, or gene disruption of the mouse. Again, given the biological variation, a statistical difference is not seen unless the mean values differ by about 0.5–0.7 log.

Simple power statistics show that four mice per group is the minimum number that should be used for *in vivo* determination of bacterial growth. Thus to follow the course of infection, one should harvest four mice at each time point. After a while one gets an intuitive sense of which time points to choose; even so, it is always sensible to have a few spare mice left over just in case.

We place organs in individual industrial-strength homogenizing tubes with a very tight pestle (keep on dry-ice before using so the pestle does not expand) and grind for 30–40 s. Do not worry about the high shearing forces; these are needed to disrupt all the cells and disperse the bacteria. Then make serial dilutions and plate 100 µl of each dilution on quadrants of petri dishes containing 7H11 agar. Keep the homogenates on ice at all times.

Incubate the plates in a 37°C cabinet containing a tray of water to keep the air humid to prevent the plates drying. Resist the temptation to count colonies when they are still small (you will undercount). Count the quadrant in which there are between 10 and 100 colonies, but also make sure the other quadrants correlate, i.e. if you have about 40 colonies, there should be about 400 on the one above and about 3–5 on the next dilution down. Then apply dilution factors and the original homogenate volume to calculate total bacteria per organ.

If you see no colonies at the 0 or −1 dilutions of the liver tissues, but there are some at higher dilutions, do not worry. The liver contains enzymes that inhibit bacterial growth; as these are diluted out the colonies start to grow.

A further 'quality control' is to do a day 1 count; i.e. 24 h after initiating infection. The numbers you obtain here can tell you if the infectious dose required was obtained. After i.v. infection expect about a 90/10/2% distribution in uptake in the liver, spleen and lungs. Look especially at the lung counts; if your uptake was 25–50% in the lungs instead of a few per cent, then your inoculum was clumped (and the rafts got stuck in the lungs).

The sensitivity of the assay is about 1.7 logs; below this we usually designate as ND (not detected). In fact, low numbers, or zero colonies, can sometimes be a problem; if you are using a computer program, do not incorporate a zero as this will undercount your values. In the same vein, be careful about very high values; a common mistake made by beginners is not to change pipettes between dilutions. As a result bacteria are 'carried over' from one dilution to the next, which can escalate the count as much as 3–4 logs. Use common sense; if the bacterial load in wet weight is larger than the weight of the mouse, you did something wrong!

◆◆◆◆◆◆ DTH AND NON-SPECIFIC RESISTANCE

The DTH reaction is the immunological reaction underlying the clinical skin test for tuberculosis. After injection of 'tuberculin' (PPD; purified protein derivative of tuberculin) dermal Langerhans cells, dendritic cells, or tissue macrophages ingest the PPD antigens and present them to recirculating T cells. These cells then generate signals (chemokines) that attract the egress from the blood of monocytes, leading to swelling of the local tissues as they fill with cells. Because the ability to mount a DTH reaction to tuberculin usually arises concomitantly with the expression of protective immunity, this reactivity is taken to indicate exposure to the infection.

In the mouse, the DTH reaction to tuberculin is very small, with a highly immune animal giving about a 0.5 mm reaction. Precision dial-gauge calipers are recommended to measure these small changes, and if possible the experiment should be carried out blind by an 'uninterested' colleague.

To elicit a DTH reaction, inject a hind footpad with 30 µl of 5–10 µg PPD (or test protein). Use a 30G needle and inject between the 'thumb' and forefinger of the foot. Inject the other footpad with the same volume of diluent only.

Measure the size of the two feet 24 and 48 h later. Support the foot on the bottom crown of the caliper and bring the upper crown down onto the foot. If the toes begin to curl up you are starting to compress the foot. Your readings should be about 0.18–0.21 mm for control feet and about 0.25 mm for the test response in 25-g mice.

A second element of the host response that can be measured *in vivo* is the development of non-specific resistance. This refers to the activation of (mostly uninfected) macrophages that have entered the site of infection and have become exposed to IFN-γ and other cytokines. This is an important parameter to measure under certain experimental conditions, i.e. in the testing of an immunomodulator. Earlier in the history of the field, great emphasis was made on certain mycobacterial materials that appeared to vaccinate mice against *M. tuberculosis* challenge; in fact, the whole thing was due to increased non-specific resistance caused by the adjuvant properties of the materials. If you think resistance is specific, rather than non-specific, then you must formally show that this resistance can be transferred by T cells (Orme, 1988).

The classical way to measure non-specific resistance is to inject the mycobacteria-infected mouse with a lethal dose of *Listeria monocytogenes* and then plate the liver and spleen 24 h later to see how much of the *Listeria* inoculum has been destroyed. To do this, infect groups of mice i.v. with 10^5 *L. monocytogenes* EDG. Harvest organs 24 h later and plate serial dilutions on trypticase-soy agar. Incubate 18 h in humidified air and count colonies. In a highly activated mouse, expect to see about a 3–4 log decrease in bacterial load versus control (naive) mice.

◆◆◆◆◆◆◆ MEASURING THE CELLULAR RESPONSE *IN VIVO*

Over the past several years, our knowledge of the systemic response to tuberculosis has advanced. More recently, our knowledge of the local response within the lung has also been improving. Several techniques have been instrumental in generating this increased knowledge.

Of particular usefulness in the past has been the determination of mRNA levels for cytokines within infected tissue. As a first broad measure therefore, we detect cytokine-specific mRNA in the infected tissue by reverse transcribing total RNA and amplifying specific sequences using cytokine-specific primers (Svetic *et al.*, 1991; Cooper *et al.*, 1995a, b; Rhoades *et al.*, 1995). This thus provides a window into the events occurring within the tissue as the infection progresses.

Although mRNA levels are informative, they do not confirm the presence of a particular protein. It is difficult, however, to determine the actual amount of protein within the tissue as levels are generally too low to be detected by ELISA. To identify which cytokines are being generated as a result of infection we are therefore obliged to culture cells *ex vivo*. There are two methods for examining the cellular response in the lung and both require the extraction of lymphocytes from the tissue. Once a single-cell suspension has been made from the lung then lymphocyte responses to antigen can be assessed via the culture methods described above or cells can be directly analysed for their ability to produce cytokine, as described below. In addition to the response in the lung, the circulating response can also be analysed by examination of splenocytes and/or cells from the mediastinal node.

While quantitation and characterization of the cells within the lung is useful in determining the role of cells in the response the location of these same cells within the architecture of the lung is also of paramount importance. To investigate this aspect of the cellular response immunohisto-chemistry has proven invaluable.

The role of the lung, as an initiator of the cellular response to tuberculosis, has been poorly addressed in the past and new techniques have served to peak our interest regarding the unique nature of antigen presentation in the lung. To address this issue we have made use of extraction and culture techniques to examine the nature of the lung dendritic cell.

RT-PCR detection of cytokine specific mRNA

For these studies, we take either lung (aerosol infection) or liver (intra-venous infection) samples from each of four mice sacrificed at each time point. Only a small piece of tissue is required and tissue for histology and for bacterial counts can be used from the same animals. Although this approach can result in 80 samples per experiment, this is still a manageable number using the 96-well plate approach.

Immediately following sacrifice the tissues should be placed in Ultraspec (a preparatory solution containing chaotropic agents and RNAse inhibitors from Biotecx TX). The sample should then be homogenized within a glove box, and immediately frozen in liquid nitrogen. The frozen samples can be stored at −70°C until all samples from the experiment have been collected. Normal precautions for handling RNA should be observed throughout the extraction procedure. The RNA extraction should be performed following basic protocols (Sambrook *et al.*, 1989) and all samples should be treated in an identical manner. The total amount of RNA should be determined by optical density readings at 260 and 280 nm and the integrity determined by gel electrophoresis. To confirm the presence of readable RNA, any PCR tests should include analysis for housekeeping genes.

The amount of RNA in each sample can be determined by a variety of semi-quantitative or quantitative analyses. We perform RT-PCR with a limited number of PCR cycles followed by probing of sample with a labelled probe. For this kind of analysis, the cycle number is kept low such that the signal obtained correlates with a dilution curve. This method is cheap and reproducible but is labour intensive. In an alternative method, a quantitative measure of RNA can be made using a real-time PCR. This procedure requires a machine capable of monitoring fluorescence signals over time (we use an Applied Biosystems 7700). The real-time PCR analysis should be performed using specifically designed primers. Primers designed for normal PCR will not perform well in these reactions. The Applied Biosystems TaqMan Primer Express Program is a straightforward tool for generating primers for cytokines.

For all PCR analyses, it is important to use at least four samples (i.e. four mice) per group and also to ensure that the amount of readable RNA is equivalent between samples. One major advantage to this technique is that old RNA samples stored in the freezer can be analysed for any newly discovered cytokine as soon as the sequence has been published.

Extraction of cells from infected organs

To obtain cells from the spleen and mediastinal lymph nodes the tissue should be pushed gently through nylon spleen screens with the plunger of a 10-ml syringe. The resultant, single-cell suspension contains a variety of cells including both mononuclear and polymorphonuclear cells. The cell suspension should be kept on ice to avoid loss of any macrophages, which will stick to warm plastic. The red blood cells are then lysed (as above) and the remaining cells counted and plated at 1×10^6 per well in 96-well plates.

To obtain cells from the lung the tissue should be perfused with PBS containing 150 Units of heparin (endotoxin free heparin sodium salt,

Sigma). This procedure clears the blood from the lungs and the lungs should turn white. Once clear of blood the lungs are removed and placed in 2 ml of cold, incomplete D-MEM (supplemented with 1% glutamine, 0.1 mM of non-essential amino acids, 50 µM 2-mercaptoethanol, 1% penicillin-streptomycin, all from Sigma-Aldrich, Ltd., St Louis, MO, USA) without fetal bovine serum (FBS). The lung is then sliced using razor blades or scissors and 2 ml of incomplete D-MEM containing collagenase type XI (Collagenase A 0.7 mg ml^{-1}, catalogue number C-7657, Sigma) and type IV bovine pancreatic DNAse I (30 µg ml^{-1}, catalogue number D5055, Sigma) is added. The lung is then digested at 37°C for 30–45 min. To stop the digestion, 10 ml of incomplete RPMI 1640 containing 10% FBS is added and the digested lungs are gently passed through a sterile nylon cell strainer (Falcon). The single cell suspension is then centrifuged at 200g for 10 min and the cells washed. Any remaining red blood cells can be lysed at this point.

Cells prepared from infected tissue can be cultured alone or, if purified cell populations are used (prepared as described earlier), with antigen-presenting cells. For antigen-related responses 10 µg ml^{-1} of mycobacterial antigen can be added or as a control, 2 µg ml^{-1} of the mitogen concanavalin A. Certain cytokines accumulate in the supernatant and can be assessed at 96 h while others are consumed and should be assessed within 24–48 h. One should be aware that cells other than antigen-specific lymphocytes can produce many of the cytokines of interest and appropriate controls should be included. Other controls include cells from uninfected mice and unstimulated cells from infected mice.

To determine the level of cytokine produced by cell cultures, simple ELISAs are the most straightforward method. Pharmingen produces matched antibody pairs and standards for many cytokines. We use a sandwich ELISA procedure as follows:

1. Primary antibody at 1 µg ml^{-1} in coating buffer, 4°C overnight in Immulon 2 plates from Dynatech.
2. Flick out primary antibody and block the non-specific protein binding sites using 1% bovine serum albumin in PBS (0.1% Tween, PBST) for two hours at room temperature.
3. Wash plate four times with PBST, add sample and standard diluted in cell culture medium (if samples are from infected animals this and the following steps should be performed under BSL3 conditions). Leave at 4°C overnight.
4. Wash plate four times as above. Add biotinylated secondary antibody at 1 µg ml^{-1} in the BSA blocking solution and leave for 45 min at room temp.
5. Wash plate four times. Add avidin-peroxidase in working buffer and allow to develop.
6. Read plate.

Analysis of cellular response by flow cytometry

Flow cytometric techniques can be used to characterize cells directly *ex vivo*. We routinely characterize cells directly from the lung, spleen or mediastinal node. In particular, cells can be analysed for expression of surface markers and for their ability to secrete cytokines.

Cell surface expression of activation markers

To analyse the expression of surface markers on cells derived from tissue, cells are prepared as above and resuspended at a concentration of 5×10^6 cells ml^{-1} in d-RPMI (lacking biotin and phenol red) (Irvine Scientific, Santa Ana, CA, USA) containing 0.1% sodium azide. Cells are incubated in the d-RPMI on ice for at least 1 h. This incubation is an important step if one is analysing the expression of cell-surface activation molecules. It is our experience that the antibody-mediated ligation of activation markers on the cell surface results in the up-regulation of several molecules that are used to determine cell activation status. For example, immediate incubation of cells with anti-CD44 and anti-CD45RB results in the cells blasting (as determined by an increase in scatter by flow cytometer analysis) and in the increased expression of CD44 on the cell surface.

Once cells have been fixed, they are dispensed into 96-well plates (200 µl per well) and centrifuged to pellet the cells. The supernatant is removed by gently inverting the plate and discarding the contents into a vessel containing a 5% Lysol solution. Antibody is added to the cells at 25 µg ml^{-1}, in 25 µl per well, and the plate incubated for 30 min at 4°C, in the dark. Following two washes in d-RPMI the cells are analysed on a dual laser flow cytometer.

Murine Model of Tuberculosis

Recommended stains include:

FITC anti-CD19 or CD45R/B220 (B cells)
PE anti-CD3e (T cells)
PerCP anti-CD4 (T-cell subset)
APC anti-CD8 (T-cell subset)

FITC anti-NK1.1 (NK cell)
PE anti-CD3e (T cell)
PerCP anti-CD4 (T-cell subset)
APC anti-CD8 (T-cell subset)

FITC anti-CD44 (T-cell activation)
PE anti-CD45RB (T-cell memory)
PerCP anti-CD3e (T cell)
APC anti-CD4 or CD8 (T-cell subset)

It is essential to include isotype controls for each individual antibody isotype that is used, in order to appreciate the contribution of non-specific binding.

Intracellular staining for cytokine production

The capacity for a T cell to secrete cytokine can be measured using an intracellular cytokine staining protocol. Cells are harvested from the lungs and incubated with anti-CD28 (1 µg ml^{-1}), anti-CD3e (0.1 µg ml^{-1}), and monensin (3 mM) for 4 h at 37°C, 5% CO_2. Cells are harvested, washed in d-RPMI, and labelled with the outer cell surface markers as described above, and fixed and permeabilized with a solution containing paraformaldehyde and saponin (BDPharmingen, Fix/Perm reagent). Cells are then incubated with antibody specific for the cytokine of interest for 30 min at 4°C, in the dark. Cells are then washed twice and analysed using a flow cytometer.

Recommended stains include:

FITC anti-IFN-γ
PE anti-CD4
PerCP anti-CD8
APC anti-β TCR (many CD3 sites have been blocked with the anti-CD3 included in the stimulation step so an alternative T-cell marker should be used).

Again, the use of isotype control antibody particularly for the cytokine reagent is essential in confirming that the (sometimes low) specific signal is real.

Immunohistochemical analysis of cellular response in the lung

In relating the location of the cells analysed by the flow cytometric techniques to the lung and granuloma architecture immunohistochemical techniques have proven invaluable. For optimum results it is very important that specimens are collected and frozen as soon as possible in order to retain the morphology of the tissue and integrity of the antigens. The lung tissue should be inflated with 30% OCT (in PBS) (Tissue-Tek, Inc., Torrance, CA, USA) and placed in a tissue embedding cassette (Peel-A-way, Polysciences, Inc., PA, USA). Care should be taken to ensure that at least 1/8 of an inch of 100% OCT is above and below the specimen. The specimen should be central within the cassette. Once samples are placed correctly the cassette should be floated in liquid nitrogen until approximately two-thirds of the OCT turns white. Alternatively, the cassettes can be placed on dry-ice. Once frozen, cassettes can be wrapped in foil to avoid drying and can be stored at −70°C.

Frozen blocks should be considered BSL3 agents as freezing does not kill the bacteria. Sections should be cut under BL3 conditions. We house our cryostat in a small room, separate from the rest of the BSL3 containment rooms. The operator should wear a protective mask and dispose of all debris as BSL3 material.

Serial sections of 7 μm are cut on a cryostat (Leica, CM 1850). A tape transfer system (Instrumedics Inc., Hackensack, NJ, USA) is particularly useful when sectioning delicate tissue such as the lung. The sections are then fixed in cold acetone for 5–10 min and air-dried. Endogenous peroxidase should be blocked with peroxidase block (Innogenex, San Ramon, CA, USA) and the sections washed in PBS containing 1% BSA (PBS-BSA) for 5 min. The sections are then incubated for 30 min at room temperature with goat serum to block non-specific binding sites (Biogenex, San Ramon, CA, USA). Primary antibody (or isotype control) diluted in PBS/BSA should be added for an overnight incubation at 4°C. Sections should be washed three times for 15–20 min each in PBS/BSA containing 0.05% Tween-20. The last wash should be carried out with PBS alone. Secondary antibody, such as (Goat F(ab)2 anti-rat IgG conjugated to horseradish peroxidase (Biosource International, Camarillo, CA, USA) should then be added for 40 min. Wash as above. Antibody is then detected using aminoethylcarbazole (AEC) (Innogenex, San Ramon, CA, USA) as substrate (red colour). Counterstain for cell structure with Meyer's hematoxilyn (colour blue).

As for all of the above analyses, the comparison of sections from at least four mice per group will add greater weight to any conclusions made from the immunohistochemistry data.

Analysis of lung dendritic cell function

Dendritic cells are potent inducers of immune responses yet their role in development of the pulmonary immune responses is, as yet, unclear. To examine their role in tuberculosis we have used the following techniques.

To isolate lung dendritic cells use 15 mice (processed in groups of five). For each group of five, harvested lungs are cut into small pieces (pooled together) and added to 10 ml of RPMI containing 0.7 mg ml^{-1} of Collagenase type XI (Sigma) and 30 μg ml^{-1} of DNase I type IV (Sigma). Digestion is allowed to proceed at 37°C for 30–45 min. The tissue is then passed through a sterile nylon cell strainer (Falcon) and the single-cell suspension is then washed. The pellet is resuspended in 400 μl of PBS containing 0.5% of FBS, without calcium or magnesium. To disrupt clumps the cells are pipetted up and down gently and 100 μl of anti-CD11c (N418) microbeads (Miltenyi Biotec) are added and the mixture is incubated for 15 minutes at 4°C. The mixture is then washed in 10–15 ml of PBS/FBS and resuspended in 5 ml of PBS/FBS. This suspension is then passed over an LS+/+ separation column (Miltenyi Biotec) which binds cells that have beads attached. The column is then washed with 10 ml of PBS/FBS three times. The magnetic field is then removed and the bound cells are collected. The purity of the CD11c positive population is assessed by flow cytometry or cytospin followed by morphometric analysis.

Cells taken directly from the lung can be analysed for functional activity or they can be matured *in vitro* as follows. Cells can be cultured in RPMI containing 10% FBS and 20 ng ml^{-1} of GM-CSF (Peprotech Inc., Rocky Hill, NJ, USA). Cells should be plated at 5×10^5 cells ml^{-1} and cultured in six well plates or petri dishes. The cultures should be fed after 48 h and on the sixth day by adding one-third of the original volume of media.

Cells cultured in this fashion can then be infected with bacteria. In order to infect the cells they should be cultured without antibiotics for 48 h prior to challenge. To collect the cells from the culture media they are centrifuged at 300g and washed once with PBS (lacking calcium and magnesium). The cells are then resuspended in 50-ml conical tubes at 2–4×10^6 cells per ml in RPMI containing 2% FBS. Mycobacteria can then be added at a ratio of 1, 5 or 10 CFU per cell and the mixture incubated at 37°C for 4 h. The cells are then washed twice with PBS and resuspended in culture media lacking antibiotics and with 10% FBS. Infection rates can be checked by acid fast staining and colony counting after 4 h, 24 h and 48 h.

Acknowledgements

We thank our colleagues in the Mycobacteria Research Laboratories for their contributions to the development of the methods described here. Several of the more modern approaches were developed using funding from NIH programs AI-40488, AI-44072 and AI-41922.

References

Bloom, B. R. (1994). *Tuberculosis – Pathogenesis, Protection, and Control*. ASM Press, Washington, DC.

Centers for Disease Control (1993). *Biological Safety in Microbiological and Biomedical Laboratories*, 3rd edn. HHS publication No. (CDC) 93-8395. US Government Printing Office, Washington.

Collins, F. M., Wayne L. G. and Montalbine, V. (1974). The effect of cultural conditions on the distribution of *Mycobacterium tuberculosis* in the spleens and lungs of specific pathogen-free mice. *Am. Rev. Respir. Dis.* **110**, 147–156.

Cooper, A. M., Dalton, D. K., Stewart, T. A., Griffin, J. P., Russell, D. G. and Orme, I. M. (1993). Disseminated tuberculosis in gamma interferon gene-disrupted mice. *J. Exp. Med.* **178**, 2243–2247.

Cooper, A. M., Roberts, A. D., Rhoades, E. R., Callahan, J. E., Getzy, D. M. and Orme, I. M. (1995a). The role of interleukin-12 in acquired immunity to *Mycobacterium tuberculosis* infection. *Immunology* **84**, 423–432.

Cooper, A. M., Callahan, J. E., Griffin, J. P., Roberts, A. D. and Orme, I. M. (1995b). Old mice are able to control low-dose aerogenic infections with *Mycobacterium tuberculosis*. *Infect. Immun.* **63**, 3259–3265.

Davis, B. D. and Dubos, R. J. (1948). The inhibitory effect of lipase on the bacterial growth in media containing fatty acid esters. *J. Bacteriol.* **55**, 11–23.

Dubos, R. J. (1950). The effect of organic acids on mammalian tubercle bacilli. *J. Exp. Med.* **92**, 319–332.

Dubos, R. J. and Middlebrook, G. (1948). The effect of wetting agents on the growth of tubercle bacilli. *J. Exp. Med.* **83**, 409–423.

Hirschfield, G. R., McNeil, M. and Brennan, P. J. (1990). Peptidoglycan-associated polypeptides of *Mycobacterium tuberculosis*. *J. Bacteriol.* **172**, 1005–1013.

Hjelemeland, L. M. (1990). Removal of detergents from membrane proteins. *Methods Enzymol.* **182**, 277–282.

Hutchinson, F. and Pollard, A. (1961). Target theory and radiation effects on biological molecules. In: *Mechanisms in Radiology* (M. Errara and A. Forssberg, Eds) pp. 71–92. Academic Press, NY.

Jensen, A. (1932). Reinzuchtung und ypenbestimmung von tuberkelbacillenstammen. Eine vereinfachung der methoden fur die praxis. *Zentralbl. Bakteriol.* **125**, 222–226.

Kubica, G. P. and Wayne, L. G. (1984). *The Mycobacteria – a Sourcebook*, vol. 15. A&B. Marcel Dekker, Inc., NY.

Lee, B., Hefta, S. A. and Brennan, P. J. (1992). Characterization of the major membrane protein of virulent *Mycobacterium tuberculosis. Infect. Immun.* **60**, 2066–2074.

Orme, I. M. (1988). Induction of nonspecific acquired resistance and delayed-type hypersensitivity, but not specific acquired resistance, in mice inoculated with nonliving mycobacterial vaccines. *Infect. Immun.* **56**, 3310–3312.

Orme, I. M. (1995). *Immunity to Mycobacteria.* RG Landes Company, Austin, TX, USA.

Orme, I. M. and Collins, F. M. (1994). Mouse model of tuberculosis. In: *Tuberculosis; Pathogenesis, Protection, and Control* (B. R. Bloom, Ed.), pp. 113–134. ASM Press, Washington, DC.

Orme, I. M., Miller, E. S., Roberts, A. D., Furney, S. K., Griffin, J. P., Dobos, K. M., Chi, D., Rivoire, B. and Brennan, P. J. (1992). T lymphocytes mediating protection and cellular cytolysis during the course of *Mycobacterium tuberculosis* infection. *J. Immunol.* **148**, 189–196.

Rhoades, E. R. and Orme, I. M. (1997). Susceptibility of a panel of virulent strains of *Mycobacterium tuberculosis* to reactive nitrogen intermediates. *Infect. Immun.* **65**, 1189–1195.

Rhoades, E., Cooper, A. M. and Orme, I. M. (1995). Chemokine response in mice infected with *Mycobacterium tuberculosis. Infect. Immun.* **63**, 3871–3877.

Roberts, A. D., Sonnenberg, M. G, Ordway, D. J, Furney, S. K, Brennan, P. J, Belisle, J. T. and Orme, I. M. (1995). Characteristics of protective immunity engendered by vaccination of mice with purified culture filtrate protein antigens of *Mycobacterium tuberculosis. Immunology* **85**, 502–508.

Rom, W. N. and Garay, S. (1995). *Tuberculosis.* Little, Brown and Company, NY.

Sambrook, J., Fritsch E. F. and Maniatis, T. (1989). *Molecular Cloning. A Laboratory Manual.* Cold Spring Harbor Press, Cold Spring Harbor.

Scopes, R. K. (1994). *Protein Purification: Principles and Practice.* Springer-Verlag, NY.

Sonnenberg, M. G. and Belisle, J. T. (1997). Definition of *Mycobacterium tuberculosis* culture filtrate proteins by two-dimensional polyacrylamide gel electrophoresis, N-terminal amino acid sequencing and electrospray mass septrometry. *Infect. Immun.* **65**, 4515–4524.

Svetic, A., Finkelman, F., Jian, Y., Dieffenbach, C., Scott, D., McCarthy, K., Steinberg, A. and Gause, W. (1991). Cytokine gene expression after *in vivo* primary immunization with goat antibody to mouse IgD antibody. *J. Immunol.* **147**, 2391–2397.

Thomas, T. C. and McNamee, M. G. (1990). Purification of membrane proteins. *Methods Enzymol.* **182**, 499–520.

Trias, J., Jariler, V. and Benz, R. (1992). Porins in the cell wall of mycobacteria. *Science* **258**, 1479–1481.

Vestal, A. L. (1975). *Procedures for the Isolation and Identification of Mycobacteria.* Centers for Disease Control, Washington, DC.

Wheeler, P. R., Besra, G. S, Minnikin, D. E, and Ratledge, C. (1993). Stimulation of mycolic acid biosynthesis by incorporation of cis-tetracos-5-enoic acid in a cell-wall preparation from *Mycobacterium smegmatis. Biochim. Biophys. Acta* **1167**, 182–188.

Youmans, G. P. (1979). *Tuberculosis.* W.B. Saunders Co., PA, USA.

Youmans, G. P. and Karlson, A. G. (1947). Streptomycin sensitivity of tubercle bacilli; studies on recently isolated tubercle bacilli and the development of resistance to streptomycin *in vivo. Am. Rev. Tuberc.* **55**, 529–534.

4.2 The Leishmaniasis Model

Pascale Kropf and Ingrid Müller
Department of Immunology, Faculty of Medicine, Imperial College of Science, Technology and Medicine, London, UK

Kevin Brunson
Washington State Caseload Forecast Council, Olympia, WA, USA

◆◆

CONTENTS

◆◆◆◆◆◆ INTRODUCTION

The study of experimental infection of mice with the intracellular protozoan parasite *Leishmania major* has not only contributed significantly to our understanding of this fascinating host–parasite relationship, but also to many basic immunological phenomena. Much has been learned about the interaction of antigen-specific T-cells and antigen-presenting cells, about cytokine and T-cell subset regulation, and co-stimulatory requirements. The immune response to experimental *L. major* infection in inbred strains of mice is the paradigm for polarized T-helper cell differentiation: in this model system, a Th1 response characterized by interleukin (IL)-2 and interferon (IFN)-γ secretion leads to self-curing disease, whereas a Th2 response (IL-4, IL-5, IL-10 and IL-13) leads to non-healing disease. Numerous manipulations, including the injection of cytokines and of neutralizing anti-cytokine antibodies, cytokine transgene expression, and more recently studies in cytokine and cytokine receptor gene knockout mice, have all provided intriguing new pieces to the as yet incomplete mosaic of our understanding of the immune response.

The leishmaniases are a group of parasitic diseases in humans caused by more than 20 different *Leishmania* species. They affect 88 countries world-wide and the disease incidence is still spreading (World Health Organization, 1995). The increase can partly be attributed to rapid

Copyright © Elsevier Science Ltd
All rights of reproduction in any form reserved

economic development and the establishment of new settlements and construction sites in forest areas of the Amazon and other regions where the disease is endemic (PAHO, 1994). Early diagnosis is difficult and treatment of leishmaniasis is costly and requires injections for weeks. Urbanization of the mostly rural diseases and the growing number of AIDS patients developing leishmaniasis as a secondary infection are becoming increasingly serious, especially since there is still no defined and efficient vaccine available for use in humans (Modabber, 1996).

The mouse model of infection with *Leishmania major*

Leishmaniasis is the general term given to at least three distinct complexes of diseases, including cutaneous, mucocutaneous and visceral leishmaniasis (Carvalho *et al.*, 1994). These different clinical manifestations observed in humans can be mimicked in various inbred strains of mice infected with *L. major* (Carvalho *et al.*, 1994; Reiner and Locksley, 1995; Etges and Müller, 1998). In this experimental system, inbred strains of mice are infected subcutaneously with infective promastigotes. Mice from the majority of inbred strains, like CBA mice, can control the replication of the parasites; they will develop small lesions, which will spontaneously heal within a few weeks. These mice become immune to subsequent challenge with parasites. On the other hand, mice from a few inbred strains, like BALB/c mice, will develop severe non-healing lesions and uncontrolled parasite growth. This mouse model of experimental cutaneous leishmaniasis is very useful to analyse the mechanisms resulting in protective immunity or in non-healing disease.

Many factors are known to influence the outcome of infection with *Leishmania* parasites:

- The developmental stage of the parasites as well as their maintenance conditions are of crucial importance. Indeed, promastigotes grown in culture medium are only infective when they reach a stationary phase and it is now possible to separate infective from non-infective parasites using lectin agglutination (peanut agglutinin: PNA) (Sacks and Perkins, 1984; Sacks et al., 1985; DaSilva and Sacks, 1987; Sacks and Da Silva, 1987). In addition, frequent *in vitro* passage of *Leishmania* promastigotes was associated with a loss of virulence and eventually, s.c. injection of these parasites did not induce any lesion in BALB/c mice (DaSilva and Sacks, 1987).
- The dose of parasites injected s.c. can also influence the development of the disease. Usually, to induce healing in CBA mice or non-healing in BALB/c mice, between 1×10^6 and 1×10^7 stationary phase *L. major* promastigotes are injected s.c. However, healing and non-healing have been shown to be dose dependent, as 1×10^4 *L. major* parasites injected s.c. induced self-healing in BALB/c mice (Bretscher et al., 1992).
- The route of infection can also cause different outcomes of the disease. Indeed, mice infected s.c. in the rump develop progressive disease, whereas they can control s.c. injection in the footpad with the same concentration of parasites (Nabors and Farrell, 1994). It has to be noted that 10–1000 metacyclic parasites are transmitted during the blood meal of a sand fly, therefore it is

likely that the common dose of parasites used to induce infection in mice causes quite a different outcome.

- The genetic background of the mice can influence the outcome of *Leishmania* infection: a comparative study of 13 inbred strains of mice showed that only BALB/c and SWR mice were unable to control infection with *L. major* parasites (De Tolla *et al.*, 1981). The exact genetic basis for susceptibility or resistance is not clear. Indeed, loci on chromosome 6, 7, 10, 11, 15 and 16 have been identified recently by back-crossing B10.D2 mice onto the BALB/c background (Beebe *et al.*, 1997). The presence of all six loci was not essential for healing and no single locus appeared to be indispensable for resistance.

The studies outline above show that the genetic background of the mice, the route of infection, the parasite strains, the maintenance and the dose of parasites can influence the outcome of infection with *Leishmania*.

Th cell responses

The healing phenotype of *L. major* infected mice is associated with a preferential expansion of Th1 cells: CD4⁺ T cells from the lymphoid organs from healer strains of mice like C57BL/6 and CBA mice have been shown to produce high levels of IFN-γ and no IL-4 (Reiner and Locksley, 1995). Macrophage-derived IL-12 is one of the principal inducers of Th1 responses. IL-12 shifts the balance of Th subsets and even redirects established immune responses (Nabors *et al.*, 1995). Both IL-12 and IFN-γ are required for the effective resolution of *L. major* infection (Belosevic *et al.*, 1989; Heinzel *et al.*, 1995). IL-12-deficient mice on a resistant background develop progressive disease, even after low-dose infection that usually induces healing in susceptible BALB/c mice (Mattner *et al.*, 1996). In addition, when IL-12 is administered in BALB/c mice at the time of *L. major* infection, they can control the replication of the parasites and switch to a Th1 phenotype (Sypek *et al.*, 1993). Even an established Th2 phenotype can be altered by injection of IL-12 and chemotherapy (Nabors *et al.*, 1995).

The mouse model of *L. major* infection offers remarkable possibilities to alter immune responses and interfere drastically with the normal development of the disease. The non-healing response of BALB/c mice to *L. major* infection can be manipulated in the early phase of infection in such a way that these mice are able to fight infection, resolve their lesions, and reduce the parasite load. Manipulations like sublethal irradiation (Howard *et al.*, 1981), repeated intravenous immunizations (Farrell *et al.*, 1989), low-dose infection (Bretscher *et al.*, 1992), anti-CD4 mAb (Titus *et al.*, 1985a) and anti-IL-4 mAb injection (Sadick *et al.*, 1990) can reverse the non-healer phenotype of BALB/c mice. The immunological manipulations such as treatment with antibodies to neutralize IL-4 or to eliminate CD4⁺ helper cells transiently must be done during the first week of infection to be effective. A lower Th2 response and a more pronounced Th1 response systematically accompanied the switch from a non-healer phenotype to a healer phenotype (Reiner and Locksley, 1995; Etges and Müller, 1998).

On the other hand, the non-healer phenotype of BALB/c mice is associated with a strong Th2 response, characterized by a sustained production of IL-4 and low levels of IFN-γ (Reiner and Locksley, 1995). Neutralizing anti-IL-4 mAb treatment reverts the non-healing phenotype of BALB/c mice and deviates the Th2 to a Th1 response (Chatelain *et al.*, 1992). Recently, the source of IL-4 thought to be responsible for the differentiation of naïve CD4$^+$ T cells into Th2 cells has been identified: 16 h after infection of BALB/c mice with *L. major*, Vβ4$^+$ Vα8$^+$ NK1.1$^-$ CD4$^+$ T cells produced a burst of IL-4 (Launois *et al.*, 1997). Depletion of this population with a mouse mammary tumour virus (MMTV) (Launois *et al.*, 1997) and the subsequent abrogation of the IL-4 peak enabled the genetically susceptible BALB/c mice to mount a Th1 response and control infection with *L. major*. Together with the experiment showing that neutralization of IL-4 *in vivo* enables BALB/c mice to control *L. major* infection, these results confirm the detrimental role of IL-4 in *L. major* infection. However, we and others have recently published that despite the absence of IL-4, IL-4-/- BALB/c mice developed non-healing lesions, similar to those observed in wild-type mice (Noben-Trauth *et al.*, 1996, 1999; Kropf *et al.*, 1997, 1999), showing that the non-healing phenotype can not be attributed to IL-4 alone. Indeed, we showed that the susceptible phenotype of BALB/c mice is associated with a loss of IL-12 responsiveness and therefore can not be exclusively ascribed to an overwhelming Th2 response (Kropf *et al.*, 1997). These results obtained both in the presence and in the absence of IL-4 indicate that the non-healer phenotype is associated with an inability of CD4$^+$ T cells to respond to IL-12 and to mount an appropriate Th1 response rather than with an uncontrolled Th2 response.

◆◆◆◆◆◆ METHODS

Parasite isolation and maintenance

In the following paragraphs we describe in detail the materials as well as the preparative steps necessary for the handling of *Leishmania* parasites.

Maintenance media

L. major promastigotes can be maintained in different media. The most commonly used media are Schneider's Drosophila Medium (Gibco), Grace's Insect Cell Culture Medium (Gibco) and a biphasic system, consisting of a liquid phase of Dulbecco's Modified Eagle Medium (DMEM, Gibco) over a solid layer of rabbit blood agar. In our laboratory we use the biphasic culture.

The DMEM used to routinely maintain the parasites in culture is supplemented with:

- 10% complement-inactivated Fetal Bovine Serum (FBS, Gibco)
- 50 IU ml^{-1} Penicillin (Gibco)
- 50 μg ml^{-1} Streptomycin (Gibco)
- 292 μg ml^{-1} L-glutamine (Gibco)
- 4.5 mg m^{-1} Glucose (Gibco)

The same medium is used for the parasite limiting dilution assay (LDA medium).

The FBS is necessary for optimal culture conditions. However, some batches of FBS can inhibit the parasite growth. Therefore, careful screening of the FBS is necessary.

Blood agar

In order to prepare 100 ml of solid blood agar medium, the following material is necessary:

- 3.0 g nutrient agar (Difco)
- 0.6 g NaCl (Sigma)
- 100 ml nanopure H_2O
- 10 ml rabbit blood with 10% sodium citrate (Merck)
- 5 ml glucose 30% in phosphate buffer (Sigma)
- 70 ml polystyrene tissue culture flasks (Corning Costar)

1. Autoclave the mixture of agar, water and NaCl for 20 min.
2. Cool the agar to about 45°C.
3. Place into a prewarmed water bath (45°C).
4. Add 10 ml of rabbit blood and 5 ml of glucose 30%.
5. Distribute 5 ml of agar mixture on the flat side of 70 ml polystyrene tissue culture flask.
6. Let solidify and cool at room temperature.
7. Tightly close the flasks.
8. Keep at 4°C for up to 4 weeks.

Parasite counting

Promastigotes are actively moving and need to be fixed for counting. Before fixing, parasites are washed as follows:

1. Centrifuge for 5 min at 500 rpm to pellet debris.
2. Transfer the supernatant in a new tube.
3. Wash the supernatant twice at 3500 rpm for 10 min.
4. Resuspend the parasites in the maintenance medium and fix them using one of the two solutions described here:

- 2% formaldehyde in phosphate-buffered saline
- Hayem's solution:

  - 0.5 g $HgCl_2$ (Sigma)
  - 1.0 g NaCl (Sigma)
  - 5.0 g Na_2SO_4 (Sigma)
  - 200 ml H_2O (Sigma).

5. After resuspending the parasites in DMEM, we usually dilute them in 2% formaldehyde in phosphate buffer and count them in an haemocytometer.

Parasite isolation

L. major is easy to handle *in vitro* and *in vivo*, however, it is important to pay attention to a few points relating to the virulence and infectivity of these organisms. The maintenance of *L. major* in culture has an innate problem in that the parasite undergoes an evolution different from that in the natural environment where it is cycled through the sandfly prior to infection in a new vertebrate host. A decrease in virulence is the dominant characteristic of a long-term maintenance *in vitro*. To maintain their virulence, a monthly passage *in vivo* is necessary. Many strains of mice can be used for this purpose. In our laboratory, we use BALB/c mice and infect them in the hind footpad with *L. major* in a final volume of 50 μl. Two to four weeks after infection, the mice are sacrificed and the parasites are isolated from the footpad lesion.

L. major are isolated from the infected footpad as follows:

1. Swab the skin of the dead mouse with 70% ethanol.
2. Make an incision of the skin around the ankle.
3. Cut the toes of the footpad.
4. Carefully remove the skin and necrotic tissue of the infected footpad.
5. Cut the footpad just above the joint.
6. Place the footpad in a sterile petri dish.
7. Cut the footpad in several small pieces.
8. Transfer them into a sterile glass homogenizer (Bellco).
9. Gently homogenize the tissue in 5 ml medium.
10. Transfer the suspension in a blood agar flask.
11. Incubate at 26°C, in a humid atmosphere, 5% CO_2 in air to allow transformation of amastigotes into promastigotes[1].
12. After 3–5 days, the suspension is washed as before.
13. Resuspend the parasites in 5 ml of maintenance medium and transfer the suspension in a new flask.

[1] The parasites can also transform and grow in tightly closed flasks in a dry incubator.

Parasite maintenance

Once isolated, the parasites should be maintained for a maximum of 4 weeks with the minimal *in vitro* dilutions possible. Like other unicellular organisms, *L. major* have lag, log and stationary phases and promastigote populations during these phases are not uniform with respect to infectivity. During the log phase, parasites are motile and dividing cells, whereas in the stationary phase, the parasites are stationary and not dividing. The cultures are started with 10^6 parasites ml^{-1}. After a few days, the parasites will reach the stationary phase. Leishmanial growth can be monitored by daily counting of the cells and by microscopic observation of the cultures. Stationary parasites are washed, counted and a new culture is started with 10^6 parasites ml^{-1} in a new blood agar flask.

Cryopreservation of parasites

L. major promastigotes can be cryopreserved by using the same techniques as for other eukaryotic cells. For successful cryopreservation, it is important to use parasites from the first *in vitro* passage after a new isolation and to freeze them in a late log phase. We cryopreserve promastigotes in the following way:

1. Wash the parasites as described before.
2. Adjust them to 1×10^8 cells ml^{-1} in FBS containing 10% of dimethyl sulfoxide (Fluka) in a final volume of one ml per tube. The tubes are placed in a styrofoam box and immediately transferred at −70°C for 24 h and are then stored in the vapor phase of liquid nitrogen until further use.

As for eukaryotic cells, the parasites have to be thawed quickly. We thaw the vial in a water bath pre-warmed at 37°C and distribute the cells drop by drop in a tube containing medium at room temperature. The cells are washed once at 3500 rpm for 10 min, resuspended in their culture medium and transferred in a blood agar flask.

It is imperative to passage the thawed batch of parasites in a mouse. In our experience, a clear reduction in virulence was noted with the time of storage.

Isolation of metacyclic promastigotes for infection: PNA agglutination

Metacyclic *L. major* parasites represent the infectious form of the parasite. Log phase procyclic and stationary phase metacyclic promastigotes of *L. major* differ in the composition of the repeating phosphorylated saccharide unit of LPG. The repeat units of LPG from log phase *L. major* contain terminal β-galactose residues and these galactose residues account for the agglutination by the lectin peanut agglutinin (PNA). The repeat units of metacyclic promastigotes terminate predominantly with α-arabinose, which do not serve as ligands for the lectin (Sacks *et al.*, 1985; Sacks and Da Silva, 1987; Turco and Descoteaux, 1992). Metacyclic promastigotes have lost their ability to bind peanut agglutinin and a technique using this differential binding has been developed by Sacks *et al.* to purify infective stages of *L. major* from cultures (Sacks *et al.*, 1985).

To isolate metacyclic *L. major* parasites, the following material is necessary:

- phosphate-buffered saline (PBS)
- Peanut agglutinin isolated from *Arachis hypogae* (PNA, Vector Laboratories or Sigma)

(contd.)

Isolation

1. Wash the parasites 3 × with PBS.
2. Adjust the parasites to 1×10^8 cells ml^{-1} in PBS.
3. Add an equal volume of PNA (100 µg ml^{-1}) in PBS.
4. Incubate one hour at room temperature.
5. To isolate the non-agglutinated parasites, two methods can be used:

1.
- Harvest the suspension and carefully layer it on top of the same volume of PBS containing 50% of FBS
- Incubate 30 min at room temperature
- Carefully harvest the parasites remaining above the interphase
- Wash 3 times in PBS and count.

2.
- Harvest the suspension and centrifuge at 600 rpm for 5 min
- Carefully harvest the supernatant, wash 3 times in PBS and count.

◆◆◆◆◆◆ INFECTION OF MICE

Factors influencing the lesion development

Usually, to induce healing in genetically resistant mice like CBA or C57BL/6 or non-healing in BALB/c mice, between 1×10^6 and 1×10^7 stationary phase *L. major* promastigotes are injected s.c. in a final volume of 50 µl PBS in one hind footpad. However, many factors can influence the lesion development. The route of entry of the *L. major* promastigotes can influence the nature of the developing immune response. Nabors and Farrell showed that SWR mice display a non-healing response to *L. major* comparable to that of BALB/c mice when the parasites are inoculated subcutaneously at the base of the tail (Nabors and Farrell, 1994). In contrast, if infected subcutaneously in the footpad with the same number of parasites, the SWR mice were able to control their lesion.

Mouse strain	Route of parasite entry	Form of disease
SWR	s.c. base of the tail	non-healing
SWR	s.c. footpad	healing

s.c., subcutaneous.

The development of progressive disease is not restricted to BALB/c mice and is not exclusively determined by the genetic composition of the infected host. C57BL/6 mice, one of the prototype healer strains, can resolve cutaneous lesions developing after subcutaneous infection. However, when infected intravenously with *L. major*, mice from this strain developed non-healing disease and were unable to mount a DTH response (Scott and Farrell, 1982). Local differences in the skin temperature might influence the growth of *L. major* amastigotes and the responsiveness of macrophages to cytokines and contribute to the expression of site specific immunity.

The dose of *L. major* parasites injected can also influence the lesion development. Indeed, resistance and susceptibility have been shown to be dose dependent. In this study, it was shown that 1×10^4 parasites induced self-healing lesions in BALB/c mice and enabled them to become immune to reinfection, whereas 3.3×10^4 parasites induced progressive disease (Bretscher *et al.*, 1992; Menon and Bretscher, 1996). This suggests that the antigen dose affects the development of T-cell subsets.

Enumeration of viable parasites by limiting dilution assay (LDA)

The most commonly used index for disease progression in *L. major* infected mice as well as for the determination of the effects of immunotherapies is the measurement of the size of cutaneous lesions which develop at the site of parasite inoculation. However, lesion size does not always correlate with the number of viable parasites within the lesion. A technique was developed by Titus *et al.* (1985b) to determine the degree of parasitism in different infected tissues. An example of an infected footpad is described here, but this same technique can be used for different organs.

For the determination of the number of viable parasites, the tissue homogenates from infected mice need to be plated on blood agar plates.

1. Prepare the blood agar as described before.
2. Distribute 50 µl per well of agar mixture in a slant (see Fig. 1) with a repeater pipetter (Eppendorf) in 96-well flat bottom tissue culture plates.
3. Let solidify and cool at room temperature and keep in a humid box at 4°C for up to 4 weeks.
4. Determine the weight of a petri dish.
5. Prepare the footpad as described for the parasite isolation.
6. Weigh the petri dish with the footpad.
7. Calculate the weight of the footpad.
8. Cut the footpad in several small pieces.
9. Transfer them into a sterile glass homogenizer.
10. Gently homogenize the tissue in 5 ml of LDA medium.
11. Prepare 8–12 serial 10-fold dilutions of the footpad homogenate in a final volume of 2.5 ml (see Fig. 1).
12. Distribute 100 µl per well of each dilution in at least 16 replicate wells, change pipettes between each dilution (see Fig. 1).
13. Incubate the plate at 26°C, in a humid atmosphere, 5% CO_2 in air to allow transformation of amastigotes into promastigotes[1].

(contd.)

14. After 10 days, the assay is read by scoring the number of positive wells (presence of motile parasites) and negative wells (absence of motile parasites). An example is illustrated in Table 1.

Note:

If the parasite load is determined in lymphoid organs, count the cells and use the number of cells instead of the ng of tissue for the frequency estimation.

[1] The parasites can also transform and grow in plates sealed with Parafilm in a dry incubator.

To determine the plating efficiency of the LDA, a control plate consisting of serial dilutions of a known number of parasites has to be set up. The model of dilutions depicted in Fig. 1 can be used. We perform six serial 10-fold dilutions, starting with a concentration of 1×10^4 *L. major* per well. After 10 days, the assay is read by scoring the number of positive wells (presence of motile parasites) and negative wells (absence of motile parasites).

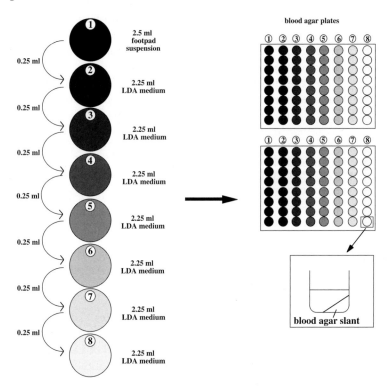

Figure 1. Plating of tissue homogenate.
1. Carefully mix well 1 with a 1-ml pipette.
2. Remove 0.25 ml, transfer it into well 2.
3. Go back to well 1 and remove 0.8 ml with the same 1-ml pipette.
4. Distribute 0.1 ml into eight replicates of one blood agar plate.
5. Repeat steps 3 and 4, but distribute into the second plate.
6. Change the pipette and repeat the procedure with wells 2 and 3.

Table 1. Results of the microscopic scoring of positive and negative wells

	1	2	3	4	5	6	7	8		1	2	3	4	5	6	7	8
A	+	+	+	+	+	+	-	-	A	+	+	+	+	+	+	-	-
B	+	+	+	+	+	+	-	-	B	+	+	+	+	+	+	-	-
C	+	+	+	+	+	-	-	-	C	+	+	+	+	+	+	-	-
D	+	+	+	+	+	+	-	-	D	+	+	+	+	+	+	-	-
E	+	+	+	+	+	+	-	-	E	+	+	+	+	+	+	+	-
F	+	+	+	+	+	+	-	-	F	+	+	+	+	+	+	-	-
G	+	+	+	+	+	+	+	-	G	+	+	+	+	+	+	+	-
H	+	+	+	+	+	+	-	-	H	+	+	+	+	+	+	-	-

Analysis of the parasite LDA results

Plating infected tissues under conditions optimal for parasite transformation and growth and determining the number of transformed, viable, motile promastigotes in these cultures permits us to determine the number of viable parasites in a given tissue. The raw data obtained will be analysed using a SAS® program to evaluate the number of parasites in the tissue. Experimental data and statistical techniques were conjoined to estimate the frequency of parasites in the tissue. Based upon the estimated frequency, simple arithmetic produced an estimate of the total number of parasites in the initial amount of tissue.

Consistent with methods detailed in Taswell (1987), a SAS® PROC IML program[1] was written to calculate frequencies, test statistics, and descriptive statistics. The single-hit Poisson model (SHPM) described was assumed to represent the distribution of organisms in the dose and the ability to detect parasites. The minimum chi-squared (MC) iterative method was chosen to estimate the frequency of parasites. Chi-squared tests were used to validate the assay and the final estimate and confidence levels were reported for completeness. The SAS® program was validated using data published in Taswell. Notation is also consistent with Taswell.

Validity models

d number of dose levels
r_d number of negatively responding wells in the dth dose level
n_d number of wells in the dth dose level
λ_d known dose of the dth dose level in nanograms
ϕ unknown relative frequency to be estimated
$p_d = r_d/n_d$ observed negative response frequency

The Assay Validity Test (AVT) was conducted by testing the estimate of the slope parameter β_m

$$\hat{\beta}_m = \Sigma w_d x_d y_d / \Sigma w_d x_d^2 \tag{1}$$

of two generalized linear models

$$\ln \hat{\phi}_d = \alpha_1 + \beta_1 \lambda_d \tag{2}$$

$$\hat{\phi}_d = \alpha_2 + \beta_2 \lambda_d \tag{3}$$

with the chi-squared statistic

$$\chi^2 = (\Sigma w_d x_d y_d)^2 / \Sigma w_d x_d^2 = \hat{\beta}_m \, \Sigma w_d x_d y_d \tag{4}$$

with 1 degree of freedom. For (2),

$$w_d = n_d p_d \, (\ln p_d)^2 / (1 - p_d), \tag{5}$$

and for (3)

$$w_d = n_d p_d \lambda_d^2 / (1 - p_d). \tag{6}$$

The Estimate Validity Test (EVT) tested the validity of $\hat{\phi}$, the final estimate of $\hat{\phi}$, by substituting

$$\hat{p}_d = \exp(-\hat{\phi}\lambda_d) \tag{7}$$

for p_d in (5) and (6) above. The larger of (4) from (2) or (3) with a correspondingly smaller p-value was then reported as the chi-squared test statistic with 1 degree of freedom for determining estimate validity.

Hypothesis testing

Two hypotheses were tested; in both cases failing to reject the null hypothesis that $\beta_m = 0$ was desired. The SHPM assumes that the presence of only a single organism is sufficient for detection, and independence between the sample frequencies and the dose level are consistent with that assumption. The AVT tests the fit between the model and the data. The EVT adds information on the estimated frequency to the AVT test.

Estimation

The MC estimate $\hat{\phi}$ of the unknown frequency ϕ was calculated as the value that minimizes

$$\chi^2 = \Sigma \left[\frac{(r_d - n_d \, e^{-\phi\lambda_d})^2}{n_d \, e^{-\phi\lambda_d} \, (1 - e^{-\phi\lambda_d})} \right] \tag{8}$$

where $\hat{\phi}$ is determined iteratively from Newton's method

$$\hat{\phi}_{i+1} = \hat{\phi}_i - \frac{(\partial \chi^2 / \partial \phi)}{(\partial^2 \chi^2 / \partial \phi^2)} \bigg|_{\hat{\phi}_i} \tag{9}$$

where $\hat{\phi}_i$ is the estimate from the ith iteration, the first partial derivative is

$$\frac{\partial \chi^2}{\partial \phi} \bigg|_{\hat{\phi}_i} = \Sigma \left[\frac{n_d \lambda_d e^{-\phi\lambda_d}(2r_d - n_d) + r_d^2 \lambda_d(e^{-\phi\lambda_d} - 2)}{n_d(1 - e^{-\phi\lambda_d})^2} \right] \bigg|_{\hat{\phi}_i} \tag{10}$$

and the second partial derivative is

$$\frac{\partial^2 \chi^2}{\partial \phi^2} \bigg|_{\hat{\phi}_i} = \Sigma \left[\frac{n_d \lambda_d^2 (n_d - 2r_d)(e^{-\phi\lambda_d} - e^{-3\phi\lambda_d}) + r_d^2 \lambda_d^2 (e^{-\phi\lambda_d} - 4 + 7e^{-\phi\lambda_d} - 4e^{-2\phi\lambda_d})}{n_d(1 - e^{-\phi\lambda_d})^4} \right] \bigg|_{\hat{\phi}_i} \tag{11}$$

The estimated variance is calculated as twice the reciprocal of the second partial derivative evaluated at the final estimate $\hat{\phi}$

$$\hat{V}(\hat{\phi}|\phi) = \frac{2}{(\partial^2 \chi^2 / \partial \phi^2)}\Big|_{\hat{\phi}} \quad (12)$$

Example

Two footpads (weight 0.165 g) were homogenized, resuspended in 10 ml of LDA medium (1 650 000 ng of tissue in the first dilution) and plated as described in Fig. 1. After 10 days, the assay was read by scoring the number of positive and negative wells. The results are illustrated in Tables 1 and 2.

Table 2. Example of responding groups

Dilutions	Negative wells	ng of tissue	
1. 1/1	0	1,650,000	
2. 1/10	0	165,000	
3. 1/100	0	16,500	
4. 1/1000	0	1,650	
5. 1/10 000	0	165	
6. 1/100 000	1	16.5	responding group
7. 1/1 000 000	13	1.65	responding group
8. 1/10 000 000	16	0.165	

Table 3. Data from a single experiment, 165 000 000 ng; $n_d = 16$

d	r_d	λ_d
1	1	16.5
2	13	1.65

Table 4. The P-values for AVT are both large, meaning that the null hypothesis is not rejected

	Model (2)	Model (3)
chi-squared AVT	0.1831529	0.2036667
P-values of AVT	0.6686776	0.6517773

Table 5. Statistics from the EVT which also fail to reject the null hypothesis

chi-squared EVT	0.2122454
P-value of EVT	0.6450132

The Leishmaniasis Model

Table 6. Estimates and descriptive statistics

Estimated frequency	0.1533245
Estimated variance	0.0023195
95% upper confidence level	0.2477212
95% lower confidence level	0.0589279
1/frequency	6.5221145
Estimated number of parasites per lesion	12,649,272

◆◆◆◆◆◆ STUDIES OF THE HOST RESPONSE TO *L. MAJOR* INFECTION

Many studies of experimental cutaneous leishmaniasis rely on the measurement of the changes of the local swelling which develops at the site of parasite inoculation as a means of following the course of infection. The replicating parasites, inflammatory responses of the infected host, and pathological changes due to infiltrating cells contribute to the expression of the lesion. The lesion size can be due to increased parasite growth or to inflammatory or pathological reactions (Hill *et al.*, 1983; Titus *et al.*, 1985b). Therefore, monitoring only the increase or decrease of the cutaneous lesion may give inaccurate or even false impressions of the true progress of the disease (Hill *et al.*, 1983; Vieira *et al.*, 1996). Similar differences in lesion sizes can be accompanied by widely varying differences in the number of viable parasites. Lesions could be present when there are no parasites and parasites could be found when there were no lesions (Hill *et al.*, 1983; Titus *et al.*, 1985b; Vieira *et al.*, 1996). The combined use of monitoring the size of the cutaneous lesions and the determination of the parasite burden in the infected tissue will reflect more accurately the status of disease, the effect of vaccines, and immunotherapies.

Evaluation of cutaneous lesions

After subcutaneous inoculation of infectious *L. major* parasites in one hind footpad, the development of the infection can be followed by determining the degree of local swelling at the site of inoculation. Infection into the footpads is widely used. The advantage of this route is that in immunocompetent individuals, the infection remains mainly restricted to the site of inoculation and the uninfected, contralateral footpad can be used as a control. To determine the size of the cutaneous lesions, the actual thickness of the infected and the non-infected, contralateral footpad are measured in regular intervals (for example, on a weekly basis). The size of the footpads can be determined with a dial caliper (Kröplin, Schlüchtern) or a digimatic caliper (Mitutoyo, Japan) or with any other metric caliper. The lesion size is calculated by substracting the thickness of the non-infected contralateral footpad from the thickness of the infected footpad.

An example of the lesion size during the course of *L. major* infection in healer and non-healer mice is given in Table 7 and Fig. 2.

Table 7. Evaluation of lesions in healer and non-healer mice

Weeks post-infection	BALB/c footpad size ± SEM	CBA footpad size ± SEM
1	0.25 ± 0.08	0.30 ± 0.04
2	0.48 ± 0.07	0.91 ± 0.04
3	1.02 ± 0.06	1.10 ± 0.06
4	1.80 ± 0.30	1.09 ± 0.09
5	2.43 ± 0.13	0.68 ± 0.06
6	3.80 ± 0.10	0.50 ± 0.02
7	4.65 ± 0.75	0.32 ± 0.04

As already mentioned, the course of experimental infection should not only be evaluated by the lesion development, the enumeration of the number of viable parasites present in the lesions should also be done at several time points during infection.

T-cell and cytokine responses

CD4+ helper T cells have been divided into the Th1 and Th2 subset based on the production of a distinct panel of cytokines: Th1 cells secrete IFN-γ,

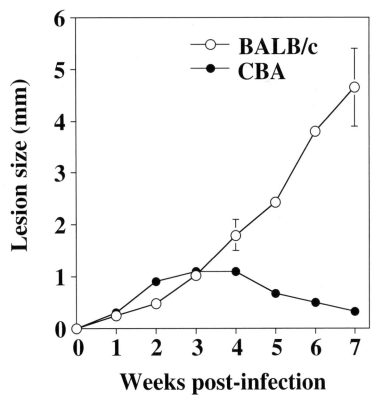

Figure 2. Lesion-size development during *L. major* infection.

IL-2 and TNF-β; Th2 cells produce IL-4, IL-5, IL-6, IL-10, IL-13. The elucidation of the steps leading to the priming of these cells is a major focus of immunology and it is widely accepted that cytokines play a key role in this priming. Cytokines can cross-regulate each other's production and this influences the function of Th1 and Th2 cells. Experimental cutaneous leishmaniasis induced by infection with *L. major* is one of the best studied models in which selective activation of Th1 or Th2 cells occurs (Liew and O'Donnell, 1993; Reiner and Locksley, 1995). The induced pattern of cytokines is mostly used as an indicator for the type of activated Th-cell subset. The cytokine pattern can be determined by different methods. The induction of mRNA by quantitative RT-PCR, the intracellular staining with fluorescent anticytokine antibodies, the estimation of the frequency of cytokine secreting cells by ELISPOT assay and the determination of the amount of cytokine secretion by ELISA or bioassays are possible methods. All of the different methods used for cytokine analysis as well as for T-cell isolation, phenotyping, stimulation and proliferation are covered in detail elsewhere in this volume and we suggest to refer to those chapters. However, we would like to describe briefly the stimulation conditions used in our laboratory for the detection of cytokines by ELISA, bioassays and ELISPOT. After infection of mice with *L. major*, T cells will be activated and migrate. Therefore, the pattern of cytokines secreted by T cells is determined in at least two distinct lymphoid organs, the spleen and the lymph nodes draining the site of infection. Suspensions of total spleen and lymph nodes cells are adjusted to $5 \times 10^6 - 1 \times 10^7$ cells ml^{-1} and stimulated in 24-well culture plates with the homologous antigen or immobilized anti-CD3 mAb (Leo *et al.*, 1987). Cells without additional stimulation *in vitro* are always used as a control. To restimulate specific T cells *in vitro*, live *L. major* promastigotes are rendered replication-incompetent by 2 min irradiation with ultra violet light and are used at a concentration of 4×10^6 parasites ml^{-1}. In order to measure the maximum T-cell response, we use plate-bound anti-CD3 mAb (Hamster mAb 145.2C11). Five μg of the antibody are added to the plates in a final volume of 200 μl of PBS and allowed to bind to the plastic for 2 h at 37°C. After 2 h, the wells are washed three times with PBS and the cells are added. IFN-γ is easily detected after 24 h, but Th2 cytokines require a longer incubation time. Therefore, we harvest the supernatants after 48 to 72 h of stimulation, aliquot them and store them at −70°C until cytokine determination.

Antibody responses and immunoglobulin isotypes

The antibody response in the cutaneous form of human leishmaniasis appears to reflect both the parasite load and the chronicity of infection (Ulrich *et al.*, 1996). Analysis of the composition of Ig subclasses in the serum of infected mice constitutes a useful readout system for the activation of Th1 or Th2 cells *in vivo* and can be used as a tool to evaluate the effect of immunomanipulation or vaccination *in vivo* in an intact animal (Ulrich *et al.*, 1996).

For the determination of Ig-isotypes, blood should be collected at different times after infection. The blood can be drawn by cardiac puncture. The choice of technique used depends in many institutions on the approval of an animal care committee and need to be selected according to their requirements. Once the blood is drawn, allow it to clot, centrifuge once at 2000 rpm, remove the clear serum and store it in small aliquots at 20°C until use.

The Ig isotypes present in the sera are determined by ELISA. We use commercially available kits or matched antibody pairs and standards (Pharmingen) and follow the protocols provided by the producers. The optimal concentration of capture and detecting antibodies need to be determined for each batch of antibodies and the appropriate dilution of the sera might vary strongly and need to be determined in preliminary titration experiments.

Macrophage activation, NO production and parasite killing

Macrophages support the growth of *Leishmania* parasites, therefore, healing of infection is clearly dependent upon the ability of macrophages to kill the intracellular parasites. Macrophages activated by exposure to cytokines acquire microbicidal activity due to the induction of the expression of iNOS and the generation of NO, the major effector molecule for the destruction of the intracellular *Leishmania* (Mauel *et al.*, 1991; Oswald *et al.*, 1994; Wei *et al.*, 1995).

Resting or activated peritoneal macrophages, splenic macrophages or bone marrow-derived macrophages can be activated, infected with *L. major* and used to determine the leishmanicidal activity of macrophages. Detailed methods for isolation and activation of macrophages of different origin are described in another chapter. We will just briefly describe the use of bone marrow derived macrophages.

Bone marrow culture derived macrophages (BM-MΦ) are obtained by differentiation of BM precursor cells *in vitro*. BM precursor cells are flushed from femurs and tibiae of normal mice and cultivated at a concentration of 5×10^5 cells ml^{-1} in hydrophobic Teflon bags (Heraeus, Hanau, Germany) (Freudenberg *et al.*, 1986). After 10 days of culture, BM-MΦ are harvested by centrifugation at 4°C, washed extensively with cold DMEM, seeded in 96-well microtitre plates at a density of 10^5 MΦ in 100 μl serum-free DMEM and cultured overnight to allow the cells to adhere. After adherence, BM-MΦ are infected at a ratio of 1:5 with *L. major* promastigotes. Six hours later, plates are washed to remove non-phagocytosed parasites and infected macrophages are acitivated with lipopolysaccharide (20 ng per well), IFN-γ (10 U per well) and TNF-α (200 U per well). After 48 h, culture supernatants are removed and the levels of NO$_2^-$ are tested by the Griess assay. The remaining macrophages can be lysed and the survival of parasites can be determined by colorimetric quantitation (see below).

Nitrite and nitrate are breakdown products of the hydrophobic gas nitric oxide and nitrite (NO$_2^-$) accumulation is used as an indicator of NO production. The Griess reaction is a fast and simple colorimetric assay for nitrite. Nitrite in cell culture supernatants is measured by the method of Ding *et al.* (1988) by adding to the culture supernatants an equal volume of Griess reagent (1% sulfanilamide and 0.1% n-(1-naphthyl)ethylenediamine dihydrochloride in 5% H_3PO_4; Sigma, St Louis, MO, USA). After 10 min at room temperature, absorbance can be measured with a spectrophotometer at 570 nm. NO_2^- concentration is determined using $NaNO_2$ dissolved in DMEM as a standard and DMEM alone as a blank.

Nitrate does not react in the Griess reaction. In the blood nitrite is rapidly converted into nitrate by haemoglobin and this is the reason why in blood or urine sample, derived from infected mice, nitrate needs to be reduced to nitrite before breakdown products of nitric oxide can be measured.

BM-MΦ infected *in vitro* are activated as described above and the survival and killing of parasites is determined (Kiderlen and Kaye, 1990). After incubation for 48–72 h, BM-MΦ are washed once with 37°C Hepes-buffered RPMI (RPMI 1640, Gibco) containing 0.008% w/v SDS (= lysis medium) and incubated with 100 μl fresh lysis medium for 7–20 min. During this time macrophage disintegration has to be monitored regularly with an inverted microscope. Once host cell lysis is complete, 150 μl per well Hepes-buffered RPMI containing 17% FCS is added to neutralize the SDS. The lysates are then incubated in a CO_2 incubator (5% CO_2 in air) for 48–72 h at 25°C to allow the parasites to transform to the promastigote stage.

The relative number of viable *Leishmania* per well is determined by a modified version of the MTT assay (Mosmann, 1983). The test is based on the ability of viable cells to metabolize water-soluble tetrazolium salt (yellow) into a water-insoluble formazan product (purple colour). The purple formazan crystals are produced by dehydrogenases in active mitochondria in viable cells. Dead cells are unable to perform this reaction. MTT stock solution (3-(4,5-Dimethylthiazol-2-yl)-2,5-diphenyltetrazolium bromide (Sigma) (10 μl 5 mg ml^{-1}) stock solution is added to each well. For the stock solution, 5 mg MTT ml^{-1} is dissolved in PBS, passed through a 0.22 μm filter and kept at 4°C for no more than 2 weeks. The parasites are incubated at 25°C for another 8–16 h before the reaction is stopped and the insoluble formazan crystals are solubilized by adding 100 μl 10% acidified SDS (pH 4.7 with acetate buffer) and incubated for a further 6–16 h at 37°C. The relative absorbance is determined with a spectrophotometer at a wavelength of 570 nm and a reference wavelength of 630 nm.

Redirection of immune responses

A variety of immunological manipulations are known that can drastically change the normal course of progressive disease in BALB/c mice. These changes clearly promote the development of Th1-type responses. For a successful redirection, the intervention has to be done early in infection and these manipulations might interfere with the normally induced pathway of T-cell activation and differentiation. We will describe two ways to redirect the non-healing response of BALB/c mice to infections with *L. major*.

Transient depletion of helper T cells *in vivo*

Rat anti-CD4 mAb GK1.5, isotype IgG 2b (Dialynas *et al.*, 1983), is used for the transient depletion of helper T cells in an intact mouse. The monoclonal antibodies are purifed by affinity chromatography from cell culture supernantants, filtered sterile and kept at concentrations ≥ 500 µg ml^{-1} in PBS at 4°C.

Before the purified antibodies are used in an experiment their potential to deplete helper T cells *in vivo* needs to be evaluated first. For this purpose, naive mice are injected i.p. with different doses of the purified anti-CD4 mAb and 48h later, the spleen is removed and the efficiency of this batch of anti-CD4 mAb in the reduction of helper cells is evaluated by flow cytometry. Non-treated mice serve as controls. Normally injection of a functional batch of purified anti-CD4 mAbs completely eliminates CD4$^+$ T cells below the detection limit at 48 h after injection.

In our hands, injection of a total of 400 to 600 µg of purified anti-CD4 mAb per mouse results in the reversal of the non-healing phenotype of BALB/c mice. These anti-CD4 treated BALB/c mice can resolve the cutaneous lesions, restrict the parasite growth and acquire immunity to reinfection (Fig. 3) (Müller *et al.*, 1991). The anti-CD4 mAb need to be injected in a time span from 10 days before to 10 days after infection to change the non-healing phenotype (Müller *et al.*, 1988). Normally we inject the anti-CD4 mAb at the time of infection or the day before. The antibodies can be injected in a single dose of 600 µg or in two doses of 300 µg in 24 h.

Neutralization of IL-4 *in vivo*

Similar to the above described reversal of the non-healing phenotype of BALB/c mice, a single injection of neutralizing anti-IL-4 mAb will also change the immune status of *L. major* infected BALB/c mice. These mice will express a Th1 response, heal the primary infection and acquire immunity to reinfection (Sadick *et al.*, 1990; Chatelain *et al.*, 1992). Anti-IL-4 mAb are purified from hybridoma culture supernatants and the most frequently used antibody is 11B11 (rat-IgG1) (Ohara and Paul, 1985). The antibodies have to be administered i.p. 2 days before or latest at the time of infection. Injection of anti-IL-4 mAb in the range of 1–10 mg will result in a stable change of the non-healing response of BALB/c mice.

◆◆◆◆◆◆ CONCLUDING REMARKS

Considerable progress has been made in the last few years in dissecting the immunological mechanisms leading to healing of *L. major* infections or progressive disease. The concept of Th1 and Th2 cells has provided a

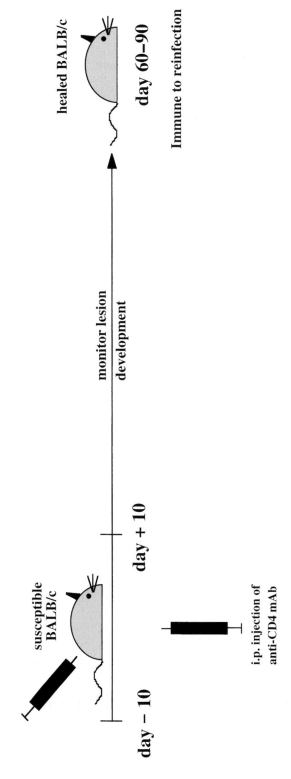

Figure 3. Transient depletion of Th cells *in vivo*.

day 0
Infection s.c. with
Leishmania major

susceptible
BALB/c

healed BALB/c

day − 10

day + 10

day 60–90

i.p. injection of
anti-CD4 mAb

monitor lesion
development

Immune to reinfection

482

useful basis to explain protection as well as pathology in experimental leishmaniasis. This knowledge can now be used to elucidate the conditions that govern the differentiation of naive Th precursor cells into mature, differentiated Th effector cells. There are still a number of critical questions that need to be answered, for example, what is the relative contribution of the parasite, the route of entry, the antigen dose and the genetic background of the infected host to the initiation, establishment and maintenance of polarized T-helper cell responses, what is the best way to change an established response, etc.

The answers to these remaining questions will not only deepen our understanding of the underlying immune mechanism but might help us to treat or prevent leishmaniasis in a more efficient way.

Note

[1] Calculations were performed on an IBM 9121 running VM/ESA 1.1; CMS Level 8, Service Level 105. SAS® was version 6.09, TS450.

The program is freely available and can be obtained from: Kevin Brunson, tel: (360) 455 1414; e-mail: brunson@tanglebank.com

References

Beebe, A. M., Mauze, S., Schork, N. J. and Coffman, R. L. (1997). Serial backcross mapping of multiple loci associated with resistance to *Leishmania major* in mice. *Immunity* **6**, 551–557.

Belosevic, M., Finbloom, D. S., van der Meide, P. H., Slayter, M. V. and Nacy, C. A. (1989). Administration of monoclonal anti-IFN-γ antibodies *in vivo* abrogates resistance of C3H/HeN mice to infection with *Leishmania major*. *J. Immunol.* **143**, 266–274.

Bretscher, P. A., Wei, G., Menon, J. N. and Bielefeldt-Ohmann, H. (1992). Establishment of stable, cell-mediated immunity that makes "susceptible" mice resistant to *Leishmania major*. *Science* **257**, 539–542.

Carvalho, E. M., Barral-Neto, M., Barral, A., Broskyn, C. I. and Bacellar, O. (1994). Immunoregulation in leishmaniasis. *Ciência e Cultura (Journal of the Brazilian Association for the Advancement of Science)* **46**, 441–445.

Chatelain, R., Varkila, K. and Coffman, R. L. (1992). IL-4 induces a Th2 response in *Leishmania major*-infected mice. *J. Immunol.* **148**, 1182–1187.

DaSilva, R. and Sacks, D. L. (1987). Metacyclogenesis is a major determinant of *Leishmania* promastigote virulence and attenuation. *Infect. Immun.* **55**, 2802–2806.

De Tolla, L. J., Scott, P. A. and Farrell, J. P. (1981). Single gene control of resistance to cutaneous leishmaniasis in mice. *Immunogenetics* **14**, 29–39.

Dialynas, D. P., Quan, Z. S., Wall, K. A., Pierres, A., Quintans, J., Loken, M. R., Pierres, M. and Fitch, F. W. (1983). Characterization of the murine T cell surface molecule, designated L3T4, identified by monoclonal antibody GK1.5: Similarity of L3T4 to the human Leu3/T4 molecule. *J. Immunol.* **131**, 2445–2451.

Ding, A. H., Nathan, C. F. and Stuehr, D. J. (1988). Release of reactive nitrogen intermediates and reactive oxygen intermediates from mouse peritoneal macrophages. Comparison of activating cytokines and evidence for independent production. *J. Immunol.* **141**, 2407–2412.

Etges, B. and Müller, I. (1998). Progressive disease or protective immunity to *Leishmania major* infection: the result of a network of stimulatory and inhibitory interactions. *J. Mol. Med.* **76**, 372–390.

The Leishmaniasis Model

Farrell, J. P., Müller, I. and Louis, J. A. (1989). A role for Lyt2⁺ T cells in resistance to cutaneous leishmaniasis in immunized mice. *J. Immunol.* **142**, 2052–2056.

Freudenberg, M. A., Keppler, D. and Galanos, C. (1986). Requirement for lipopolysaccharide-responsive macrophages in galactosamine-induced sensitization to endotoxin. *Infect. Immun.* **59**, 891–895.

Heinzel, F. P., Rerko, R. M., Ahmed, F. and Pearlman, E. (1995). Endogenous IL-12 is required for control of Th2 cytokine responses capable of exacerbating leishmaniasis in normally resistant mice. *J. Immunol.* **155**, 730–739.

Hill, J. O., North, R. J. and Collins, F. M. (1983). Advantages of measuring changes in the number of viable parasites in murine models of experimental cutaneous leishmaniasis. *Infect. Immun.* **39**, 1087–1094.

Howard, J. C., Hale, G. and Liew, F. Y. (1981). Immunological regulation of experimental cutaneous leishmaniasis. IV. Prophylactic effect of sublethal irradiation as a result of abrogation of suppressor T cell generation in mice genetically susceptible to *Leishmania tropica*. *J. Exp. Med.* **153**, 557–568.

Hsieh, C.-S., Macatonia, S. E., Tripp, C. S., Wolf, S. F., O'Garra, A. and Murphy, K. M. (1993). Development of T_H1 CD4⁺ T cells through IL-12 produced by *Listeria*-induced macrophages. *Science* **260**, 547–549.

Kiderlen, A. F. and Kaye, P. M. (1990). A modified colorimetric assay of macrophage activation for intracellular cytotoxicity against *Leishmania* parasites. *J. Immunol. Methods* **127**, 11–18.

Kropf, P., Etges, R., Schopf, L., Chung, C., Sypek, J. and Müller, I. (1997). Characterization of T cell-mediated responses in non-healing and healing *Leishmania major* infections in the absence of endogenous interleukin-4. *J. Immunol.* **159**, 3434–3443.

Kropf, P., Schopf, L. R., Chung, C. L., Xu, D., Lieu, F. Y., Sypek, J. P. and Müller, I. (1999). Expression of Th2 cytokines and the stable Th2 marker ST2L in the absence of IL-4 during *Leishmania major* infection. *Eur. J. Immunol.* **29**, 3621–3628.

Launois, P., Maillard, I., Pingel, S., Swihart, K. G., Xenarios, I., Acha-Orbea, H., Diggelmann, H., Locksley, R. M., MacDonald, R. and Louis, J. A. (1997). IL-4 rapidly produced by Vβ4Vα8 CD4⁺ T cells instructs Th2 development and susceptibility to *Leishmania major* in BALB/c mice. *Immunity* **6**, 541–549.

Leo, O., Foo, M., Sachs, D. H., Samelson, L. E. and Bluestone, J. A. (1987). Identification of a monoclonal antibody specific for a murine T3 polypeptide. *Proc. Natl. Acad. Sci. USA* **84**, 1374–1378.

Liew, F. Y. and O'Donnell, C.A. (1993). Immunology of Leishmaniasis. *Adv. Parasitol.* **32**, 162–259.

Mattner, F., Magram, J., Ferrante, J., Launois, P., Di Padova, K., Behin, R., Gately, M. K., Louis, J. A. and Alber, G. (1996). Genetically resistant mice lacking interleukin-12 are susceptible to infection with *Leishmania major* and mount a polarized Th2 cell response. *Eur. J. Immunol.* **26**, 1553–1559.

Mauel, J., Betz-Corradin, S. and Buchmüller Rouiller, Y. (1991). Nitrogen and oxygen metabolites and the killing of *Leishmania* by activated murine macrophages. *Res. Immunol.* **142**, 577–580.

Menon, J. N. and Bretscher, P. A. (1996). Characterization of the immunological state generated in mice susceptible to *Leishmania major* following exposure to low doses of *L. major* and resulting in resistance to a normally pathogenic challenge. *Eur. J. Immunol.* **26**, 243–249.

Modabber, F. (1996). Vaccine: The only hope to control leishmaniasis. In: *Molecular and Immune Mechanisms in the Pathogenesis of Cutaneous Leishmaniasis* (F. J. Tapia, G. Caceres-Dittmar, and M. A. Sanchez, eds) pp. 222–236. Texas, USA, R.G. Landes Company.

Mosmann, T. (1983). Rapid colorimetric assay for cellular growth and survival: application to proliferation and cytotoxicity assays. *J. Immunol. Methods* **65**, 55–63.

Müller, I., Pedrazzini, T. and Louis, J. A. (1988). Experimentally induced cutaneous Leishmaniasis: are L3T4+ T cells that promote parasite growth distinct from those mediating resistance? *Immunol. Lett.* **19**, 251–260.

Müller, I., Pedrazzini, T., Kropf, P., Louis, J. and Milon, G. (1991). Establishment of resistance to *Leishmania major* infection in susceptible BALB/c mice requires parasite-specific CD8+ T cells. *Int. Immunol.* **3**, 587–597.

Nabors, G. S. and Farrell, J. P. (1994). Site-specific immunity to *Leishmania major* in SWR mice: the site of infection influences susceptibility and expression of the antileishmanial immune response. *Infect. Immun.* **62**, 3655–3662.

Nabors, G. S., Afonso, L. C. C., Farrell, J. P. and Scott, P. (1995). Switch from a type 2 to a type 1 T helper cell response and cure of established *Leishmania major* infection in mice is induced by combined therapy with interleukin 12 and pentostam. *Proc. Natl. Acad. Sci.* **92**, 3142–3146.

Noben-Trauth, N., Kropf, P. and Müller, I. (1996). Susceptibility to *Leishmania major* infection in interleukin-4 deficient mice. *Science* **271**, 987–990.

Noben-Trauth, N., Paul, W. E. and Sacks, D. L. (1999). IL-4 and IL-4 receptor-deficient BALB/c mice reveal differences in susceptibility to *Leishmania major* parasite substrains. *J. Immunol.* **162**, 6132–6140.

Ohara, J. and Paul, W. E. (1985). Production of a monoclonal antibody to and molecular characterization of B-cell stimulatory factor-1. *Nature* **315**, 333–336.

Oswald, I. P., Wynn, T. A., Sher, A. and James, S. L. (1994). NO as an effector molecule of parasite killing: modulation of its synthesis by cytokines. *Comp. Biochem. Physiol.* **108C**, 11–18.

PAHO, C. D. P. (1994). Leishmaniasis in the Americas. *Epidem. Bull/PAHO* **15**(3), 8–13.

Preston, P. M. and Dumonde, D. C. (1976). Experimental cutaneous leishmaniasis. V. Protective immunity in subclinical and self-healing infection in the mouse. *Clin. Exp. Immunol.* **23**, 126–138.

Reiner, S. L. and Locksley, R. M. (1995). The regulation of immunity to *Leishmania major*. *Annu. Rev. Immunol.* **13**, 151–177.

Sacks, D. L. and Da Silva, R. P. (1987). The generation of infective stage *Leishmania major* promastigotes is associated with the cell-surface expression and release of a developmentally regulated glycolipid. *J. Immunol.* **139**, 3099–3106.

Sacks, D. L. and Perkins, P. V. (1984). Identification of an infective stage of *Leishmania* promastigotes. *Science* **223**, 1417–1419.

Sacks, D. L., Hieny, S. and Sher, A. (1985). Identification of cell surface carbohydrate and antigenic changes between noninfective and infective developmental stages of *Leishmania major* promastigotes. *J. Immunol.* **135**, 564–569.

Sadick, M. D., Heinzel, F. P., Holaday, B. J., Pu, R. T., Dawkins, R. S. and Locksley, R. M. (1990). Cure of murine leishmaniasis with anti-interleukin-4 monoclonal antibody. Evidence for a T-cell-dependent, interferon-γ-independent mechanism. *J. Exp. Med.* **171**, 115–127.

Scott, P. A. and Farrell, J. P. (1982). Experimental cutaneous leishmaniasis: disseminated leishmaniasis in genetically susceptible and resistant mice. *Am. J. Trop. Med. Hyg.* **31**, 230–238.

Sypek, J. P., Chung, C. L., Mayor, S. E. H., Subramanyam, J. M., Goldman, S. J., Sieburth, D. S., Wolf, S. F. and Schaub, R. G. (1993). Resolution of cutaneous leishmaniasis: Interleukin 12 initiates a protective T helper type 1 immune response. *J. Exp. Med.* **177**, 1797–1802.

Taswell, C. (1987). Limiting dilution assays for the separation, characterization, and quantitation of biologically active particles and their clonal progeny. In: *Cell Separation: Methods and Selected Applications*, Vol. 4. (T. G. Pretlow and T. P. Pretlow, Eds), pp. 109–145. Academic Press, London.

Titus, R. G., Ceredig, R., Cerottini, J. C. and Louis, J. A. (1985a). Therapeutic effect of anti-L3T4 monoclonal antibody GK1.5 on cutaneous leishmaniasis in genetically susceptible BALB/c mice. *J. Immunol.* **135**, 2108–2114.

Titus, R. G., Marchand, M., Boon, T. and Louis, J. A. (1985b). A limiting dilution assay for quantifying *Leishmania major* in tissues of infected mice. *Parasite Immunol.* **7**, 545–555.

Turco, S. and Descoteaux, A. (1992). The lipophosphoglycan of *Leishmania* parasites. *Annu. Rev. Microbiol.* **46**, 65–94.

Ulrich, M., Rodriguez, V. and Centeno, M. (1996). The humoral response in leishmaniasis. In: *Molecular and Immune Mechanisms in the Pathogenesis of Cutaneous Leishmaniasis* (F. J. Tapia, G. Caceres-Dittmar, and M. A. Sanchez, Eds), pp. 189–202. Austin, Texas, R.G. Landes Company.

Vieira, L. Q., Goldschmidt, M., Nashleanas, M., Pfeffer, K., Mak, T. and Scott, P. (1996). Mice lacking the TNF receptor p55 fail to resolve lesions caused by infection with *Leishmania major*, but control parasite replication. *J. Immunol.* **157**, 827–835.

Wei, X., Charles, I. G., Smith, A., Ure, J., Feng, G., Huang, F., Xu, D., Müller, W., Moncada, S. and Liew, F. Y. (1995). Altered immune responses in mice lacking inducible nitric oxide synthase. *Nature* **375**, 408–411.

World Health Organization (1995). Leishmaniasis. In: *Tropical Disease Research. 20 Years of Progress*. World Health Organization, pp. 135–146, Geneva.

◆◆◆◆◆◆◆ THE SAS® SHPM PROGRAM

The SAS® program that follows was developed for the first edition of this book to provide an alternative to functional, although idiosyncratic, PC/Mac software based on Taswell's method. The utility of the program is diminished because SAS® software, while popular worldwide in academics, is beyond the means of many research centres, mainly because of cost.

To correct this shortcoming, an alternative is under development, but it was not ready before the publication deadline of this edition. The progress of the project can be followed by choosing the LDA hyperlink at the bottom of the following URL:

http://www.tangledbank.com/

```
/*** Limited Dilution Assay (LDA)                              ***/
/*** Single-Hit Poisson Model, Minimum Chi-squared Method      ***/
options nocenter ls=78;

proc iml;
fw = 20;
                    /*** Enter data here... ***/
/* SAS(r) PROC IML is a matrix programming language. Some familiarity
/*  with it is very helpful but not essential.
/* The program needs the following information:
/* The known dose and the number of negatively responding wells
/*  for each dose level is entered in the matrix assay1. The values
/*  are entered in pairs (with the dose first and number of negative
/*  responses second) that correspond to the columns of a matrix.
/*  A comma separates the pairs into matrix rows. Only the values
/*  for dose levels greater than 0 and less than the number of wells
/*  per dose level are needed.
```

```
/* The number of footpads is the variable nfp.
/* The total amount of tissue in nanograms is the variable tissue.
/* The number of wells per dose level is the variable nwell.
/* The p-value for performing hypothesis tests is the variable rejl.
/* Example: Assume each dose level has 16 wells, 2 footpads =
/*   165,000,000 ng of tissue, dose level 16.5 had 1 negative
/*   response and dose level 1.65 had 13 negative responses. Then
/*     assay1 = {16.5 1,1.65 13};
/*     nfp = 2;
/*     tissue = 1.65 * 10**8;
/*     nwell = 16;
/*     rejl = .05;
/*   If there had been a third dose level .165 with 14 negative
/*     responses then assay1 would have appeared as
/*     assay1 = {16.5 1,1.65 13,.165 14};
*/
assay1 = {1200 14,12000 5 };
nfp = 2 ; /* the number of footpads */;
tissue = 120000000 ; /* total ng/100ul of tissue */;
nwell = 16 ; /* number of replicate wells */;
rejl = .05 ; /* p-value to reject hypothesis */;

/*** the next 4 lines prevent accidentally using negative well values
  of 0 and nwell since they contribute nothing to the analysis ***/;
excmax = (loc(assay1[,2] ^= nwell))`;
assay = assay1[excmax,];
exc0 = (loc(assay[,2] ^= 0))`;
assay = assay[exc0,];

/*** the next 5 lines calculate values needed for computations ***/;
dd = nrow(assay);
l = assay[,1];
r = assay[,2];
n = repeat(nwell,dd,1);
p = r / n;

cnames = {'l' 'r' 'n'}; /*** column labels ***/;

/*** Assay Validity Test ***/
w1 = (n # p # (log(p))##2) / (1 - p);
w2 = (n # p # l##2) / (1 - p);
w = w1||w2;
x1 = l;
x2 = 1 / l;
xx = x1||x2;
wxx = w # xx;
sumwxx = wxx[+,];
sumw = w[+,];
xbar = sumwxx / sumw;
x = xx - (repeat(xbar,dd,1));
phihat = -(log(p) / l);
yy = log(phihat)||phihat;
wyy = w # yy;
sumwyy = wyy[+,];
ybar = sumwyy / sumw;
y = yy - repeat(ybar,dd,1);
bhat = ((w # x # y)[+,] / (w # x##2)[+,]);
vbhat = abs(((w # y##2)[+,] - (w # x # y)[+,])
        / (dd - (w # x##2)[+,]));
chi = bhat # (w # x # y)[+,];

/*** Weighted Least Squares Estimates ***/
```

```
w = w2;
sumw = w[+,];
x = x2;
y = y[,2];
phiwm = (w # phihat)[+,] / sumw;
vphiwm = 1 / sumw;
phiest = phiwm;

  /*** Estimate Validity Test ***/
start validest;
 p = exp(-1 * phiest * 1);
 w1 = (n # p # (log(p))##2) / (1 - p);
 w2 = (n # p # 1##2) / (1 - p);
 w = w1||w2;
 x1 = 1;
 x2 = 1 / 1;
 xx = x1||x2;
 wxx = w # xx;
 sumwxx = wxx[+,];
 sumw = w[+,];
 xbar = sumwxx / sumw;
 x = xx - (repeat(xbar,dd));
 yy = log(phihat)||phihat;
 wyy = w # yy;
 sumwyy = wyy[+,];
 ybar = sumwyy / sumw;
 y = yy - repeat(ybar,dd,1);
 bhat = ((w # x # y)[+,] / (w # x##2)[+,]);
 vbhat = abs((((w # y##2)[+,] - (w # x # y)[+,])
         / (dd - (w # x##2)[+,])));
 chisl = bhat # (w # x # y)[+,];
 chisl = max(chisl);
finish;

  /*** Newton's Iterative Method ***/
start newton;
 diff = -.000001; /*** convergence value ***/
 phimc = phiwm;
 iter = 0;
 check = 0;
 run chisq;
 do until(check = 1);
  iter = iter + 1;
  run first;
  run second;
  delta = .001 * (jacob / hess);
  converge = phimc - delta;
  phimc = converge;
  check = (delta > diff);
  run chisq;
 end;
finish newton;

  /*** Function to Evaluate ***/
start chisq;
 expon = phimc # 1;
 negxpon = exp(-1 * expon);
 negxpon2 = exp(-2 * expon);
 negxpon3 = exp(-3 * expon);
```

488

```
 posxpon = exp(phimc # 1);
 denom = n # (1 - negxpon);
finish chisq;

 /*** First Derivative ***/
start first;
 numerf = n # 1 # negxpon # (2 # r - n) + r##2 # 1 # (posxpon - 2);
 jacob = (numerf / (n # (1 - negxpon)##2))[+,];
finish first;


 /*** Second Derivative ***/
start second;
 numers = n # 1##2 # (n - 2 # r) # (negxpon - negxpon3)
          + r##2 # 1##2 # (posxpon - 4 + 7 # negxpon
                           - 4 # negxpon2);
 hess = (numers / (n # (1 - negxpon)##4))[+,];
finish second;


 /*** Mop-up Work ***/
do;
 run validest;
  chiwm = chisl;
  varwm = vbhat;
 run newton;
  phiest = phimc;
 run validest;
reset noname;
 varmc = 2 * 1/hess;
 u95 = phimc + 1.96 * sqrt(varmc);
 l95 = phimc - 1.96 * sqrt(varmc);
 chimc = chisl;
 pv = 1 - probchi(chi,1);
 pvmc = 1 - probchi(chimc,1);
 parasite = tissue/nfp * phimc;
 inv = 1 / phimc;
 print '   MINIMUM CHI-SQUARED ESTIMATION WITH NEWTONS METHOD';
 run banner;
 print '                       ASSAY VALIDITY TEST',
       '                                     Model 1    Model2',
       'chi-squared AVT                      ' chi,
       'p-values of AVT                      ' pv,
       '                       ESTIMATE VALIDITY TEST',
       'estimated frequency                  ' phimc,
       'chi-squared EVT                      ' chimc,
       'p-value of EVT                       ' pvmc,
       'estimated variance                   ' varmc,
       '95% upper confidence level           ' u95,
       '95% lower confidence level           ' l95,
       '1/frequency                          ' inv,
       'estimated parasites per lesion       ' parasite,
       'nanograms of tissue                  ' tissue,
       'number of iterations to converge     ' iter;
start banner;
if all(pv > rejl) then
 print
       '    *********************************************',
       '    ***            CONGRATULATIONS            ***',
       '    ***       THE ASSAY IS VALID BECAUSE       ***',
       '    ***   THE NULL HYPOTHESIS WAS NOT REJECTED ***',
```

489

```
'              ***        AT THE 'rejl[format=3.2]
' ALPHA LEVEL    ***',
'              ***********************************************';
else
 print
'              ***********************************************',
'              ***               SO SORRY              ***',
'              ***      THE ASSAY IS NOT VALID BECAUSE     ***',
'              ***      THE NULL HYPOTHESIS WAS REJECTED    ***',
'              ***          AT THE 'rejl[format=3.2]
' ALPHA LEVEL        ***',
'              ***********************************************';
if pvmc > rejl then
 print
'              ***********************************************',
'              ***             CONGRATULATIONS          ***',
'              ***      THE ESTIMATE IS VALID BECAUSE      ***',
'              ***   THE NULL HYPOTHESIS WAS NOT REJECTED   ***',
'              ***          AT THE 'rejl[format=3.2]
' ALPHA LEVEL        ***',
'              ***********************************************';
else
 print
'              ***********************************************',
'              ***               SO SORRY              ***',
'              ***      THE ESTIMATE IS NOT VALID BECAUSE   ***',
'              ***      THE NULL HYPOTHESIS WAS REJECTED    ***',
'              ***          AT THE 'rejl[format=3.2]
' ALPHA LEVEL        ***',
'              ***********************************************';
finish banner;
end;

quit;
```

List of suppliers

Bellco Glass, Inc.
P.O. Box B, 340
Edrudo Road
Vineland, NJ 08360, USA

Tel: 001-609-691-1075
Fax: 001-609-691-3247
sales@bellcoglass.com
Tenbroeck glass homogenizers

Corning Costar Corp.
45 Nagog Park
Acton, MA 01720, USA

Tel.: 001-508-635-2200
Fax: 001-508-635-2476
http://www.corningcostar.com
Plastic ware, tissue culture

Difco Laboratories
P.O. Box 331058
Detroit, MI 48232-7058, USA

Tel.: 001-313-462-8500
Fax: 001-313-462-8517
Nutrient agar

Endogen
30 Commerce Way
Woburn, MA 081801-1059, USA

Tel.: 001-617-937-0890
Fax: 001-617-937-3096
http://www.endogen.com
ELISA-minikits

Gibco, Life Technologies, Inc.
P.O. Box 68
Grand Island, NY 14072-0068, USA

Tel.: 001-716-774-6700
Fax: 001-716-774-6694
http://www.lifetech.com
Cell culture medium, medium supplements

Heraeus Bereich Thermotech
P.O. Box 1563
D-63450 Hanau/Germany

Tel.: 0049-6181-350
Fax: 0049-6181-35880
Teflon for bone marrow, macrophage cultures

Kroeplin GmbH
Messzeugfabrik
P.O. Box 12555
D-6490 Schlüchtern/Germany

Tel.: 001-6661-860
Fax: 001-6661-8639
Dial caliper

PharMingen
10975 Torreyana Road
San Diego, CA 92121-111, USA

Tel.: 001-619-677-7737
Fax: 001-619-677-7749
http://www.pharmingen.com
Antibodies, cytokines

SAS Institute, Inc.
SAS Campus Drive
Cary, NC 27513-2414, USA

Tel.: 001-919-677-8000
Fax: 001-919-677-8123
Statistic software

Sigma
P.O. Box 14508
St Louis, MO 63178, USA

Tel.: 001-314-771-5765
Fax: 001-314-771-5757
http://www.sigma.sial.com
Chemicals

Southern Biotechnology Associates, Inc.
160A Oxmoor Boulevard
Birmingham, AL 35209, USA

Tel.: 001-205-945-1774
Fax: 001-205-945-8768
http://www.southernbiotech.com
Antibodies

Vector Laboratories, Inc.
30 Ingold Road
Burlingame, CA 94010, USA

Tel.: 001-650-697-3600
Fax: 001-650-697-0339
Peanut agglutinin

4.3 Animal Models: Murine Cytomegalovirus

Jürgen Podlech, Rafaela Holtappels, Natascha K A Grzimek and Matthias J Reddehase
Institute for Virology, Johannes Gutenberg University, Mainz, Germany

CONTENTS

◆◆◆◆◆◆ INTRODUCTION

Multiple-organ cytomegalovirus (CMV) disease, interstitial pneumonia in particular, is a major concern in the therapy of haematopoietic malignancies by haematoablative treatment and bone marrow transplantation (BMT). Human CMV (hCMV) is the prototype member of the subfamily *Betaherpesvirinae* of the virus family *Herpesviridae*. Its genome is a linear, double-stranded DNA with a coding capacity of ca. 165 open reading frames. Primary infection occurs usually in early childhood and does not cause symptoms of disease in immunocompetent persons, except a mild mononucleosis that is rarely diagnosed. The productive, acute infection is efficiently controlled by the immune system, but the viral genome is not eliminated and persists in a dormant state, referred to as viral latency. Disease occurs in immunodeficient patients after primary infection or after reactivation of latent endogenous virus.

The replication of CMVs is highly species-specific; accordingly, hCMV cannot be studied in an animal model. This is not really a disadvantage. During an aeon of co-evolution, CMVs have adapted themselves to their respective hosts; therefore, CMV biology is most reliably studied in a natural virus–host combination. Even though hCMV and murine CMV (mCMV) differ molecularly, basic principles in viral pathogenesis are the same. The murine model has been paradigmatic for many aspects of the immune control of CMVs (Reddehase, 2000). Specifically, the importance of CD8 T cells and of the viral immediate-early protein in the protective immune response, as well as immune evasion strategies of CMVs have

METHODS IN MICROBIOLOGY, VOLUME 32
ISBN 0–12–521532–0

Copyright © Elsevier Science Ltd
All rights of reproduction in any form reserved

first been documented in this model. Importantly, regarding clinical trials of pre-emptive cytoimmunotherapy of CMV disease (Riddell *et al.*, 1992), the murine model has provided the proof of principle.

◆◆◆◆◆◆ BONE MARROW TRANSPLANTATION AND INFECTION

Since hCMV infection of BMT patients is a major clinical problem, studies by our group were focused on the murine model of mCMV infection in an experimental BMT setting (Fig. 1). The basal model uses syngeneic BMT with female BALB/c (H-2^d haplotype) mice as donors and recipients. For tracking donor-derived haematopoietic cells in the recipient, the male sex (testes determining) gene *tdy* on the Y chromosome of male donors can be used as a reporter gene (Mayer *et al.*, 1997; Steffens *et al.*, 1998b). The model can, of course, be extended by introducing genetic differences between donor and recipient for studying an interference between the virus and graft-versus-host (GvH) or host-versus-graft (HvG) disease (Alterio de Goss *et al.*, 1998; Podlech *et al.*, 1998b). Recipients die of multiple-organ CMV disease, including interstitial pneumonia (Reddehase *et al.*, 1985; Podlech *et al.*, 1998a, 2000), unless BMT leads to an efficient haematopoietic reconstitution. Reconstituted antiviral CD8 T cells terminate primary infection (Holtappels *et al.*, 1998; Podlech *et al.*, 1998a, 2000)

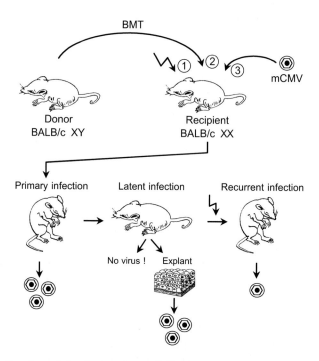

Figure 1. Basal model of experimental BMT and mCMV infection. Step (1): immunoablative conditioning of the BMT recipients by total-body γ-irradiation. Step (2): intravenous transfer of donor-derived BM cells. Step (3): intraplantar infection with mCMV.

and allow viral latency to establish (Polic *et al.*, 1998; Steffens *et al.*, 1998a; Holtappels *et al.*, 2000c). By definition, infectious virions are absent during latency (Kurz *et al.*, 1997), but viral DNA persists in organs. Upon tissue explantation or transplantation, or by secondary immunoablative treatment of the recipients, viral gene expression is reactivated and results in recurrent productive infection (Reddehase *et al.*, 1994; Polic *et al.*, 1998; Kurz and Reddehase, 1999).

Murine cytomegalovirus (mCMV)

General information

The Smith strain of mCMV (Smith, 1954) was originally isolated as *Mouse Salivary Gland Virus* from salivary gland tissue of naturally infected mice. The virus was distributed by the American Type Culture Collection (http://www.atcc.org/home.cfm) as ATCC number VR-194 and was recently re-accessioned as VR-1399. Experiments in our laboratory refer to mCMV ATCC VR-194 purchased in 1981 and propagated in murine fetal fibroblasts (see below). The VR-194 genome was cloned as *Hind* III fragments by Ebeling *et al.* (1983) and sequenced by Rawlinson *et al.* (1996). The double-stranded linear viral DNA encompasses 170 open reading frames (ORFs). Some work in the literature refers to mCMV Smith strain variant K181, for which *Hind* III fragments were cloned by Mercer *et al.* (1983). More recently, the contiguous mCMV genome was cloned as an infectious bacterial artificial chromosome (BAC) by Messerle *et al.* (1997). mCMV is a member of the herpesvirus family (*Herpesviridae*), subfamily β-herpesviruses (*β-Herpesvirinae*). Infection by β-herpesviruses is strictly species-specific. Accordingly, working with mCMV is not associated with any health risk. The US BioSafety Level is 2. The German GenTG BioSafety Level is 1, according to a decision of the ZKBS No. A VI-6782-12-96.

- Work with cell culture-propagated virus, sucrose-gradient purified as described later.
- For infection of mice (see Box below), dilute the virus stock with physiological saline.

Commentary

Much work in the literature, in particular in older literature, was performed with mCMV *Salivary Gland Virus* (SGV), which is a crude preparation consisting of diluted and only partially purified homogenate of salivary gland tissue from infected mice. This material is said to be of higher virulence. In fact, lethality in mice is higher for SGV than for cell-culture propagated, purified virus. However, SGV is an undefined mixture of virus, cytokines, hormones and cellular components. Therefore, some differences between SGV and purified virus may be attributed to components other than virus. The frequently
(contd.)

reported control experiment using salivary gland homogenate from uninfected mice is inappropriate, as cytokines derived from inflammatory infiltrates in infected salivary glands are absent in uninfected salivary glands. In fact, SGV suppresses haematopoiesis in long-term bone marrow cell cultures even after inactivation of viral infectivity (Busch *et al.*, 1991) and it causes rapid liver necrosis after intraperitoneal or intravenous inoculation independent of virus replication in the liver. It should be noted that unpurified supernatant from infected cell cultures also shows immediate liver toxicity after intravenous inoculation. Therefore, the use of purified virus is mandatory for reliable experiments.

Bone marrow transplantation (BMT) and infection

Procedure: preparation of donor BMC

For the preparation of bone marrow cells (BMC), use donor mice between 10 and 16 weeks of age. Sacrifice mice by cervical dislocation or by CO_2 atmosphere. All subsequent procedures are performed under a clean bench.

1. Sterilize the hind limbs with 70% ethanol, and dissect both with strong scissors.
2. Prepare the femurs and tibiae by removing muscle tissue.
3. Cut off the epiphysis on both ends of each bone.
4. Flush the BM out by gently rinsing the bone cavity with medium RPMI-1640 (Gibco-BRL, cat. no. 31870-025), supplemented with 5% (vol/vol) FCS. Use a syringe with a 0.45×23 mm cannula. Collect the extruded cylindrical BM tissue in a 50-ml Falcon tube.
5. Prepare a single cell suspension by gently disrupting the tissue with a pipette.
6. Wash the cells twice by sedimentation (8 min at ca. $680g$ at $4°C$) and by resuspension in 25 ml of RPMI with no FCS (!).
7. Filtrate through filtration gauze, wash once under cold conditions, count the cells, and keep them at $4°C$ in a low volume for a period as short as possible.
8. Immediately before use: adjust the required cell number with physiological saline.

Note: According to cytofluorometric analysis, the contamination of BMC with mature intravascular T cells is <1%, actually undetectable. In our experiments on the reconstitution of donor-type antiviral CD8 T cells after BMT (Alterio de Goss *et al.*, 1998; Holtappels *et al.*, 1998), depletion of CD8 T cells from the donor-derived BMC has had no effect on the reconstitution of the recipients. However, if required, CD8 T cells can be removed by using established immunomagnetical methods. By contrast, never deplete CD4 T cells, as this results in an impaired engraftment of the donor BMC.

Procedure: BMT and infection

1. Subject recipient mice to haematoablative treatment by total-body γ-irradiation with a single dose of 6 to 7 Gy, delivered by an appropriate γ-ray source. We use a ^{137}Cs γ-ray source (model OB58 Buchler, now STS Steuerungstechnik & Strahlenschutz GmbH, Braunschweig, Germany) with a dose rate of ca. 0.6 Gy per min that is adjusted monthly. This treatment is usually performed at ca. 6 h prior to BMT, in some cases also on the day before BMT.
2. Warm-up the tail of a recipient mouse gently for ca. 5 min with an infrared light bulb (e.g. Philips R95 infraphil, 100 W; cat. no. 136916) to dilate the tail veins.
3. Sterilize the inoculation site with 70% ethanol.
4. Inject 0.5 ml of donor BMC suspension (warmed up to body temperature immediately before use) very slowly into the dorsal tail vein (i.v.; intravenous cell transfer), while mice are fixed in a special V-shaped device. Use a 2-ml syringe with a 0.45 × 23 mm cannula.
5. If intended, infect recipient mice at one of the hind footpads (intraplantar infection). In detail, fix the recipient mouse in a V-shaped device, sterilize the footpad with 70% ethanol, and inject, very slowly, 25 µl of virus suspension in an appropriate dilution (10⁵ PFU was mostly used in our experiments) with physiological saline. Use a 1-ml syringe with a 0.4 × 12 mm cannula. Infection is usually performed at ca. 2 h after BMT.

Note: Volumes >25 µl as well as inoculation of more than one footpad are not allowed in Germany according to a decision of the Ethics commission at the *Regierungspraesidium Tuebingen*, decision no. AZ 37-9553. In this context, it also should be noted that unilateral intraplantar infection, rather than bilateral infection, results in an optimal priming for the antiviral T-cell response in the respective draining popliteal lymph node of immunocompetent mice. Intraplantar infection proved to give well reproducible results and low variance of virus titres in various organs of immunocompromised mice.

Bone marrow histology

Initial steps in bone marrow histology differ from standard histology by the need of bone decalcification. Principally, there are two common methods of decalcification: the EDTA method and the TCA method. Decalcification with EDTA is superior in preserving the tissue and is therefore the method of choice for subsequent haematoxylin and eosin (HE) staining and other standard methods of BM histology. It is the routine method for subsequent *in situ* hybridization (ISH), because treatment with TCA would destroy nucleic acids. Decalcification with

(*contd.*)

TCA was our method of choice for the immunohistochemical (IHC) detection of the viral IE1 protein (see later), as unspecific background staining is reduced as compared with the EDTA method. It should be noted, however, that not all epitopes withstand acid treatment.

For both methods:

1. Fix the femoral and tibial bones for 18 h at 20°C in phosphate buffered saline (PBS, pH 7.4) containing 4% (vol/vol) formalin.
2. Rinse for ca. 2 h at ca. 20°C with water to remove the fixative.

EDTA method

1. Incubate the bones with gentle stirring for 3 days at ca. 20°C in 0.53 M ethylenediamine-tetraacetic acid (EDTA) solution.
   - Dissolve 200 g EDTA (Carl Roth GmbH & Co., Karlsruhe, Germany; cat. no. 3619.1) in 800 ml aqua dist., add ca. 50 ml 40% (wt/vol) NaOH to adjust to pH 7.4, and fill up to 1 l with aqua dist.
2. Rinse for ca. 2 h at ca. 20°C with water.
3. Dehydrate with graded concentrations of isopropanol followed by xylene according to standard histological procedures.

TCA method

1. Incubate the bones with gentle stirring for 1 day at ca. 20°C in 0.3 M trichloroacetic acid (TCA) solution.
   - Dissolve 50 g TCA (Carl Roth GmbH & Co., Karlsruhe, Germany; cat. no. 8789.1) in 960 ml aqua dist. and add 40 ml of 4% (vol/vol) formalin.
2. Start the dehydration process with 90% (vol/vol) isopropanol in aqua dist., followed by 100% isopropanol and xylene.

For both methods:
1. Proceed according to standard histological methods with embedding of the decalcified and dehydrated bones in paraffin.
2. Cut 2-μm sections (e.g. with rotation microtome model HM 355, Microm, Walldorf, Germany), deparaffinize, and rehydrate according to standard procedures.
3. Perform standard histological stainings (e.g. HE) and IHC detection of viral proteins (after decalcification by EDTA) or ISH of viral DNA (after decalcification with TCA) as described for other tissues in the Boxes below.

Fig. 2A illustrates the repopulation of the bone marrow after BMT with a low dose of BMC, and Fig. 2B shows BM aplasia caused by mCMV infection of the BM stroma (Mayer *et al.*, 1997; Steffens *et al.*, 1998b) (HE staining after bone decalcification using the EDTA method).

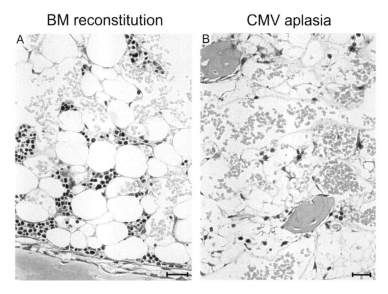

BM reconstitution   CMV aplasia

Figure 2. Bone marrow aplasia caused by mCMV. (A) Repopulation of the BM with myelomonocytic and erythroblastic colonies after BMT in absence of mCMV infection. Section of a femoral diaphysis. (B) Graft failure in the presence of mCMV infection. Section of a femoral epiphysis. HE staining after decalcification with the EDTA method. Bar marker: 25 µm. (From Steffens *et al.*, 1998b).

◆◆◆◆◆◆ DETECTION OF VIRUS REPLICATION IN ORGANS

The productive replication cycle of mCMV is cytopathogenic and actually cytolytic for most permissive cell types. In an immunocompromised and genetically susceptible murine host (for instance in BALB/c mice), mCMV shows a very broad cell-type tropism (Podlech *et al.*, 1998a). Specifically, virus replication has been documented *in vivo* for many different cell types, including various types of epithelial cells (e.g. glandular epithelial cells of the salivary glands, pneumocytes, enterocytes, hepatocytes, cortical and medullary cells of the suprarenal glands, and ependymal cells lining the brain ventricles). Further cell types include endothelial cells, heart muscle myocytes, brown fat adipocytes, dendritic cells (Andrews *et al.*, 2001), mature macrophages (Stoddart *et al.*, 1994; Hanson *et al.*, 2001) including alveolar macrophages in the lungs and Kupffer cells in the liver, connective tissue fibrocytes, and bone marrow stromal cells (Mayer *et al.*, 1997). Notably, the *in vivo* cell-type tropism of hCMV is very similar (Plachter *et al.*, 1996). Accordingly, involvement of multiple organs is a common typical feature of human and murine CMV disease.

In tissues composed of uniform parenchymal cells permissive for productive mCMV infection, viral cytolysis results in three-dimensional lesions, literally holes in the tissue, so-called virus plaques. In the histological section, plaques are characterized by a necrotic centre and by a corona of infected cells (Fig. 3). The centre is devoid of nuclei and contains remnants of lysed cells, cytoskeleton components and extracellular matrix components in particular. The plaque grows centrifugally. Accordingly, more recently infected cells form the outer rim of the corona adjacent to intact surrounding tissue, while cells in late stages of the virus replication cycle form the inner rim. This type of viral histopathology is particularly pronounced in the liver and in the suprarenal gland (Grzimek *et al.*, 1999).

Principally, virus replication in organs can be quantitated either *in situ* by histological methods (IHC for detecting viral proteins and ISH for detecting viral nucleic acids in infected cells) or *in vitro* in cell cultures of permissive indicator cells infected with organ homogenate.

Virus plaque assay in cell culture

The multitude of cell types infected *in vivo* contrasts with the limited number of cell culture systems currently available for propagating the virus *in vitro*. mCMV replicates in fibroblast cell lines, such as NIH 3T3 (ATCC CRL-1658) and, albeit with lower productivity, in macrophage cell lines, such as J774.A.1 (ATCC TIB-67) and IC-21 (ATCC TIB-186) (Hengel *et al.*, 2000; Hanson *et al.*, 2001). It has been reported that mCMV replicates to high titres in M2-10B4 cells, a continuous line of murine bone marrow stromal cells (Lutarewych *et al.*, 1997). However, the most established way to propagate mCMV is the infection of primary fetal fibroblasts, usually referred to as mouse embryo(nal) fibroblasts (MEF). Because the permissive cell lines tend to overgrow virus plaques in the infected cell monolayer, MEF remain the cell type used for quantitation of infectivity (given as plaque forming units, PFU) by the virus plaque assay.

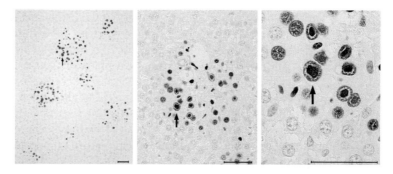

Figure 3. Virus plaques in the liver. Black IHC staining of the viral IE1 protein accumulated in an intranuclear inclusion body (so-called owl's eye) in the late (L) phase of the virus replication cycle. ABC-peroxidase/DAB-nickel method of staining with light haematoxylin counterstaining. Bar markers represent 50 μm.

Preparation of MEF

Equipment and reagents

- Sterile 300-ml Erlenmeyer glass bottle, filled with 2–3 mm diameter glass beads
- Magnetic stirrer
- 100-μm pore Nylon mesh (cell strainer, Becton Dickinson, cat. no. 352360)
- Stainless steel meshes
- 13.5-cm diameter tissue culture dishes
- DPBS solution with no Ca^{2+} and Mg^{2+} (Bio Whittaker Europe, cat. no. BE17-512F)
- Trypsin/EDTA solution-I, pH 6.4:

NaCl	8 g
$Na_2HPO_4 \times 2\,H_2O$	1.42 g
KCl	0.2 g
KH_2PO_4	0.2 g
Trypsin 1:250	1.25 g (2.2 U/mg, Serva, cat. no. 37292)
EDTA	1.25 g

  ad 1 l H_2O bidist.; adjust the pH with 1 N NaOH. Before use, dilute 1:2 with DPBS and filtrate sterile.

- Trypsin/EDTA solution-II (10×; Gibco, cat. no. 35400-027), dilute 1:10 with DPBS
- Medium MEM-FCS is MEM (Gibco, cat. no. 21090-022) supplemented with 5% (vol/vol) heat-inactivated (1 h, 56°C) FCS, 2 mM L-glutamine, and antibiotics (usually 100 U of penicillin and 0.1 mg of streptomycin per ml).

Procedure

Work under sterile conditions throughout. Cell culture is routinely performed under standard conditions (37°C, 95%-humidified atmosphere, 5% CO_2).

1. Remove the uterus on day 17 of pregnancy, dissect the fetuses (usually 5–10 per mouse), and place them immediately on ice in a tissue culture dish.
2. Remove and discard the eyes and inner organs.
3. Cut the remaining tissue into very small pieces.
4. Take up the minced tissue with cold DPBS, put it on the steel mesh, and rinse extensively with DPBS to remove the erythrocytes.
5. Transfer the minced tissue with a cell scraper to the Erlenmeyer bottle.
6. Add 30 ml Trypsin/EDTA solution-I and stir gently for 30 min at 37°C. Repeat this procedure twice, adding fresh Trypsin/EDTA in every round.

(contd.)

7. Collect the cell suspension and pass it through steel mesh to remove clumps. Wash the glass beads with MEM-FCS and collect the cell suspension accordingly.
8. Sediment the cells by centrifugation for 10 min at ca. 300g.
9. Resuspend the cell pellet in MEM-FCS, wash twice, filter the cell suspension through a Nylon mesh, and determine the cell number.
10. Seed the cells into tissue culture dishes (per dish: $2–3 \times 10^7$ cells in 25–30 ml MEM-FCS).
11. Important: on the next day, rinse the monolayer three times with DPBS to remove erythrocytes and cell debris, and add 30 ml of fresh MEM-FCS.
12. After 3–5 days of cultivation, the cells form a confluent monolayer and can either be split for a first passage or harvested for storage in liquid nitrogen. Discard the culture medium, wash the cell monolayer with DPBS to remove FCS, add 3 ml of Trypsin/EDTA solution-II, and incubate for a few minutes until the cells detach (visual control under an inverted microscope).
13. Resuspend and wash the cells twice in MEM-FCS.
14. For freezing, resuspend the cells in 10% DMSO–90% FCS. For seeding, resuspend the cells in MEM-FCS. Cells can be split three-fold or four-fold every 3–5 days.
15. The virus plaque assay (see below) can be performed with cells in the second or third passage, but not in later passages.

Note: The cell yield is variable. Usually, 5–10 fetuses from one pregnant mouse give between five and 10 culture dishes.

MEF are used for propagating the virus and for quantitating virus infectivity by the virus plaque assay. Infected MEF are used as target cells in cytolytic assays and as stimulator cells in the enzyme-linked immunospot (ELISPOT) assay.

Virus propagation and purification

1. Prepare MEF in the second passage in 13.5-cm tissue culture dishes (see above).
2. Infect the cells with $1–2 \times 10^5$ plaque forming units (PFU, see below) of mCMV per dish shortly before the cell monolayer is confluent. As a dish contains at that stage ca. 1×10^7 cells, the multiplicity of infection (MOI) is ca. 0.01–0.02 PFU per cell. In detail, remove the culture medium and add the virus (in MEM-FCS) in a volume of 5 ml. After 30 min at ca. 20°C and gentle occasional shaking, add 20 ml of fresh medium.
3. Incubate the dishes under cell culture conditions for 4–5 days, until most cells show a cytopathic effect (CPE), i.e. get rounded and start to detach.

(contd.)

4. Collect supernatants and cells in sterile 500-ml centrifugation tubes. Scrape adherent cells off with a cell scraper (do not use trypsin!).
5. From here on, perform all virus purification steps at 4°C or on ice.
6. Centrifugate for 20 min at ca. 6400*g* in order to sediment cell debris and cell nuclei.
7. Collect the supernatants in sterile 250-ml centrifugation tubes.

Note: If it is the intention to isolate pure virions (as in Kurz *et al.*, 1997), one should discard the pellets, because nuclei contain a high number of non-enveloped, non-infectious nucleocapsids. If it is the intention to get an optimal virus yield, one should proceed as follows:

8. Resuspend the pellets in altogether ca. 10 ml of MEM-FCS and homogenize the material with a Dounce homogenizer (ca. 20 times). Transfer the homogenate to a 50-ml centrifugation tube and centrifugate for 20 min at ca. 3600*g*. Discard the pellet, collect the supernatant, and pool the supernatants of both centrifugations.
9. Centrifugate for 3 h at ca. 26 000*g*. Discard the supernatants, except some fluid covering the pellets, and store the tubes on ice for 4 h or overnight.
10. Resuspend the pellets in the residual medium, homogenize again (on ice), and carefully load 2 ml of the homogenate on an 18-ml pre-cooled sucrose/VBS density cushion in 20-ml polyallomer ultracentrifugation tubes.
    - 15% (wt/vol) sucrose/VSB (virus standard buffer): 15 g sucrose in 100 ml VSB. VSB: 50 ml of 1 M Tris (tris-hydroxymethyl-aminomethan), 12 ml of 1 M KCl, and 50 ml of 0.1 M Na-EDTA and 1 l aqua bidist., pH 7.8.
11. Centrifugate for 1 h at ca. 53 000*g* (swing-out rotor).
12. Discard the supernatant, cover the pellet with 4 ml sucrose-VSB, and incubate for 4–12 h on ice.
13. Resuspend the pellet thoroughly with a Pasteur pipette, homogenize with a Dounce homogenizer, and store 20–100 µl aliquots at –70°C or in liquid nitrogen.

Note: In earlier publications, DPBS (also referred to as PBS-A) was used instead of VSB for virus purification and storage. It is essential that the buffer used is devoid of Ca^{2+} and Mg^{2+} to preclude aggregation of the virions.

Virus yield

Starting with 50 tissue culture plates, one can expect 3–4 ml of virus suspension with an infectivity titre ranging from 5×10^8 to 2×10^9 PFU per ml.

A highly purified virion population, as revealed by electron microscopy, consists of monocapsid virions and multicapsid virions in a ratio of ca. 3 : 2 (Kurz *et al.*, 1997). The formation of multicapsid virions is a special feature of mCMV morphogenesis, as it occurs *in vitro* and *in vivo* in most cell types, with the exception of salivary gland epithelial cells. In a multicapsid virion, 2–8 capsids (3.4 on average) are embedded in tegument proteins and share one envelope (Weiland *et al.*, 1986; Kurz *et al.*, 1997).

Virus plaque assay

By definition, one plaque-forming unit (PFU) is the amount of virus required to cause the formation of a single plaque in a monolayer of permissive cells. The virus plaque assay can be performed with purified virions, with supernatant of infected cell cultures, and with homo-genates of infected cells or organs. There exist two versions of the assay: the standard plaque assay and the plaque assay with 'centrifugal enhancement of infectivity, *CEI*' (Hudson *et al.*, 1976; Kurz *et al.*, 1997).

1. Seed MEF in 48-well flat-bottom culture plates for the second or third passage.
2. Infect the MEF when the cell monolayer has just reached confluence. For this, remove most of the cell culture medium such that the cell monolayer remains covered with fluid, and add 100 µl of test sus-pension in an appropriate dilution.
3. Incubate for ca. 1 h at 37°C for virus adsorption and penetration. In the case of *CEI*, incubate for ca. 30 min at 37°C for adsorption, and centrifugate the plates for another 30 min at ca. 1000g at ca. 20°C for enhanced penetration.
4. Cover the cultures with 0.5 ml of semi-solid methylcellulose medium to prevent the formation of secondary plaques.
   a. 8.8 g of methylcellulose (Fluka, Buchs, CH) suspended in 360 ml of aqua bidist.
   b. Sterilize the suspension at 121°C.
   c. Dissolve the methylcellulose at 20°C under permanent stirring for ca. 3 h and keep the solutioin at 4°C until it becomes clear (usually overnight). The solution can be stored for several months.
   d. Before use, add 40 ml of 10x MEM, 20 ml FCS, 3.5 ml L-glutamine (50 mg ml^{-1}), antibiotics, and 8–16 ml NaHCO$_3$ (55 g l^{-1}) until a pH of 7.4 is reached.
5. Count virus plaques under an inverted microscope after 3–4 days of cultivation.

Note: Usually, the test suspension is serially diluted in factor-10 steps with two or three replicate cultures for each dilution. Plaques are counted for the dilution that gives >10 and <100 plaques. In the case of high plaque numbers, round plaques may fuse to form doublet, triplet or multiple-leaved structures, which have to be counted accordingly as two, three or more plaques. The virus titre (*VT*) of the test suspension is calculated according to the formula: *VT* [PFU/ml] = $M \times 10 \times$ *Dilution Factor*, where M is the mean value of the plaque numbers counted for the replicate cultures.

Animal Models:
Murine Cytomegalovirus

> ### Virus titre in organs
>
> 1. Dissect organs, transfer them into MEM-FCS, and freeze at −20°C. Note: Frozen organs can be stored for a longer period at −70°C.
> 2. Thaw slowly at ca. 4°C and pass the freeze–thaw disrupted tissue through a steel mesh. Rinse the mesh with MEM-FCS so that the organ homogenate reaches a final volume of 2 ml (e. g. for spleen, lungs, salivary glands, and suprarenal glands) or 20 ml in the case of the liver.
> 3. Prepare dilutions of the homogenate and continue as described above for the *CEI* method.
>
> *Note 1:* Organ homogenate may be toxic for the MEF indicator cell monolayer. Usually, the first suspension to be tested is the 1:10 dilution of the homogenate (1/200 and 1/2000 aliquot for diverse organs and liver, respectively), which defines the detection limit per culture accordingly. The virus titre *VT** (with the asterisk indicating *CEI*) is here given as PFU* per organ.
>
> *Note 2:* Undiluted, purified virus stored in 15% sucrose–VSB solution can be frozen and thawed up to five times with no loss of infectivity. By contrast, infectivity in organ homogenates is reduced by a factor of ca. 5 with every freeze-and-thaw cycle.

Purified virion preparations have a titre in the order of 10^8 to 10^9 PFU per ml (see earlier). In the case of *CEI*, the infectivity enhancement factor PFU*:PFU is ca. 20. The molecular equivalent of a PFU has been defined only recently (Kurz *et al.*, 1997). Specifically, phosphorimaging analysis of a Southern blot performed with purified virion DNA revealed a genome-to-PFU ratio of 500, that is a genome-to-PFU* ratio of 25. For special purposes, for instance for excluding virus replication during latent infection, the sensitivity of the infectivity assay can be further enhanced by a factor of 5 by using *ie1* gene specific RT PCR as the read-out instead of visible plaques. The genome-to-infectivity ratio was thereby defined as being 4.5 (95% confidence limits 2–9). Figure 7 illustrates the result of a typical virus plaque assay using the *CEI* method for determining the amount of infectious virus in the lungs after cytoimmunotherapy with graded doses of virus peptide-specific CTL (Holtappels *et al.*, 2001).

While the virus plaque assay determines infectivity, histological methods determine the number of infected cells in a tissue and the degree of tissue destruction. Principally, one can detect viral proteins by immuno-histochemistry (IHC) or viral nucleic acids, DNA or RNA, by *in situ* hybridization (ISH).

> ### Staining of viral antigen by single-colour IHC
>
> The most sensitive method for the *in situ* detection of mCMV-infected cells is the staining of the intranuclear viral immediate-early (IE) phase
> *(contd.)*

protein IE1 (pp89). This protein is expressed at ca. 2 h after infection and is maintained throughout the virus replication cycle. In the late (L) phase of the cycle, IE1 accumulates in an intranuclear inclusion body, which is the site of nucleocapsid morphogenesis and viral DNA packaging.

Reagents

- Trypsin solution, pH 7.4:
Dissolve 1.25 g Trypsin (Sigma, cat. no. T 7409) in 1 l of trypsin buffer:

a. NaCl	8 g	(137 mM)
b. KCl	0.2 g	(2.7 mM)
c. KH_2PO_4	0.2 g	(1.5 mM)
d. Na_2HPO_4	1.15 g	(6.5 mM)
e. EDTA	1.25 g	(3.4 mM)

dissolved in 1 l aqua dist., and adjusted to pH 7.4
- Blocking solution (0.3% H_2O_2 in methanol):
Add 0.6 ml of a 30% (vol/vol) H_2O_2 solution (in aqua dist.) to 59.4 ml methanol.
- TBS (Tris buffered saline)

a. Tris	12.1 g	(100 mM)
b. NaCl	8.8 g	(150 mM)

ad 1 l aqua dist., adjust to pH 7.4
- Normal rabbit serum (Sigma, cat. no. R 4505)
- Virus-specific antibody:
MAb mouse IgG1, clone CROMA 101, directed against mCMV-IE1 (from Prof. S. Jonjic, University of Rijeka, Rijeka, Croatia).
- Further antibodies:
a. Polyclonal goat Ab directed against mouse-IgG (Fab), biotin-conjugated and affinity-purified (Sigma, cat. no. B 0529)
b. Mouse IgG1 (Dako, cat. no. X 0931) serving as an isotype control
- ABC-enzyme kits:
a. Vectastain ABC-peroxidase kit PK-4000 Standard or (alternatively)
b. Vectastain ABC-alkaline phosphatase kit AK-5000 Standard
both from Vector Laboratories Inc. (Burlingame, CA 94010, USA)
- Peroxidase staining substrate:
10 mg 3,3' diaminobenzidine tetrahydrochloride (DAB) (Sigma, cat. no. D 5637) and 75 mg ammonium-nickel-sulfate hexahydrate (cat. no. 09885; Fluka, Buchs, CH) dissolved in 50 ml of Tris (50 mM, pH 7.5) and supplemented with 17 µl of 30% (vol/vol) H_2O_2 (in aqua dist.).
- Alkaline-phosphatase staining substrate:
Fuchsin substrate system (Dako, cat. no. K 0625), consisting of the chromogen Fuchsin, activating reagent, and substrate buffer.
- Mayer's hematoxylin (Sigma, cat. no. MHS-16)
- Embedding medium:
PARAmount (Earth Safe Industries, Inc.; Belle Mead, NJ 08502, USA).

Procedure

1. Prepare 2-μm sections of paraffin-embedded tissue, deparaffinize, and rehydrate according to established histological methods.
2. Partially digest proteins with trypsin solution for 15 min at 37°C.
3. Wash with aqua dist. for 3 min.
4. Block endogenous peroxidase with blocking solution for 30 min at 20°C.
5. Wash with aqua dist. for 3 min.
6. Incubate the slides for 20 min at 20°C in a 1:10 dilution of normal rabbit serum in TBS. Do not wash after this step, but let the serum run off. Perform all further steps in a humid chamber to avoid drying of the tissue sections.
7. Incubate for 18 h at 4°C with MAb CROMA 101 in appropriate dilution in TBS. For specificity control, incubate sections accordingly with an isotype-matched Ab.
8. Wash twice with TBS for 3 min.
9. Incubate for 30 min at 20°C with Sigma B 0529, diluted 1:200 in TBS.
10. Wash twice with TBS for 3 min.
11. Prepare ABC solution from the PK-4000 (alternatively, the AK-5000) kit 30 min before use according to the product instructions.
12. Incubate the slides for 30 min at 20°C with the ABC-solution.
13. Wash three times for 10 min with TBS.
14. Incubate for 5–10 min at ca. 20°C with peroxidase or alkaline-phosphatase staining substrate until the coloured precipitate becomes visible.
15. Wash three times for 1 min with aqua dist.
16. Counterstain with haematoxylin solution for 5 s.
17. Let the slides dry, and seal them with PARAmount and a coverslip for storage.

Three colour options are used routinely in our lab:
1. **Black:**   Peroxidase combined with DAB-nickel.
2. **Brown:**   Peroxidase combined with DAB (with no nickel).
3. **Red:**   Alkaline phosphatase combined with Fuchsin.

Note: the black staining of IE1 with light haematoxylin counterstaining gives the best contrast, and the microphotographs can be documented as halftone prints. Examples are shown in Figs 3 and 8.

Staining of viral DNA by two-colour ISH

The DNA-ISH detects the viral genomic DNA within an intranuclear inclusion body formed in the late (L) phase of the virus replication cycle. Viral DNA is highly accumulated in the inclusion body, as this is

(contd.)

the site of viral DNA packaging. Cytoplasmic inclusion bodies can be detected by ISH in infected salivary gland epithelial cells (Podlech *et al.*, 1998a), as these cells, unlike other cell types, form vacuoles filled with numerous monocapsid virions for secretion into the salivary duct. As with IHC, there are different colour options for single-colour ISH. Two-colour (black and red) ISH is a method that allows to distinguish between two viruses, e.g. between two different mCMV recombinants, after co-infection. It can be used to compare the '*in vivo* replication fitness' of two viruses in the same individual host, avoiding other parameters of variance (Grzimek *et al.*, 1999).

Reagents (except those already listed in the previous Box)

- Proteinase K buffer:

a. NaCl	0.584 g	(10 mM)
b. Tris	6.057 g	(50 mM)
c. EDTA	3.722 g	(10 mM)

dissolved in 1 l aqua dist., and adjusted to pH 7.4
- Proteinase K solution:

Dissolve proteinase K (Sigma, cat. no. P 5056) in proteinase K buffer. A stock solution of 50 µg in 20 µl is diluted to 1 ml before use.
- Hybridization buffer:

HybriBuffer ISH (cat. no. R 012-050; Biognostik, Göttingen, Germany)
- Washing solution (SSC, sodium salt citrate); 20-fold stock solution:

a. NaCl	175.32 g	(3 M)
b. TSCD	88.23 g	(0.3 M)

(Tri-sodium-citrate-dihydrate, cat. no. 3580.1; Roth, Karlsruhe, Germany) ad 1 l aqua dist., adjusted to pH 7.0.
- Antibodies:
a. anti-Fluorescein, conjugated with alkaline phosphatase (Roche, cat. no. 1426338)
b. anti-Digoxigenin, conjugated with peroxidase (Roche, cat. no. 1207733)
- Conjugated nucleotides:
a. Fluorescein-12-dUTP (Roche, cat. no. 1373242)
b. Digoxigenin-11-dUTP (Roche, cat. no. 1093088)
- Rubber cement 'Fixogum' (Marabu Company, Tamm, Germany).

Generation of hybridization probes

Labelled hybridization probes are synthesized by polymerase chain reaction (PCR) with specific template and primers, using Fluorescein-12-dUTP or Digoxigenin-11-dUTP in the dNTP mix for incorporation into the DNA amplificate. For optimal staining in ISH, the PCR should be designed so as to result in a probe of >0.5 kbp (for an example and for more PCR details, see Grzimek *et al.*, 1999). Probes >10 kbp can be

(*contd.*)

derived from labelled plasmids and result in a very strong and specific signal after just a few minutes of development in substrate solution. This is the method of choice for single-colour ISH detecting one virus with highest sensitivity (Podlech *et al.*, 1998a). Shorter probes are advantageous for the discrimination of virus mutants that differ in single genes or regulatory sequences. Oligonucleotide probes give only faint staining even after an extended period of development in substrate solution. If applicable, several oligonucleotides should therefore be used as a probe cocktail.

Note: Fluorescein-labelled hybridization probes do not withstand repeated freezing and thawing and should therefore always be frozen in ready-to-use aliquots.

Procedure

1. Prepare 2-µm sections of paraffin-embedded tissue, deparaffinize, and rehydrate.
2. Digest proteins with proteinase K solution (use ca. 0.1 ml per slide) for ca. 18 min at 50°C in a humid chamber.
3. Wash with aqua dist. for 3 min.
4. Block endogenous peroxidase with blocking solution for 30 min at 20°C.
5. Wash with aqua dist. for 3 min.
6. Dehydrate with graded alcohol (50, 70, 90 and 100% isopropanol), 30 s each.
7. Prepare the hybridization solution by adding both labelled probes to the hybridization buffer (of each probe, 0.5–1 µg per ml).
8. Give ca. 5 µl of the hybridization solution onto the tissue specimen, cover it with an autoclaved coverslip, seal the edges with rubber cement, and let it dry for 1 h.
9. Denature at 95°C for 6 min, hybridize at 37°C for 18 h, and remove the sealing.
10. Wash the slides in washing solution, 5 min each in 2x SSC, 1x SSC, 0.5x SSC, and 0.1x SSC.
11. Incubate the slides for 30 min at 20°C with a mixture of the anti-Fluorescein (1 : 200 in TBS) and anti-Digoxigenin (1 : 25 in TBS) antibodies. Perform all steps in a humid chamber to avoid drying of the tissue sections.
12. Wash three times with TBS for 5 min.
13. Stain with Dako K 0625 (Fuchsin) for 10–50 min, until a red precipitate appears.
14. Wash with TBS for 5 min.
15. Stain with DAB-nickel for 5–10 min, until a black precipitate appears.
16. Wash with TBS for 5 min.
17. Counterstain with haematoxylin solution for 5 s, let the slides dry, and seal them with PARAmount and a coverslip for storage.

A particularly instructive example for the scientific application of two-colour ISH is shown in Plate 7. Enhancer swap mutants of mCMV (virus A) were generated in which the major immediate-early promoter-enhancer (MIEPE) of mCMV is replaced either by the paralogous region of hCMV (virus B) (Grzimek et al., 1999) or the analogous regulatory unit of SV40 (virus C), a virus unrelated to mCMV (Grzimek, unpublished data). The question was whether the mutants B and C can autonomously replicate in vivo in the murine host. Mice were co-infected with viruses A & B or A & C, and virus replication in the liver was analysed by two-colour ISH. In the case of A & B co-infection, intranuclear inclusion bodies in infected hepatocytes stained either red (virus A) or black (virus B), indicating that both viruses replicated autonomously. This finding allowed us to conclude that the hCMV enhancer can substitute for the mCMV enhancer for virus growth in the liver (Grzimek et al., 1999). By contrast, after A & C co-infection, intranuclear inclusion bodies that stained black (virus C) always also stained red (virus A), indicating that virus C needs virus A as a helper virus for replicating in hepatocytes in vivo.

◆◆◆◆◆◆ INTERSTITIAL CYTOMEGALOVIRUS PNEUMONIA

Interstitial pneumonia is the most frequent and most severe clinical complication of hCMV disease in BMT patients. Accordingly, much work in the mouse model of CMV disease after experimental BMT was focused on the lungs as a target organ of mCMV (Reddehase et al., 1985; Alterio de Goss et al., 1998; Holtappels et al., 1998; Holtappels et al., 2000c; Podlech et al., 2000). In addition to being a major site of acute disease, the lungs also proved to be a major site of mCMV latency and of recurrent infection after a secondary immunoablative treatment (Balthesen et al., 1993; Kurz et al., 1997; Kurz and Reddehase, 1999).

Staining of T cells and infected lung cells by two-colour IHC

Preparing the lungs for histology

1. Open the thoracic cavity and prepare the right atrium with the vena cava superior and vena cava inferior. Incise the abdominal aorta, and fix the lung tissue via the vascular tree by *in situ* perfusion with PBS, pH 7.4, containing 4% (vol/vol) formalin.
2. Dissect the lungs and distend alveolar spaces by careful instillation of the fixative into the trachea.
3. Fix the organ by incubation in the fixative for 18 h at 20°C.
4. Proceed as usual for the preparation of 2-μm paraffin sections.

Reagents (except those already listed earlier)

- Unmasking solution:
10 mM Tri-sodium-citrate-dihydrate (2.941 g ad 1 l aqua dist.), adjusted to pH 6.0.
- Dako Biotin Blocking System:
Blocking solution (cat. no. X 0590; Dako Corporation, Carpinteria, CA 93013, USA).
- Antibodies and sera:
    a. MAb rat IgG1 directed against murine CD3ε conserved cytoplasmic epitope ERPPPVPNPDYEPC (cat. no. NCL-CD3-12; Novocastra, Newcastle upon Tyne, UK).
    b. Biotinylated polyclonal goat Ig directed against rat Ig (BD Pharmingen; cat. no. 554014, formerly known as 12112D)
    c. Polyclonal goat antiserum directed against mouse IgG (Sigma, cat. no. M 5899).
    d. Alkaline phosphatase anti alkaline phosphatase (APAAP) soluble complex: mouse Mab, clone AP1B9 (IgG1), conjugated with calf intestine alkaline phosphatase (Sigma, cat. no. A 7827).

Procedure

1. Prepare 2-μm sections of paraffin-embedded tissue, deparaffinize, rehydrate, wash (aqua dist., 3 min), and partially digest proteins with trypsin solution (see earlier).
2. Perform *Heat Induced Epitope Retrieval*: boil the slides (in a microwave oven) for 5 min in unmasking solution. Let it cool down to ca. 20°C (which takes ca. 45 min) and wash with water.
3. Block endogenous peroxidase and wash with water.

Note: From here on, work in a humid chamber to avoid drying of the sections.

4. Block endogenous biotin with the Dako kit according to the suppliers' instructions.
5. Incubate the slides at 4°C for 18 h with MAb anti-CD3ε, diluted 1 : 20 in TBS.
6. Wash twice with TBS for 3 min.
7. Incubate at 20°C for 30 min with biotinylated anti-rat Ab, diluted 1 : 100 in TBS.
8. Wash twice with TBS for 3 min.
9. Incubate for 30 min at 20°C with the ABC solution of the ABC-peroxidase kit PK-4000 Standard, freshly prepared 30 min before use as described by the supplier.
10. Wash three times for 10 min with TBS.
11. Stain for 5–20 min at ca. 20°C with the peroxidase staining substrate (DAB-nickel), until a black precipitate appears.
12. Wash three times for 10 min with TBS.

(contd.)

**Animal Models:
Murine Cytomegalovirus**

13. Incubate the slides for 20 min at 20°C in a 1:10 dilution of normal rabbit serum in TBS. Do not wash after this step, but let the serum run off.
14. Incubate for 18 h at 4°C with MAb CROMA 101 (see above).
15. Wash twice with TBS for 3 min.
16. Incubate for 30 min at 20°C with Sigma M 5899, diluted 1:20 in TBS.
17. Wash twice with TBS for 3 min.
18. Incubate for 30 min at ca. 20°C with the APAAP complex, diluted 1:50 in TBS.
19. Wash twice with TBS for 3 min.
20. Stain for 10–30 min at ca. 20°C with the Fuchsin substrate kit, until a red precipitate appears.
21. Wash with TBS for 3 min, counterstain with haematoxylin for 5 s, allow the slides to dry, and seal them with PARAmount and a coverslip for storage.

Plate 8 illustrates the key results obtained by the histological analysis of mCMV pneumonia in the BMT setting. (A1, A2) compares the lung histology (standard HE staining) after BMT in the absence of infection (A1) with mCMV pneumonia characterized by widening of the alveolar septae and massive infiltration of leukocytes (A2) (from Holtappels *et al.*, 1998). (B1, B2) show an inflammatory focus with CD8 T cells (black IHC membrane staining of CD3ε in mice depleted of CD4 T cells), which confine the infection to a few infected lung cells (red IHC staining of IE1) and eventually resolve productive infection. (C1, C2) illustrate disseminated, lethal viral pneumonia with numerous infected lung cells after *in vivo* depletion of CD8 T cells with MAb YTS 169-4 and of CD4 T cells with MAb YTS 191.1 (Cobbold *et al.*, 1984). Note that disseminated pneumonia occurred also after selective depletion of CD8 T cells. Apparently, in this model, CD4 T cells are not antiviral effector cells and are not required for recruitment and antiviral function of the CD8 T cells (data from Podlech *et al.*, 2000).

◆◆◆◆◆◆ SPECIFICITY OF T CELLS IN PULMONARY INFILTRATES

Three-colour cytofluorometric analysis of pulmonary infiltrate cells (staining for T-cell receptor (TCR) α/β, CD4 and CD8) revealed a preferential recruitment of CD8 T cells to the infected lungs as well as a correlation between CD8 T-cell infiltration and control of virus replication in the lungs (Alterio de Goss *et al.*, 1998; Holtappels *et al.*, 1998). Thus, in agreement with the histological data, the infiltrating CD8 T cells were apparently protective. CD8 T cells were purified from the pulmonary infiltrates for analysing their viral antigen specificity and their functional properties.

Isolation of pulmonary CD8 T lymphocytes

According to Holt *et al.* (1985), with some modification.

Equipment and reagents

- Sterile 100-ml Erlenmeyer glass bottle with magnetic stirrer
- Three-way stopcock with a very fine tube fixed
- 50-ml syringe with infusion tube and a 0.6×30 mm cannula
- 100-μm pore Nylon mesh (cell strainer, Becton Dickinson, cat. no. 352360)
- Heparin solution:

DPBS with no Ca^{2+} and Mg^{2+} (Bio Whittaker Europe, cat. no. BE17-512F), supplemented with 10 units of Heparin per ml.

- Culture medium:
  DMEM (Gibco, cat. no. 41965-039), supplemented with 10% (vol/vol) FCS and with antibiotics.
- Enzymes:
a. DNase I (Sigma, cat. no. DN-25); stock solution: 3 mg per ml of aqua bidist., filtered sterile (0.2 μm). Concentration for use: 0.05 mg per ml.
b. Collagenase type II (Sigma, cat. no. C-6885); stock solution: 10 mg per ml in DPBS, filtered sterile (0.8 μm followed by 0.2 μm). Concentration for use: 0.7 mg per ml.
- Ficoll, with a density of 1.077 g per ml.

Procedure: perfusion of the lung

1. Fix the syringe with tube and cannula at the side wall of the clean bench, and fill it with warmed Heparin solution.
2. Sacrifice the mouse in CO_2 atmosphere, and open the thoracic and abdominal cavities: take care not to injure the lung.
3. Open the arteriae subclaviae and the aorta abdominalis.
4. Puncture the right heart chamber with the cannula and flush the Heparin solution through the lung, until its colour turns into white.

Note: Bronchoalveolar lavage (BAL) is optional, if it is intended either to collect or discard alveolar leukocytes. In that case, insert the tube of the three-way stopcock into the trachea and rinse the alveolar spaces 10 times with 1 ml of appropriate medium.

5. Excise the lungs (preferentially under stereomicroscopic control to exclude lymph nodes), and transfer them into DMEM in 50-ml tubes.

Procedure: isolation of interstitial mononuclear leukocytes

1. Transfer the lungs on a Petri dish and select parenchymal lung tissue, discarding everything else, including trachea, bronchi, and bronchioles.

(contd.)

Animal Models:
Murine Cytomegalovirus

2. Cut the lung parenchyma into very small pieces, and transfer the minced tissue to the Erlenmayer bottles (material from 3–5 lungs per bottle).
3. Per bottle, add 15 ml DMEM, 300 µl DNase and 1 ml collagenase. Stir the suspension for 1 to 1.5 h at 37°C.
4. Filter the suspension through a stainless steel mesh, resolve clumps and grind the remaining tissue with the stamp of a syringe, and swill with DMEM.
5. Centrifugate the cell suspension for 8 min at ca. 680g, and wash with DMEM by resuspension and centrifugation.
6. Resuspend the cell pellet in DMEM (material from 2–3 lungs per 8 ml) and filter the suspension through the nylon cell strainer or gauze.
7. Carefully overlay 4 ml of Ficoll in a 14-ml polystyrene tube with 8 ml of the filtered cell suspension, and centrifugate for 30 min at 760g.
8. Carefully recover the mononuclear leukocyte band at the Ficoll-medium interphase with a Pasteur pipette.
9. Wash the cells, count, and resuspend in appropriate medium for functional analyses.

Yield

The yield varies greatly depending on the experimental conditions. One can expect $0.5–1 \times 10^6$ cells from a normal lung, and up to ca. 5×10^6 cells at 1 month after BMT and infection, which is at the peak of lung infiltration.

Important notes

1. All glassware must be very clean.
2. The specific activity and batch of collagenase is a very critical parameter. Therefore, every batch needs to be tested for its suitability.

Purification of CD8 T lymphocytes

The functional assays are usually performed with CD8 T lymphocytes purified from the pulmonary mononuclear leukocyte population (Alterio de Goss et al., 1998; Holtappels et al., 2000c). Techniques of cytofluorometric or immunomagnetic cell sorting are not described here, as they are covered in detail in other chapters of this book.

Infected target cells for cytolytic assays

Cytolytic assays and the generation of target cells presenting synthetic peptides or acid-extracted naturally processed peptides present in

(contd.)

HPLC fractions were described previously (Holtappels *et al.*, 1998, 2000a, 2002a) and are dealt with in other chapters of this book. Gene expression of herpesviruses (including CMVs) is subdivided in three coordinately regulated kinetic classes, referred to as 'immediate-early' (IE), 'early' (E), and 'late' (L). In addition, exogenous loading of the MHC class-I pathway with virion proteins during the process of virus entry generates target cells in absence of viral gene expression (Reddehase and Koszinowski, 1984; Reddehase *et al.*, 1984).

- Prepare MEF in the second passage in tissue culture dishes of appropriate size.
- Infect the cells under conditions of *CEI* (see earlier Box) with 0.2 PFU per cell, which equals a multiplicity of infection of 4 PFU* per cell, so that every cell is infected.

Metabolic inhibitors

- Actinomycin D (ActD); irreversible inhibitor of transcription:
prepare a stock solution of 1 mg ActD (Sigma, cat. no. A 9415) per ml in ethanol (analytical grade). Store at −20°C in the dark.
- Cycloheximide (CH); reversible inhibitor of protein synthesis:
prepare a stock solution of 2 mg CH (Sigma, cat. no. C 6679) per ml aqua bidist, filtrate, sterile and store in aliquots at −20°C. Do not freeze again.
- Phosphonoacetic acid (PAA); irreversible inhibitor of DNA synthesis:
prepare a stock solution of 50 mg PAA (Sigma, cat. no. P 6909) per ml in aqua bidist., filtrate sterile, and store in aliquots at −20°C.

IE phase target cells (IE)

1. Replace the culture medium by fresh medium containing 50 μg CH per ml.
2. Infect the cells 15 min later.
3. At 3 h after infection, discard the CH medium and wash the cell monolayer thoroughly in medium containing 5 μg of ActD per ml.
4. After 2 h of incubation in ActD medium, trypsinize with Trypsin/EDTA solution-II (see earlier) just until the cells detach, wash the cells, and label them with $Na_2^{51}CrO_4$ for the chromium release assay. Note: It is not necessary to keep the cells in the presence of ActD during radioactive labelling and cytolytic assay.

E phase window target cells (EW)

After removal of the CH medium, cultivate the cells for a defined period in inhibitor-free culture medium before adding the ActD. This procedure allows to study the influence of E-gene expression on the presentation of IE antigens, and has led to the discovery of E-phase immune evasion genes (Del Val *et al.*, 1989; for a review, see Hengel *et al.*, 1999).

> *E phase target cells (E)*
>
> At 15 min prior to infection, replace the culture medium by medium containing 250 µg of PAA per ml. Harvest the cells 16 h after infection.
>
> *L phase target cells (L)*
>
> Infect the cells for ca. 22–24 h in absence of metabolic inhibitors.
>
> *Virion protein target cells (VP)*
>
> These are target cells that present still unidentified antigenic peptides processed from virion proteins (VP) in absence of viral gene expression.
>
> Mock-infect the cells with UV-light (254 nm, ca. 4500 J per m²; use purified virus in VSB; see earlier) inactivated virions in doses (PFU^{UV}) of up to 100-fold the dose used for infection. Alternatively, infect cells in the presence of ActD.

CD3ε-redirected ELISPOT assay

Stimulator cells in the ELISPOT assay can be infected target cells (see Box above) as well as cells pulsed with synthetic antigenic peptides or with acid-extracted naturally processed peptides present in HPLC fractions (Holtappels *et al.*, 2002a). The technique of the ELISPOT assay is covered in other chapters of this book, and our modifications of the IFN-γ-based ELISPOT assay were described in detail previously (Holtappels *et al.*, 2000a, c, 2001). An important positive control for determining the total frequency of cells in a population that is capable of responding with IFN-γ secretion is the recently described *CD3ε-redirected ELISPOT assay* (Holtappels *et al.*, 2000c, 2001). In this assay, recognition of MHC class-I presented peptide by TCR α/β is bypassed by polyclonal signalling via CD3ε. In the case of a monospecific CTL line (CTLL), the number of cells responding to the presentation of the cognate peptide and the number of cells responding to signalling via CD3ε should be the same. For the example of a CTLL specific for the IE1 peptide, referred to as IE1-CTLL, this condition was found to be fulfilled (Fig. 4.; data from Holtappels *et al.*, 2002a).

Stimulator cells

- Incubate P815-B7 transfectants (Azuma *et al.*, 1992) for 20 min at 20°C in culture medium containing MAb directed against murine CD3ε (clone 145-2C11; cat. no. 1530-14; Dianova, Hamburg, Germany). The MAb binds to Fc receptors on P815-B7. A plateau response is reached at ca. 0.6 µg per 10⁵ cells (Holtappels *et al.*, 2001), but is somewhat batch-dependent. Wash out excess antibody before use.

Alternative (Holtappels *et al.*, 2001):
- Directly use the 145-2C11 hybridoma (Leo *et al.*, 1987; ATCC no. CRL-1975).

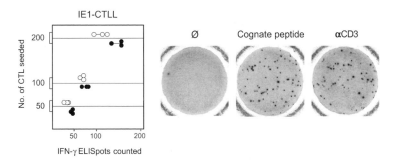

Figure 4. Peptide-specific and CD3ε-redirected IFN-γ-based ELISPOT assays. IE1-CTLL cells specific for mCMV IE1 peptide [168]YPHFMPTNL[176] (see Table 1) were stimulated either with P815-B7 cells pulsed with the cognate peptide (solid circles) or with 145-2C11 hybridoma cells that produce MAb directed against murine CD3ε (open circles). Left: data of triplicate assay cultures for three cell numbers seeded. Right: photodocumentation of representative ELISPOT filters with 100 IE1-CTL seeded. Ø, P815-B7 cells with no peptide added (from Holtappels *et al.*, 2002a).

In acute mCMV infection of immunocompetent BALB/c mice, CD8 T cells present in draining lymph nodes are not cytolytically active, unless they are expanded by a 5–7 day period of cultivation in the presence of IL-2 (Reddehase *et al.*, 1984). By contrast, CD8 T cells isolated from infected lungs after BMT include *ex vivo* cytolytic effector CTL (Fig. 5A). With a

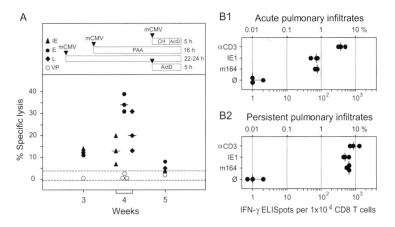

Figure 5. Specificity of CD8 T cells in pulmonary infiltrates. (A) *Ex vivo* cytolytic activity of lung infiltrate cells isolated at the indicated time points after BMT and mCMV infection. Triplicate data for an effector/target cell ratio of 200, with the median value marked. For the target cells, see Box above (from Holtappels *et al.*, 1998). (B1 and B2) ELISPOT frequencies of CD8 T cells purified from acute (1 month) and persistent (3 months) pulmonary infiltrates, respectively. Dots represent triplicate data (median value marked) from the linear portion of a titration, extrapolated to 10 000 CD8 T cells. αCD3, polyclonal stimulation with the 145-2C11 hybridoma. IE1 and m164, stimulation with P815-B7 cells pulsed with the respective peptides (see Table 1). Ø, P815-B7 cells with no peptide added (from Holtappels *et al.*, 2002a).

peak at 4 weeks after BMT and infection, these CTL preferentially lysed target cells in the E phase of the virus replication cycle (data from Holtappels *et al.*, 1998), in spite of the 'immune evasion functions' of mCMV that are expressed in the E phase (Hengel *et al.*, 1999). ELISPOT assays revealed similar frequencies of CD8 T cells specific for the immunodominant IE1 peptide (Reddehase *et al.*, 1989) and for the more recently identified co-dominant E-phase peptide m164 (Fig. 5B1; data from Holtappels *et al.*, 2002a). Both specificities were found to be enriched among CD8 T cells isolated from persisting pulmonary infiltrates during latent infection of the lungs (Fig. 5B2; Holtappels *et al.*, 2000c, 2002a).

◆◆◆◆◆◆ CYTOIMMUNOTHERAPY WITH CTL-LINES

All evidence presented so far has pointed to a protective role of the CD8 T cells in the pulmonary infiltrates. The most direct approach to demonstrate protective antiviral activity is the adoptive transfer of antiviral immunity by transfer of the cells in question. CD8 T cells isolated from the pulmonary infiltrates during acute as well as during latent infection indeed proved to be capable of resolving CMV organ infection in indicator recipients (Alterio de Goss *et al.*, 1998; Podlech *et al.*, 2000). Indicator recipients are mice that have received immunoablative treatment and intraplantar infection (see earlier Box), but no reconstitution by BMT. Under such a condition, mice die of multiple-organ CMV disease between days 12 and 18, unless intravenously transferred cells protect against infection (Reddehase *et al.*, 1985). This assay provides a model for a pre-emptive cytoimmunotherapy of CMV disease with peptide-specific CTLL, and can verify the *in vivo* relevance of a particular antigenic peptide in antiviral immunity.

Table 1 gives a list of all currently known antigenic peptides of mCMV. All peptides were used to generate CTLL by repetitive *in vitro* restimulation of memory CD8 T cells present in the spleen during viral latency after clearance of the acute infection. The methods for generating clonal and polyclonal, peptide-specific CTLL were described in detail previously (Reddehase *et al.*, 1987; Holtappels *et al.*, 2001, 2002a). Figure 6 emphasizes

Table 1. List of currently known antigenic peptides of mCMV (January 2002)

ORF	Replication phase	Peptide sequence	MHC class-I restriction	Reference
m04	E	[243]YGPSLYRRF[251]	D^d	Holtappels *et al.* (2000a)
m18	E	[346]SGPSRGRII[254]	D^d	Holtappels *et al.* (2002b)
M83	E/L	[761]YPSKEPFNF[769]	L^d	Holtappels *et al.* (2001)
M84	E	[297]AYAGLFTPL[305]	K^d	Holtappels *et al.* (2000b)
m123ex4 (ie1)	IE	[168]YPHFMPTNL[176]	L^d	Reddehase *et al.* (1989)
m164	E	[257]AGPPRYSRI[265]	D^d	Holtappels *et al.* (2002a)

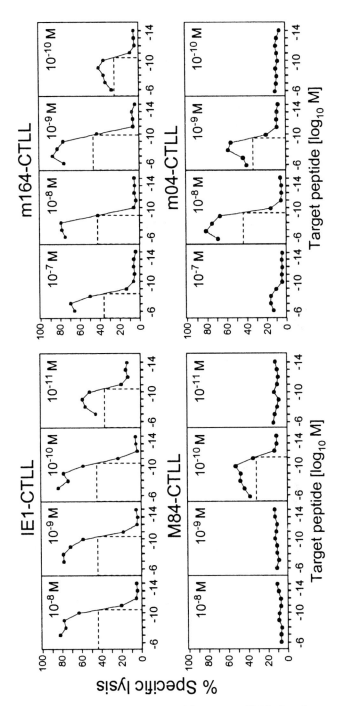

**Animal Models:
Murine Cytomegalovirus**

Figure 6. Conditions for the generation of long-term CTLL. For the peptides, see Table 1. Molar concentrations of peptides used for the restimulation of memory CD8 T cells are indicated in each panel. After three rounds of restimulation, the cytolytic assay was performed at an effector/target cell ratio of 15, with P815 target cells that were pulsed with the indicated molar concentrations (abscissa) of the corresponding peptides. Dashed lines show the target peptide concentration at which the lysis was half-maximal (from Holtappels *et al.*, 2000a, 2002a).

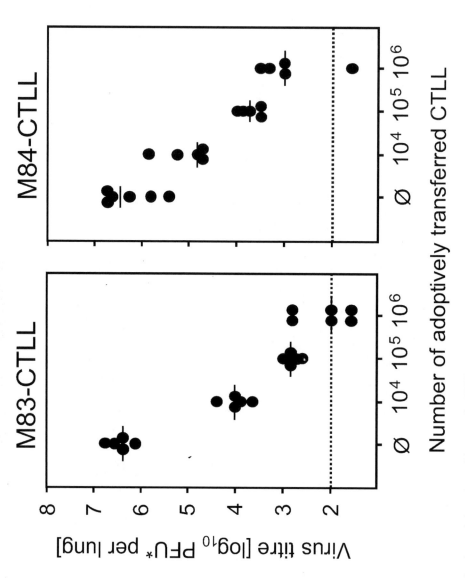

Figure 7. Antiviral *in vivo* activity of peptide-specific CTLL upon cell transfer. Graded numbers of cells from M83-CTLL and M84-CTLL were transferred intravenously to immunocompromised indicator recipients on the day of infection. $\varnothing$, no cell transfer. Virus titres in the lungs were determined on day 12 by the virus plaque assay under conditions of *CEI*. Dots represent titres in individual mice, with the median value marked. The dashed line indicates the detection limit (from Holtappels *et al.*, 2001).

the importance of the molar concentration of the peptide during the restimulations. Notably, the peptide concentration can be not only sub-optimal but also supraoptimal. IE1-CTLL and m164-CTLL tolerated a wide range of peptide concentrations, whereas the generation of M84-CTLL and m04-CTLL was delicately dependent upon peptide concentration. This is an important note, as it may explain failures in raising CTLL. As shown by the molar concentrations of peptide required for target cell formation, this feature of CTLL does not reflect peptide affinity to the presenting MHC class-I molecule, but rather the affinity of the TCR to the MHC–peptide complex.

All mCMV peptide-specific CTLL that we have tested so far were efficient in controlling mCMV infection in indicator recipients upon cell transfer. Figure 7 shows a virus plaque assay (see earlier Box) comparing

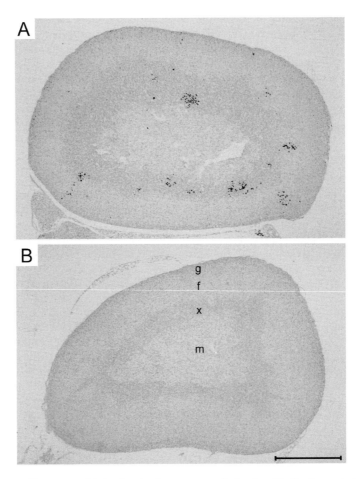

Animal Models:
Murine Cytomegalovirus

Figure 8. Clearance of infection in the suprarenal glands of indicator recipients by pre-emptive cytoimmunotherapy with the M83-CTLL. (A) No cell transfer. (B) transfer of 100 000 CTL. Shown are whole-organ sections with black IHC (ABC-peroxidase/DAB-nickel) staining of the IE1 protein in the nuclei of infected cells (see Fig. 3). (g) zona glomerulosa, (f) zona fasciculata, (x) X zone or juxtamedullary cortex, (m) medulla. Bar marker: 0.5 mm (from Holtappels *et al.*, 2001).

the capacities of M83-CTLL and M84-CTLL in reducing virus replication in the lungs. In Fig. 8, IE1 protein-specific IHC (see earlier Box) is used to illustrate the antiviral effect of M83-CTLL in whole-organ sections of the suprarenal glands (data from Holtappels *et al.*, 2001).

References

Alterio de Goss, M., Holtappels, R., Steffens, H. P., Podlech, J., Angele, P., Dreher, L., Thomas, D. and Reddehase, M. J. (1998). Control of cytomegalovirus in bone marrow transplantation chimeras lacking the prevailing antigen-presenting molecule in recipient tissues rests primarily on recipient-derived CD8 T cells. *J. Virol.* **72**, 7733–7744.

Andrews, D. M., Andoniou, C. D., Granucci, F., Ricciardi-Castagnoli, P. and Degli-Esposti, M. A. (2001). Infection of dendritic cells by murine cytomegalovirus induces functional paralysis. *Nat. Immunol.* **2**, 1077–1084.

Azuma, M., Cayabyab, M., Buck, D., Philipps, J. H. and Lanier, L. L. (1992). CD28 interaction with B7 costimulates primary allogeneic proliferative responses and cytotoxicity mediated by small, resting T lymphocytes. *J. Exp. Med.* **175**, 353–360.

Balthesen, M., Messerle, M. and Reddehase, M. J. (1993). Lungs are a major organ site of cytomegalovirus latency and recurrence. *J. Virol.* **67**, 5360–5366.

Busch, F. W., Mutter, W., Koszinowski, U. H. and Reddehase, M. J. (1991). Rescue of myeloid-lineage-committed preprogenitor cells from cytomegalovirus-infected bone marrow stroma. *J. Virol.* **65**, 981–984.

Cobbold, S. P., Jayasuriya, A., Nash, A., Prospero, T. D. and Waldmann, H. (1984). Therapy with monoclonal antibodies by elimination of T-cell subsets in vivo. *Nature (Lond.)* **312**, 548–550.

Del Val, M., Münch, M., Reddehase, M. J. and Koszinowski, U. H. (1989). Presentation of CMV immediate-early antigen to cytolytic T lymphocytes is selectively prevented by viral genes expressed in the early phase. *Cell* **58**, 305–315.

Ebeling, A., Keil, G. M., Knust, E. and Koszinowski, U. H. (1983). Molecular cloning and physical mapping of murine cytomegalovirus DNA. *J. Virol.* **47**, 421–433.

Grzimek, N. K. A., Podlech, J., Steffens, H. P., Holtappels, R., Schmalz, S. and Reddehase, M. J. (1999). In vivo replication of recombinant murine cytomegalovirus driven by the paralogous major immediate-early promoter-enhancer of human cytomegalovirus. *J. Virol.* **73**, 5043–5055.

Hanson, L. K., Slater, J. S., Karabekian, Z., Ciocco-Schmitt, G. and Campbell, A. E. (2001). Products of US22 genes M140 and M141 confer efficient replication of murine cytomegalovirus in macrophages and spleen. *J. Virol.* **75**, 6292–6302.

Hengel, H., Reusch, U., Gutermann, A., Ziegler, H., Jonjic, S., Lucin, P. and Koszinowski, U. H. (1999). Cytomegaloviral control of MHC class I function in the mouse. *Immunol. Rev.* **168**, 167–176.

Hengel, H., Reusch, U., Geginat, G., Holtappels, R., Ruppert, T., Hellebrand, E. and Koszinowski, U. H. (2000). Macrophages escape inhibition of major histocompatibility complex class I-dependent antigen presentation by cytomegalovirus. *J. Virol.* **74**, 7861–7868.

Holt, P. G., Degebrodt, A., Venaille, T., O'Leary, C., Krska, K., Flexman, J., Farrell, H., Shellam, G., Young, P., Penhale, J., Robertson, T. and Papadimitriou, J. M. (1985). Preparation of interstitial lung cells by enzymatic digestion of tissue slices: preliminary characterization by morphology and performance in functional assays. *Immunology* **54**, 139–147.

Holtappels, R., Podlech, J., Geginat, G., Steffens, H. P., Thomas, D. and Reddehase, M. J. (1998). Control of murine cytomegalovirus in the lungs: relative but not absolute immunodominance of the immediate-early 1 nonapeptide during the antiviral cytolytic T-lymphocyte response in pulmonary infiltrates. *J. Virol.* **72**, 7201–7212.

Holtappels, R., Thomas, D., Podlech, J., Geginat, G., Steffens, H. P. and Reddehase, M. J. (2000a). The putative natural killer decoy early gene m04 (gp34) of murine cytomegalovirus encodes an antigenic peptide recognized by protective antiviral CD8 T cells. *J. Virol.* **74**, 1871–1884.

Holtappels, R., Thomas, D. and Reddehase, M. J. (2000b). Identification of a K(d)-restricted antigenic peptide encoded by murine cytomegalovirus early gene M84. *J. Gen. Virol.* **81**, 3037–3042.

Holtappels, R., Pahl-Seibert, M. F., Thomas, D. and Reddehase, M. J. (2000c). Enrichment of immediate-early 1 (m123/pp89) peptide-specific CD8 T cells in a pulmonary CD62L(lo) memory-effector cell pool during latent murine cytomegalovirus infection of the lungs. *J. Virol.* **74**, 11495–11503.

Holtappels, R., Podlech, J., Grzimek, N. K. A., Thomas, D., Pahl-Seibert, M. F. and Reddehase, M. J. (2001). Experimental preemptive immunotherapy of murine cytomegalovirus disease with CD8 T cell lines specific for ppM83 and pM84, the two homologs of human cytomegalovirus tegument protein ppUL83 (pp65). *J. Virol.* **75**, 6584–6600.

Holtappels, R., Thomas, D., Podlech, J. and Reddehase, M. J. (2002a). Two antigenic peptides from genes m123 and m164 of murine cytomegalovirus quantitatively dominate CD8 T-cell memory in the H-2(d) haplotype. *J. Virol.* **76**, 151–164.

Holtappels, R., Grzimek, N. K. A., Thomas, D. and Reddehase, M. J. (2002b). Early gene m18, a novel player in the immune response to murine cytomegalovirus. *J. Gen. Virol.* **83**, 311–316.

Hudson, J. B., Misra, V. and Mosmann, T. R. (1976). Cytomegalovirus infectivity: analysis of the phenomenon of centrifugal enhancement of infectivity. *Virology* **72**, 235–243.

Kurz, S. K. and Reddehase, M. J. (1999). Patchwork pattern of transcriptional reactivation in the lungs indicates sequential checkpoints in the transition from murine cytomegalovirus latency to recurrence. *J. Virol.* **73**, 8612–8622.

Kurz, S. K., Steffens, H. P., Mayer, A., Harris, J. R. and Reddehase, M. J. (1997). Latency versus persistence or intermittent recurrences: evidence for a latent state of murine cytomegalovirus in the lungs. *J. Virol.* **71**, 2980–2987.

Leo, O., Foo, M., Sachs, D. H., Samelson, L. E. and Bluestone, J. A. (1987). Identification of a monoclonal antibody specific for a murine T3 polypeptide. *Proc. Natl. Acad. Sci. USA* **84**, 1374–1378.

Lutarewych, M. A., Quirk, M. R., Kringstad, B. A., Li, W., Verfaillie, C. M. and Jordan, M. C. (1997). Propagation and titration of murine cytomegalovirus in a continuous bone marrow-derived stromal cell line (M2-10B4). *J. Virol. Meth.* **68**, 193–198.

Mayer, A., Podlech, J., Kurz, S. K., Steffens, H. P., Maiberger, S., Thalmeier, K., Angele, P., Dreher, L. and Reddehase, M. J. (1997). Bone marrow failure by cytomegalovirus is associated with an in vivo deficiency in the expression of essential stromal hemopoietin genes. *J. Virol.* **71**, 4589–4598.

Mercer, J. A., Marks, J. R. and Spector, D. H. (1983). Molecular cloning and restriction endonuclease mapping of the murine cytomegalovirus genome (Smith strain). *Virology* **129**, 94–106.

Messerle, M., Crnkovic, I. Hammerschmidt, W., Ziegler, H. and Koszinowski, U. H. (1997). Cloning and mutagenesis of a herpesvirus genome as an infectious bacterial artificial chromosome. *Proc. Natl. Acad. Sci. USA* **94**, 14759–14763.

Plachter, B., Sinzger, C. and Jahn, G. (1996). Cell types involved in replication and distribution of human cytomegalovirus. *Adv. Virus Res.* **46**, 195–261.

Podlech, J., Holtappels, R., Wirtz, N., Steffens, H. P. and Reddehase, M. J. (1998a). Reconstitution of CD8 T cells is essential for the prevention of multiple-organ cytomegalovirus histopathology after bone marrow transplantation. *J. Gen. Virol.* **79**, 2099–2104.

Podlech, J., Steffens, H. P., Holtappels, R., Mayer, A., Alterio de Goss, M., Oettel, O., Wirtz, N., Maiberger, S. and Reddehase, M. J. (1998b). Cytomegalovirus pathogenesis after experimental bone marrow transplantation. *Monogr. Virol.* **21**, 119–128.

Podlech, J., Holtappels, R., Pahl-Seibert, M. F., Steffens, H. P. and Reddehase, M. J. (2000). Murine model of cytomegalovirus interstitial pneumonia in syngeneic bone marrow transplantation: persistence of protective pulmonary CD8-T-cell infiltrates after clearance of acute infection. *J. Virol.* **74**, 7496–7507 (includes cover photograph).

Polic, B., Hengel, H., Krmpotic, A., Trgovcich, J., Pavic, I., Lucin, P., Jonjic, S. and Koszinowski, U. H. (1998). Hierarchical and redundant lymphocyte subset control precludes cytomegalovirus replication during latent infection. *J. Exp. Med.* **188**, 1047–1054.

Rawlinson, W. D., Farrell, H. E. and Barrell, B. G. (1996). Analysis of the complete DNA sequence of murine cytomegalovirus. *J. Virol.* **70**, 8833–8849.

Reddehase, M. J. (2000). The immunogenicity of human and murine cytomegaloviruses. *Curr. Opin. Immunol.* **12**, 390–396, 738.

Reddehase, M. J. and Koszinowski, U. H. (1984). Significance of herpesvirus immediate early gene expression in cellular immunity to cytomegalovirus infection. *Nature (Lond.)* **312**, 369–371.

Reddehase, M. J., Keil, G. M. and Koszinowski, U. H. (1984). The cytolytic T lymphocyte response to the murine cytomegalovirus. II. Detection of virus replication stage-specific antigens by separate populations of in vivo active cytolytic T lymphocyte precursors. *Eur. J. Immunol.* **14**, 56–61.

Reddehase, M. J., Weiland, F., Münch, K., Jonjic, S., Lüske, A. and Koszinowski, U. H. (1985). Interstitial murine cytomegalovirus pneumonia after irradiation: characterization of cells that limit viral replication during established infection of the lungs. *J. Virol.* **55**, 264–273.

Reddehase, M. J., Zawatzky, R., Weiland, F., Bühring, H.-J., Mutter, W. and Koszinowski, U. H. (1987). Stable expression of clonal specificity in murine cytomegalovirus-specific large granular lymphoblast lines propagated long-term in recombinant interleukin 2. *Immunobiol.* **174**, 420–431.

Reddehase, M. J., Rothbard, J. B. and Koszinowski, U. H. (1989). A pentapeptide as minimal antigenic determinant for MHC class I-restricted T lymphocytes. *Nature (Lond.)* **337**, 651–653.

Reddehase, M. J., Balthesen, M., Rapp, M., Jonjic, S., Pavic, I. and Koszinowski, U. H. (1994). The conditions of primary infection define the load of latent viral genome in organs and the risk of recurrent cytomegalovirus disease. *J. Exp. Med.* **179**, 185–193.

Riddell, S. R., Watanabe, K. S., Goodrich, J. M., Li, C. R., Agha, M. E. and Greenberg, P. D. (1992). Restoration of viral immunity in immunodeficient humans by adoptive transfer of T cell clones. *Science* **257**, 238–241.

Smith, M. G. (1954). Propagation of salivary gland virus of the mouse in tissue cultures. *Proc. Soc. Exp. Biol. Med.* **86**, 435–440.

Steffens, H. P., Kurz, S. K., Holtappels, R. and Reddehase, M. J. (1998a).

Preemptive CD8 T-cell immunotherapy of acute cytomegalovirus infection limits the burden of latent viral genomes and reduces the risk of virus recurrence. *J. Virol.* **72**, 1797–1804.

Steffens, H. P., Podlech, J., Kurz, S. K., Angele, P., Dreis, D. and Reddehase, M. J. (1998b). Cytomegalovirus inhibits the engraftment of donor bone marrow cells by downregulation of hemopoietin gene expression in recipient stroma. *J. Virol.* **72**, 5006–5015.

Stoddart, C. A., Cardin, R. D., Boname, J. M., Manning, W. C., Abenes, G. B. and Mocarski, E. S. (1994). Peripheral blood mononuclear phagocytes mediate dissemination of murine cytomegalovirus. *J. Virol.* **68**, 6243–6253.

Weiland, F., Keil, G. M., Reddehase, M. J. and Koszinowski, U. H. (1986). Studies on the morphogenesis of murine cytomegalovirus. *Intervirology* **26**, 192–201.

Copyright declaration

Data were reproduced, mostly in modified form and rearranged, from previously published work of the authors with the permission of the *American Society for Microbiology* (Journal of Virology).

Contact

Matthias.Reddehase@uni-mainz.de   (for general questions)
Podlech@mail.uni-mainz.de   (for histology and pathology)
Holtappe@mail.uni-mainz.de   (for cellular immunology)
Grzimek@mail.uni-mainz.de   (for molecular virology)

Animal Models: Murine Cytomegalovirus

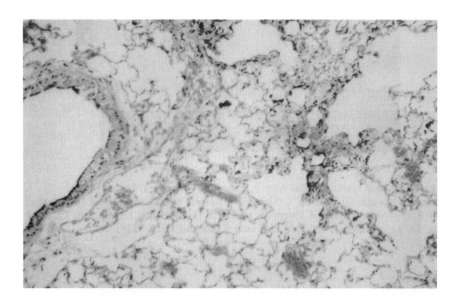

Plate 1. Murine lung showing the distribution of intranasally administered India ink. H & E, ×320.

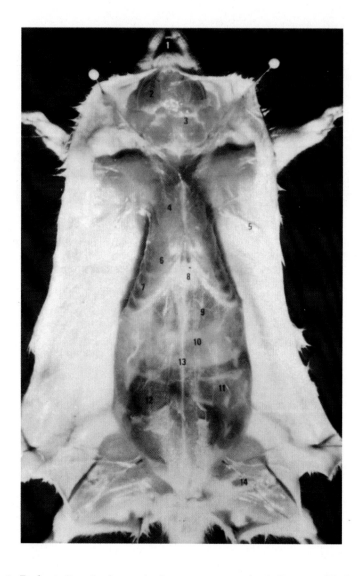

Plate 2. Rodent sites: 1, rima oris; 2, masseter muscle; 3, Lc. mandibulare, Gl. mandibularis and Gl. parotis; 4, pectoral muscle; 5, cutaneous nerves; 6, thorax; 7, ribs; 8, xiphoid cartilage; 9, liver; 10, stomach; 11, intestine; 12, caecum; 13, linea alba; 14, saphenus nerve, artery and vein (Olds and Olds, 1979).

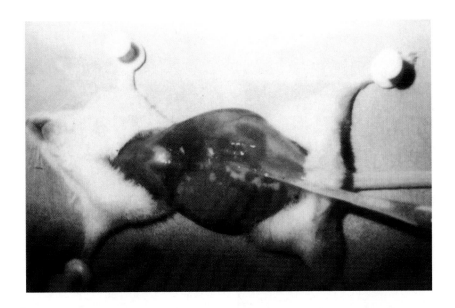

Plate 3. Lavage of the peritoneal cavity (Versteeg, 1985).

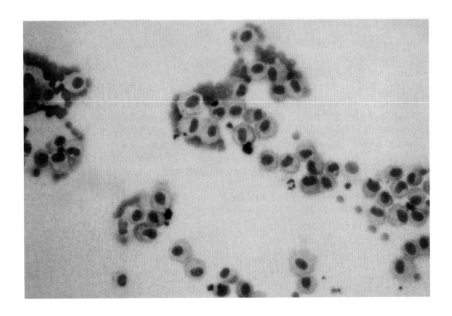

Plate 4. Typical cellular composition of bronchoalveolar fluid. Diff-Quick stain, ×640.

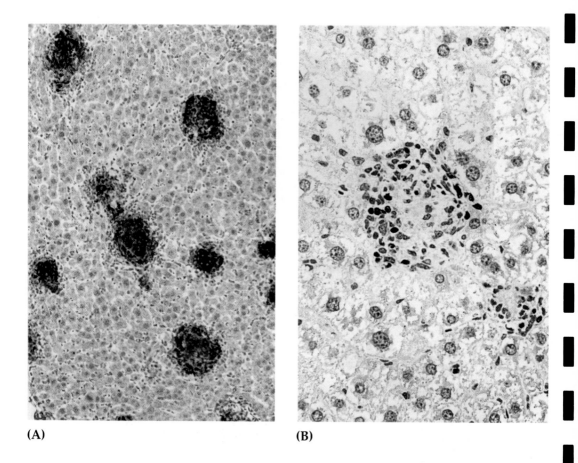

(A) (B)

Plate 5. (A) Expression of inducible nitric oxide synthase (iNOS) in granulomas of C57BL/6 mice intravenously infected with 2×10^5 cfu *Mycobacterium avium* 6 weeks previously. As primary antibody, a polyclonal rabbit-anti-NOS antiserum (Genzyme) was used at a 1:1000 dilution. Pressure cooking of paraffin-embedded sections was performed for 1 min. (B) Expression of Ki-67 (proliferation-associated) antigen in granulomas of C57BL/6 mice intravenously infected with 2×10^5 cfu *M. avium* 4 weeks previously. As primary antibody, a polyclonal rabbit-anti-mouse-Ki-67-equivalent antiserum (kindly provided by J. Gerdes, Borstel) was used at a 1:1000 dilution. Microwaving of paraffin-embedded sections was performed five consecutive times for 5 min.

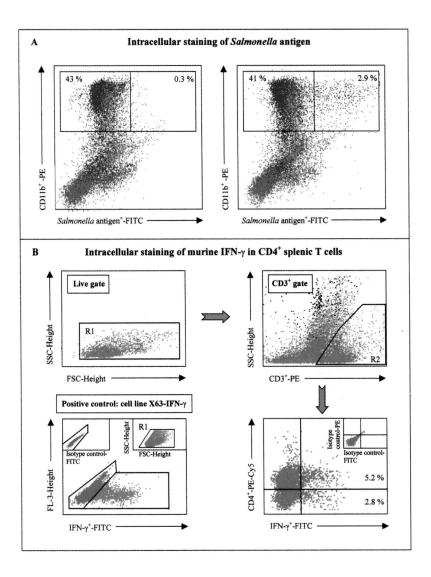

Plate 6. Intracellular fluorochrome staining

A. Intracellular *Salmonella* antigen (comprising viable and killed/processed intracellular salmonellae) was stained using a polyclonal rabbit anti-*Salmonella* Enteritidis antiserum as primary antibody and a FITC-labelled goat anti-rabbit immunoglobulin F(ab)$_2$ fragment (Jackson ImmunoResearch) as secondary antibody. Phagocytes, including macrophages and granulocytes, were identified by staining with the PE-labelled anti-CD11b mAb M1/70.15 (Caltag) simultaneously to the secondary staining of the *Salmonella* antigen.

B. Isolated splenocytes were stained simultaneously with mAb against IFN-γ (FITC-labelled), CD3 (PE-labelled), and a biotinylated mab against CD4 (Streptavidin-PE-Cy5-labelled; all reagents from BD-Pharmingen). T cells were discriminated via CD3$^+$ gating (upper right). The IFN-γ-transfected cell line X63-IFN-γ was used as a positive control (lower left).

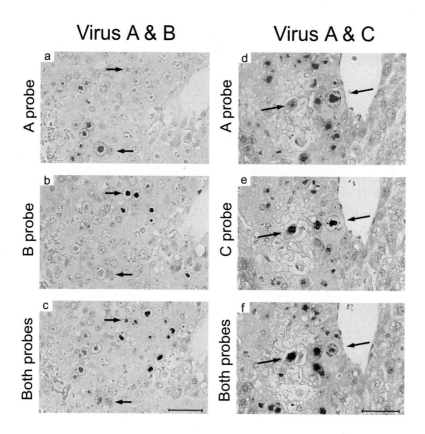

Plate 7. Analysis of *in vivo* virus replication by two-colour *in situ* DNA hybridization (DNA-ISH). Virus A, wildtype mCMV. Virus B, enhancer swap mutant with the mCMV enhancer replaced by the hCMV enhancer. Virus C, enhancer swap mutant with the mCMV enhancer replaced by the SV40 enhancer. Mice were co-infected with viruses A&B or A&C. Shown are serial (2-µm distance) sections of the co-infected livers, hybridized with labelled probes specific for the respective enhancer elements in the three viruses. (A probe) red. (B and C probes) black. Counterstaining with haematoxylin. Arrows point to infected hepatocytes of interest. Bar markers represent 50 µm. For the scientific interpretation see main body of the text in the chapter 'Animal models: murine cytomegalovirus' and accompanying Box for the technique (in part from Grzimek *et al.*, 1999).

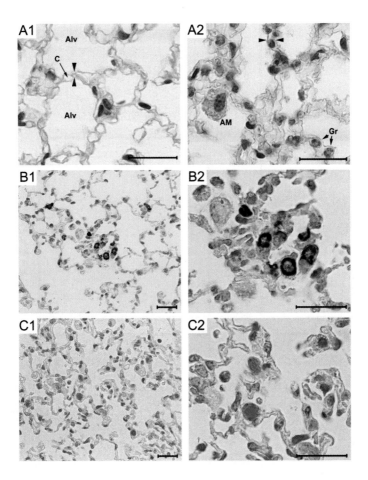

Plate 8. Analysis of interstitial cytomegalovirus pneumonia by two-colour immunohistochemistry (IHC). (A1) HE histology after BMT. Alv, alveoli; C, capillary. Arrowheads mark an alveolar septum. (A2) HE histology after BMT and mCMV infection. AM, alveolar macrophage; Gr, granulocytes. Arrowheads highlight a widened alveolar septum with interstitial mononuclear leukocytes (from Holtappels *et al.*, 1998). (B1) Inflammatory focus with CD8 T cells (black IHC membrane staining of CD3ε) and infected lung cells (red IHC staining of intranuclear viral IE1), seen in BMT recipients depleted of CD4 T cells. (B2) B1 resolved to greater detail. (C1) Disseminated viral pneumonia with numerous infected lung cells, seen after depletion of CD4 and CD8 T cells. (C2) C1 resolved to greater detail. Bar markers represent 25 μm (from Podlech *et al.*, 2000). For the scientific interpretation see main body of the text in the chapter 'Animal models: murine cytomegalovirus' and accompanying Box for the technique.

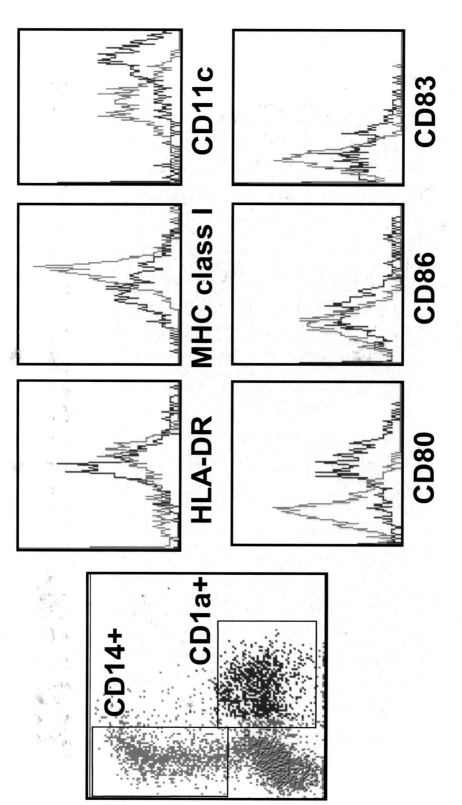

Plate 9. CD34-HPC derived DC vaccine. G-CSF mobilized blood CD34+ HPC cultured under clinical grade conditions with GM-CSF, TNF and FLT3 ligand give rise to two subsets: CD1a+ DC (blue lines) which contain Langerin + cells and CD14+ DC (red lines) which contain DC-SIGN+ cells.

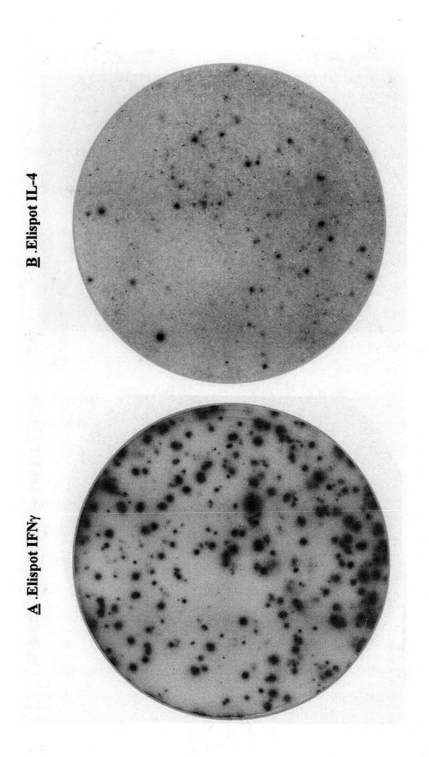

Plate 10. Single cytokine ELISPOT. 10^5 PBMC were stimulated with 1 ng ml^{-1} PMA and 500 ng ml^{-1} ionomycin for 15 h. IFN-γ (A) and IL-4 (B) spots were revealed by immuno-enzymatic method using HRP and alkline phosphatase, respectively. Spots were read on a Carl Zeiss vision system.

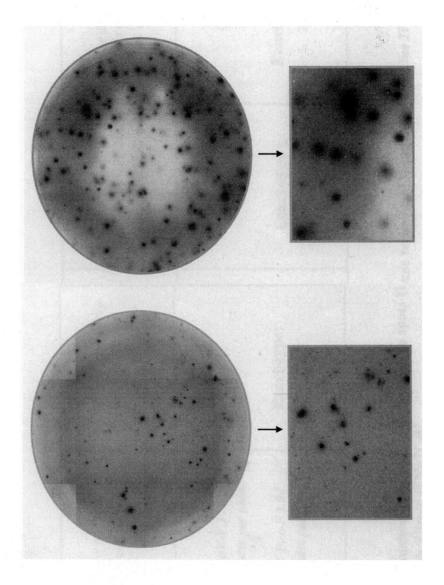

Plate 11. Dual colour Eli-spot.
Top panel. 5.10⁴ PBMC were mixed with 2.5.10⁴ Th2 cell clone. Cells were stimulated for 15 h in the presence of 1 ng ml⁻¹ PMA and 500 ng ml⁻¹ ionomycin. Red and blue/purple spots indicate IFN-γ-producing and IL-4-producing cells, respectively.
Bottom panel. 5.10⁴ PBMC were stimulated for 15 h in the presence of 1 ng ml⁻¹ PMA and 500 ng ml⁻¹ ionomycin. Red and blue/purple spots indicate IFN-γ/IL-10-producing cells, respectively. Spots were read on a Carl Zeiss vision system.

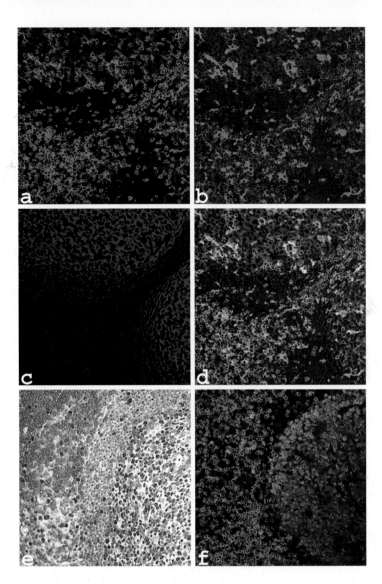

Plate 12. Immunofluorescent and immunoenzymatic detection of antigens in human tissue

(a)–(d) Multicolour immunofluorescence staining of tonsillar tissue. A cryosection of human tonsillar tissue was stained in a combination of indirect and direct immuonfluorescence staining procedures. The analysis was performed by confocal laser scanning microscopy. (a) Detection of CD4 (red); (b) detection of CD11c (green); (c) detection of DRC1 (blue) and (d) merged image of red, green and blue fluorescence. Two adjacent germinal centres become apparent by staining the network built by follicular dendritic cells appearing in blue (c). The dense network of the germinal centre is surrounded by the less dense mantle zone which forms the boundary to the interdigitating regions. The small CD4-positive T-lymphocytes appear in red (d). Cells positive for both CD4 (a) and CD11c (b) (macrophages and dendritic cells) appear in yellow due to super-position of the colours red and green (d).

(e)–(f) Comparison of double-staining techniques for the proliferation marker Ki-67 and the T-cell marker CD3 in human hyperplastic tonsil tissue

(e) Expression of the Ki-67 antigen is detected by a Ki-67 monoclonal antibody (MIB-1; Dianova) and PO-conjugated secondary reagents. CD3 expression is shown as detected by a rabbit antiserum (DAKO) and visualization with AP-conjugated secondary antibody and enzyme reagents. The highest expression of the Ki-67 antigen is seen in the germinal centre and CD3 positive T cells are found in great abundance in the marginal T-cell zone. Expression of these two antigens appears to be exclusive.

(f) Expression of the Ki-67 antigen in red fluorescence as detected by the Alexa-568-conjugated MIB-1 antibody (prepared in this laboratory) and CD3 expression in green fluorescence as detected by the CD3-antiserum (DAKO) and FITC-conjugated goat-anti-rabbit antiserum (Dianova). Analysis with confocal laser scanning microscopy demonstrates the proliferating cell in the germinal centre with a few non-proliferating T cells, and the majority of the T cells in the marginal zone demonstrating the absence of CD3 in the majority of the proliferating cells.

4.4 DNA Vaccines: Fundamentals and Practice

Maripat Corr, Delphine J Lee and Eyal Raz
Division of Rheumatology, Allergy and Immunology, and The Sam and Rose Stein Institute for Research on Aging, University of California, USA

◆◆◆

CONTENTS

The science of gene immunization
The principles of DNA immunization
The pros and cons of DNA vaccination
Applications
Concluding remarks

◆◆◆◆◆◆ THE SCIENCE OF GENE IMMUNIZATION

Introduction

The surprising finding that a gene encoded on plasmid (pDNA) injected intramuscularly (i.m.) could be expressed *in vivo* (Wolff *et al.*, 1990) opened an entirely new approach to vaccine development. Since that discovery, an explosion of studies has embarked upon a number of applications using direct plasmid injection including prevention of infectious disease, and treatment of cancer and allergic diseases. The potent T-cell mediated responses to DNA immunization can be attributed to the *in vivo* expression of the encoded antigen as well as the specific immunostimulatory sequences (ISS) in the non-coding regions of the vector. The cellular and molecular mechanisms of these effects on the immune response to DNA vaccination followed by practical considerations of DNA immunization will be discussed.

The use of injecting pDNA encoding a protein antigen as an immunogen is referred to as genetic vaccination, genetic immunization, and DNA or gene vaccination. The pDNA is circular, with several elements (Fig. 1). An origin of replication, along with an antibiotic resistance gene allows the plasmid to be replicated and produced in high quantities in *E. coli* under selective pressure. Expression of the gene encoding the antigen is usually under the control of a strong viral or eukaryotic promoter, such as the human cytomegalovirus (HCMV) E1 promoter. The appropriate termination and polyadenylation sequences follow the antigen gene

METHODS IN MICROBIOLOGY, VOLUME 32
ISBN 0–12–521532–0

Copyright © Elsevier Science Ltd
All rights of reproduction in any form reserved

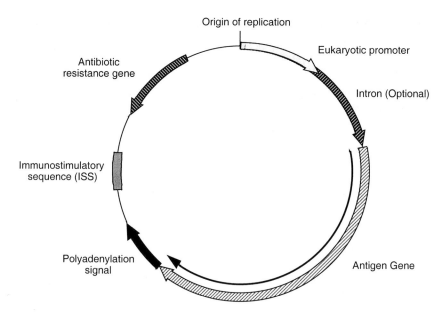

Figure 1. Schematic drawing of a typical plasmid used for DNA vaccines. The plasmid contains an origin of replication, an antibiotic resistance gene, and a sequence that encodes the antigen of interest. The gene encoding the antigen is preceded by a strong eukaryotic promoter and an optional intron sequence and is followed by termination and polyadenylation signal sequences.

sequence so that any mammalian cell that takes up the plasmid can express the protein.

Delivery of the DNA can be performed in a variety of ways. While most studies have used i.m. and intradermal (i.d.) routes of immunization, other routes of injection such as intravenous, intranasal, subcutaneous and intraperitoneal methods have been tested (Fynan *et al.*, 1993). The plasmid may simply be dissolved in normal saline and injected intradermally (Raz *et al.*, 1994) or intramuscularly (Wolff *et al.*, 1990). Alternatively, a 'gene gun' can be used to transduce skin cells with gold microprojectiles coated with pDNA encoding the antigen (also known as particle-mediated gene delivery technology) (Tang *et al.*, 1992; Haynes *et al.*, 1994; Hui *et al.*, 1994; Johnston and Tang, 1994; Pertmer *et al.*, 1995). The delivery of pDNA has also been enhanced by encapsulation with liposomes (Felgner *et al.*, 1995; Wheeler *et al.*, 1996; Ishii *et al.*, 1997; Okada *et al.*, 1997; Bennett *et al.*, 1998). These different methods of injection elicit slightly different immune responses and will be discussed further below (Table 1).

The immune response to pDNA vaccination

Immunization with plasmid DNA results in potent humoral and cellular immune responses that are specific to the expressed antigen. Not only is the adaptive immune system responsive to plasmid expression, but the

Table I. Comparison of immune responses of directly injected DNA immunization versus biolistic DNA immunization.

| | Mode of delivery | |
	Direct injection	Biolistic (i.e. 'gene-gun') delivery
T-helper cell response	Th1	Th0 or Th2
Cytokine response	IFNγ	IL4, sometimes also IFNγ
CTL response	Potent CTL	Potent CTL
Antibody response	IgG2a > IgG1	IgG1 > IgG2a

innate immune system is activated by the presence of unmethylated DNA or immunostimulatory sequences (ISS) found in the backbone of the plasmid vector (Fig. 2). There is an interplay between the innate and the adaptive immune responses, whereby the innate response augments the adaptive response through cytokine secretion and activation of antigen-presenting cells.

An important advantage of DNA vaccination is the type of immune response induced in CD4$^+$ T helper (Th) cells. Th cells differentiate from Th0 precursors into two readily discernible populations, Th1 and Th2, based on the types of cytokines they produce (Mosmann *et al.*, 1986). Th1 cells produce interleukin (IL)-2, interferon (IFN)γ, and tumour necrosis factor (TNF)β while Th2 cells produce IL-4, IL-5, IL-6 and IL-13 (Abbas *et al.*, 1996; O'Garra, 1998). Such differentiation is determined by the cytokines present during the priming period (Seder and Paul, 1994). Th1 cells mediate Delayed Type Hypersensitivity (DTH) reactions, increase IgG2a and IgG3 isotype synthesis via IFNγ secretion (in the mouse) (Street and Mosmann, 1991), and are associated with a strong cytotoxic T-lymphocyte (CTL) response. In contrast, Th2 cells activate B cells to produce IgG1 and IgE subclasses, and also stimulate eosinophils via IL-5 (Drazen *et al.*, 1996). These two polarized responses are cross-regulated by the cytokines they secrete *in vivo*.

Plasmid DNA immunization by direct i.d. injection induces T-helper cell differentiation to the Th1 type. Subsequently, antigen specific T cells secrete high levels of IFNγ which leads to the production of IgG2a antibodies, as well as the activation of strong MHC class I-restricted CTL responses (Ulmer *et al.*, 1993; Davis *et al.*, 1995; Manickan *et al.*, 1995; Raz *et al.*, 1996; Feltquate *et al.*, 1997). Not only can an ongoing antigen-specific Th2 response induced by protein in alum be 'switched' to an antigen-specific Th1 response by naked pDNA immunization, but the Th1 response induced by pDNA immunization also prevails over a later attempt to induce a Th2 response by protein in alum immunization (Raz *et al.*, 1996).

Plasmid DNA administered by gene gun immunization (also termed biolistic DNA injection) results in the induction of antigen-specific humoral and cytotoxic cellular immune responses (Haynes *et al.*, 1994). However, unlike intradermal (or intramuscular) delivery by direct injection, the gene gun delivery method gives rise to less clearly polarized T-helper responses

**DNA Vaccines:
Fundamentals and Practice**

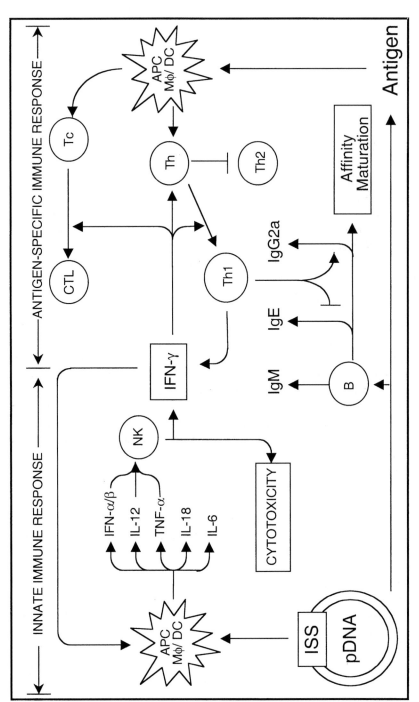

Figure 2. A proposed model of the effects of ISS on the immune system. Plasmid DNA contains immunostimulatory DNA sequences, which directly stimulate the innate immune system enhancing surface ligand expression and cytokines secretion by antigen-presenting cells. The local antigen-presenting environment is then preconditioned with cytokines. The antigen-presenting cells also become loaded with the antigenic protein expressed by the plasmid and stimulate antigen specific CD4+ T-helper cells and CD8+ cytotoxic T lymphocytes in the adaptive immune response.

(Pertmer *et al.*, 1996) or Th2 responses (Feltquate *et al.*, 1997). Pertmer and colleagues found that immunization with the gene gun resulted in CTL responses and IFNγ production as well as IgG1 antibodies. Over time the IFNγ production waned as the mice showed a marked increase in IL-4 production following repeated immunizations (Pertmer *et al.*, 1996). In contrast, Feltquate and colleagues compared the immune response induced by direct injection and gene gun immunization in skin and muscle. Biolistic immunization in mice with plasmid encoding the haemagglutinin protein resulted in Th2 responses with predominantly IgG1 and IL-4 production, while the same plasmid in saline injected i.d. gave Th1 responses with IgG2a and IFNγ. The difference in T-cell responses elicited by gene gun and direct injection may be attributed to the smaller amount of DNA and ISS exposure in the biolistic method. However, as little as 1 μg of pDNA delivered by direct injection still resulted in IgG2a antibodies while 1.5 μg delivered by gene gun produced mostly IgG1.

The bias in the T-cell cytokine profile and the potent response relative to the quantitatively small amount of protein produced by pDNA injection suggested that the DNA was acting as an adjuvant to enhance the response. The elements that augment the immune response were localized to the non-coding region in the backbone of the plasmid vector and are referred to as immunostimulatory sequences (ISS). ISS are typically palindromic hexamer motifs containing central CpG dinucleotides. In microbial DNA these sequences are unmethylated and trigger a response from the innate branch of immunity in a mammalian host (Fig. 2). The immunogenicity of the pDNA vaccine is significantly reduced by methylating its CpG motifs and increased by co-administering exogenous CpG-containing DNA (Klinman *et al.*, 1997b). Altering the number of ISS within the pDNA can change the magnitude of the Th1 response (Sato *et al.*, 1996). In addition, co-injection of protein and oligonucleotides containing the ISS (Roman *et al.*, 1997), or antigen combined with incomplete Freund's adjuvant and ISS containing oligonucleotides (Chu *et al.*, 1997), can similarly elicit an antigen-specific Th1 response. Oligonucleotides covalently linked to protein antigens are more potent in raising a Th1 response than immunization with mixtures of antigen and ISS containing oligonucleotides (Cho *et al.*, 2000; Tighe *et al.*, 2000a).

T cells, however, are not directly stimulated by ISS and the mechanism by which ISS induces Th1 differentiation has not been clearly delineated. Fresh human monocytes/macrophages upregulate expression of IFNα, IFNβ, IL-12 and IL-18 mRNA *in vitro* upon transfection with ISS (Roman *et al.*, 1997) suggesting a role for soluble factors, particularly the type 1 interferons (Sun *et al.*, 1998). All of these cytokines have been established as inducers of IFNγ and promote the differentiation of naive Th0 cells to Th1 (Brinkmann *et al.*, 1993; Okamura *et al.*, 1995; Trinchieri, 1995; Yaegashi *et al.*, 1995). Other cytokines induced by murine or human cells include TNFα, IL-6, IL1β, IL-1RA, MIP-1β, MCP-1 (Yamamoto *et al.*, 1988, 1994; Yi *et al.*, 1996a, b; Roman *et al.*, 1997; Schwartz *et al.*, 1997; Sun *et al.*, 1998; Hartmann and Krieg, 1999).

However, despite the suggestive *in vitro* evidence of a role for cytokines from antigen-presenting cells such as monocytes/macrophages in the promotion of Th1 differentiation in DNA vaccination, the mechanism by which DNA vaccines induce an antigen-specific Th1 response *in vivo* is still not precisely defined. Antigen is expressed at the site of injection by myocytes (Wolff *et al.*, 1990), keratinocytes, fibroblasts and cells with the morphologic appearance of dendritic cells (Raz *et al.*, 1994). Additional antigen expressing dendritic cells have been detected in the draining lymph nodes of mice immunized with pDNA (Condon *et al.*, 1996; Casares *et al.*, 1997; Porgador *et al.*, 1998; Akbari *et al.*, 1999). If the antigen-presenting cells (APCs) are directly transfected, they may be stimulated by ISS, make Th1-inducing cytokines (like IL-12), and prime naive antigen-specific T-helper precursor cells to become Th1 cells. However, Th cell priming may occur by a professional APC, which takes up exogenous protein synthesized by other transfected cells. APCs such as monocytes/macrophages do make several Th1-inducing cytokines and upregulate several costimulators in response to the ISS present on the pDNA (Martin-Orozco *et al.*, 1999). Perhaps ISS affect the APC in its ability to process extracellular antigens in a more efficient manner to present on MHC class I and II, or the ISS may change the overall environment of injected tissues to promote Th1 differentiation.

◆◆◆◆◆◆ THE PRINCIPLES OF DNA IMMUNIZATION

This section will provide an overview of some of the basic techniques that are employed in the generation of plasmid or oligonucleotide based vaccines. For detailed experimental protocols please refer to Ausubel *et al.* (2000). Several factors should be taken into consideration in the choice of vector for plasmid injection (Fig. 1). Either a tissue specific promoter can be selected or a promoter with a high level of expression such as the human cytomegalovirus (HCMV) E1 promoter can be used. If the gene product may be toxic to the transduced cells *in vivo*, a lower expression promoter may be employed. The stability of the transcript may depend on the transcription terminator used. The presence of one or two ISS sequences in the vector provides an adjuvant like effect. A greater number of ISS sequences have not been found further to enhance this property. Lastly the stability of the vector and its ability to amplify in *E. coli* will facilitate the production of high yields of purified plasmid DNA.

The expressed gene can be generated as a PCR fragment or subcloned from a different plasmid. The PCR fragment can be digested with restriction enzymes that are specifically engineered into the primers used for amplification. The enzyme-digested vector and insert should be gel purified. These fragments should then be ligated together at a high insert to vector ratio (5:1). Competent *E. coli* should be transformed with the ligation product and selected for growth usually by antibiotic resistance. The resulting colonies should be grown up in small 1-ml cultures under selection conditions, the plasmid construct isolated and verified by restriction digest or sequence analysis. Expression should also be verified by transient transfection in a suitable cell line.

Once the plasmid vector has been validated a large-scale preparation of purified plasmid is made for further experiments. Transformed *E. coli* are grown to saturation in selecting media with vigorous aeration by shaking 500 cc to 1 l in large 2–4 l flasks. After harvesting the bacterial pellet is lysed in 100 mM NaOH, 0.5%SDS in the presence of RNase. The protein and genomic DNA is removed by precipitation with 3 M potassium acetate and centrifugation. The addition of one-tenth volume of 10% Triton X 100 to the transferred supernatant will help diminish the amount of bound endo-toxin. The DNA can then be passed over a DNA binding resin and eluted. Alternatively DNA can be purified by caesium chloride gradient. How-ever, this methodology requires the use of ethidium bromide, which needs to be removed by 1-butanol extraction. The concentration and purity of the DNA can be determined by optical density (OD) at 260 and 280 nm. The OD_{260}/OD_{280} ratio optimally should be ≥ 1.8. To minimize degradation DNA should be stored at $-70°C$ in 10 mM Tris Cl, 1 mM EDTA (TE). The level of bound endotoxin should be checked by *Limulus* amebocyte lysate assay which can be performed using commercial kits.

Protocol for DNA isolation

1. Transformed bacteria (usually *E. coli*) are grown shaking at 37°C overnight in Luria Broth under selecting conditions (e.g. ampicillin 0.1 mg ml^{-1}).
2. Bacteria are harvested by centrifugation at 6000g for 20 min. The supernatant is discarded and the pellet is resuspended in 40 ml 50 mM Tris•Cl, 10 mM EDTA. If RNA contamination is a problem then 50–100 mcg ml^{-1} RNase A can be added.
3. The bacteria are lysed in 40 ml 100 mM NaOH, 0.5%SDS, slowly rotating the bottle.
4. Genomic DNA and protein is precipitated by adding 40 ml cold with 3 M potassium acetate and then pelleted by centrifugation at 20000g for 10 min. The supernatant may be further clarified by pass-ing it through cheesecloth or a filter.
5. To diminish the amount of endotoxin one-tenth volume of 10% Triton X 100 is added and the tube is inverted 10 times.
6. To separate the plasmid DNA from contaminants we recommend using a commercially prepared anion exchange resin.
7. The eluted DNA can be precipitated using 0.7 volume of iso-propanol and centrifugation at 15000g for 30 min.
8. Resuspend the DNA pellet in 10 mM Tris Cl, 1 mM EDTA (TE). Store frozen.

Other forms of DNA based vaccines include oligodeoxynucleotides, which contain ISS motifs. Phosphorthioate oligodeoxynucleotides are more nuclease resistant than their phosphodiester counterparts and are more widely used. These can be administered as adjuvants mixed with proteins or chemically conjugated to proteins (Cho *et al.*, 2000; Tighe *et al.*, 2000a). To conjugate purified protein to ISS-ODN, the protein is treated with N-ethyl-maleimide and activated with sulfosuccinimidyl-4-(N maleimidomethyl) cyclohexane-1-carboxylate. 5′ Disulfide ISS-ODN is reduced with tris(2-carboxyethyl)phosphine. The activated 5′ thio ISS-ODN and the protein are individually separated on G-25 desalting columns and then mixed together at room temperature to allow conjugation. This product is then purified by Superdex HR 200 gel filtration chromatography (Chu *et al*, 2000; Tighe *et al*., 2000b). The extent of conjugation can be assessed by acrylamide gel electrophoresis, and Coomassie and DNA silver stains.

Protocol for DNA conjugation to protein

1. The protein should be assessed for a free cysteine. If none are available then the protein is treated with a 20-fold molar excess of sulfo-succinimidyl-4-(*N*-maleimidomethyl) cyclohexane-1-carboxylate (sulfo-SMCC) at room temperature for 1 h. This modifies the amino side-chains of L-lysine residues by the addition of maleimide groups.
2. Reduce 5′-disulfide-ISS–ODN with 200 mM Tris-(2-carboxyethyl) phosphine (TCEP) at room temperature for 1 h.
3. Remove residual reagents from the protein and activated oligonucleotides by chromatography on a G-25 desalting column.
4. The resulting 5′-thio-ISS–ODN are mixed with the modified protein at a 5:1 molar ratio and incubated overnight at room temperature. This product is then purified by Superdex HR 200 gel filtration chromatography. This procedure usually results in one to three oligonucleotides conjugated per molecule of protein.

Prior to injection the DNA should be precipitated and resuspended in isotonic saline. The DNA can be resuspended easily up to 2 mg ml^{-1}. For intramuscular or intradermal injection a tuberculin or insulin syringe with a 28-gauge needle is typically employed and a single injection site is usually sufficient. Another method of application uses a biolistic system which can propel DNA-coated gold microprojectiles into living tissue with a 'gene gun' (Tang *et al*., 1992). More recently *in vivo* electroporation has been successfully applied (Smith and Nordstrom, 2000; Drabick *et al*., 2001; Lucas *et al*., 2001).

Protocol for biolistic plasmid administration adapted from Tang *et al*. (1992) and Porgador *et al*. (1998)

1. Ten to 50 mg of gold powder (diameter: 0.6 or 1.0 μm) and 100 μl of 0.05 M spermidine are vortexed for 5 s and sonicated for 10 s to break up the gold particles.

(contd.)

2. Add 100 µg plasmid DNA in 100 µl distilled H_2O.
3. The DNA and gold are precipitated by adding 100 µl 1 M $CaCl_2$ drop-wise and vortexing the mixture. Incubate the mixture for 10 min at room temperature, then centrifuge for 15 s at 14 000 rpm.
4. Wash the pellet three times in 1 ml 100% ethanol.
5. Resuspend the pellet in 200 µl ethanol containing 0.1 mg ml^{-1} polyvinylpyrrolidone (PVP). Adjust to the desired concentration with the ethanol/PVP solution.
6. Mice are anaesthetized with Avertin or sodium pentobarbital. The abdominal area is shaved and 2–10 non-overlapping doses are administered to the dermis with a hand held gene delivery system (usually helium driven, using discharge pressures of 350–450 psi).

The method selected for immunization will in part determine the schedule of immunizations. A single injection of 50 µg of conjugated protein and ISS containing oligonucleotide will give a robust reproducible response. The response levels using 50 µg of naked DNA injection can be variable and to minimize the differences between individual animals two injections are administered on days 0 and 14. A response to a single injection of naked plasmid DNA can be detected within 5 days at the T-cell level by cytokine secretion. Generally 4 weeks after the initial injection antibody and cellular responses are readily measurable.

◆◆◆◆◆◆ THE PROS AND CONS OF DNA VACCINATION

Pros

DNA vaccines have a number of advantages as well as some disadvantages compared with other forms of immunization such as protein vaccines or viral vector delivery systems (Table 2). One of the greatest advantages of genetic vaccines is the relative ease with which they can be constructed. Furthermore, simplicity of molecular biology techniques with which DNA can be manipulated also allows one to modify the antigenicity of a protein. For example, certain antigenic epitopes can be re-introduced or deleted from a given gene of interest. Alternatively, antigenic proteins may be fused to other biologically active molecules to alter their immunogenicity. In addition, more than one plasmid may be co-injected at a time to manipulate the desired immune response to the antigen. Strategies to influence the immune response have included the co-delivery of plasmids expressing cytokines, chemokines and costimulatory molecules (for an extensive list see Gurunathan et al. (2000)).

Synthesis and purification of pDNA are also relatively simple compared with the conventional vaccines using attenuated pathogens or recombinant proteins. DNA is relatively heat stable which is especially beneficial for developing nations where refrigeration is not readily available.

Table 2. Pros and cons of DNA vaccines.

DNA vaccines	
Pros	Cons
Strong immune responses in animal models: humoral, CTL, Th1	Strong Th1 environment may not always be desirable
Inherent adjuvant properties	Questionable efficacy in widely outbred species such as humans
Relative ease of construction, and manipulation of antigenicity of protein	Possible tolerance rather than immunity in some number of individuals
Simplicity of DNA synthesis and purification procedures	Theoretical risk of triggering autoimmunity
Stable for transport and storage	Potential integration into host genome
Extended antigen expression	
Simultaneous gene delivery expressing multiple proteins such as cytokines or co-stimulatory molecules	
Antigen presentation by both MHC class I and class II	
No risk of infection relative to live attenuated viruses	
Less potential risk of recombination than viral vectors	
Repeated immunizations are possible unlike viral vectors	

Furthermore, unlike protein vaccination, immunization with pDNA provides both an extended period of antigenic expression and the adjuvant effects of the immunostimulatory sequences to continuously stimulate the immune system. In addition, unlike other viral DNA delivery strategies, pDNA immunization has no risk of infection, and there is less potential risk of recombination associated with replication-deficient viral vectors. Moreover, repeated immunizations are possible since no other viral proteins or carrier molecules are associated.

While protein vaccines may be degraded or cleared by antibodies, DNA vaccines allow host cells to take up the DNA and synthesize antigenic protein. Antigens synthesized by host cells via DNA immunization can also be released from cells to be endocytosed by professional antigen-presenting cells and then enter the MHC class II presentation pathway to stimulate CD4$^+$ T-helper lymphocytes. CD4$^+$ Th lymphocytes which recognize the antigenic peptide bound by MHC class II become activated and then secrete cytokines to aid in the activation of other immune cells, such as B cells and cytotoxic T cells. Unlike standard protein-based vaccines, in addition to being presented on MHC class II, the intracellular source of antigen then has access to the MHC class I antigen presentation pathway. Antigens encoded by pDNA vaccines may utilize this intracellular pathway subsequently to elicit CD8$^+$ CTL responses. However, the mechanisms by which plasmid encoded antigens are presented via MHC class I molecules to naive CD8$^+$ T lymphocytes may be more complex.

Cons

One concern with the use of DNA vaccines, as with viral vector vaccines, is the possibility of integration into the host genome. Depending on the site of integration, there would be a risk of affecting the expression of genes controlling cell growth and increasing the potential risk of malignancy. In fact, the vaccination may induce tolerance to the antigen rather than immunity in some number of individuals depending on genetic make-up or simply maturity of the immune system. For example, DNA immunization against the circumsporozoite protein of *Plasmodium yoelii* in neonatal mice resulted in persistent tolerance (over 9 months), and in aged mice produced significantly lower humoral and cell-mediated immunity (and provided less protection) than young adult mice (Klinman *et al.*, 1997a). However, not all DNA vaccines induce tolerance in neonates (Siegrist and Lambert, 1997).

The possibility of DNA vaccines influencing the immune system (either by generating potent antigen-specific immune responses or through the innate adjuvanticity of the DNA molecule itself) to induce responses to self-antigens and consequently trigger autoimmunity is another uncertainty. Studies by Klinman and colleagues (Klinman *et al.*, 1997a) demonstrate a threefold rise in the number of splenocytes producing anti-DNA antibodies after DNA booster immunizations compared with mice who only received primary immunizations. Despite these increases, vaccination did not significantly increase the production of autoantibodies to DNA or myosin in mice prone to spontaneous overproduction of pathogenic anti-DNA antibodies. Although the possibility of autoimmune disease induction by DNA vaccination exists, DNA immunized mice did not develop autoimmunity. At the present time, the advantages of DNA vaccines outweigh the disadvantages to justify their use in the development of viable immunization protocols for infectious diseases as well as immunotherapy for cancer and allergies.

◆◆◆◆◆◆ APPLICATIONS

Plasmid DNA immunization is being widely investigated as a potential prophylactic or therapeutic intervention for a variety of applications. The ease with which plasmid constructs can be manipulated allows the use of different forms of an antigen such as secreted, membrane bound, or even one with deleted sequences to avoid the expression of unwanted antigenic epitopes. Furthermore, gene immunization gives the added advantage of ectopically expressing membrane bound molecules such as potential antigen-presenting molecules or co-stimulatory ligands which may alter the overall immune response. Gene immunization with combinations of plasmids expressing antigen and co-stimulators has proved to be effective in enhancing different arms of the immune system. These experiments found B7.1 to be useful for cytotoxic T-cell priming and B7.2 to enhance antibody responses (Corr *et al.*, 1997; Iwasaki *et al.*, 1997b; Kim *et al.*, 1997). Other studies have also shown enhanced immune responses using co-linear expression of co-stimulatory molecules (Iwasaki *et al.*, 1997a). The expression of the co-stimulator ligand appears to act locally and is dependent on the presence of the plasmid expressing the co-stimulator molecule at the same site as the antigen-encoding plasmid (Corr *et al.*, 1997). In addition, plasmid encoding CD40 ligand, a co-stimulatory molecule upregulated on activated T cells to stimulate APCs, enhances antibody and CTL activity when co-immunized with plasmid encoding antigen (Mendoza *et al.*, 1997; Gurunathan *et al.*, 1998).

The immune response may also be modulated by simply mixing different plasmids co-expressing soluble cytokines to skew toward a desired immune response. Granulocyte-macrophage-colony stimulating factor (GM-CSF), which may affect antigen presentation, has been shown to stimulate both T-helper and B-cell responses in a number of systems (Xiang and Ertl, 1995; Pasquini *et al.*, 1997; Svanholm *et al.*, 1997). Chow and colleagues demonstrated that co-immunization of the hepatitis B virus (HBV) DNA vaccine with IL-12 or IFN-gamma gene exhibited a significant enhancement of Th1 responses while maintaining inhibition of Th2 responses. On the other hand, co-injection with the IL-4 gene significantly enhanced the development of specific Th2 cells while Th1 differentiation was suppressed. IL-2 or GM-CSF enhanced the development of Th1 cells without affecting the development of Th2 cells (Chow *et al.*, 1998). The co-delivery of antigen with cytokine can also be achieved by expressing the antigen and the cytokine as a fusion product (Maecker *et al.*, 1997). Variations of these immunization strategies show promise in a number of areas including infectious diseases, allergy and cancer immunotherapies.

Infectious diseases

An optimal immunization strategy to an invading pathogen would generate both cellular and humoral immune responses. For example, influenza viruses mutate in their envelope genes and as a result are able to evade the previous year's vaccine containing protein subunits that are directed at the envelope glycoprotein. The nucleoprotein of influenza is an internal viral protein and is less subject to the antibody-induced antigenic drift than the surface glycoproteins. The cellular immune response may be primed to the internal proteins whereas the initial humoral response may primarily target the outer proteins. Early studies of naked DNA vaccine applications utilized plasmids encoding the influenza internal core proteins and/or the surface glycoproteins and have demonstrated protection among different viral strains with varying degrees of efficacy in several animal models including mice (Ulmer *et al.*, 1993; Justewicz *et al.*, 1995), chickens (Fynan *et al.*, 1993) and ferrets (Donnelly *et al.*, 1995).

DNA vaccines have also been developed for the production of *in vivo* immunity against HIV-1 (Cohen *et al.*, 1998). Chimpanzees can maintain protective immune responses to HIV-1 up to 48 weeks after challenge (Boyer *et al.*, 1997); however other challenge models in macaques have produced partial protection at best (Kim and Weiner, 1997). Combinations of pDNA and protein immunization strategies have also been tested. The most effective method in rhesus macaques to protect from subsequent challenge infections with immunodeficiency virus was direct intradermal injection of plasmid followed by boosting with recombinant pox virus (Robinson *et al.*, 1999).

A few examples of infectious diseases for which DNA vaccines are being developed include malaria (Hedstrom *et al.*, 1997), tuberculosis (Tascon *et al.*, 1996; Lowrie *et al.*, 1997), hepatitis B virus (Davis and Brazolot Millan, 1997; Prince *et al.*, 1997) and hepatitis C virus (Inchauspe, 1997). In addition, a number of other preclinical animal models have demonstrated protective immune responses to viruses such as bovine herpes virus (Cox *et al.*, 1993), herpes simplex virus in rodents (Manickan *et al.*, 1995; Bourne *et al.*, 1996; McClements *et al.*, 1996, 1997), rabies virus (Xiang *et al.*, 1994; Lodmell *et al.*, 1998), lymphocytic choriomeningitis virus (Martins *et al.*, 1995; Yokoyama *et al.*, 1995), and cottontail rabbit papilloma virus (Donnelly *et al.*, 1996). In fact, immunization of genetically susceptible BALB/c mice using plasmid that expressed an antigen from the pathogen protected virtually all of the mice against progressive, non-healing infections with *L. major* (Gurunathan *et al.*, 1997).

Recently, Amara *et al.* have shown combination vaccine strategies utilizing DNA priming followed by a recombinant modified vaccinia Ankara booster controlled a highly pathogenic immunodeficiency virus challenge in a rhesus macaque model (Amara *et al.*, 2001). As the adjuvanticity of ISS persists for up to 2 weeks *in vivo* after intradermal or intranasal delivery, ISS prepriming has been shown to induce immune responses that are significantly stronger than with ISS/antigen co-administration. Preinjection with ISS offers an alternative approach to the traditional use of adjuvants (i.e., antigen/adjuvant co-injection) and expands the potential clinical applications for ISS (Kobayashi *et al.*, 1999). This strategy of combining previous vaccination strategies with DNA vaccination will likely continue to be used in the development of future vaccines (Cohen, 2001).

Allergic diseases

The induction of true allergen desensitization remains an elusive therapeutic goal despite a number of effective therapeutic options for the prevention and treatment of the pathophysiologic responses, which characterize allergic diseases. Only immunotherapy (IT) has been shown to have any effect on the underlying hypersensitivities, which mediate allergic reactions. Traditional protein allergen based IT has a limited scope of efficacy.

Allergic diseases are the result of enhanced allergen specific Th2 responses (Wierenga *et al.*, 1990; Parronchi *et al.*, 1991). The deleterious response is triggered by allergen-specific IgE antibodies bound to IgE receptors at the surface of mast cells and basophils. The presence of allergen causes cross-linking of the bound IgE and results in the immediate release of histamine, IL-4 and IL-5 (Mygind *et al.*, 1996) as well as the subsequent production of proinflammatory leukotrienes and platelet-activating factor. Since antigen-encoding pDNA injected i.m. or i.d. into mice elicits a long-lasting antigen-specific cellular and humoral response with a Th1 phenotype (Manickan *et al.*, 1995; Raz *et al.*, 1996), this may provide a novel method of immunotherapy for the treatment of allergic diseases. In fact, pDNA encoding the major allergen of birch pollen has been used to immunize mice (Hartl *et al.*, 1999). These studies showed that DNA immunization induces a strong Th1 immune response against a relevant inhalant allergen.

Different strategies using ISS and pDNA have emerged in designing potential immunotherapeutics. Immunization with pDNA vaccines, protein allergens mixed with ISS-ODN, allergens chemically conjugated to ISS-ODN (AIC), and ISS-ODN alone have been highly effective in the prevention and reversal of Th2 mediated hypersensitivity states in mouse models of allergic disease (Horner *et al.*, 2000; Tighe *et al.*, 2000b). Administration of ISS prior to antigen exposure has been shown to be effective in a mouse model of asthma. The protective effect of ISS lasted at least 4 weeks but was not sustained up to 8 weeks (Broide *et al.*, 2001). Mucosal (intranasal and intratracheal) delivery of ISS was as effective as systemic (intraperitoneal) ISS delivery in inhibiting airway eosinophilia and switching cytokine responses from Th2 to Th1 responses (Broide *et al.*, 1998). Thus, ISS administration prior to or in combination with allergen should be considered as a method of allergen-based immunotherapy (Broide and Raz, 1999). Several clinical trials have already been initiated for the reversal of allergic hypersensitivity states in humans.

Neoplastic disorders

Stimulating the host's immune system to combat a cancerous growth as 'foreign' cells with aberrant protein expression has been an attractive approach to the treatment of micrometastases (Yang and Sun, 1995; Durrant, 1997). Mutations in tumour oncogenes or suppressor genes, which lead to malignant transformation, may also be recognized as 'foreign' tumour-specific antigens. Studies by Conry and colleagues (Conry et al., 1995) have demonstrated the induction of a human carcinoembryonic antigen (CEA) response by the injection of pDNA encoding CEA and the subsequent protection of mice from syngeneic CEA-expressing tumour cell lines. CEA is expressed at high levels in human colon, breast and non-small-cell lung cancers. Likewise, studies by Graham et al. (1996) using pDNA encoding the polymorphic epithelial mucin (PEM) associated with breast, pancreatic and colon cancers, protected mice from challenge with syngeneic PEM-expressing tumour cells.

The use of directly expressing ectopic cytokine genes in vivo has also been utilized in anti-neoplastic therapy. Sun and colleagues inhibited growth of a renal carcinoma tumour model with a combination of murine tumour necrosis factor alpha and interferon gamma genes via biolistic injection (Sun et al., 1995). In addition, treatment with murine IL-2 and interferon gamma genes prolonged the survival of tumour-bearing mice and resulted in tumour eradication in a quarter of these animals. The development of DNA vaccines which can elicit an inflammatory response from cells of the innate immune system, along with a strong cell-mediated immune response to a specific tumour antigen, offers great potential for the future.

Clinical trials

The first demonstration in healthy naive humans of the induction of CD8+ CTLs by DNA vaccines used a malarial antigen (Wang et al., 1998). A Phase I safety and tolerability clinical study (Le et al., 2000) in which three i.m. injections of different dosages (20, 100, 500 and 2500 µg) of pDNA encoding Plasmodium falciparum circumsporozoite protein (PfCSP) caused only local reactogenicity. Systemic symptoms were few and mild, and no severe or serious adverse events, clinically significant biochemical or haematologic changes, nor anti-dsDNA antibodies were observed. Although CTL were induced, antigen-specific antibodies were not detected. Another Phase 1 trial in humans for the safety and immunogenicity of human papillomavirus (HPV)-11 L1 virus-like particle showed the injection was well tolerated and induced high levels of both binding and neutralizing antibodies (Evans et al., 2001).

Clinical trials for pDNA vaccines against HIV-1 have also been reported. In a chimpanzee model system DNA vaccination with constructs which express the env, rev, gag and pol proteins led to a CD8 cytolytic response (Boyer *et al.*, 1997). In humans, pDNA encoding HIV-1 ENV and REV genes was tested for safety and host immune response in 15 asymptomatic HIV-infected patients, who were not using antiviral drugs and had CD4+ lymphocyte counts of > or = 500 per microlitre of blood (MacGregor *et al.*, 1998). No anti-DNA antibody or muscle enzyme elevations were detected, and no consistent change occurred in CD4 or CD8 lymphocyte counts or plasma HIV concentration. Antibody against gp120 increased in individual patients who received higher doses of pDNA (100, 300 µg). Some increases were noted in cytotoxic T-lymphocyte activity against gp160-bearing targets and in lymphocyte proliferative activity (MacGregor *et al.*, 1998) and three of three patients in the 300 µg dose group also developed increased MIP-1alpha levels detectable in sera (Boyer *et al.*, 1999). Studies in HIV-1-seronegative persons have also demonstrated the immunogenicity of an HIV env/rev DNA vaccine with vaccine-induced antigen-specific lymphocyte proliferative responses and antigen-specific production of IFN-γ and β-chemokine (Boyer *et al.*, 2000).

Studies in Hepatitis B pDNA vaccines have also entered clinical trials, showing safety and immunogenicity in volunteers using a plasmid-encoding hepatitis B surface antigen delivered by the PowderJect XR1 gene delivery system into human skin (Tacket *et al.*, 1999). The vaccine has been reported to be safe and well tolerated, causing only transient inflammatory responses at the site of administration. Protective antibody titres and both humoral and cell-mediated immune responses were measured (Roy *et al.*, 2000).

DNA vaccination has also entered clinical trials for immunotherapy of prostate cancer (Mincheff *et al.*, 2000). Using the prostate-specific membrane antigen (PSMA) as a target molecule, patients with prostate cancer received i.d. immunizations with no immediate or long-term side effects recorded. However, patients who received initial injection with a replication-deficient adenoviral expression vector followed by PSMA-encoded pDNA boosts responded by delayed-type hypersensitivity (DTH). However, only 50% of patients who received a combination of PSMA-encoded pDNA combined with CD86-encoded pDNA and 67% of patients who received PSMA-encoded pDNA and soluble GM-CSF (sGM-CSF) showed signs of successful immunization. These trials indicate that DNA based immunizations are well tolerated and may yield new therapeutic modalities.

◆◆◆◆◆◆ CONCLUDING REMARKS

DNA vaccines are proving to be an attractive modality for use in research and clinical treatments. Future advances in gene expression technology and in enhancing the adjuvanticity may make this methodology a more potent investigational tool with broader applications. The transition of DNA vaccine to clinical trials was in part facilitated by the relative ease of development and production, as well as their efficacy in animal models. While not all safety concerns have been completely addressed, human trial data are encouraging.

Acknowledgements

The work was supported in part by grants AI-40682, AR40770, and AR44850 from the National Institutes of Health. We would like to thank N. Noon and J. Uhle for their assistance.

References

Abbas, A. K., Murphy, K. M. and Sher, A. (1996). Functional diversity of helper T lymphocytes. *Nature* **383**, 787–793.

Akbari, O., Panjwani, N., Garcia, S., Tascon, R., Lowrie, D. and Stockinger, B. (1999). DNA vaccination: transfection and activation of dendritic cells as key events for immunity. *J. Exp. Med.* **189**, 169–178.

Amara, R. R., Villinger, F., Altman, J. D., Lydy, S. L., O'Neil, S. P., Staprans, S. I., Montefiori, D. C., Xu, Y., Herndon, J. G., Wyatt, L. S., Candido, M. A., Kozyr, N. L., Earl, P. L., Smith, J. M., Ma, H., Grimm, B. D., Hulsey, M. L., Miller, J., McClure, H. M., McNicholl, J. M., Moss, B. and Robinson, H. L. (2001). Control of a mucosal challenge and prevention of AIDS by a multiprotein DNA/MVA Vaccine. *Science* **292**, 69–74.

Ausubel, F. M., Brent, R., Kingston, R. E., Moore, D. D., Seidman, J. G., Smith, J. A. and Struhl, K. (eds). (2000). *Current Protocols in Molecular Biology*. John Wiley & Sons, Inc.

Bennett, C. F., Mirejovsky, D., Crooke, R. M., Tsai, Y. J., Felgner, J., Sridhar, C. N., Wheeler, C. J. and Felgner, P. L. (1998). Structural requirements for cationic lipid mediated phosphorothioate oligonucleotides delivery to cells in culture. *J. Drug Target* **5**, 149–162.

Bourne, N., Stanberry, L. R., Bernstein, D. I. and Lew, D. (1996). DNA immunization against experimental genital herpes simplex virus infection. *J. Infect. Dis.* **173**, 800–807.

Boyer, J. D., Ugen, K. E., Wang, B., Agadjanyan, M., Gilbert, L., Bagarazzi, M. L., Chattergoon, M., Frost, P., Javadian, A., Williams, W. V., Refaeli, Y., Ciccarelli, R. B., McCallus, D., Coney, L. and Weiner, D. B. (1997). Protection of chimpanzees from high-dose heterologous HIV-1 challenge by DNA vaccination. *Nat. Med.* **3**, 526–532.

Boyer, J. D., Chattergoon, M. A., Ugen, K. E., Shah, A., Bennett, M., Cohen, A., Nyland, S., Lacy, K. E., Bagarazzi, M. L., Higgins, T. J., Baine, Y., Ciccarelli, R. B., Ginsberg, R. S., MacGregor, R. R. and Weiner, D. B. (1999). Enhancement of cellular immune response in HIV-1 seropositive individuals: A DNA-based trial. *Clin. Immunol.* **90**, 100–107.

Boyer, J. D., Cohen, A. D., Vogt, S., Schumann, K., Nath, B., Ahn, L., Lacy, K., Bagarazzi, M. L., Higgins, T. J., Baine, Y., Ciccarelli, R. B., Ginsberg, R. S., MacGregor, R. R. and Weiner, D. B. (2000). Vaccination of seronegative volunteers with a human immunodeficiency virus type 1 env/rev DNA vaccine induces antigen-specific proliferation and lymphocyte production of beta-chemokines. *J. Infect. Dis.* **181**, 476–483.

Brinkmann, V., Geiger, T., Alkan, S. and Heusser, C. H. (1993). Interferon alpha increases the frequency of interferon gamma-producing human CD4+ T cells. *J. Exp. Med.* **178**, 1655–1663.

Broide, D. and Raz, E. (1999). DNA-based immunization for asthma. *Int. Arch. Allergy Immunol.* **118**, 453–456.

Broide, D., Schwarze, J., Tighe, H., Gifford, T., Nguyen, M. D., Malek, S., Van Uden, J., Martin-Orozco, E., Gelfand, E. W. and Raz, E. (1998). Immunostimulatory DNA sequences inhibit IL-5, eosinophilic inflammation, and airway hyperresponsiveness in mice. *J. Immunol.* **161**, 7054–7062.

Broide, D. H., Stachnick, G., Castaneda, D., Nayar, J., Miller, M., Cho, J. Y., Roman, M., Zubeldia, J., Hyashi, T. and Raz, E. (2001). Systemic administration of immunostimulatory DNA sequences mediates reversible inhibition of Th2 responses in a mouse model of asthma. *J. Clin. Immunol.* **21**, 175–182.

Casares, S., Inaba, K., Brumeanu, T. D., Steinman, R. M. and Bona, C. A. (1997). Antigen presentation by dendritic cells after immunization with DNA encoding a major histocompatibility complex class II-restricted viral epitope. *J. Exp. Med.* **186**, 1481–1486.

Cho, H. J., Takabayashi, K., Cheng, P. M., Nguyen, M. D., Corr, M., Tuck, S. and Raz, E. (2000). Immunostimulatory DNA-based vaccines induce cytotoxic lymphocyte activity by a T-helper cell-independent mechanism. *Nat. Biotechnol.* **18**, 509–514.

Chow, Y. H., Chiang, B. L., Lee, Y. L., Chi, W. K., Lin, W. C., Chen, Y. T. and Tao, M. H. (1998). Development of Th1 and Th2 populations and the nature of immune responses to hepatitis B virus DNA vaccines can be modulated by codelivery of various cytokine genes. *J. Immunol.* **160**, 1320–1329.

Chu, R. S., Targoni, O. S., Krieg, A. M., Lehmann, P. V. and Harding, C. V. (1997). CpG oligodeoxynucleotides act as adjuvants that switch on T helper 1 (Th1) immunity. *J. Exp. Med.* **186**, 1623–1631.

Cohen, A. D., Boyer, J. D. and Weiner, D. B. (1998). Modulating the immune response to genetic immunization. *Faseb J.* **12**, 1611–1626.

Cohen, J. (2001). Merch reemerges with a bold AIDS vaccine effort. *Science* **292**, 24–25.

Condon, C., Watkins, S. C., Celluzzi, C. M., Thompson, K. and Falo, L. D., Jr. (1996). DNA-based immunization by in vivo transfection of dendritic cells. *Nat. Med.* **2**, 1122–1128.

Conry, R. M., LoBuglio, A. F., Loechel, F., Moore, S. E., Sumerel, L. A., Barlow, D. L., Pike, J. and Curiel, D. T. (1995). A carcinoembryonic antigen polynucleotide vaccine for human clinical use. *Cancer Gene Ther.* **2**, 33–38.

Corr, M., Tighe, H., Lee, D., Dudler, J., Trieu, M., Brinson, D. C. and Carson, D. A. (1997). Costimulation provided by DNA immunization enhances antitumor immunity. *J. Immunol.* **159**, 4999–5004.

Cox, G. J., Zamb, T. J. and Babiuk, L. A. (1993). Bovine herpesvirus 1: immune responses in mice and cattle injected with plasmid DNA. *J. Virol.* **67**, 5664–5667.

Davis, H. L. and Brazolot Millan, C. L. (1997). DNA-based immunization against hepatitis B virus. *Springer Semin. Immunopathol.* **19**, 195–209.

Davis, H. L., Schirmbeck, R., Reimann, J. and Whalen, R. G. (1995). DNA-mediated immunization in mice induces a potent MHC class I-restricted cytotoxic T lymphocyte response to the hepatitis B envelope protein. *Hum. Gene. Ther.* **6**, 1447–1456.

Donnelly, J. J., Friedman, A., Martinez, D., Montgomery, D. L., Shiver, J. W., Motzel, S. L., Ulmer, J. B. and Liu, M. A. (1995). Preclinical efficacy of a prototype DNA vaccine: enhanced protection against antigenic drift in influenza virus. *Nat. Med.* **1**, 583–587.

Donnelly, J. J., Martinez, D., Jansen, K. U., Ellis, R. W., Montgomery, D. L. and Liu, M. A. (1996). Protection against papillomavirus with a polynucleotide vaccine. *J. Infect. Dis.* **173**, 314–320.

Drabick, J. J., Glasspool-Malone, J., King, A. and Malone, R. W. (2001). Cutaneous transfection and immune responses to intradermal nucleic acid vaccination are significantly enhanced by in vivo electropermeabilization. *Mol. Ther.* **3**, 249–255.

Drazen, J. M., Arm, J. P. and Austen, K. F. (1996). Sorting out the cytokines of asthma. *J. Exp. Med.* **183**, 1–5.

Durrant, L. G. (1997). Cancer vaccines. *Anticancer Drugs* **8**, 727–733.

Evans, T. G., Bonnez, W., Rose, R. C., Koenig, S., Demeter, L., Suzich, J. A., O'Brien, D., Campbell, M., White, W. I., Balsley, J. and Reichman, R. C. (2001). A Phase 1 study of a recombinant viruslike particle vaccine against human papillomavirus type 11 in healthy adult volunteers. *J. Infect. Dis.* **183**, 1485–1493.

Felgner, P. L., Tsai, Y. J., Sukhu, L., Wheeler, C. J., Manthorpe, M., Marshall, J. and Cheng, S. H. (1995). Improved cationic lipid formulations for in vivo gene therapy. *Ann. NY Acad. Sci.* **772**, 126–139.

Feltquate, D. M., Heaney, S., Webster, R. G. and Robinson, H. L. (1997). Different T helper cell types and antibody isotypes generated by saline and gene gun DNA immunization. *J. Immunol.* **158**, 2278–2284.

Fynan, E. F., Webster, R. G., Fuller, D. H., Haynes, J. R., Santoro, J. C. and Robinson, H. L. (1993). DNA vaccines: protective immunizations by parenteral, mucosal, and gene-gun inoculations. *Proc. Natl. Acad. Sci. USA* **90**, 11478–11482.

Graham, R. A., Burchell, J. M., Beverley, P. and Taylor-Papadimitriou, J. (1996). Intramuscular immunisation with MUC1 cDNA can protect C57 mice challenged with MUC1-expressing syngeneic mouse tumour cells. *Int. J. Cancer* **65**, 664–670.

Gurunathan, S., Sacks, D. L., Brown, D. R., Reiner, S. L., Charest, H., Glaichenhaus, N. and Seder, R. A. (1997). Vaccination with DNA encoding the immunodominant LACK parasite antigen confers protective immunity to mice infected with *Leishmania major. J. Exp. Med.* **186**, 1137–1147.

Gurunathan, S., Irvine, K. R., Wu, C. Y., Cohen, J. I., Thomas, E., Prussin, C., Restifo, N. P. and Seder, R. A. (1998). CD40 ligand/trimer DNA enhances both humoral and cellular immune responses and induces protective immunity to infectious and tumor challenge. *J. Immunol.* **161**, 4563–4571.

Gurunathan, S., Klinman, D. M. and Seder, R. A. (2000). DNA vaccines: immunology, application, and optimization. *Annu. Rev. Immunol.* **18**, 927–974.

Hartl, A., Kiesslich, J., Weiss, R., Bernhaupt, A., Mostböck, S., Scheiblhofer, S., Ebner, C., Ferreira, F. and Thalhamer, J. (1999). Immune responses after immunization with plasmid DNA encoding Bet v 1, the major allergen of birch pollen. *J. Allergy Clin. Immunol.* **103**, 107–113.

Hartmann, G. and Krieg, A. M. (1999). CpG DNA and LPS induce distinct patterns of activation in human monocytes. *Gene Ther.* **6**, 893–903.

Haynes, J. R., Fuller, D. H., Eisenbraun, M. D., Ford, M. J. and Pertmer, T. M. (1994). Accell particle-mediated DNA immunization elicits humoral, cytotoxic, and protective immune responses. *AIDS Res. Hum. Retroviruses* **10** (Suppl 2), S43–45.

Hedstrom, R. C., Doolan, D. L., Wang, R., Gardner, M. J., Kumar, A., Sedegah, M., Gramzinski, R. A., Sacci, J. B., Jr., Charoenvit, Y., Weiss, W. R., Margalith, M., Norman, J. A., Hobart, P. and Hoffman, S. L. (1997). The development of a multivalent DNA vaccine for malaria. *Springer Semin. Immunopathol.* **19**, 147–159.

Horner, A. A., Nguyen, M. D., Ronaghy, A., Cinman, N., Verbeek, S. and Raz, E. (2000). DNA-based vaccination reduces the risk of lethal anaphylactic hypersensitivity in mice. *J. Allergy Clin. Immunol.* **106**, 349–356.

Hui, K. M., Sabapathy, T. K., Oei, A. A. and Chia, T. F. (1994). Generation of allo-reactive cytotoxic T lymphocytes by particle bombardment-mediated gene transfer. *J. Immunol. Methods* **171**, 147–155.

Inchauspe, G. (1997). Gene vaccination for hepatitis C. *Springer Semin. Immunopathol.* **19**, 211–221.

Ishii, N., Fukushima, J., Kaneko, T., Okada, E., Tani, K., Tanaka, S. I., Hamajima, K., Xin, K. Q., Kawamoto, S., Koff, W., Nishioka, K., Yasuda, T. and Okuda, K. (1997). Cationic liposomes are a strong adjuvant for a DNA vaccine of human immunodeficiency virus type 1. *AIDS Res. Hum. Retroviruses* **13**, 1421–1428.

Iwasaki, A., Stiernholm, B. J., Chan, A. K., Berinstein, N. L. and Barber, B. H. (1997a). Enhanced CTL responses mediated by plasmid DNA immunogens encoding costimulatory molecules and cytokines. *J. Immunol.* **158**, 4591–4601.

Iwasaki, A., Torres, C. A., Ohashi, P. S., Robinson, H. L. and Barber, B. H. (1997b). The dominant role of bone marrow-derived cells in CTL induction following plasmid DNA immunization at different sites. *J. Immunol.* **159**, 11–14.

Johnston, S. A. and Tang, D. C. (1994). Gene gun transfection of animal cells and genetic immunization. *Methods Cell Biol.* **43**, 353–365.

Justewicz, D. M., Morin, M. J., Robinson, H. L. and Webster, R. G. (1995). Antibody-forming cell response to virus challenge in mice immunized with DNA encoding the influenza virus hemagglutinin. *J. Virol.* **69**, 7712–7717.

Kim, J. J. and Weiner, D. B. (1997). DNA gene vaccination for HIV. *Springer Semin. Immunopathol.* **19**, 175–194.

Kim, J. J., Bagarazzi, M. L., Trivedi, N., Hu, Y., Kazahaya, K., Wilson, D. M., Ciccarelli, R., Chattergoon, M. A., Dang, K., Mahalingam, S., Chalian, A. A., Agadjanyan, M. G., Boyer, J. D., Wang, B. and Weiner, D. B. (1997). Engineering of in vivo immune responses to DNA immunization via codelivery of costimu-latory molecule genes. *Nat. Biotechnol.* **15**, 641–646.

Klinman, D. M., Takeno, M., Ichino, M., Gu, M., Yamshchikov, G., Mor, G. and Conover, J. (1997a). DNA vaccines: safety and efficacy issues. *Springer Semin. Immunopathol.* **19**, 245–256.

Klinman, D. M., Yamshchikov, G. and Ishigatsubo, Y. (1997b). Contribution of CpG motifs to the immunogenicity of DNA vaccines. *J. Immunol.* **158**, 3635–3639.

Kobayashi, H., Horner, A. A., Takabayashi, K., Nguyen, M. D., Huang, E., Cinman, N. and Raz, E. (1999). Immunostimulatory DNA pre-priming: a novel approach for prolonged Th1-biased immunity. *Cell. Immunol.* **198**, 69–75.

Le, T. P., Coonan, K. M., Hedstrom, R. C., Charoenvit, Y., Sedegah, M., Epstein, J. E., Kumar, S., Wang, R., Doolan, D. L., Maguire, J. D., Parker, S. E., Hobart, P., Norman, J. and Hoffman, S. L. (2000). Safety, tolerability and humoral immune responses after intramuscular administration of a malaria DNA vaccine to healthy adult volunteers. *Vaccine* **18**, 1893–1901.

Lodmell, D. L., Ray, N. B., Parnell, M. J., Ewalt, L. C., Hanlon, C. A., Shaddock, J. H., Sanderlin, D. S. and Rupprecht, C. E. (1998). DNA immunization protects nonhuman primates against rabies virus. *Nat. Med.* **4**, 949–952.

Lowrie, D. B., Silva, C. L. and Tascon, R. E. (1997). Genetic vaccination against tuberculosis. *Springer Semin. Immunopathol.* **19**, 161–173.

Lucas, M. L., Jaroszeski, M. J., Gilbert, R. and Heller, R. (2001). In vivo electropo-ration using an exponentially enhanced pulse: a new waveform. *DNA Cell Biol.* **20**, 183–188.

MacGregor, R. R., Boyer, J. D., Ugen, K. E., Lacy, K. E., Gluckman, S. J., Bagarazzi, M. L., Chattergoon, M. A., Baine, Y., Higgins, T. J., Ciccarelli, R. B., Coney, L. R., Ginsberg, R. S. and Weiner, D. B. (1998). First human trial of a DNA-based vac-cine for treatment of human immunodeficiency virus type 1 infection: safety and host response. *J. Infect. Dis.* **178**, 92–100.

Maecker, H. T., Umetsu, D. T., DeKruyff, R. H. and Levy, S. (1997). DNA vaccination with cytokine fusion constructs biases the immune response to ovalbumin. *Vaccine* **15**, 1687–1696.

Manickan, E., Rouse, R. J., Yu, Z., Wire, W. S. and Rouse, B. T. (1995). Genetic immunization against herpes simplex virus. Protection is mediated by CD4+ T lymphocytes. *J. Immunol.* **155**, 259–265.

Martin-Orozco, E., Kobayashi, H., Van Uden, J., Nguyen, M., Kornbluth, R. S. and Raz, E. (1999). Enhancement of antigen-presenting cell surface molecules involved in cognate interactions by immunostimulatory DNA sequences. *Int. Immunol.* **11**, 1111–1118.

Martins, L. P., Lau, L. L., Asano, M. S. and Ahmed, R. (1995). DNA vaccination against persistent viral infection. *J. Virol.* **69**, 2574–2582.

McClements, W. L., Armstrong, M. E., Keys, R. D. and Liu, M. A. (1996). Immunization with DNA vaccines encoding glycoprotein D or glycoprotein B, alone or in combination, induces protective immunity in animal models of herpes simplex virus-2 disease. *Proc. Natl. Acad. Sci. USA* **93**, 11414–11420.

McClements, W. L., Armstrong, M. E., Keys, R. D. and Liu, M. A. (1997). The prophylactic effect of immunization with DNA encoding herpes simplex virus glycoproteins on HSV-induced disease in guinea pigs. *Vaccine* **15**, 857–860.

Mendoza, R. B., Cantwell, M. J. and Kipps, T. J. (1997). Immunostimulatory effects of a plasmid expressing CD40 ligand (CD154) on gene immunization. *J. Immunol.* **159**, 5777–5781.

Mincheff, M., Tchakarov, S., Zoubak, S., Loukinov, D., Botev, C., Altankova, I., Georgiev, G., Petrov, S. and Meryman, H. T. (2000). Naked DNA and adenoviral immunizations for immunotherapy of prostate cancer: a phase I/II clinical trial. *Eur. Urol.* **38**, 208–217.

Mosmann, T. R., Cherwinski, H., Bond, M. W., Giedlin, M. A. and Coffman, R. L. (1986). Two types of murine helper T cell clone. I. Definition according to profiles of lymphokine activities and secreted proteins. *J. Immunol.* **136**, 2348–2357.

Mygind, N., Dahl, R., Pederson, S. and Thestrup-Pederson, K. (1996). *Essential Allergy*, 2nd edn. Cambridge, MA, Blackwell Science.

O'Garra, A. (1998). Cytokines induce the development of functionally heterogeneous T helper cell subsets. *Immunity* **8**, 275–283.

Okada, E., Sasaki, S., Ishii, N., Aoki, I., Yasuda, T., Nishioka, K., Fukushima, J., Miyazaki, J., Wahren, B. and Okuda, K. (1997). Intranasal immunization of a DNA vaccine with IL-12- and granulocyte-macrophage colony-stimulating factor (GM-CSF)-expressing plasmids in liposomes induces strong mucosal and cell-mediated immune responses against HIV-1 antigens. *J. Immunol.* **159**, 3638–3647.

Okamura, H., Tsutsi, H., Komatsu, T., Yutsudo, M., Hakura, A., Tanimoto, T., Torigoe, K., Okura, T., Nukada, Y., Hattori, K. *et al.* (1995). Cloning of a new cytokine that induces IFN-gamma production by T cells. *Nature* **378**, 88–91.

Parronchi, P., Macchia, D., Piccinni, M. P., Biswas, P., Simonelli, C., Maggi, E., Ricci, M., Ansari, A. A. and Romagnani, S. (1991). Allergen- and bacterial antigen-specific T-cell clones established from atopic donors show a different profile of cytokine production. *Proc. Natl. Acad. Sci. USA* **88**, 4538–4542.

Pasquini, S., Xiang, Z., Wang, Y., He, Z., Deng, H., Blaszczyk-Thurin, M. and Ertl, H. C. (1997). Cytokines and costimulatory molecules as genetic adjuvants. *Immunol. Cell Biol.* **75**, 397–401.

Pertmer, T. M., Eisenbraun, M. D., McCabe, D., Prayaga, S. K., Fuller, D. H. and Haynes, J. R. (1995). Gene gun-based nucleic acid immunization: elicitation of humoral and cytotoxic T lymphocyte responses following epidermal delivery of nanogram quantities of DNA. *Vaccine* **13**, 1427–1430.

DNA Vaccines: Fundamentals and Practice

Pertmer, T. M., Roberts, T. R. and Haynes, J. R. (1996). Influenza virus nucleoprotein-specific immunoglobulin G subclass and cytokine responses elicited by DNA vaccination are dependent on the route of vector DNA delivery. *J. Virol.* **70**, 6119–6125.

Porgador, A., Irvine, K. R., Iwasaki, A., Barber, B. H., Restifo, N. P. and Germain, R. N. (1998). Predominant role for directly transfected dendritic cells in antigen presentation to CD8+ T cells after gene gun immunization. *J. Exp. Med.* **188**, 1075–1082.

Prince, A. M., Whalen, R. and Brotman, B. (1997). Successful nucleic acid based immunization of newborn chimpanzees against hepatitis B virus. *Vaccine* **15**, 916–919.

Raz, E., Carson, D. A., Parker, S. E., Parr, T. B., Abai, A. M., Aichinger, G., Gromkowski, S. H., Singh, M., Lew, D., Yankauckas, M. A., *et al.* (1994). Intradermal gene immunization: the possible role of DNA uptake in the induction of cellular immunity to viruses. *Proc. Natl. Acad. Sci. USA* **91**, 9519–9523.

Raz, E., Tighe, H., Sato, Y., Corr, M., Dudler, J. A., Roman, M., Swain, S. L., Spiegelberg, H. L. and Carson, D. A. (1996). Preferential induction of a Th1 immune response and inhibition of specific IgE antibody formation by plasmid DNA immunization. *Proc. Natl. Acad. Sci. USA* **93**, 5141–5145.

Robinson, H. L., Montefiori, D. C., Johnson, R. P., Manson, K. H., Kalish, M. L., Lifson, J. D., Rizvi, T. A., Lu, S., Hu, S. L., Mazzara, G. P., Panicali, D. L., Herndon, J. G., Glickman, R., Candido, M. A., Lydy, S. L., Wyand, M. S. and McClure, H. M. (1999). Neutralizing antibody-independent containment of immunodeficiency virus challenges by DNA priming and recombinant pox virus booster immunizations. *Nat. Med.* **5**, 526–534.

Roman, M., Martin-Orozco, E., Goodman, J. S., Nguyen, M. D., Sato, Y., Ronaghy, A., Kornbluth, R. S., Richman, D. D., Carson, D. A. and Raz, E. (1997). Immunostimulatory DNA sequences function as T helper-1-promoting adjuvants. *Nat. Med.* **3**, 849–854.

Roy, M. J., Wu, M. S., Barr, L. J., Fuller, J. T., Tussey, L. G., Speller, S., Culp, J., Burkholder, J. K., Swain, W. F., Dixon, R. M., Widera, G., Vessey, R., King, A., Ogg, G., Gallimore, A., Haynes, J. R. and Heydenburg Fuller, D. (2000). Induction of antigen-specific CD8+ T cells, T helper cells, and protective levels of antibody in humans by particle-mediated administration of a hepatitis B virus DNA vaccine. *Vaccine* **19**, 764–778.

Sato, Y., Roman, M., Tighe, H., Lee, D., Corr, M., Nguyen, M. D., Silverman, G. J., Lotz, M., Carson, D. A. and Raz, E. (1996). Immunostimulatory DNA sequences necessary for effective intradermal gene immunization. *Science* **273**, 352–354.

Schwartz, D. A., Quinn, T. J., Thorne, P. S., Sayeed, S., Yi, A. K. and Krieg, A. M. (1997). CpG motifs in bacterial DNA cause inflammation in the lower respiratory tract. *J. Clin. Invest.* **100**, 68–73.

Seder, R. A. and Paul, W. E. (1994). Acquisition of lymphokine-producing phenotype by CD4+ T cells. *Annu. Rev. Immunol.* **12**, 635–673.

Siegrist, C. A. and Lambert, P. H. (1997). Immunization with DNA vaccines in early life: advantages and limitations as compared to conventional vaccines. *Springer Semin. Immunopathol.* **19**, 233–243.

Smith, L. C. and Nordstrom, J. L. (2000). Advances in plasmid gene delivery and expression in skeletal muscle. *Curr. Opin. Mol. Ther.* **2**, 150–154.

Street, N. E. and Mosmann, T. R. (1991). Functional diversity of T lymphocytes due to secretion of different cytokine patterns. *Faseb J.* **5**, 171–177.

Sun, S., Zhang, X., Tough, D. F. and Sprent, J. (1998). Type I interferon-mediated stimulation of T cells by CpG DNA. *J. Exp. Med.* **188**, 2335–2342.

Sun, W. H., Burkholder, J. K., Sun, J., Culp, J., Turner, J., Lu, X. G., Pugh, T. D., Ershler, W. B. and Yang, N. S. (1995). In vivo cytokine gene transfer by gene gun reduces tumor growth in mice. *Proc. Natl. Acad. Sci. USA* **92**, 2889–2893.

Svanholm, C., Lowenadler, B. and Wigzell, H. (1997). Amplification of T-cell and antibody responses in DNA-based immunization with HIV-1 Nef by co-injection with a GM-CSF expression vector. *Scand. J. Immunol.* **46**, 298–303.

Tacket, C. O., Roy, M. J., Widera, G., Swain, W. F., Broome, S. and Edelman, R. (1999). Phase 1 safety and immune response studies of a DNA vaccine encoding hepatitis B surface antigen delivered by a gene delivery device. *Vaccine* **17**, 2826–2829.

Tang, D. C., DeVit, M. and Johnston, S. A. (1992). Genetic immunization is a simple method for eliciting an immune response. *Nature* **356**, 152–154.

Tascon, R. E., Colston, M. J., Ragno, S., Stavropoulos, E., Gregory, D. and Lowrie, D. B. (1996). Vaccination against tuberculosis by DNA injection. *Nat. Med.* **2**, 888–892.

Tighe, H., Takabayashi, K., Schwartz, D., Marsden, R., Beck, L., Corbeil, J., Richman, D. D., Eiden, J. J., Jr., Spiegelberg, H. L. and Raz, E. (2000a). Conjugation of protein to immunostimulatory DNA results in a rapid, long-lasting and potent induction of cell-mediated and humoral immunity. *Eur. J. Immunol.* **30**, 1939–1947.

Tighe, H., Takabayashi, K., Schwartz, D., Van Nest, G., Tuck, S., Eiden, J. J., Kagey-Sobotka, A., Creticos, P. S., Lichtenstein, L. M., Spiegelberg, H. L. and Raz, E. (2000b). Conjugation of immunostimulatory DNA to the short ragweed allergen amb a 1 enhances its immunogenicity and reduces its allergenicity. *J. Allergy Clin. Immunol.* **106**, 124–134.

Trinchieri, G. (1995). Interleukin-12: a proinflammatory cytokine with immunoregulatory functions that bridge innate resistance and antigen-specific adaptive immunity. *Annu. Rev. Immunol.* **13**, 251–276.

Ulmer, J. B., Donnelly, J. J., Parker, S. E., Rhodes, G. H., Felgner, P. L., Dwarki, V. J., Gromkowski, S. H., Deck, R. R., DeWitt, C. M. and Friedman, A. (1993). Heterologous protection against influenza by injection of DNA encoding a viral protein. *Science* **259**, 1745–1749.

Wang, R., Doolan, D. L., Le, T. P., Hedstrom, R. C., Coonan, K. M., Charoenvit, Y., Jones, T. R., Hobart, P., Margalith, M., Ng, J., Weiss, W. R., Sedegah, M., de Taisne, C., Norman, J. A. and Hoffman, S. L. (1998). Induction of antigen-specific cytotoxic T lymphocytes in humans by a malaria DNA vaccine. *Science* **282**, 476–480.

Wheeler, C. J., Felgner, P. L., Tsai, Y. J., Marshall, J., Sukhu, L., Doh, S. G., Hartikka, J., Nietupski, J., Manthorpe, M., Nichols, M., Plewe, M., Liang, X., Norman, J., Smith, A. and Cheng, S. H. (1996). A novel cationic lipid greatly enhances plasmid DNA delivery and expression in mouse lung. *Proc. Natl. Acad. Sci. USA* **93**, 11454–11459.

Wierenga, E. A., Snoek, M., de Groot, C., Chretien, I., Bos, J. D., Jansen, H. M. and Kapsenberg, M. L. (1990). Evidence for compartmentalization of functional subsets of CD4+ T lymphocytes in atopic patients. *J. Immunol.* **144**, 4651–4656.

Wolff, J. A., Malone, R. W., Williams, P., Chong, W., Acsadi, G., Jani, A. and Felgner, P. L. (1990). Direct gene transfer into mouse muscle in vivo. *Science* **247**, 1465–1468.

Xiang, Z. and Ertl, H. C. (1995). Manipulation of the immune response to a plasmid-encoded viral antigen by coinoculation with plasmids expressing cytokines. *Immunity* **2**, 129–135.

Xiang, Z. Q., Spitalnik, S., Tran, M., Wunner, W. H., Cheng, J. and Ertl, H. C. (1994). Vaccination with a plasmid vector carrying the rabies virus glycoprotein gene induces protective immunity against rabies virus. *Virology* **199**, 132–140.

Yaegashi, Y., Nielsen, P., Sing, A., Galanos, C. and Freudenberg, M. A. (1995). Interferon beta, a cofactor in the interferon gamma production induced by gram-negative bacteria in mice. *J. Exp. Med.* **181**, 953–960.

Yamamoto, S., Kuramoto, E., Shimada, S. and Tokunaga, T. (1988). In vitro augmentation of natural killer cell activity and production of interferon-alpha/beta and -gamma with deoxyribonucleic acid fraction from *Mycobacterium bovis* BCG. *Jpn J. Cancer Res.* **79**, 866–873.

Yamamoto, T., Yamamoto, S., Kataoka, T., Komuro, K., Kohase, M. and Tokunaga, T. (1994). Synthetic oligonucleotides with certain palindromes stimulate interferon production of human peripheral blood lymphocytes in vitro. *Jpn J. Cancer Res.* **85**, 775–779.

Yang, N. S. and Sun, W. H. (1995). Gene gun and other non-viral approaches for cancer gene therapy. *Nature Med.* **1**, 481–483.

Yi, A. K., Chace, J. H., Cowdery, J. S. and Krieg, A. M. (1996a). IFN-gamma promotes IL-6 and IgM secretion in response to CpG motifs in bacterial DNA and oligodeoxynucleotides. *J. Immunol.* **156**, 558–564.

Yi, A. K., Klinman, D. M., Martin, T. L., Matson, S. and Krieg, A. M. (1996b). Rapid immune activation by CpG motifs in bacterial DNA. Systemic induction of IL-6 transcription through an antioxidant-sensitive pathway. *J. Immunol.* **157**, 5394–5402.

Yokoyama, M., Zhang, J. and Whitton, J. L. (1995). DNA immunization confers protection against lethal lymphocytic choriomeningitis virus infection. *J. Virol.* **69**, 2684–2688.

4.5 Preparation and Use of Adjuvants

Karin Lövgren-Bengtsson and Caroline Fossum
Department of Veterinary Microbiology, Swedish University of Agricultural Sciences, Uppsala, Sweden

◆◆

CONTENTS

◆◆◆◆◆◆ INTRODUCTION

The immunogenicity of an antigen–adjuvant formulation, in terms of humoral and cellular immune responses, is dependent on a range of different factors. Among these are the physical characteristics of the antigen, in particular its size, charge and degree of glycosylation, the antigenicity and innate adjuvanticity, and the choice of adjuvant. Most of these factors regarding the physical and biological characteristics of the antigen cannot easily be altered, but the adjuvant part can be selected to suit a particular protein and to modulate the immune responses in a desired direction, for example with an emphasis on an antibody response dominated by certain isotypes or subclasses, or a need for cellular effector mechanisms such as cytotoxic T cells.

In the literature there are numerous examples of different antigens and their performance with various adjuvants, review articles and book chapters on adjuvants and adjuvant characteristics. However, despite all this information, it is difficult to predict the immunological outcome of a previously untested adjuvant–antigen combination. Adjuvant research is far less empirical today than 20 years ago. Many more of the mechanisms behind adjuvant activity are known, helping us to make rational decisions in our choice of adjuvant formulation for a certain antigen preparation.

This chapter focuses on the interactions between antigens and a few commonly used adjuvants, identifying some key factors affecting the

Preparation and Use of Adjuvants

Copyright © Elsevier Science Ltd
All rights of reproduction in any form reserved

success of the antigen–adjuvant formulation. It is not possible to cover here all the types of adjuvants available today, but a selected number of adjuvants and adjuvant formulations will serve as examples of the parameters involved.

◆◆◆◆◆◆ ADJUVANT ACTIVITY AND ANTIGEN–ADJUVANT FORMULATION

The purpose of an adjuvant is simply to increase the immunogenicity of an antigen and to modulate the immune response to the antigen in a desired direction. Although we cannot generalize, considering the mode of action of the hundreds of adjuvant formulations available today (Cox and Coulter, 1992, 1997; Vogel and Powell, 1995) there are four major areas in which adjuvants exert their adjuvant activities or, alternatively, where antigens may require help to increase their immunogenicity (Morein *et al.*, 1996). These areas are:

- *Physical presentation of the antigen.* This involves the physical appearance and the antigenicity of the antigen in an adjuvant formulation. Factors included here are stabilization and exposure of native conformational epitopes in the antigen, adjuvant-mediated formulation of the antigen into small soluble particles or aggregates, and other mechanisms bringing the antigen into an organized multimeric formation and hence increasing the surface area of the antigen.
- *Antigen and adjuvant uptake and distribution (targeting).* This covers a whole range of activities, including slow release of antigen from a depot at the injection site, the initiation of the immune response by attracting appropriate antigen-presenting cells, and other activities leading to increased antigen uptake and transport of antigen to relevant lymphatic organs. A subset of these activities covers the intracellular handling of antigens, especially their proteolytic processing and association with major histocompatibility complex (MHC) class I or II molecules.
- *Immune potentiation and modulation.* These include activities that regulate both quantitative and qualitative aspects of the ensuing immune responses, generally as a result of modulation of cytokine networks. This will usually result in expansion of T-cell clones with different profiles of cytokine production.
- *Activation of innate immunity.* Recognition of PAMP (pathogen associated molecular pattern) by PRR (pathogen recognizing receptors) and upregulation of co-stimulatory molecules on APC (antigen-presenting cells). Innate immunity has for long been described as a non-specific immune response characterized by cellular engulfment and destruction of micro-organisms. According to more recent data however, the innate immunity, encompassing macrophages, NK cells, complement factors, type I interferons and other early cytokines as well as acute phase reactants, has considerable specificity and is capable of discriminating between pathogens and self (Medzhitov and Janeway, 1997; Hoffman *et al.*, 1999; Aderem and Ulevitch, 2000) or dangerous from harmless (Matzinger, 1994; Pennisi, 1996; Ridge *et al.*, 1996). In addition the activation of the innate immune response is likely a prerequisite for

triggering of the acquired immunity and hence a key event for successful immunization (Akira *et al.*, 2001)

Very few adjuvants exert strong activities in all areas above and for poor antigens that require extensive adjuvant help a combination of adjuvants may be required for appropriate immune stimulation.

The term 'adjuvant' or 'adjuvant formulation' covers activities within one or more of the four areas mentioned above and possibly other mechanisms leading to increased immunogenicity of an antigen. Allison and Byars (1990) introduced some structure into the terminology by defining an adjuvant as an agent that augments specific immune stimulation to antigens, a vehicle as the substance used for the delivery of the antigen, and an adjuvant formulation as the combination of adjuvants in a suitable vehicle. However, since vehicles also may exert adjuvant activities, especially in the area of antigen presentation, the concept of adjuvant activity remains unclear.

In order to discuss the practical aspects of adjuvants, vehicles and adjuvant formulations, this chapter will not strictly follow this definition, but rather uses the term 'adjuvant' in a broader sense to mean to help (increase immunogenicity), as from its Latin origin 'adjuvare' (to help). The only clear distinction that will be made is that between particulate and non-particulate adjuvants (Cox and Coulter, 1992).

◆◆◆◆◆◆ ANTIGENS

The physical and biological variability of antigens forms a continuum. In practice it is useful to make the following categorization.

Particulate antigens

Among the particulate antigens we find whole cells and micro-organisms or various sized parts thereof. Smaller particles (of the order of nanometres) are protein micelles (or rosettes) and other small aggregates spontaneously formed in aqueous solution by hydrophobic and amphipathic antigens. Particulate antigens are generally good immunogens and it is comparatively easy to increase their immunogenicity further by the addition of an immunomodulatory adjuvant.

Monomeric antigens

Monomeric antigens are proteins that do not spontaneously form particles or aggregates in solution. These antigens are generally poor immunogens and require help from potent adjuvants. Since the antigens themselves do not form particles, these antigens often perform well when administered with particulate adjuvants. Unfortunately, most recombinant-DNA products belong to this group of antigens, and this has so far limited their use in vaccines.

Biological activities: antigen-induced immunomodulation

Many antigens, especially those derived from micro-organisms, have immunomodulatory effects. The immunomodulation can be both positive and negative from immunization points of view, and cannot easily be overcome by addition of adjuvants. For example, the flagellar fraction of *Trypanosoma cruzi* contains an antigen (Ag 123) that has immunosuppressive activity (Hansen *et al.*, 1997). This antigen possibly prevents the development or the recall of memory cells secreting γ-interferon (IFN-γ) (Hansen *et al.*, 1996). Many microbial antigens have useful affinities, e.g., the sialic acid binding property of haemagglutinin from influenza virus. This enables mucosal uptake of haemagglutinin as well as other antigens lacking binding properties, provided that they are physically associated with the haemagglutinin (Hu *et al.*, 2001). Likewise, the ADP-ribosylating ability of cholera toxin derived hybrid molecule CTA1-DD has a mucosal adjuvant effect mediated by the enzymatic activity of the protein (Lycke, 2001). Contrary to many adjuvants CTA1-DD induces an immune response to itself as well as to admixed antigens (Lycke, 2001; Lycke and Schön, 2001). Formulated in ISCOMs the adjuvant effect was further potentiated due to a synergistic effect of the two adjuvant systems (Mowat *et al.*, 2001).

The fact that some antigens are immunogenic at 100-fold smaller doses than others and that each antigen–adjuvant combination has its optimum of both adjuvant and antigen dose should also be taken into consideration in the search of adjuvant for a particular antigen.

Modulation by antigen dose and mode of administration

It has been shown that not only the physical and biological activities of an antigen influence its immunogenicity, but also the dose of the antigen may modulate the immune responses. For example, a medium dose of peptide antigen induced a T-helper 0/T-helper 1 cell (Th0/Th1) type response, while high and low doses of the same peptide induced predominantly Th2 responses (Mosmann and Coffman, 1989; Bretscher *et al.*, 1992; Hosken *et al.*, 1995). It was also shown that neonatal tolerance may not be an intrinsic property of the newborn immune system. Instead the dose of antigen had to be balanced against the low number of virgin T cells present in the newborn spleen, and consequently the unresponsiveness of newborn mice could be avoided simply by lowering the antigen dose (Forsthuber *et al.*, 1996; Ridge *et al.*, 1996; Sarzotti *et al.*, 1996).

Several adjuvants such as oil emulsions, the early non-ionic block polymers (e.g., L121) and to a certain extent Al(OH)$_3$ gel employ a sustained release of antigen as an important mode of action. Typically these adjuvants are very strong in priming for an immune response but weaker in the booster situation. An adjuvant like ISCOMs that are efficient for priming but particularly strong in the booster situation is hampered by a sustained release or frequent administrations (Johansson *et al.*, 2000). The resulting immune response is not only decreased but also shifted towards TH2 reactions.

◆◆◆◆◆◆ PARTICULATE ADJUVANTS

Particulate adjuvants are adjuvants that are capable of arranging antigens into particle-like organizations, small aggregates, droplets or precipitates, where such activity is an important component of their adjuvant activity (Cox and Coulter, 1992). It is among the particulate adjuvants we find the most frequently used adjuvants, such as Freund's complete and incomplete adjuvants, the aluminium salt gels but also liposomes and ISCOMs.

Aluminium salts

The history of aluminium adjuvants begins early in the 20th century with the observation that alum precipitation significantly increased the immunogenicity of toxoids (Glenny et al., 1926). Alum-precipitated antigens were prepared by mixing antigens in solution with potassium alum to form protein aluminates, a product that could be highly heterogeneous depending on which anions (bicarbonate, sulphate or phosphate) were present in the mixture. This heterogeneity was later overcome by the use of preformed aluminium hydroxide and aluminium phosphate gels to which antigens were adsorbed. However, preformed gels also differ in terms of physico-chemical characteristics, stability and protein adsorption. To permit comparisons between research activities, an international workshop elected to use one good commercial preparation as the research standard (Stewart-Tull, 1989). Although aluminium adjuvants have immunostimulatory activities such as attraction of eosinophils to the injection site (Walls, 1977) and complement activation (Ramanathan et al., 1979), most evidence indicates the absorption of antigens to the gel as the key adjuvant activity. The mechanisms responsible for adsorption of antigens include intramolecular forces such as electrostatic attraction, hydrophilic and hydrophobic interactions, hydrogen bonds and van der Waals forces (Seeber et al., 1991).

The antigen is adsorbed to the gel by incubation (slow agitation) at a preselected pH for a few hours or overnight. The pH optimum for adsorption varies between antigens. In a mixture of antigens some may bind better than others, and to obtain a reproducible and reliable binding different pH values for both the hydroxide and phosphate gels need to be tested. For example, Seeber et al. (1991) showed that proteins with an acid isoelectric point adsorb better to aluminium hydroxide and that proteins with an alkaline isoelectric point adsorb better to aluminium phosphate.

The stability of the antigen is also of vital importance. Some antigens are highly sensitive to pH alterations and the final pH chosen may very well be a compromise between antigen stability and antigen adsorption to the gel. The degree of antigen adsorption to the gel can readily be tested. After incubation of the antigen with the gel at different pH, the gel is spun in a simple low speed centrifuge and the clear supernatant is tested for presence of antigen by enzyme-linked immunosorbent assay (ELISA), electrophoresis, bioassay or any other suitable technique.

Preparation and Use of Adjuvants

To date Al(OH)$_3$ is the most used adjuvant in commercial vaccines, until recently when MF59 was approved, aluminium salts were the only approved adjuvants for use in human vaccines.

Type of immune modulation

Adsorption of antigens to Al(OH)$_3$ increases the efficiency of antigen presentation to antigen-presenting cells and upregulates the latter as measured by an increased interleukin-1 (IL-1) production (Mannhalter *et al.*, 1985). Al(OH)$_3$ activate complement (Ramanathan *et al.*, 1979), improves antigen trapping in lymph nodes, and acts as a short-term (1–2 weeks) depot for antigen release. Thus the antigen is presented in a 'particulate' manner and the rate of antigen targeting is increased (Mannhalter *et al.*, 1985). These properties of Al(OH)$_3$ are probably the major reasons why it works so well with small soluble antigens such as toxoids. Experiments done with Al(OH)$_3$ and different antigens suggest that Al(OH)$_3$ is superior with poorly immunogenic antigens, but of less value in potentiating the immunogenicity of strong antigens (Bomford, 1984). The resulting immune modulation afforded by Al(OH)$_3$ is moderate and mainly affects the antibody response, characterized by an increased production of IL-4 leading to a Th2-type antibody response dominated by murine IgG1 antibodies (IgG2 in humans) but poor cell-mediated responses (CMI) (Bomford, 1980a). Whether the Th2-type response (Mosmann and Coffman, 1989; Hu and Kitagawa, 1990) is an immunomodulatory effect of Al(OH)$_3$ or merely a potentiation of the response generated by the antigen itself is not clear, since monomeric, poorly immunogenic antigens will themselves induce a Th2-type response. However, the induction of IgE, at least in rodents (Hamamoka *et al.*, 1973; Uede *et al.*, 1982; Uede and Ishizaka, 1982), may imply the former. The value of Al(OH)$_3$ in the induction of secondary responses is often low after a strong priming (Hu and Kitagawa, 1990).

Dosing

The amount of Al(OH)$_3$ gel used for injection has to be tested for each antigen and species, but the concentration normally varies in the range 15–40% v/v. No more than 1.25 mg aluminium is allowed in one dose of human vaccine.

Oil emulsions

Water-in-oil emulsions

The aluminium salts adsorb antigen to a gel, while in oil adjuvants the antigens are in aqueous-phase droplets in water-in-oil (w/o) emulsions, best exemplified by Freund's complete (FCA) and incomplete (FIA) adjuvants. A w/o emulsion is an even dispersion of microdroplets of water in a continuous phase of oil. As a lot of energy is required for its production, the resulting emulsion becomes unstable unless an emulsifier (an agent consisting of polar and non-polar regions) is added to reduce the interfacial tension.

The most common oils used in w/o adjuvants are light mineral oils such as Drakeol 6 VR, Bayol F and Marcol 52. These are complicated mixtures, but usually contain varying proportions of paraffin and naphthene derivatives (Dalsgaard *et al.*, 1990). Arlacel A (mannide monooleate) was one of the first emulsifiers commonly used, and is still in use in a more purified form (Arlacel A Special, ICI, Montanide 80, Seppic). Objection to the use of mineral oils for vaccines for food animals and, in some countries, all animals, has led to the search for metabolizable alternatives. Vegetable oils such as peanut and sesame oils are reported to be useful, but are less efficient than FIA (Hilleman, 1966; Fukumi, 1967; Kimura *et al.*, 1978; Brugh *et al.*, 1983). Squalene, an intermediate in the biosynthesis of cholesterol and its hydrogenated form, squalane, are preferred options.

A report that Arlacel A is carcinogenic in Swiss Webster mice and co-carcinogenic in all strains of mice tested (Murray *et al.*, 1972) increased the importance of alternative biodegradable emulsifiers. A blend of Tween 85 and Span 85 (Bokhout *et al.*, 1981) is reported to form stable emulsions with higher adjuvant activity than FIA (Ott *et al.*, 1995). The non-ionic Pluronic 122 (see below) has been used as an emulsifier for a range of fatty acid esters, with good results (Bomford, 1981).

Type of immune modulation

Water-in-oil formulations are generally efficient adjuvants and their main adjuvant effects rely on their ability to target the antigen to the immune system. The antigen must be located in microdroplets of water which are in the oil phase, but such an emulsion can be formed with almost any antigen, regardless of its size, charge or other physical characteristics, provided it is water soluble. However, it is argued that antigens may denature during the emulsification process (Kenney *et al.*, 1989), and w/o emulsions are possibly less well suited for antigens where preservation of conformational epitopes is vital. Water-in-oil formulations are highly efficient for inducing T-cell responses and antibody responses to linear B-cell epitopes. After injection, the antigen is continuously released from a depot formed at the injection site, attracting mononuclear cells. However, the usefulness of this depot is debated, since excision of the depot 6–8 weeks (Lascelles *et al.*, 1989) after administration failed to affect the immune responses. Like $Al(OH)_3$, w/o emulsions without additional immunomodulators such as mycobacteria in the case of FCA, induce mouse antibodies predominantly of the IgG1 isotype, and do not stimulate delayed-type hypersensitivity (White, 1976).

Oil-in-water emulsions

Oil-in-water emulsions consist of microdroplets of oil in a continuous phase of water, and were produced in attempts to prepare emulsions of reduced viscosity compared with w/o emulsions (Meyer *et al.*, 1974). In addition, emulsions with reduced amounts of oil would be expected to be less reactogenic. The oil phase of oil-in-water (o/w) emulsions is typically 1–5%, compared with 25–50% in w/o emulsions.

An o/w emulsion recently approved for human use is MF59 (Ott *et al.*, 2000). The original MF59 emulsion was composed of 5% v/v squalene, 0.5% v/v polysorbate 80 (Tween 80) and 0.5% v/v sorbitan trioleate (Span 85), emusified under high pressure conditions in a microfluidizer and transformed in small uniformed droplets. A second generation emulsion with enhanced stability designated MF59C.1 has been developed (Ott *et al.*, 2000). This formulation has proven effective for a large variety of antigens and will most likely constitute a serious competitor to the aluminium salt adjuvants (Dupuis *et al.*, 2001; Drulak *et al.*, 2001; Gasparini *et al.*, 2001; Nicholson *et al.*, 2001, Stowers *et al.*, 2001).

Type of immune modulation

Oil-in-water emulsions are generally less potent than w/o formulations, at least with regard to the duration of the immune response (Dalsgaard *et al.*, 1990). They are generally well tolerated and have a viscosity which is low compared with w/o formulations. Also, in o/w emulsions, hydrophobic and amphipathic antigens are presented in an organized particulate form (Dalsgaard, 1987). However, it is generally believed that addition of surface-active molecules such as non-ionic block co-polymers (Allison and Byars, 1986), sorbitan or glycerol trioleate (Woodard, 1989) or trehalose dimycolate (TDM) (McLaughlin *et al.*, 1978; Hunter *et al.*, 1981; Lemaire *et al.*, 1986) is essential for optimum presentation of most antigens. Oil-in-water emulsions are generally not considered to have immunomodulatory activity, but they offer a suitable vehicle for a range of immunomodulatory substances.

Water/oil/water emulsions

Water/oil/water emulsions (i.e. a w/o emulsion re-emulsified in water) have also been made in attempts to reduce the viscosity of w/o emulsions (Hunter *et al.*, 1995). These are technically difficult to prepare and are less stable than w/o and o/w emulsions.

Non-ionic block polymers

The finding that many adjuvants are surface-active molecules led Hunter and co-workers to synthesize and study a new group of surface-active molecules with an apparent adjuvant effect – the non-ionic block polymers (NBPs) (Hunter *et al.*, 1981; Hunter and Bennett, 1984, 1986). NBPs are co-polymers of polyoxyethylene and polyoxypropylene of various chain length and different hydrophilic/lipophilic balance (HLB). The adjuvant active NBPs tend to have a lower HLB, i.e. they are more lipophilic. Many of these polymers also act as emulsifiers, forming stable w/o, o/w and w/o/w emulsions.

Oil-in-water co-polymer adjuvants have been reported to have induced potent immune responses and protection in many studies (Byars and Allison, 1987; Millet *et al.*, 1992). The co-polymers used in these studies have one severe disadvantage; they are insoluble at room temperature despite being soluble when refrigerated (Hunter *et al.*, 1995). Consequently, the emulsion is destabilized when refrigerated. This problem can be solved by microfluidization of the emulsion; however, even then the emulsions are less effective adjuvants than classical o/w emulsions (Hunter *et al.*, 1981).

The largest co-polymers are particularly effective in stabilizing w/o emulsions. For example, water-in-squalene emulsions containing up to 90% water were stable at room temperature for months, and can even be frozen and thawed (Hunter *et al.*, 1995).

Large block polymers can also stabilize w/o/w emulsions. Such double emulsions have been used with Simian immunodeficiency virus (SIV) (Hunter *et al.*, 1995) and malaria (Millet *et al.*, 1992) in monkeys with good efficacy, but with some severe local reactions.

The new larger co-polymers form stable microparticulate suspensions in saline without oil. The efficacy of these preparations is largely dependent on the physical properties of the antigens, since antigens that fail to bind to the polymer give poorer responses (Hunter *et al.*, 1995), while antigens that bind become highly immunogenic (Todd and Newman, 2000). Promising results have been reported with a range of different antigens in mice (Newman *et al.*, 1997; Todd *et al.*, 1997; McNicholl *et al.*, 1998; Katz *et al.*, 2000) monkeys (Collins *et al.*, 2000) and man (Triozzi *et al.*, 1997).

Some of these block polymers have been commercialized as research adjuvants under the names TiterMax and TiterMax Gold (CytRx Corporation). TiterMax emulsions consist of 80% saline in squalene containing block copolymer CRL-8941, sorbitan monooleate and co-polymer-coated silica particles (Hunter *et al.*, 1995). TiterMax Gold contains another block co-polymer, CRL-8300.

Type of immune modulation

The precise mode of action of this class of adjuvant formulation is unknown, but the binding of antigen to the polymer seems to be essential, since co-polymers with low antigen-binding capacity have a low adjuvant activity. The co-polymers also augment the expression of class II major histocompatibility antigen by macrophages (Howerton *et al.*, 1990) and it is likely that the major activities exerted by the block co-polymers relate to a favourable physical presentation of the antigen in a condensed three-dimensional matrix in a milieu of activated antigen-presenting cells (Hunter *et al.*, 1994). CRL-1005 is reported to induce a balanced Th1/Th2 response that can be altered by modifying the hydrophobicity/hydrophilicity of the co-polymer (Newman *et al.*, 1998; Todd and Newman, 2000).

Liposomes

Lipid vesicles or liposomes are vehicles for vaccine adjuvant formulations that have attracted the interest of many groups due to their versatility and liposomes are used as vaccine carriers with both protein antigens and DNA (Gregoriadis *et al.*, 2000). Their physical properties can be varied to meet a variety of demands and liposomes are on the market and in advanced clinical studies for drug or antigen delivery (Lasic, 1998). Liposomes with physically diverse properties can be made by altering the composition of the lipids and the method used for production (Alving, 1992). Liposomes range in size from 20 nm up to more than 10 μm, and they can be uni- or multilamellar, rigid or fusogenic, charged or non-charged. Hydrophilic antigens can be contained in the interior or between the lipid bilayers, while hydrophobic or amphipathic antigens can be integrated in the lipid membrane during the preparation. Due to the great number of possible variations of both the composition of liposomes and their interaction with different antigens, it is difficult to generalize their effects. It is clear that liposomes are potent in presenting antigens, especially hydrophobic or amphipathic antigens presented on the liposomal surface. Liposomes by themselves are not generally regarded as particles that efficiently enhance or modulate the immune response, and require supplementation with immunomodulators. However, a paper by Phillips *et al.* (1996) indicates that the phospholipid composition influences the antibody response to encapsulated antigens. Almost every possible combination of liposomes and other adjuvants has been tested (Alving, 1991; Kersten and Crommelin, 1995).

Variants of liposomes, a form of antigen presentation related to liposomes are the so-called protein cochleate formulations. These are protein–lipid (cholesterol/phosphatidyl serine)–calcium (Ca^{2+}) precipitates consisting of a large continuous solid lipid bilayer sheet rolled up in a 'jelly roll' fashion (Gould-Fogerite *et al.*, 1994). Amphipathic or hydrophobic antigens can be integrated into the lipid bilayer of the cochleates. The presence of calcium maintains the cochleates in their rolled-up form, and removal of the calcium by diffusion or addition of a chelating agent such as ethylene diamine tetraacetic acid (EDTA) allows the cochleate to unroll and form large, mainly unilaminar, liposomes.

Trehalose dimycolate

Trehalose-6,6'-dimycolate (TDM) is a high-molecular-weight glycolipid isolated from mycobacteria (Bloch, 1950; Noll *et al.*, 1956). TDM is an amphipathic surface-active molecule frequently employed in o/w emulsions in the presence of Tween 80 (Lemaire *et al.*, 1986). The orientation of TDM in o/w emulsions promotes retention of soluble antigens (McLaughlin *et al.*, 1978) and is able to concentrate soluble antigens on the surface of the oil droplets. The efficacy of TDM as an adjuvant in emulsions depends on the physical characteristics of the emulsion. The most potent emulsions of TDM comprise small oil droplets with concomitant large surface areas (Rudbach *et al.*, 1995).

Cell-wall skeleton

The mycobacterium cell-wall skeleton (CWS) is defined as the material remaining after nucleic acid, protein and free fatty acid have been removed from the cell wall. The CWS is a particulate complex consisting of peptidoglycan, a polymerized form of muramyl dipeptide (MDP), arabinogalactane and mycolic acid (Rudbach *et al.*, 1995). When formulated in o/w emulsions, the CWS tends to coat the surface of the oil droplets (McLaughlin *et al.*, 1978).

The immunostimulatory effect of the CWS is probably mediated by several parts of the complex, since both peptidoglycan and mycolic acid are immunostimulatory.

The CWS is one of the constituents in Ribi adjuvant systems (see below).

◆◆◆◆◆◆ NON-PARTICULATE ADJUVANTS

Saponins

Data on saponin derivation are often not available. However, most adjuvant-active saponin preparations are probably derived from the South American soap tree *Quillaja saponaria Molina*. Saponins from other plants may also be adjuvant active, and saponins in general are often mentioned as the active component in medicinal plants (Campbell, 1995).

The adjuvant activity of *Quillaja* saponins was recognized in the early literature on adjuvants (Ramon, 1926; Thibault and Richou, 1936) and their commercial use in vaccines dates back to 1951 (Espinet, 1951). Since the pioneer work of Dalsgaard (1974, 1978), much effort has been directed at the separation of *Quillaja* saponin extracts in order to identify which components are adjuvant active and which components are responsible for undesired effects such as toxicity. The goal for this activity has been to find a component or a fraction of *Quillaja* saponin that is well characterized and suitable for use in human vaccines. Currently, companies like Antigenics Inc, Galenica Pharmaceuticals Inc. CSL Ltd and Isconova AB are working with characterized preparations of saponins for vaccine use. Antigenics use a highly purified fraction of *Quillaja* saponin, 21-QS-21 (Kensil *et al.*, 1991; Kensil, 2000), Galenica has developed semisynthetic triterpenoid saponin derivatives (GPI-0100) with increased tolerability and prolonged stability in aqueous solution (Marciani *et al.*, 2000, 2001). The immunostimulatory capacities of these derivatives seem in essence to parallel that of the parent molecules. CSL and Isconova AB are developing the use of *Quillaja* saponins (Iscoprep 703 and Quil A) for use in ISCOM and ISCOMMATRIX for human and veterinary applications (Vogel and Powell, 1995; Barr and Mitchell, 1996).

QS-21 and GPI-0100 are used in addition to an antigen(s) in solution or mixed with $Al(OH)_3$ while Iscoprep 703 is intended for production of iscoms or iscom-matrix (see below). QS-21 is a substantially pure saponin component, whereas Iscoprep 703 consists of two fractions of *Quillaja*

Preparation and Use of Adjuvants

saponins. Iscoprep 703 is designed not only for optimum adjuvant activity with minimum toxicity, but also efficiently to form iscoms with a range of different antigens and lipids (see below).

Type of immune modulation

In general, saponins induce potent antibody and cell-mediated immune responses, especially with antigens in cell membranes (Bomford, 1980a,b; Scott et al., 1984) and other particulate antigens. Monomeric or small antigens, i.e. weak antigens, often perform poorly when adjuvanted with saponins (Bomford, 1984), most likely because saponins as such do not provide good physical antigen presentation. However, saponin used in iscoms (see below) or combined with $Al(OH)_3$ (Dalsgaard, 1978; Ma et al., 1994) or liposomes (Lipford et al., 1994), to confer a good physical presentation of antigen, often yield very efficient adjuvant formulations, inducing both cellular and humoral responses. Scott et al. (1984) noted an acute local inflammation after subcutaneous injection of unfractionated *Quillaja* saponins. Although antigen was retained at the injection site, increased amounts also reached the spleen. The local inflammation was probably caused by cell and/or tissue damage due to saponin binding to cell-membrane cholesterol. Both the inflammation and the splenic localization were blocked by added liposomal cholesterol, suggesting that these effects were mediated through the local inflammation. In contrast, iscoms and iscom-matrix (see below) comprise *Quillaja* saponins bound to cholesterol, and hence a local inflammation is less likely to occur and the antigen is rapidly removed from the injection site and focused to the spleen (Watson et al., 1989; Lövgren-Bengtsson and Sjölander, 1996; Sjölander et al., 1996a, 1997a,b). The immune potentiation of *Quillaja* saponins is probably mediated by several mechanisms and influenced not only by the antigen but also by the antigen presentation form.

There is conflict about which subclasses of IgG are enhanced (Scott et al., 1984; Kenney et al., 1989; Karagouni and Hadjipetrou-Kourounakis, 1990; Kensil et al., 1991). However, oral ingestion of saponins resulted in a non-specific resistance to intracerebral challenge with rabies virus (Chavali and Campbell, 1987), suggesting a stimulation of IL-2 and IFN-γ. This is supported by the results of Heath et al. (1991), which showed that saponin seems to mimic the effects of IFN-γ, suggesting that one property of saponin is release of IFN-γ.

Dosing

As with most adjuvants, the dose of saponin should be adjusted to the antigen and the animal species. Attention must also be focused on the purity of the saponins, since crude saponin preparations may need to be used in considerably higher doses. As a guideline, Dalsgaard et al. (1990) gave the following dose recommendations for Quil A (a semipurified and characterized preparation of adjuvant active saponins): 10 μg for use in mice, 50 μg in guinea-pigs, 200 μg in rabbits, 500 μg in pigs and 1000 μg in cattle. The amount of antigen can vary from microgram up to milligram quantities, depending on its intrinsic immunogenicity, although a range of 10–100 µg is typical for most vaccines. The tolerability of purified or modified saponins mentioned above are generally higher.

Iscom-matrix

The iscom-matrix is a particulate complex with identical composition, shape and appearance as the iscom (see below), except that it lacks inserted antigens (Lövgren and Morein, 1988). Since the iscom-matrix contains adjuvant-active Quil A, it can be used as an adjuvant simply mixed with antigens. A major advantage of using the iscom-matrix, compared with free saponin, is that the haemolytic activity of saponins in iscom-matrix is absent or drastically reduced (Kersten et al., 1991; Rönnberg et al., 1995). The saponins are bound to cholesterol within the complex, and therefore do not bind to tissue cholesterol at the site of injection and thereby cause local reactions (Rönnberg et al., 1995; B. Sundquist, personal communication). However, to date only limited information is available concerning the mechanism of adjuvant activity of iscom-matrix, especially in comparison to iscoms (Lövgren-Bengtsson and Sjölander, 1996; Cox et al., 1997). However, the immunogenicity of iscom-matrix antigen is more dependent on the properties of the antigen than that of iscoms.

Type of immune modulation

The subclass distribution of antigen-specific serum antibodies in mice immunized with antigen mixed with iscom-matrix roughly parallel that in mice immunized with iscoms. Likewise, antigen-specific spleen cells from mice immunized with iscoms or flu-Ag adjuvanted with iscom-matrix produce high levels of IL-2 and IFN-γ after restimulation (Lövgren-Bengtsson and Sjölander, 1996). Another feature of iscoms, the activation of cytotoxic T lymphocytes (CTLs), is reported after immunization with antigen adjuvanted with iscom-matrix (Cox et al., 1997). However, these CTL responses were substantially weaker (about three-fold) than those induced by iscoms.

Iscom-matrix can lack the superior antigen-presenting ability of iscoms, because an antigen mixed with iscom-matrix may not be physically associated with the iscom-matrix. Like saponins in general, iscom-matrix is a potent and useful adjuvant for particulate or strong antigens (Jones et al., 1995; Snodgrass et al., 1995), but inferior with poor antigens (Lövgren and Sundquist, 1994). Recent data however, suggest that iscom-matrix is a potent adjuvant with many antigens after local and systemic administrations (Ugozzoli et al., 1998; Windon et al., 2001; Polakos et al., 2001).

Dosing

The dose recommendations for *Quillaja* saponins (see above) are likely to be valid also for iscom-matrix.

Lipid A

The outer cell membrane of Gram-negative bacteria comprises an amphipathic molecule known as lipopolysaccharide (LPS) or endotoxin. LPS is a potent adjuvant (Johnsson *et al.*, 1956), but also induces toxic reactions such as fever and lethal shock. It has been shown (Takada and Kotani, 1989) that most of the biological activities of LPS are exerted by lipid A, a structural component of LPS. Variants of lipid A can be obtained from a variety of bacterial species, varying in their degree of toxicity and immunomodulating activity (Alving, 1993). Lipid A is a disaccharide of glucosamine with two phosphate groups and five or six fatty acid chains of variable length, usually C12 to C16 (Rietschel *et al.*, 1985). Attempts to separate the toxic properties from the immunomodulating activities have resulted in molecules with substantially reduced toxicity, the best known being monophosphoryl lipid A (MPL). Other detoxified derivatives of lipid A have been obtained by succinylation and phthalyation (Schenk *et al.*, 1969; Chedid *et al.*, 1975).

Type of immune modulation

Lipid A is a potent immunomodulator that can be formulated in an aqueous mixture with an antigen, but due to its amphipathic properties it is often contained in liposomes or oil emulsions.

Macrophages stimulated by lipid A increase their phagocytic capacity facilitating uptake, processing and presentation of antigen (Unanue and Allen, 1987) and release of IL-1, IL-6, IL-8, granulocyte–macrophage colony stimulating factor (GM-CSF) and tumour necrosis factor (TNF) (Arend and Massoni, 1986; Henricson *et al.*, 1990; Nowotny, 1990; Astiz *et al.*, 1995). In addition, lymphocytes produce IFN-γ and IL-2 (Carozzi *et al.*, 1989; Odean *et al.*, 1990). In contrast to most adjuvants, MPL (like its parent molecule LPS) is not dependent on physical association with the antigen to exert adjuvant effects. This property allows MPL to be used alone or together with other adjuvant/antigen formulations depending on the physical and immunological characteristics of the antigen (Ulrich, 2000).

Dosing

In mice, the dose of lipid A derivatives may vary from about 10–300 μg given in saline, w/o emulsion, o/w emulsion, in liposomes alone or in liposomes mixed with other immunomodulators. Rats, guinea-pigs and rabbits are given about 100–300 μg.

MDP derivatives

Muramyl dipeptide (MDP) is the minimum adjuvant-active component obtained from the cell wall of certain Gram-positive bacteria, particularly mycobacteria. MDP has diverse biological activities, from modulation of acquired and innate immune responses to modulatory effects on the central nervous system (Takada and Kotani, 1995). It is a strong but reactive immunomodulator in w/o emulsions. In order to reduce the side-effects and to improve the physical characteristics of the molecule, a large number of MDP derivatives, including lipophilic derivatives, have been prepared by chemical synthesis (for references, see Takada and Kotani (1995)). 6-O-Stearoyl-MDP (L18-MDP) (Azuma, 1992), 6-O-(tetra-decylhexadecanoyl)-MDP (B30-MDP) (Azuma, 1992) and murabutide (N-acetylmuramyl-L-alanyl-D-glutaminyl-n-butyl ester) (Chedid *et al.*, 1982) are MDP derivatives reported to exert powerful adjuvant activity in experimental vaccines (Takada and Kotani, 1995).

MTP-PE (muramyl tripeptide–phosphatidylethanolamine, Ciba-Geigy Ltd, Chiron Corp.) is a lipophilic derivative (Braun *et al.*, 1995), which in liposomes is superior to its aqueous form in terms of both immuno-stimulation and toxicity (Schumann *et al.*, 1989).

CpG-ODN

Ten years ago Yamamoto *et al.* (1992a) described that DNA from bacteria, but not from vertebrates, can induce production of type 1 and 2 inter-ferons and inhibit tumour growth in mice. These findings were preceded by the description of an antitumour activity of DNA from *Mycobacterium bovis* BCG (Tokunaga *et al.*, 1984; Shimada *et al.*, 1985) and the notion that such DNA could promote NK cell activity *in vitro* via interferon production (Yamamoto *et al.*, 1988). Using synthetic oligodinucleotides (ODN), the interferon inducing capacity of BCG DNA was ascribed to certain palindromic hexamers, all with a central unmethylated CpG motif (Yamamoto *et al.*, 1992b). Earlier studies (Messina *et al.*, 1991) had revealed that bacterial DNA could induce proliferation of murine B cells and some years later a number of immune stimulatory activities of bacterial DNA, that could be mimiced by CpG-ODNs (proliferation, Ig secretion, and resistance to apoptosis in B cells, production of IL-6, IL-12, IFN-γ, IFN α/β by macrophages, NK cells and T cells) were reviewed (Krieg, 1996; Pisetsky, 1996). These features all pointed towards adjuvant properties and during the next 5 years a substantial body of evidence has accumulated concerning the applicability of immunostimulatory-DNA, e.g. in the form of CpG-ODNs for immune modulation.

In addition to those listed above, the immunomodulatory effects described for CpG-DNA today also include increased expression of MHC class II molecules, co-stimulatory molecules (B7-1, B7-2), as well as production of TNF-α, IL-10, IL-18 (Krieg, 2000) and chemokines (Stan *et al.*, 2001). Initial work using CpG-ODN and aiming to identify optimal 'immunostimulatory' motifs and flanking bases soon revealed that these not only differed between species but also for parameter studied (Verthelyi *et al.*, 2001). Further, it became clear that ODN backbones of phosphothioates are more resistant to nucleases and therefore in many cases preferable to the natural phosphodiesters (Pisetsky and Reich, 1999). It is now generally agreed that GACGTT is an optimal motif in mice whereas GTCGTT preferentially stimulates human cells (reviewed in

Krieg and Wagner, 2000). For other species no consensus is accepted but various motifs have been tested in cultures of leukocytes from pigs (Kamstrup et al., 2001), cattle (Shoda et al., 2001; Zhang et al., 2001), fish (Kanellos et al., 1999; Jørgensen et al, 2001), dogs and cats (Wernette et al., 2002). The indicated difference in species specificities could be a result of dissimilarities between cellular receptors used for interaction with CpG motifs. Indeed, transfection with human TLR9 renders cells responsive to GTCGTT motifs whereas transfection with the murine TLR9 confers responsiveness to GACGTT (Bauer et al., 2001). Although TLR9 today is approved as the cellular receptor for CpG-DNA (Hemmi et al., 2000; Akira et al., 2001; Kadowaki et al., 2001) it is likely that also other induction pathways will be revealed. In this context it is notable that encapsulation of CpG-DNA in liposomes can by-pass the receptor-dependent uptake of CpG-DNA (Mui et al., 2001).

The adjuvant activity of CpG-motifs is a feature that most DNA vaccines benefit from. In addition, CpG-ODN as such are applied as immune modulators, not only in experimental vaccines but also as potential immunotherapeutic agents. Experimental studies in mice clearly demonstrate that CpG-ODN can be effective against tumours, but again the need to design CpG-motifs that promote the desired type of immune response is emphasized (Ballas et al., 2001). In the case of allergic disorders, the inbuilt capacity of certain CpG-ODN to induce production of cytokines that promote a Th-1 type of response is utilized for re-direction of Th-2 driven hypersensitivities (Krieg and Kline, 2000; Hochreiter et al., 2001; Peng et al., 2001).

The potential to use CpG-ODN in vaccine formulations for priming of naive cells and boosting of memory cells was first demonstrated in mice. Using various ODNs in combination with liposomes containing ovalbumin a strong antibody response, as well as a cytotoxic T-cell response, was demonstrated (Lipford et al., 1997). Interestingly, immunization of mice with CpG-ODN in combination with hen egg lysozyme and IFA, induced high titres of IgG2a antibodies and production of IFN-γ by antigen-stimulated T cells despite that HEL in IFA alone induced a strong Th-2 type of response (Chu et al., 1997). The same Th-1 promoting effect was obtained in BALB/c mice (Th-2 biased) as well as in B10.D2 mice (Th-1 biased). Thus, the results of these two elegant studies encouraged the continued research on the practical use of CpG-ODNs in vaccines against micro-organisms. Promising results from challenge experiments in mice vaccinated with recombinant proteins and CpG-ODN have, for example, been reported for dengue virus (Porter et al., 1998), influenza virus (Moldoveanu et al., 1998), lymphocytic choriomeningitis virus, vaccinia virus and Listeria monocytogenes (Oxenius et al., 1999), but still results from other species are scanty. It is however notable that CpG-ODN can improve the antibody response to a synthetic peptide malaria vaccine in Aotus monkeys (Jones et al., 1999) and increase the antibody titres of orangutans to a commercial hepatitis B vaccine (Davis et al., 2000). Indeed, it was recently reported that CpG-ODNs also in humans could improve the efficacy of a commercial hepatitis B virus vaccine that contains HbsAg in Al(OH)$_3$ (Davis et al., 2002). It is also of great interest that CpG-ODN seems to be an efficient adjuvant for induction of mucosal immunity (for review see McCluskie *et al.*, 2001). Further, CpG-ODN can contribute to a Th-1 type of response in young mice (Brasolot Millan *et al.*, 1998; Kovarik *et al.*, 1999; Weeratna *et al.*, 2001). Currently, liposomes seem to be a suitable carrier system for CpG-ODN in vaccines (Li *et al.*, 2002) but certainly other modes for targeting of CpG-ODNs and the antigen (Shirota *et al.*, 2001) will, together with the continued identification of optimal motifs (Hartmann *et al.*, 2001), contribute to future improvements.

◆◆◆◆◆◆ ADJUVANT FORMULATIONS: COMBINATIONS

Since different antigens require different adjuvant activities from an adjuvant, mixed formulations (or combinations) have been designed to combine several areas of adjuvant activity and to increase adjuvant versatility. Addition of immunomodulators such as lipid A or MDP derivatives, saponins or γ-inulin, along with the proper presentation of antigens afforded by co-polymers, o/w emulsions, $Al(OH)_3$, liposomes or the iscom structure, further increase the adjuvant activity. Although physically different, the adjuvant formulations described below all function by providing increased antigen presentation and targeting, together with immunomodulation.

Ribi adjuvant system

The Ribi adjuvant system (RAS) is a preformed o/w emulsion containing various combinations of MPL, TDM and CWS, which like SAF only requires mixing with aqueous antigen. Aqueous forms of MPL generally stimulate antibody responses, and TDM and CWS tend to enhance cell-mediated responses. Consequently, a combination of MPL and TDM or CWS is recommended to obtain a mixed response (Rudbach *et al.*, 1995).

Dosing

The final emulsion contains 2% squalene, 0.2% Tween 80, and 0.25 mg ml^{-1} of each of MPL, CWS or TDM. For mice, 50 µg of each is recommended. The corresponding dose for rabbits and goats is 250 µg. The concentration of antigen should be in the range 50–250 µg ml^{-1}; weak antigens can be used at concentrations up to 1.0 mg ml^{-1}.

ISCOM (Immune stimulating complexes)

The complex

Unlike the classical adjuvant formulations such as Freund's adjuvants and aluminium salts, which are blended with an antigen to formulate an emulsion or the antigen is adsorbed onto a three-dimensional gel, the iscom (immune stimulating complex, Iscotec AB) is a complex consisting of lipid, saponin and antigen. The complex will form only if the right constituents are allowed to interact in the correct stoichiometry.

The iscom complex is typically a 40-nm cage-like structure combining a multimeric presentation of antigen with a built in saponin adjuvant (Morein *et al.*, 1984), e.g. Quil A (Dalsgaard, 1978). The physical three-dimensional structure of the iscom is built up from 10- to 12-nm subunits formed by the interaction of Quil A with cholesterol (Özel *et al.*, 1989; Kersten *et al.*, 1991). By the addition of phospholipids, hydrophobic and amphipathic antigens are incorporated into iscoms by hydrophobic interactions during the assembly of the subunits (Lövgren and Morein, 1988).

In practice, iscoms are constructed by mixing antigens and saponin with detergent-solubilized cholesterol and phospholipid (Lövgren *et al.*, 1987). The detergent is removed by dialysis, ultrafiltration or ultracentrifugation, and iscoms form spontaneously during this process. To date, a large number of reports have been published describing different procedures for constructing iscom and the immune-enhancing properties of iscom preparations (Morein *et al.*, 1995; Rimmelzwaan and Osterhaus, 1995; Barr and Mitchell, 1996).

The experimental conditions for incorporation of different antigens and saponins into iscoms have mostly been empirical. Furthermore, iscom preparations are mostly characterized by their protein content, largely neglecting the amount of incorporated saponin. As a consequence, little consideration has been given to the importance of the ratio between antigen and saponin so that the iscom has sufficient adjuvanticity but minimal toxicity (Lövgren-Bengtsson and Sjölander, 1996).

The iscom resulted from efforts to present antigens and adjuvant together in the same particle. Such an approach was expected to increase the immunogenicity of incorporated proteins and thereby decrease the required dose of both antigen and adjuvant. However, not all antigens spontaneously incorporate into iscoms using standard procedures, even if they are hydrophobic, and many antigens are too hydrophilic to incorporate into iscoms.

Basic constituents and prerequisites for iscom formation

Quillaja saponin

The basis for iscom formation is the interaction of Quillaja saponins with cholesterol. The action of saponins on biological membranes and their specific affinity for cholesterol has been known for many years (Dourmashkin *et al.*, 1962; Lucy and Glauert, 1964), but its practical application to the creation of the iscom is much more recent (Morein *et al.*, 1984). Not all saponins in a crude or semipurified extract form 'classical' 40-nm iscoms, even though some type of complex or aggregate is often formed when Quillaja saponins (Kersten *et al.*, 1991) or some other saponins (Bomford *et al.*, 1992) are incubated with cholesterol. There are preparations of Quillaja saponins available commercially that have been tested and selected for their iscom-forming ability: Quil-A (Superfos AS), Spikoside (Isconova AB) and Iscoprep 703 (Iscotec AB).

Lipids

Cholesterol is indispensable for iscom formation, since the unique affinity of Quillaja saponins for cholesterol is the cause of the complex formation. Many other sterols, such as cholesterol derivatives and plant sterols, also form complexes with saponins, with structures similar or identical to that of iscoms (K. Lövgren and B. Morein, unpublished), and saponins other than Quillaja saponins interact with cholesterol to form some sort of structure (Bomford et al., 1992). However, so far, cholesterol and Quillaja saponin is the only suitable combination for the production of iscoms, at least for vaccine use.

Phospholipids are used as a supplementary lipid, even though other lipids can replace them. In terms of constructional aspects, phospholipids have different characteristics.

Iscom formation

The basic instructions for making iscoms have been to take 1 mg each of detergent-solubilized cholesterol and phospholipid per milligram of antigen and to mix this with 5 mg saponin before extensive dialysis against, for example, PBS (Lövgren et al., 1990).

The first method described for iscom formation was based on ultra-centrifugation (Morein et al., 1984). Detergent-solubilized membrane derived antigens were placed on the top of a 20–60% w/w sucrose gradient containing 0.05% saponin. Between the gradient and the antigen was a thin layer of 10% sucrose containing 0.1% Triton X100. This detergent-containing zone retains solubilized membrane lipids, but not those lipids associated with the membrane-anchoring parts of the antigen. Later, when the antigens carrying lipids enter the saponin-containing zone, iscoms are formed. Both the centrifugation method and the methods based on dialysis/ultrafiltration have advantages and disadvantages. The method of choice should be selected to suit the antigen and the desired composition of the iscom. Since the centrifugation method relies on the presence of lipids already associated with the antigen, this method is most suited for membrane antigens (native or recombinants) extracted from the membrane with a suitable detergent. Since the antigen is removed from the detergent-containing sample zone during the centrifugation, there are no other demands on the detergent than to solubilize the antigen gently without replacing the lipids and without denaturation of conformational epitopes stabilized by hydrophobic interactions (Merza et al., 1989). Iscoms prepared using the centrifugation method will contain a minimum of lipid, and therefore also low amounts of saponin. Iscoms made from measles virus glycoproteins using the centrifugation method contained only 0.2 µg saponin per microgram of antigen (Morein et al., 1984).

Compared with the dialysis/ultrafiltration methods, iscoms produced by the centrifugation technique are probably less toxic.

If the antigens carry too few lipids there will not be enough lipid to construct complexes with saponin, and the preparation will contain iscoms, iscom fragments, protein micelles or protein aggregates.

Many antigens are not membrane proteins, and even membrane-derived antigens need purification that might remove too much membrane lipid before incorporation into iscoms. Other antigens do not bind very well to lipids in the presence of detergent, and will not bring their lipids down to the saponin-containing zone during centrifugation. For these antigens, the dialysis/ultrafiltration methods are most useful. As mentioned above, the antigens are mixed and incubated with a suitable amount of lipid and an adequate amount of saponin before the detergent is removed. During removal of the detergent, hydrophobic and amphipathic antigens associate with the iscom constituents and incorporate into the iscom particles. Even if the incorporation of the antigen fails, the iscom backbone consisting of saponin and lipids, the so-called iscom-matrix, will form.

Antigen handling

The composition of an iscom should be chosen to tolerate the incorporation of an antigen into the structure without weakening the physical stability of the complex. In practice this has not been much of a problem, and most antigens, large or small, incorporate without disruption of the complex.

The handling of the antigen before incorporation is important. Since the antigens most suited for iscom formulation are hydrophobic or amphipathic, these must be purified in the presence of a detergent in such a way that they will not form protein micelles or aggregates. This is because pure protein micelles or aggregates are very difficult to dissociate, so that their hydrophobic part is accessible for hydrophobic interactions. There are many examples where antigens which theoretically should incorporate well in fact do not as a result of such aggregates. A comprehensive review of the methods used for iscom preparation has recently been published (Lövgren-Bengtsson and Morein, 2000).

Amphipathic/hydrophobic antigens

The iscom technology was originally developed for amphipathic/hydrophobic antigens, which generally incorporate nicely into iscoms. For many of these antigens, the choice of methodology is not important.

Some hydrophobic antigens are poorly soluble, even in detergents. Such antigens can be solubilized in, for example, dimethyl sulfoxide (DMSO), urea, ethylene glycol or guanidine hydrochloride (GuaHCl), and then lipid in detergent stock solution and saponin is added, as for other antigens. Initial dialysis/ultrafiltration is performed against a solution containing a lower concentration of solubilizer (DMSO, urea, ethylene glycol or GuaHCl), followed by a physiological buffer (e.g. PBS).

Non-amphipathic antigens

Hydrophilic antigens do not incorporate into iscoms. To make iscoms containing hydrophilic antigen, hydrophobic regions must be introduced by, for example, binding lipids to them (Mowat et al., 1991; Browning et al., 1992; Reid, 1992; Scheepers and Becht, 1994) or exposed by, for example, low pH treatment. At first glance, one may think that such a treatment will irreversibly denature the antigens, and for some antigens this is true. In fact, low pH treatment is often less denaturing than chemical modification (K. Lövgren-Bengtsson, B. Morein, J. Ekström, L. Akerblom, M. Villacrés-Eriksson, unpublished).

Some antigens, particularly basic antigens with high isoelectric points but also many E. coli-expressed recombinant antigens (glycoproteins in their native form), do adsorb spontaneously to iscom-matrix (Lövgren-Bengtsson and Morein, 2000; Polakos et al., 2001).

Low pH method (Pyle et al., 1989; Morein et al., 1990; Sjölander et al., 1996b). The protein is mixed with lipid and the pH lowered generally to between 2.5 and 5, depending on the nature of the antigen. After a short incubation (about 1 h) at room temperature, the saponin is added and the mixture is incubated for 1–2 h before dialysis/filtration. The first volumes of buffer should have the same low pH as the initial mixture; the buffer is then changed to, for example, PBS. Using the detergent MEGA-10, a white precipitate may form at low pH. The precipitate consists of MEGA-10 and possibly some lipid, and will dissolve as soon as the pH is brought back to neutral.

Conjugation of antigens to preformed iscom-matrix. As an alternative to incorporation, the antigen(s) can be covalently linked to preformed iscom-matrix. The iscom-matrix used for conjugation of antigens is the same as that used as adjuvant (see above), with the exception that it contains a phospholipid with a functional polar head group suitable for conjugation (e.g. phosphatidyl ethanolamine) (Sjölander et al., 1996b; Lundén et al., 1997). The same chemistry as that described below for lipid modifications (below) is applicable.

Modification of antigens

There is a variety of protein modification and cross-linking reagents (e.g. Pierce) and activated lipids (e.g. Northern Lipids) commercially available. The vast majority of these reagents rely on the presence of amino groups ($-NH_2$) reagents (Larsson et al., 1993; Lövgren and Larsson, 1994; Sjölander et al., 1996b) that can be used for binding. Some reagents react with sulfhydryl groups ($-SH$) (Mowat et al., 1991; Browning et al., 1992; Reid, 1992) and a few methods are available for directed reaction involving carboxyl groups ($-COOH$) (Sjölander et al. 1997b) and carbohydrates (Wilson et al., 1999). Alternative strategies for efficient lipid tagging of recombinant vaccine antigens, *in vivo* during their production or *in vitro* after purification, have been suggested (Andersson et al., 1999, 2000, 2001).

Peptides/haptens

Synthetic oligopeptides and other low-molecular-weight haptens (small B-cell epitopes) that require a carrier protein for T-cell help, can be covalently linked to pre-formed carrier iscoms using the techniques for covalent conjugation described above.

Oligopeptides or small proteins consisting of both T- and B-cell epitopes can, if they are amphipathic or hydrophobic, be incorporated using the same general methods described above or can be conjugated to a lipid prior to incorporation into iscoms (Pedersen et al., 1992; Weijer et al., 1993). Hydrophilic peptides can be conjugated to preformed iscoms containing carrier proteins (Sjölander et al., 1991, 1993; Larsson et al., 1993; Lövgren and Larsson, 1994).

Characterization of the final product: composition of iscoms

Whichever method is used for incorporating antigens into iscoms, the goal is the same – a highly immunogenic particle of low toxicity, containing an exposed antigen with antigenicity of the native molecule. The composition of the 'ideal' iscom is, of course, dependent on several factors, such as antigen immunogenicity, grade and purity of Quillaja saponin, the species to be immunized and the antigen dose to be used for immunization. Even with a strong antigen such as influenza virus membrane glycoproteins, an increased ratio ($\mu g/\mu g$) of Quillaja saponin to protein can increase the immunogenicity in such a way that a 10-μg dose of antigen in iscoms with a low proportion of Quil A (1 μg Quil A per μg antigen) is as immunogenic as a 1-μg dose of antigen in iscoms with a high proportion of Quil A (5 μg Quil A per mg antigen) (Lövgren-Bengtsson and Sjölander, 1996).

Problems

With proteins that spontaneously incorporate into iscoms there are generally no problems encountered during iscom preparation. The antigens incorporate quantitatively and it is easy to predict the composition of the iscoms produced. Most of the problems concerning iscoms arise when antigens that incorporate poorly (or not at all) are used and the resulting preparation is tested in animals without a proper analysis of the final product. There are several examples in the literature of poor or toxic iscoms resulting from problems with insufficient antigen incorporation.

In the study by Kersten et al. (1988) in which a strongly hydrophobic bacterial antigen was used, the iscoms were toxic at a dose of 2.5 μg antigen with a Quil A/protein ratio (w/w) of 20:1. In another study (Stienecker et al., 1995) gp120 iscoms were not at all immunogenic, according to the authors, probably due to the fact that gp120 was not incorporated into the iscoms. However, by analysing the final iscoms to establish the antigen/Quil A ratio, suboptimal iscom preparations can be safety tested in laboratory animals. Special attention must be given to preparations for use in mice. Mice in general are sensitive to saponins, but there is a large strain variation. Also, the mode of administration to mice plays a substantial role. Intravenous and intraperitoneal administration should preferably be avoided. If necessary, the dose of iscoms should be reduced to about 10% of that suitable for use subcutaneously or intramuscularly, i.e. 0.1–0.5 mg (antigen) or a dose of iscoms containing <1 μg Quil A. The increased toxicity of intravenous and intraperitoneal administration is most probably related to a highly efficient and rapid uptake of the iscoms. In a dose–response study of iscoms administered intraperitoneally and subcutaneously, about a 10-fold lower dose was required intraperitoneally to generate the same magnitude of response as that of iscoms given subcutaneously (K. Lövgren-Bengtsson, unpublished).

Type of immune modulation

Iscoms enhance immune responses in various ways, including increased MHC class II expression on antigen-presenting cells (Bergström-Mollauglu et al., 1992; Watson et al., 1992), induction of IL-1 production (Villacrés-Eriksson et al., 1993; Valensi et al., 1994), activation of helper (Fossum et al., 1990; Villacrés-Eriksson et al., 1992) and cytotoxic T cells (Jones et al., 1988; Takahashi et al., 1990; Ennis et al., 1999; da Fonseca et al., 2000; Le et al., 2001; Sambhara et al., 2001), and generation of potent long-lasting antibody responses (Lövgren, 1988; Sundquist et al., 1988; Hannant et al., 1993; Mumford et al., 1994a,b), involving all subclasses and isotypes (Lövgren, 1988; Villacrés-Eriksson et al., 1993). In the mouse a typical antibody response consists of equal proportions of IgG1 and IgG2, a medium proportion of IgG2b and minor but substantial amounts of IgG3.

The cellular response includes CD8+ class I dependent CTLs and proliferating lymphocytes secreting IL-2 and high amounts of IFN-γ (Morein et al., 1995) and IL-12 (Villacrés-Eriksson et al., 1997). Antigen-presenting cells such as macrophages and dendritic cells produce IL-1 and IL-6 after in vitro stimulation with iscoms (Villacrés-Eriksson, 1995).

Iscom-borne antigen is rapidly taken up from the site of injection and transported to draining lymph nodes where there is a potent but transient response of proliferating T cells producing IL-2 and IFN-γ at days 4–8 after immunization; this is then transferred to the spleen (Sjölander et al., 1997a). Iscoms also induce good immune responses following intranasal and oral administration (Lövgren, 1988; Mowat et al., 1991, 2001; Hu et al., 2001).

Dosing

In general, the same maximum dose limitations as those mentioned for free Quil A (see above) are valid also for iscoms, although the dose of Quil A in iscoms required for a potent immune response is by far lower. The dose of antigen required, especially with monomeric antigens, may be one-tenth of that required with other adjuvant formulations (Morein et al., 1995).

◆◆◆◆◆◆ SUMMARY

To choose the optimum adjuvant for a particular antigen preparation is not easy and there are many factors to consider. Of major importance is whether the purpose of the immunization is to develop a vaccine or to raise antibodies for use in immunoassays or affinity chromatography. Antigen availability and choice of animal model are also often issues of primary concern. Secondly, we must consider the physical characteristics of the antigen: is it likely to be immunogenic or is it one of those small hydrophilic antigens that are usually poorly immunogenic?

If there is only a small quantity of a hydrophilic antigen available and the purpose is to raise antibodies for use in immunoassay, one of the most reliable adjuvants is FCA. Although efficient, FCA does have disagreeable side-effects, and for the well-being of laboratory animals the use of FCA is discouraged or forbidden in many countries. The modern commercially available alternatives that are equally simple to use are the oil emulsion type adjuvants such as TiterMax and Ribi adjuvants. If the antigen is not in short supply and there is time for some development work, the incorporation into iscoms will generate a highly immunogenic formulation. With particulate antigens there is a wider choice of adjuvants, and the choice and efforts can be focused on obtaining the desired immune response.

For vaccine development many factors must be considered, and it is beyond the scope of this chapter to give a rationale for the choice of adjuvant, although several of the adjuvants discussed are either in clinical trials in humans or already in use in commercially available animal vaccines.

The market for commercially available and experimental adjuvants is increasing. Adjuvants are becoming more and more efficient and versatile. Modern immunotechnology and immunology has brought new demands and new possibilities into the vaccine adjuvant area. The modes of action of immunological adjuvants/immunomodulators are revealed and a deeper understanding of mechanisms is replacing the previous more empiric 'sense' of adjuvanticity. Antigen preparations are different and the goals of immunization are disperse (a) immune prophylaxis, (b) immune therapy for cancer, chronic infections, autoimmunity and allergy, (c) generations of antibodies for therapy or diagnostics, therefore we will make use of a collection of well-defined adjuvant systems that we can use for safe and appropriate immune modulation.

References

Aderem, A., and Ulevitch R. J. (2000) Toll-like receptors in the induction of the innate immune response. *Nature* **406**, 782–787.

Akira, S., Takeda, K. and Kaisho, T. (2001). Toll-like receptors: critical proteins linking innate and acquired immunity. *Nat. Immunol.* **2**, 675–680.

Allison, A. C. and Byars, N. E. (1986). An adjuvant formulation that selectively elicits the formation of antibodies of protective isotype and of cell-mediated immunity. *J. Immunol. Methods* **95**, 157–168.

Allison, A. C. and Byars, N. E. (1990). Adjuvant formulations and their mode of action. *Semin. Immunol.* **2**, 369–374.

Alving, C. R. (1991). Liposomes as carriers of antigens and adjuvants. *J. Immunol. Methods* **140**, 1–13.

Alving, C. R. (1992). Immunological aspects of liposomes: presentation and processing of liposomal protein and phospholipid antigens. *Biochim. Biophys. Acta* **1113**, 307–322.

Alving, C. R. (1993). Lipopolysaccharide, lipid A and liposomes containing lipid A as immunologic adjuvants. *Immunobiology* **187**, 430–446.

Andersson, C., Sandberg, L., Murby, M., Sjölander, A., Lövgren-Bengtsson, K. and Ståhl, S. (1999). General expression vectors for production of hydrophobically tagged immunogens for direct iscom incorporation. *J. Immunol. Methods* **222**, 71–82.

Andersson, C., Sandberg, L., Wernerus, H., Johansson, M., Lövgren-Bengtsson, K. and Ståhl, S. (2000) Improved systems for hydrophobic tagging of recombinant immunogens for efficient iscom incorporation. *J. Immunol. Methods* **238**, 181–193.

Andersson, C., Wikman, M., Lövgren-Bengtsson, K., Lundén A. and Ståhl, S. (2001). In vivo and in vitro lipidation of recombinant immunogens for direct iscom incorporation. *J. Immunol. Methods* **255**, 135–148.

Arend, W. P. and Massoni, R. J. (1986). Characteristics of bacterial lipopoly-saccharide induction of interleukin 1 synthesis and secretion by human monocytes. *Clin. Exp. Immunol.* **64**, 656–664.

Astiz, M. E., Rackow, E. C., Still, J. G., Howell, S. T., Cato, A., Von Eschen, K. B., Ulrich, J. T., Rudbach, J. A., McMahon, G. and Vargas, R. (1995). Pretreatment of normal humans with monophosphoryl lipid A induces tolerance to endotoxin. *J. Crit. Care Med.* **23**, 9–17.

Azuma, I. (1992). Synthetic immunoadjuvants: application to non-specific host stimulation and potentiation of vaccine immunogenicity. *Vaccine* **10**, 1000–1006.

Ballas, Z. K., Krieg, A. M., Warren, T., Rasmussen, W., Davis, H. L., Waldschmidt, M. and Weiner, G. J. (2001). Divergent therapeutic and immunologic effects of oligodeoxynucleotides with distinct CpG motifs. *J. Immunol.* **167**, 4878–4886.

Barr, I. G. and Mitchell, G. F. (1996). ISCOMs (immunostimulating complexes): the first decade. *Immunol. Cell. Biol.* **74**, 8–25.

Bauer, S., Kirschning, C. J., Hacker, H., Redecke, V., Hausmann, S., Akira, S., Wagner, H. and Lipford, G. B. (2001). Human TLR9 confers responsiveness to bacterial DNA via species-specific CpG motif recognition. *Proc. Natl. Acad. Sci.* **98**, 9237–2342.

Bergström-Mollauglu, M., Lövgren, K., Åkerblom, L., Fossum, C. and Morein, B. (1992). Antigen-specific increases in the number of splenocytes expressing MHC class II molecules following restimulation with antigen in various physical forms. *Scand. J. Immunol.* **36**, 565–574.

Bloch, H. (1950). Studies on the virulence of tubercle bacilli: isolation and biological properties of a constituent of virulent organisms. *J. Exp. Med.* **91**, 197–217.

Bokhout, B. A., van Gaalen, C. and Van Der Heijden, Ph. J. (1981). A selected water-in-oil emulsion; composition and usefulness as an immunological adjuvant. *Vet. Immunol. Immunopathol.* **2**, 491–500.

Bomford, R. (1980a). The comparative selectivity of adjuvants for humoral and cell mediated immunity. I. Effect on the antibody response to bovine serum albumin and sheep red blood cells of Freund's incomplete and complete adjuvants, allhydrogel, *Corynebacterium parvum*, *Bordetella pertussis*, muramyldipeptide and saponin. *Clin. Exp. Immunol.* **39**, 426–434.

Bomford, R. (1980b). The comparative selectivity of adjuvants for humoral and cell mediated immunity. II. Effect on delayed-type hypersensitivity in the mouse and guinea pigs, and cell-mediated immunity to tumour antigens in the mouse of Freund's incomplete and complete adjuvants, allhydrogel, *Coryne-bacterium parvum*, *Bordetella pertussis*, muramyldipeptide and saponin. *Clin. Exp. Immunol.* **39**, 435–441.

Bomford, R. (1981). The adjuvant activity of fatty-acid esters. The role of acyl chain length and degree of saturation. *Immunology* **44**, 187–192.

Bomford, R. (1984). Relative adjuvant efficacy of Al(OH)$_3$ and saponin is related to the immunogenicity of the antigen. *Int. Arch. Allergy Appl. Immunol.* **75**, 280–281.

Bomford, R., Stapleton, M., Winsor, S., Beesley, J. E., Jessup, E. A., Price, K. R. and Fenwick, G. R. (1992). Adjuvanticity and iscom formation by structurally diverse saponins. *Vaccine* **10**, 572–577.

Brasolot Millan, C. L., Weeratna, R., Krieg, A. M., Siegrist, C. A. and Davis, H. L. (1998). CpG DNA can induce strong Th1 humoral and cell-mediated immune responses against hepatitis B surface antigen in young mice. *Proc. Natl. Acad. Sci.* **95**, 15553–15558.

Braun, D. G., van Hoogevest, P. and Schumann, G. (1995). Muramyl tripeptide–phosphatidylethanolamine: a muramyl dipeptide derivative with lipophilic properties. In: *The Theory and Practical Application of Adjuvants* (D. E. S. Stewart-Tull, Ed.), pp. 213–237. Wiley, New York.

Bretscher, P. A., Wei, G., Menon, J. N. and Bielefeldt-Ohmann, H. (1992). Establishment of a stable cell-mediated immunity that makes 'susceptible' mice resistant to *Leishmania major*. *Science* **257**, 539–542.

Browning, M., Reid, G. R., Osborne, R. and Jarret, O. (1992). Incorporation of soluble antigens into iscoms: HIV gp120 iscoms induce virus neutralizing anti-bodies. *Vaccine* **10**, 585–590.

Brugh, M., Stone, H. D. and Lupton, H. W. (1983). Comparison of inactivated Newcastle disease viral vaccines containing different emulsion adjuvants. *Am. J. Vet. Res.* **44**, 72–75.

Byars, N. E. and Allison, A. C. (1987). Adjuvant formulation for use in vaccines to elicit both cell-mediated and humoral immunity. *Vaccine* **5**, 223–228.

Campbell, J. B. (1995) Saponins. In: *The Theory and Practical Application of Adjuvants* (D. E. S. Stewart-Tull, Ed.), pp. 95–127. Wiley, New York.

Carozzi, S., Salit, M., Cantaluppi, A., Nasini, M. G., Barocci, S., Cantarella, S. and Lamperi, S. (1989). Effect of monophosphoryl lipid A on the in vitro function of peritoneal leukocytes from uremic patients on continuous ambulatory peritoneal dialysis. *J. Clin. Microbiol.* **27**, 1748–1753.

Chavali, S. R. and Campbell, J. B. (1987). Immunomodulatory effects of orally-administered saponins and nonspecific resistance against rabies infection. *Int. Arch. Allergy Appl. Immunol.* **84**, 129–134.

Chedid, L., Audibert, F., Bona, C., Damais, C., Parant, F. and Parant, M. (1975). Biological activities of endotoxins detoxified by alkylation. *Infect. Immun.* **12**, 714–721.

Chedid, L., Parant, M., Audibert, F., Riveau, G., Parant, F., Lederer, E., Choay, J. and Lefrancier, P. (1982). Biological activity of a new synthetic muramyl peptide adjuvant devoid of pyrogenicity. *Infect. Immun.* **35**, 417–424.

Chu, R. S., Targoni, O. S., Krieg, A. M., Lehmann, P. V. and Harding, C. V. (1997). CpG oligodeoxynucleotides act as adjuvants that switch on T helper 1 (Th1) immunity. *J. Exp. Med.* **186**, 1623–1631.

Collins, W. E., Walduck, A., Sullivan, J. S., Andrews, K., Stowers, A., Morris, C. L., Jennings, V., Yang, C., Kendall, J., Lin, Q., Martin, L. B., Diggs, C. and Saul, A. (2000). Efficacy of vaccines containing rhoptry-associated proteins RAP1 and RAP2 of *Plasmodium falciparum* in *Saimiri boliviensis* monkeys. *Am. J. Trop. Med. Hyg.* **62**, 466–479.

Cox, J. and Coulter, A. (1992). Advances in adjuvant technology and application. In: *Animal Parasite Control Utilizing Biotechnology* (W. K. Yong, Ed.), pp. 51–112. CRC Press, Boca Raton, FL, USA.

Cox, J. and Coulter, A. (1997). Adjuvants – a classification and review of their mode of action. *Vaccine* **15**, 248–256.

Cox, J., Coulter, A., Macfarlane, R., Beezum, L., Bates, J., Wong, T. and Drane, D. (1997). Development of an influenza-izcom vaccine. In: *Vaccine Design: The Role of Cytokine Networks* (Gregoriadis, G., McCormack, B., Allison, A. C., Eds), pp. 33–49. Plenum Publishing, New York.

da Fonseca, D. P., Frerichs, J., Singh, M., Snippe, H. and Verheul, A. F. (2000) Induction of antibody and T-cell responses by immunization with ISCOMS containing the 38-kilodalton protein of *Mycobacterium tuberculosis*. *Vaccine* **9**, 122–131.

Dalsgaard, K. (1974). Saponin adjuvants III. Isolation of a substance from *Quillaja saponaria* Molina with adjuvant activity in foot-and-mouth disease vaccines. *Arch. Ges. Virusforsch.* **44**, 243–254.

Dalsgaard, K. (1978). A study on the isolation and characterisation of the saponin Quil A. Evaluation of its adjuvant activity with special reference to the application in foot-and-mouth disease. *Acta Vet. Scand.* **69** (Suppl.), 1–40.

Dalsgaard, K. (1987). Adjuvants. *Vet. Immunol. Immunopathol.* **17**, 145–152.

Dalsgaard, K., Hilgers, L. and Trouvé, G. (1990). Classical and new approaches to adjuvant use in domestic animals. *Adv. Vet-Sci. Comp. Med.* **35**, 121–160.

Davis, H. L. (2000). Use of CpG DNA for enhancing specific immune responses. In: *Immunobiology of bacterial CpG-DNA* (H. Wagner Ed.), pp. 171–182. Springer-Verlag, Berlin.

Davis, H. L., Suparto, I. I., Weeratna, R. R., Jumintarto, Iskandriati, D. D., Chamzah, S. S., Ma'ruf, A. A., Nente, C. C., Pawitri, D. D., Krieg, A. M., Heriyanto, Smits, W. and Sajuthi, D. D. (2000). CpG DNA overcomes hyporesponsiveness to hepatitis B vaccine in orangutans. *Vaccine* **18**, 1920–1924.

Davis, H. L., Krieg, A. M., Cooper, C. L., Morris, M. L., Cameron, D. W. and Heathcote, J. (2002). CpG ODN is generally well tolerated and highly effective in humans as adjuvant to HBV vaccine: Preliminary results of phase 1 trial with CpG ODN 7909. Proceedings 2ⁿᵈ International Symposium 'Activating Immunity with CpG Oligos'. 7–10 October 2001. Amelia Island, FL, USA.

Dourmashkin, R. R., Dougherty, R. M. and Harris, R. J. (1962). Electron microscopic observation on rous sarcoma virus and cell membranes. *Nature* **194**, 1116–1119.

Drulak, M. W., Malinoski, F. J., Fuller, S. A., Stewart, S. S., Hoskin, S., Duliege, A. M., Sekulovich, R., Burke, R. and Winston, S. (2001). Vaccination of seropositive subjects with CHIRON CMV gB subunit vaccine combined with MF59 adjuvant for production of CMV immune globulin. *Viral Immunol.* **13**, 49–56.

Dupuis, M., Denis-Mize, K., LaBarbara, A., Peters, W., Charo, I. F., McDonald, D. M. and Ott, G. (2001). Immunization with the adjuvant MF59 induces macrophage trafficking and apoptosis. *Eur. J. Immunol.* **31**, 2910–2918.

Ennis, F. A., Cruz, J., Jameson, J., Klein, M., Burt, D. and Thipphawong, J. (1999). Augmentation of human influenza A virus-specific cytotoxic T lymphocyte memory by influenza vaccine and adjuvanted carriers (ISCOMS). *Virology* **259**, 256–261.

Espinet, R. G. (1951). Nuevo tipo de vacuna antiafosa a complejo glucovirico. *Gac. Vet.* **74**, 1–13.

Forsthuber, T., Yip, H. C. and Lehmann, P. V. (1996). Induction of Th1 and Th2 immunity in neonatal mice. *Science* **271**, 1728–1730.

Fossum, C., Bergström, M., Lövgren, K., Watson, D. L. and Morein, B. (1990). Effect of iscom and/or their adjuvant moiety (matrix) on the initial proliferative and IL-2 responses in vitro. Comparison of spleen cells from mice inoculated with iscoms and/or matrix. *Cell. Immunol.* **129**, 414–425.

Fukumi, H. (1967). Effectiveness and untoward reactions of oil adjuvant influenza vaccines. *Symp. Ser. Immunobiol.*, Standard, 6.

Gasparini, R., Pozzi, T., Montomoli, E., Fragapane, E., Senatore, F., Minutello, M. and Podda, A. (2001) Increased immunogenicity of the MF59-adjuvanted influenza vaccine compared to a conventional subunit vaccine in elderly subjects. *Eur. J. Epidemiol.* **17**, 135–140.

Glenny, A. T., Pope, C. G., Waddington, H. and Wallas, U. (1926). Immunological notes XVII to XXIV. *J. Pathol.* **29**, 31–40.

Gould-Fogerite, S., Edghill-Smith, Y., Kheiri, M., Wang, Z., Das, K., Feketova, E., Canki, M. and Mannio, R. J. (1994). Lipid matrix-based subunit vaccines: A structure-function approach to oral and parenteral immunization. *AIDS Res. Human Retrovirus* **10** (Suppl. 2), S99–S103.

Gregoriadis, G., McCormack, B., Obrenovic, M., Perrie, Y. and Saffie, R. (2000). *Methods in Molecular Medicine, Vol. 42: Vaccine Adjuvants: Preparation Methods and Research Protocols* (D. T. O'Hagan Ed.), pp 137–150. Humana Press, Inc., Totowa, NJ, USA.

Hamamoka, T., Katz, D. H., Bloch, K. J. and Benacerraf, B. (1973) Hapten-specific IgE antibody response in mice. I. Secondary IgE response in irradiated recipients of synergic primed spleen cells. *J. Exp. Med.* **138**, 306–311.

Hannant, D., Jessett, D. M., O'Neill, T., Dolby, C. A., Cook, R. F. and Mumford, J. A. (1993). Responses of ponies to equid herpesvirus-1 ISCOM vaccination and challenge with virus of the homologous strain. *Res. Vet. Sci.* **54**, 299–305.

Hansen, D. S., Alievi, G., Segura, E. L., Carlomagni, M., Morein, B. and Villacrés-Eriksson, M. (1996). The flagellar fraction of *Trypanosoma cruzi* depleted of an immunosuppressive antigen enhances protection to infection and elicits spontaneous T cell responses. *Parasit. Immunol.* **18**, 607–615.

Hansen, D., Alievi, G., Segura, E., Carlomagno, M., Morein, B. and Villacres-Eriksson, M. (1966). The flaggelar fraction of *Trypanosoma cruzi* depleted of an immunosuppressive antigen enhances protection to infection and elicits spontaneous T cell responses. *Parasite Immunol.* **18**, 607–615.

Hartmann, G., Weeratna, R. D., Ballas, Z. K., Payette, P., Blackwell, S., Suparto, I., Rasmussen, W. L., Waldschmidt, M., Sajuthi, D., Purcell, R. H., Davis, H. L. and Krieg, A. M. (2001). Delineation of a CpG phosphorothioate oligodeoxy-nucleotide for activating primate immune responses in vitro and in vivo. *J. Immunol.* **164**, 1617–1624.

Heath, A. W., Nyan, O., Richards, C. E. and Playfair, J. H. L. (1991). Effects of interferon-γ and saponin on lymphocyte traffic are inversely related to adjuvanticity and enhancement of MHC class II expression. *Int. Immunol.* **3**, 285–292.

Hemmi, H., Takeuchi, O., Kawai, T., Kaisho, T., Sato, S., Sanjo, H., Matsumoto, M., Hoshino, K., Wagner, H., Takeda, K. and Akira, S. (2000). A Toll-like receptor recognizes bacterial DNA. *Nature* **408**, 740–755.

Henricson, B. E., Benjamin, W. R. and Vogel, S. N. (1990). Differential cytokine induction by doses of lipopolysaccharide and monophosphoryl lipid A that result in equivalent early endotoxin tolerance. *Infect. Immun.* **58**, 2429–2437.

Hilleman, M. R. (1966). Critical appraisal of emulsified oil adjuvants to viral vaccines. In: *Progress in Medical Virology*, Vol. 8, pp. 131–182. Krager, New York.

Hochreiter, R., Hartl, A., Freund, J., Valenta, R., Ferreira, F. and Thalhamer, J. (2001). The influence of CpG motifs on a protein or DNA-based Th2-type immune response against major pollen allergens Bet v 1a, Phl p 2 and *Escherichia coli*-derived beta-galactosidase. *Int. Arch. Allergy Immunol.* **124**, 406–410.

Hoffmann, J. A., Kafatos, F. C., Janeway, C. A. and Ezekowitz, R. A. (1999) Phylogenetic perspectives in innate immunity. *Science* **284**, 1313–1318.

Hosken, N. A., Shibuya, K., Heath, A. W., Murphy, K. M. and O'Garra, A. (1995). The effect of antigen dose on CD4+ T helper cell phenotype development in a T cell receptor-ab-transgenic model. *J. Exp. Med.* **182**, 1579–1584.

Howerton, D. A., Hunter, R. L., Ziegler, H. K. and Check, I. J. (1990). Induction of macrophage Ia expression in vivo by a synthetic block copolymer, L81. *J. Immunol.* **144**, 1578–1584.

Hu, J. G. and Kitagawa, T. (1990). Studies on the optimal immunization schedule of experimental animals. VI. Antigen dose–response of aluminium hydroxide-aided immunization and booster effect under low antigen dose. *Chem. Pharm. Bull.* **38**, 2775–2779.

Hu K.-F., Lövgren-Bengtsson, K. and Morein, B. (2001) Immunostimulating complexes (ISCOMs) for nasal vaccination. *Adv. Drug Delivery Rev.* **51**, 149–159.

Hunter, R. L. and Bennett, B. (1984). The adjuvant activity of nonionic block polymer surfactants. II. Antibody formation and inflammation related to the structure of tri-block and octablock copolymers. *J. Immunol.* **133**, 3167–3175.

Hunter, R. L. and Bennett, B. (1986). The adjuvant activity of nonionic block polymer surfactants. II. Characterization of selected biologically active surfaces. *Scand. J. Immunol.* **23**, 287–300.

Hunter, R. L., McNicholl, J. and Lal, A. (1994). Mechanism of action of nonionic block polymer adjuvants. *AIDS Res. Human Retrovirus* **10** (Suppl. 2), S95–S98.

Hunter, R., Strickland, F. and Kézdy, F. (1981). The adjuvant activity of nonionic block polymer surfactants. I. The role of hydrophile–lipophile balance. *J. Immunol.* **127**, 1244–1250.

Hunter, R. L., Olsen, M. R. and Bennett, B. (1995). Copolymer adjuvants and TiterMax. In: *The Theory and Practical Application of Adjuvants* (D. E. S. Stewart-Tull, Ed.), pp. 51–94. Wiley, New York.

Johnsson, A. G., Gaines, S. and Landy, M. (1956). Studies on the *O antigen of *Salmonella typhosa*. V. Enhancement of antibody responses to protein antigens by the purified lipopolysaccharide. *J. Exp. Med.* **103**, 225–246.

Jones, G. E., Jones, K. A., Machell, J., Brebner, J., Anderson, I. E. and How, S. (1995). Efficacy trials with tissue-culture grown, inactivated vaccines against chlamydial abortion in sheep. *Vaccine* **13**, 715–723.

Jones, P. D., Tha Hla, R., Morein, B., Lövgren, K. and Ada, G. L. (1988). Cellular immune responses in the murine lung to local immunization with influenza A virus glycoproteins in micelles and immunostimulating complexes (iscoms). *Scand. J. Immunol.* **27**, 645–652.

Jones, T. R., Obaldia, N. 3rd, Gramzinski, R. A., Charoenvit, Y., Kolodny, N., Kitov, S., Davis, H. L., Krieg, A. M. and Hoffman, S. L. (1999). Synthetic oligodeoxynucleotides containing CpG motifs enhance immunogenicity of a peptide malaria vaccine in *Aotus* monkeys. *Vaccine* **17**, 3065–3071.

Jørgensen, J. B., Johansen, A., Stenersen, B. and Sommer, A. I. (2001). CpG oligodeoxynucleotides and plasmid DNA stimulate Atlantic salmon (*Salmo salar* L.) leucocytes to produce supernatants with antiviral activity. *Dev. Comp. Immunol.* **25**, 313–321.

Kadowaki, N., Ho, S., Antonenko, S., Malefyt, R. W., Kastelein, R. A., Bazan, F. and Liu, Y. J. (2001). Subsets of human dendritic cell precursors express different toll-like receptors and respond to different microbial antigens. *J. Exp. Med.* **194**, 863–869.

Kamstrup, S., Verthelyi, D. and Klinman, D. M. (2001). Response of porcine peripheral blood mononuclear cells to CpG-containing oligodeoxynucleotides. *Vet. Microbiol.* **78**, 353–362.

Kanellos, T. S., Sylvester, I. D., Butler, V. L., Ambali, A. G., Partidos, C. D., Hamblin, A. S. and Russell, P. H. (1999). Mammalian granulocyte-macrophage colony-stimulating factor and some CpG motifs have an effect on the immunogenicity of DNA and subunit vaccines in fish. *Immunology* **96**, 507–510.

Karagouni, E. E. and Hadjipetrou-Kourounakis, L. (1990). Regulation of immunoglobulin production by adjuvants in vivo. *Scand. J. Immunol.* **31**, 745–754.

Katz, J. M., Lu, X., Todd, C. W. and Newman, M. J. (2000). A nonionic block copolymer adjuvant (CRL1005) enhances the immunogenicity and protective efficacy of inactivated influenza vaccine in young and aged mice. *Vaccine* **18**, 2177–2187.

Kenney, J. S., Hughes, B. W., Masada, M. P. and Allison, A. C. (1989). Influence of adjuvants on the quantity, affinity, isotype and epitope specificity of murine antibodies. *J. Immunol. Methods* **121**, 157–166.

Kensil, C. (2000). QS-21 adjuvant. In: *Methods in Molecular Medicine, Vol. 42: Vaccine Adjuvants: Preparation Methods and Research Protocols* (D. T. O'Hagan Ed.), pp. 259–271. Humana Press, Inc., Totowa, NJ, USA.

Kensil, C., Patel, C., Lennick, M. and Marciani, D. (1991). Separation and characterization of saponins with adjuvant activity from *Quillaja saponaria* Molina cortex. *J. Immunol.* **146**, 431–437.

Kersten, G. F. A. and Crommelin, D. J. A. (1995). Liposomes and iscoms as vaccine formulations. *Biochim. Biophys. Acta* **1241**, 117–138.

Kersten, G. F. A., Teerlink, T., Derks, H. J. G. M., Verkleij, A. J., van Wesel, T. L., Crommelin, D. J. A. and Beuvery, E. C. (1988). Incorporation of the major outer membrane protein of *Neisseria gonorrhoeae* in saponin–lipid complexes (iscoms): chemical analysis, some structural features and comparison of their immunogenicity with other antigen delivery systems. *Infect. Immun.* **56**, 432–438.

Kersten, G. F. A., Spiekstra, A., Beuvery, E. C. and Crommelin, D. J. A. (1991). On the structure of immune-stimulating saponin–lipid complexes (iscoms). *Biochim. Biophys. Acta* **1062**, 165–171.

Kimura, J., Nariuchi, H., Wantanabe, T., Matuhashi, T., Okayasu, I. and Hatakeyama, S. (1978). Studies of the adjuvant effect of water-in-oil-in-water (w/o/w) emulsion of sesame oil. I. Enhanced and persistent antibody formation by antigen incorporated into the water-in-oil-in-water emulsion. *Jpn. J. Exp. Med.* **48**, 149–154.

Kovarik, J., Bozzotti, P., Love-Homan, L., Pihlgren, M., Davis, H. L., Lambert, P. H., Krieg, A. M. and Siegrist, C. A. (1999). CpG oligodeoxynucleotides can circumvent the Th2 polarization of neonatal responses to vaccines but may fail to fully redirect Th2 responses established by neonatal priming. *J. Immunol.* **162**, 1611–1617.

Krieg, A. M. (1996). Lymphocyte activation by CpG dinucleotide motifs in procaryotic DNA. *Trend Microbiol.* **4**, 73–77.

Krieg, A. M. (2000). The role of CpG motifs in innate immunity. *Curr. Opin. Immunol.* **12**, 35–43.

Krieg, A. M. and Kline, J. N. (2000). Immune effects and therapeutic applications of CpG motifs in bacterial DNA. *Immunopharmacology* **48**, 303–305.

Krieg, A. M. and Wagner, H. (2000). Causing a commotion in the blood: immunotherapy progresses from bacteria to bacterial DNA. *Immunol. Today* **21**, 521–526.

Larsson, M., Lövgren, K. and Morein, B. (1993). Immunopotentiation of synthetic oligopeptides by chemical conjugation to ISCOMs. *J. Immunol. Methods* **162**, 257–260.

Lascelles, A. K., Eagleson, G., Beh, K. J. and Watson, D. L. (1989). Significance of Freund's adjuvant/antigen injection granuloma in the maintenance of serum antibody response. *Vet. Immunol. Immunopathol.* **22**, 15–27.

Lasic, D. D. (1998). Novel application of liposomes. *TIBTECH* **16**, 307–321.

Le, T. T., Drane, D., Malliaros, J., Cox, J. C., Rothel, L., Pearse, M., Woodberry, T., Gardner, J. and Suhrbier, A. (2001). Cytotoxic T cell polyepitope vaccines delivered by ISCOMs. *Vaccine* **19**, 4669–4675.

Lemaire, G., Tenu, J.-C. and Lacave, C. S. (1986). Natural and synthetic trehalose diesters as immunomodulators. *Med. Res. Rev.* **6**, 243–274.

Li, W. M., Bally, M. B. and Schutze-Redelmeier, M. P. (2002). Enhanced immune response to T-independent antigen by using CpG oligodeoxynucleotides encapsulated in liposomes. *Vaccine* **20**, 148–157.

Lipford, G. B., Wagner, H. and Heeg, K. (1994). Vaccination with immuno-dominant peptides encapsulated in Quil A-containing liposomes induces peptide-specific primary CD8+ cytotoxic T cells. *Vaccine* **12**, 73–80.

Lipford, G. B., Bauer, M., Blank, C., Reiter, R., Wagner, H. and Heeg, K. (1997). CpG-containing synthetic oligonucleotides promote B and cytotoxic T cell responses to protein antigen: a new class of vaccine adjuvants. *Eur. J. Immunol.* **27**, 2340–2344.

Lucy, J. A. and Glauert, A. M. (1964). Structure and assembly of macromolecular lipid complexes composed of globular micelles. *J. Mol. Biol.* **8**, 727–748.

Lundén, A., Parmely, S. F., Lövgren Bengtsson, K. and Araujo, F. G. (1997) Use of recombinant antigen (SAG2) expressed as a glutathione-S-transferase fusion protein to immunize mice against *Toxoplasma gondii*. *Parasitol. Res.* **83**, 6–9.

Lycke, N. (2001). The B-cell targeted CTA1-DD vaccine adjuvant is highly effective at enhancing antibody as well as CTL responses. *Curr. Opin. Mol. Ther.* **3**, 37–44.

Lycke, N. and Schön K. (2001). The B cell targeted adjuvant, CTA1-DD, exhibits potent mucosal immunoenhancing activity despite pre-existing anti-toxin immunity. *Vaccine* **19**, 2542–2548.

Lövgren, K. (1988). The serum antibody responses distributed in subclasses and isotypes following intranasal and subcutaneous immunization with influenza virus iscoms. *Scand. J. Immunol.* **27**, 241–245.

Lövgren, K. and Larsson, M. (1994). Conjugation of synthetic peptides to carrier ISCOMs: factors affecting the immunogenicity of the conjugate. *J. Immunol. Methods* **173**, 237–243.

Lövgren, K. and Morein, B. (1988). The requirement of lipids for the formation of immunostimulating complexes (iscoms). *Biotechnol. Appl. Biochem.* **10**, 161–172.

Lövgren, K. and Sundquist, B. (1994). Comparison of four different adjuvant formulations with two types of antigens; influenza virus protein micelles and ovalbumin. IBC Conference on Novel Vaccine Strategies, Washington, 23–25 February 1994.

Lövgren, K., Lindmark, J., Pipkorn, R. and Morein, B. (1987). Antigenic presentation of small molecules and synthetic peptides conjugated to a preformed iscom as carrier. *J. Immunol. Methods* **98**, 137–143.

Lövgren, K., Kåberg, H. and Morein, B. (1990). An experimental influenza subunit vaccine (ISCOM) – induction of protective immunity to challenge infection in mice after intranasal or subcutaneous administration. *Clin. Exp. Immunol.* **82**, 435–439.

Lövgren-Bengtsson, K. and Morein, B. (2000). The iscom™ technology. Methods in Molecular Medicine. *Vaccine adjuvants: Preparation Methods and Study Protocols.* (O'Hagan, D. Ed.) Humana Press, pp. 239–258.

Lövgren-Bengtsson, K. and Sjölander, A. (1996). Adjuvant activities of iscoms and iscom matrix; two different formulations of Quillaja saponin and antigen. *Vaccine* **14**, 753–760.

Ma, J., Bulger, P. A., Davis, D. R., Perilli-Palmer, P., Bedore, D. A., Kensil, C. R., Young, E. M., Hung, C. H., Seals, J. R. and Pavia, C. S. (1994). Impact of the saponin adjuvant QS-21 and aluminium hydroxide on the immunogenicity of recombinant OspA and OspB of *Borrelia burgdorferi*. *Vaccine* **12**, 925–932.

Mannhalter, J. W., Neychev, H. O., Zlabinger, G. J., Ahmad, R. and Eibl, M. M. (1985). Modulation of the human immune response by the non-toxic and non-pyrogenic adjuvant aluminium hydroxide: effect on antigen uptake and antigen processing. *Clin. Exp. Immunol.* **61**, 143–151.

Marciani, D. J., Press, J. B., Reynolds, R. C., Pathak, A. K., Pathak, V., Gundy, L. E., Farmer, J. T., Koratich, M. S. and May, R. D. (2000). Development of semi-synthetic triterpenoid saponin derivatives with immune stimulating activity. *Vaccine* **18**, 3141–3151.

Marciani, D. J., Pathak, A. K., Reynolds, R. C., Seitz, L. and May, R. D. (2001). Altered immunomodulating and toxicological properties of degraded *Quillaja saponaria* Molina saponins. *Int. Immunopharmacol.* **4**, 813–818.

Matzinger, P. (1994). Tolerance, danger and the extended family. *Annu. Rev. Immunol.* **12**, 991–1045.

McCluskie, M. J., Weeratna, R. D., Payette, P. J. and Davis, H. L. (2001). The potential of CpG oligodeoxynucleotides as mucosal adjuvants. *Crit. Rev. Immunol.* **21**, 103–120.

McLaughlin, C. A., Ribi, E. E., Goren, M. B. and Toubiana, R. (1978). Tumor regression induced by defined microbial components in an oil-in-water emulsion is mediated through their binding to oil droplets. *Cancer Immunol. Immunother.* **4**, 109–113.

McNicholl, J. M., Bond, K. B., Ruhadze, E. R., Olsen, M. R., Takayama, K. and Hunter, R. L. (1998). Enhancement of HIV type 1 vaccine immunogenicity by block copolymer adjuvants. I. Induction of high-titer, long-lasting, cross-reactive antibodies of broad isotype. *AIDS Res. Hum. Retroviruses* **14**, 1457–1471.

Medzhitov, R. and Janeway, C. A. Jr. (1997). Innate immunity: the virtues of a non-clonal system of recognition. *Cell* **91**, 295–298.

Merza, M., Linné, T., Höglund, S., Portetelle, D., Burny, A. and Morein, B. (1989). Bovine leukemia virus iscoms: biochemical characterization. *Vaccine* **7**, 22–28.

Messina, J. P., Gilkeson, G. S. and Pisetsky, D. S. (1991). Stimulation of in vitro murine lymphocyte proliferation by bacterial DNA. *J Immunol.* **147**, 1759–1764.

Meyer, T. J., Ribi, E. E., Azuma, I. and Zbar, B. (1974). Biologically active components from mycobacterial cell walls. II. Suppression and regression of strain-2 guinea pig hepatoma. *J. Natl Cancer Inst.* **52**, 103–111.

Millet, P., Kalish, M., Collins, W. E. and Hunter, R. L. (1992). Effect of adjuvant formulation on the selection of B-cell epitopes expressed by a malaria peptide vaccine. *Vaccine* **10**, 547–550.

Moldoveanu, Z., Love-Homan, L., Huang, W. Q. and Krieg, A. M. (1998). CpG DNA, a novel immune enhancer for systemic and mucosal immunization with influenza virus. *Vaccine* **16**, 1216–1224.

Morein, B., Sundquist, B., Höglund, S., Dalsgaard, K. and Osterhaus, A. (1984). Iscom, a novel structure for antigenic presentation of membrane proteins from enveloped viruses. *Nature* **308**, 457–460.

Morein, B., Ekström, J. and Lövgren, K. (1990). Increased immunogenicity of a non-amphipathic protein (BSA) after inclusion into iscoms. *J. Immunol. Methods* **128**, 177–181.

Morein, B., Lövgren, K., Rönnberg, B., Sjölander, A. and Villacrés-Eriksson, M. (1995). Immunostimulating complexes. Clinical potential in vaccine development. *Clin. Immunother.* **3**, 409–487.

Morein, B., Lövgren-Bengtsson, K. and Cox, J. (1996). Modern adjuvants. Functional aspects. In: *Concepts in Vaccine Development* (S. H. E. Kaufmann, Ed.), pp. 243–263. Walter de Gruyter, Berlin.

Mosmann, T. R. and Coffman, R. L. (1989). Th1 and Th2 cells: different patterns of lymphokine secretion lead to different functional properties. *Annu. Rev. Immunol.* **7**, 145–173.

Mowat, A. MCl, Donachie, A. M., Reid, G. and Jarret, O. (1991). Immune-stimulating complexes containing Quil A and protein antigen prime class 1 MHC-restricted T lymphocytes in vivo and are immunogenic by the oral route. *Immunology* **72**, 317–322.

Mowat, A. M., Donachie, A. M., Jagewall, S., Schön, K., Löwenadler, B., Dalsgaard, K., Kaastrup, P. and Lycke, N. (2001). CTA1-DD-immune stimulating complexes: a novel, rationally designed combined mucosal vaccine adjuvant effective with nanogram doses of antigen. *J. Immunol.* **167**, 3398–3405.

Mui, B., Raney, S. G., Semple, S. C. and Hope, M. J. (2001). Immune stimulation by a CpG-containing oligodeoxynucleotide is enhanced when encapsulated and delivered in lipid particles. *J. Pharmacol. Exp. Ther.* **298**, 1185–1189.

Mumford, J. A., Jesset, D., Dunleavy, U., Wood, J., Hannant, D., Sundquist, B. and Cook, R. F. (1994a). Antigenicity and immunogenicity of experimental equine influenza ISCOM vaccines. *Vaccine* **12**, 857–863.

Mumford, J. A., Jesset, D., Rollinson, E. A., Hannant, D. and Draper, M. E. (1994b). Duration of protective efficacy of equine influenza immunostimulating complex/tetanus vaccines. *Vet. Record* Feb. **12**, 158–162.

Murray, R., Cohen, P. and Hardegree, M. C. (1972). Mineral oil adjuvants: biological and chemical studies. *Ann. Allerg.* **30**, 146–151.

Newman, M. J., Todd, C. W., Lee, E. M., Balusubramanian, M., Didier, P. J. and Katz, J. M. (1997). Increasing the immunogenicity of a trivalent influenza virus vaccine with adjuvant-active nonionic block copolymers for potential use in the elderly. *Mech. Ageing Dev.* **93**, 189–203.

Newman, M. J., Actor, J. K., Balusubramanian, M. and Jagannath, C. (1998). Use of nonionic block polymers in vaccines and therapeutics. *Critical Rev. Therap. Drug Carrier Sys.* **15**, 89–142.

Nicholson, K. G., Colegate, A. E., Podda, A., Stephenson, I., Wood, J., Ypma, E. and Zambon, M. C. (2001). Safety and antigenicity of non-adjuvanted and MF59-adjuvanted influenza A/Duck/Singapore/97 (H5N3) vaccine: a randomised trial of two potential vaccines against H5N1 influenza. *Lancet* **357**, 1937–1943.

Noll, H., Bloch, H., Asselineau, J. and Lederer, E. (1956). The chemical structure of the cord factor of *Mycobacterium tuberculosis*. *Biochim. Biophys. Acta* **20**, 299–309.

Nowotny, A. (1990). Immune reactions elicited or modulated by endotoxin: introduction. In: *Cellular and Molecular Aspects of Endotoxins* (A. Nowotny, J. J. Spitzer and E. J. Ziegler, Eds), pp. 329–338. Elsevier, Amsterdam.

Odean, M. V., Frane, C. M., Van der Vieren, M., Tomai, M. A. and Johnson, A. G. (1990). Involvement of gamma interferon in antibody enhancement by adjuvants. *Infect. Immun.* **58**, 427–432.

Ott, G., Barchfeld, G. L., Chernoff, D., Radhakrishnan, R., van Hoogevest, P. and Van Nest, G. (1995). Design and evaluation of a safe and potent adjuvant for human vaccines. In: *Vaccine Design: The Subunit Adjuvant Approach* (M. F. Powell and M. J. Newman, Eds), pp. 277–296. Plenum Press, New York.

Ott, G., Radhakrishnan, R., Fang, J.-H. and Hora, M. (2000). The adjuvant MF59: A 10-year perspective. *Methods in Molecular Medicine, Vol. 42: Vaccine Adjuvants: Preparation Methods and Research Protocols* (D. T. O'Hagan Ed.), pp. 211–228. Humana Press, Inc., Totowa, NJ.

Oxenius, A., Martinic, M. M., Hengartner, H. and Klenerman P. (1999). CpG-containing oligonucleotides are efficient adjuvants for induction of protective antiviral immune responses with T-cell peptide vaccines. *J. Virol.* **73**, 4120–4126.

Özel, M., Höglund, S., Gelderblom, H. and Morein, B. (1989). Quaternary structure of the immunostimulating complex (ISCOM). *J. Ultrastruct. Mol. Struct. Res.* **102**, 240–248.

Preparation and Use of Adjuvants

Pedersen, I. R., Bøg-Hansen, T. C., Dalsgaard, K. and Heegaard, P. M. H. (1992). Iscom immunization with synthetic peptides representing measles virus hemagglutinin. *Virus Res.* **24**, 145–169.

Peng, Z., Wang, H., Mao, X., HayGlass, K. T. and Simons, F. E. (2001). CpG oligodeoxynucleotide vaccination suppresses IgE induction but may fail to down-regulate ongoing IgE responses in mice. *Int. Immunol.* **13**, 3–11.

Pennisi, E. (1996) Teetering on the brink of danger. *Science* **271**, 1665–1667.

Phillips, N. C., Gagné, L., Ivanoff, N. and Riveau, G. (1996). Influence of phospholipid composition on antibody response to liposome encapsulated protein and peptide. *Vaccine* **14**, 898–904.

Pisetsky, D. S. (1996). Immune activation by bacterial DNA: a new genetic code. *Immunity* **5**, 303–310.

Pisetsky, D. S. and Reich, C. F. III. (1999). Influence of backbone chemistry on immune activation by synthetic oligonucleotides. *Biochem. Pharmacol.* **58**, 1981–1988.

Polakos, N. K., Drane, D., Cox, J., Ng, P., Selby, M. J., Chien, D., O'Hagan, D. T., Houghton, M. and Paliard, X. (2001) Characterization of hepatitis C virus core-specific immune responses primed in rhesus macaques by a nonclassical ISCOM vaccine. *J. Immunol.* **166**, 3589–3598.

Porter, K. R., Kochel, T. J., Wu, S. J., Raviprakash, K., Phillips, I. and Hayes, C. G. (1998). Protective efficacy of a dengue 2 DNA vaccine in mice and the effect of CpG immuno-stimulatory motifs on antibody responses. *Arch. Virol.* **143**, 997–1003.

Pyle, S., Morein, B., Bess, J., Åkerblom, L., Nara, P., Nigida, S., Lerche, N., Fishinger, P. and Arthur, L. (1989). Immune response to immunostimulatory complexes (ISCOMs) prepared from human immunodeficiency virus type 1 (HIV-1) or the HIV-1 external envelope glycoprotein. *Vaccine* **7**, 465–473.

Ramanathan, V. D., Badenoch-Jones, P. and Turk, J. L. (1979). Complement activation by aluminium and zirconium compounds. *Immunology* **37**, 881–888.

Ramon, G. (1926). Procédés pour accroitre la production des antitoxines. *Ann. Inst. Pasteur, Paris* **40**, 1–10.

Reid, G. (1992). Soluble proteins incorporate into iscoms after covalent attachment of fatty acid. *Vaccine* **10**, 597–602.

Ridge, J. P., Fuchs, E. J. and Matzinger, P. (1996) Neonatal tolerance revisited: turning on newborn T cells with dendritic cells. *Science* **271**, 1723–1726.

Rietschel, E. T., Brade, H., Brade, L., Kaca, W., Kawahara, K., Lindner, B., Luderitz, T., Tomita, T., Schade, U. and Seydel, U. (1985). Newer aspects of the chemical structure and biological activity of bacterial endotoxins. *Prog. Clin. Biol. Res.* **189**, 31–51.

Rimmelzwaan, G. F. and Osterhaus, A. D. (1995). A novel generation of viral vaccines based on the ISCOM matrix. *Pharm. Biotechnol.* **6**, 543–558.

Rönnberg, B., Fekadu, M. and Morein, B. (1995). Adjuvant activity of non-toxic *Quillaja saponaria* Molina components for use in iscom-matrix. *Vaccine* **13**, 1375–1382.

Rudbach, J. A., Johnson, D. A. and Ulrich, J. T. (1995). Ribi adjuvants: chemistry, biology and utility in vaccines for human and veterinary medicine. In: *The Theory and Practical Application of Adjuvants* (D. E. S. Stewart-Tull, Ed.), pp. 287–313. Wiley, New York.

Sambhara, S., Kurichh, A., Miranda, R., Tumpey, T., Rowe, T., Renshaw, M., Arpino, R., Tamane, A., Kandil, A., James, O., Underdown, B., Klein, M., Katz, J. and Burt, D. (2001) Heterosubtypic immunity against human influenza A viruses, including recently emerged avian H5 and H9 viruses, induced by FLU-ISCOM vaccine in mice requires both cytotoxic T-lymphocyte and macrophage function. *Cell Immunol.* **211**, 143–153.

Sarzotti, M., Robbins, D. S. and Hoffman, P. M. (1996). Induction of protective CTL responses in newborn mice by a murine retrovirus. *Science* **271**, 1726–1728.

Scheepers, K. and Becht, H. (1994). Protection of mice against an influenza virus infection by oral vaccination with viral nucleoprotein incorporated into immunostimulating complexes. *Med. Microbiol. Immunol.* **183**, 265–278.

Schenck, J. R., Hargie, M. P., Brown, M. S., Ebert, D. S., Yoo, A. L. and McIntire, F. C. (1969). The enhancement of antibody formation by *Escherichia coli* lipopolysaccharide and detoxified derivative. *J. Immunol.* **102**, 1411–1422.

Schumann, G., van Hoogevest, P. and Frankhauser, P. (1989). Comparison of free and liposomal MTP-PE: pharmacological, toxicological and pharmakokinetic aspect. In: *Liposomes in the Therapy of Infectious Disease and Cancer* (G. Lopez-Berestein and I. J. Fidler, Eds), pp. 191–203. A. R. Liss, New York.

Scott, M. T., Bahr, G., Moddaber, F., Afchain, D. and Chedid, L. (1984). Adjuvant requirements for protective immunization of mice using a *Trypanosoma cruzi* 90 kilodalton cell surface glycoprotein. *Int. Arch. Allergy Appl. Immunol.* **74**, 373–377.

Seeber, S. J., White, J. L. and Hem, S. L. (1991). Predicting the adsorbtion of proteins by aluminium-containing adjuvants. *Vaccine* **9**, 201–203.

Shimada, S., Yano, O., Inoue, H., Kuramoto, E., Fukuda, T., Yamamoto, H., Kataoka, T. and Tokunaga, T. (1985). Antitumor activity of the DNA fraction from *Mycobacterium bovis* BCG. II. Effects on various syngeneic mouse tumors. *J. Natl. Cancer Inst.* **74**, 681–688.

Shirota, H., Sano, K., Hirasawa, N., Terui, T., Ohuchi, K., Hattori, T., Shirato, K. and Tamura, G. (2001). Novel roles of CpG oligodeoxynucleotides as a leader for the sampling and presentation of CpG-tagged antigen by dendritic cells. *J. Immunol.* **167**, 66–74.

Shoda, L. K., Kegerreis, K. A., Suarez, C. E., Mwangi, W., Knowles, D. P. and Brown, W. C. (2001). Immunostimulatory CpG-modified plasmid DNA enhances IL-12, TNF-alpha, and NO production by bovine macrophages. *J. Leukoc. Biol.* **70**, 103–112.

Sjölander, A., Lövgren, K., Ståhl, S., Åslund, L., Hamsson, M., Nygren, P.-Å, Larsson, M., Hagstedt, M., Wåhlin, B., Berzins, K., Uhlén, M., Morein, B. and Perlmann, P. (1991). High antibody responses in rabbits immunized with influenza virus ISCOMs containing a repeated sequence of the *Plasmodium falciparum* antigen Pf155/RESA. *Vaccine* **9**, 443–450.

Sjölander, A., Hansson, M., Lövgren, K., Wåhlin, B., Berzins, K. and Perlmann, P. (1993). Immunogenicity in rabbits and monkeys of influenza ISCOMs conjugated with repeated sequences of the *Plasmodium falciparum* antigen Pf 155/RESA. *Parasite Immunol.* **15**, 355–359.

Sjölander, A., Lövgren-Bengtsson, K., Johansson, M. and Morein, B. (1996a). Kinetics, localization and isotype profile of antibody responses to immune stimulating complexes (iscoms) containing human influenza virus envelope glycoproteins. *Scand. J. Immunol.* **43**, 164–172.

Sjölander, S., Hansen, J.-E., Lövgren-Bengtsson, K., Åkerblom, L. and Morein, B. (1996b). Induction of homologous virus neutralizing antibodies in guinea-pigs immunized with two human immunodeficiency virus type 1 glycoprotein go120-iscom preparations. A comparison with other adjuvant systems. *Vaccine* **14**, 344–352.

Sjölander, A., Lövgren-Bengtsson, K. and Morein, B. (1997a). Kinetics, localization and cytokine profile of T cell responses to immune stimulating complexes (iscoms) containing human influenza virus envelope glycoproteins. *Vaccine* **15**, 1030–1038.

Sjölander, A., van't Land, B. and Lövgren-Bengtsson, K. (1997b). Iscoms containing purified Quillaja saponins upregulate both Th1-like and Th2-like immune responses. *Cell. Immunol.* **177**, 69–76.

Snodgrass, D. R., Campbell, I., Mwenda, J. M., Chege, G., Seleman, M. A., Morein, B. and Hart, C. A. (1995). Stimulation of rotavirus IgA, IgG and neutralising antibodies in baboon milk by parenteral vaccination. *Vaccine* **13**, 408–413.

Stan, A. C., Casares, S., Brumeanu, T. D., Klinman, D. M. and Bona, C. A. (2001). CpG motifs of DNA vaccines induce the expression of chemokines and MHC class II molecules on myocytes. *Eur. J. Immunol.* **31**, 301–310.

Stewart-Tull, D. E. S. (1989). Recommendations for the assessment of adjuvants (immunomodulators). In: *Immunological Adjuvants and Vaccines* (G. Gregoriadis, A. C. Allison and G. Poste, Eds), pp. 213–226. Plenum Press, New York.

Stienecker, F., Kersten, G., van Bloois, L., Crommelin, D., Hem, S., Löwer, J. and Kreuter, J. (1995). Comparison of 24 different adjuvants for inactivated HIV-2 split whole virus as antigen in mice. Induction of titres of binding antibodies and toxicity of the formulations. *Vaccine* **13**, 45–53.

Stowers, A. W., Cioce, V., Shimp, R. L., Lawson, M., Hui, G., Muratova, O., Kaslow, D. C., Robinson, R., Long, C. A. and Miller, L. H. (2001). Efficacy of two alternate vaccines based on *Plasmodium falciparum* merozoite surface protein 1 in an Aotus challenge trial. *Infect. Immun.* **69**, 1536–1546.

Sundquist, B., Lövgren, K. and Morein, B. (1988). Influenza virus iscoms: antibody response in animals. *Vaccine* **6**, 49–52.

Takada, H. and Kotani, S. (1989). Structural requirements of lipid A for endotoxicity and other biological activities. *Crit. Rev. Microbiol.* **16**, 477–523.

Takada, H. and Kotani, S. (1995). Muramyl dipeptide and derivatives. In: *The Theory and Practical Application of Adjuvants* (D. E. S. Stewart-Tull, Ed.), pp. 51–94. Wiley, New York.

Takahashi, H., Takeshita, T., Morein, B., Putney, S., Germain, R. N. and Berzofsky, J. (1990). Induction of CD8+ cytotoxic T cells by immunization with purified HIV-1 envelope proteins in iscoms. *Nature* **344**, 873–875.

Thibault, P. and Richou, R. (1936). Sur l'accroissement de l'immunité antitoxique sous influence de l'addition de diverses substances à l'antigène (anatoxines diphtéroque et tétanique). *CR Soc. Biol.* **121**, 718–721.

Todd, C. W. and Newman, M. J. (2000). Aqueous formulation of adjuvant-active nonionic block copolymers. In: *Methods in Molecular Medicine, Vol. 42: Vaccine Adjuvants: Preparation Methods and Reseach Protocols* (D. T. O'Hagan Ed.), pp. 121–136. Humana Press, Inc., Totowa, NJ.

Todd, C. W., Pozzi, L. A., Guarnaccia, J. R., Balasubramanian, M., Henk, W. G., Younger, L. E. and Newman, M. J. (1997). Development of an adjuvant-active nonionic block copolymer for use in oil-free subunit vaccines formulations. *Vaccine* **15**, 564–570.

Tokunaga, T., Yamamoto, H., Shimada, S., Abe, H., Fukuda, T., Fujisawa, Y., Furutani, Y., Yano, O., Kataoka, T., Sudo, T. *et al.* (1984). Antitumor activity of deoxyribonucleic acid fraction from *Mycobacterium bovis* BCG. I. Isolation, physicochemical characterization, and antitumor activity. *J. Natl. Cancer Inst.* **72**, 955–962.

Triozzi, P. L., Stevens, V. C., Aldrich, W., Powell, J., Todd, C. W. and Newman, M. J. (1997). Effects of a b-human chorionic gonadotrophin subunit immunogen administered in aqueous solution with a novel nonionic block copolymer adjuvant in patients with advanced cancer. *Clinical Cancer Research* **3**, 2355–2362.

Uede, T. and Ishizaka, K. (1982). Formation of IgE binding factors by rat T lymphocytes. VI. Cellular mechanisms for the formation of IgE-potentiating factor and IgE-suppressive factor by antigenic stimulation of antigen primed spleen cells. *J. Immunol.* **129**, 1391–1397.

Uede, T., Huff, T. F. and Ishizaka, K. (1982). Formation of IgE binding factors by rat T lymphocytes. V. Effect of adjuvant for the priming immunization on the nature of IgE binding factors formed by antigenic stimulation. *J. Immunol.* **129**, 1384–1390.

Ugozzoli, M., O'Hagan, D. T. and Ott, G. S. (1998). Intranasal immunization of mice with herpes simplex virus type 2 recombinant gD2: the effect of adjuvants on mucosal and serum antibody responses. *Immunology* **93**, 563–571.

Ulrich, J. T. (2000). MPL® Immunostimulant: adjuvant formulations. In: *Methods in Molecular Medicine, Vol. 42: Vaccine Adjuvants: Preparation Methods and Research Protocols* (D. T. O'Hagan Ed.), pp. 273–282. Humana Press, Inc., Totowa, NJ.

Unanue, E. R. and Allen, P. M. (1987). The basis for the immunoregulatory role of macrophages and other accessory cells. *Science* **236**, 551–557.

Valensi, J. P., Carlsson, J. R. and Van Nest, G. A. (1994). Systemic cytokine profiles in BALB/c mice immunized with trivalent influenza vaccine containing MF59 oil emulsion and other adjuvants. *J. Immunol.* **153**, 4029–4039.

Verthelyi, D., Ishii, K., Gursel, M., Takeshita, F. and Klinman, D. (2001). Human peripheral blood cells differentially recognize and respond to two distinct CpG motifs. *J. Immunol.* **166**, 2372–2377.

Villacrés-Eriksson, M. (1995). Antigen presentation by naive macrophages, dendritic cells and B cells to primed T lymphocytes and their cytokine production following exposure to immunostimulating complexes. *Clin. Exp. Immunol.* **102**, 46–52.

Villacrés-Eriksson, M., Bergström-Mollauglu, M., Kåberg, H. and Morein, B. (1992). Involvement of interleukin-2 and interferon-gamma in the immune response induced by influenza virus iscoms. *Scand. J. Immunol.* **36**, 421–426.

Villacrés-Eriksson, M., Bergström-Mollauglu, M., Kåberg, H., Lövgren, K. and Morein B. (1993). The induction of cell-associated and secreted interleukin-1 by iscoms, matrix or micelles in murine splenic cells. *Clin. Exp. Immunol.* **93**, 120–125.

Villacrés-Eriksson, M., Behboudi, S., Morgan, A. J., Trichieri, G. and Morein, B. (1997). Immunomodulation by *Quillaja saponaria* adjuvant formulations: in vivo stimulation of interleukin 12 and its effects on the antibody response. *Cytokine* **9**, 73–82.

Vogel, F. R. and Powell, M. F. (1995). A compendium of vaccine adjuvants and excipients. *Pharm. Biotechnol.* **6**, 141–228.

Walls, R. S. (1977) Eosinophil response to alum adjuvants: involvements of T cells in non-antigen-dependent mechanisms. *Proc. Soc. Exp. Biol. Med.* **156**, 431–435.

Watson, D. L., Lövgren, K., Watson, N. A., Fossum, C., Morein, B. and Höglund, S. (1989). The inflammatory response and antigen localization following immunization with influenza virus iscoms. *Inflammation* **13**, 641–649.

Watson, D., Watson, N., Fossum, C., Lövgren, K. and Morein, B. (1992). Interactions between immune-stimulating complexes (ISCOMS) and peritoneal mononuclear leucocytes. *Microbiol. Immunol.* **36**, 199–203.

Weeratna, R. D., Brazolot, Millan, C. L., McCluskie, M. J. and Davis, H. L. (2001). CpG ODN can re-direct the Th bias of established Th2 immune responses in adult and young mice. *FEMS Immunol. Med. Microbiol.* **32**, 65–67.

Weijer, K., Pfauth, A., van Herwijnen, R., Jarret, O., Meloen, R. H., Tomee, C. and Osterhaus, A. (1993). Induction of feline leukaemia virus-neutralizing antibodies by peptides derived from the FeLv env gene. *Vaccine* **11**, 946–956.

Wernette, C. M., Smith, B. F., Barksdale, Z. L., Hecker, R. and Baker, H. J. (2002). CpG oligodeoxynucleotides stimulate canine and feline immune cell proliferation. *Vet. Immunol. Immunopathol.* **84**, 223–236.

Preparation and Use of Adjuvants

White, R. G. (1976). The adjuvant effect of microbial products on the immune response. *Annu. Rev. Microbiol.* **30**, 579–600.

Wilson, A. D., Lövgren-Bengtsson, K., Villacres-Ericsson, M., Morein, B. and Morgan, A. J. (1999) The major Epstein-Barr virus (EBV) envelope glycoprotein gp340 when incorporated into Iscoms primes cytotoxic T-cell responses directed against EBV lymphoblastoid cell lines. *Vaccine* **17**, 1282–1290.

Windon, R. G., Chaplin, P. J., McWaters, P., Tavarnesi, M., Tzatzaris, M., Kimpton, W. G., Cahill, R. N., Beezum, L., Coulter, A., Drane, D., Sjölander, A., Pearse, M., Scheerlinck, J. P. and Tennent, J. M. (2001). Local immune responses to influenza antigen are synergistically enhanced by the adjuvant ISCOMATRIX. *Vaccine* **20**, 490–497.

Woodard, L. F. (1989). Adjuvant activity of water-insoluble surfactants. *Lab. Anim. Sci.* **39**, 222–225.

Yamamoto, S., Kuramoto, E., Shimada, S. and Tokunaga, T. (1988). In vitro augmentation of natural killer cell activity and production of interferon-alpha/beta and -gamma with deoxyribonucleic acid fraction from *Mycobacterium bovis* BCG. *Jpn. J. Cancer Res.* **79**, 866–873.

Yamamoto, S., Yamamoto, T., Shimada, S., Kuramoto, E., Yano, O., Kataoka, T. and Tokunaga, T. (1992a). DNA from bacteria, but not from vertebrates induces interferon, activates Natural Killer cells and inhibits tumor growth. *Microbiol. Immunol.* **36**, 983–997.

Yamamoto, S., Yamamoto, T., Kataoka, T., Kuramoto, E., Yano, O. and Tokunaga, T. (1992b). Unique palindromic sequences in synthetic oligonucleotides are required to induce IFN and augment IFN-mediated natural killer activity. *J. Immunol.* **148**, 4072–4076.

Zhang, Y., Shoda, L. K., Brayton, K. A., Estes, D. M., Palmer, G. H. and Brown, W. C. (2001). Induction of interleukin-6 and interleukin-12 in bovine B lymphocytes, monocytes, and macrophages by a CpG oligodeoxynucleotide (ODN 2059) containing the GTCGTT motif. *J. Interferon Cytokine Res.* **21**, 871–881.

PART III

Human Systems

◆◆

1 Isolation and Propagation of Human Dendritic Cells

A Karolina Palucka and Jacques Banchereau
Baylor Institute for Immunology Research, Dallas, USA

Christophe Caux
Schering-Plough, Laboratory for Immunological Research, Dardilly, France

Colette Dezutter-Dambuyant
INSERM, U346, Hopital Edouard Herriot, 69374 Lyon, France

Yong-Jun Liu
Department of Immunobiology, DNAX Research Institute, Palo Alto, California, USA

◆◆◆

CONTENTS

Isolation and Propagation of Human Dendritic Cells

◆◆◆◆◆◆ INTRODUCTION

Dendritic cells (DCs) are unique antigen-presenting cells (APCs), which induce primary immune responses thus permitting establishment of immunological memory (Steinman, 1991; Hart, 1997; Steinman *et al.*, 1997; Banchereau *et al.*, 2000). DC progenitors in the bone marrow give rise to circulating precursors that home to the tissue where they reside as immature cells with high phagocytic capacity. Immature, antigen-capturing DC in peripheral tissues sense pathogens, tissue necrosis, and local inflammation. These 'danger' signals induce DCs to undergo a maturation process that includes their migration into T-cell areas of draining lymph nodes. There, they present processed antigen(s) (Ags) to T cells

METHODS IN MICROBIOLOGY, VOLUME 32
ISBN 0–12–521532–0

Copyright © Elsevier Science Ltd
All rights of reproduction in any form reserved

via antigen presenting molecules such as MHC and CD1, thereby inducing T-cell proliferation and differentiation into Type 1 (IFN-γ), Type 2 (IL-4/IL-13) or regulatory T cells (IL-10). DC present Ag to CD4+ T cells, which in turn regulate the immune effectors including Ag specific CD8+ T cells and B cells as well as non-Ag-specific macrophages, eosinophils and NK cells. DCs can also bypass CD4 T cells and directly activate CD8+ T cells, B cells and, as shown most recently, NK cells (Gerosa *et al.*, 2000; Crotta *et al.*, 2002; Ferlazzo *et al.*, 2002; Piccioli *et al.*, 2002). There is now evidence that immature DCs that capture self-antigens (e.g. apoptotic cells) induce tolerance (Huang *et al.*, 2000; Steinman *et al.*, 2000). In contrast, if antigen-loaded DC undergo maturation, they induce antigen-specific immunity (Finkelman *et al.*, 1996; Heath and Carbone, 2001).

DCs can be found in many tissues including: (a) peripheral non-lymphoid tissues with the Langerhans cells (LCs) of epithelium (skin, mucosa, lung), characterized by the expression of CD1a and Birbeck granules, and the interstitial (dermal) DC (intDCs) of heart, kidney as well as virtually all other organs, lacking Birbeck granules but expressing coagulation factor XIIIa; (b) circulation with the afferent lymph veiled cells and the peripheral blood DC. The latter ones are composed of two subsets identified by differential expression of CD11c and CD123 {IL3-Rα}, each representing a small fraction {~0.3%} of the mononuclear cells; (c) secondary lymphoid organs with interdigitating cells (IDC) within the T cell rich areas and germinal centre DC (GCDC) in B cell follicles, and (d) the thymic medulla (Banchereau *et al.*, 2000).

For nearly 20 years after their discovery (Steinman and Cohn, 1973), DCs had to be painstakingly isolated from tissues to be studied. In 1992, *in vitro* culture systems were identified to generate large numbers of mouse (Inaba *et al.*, 1992) and human DCs (Caux *et al.*, 1992). More recently, Flt3 ligand (Flt3-L), a stromal cell product, was found to induce a massive expansion of DCs, *in vivo* in mice (Maraskovsky *et al.*, 1996; Pulendran *et al.*, 1997) and in humans (Maraskovsky *et al.*, 2000; Pulendran *et al.*, 2000). Currently, the standard methods to generate human DC *in vitro* include (a) culturing bone marrow progenitors in GM-CSF and TNF (Caux *et al.*, 1992), and (b) culturing blood monocytes with GM-CSF and IL-4 (Peters *et al.*, 1993; Romani *et al.*, 1994; Sallusto and Lanzavecchia, 1994). This chapter summarizes methods for making human DCs *ex vivo* as well as for *in vivo* DC propagation. Bibliography provides the key references. The reader is invited to consult the earlier book edition as well as most recent reviews.

◆◆◆◆◆◆ ISOLATION OF DENDRITIC CELLS

From human skin

Characteristic features

The stratified squamous epithelia (skin and mucosa) is the first line of defence of the organism against external agents, not only as a physical barrier between the body and the environment but also as the site of initiation of immune reactions. Human epidermis is a heterogeneous epithelium composed of keratinocytes, melanocytes, Merkel cells and bone marrow derived Langerhans cells (LC) (Stingl and Shevach, 1991). LC represent 1–6% of the epidermal cells (EC) (Romani et al., 1991) and can be distinguished from the surrounding EC by the presence of rod-shaped cytoplasmic organelles, termed Birbeck granules and by the expression of Langerin, a C-type lection uniquely expressed on LC and possibly involved in the formation of Birbeck granules (Valladeau et al., 1998, 2000). Other molecules expressed on LC include classical and non-classical major histocompatibility complex molecules such as MHC class II and CD1a, respectively; β-integrins such as CD11b and CD11c; CD32 (Fc_RII), CD86 (B7-2) and CCR6, a chemokine receptor mediating LC migration to MIP3-α (Banchereau et al., 2000; Caux et al., 2000). Freshly isolated epidermal LCs are weak stimulators of naive T cells, but are effective processors of soluble protein antigens. After short-term culture LCs down-regulate the expression of some surface molecules (CD1a, CD32, E-Cadherin), lose Birbeck granules and their antigen-processing capacity while they upregulate MHC class I and II, accessory molecules (CD58, CD80, CD86) and acquire activation/maturation markers (CD25, CD83). Cultured LCs become potent immunostimulatory cells. This *in vitro* phenotypical and functional maturation process mimics what occurs *in vivo* when perturbation of the skin milieu causes migration of epidermal LC via dermal lymphatic vessels to draining lymph nodes and maturation of the migrating LC into potent immunostimulatory APC. *In vivo* they are responsible for the induction of contact hypersensitivity to haptens (Kripke et al., 1990; Enk and Katz, 1992) and take part in skin graft rejection and in presentation of bacterial, viral and tumour antigens to T cells (Dezutter-Danbuyant et al., 1991; Grabbe et al., 1995) and reviewed in (Steinman, 1991; Banchereau et al., 2000).

Technical aspect

Isolation of epidermal cells

Epidermal cell suspensions are obtained from fresh normal skin. Trypsinization allows the separation of epidermis from dermis and disrupts the epidermal sheets into a single-cell suspension (Ray et al., 1989). Trimmed skin is split-cut with a keratotome set. The dermo–epidermal slices are treated for 18 h at 4°C or for 1 h at 37°C with 0.05% trypsin in Hank's balanced salt solution without Ca^{2+} and Mg^{2+}. Such a low trypsin concentration provides a good balance between tissue digestion and surface molecule damage. The epidermis is then detached from the dermis with fine forceps. The epidermal sheets are pooled in medium 199 Hank's supplemented with 20% fetal calf serum (TC199/FCS) to stop trypsin action, then they are vigorously and repeatedly blown in and out of a 5-ml pipette. After filtration through sterile gauze, the suspended cells are washed three times with TC199/FCS. Cells are counted and the viability is estimated by trypan blue exclusion. Viability after trypsinization must range from 80–95%.

Enrichment of epidermal dendritic cells or Langerhans cells

LC enrichment is achieved by successive density gradient centrifugation steps and depletion of basal keratinocytes. Total EC suspensions (4.10^6 cells ml^{-1}) in TC199/FCS are layered on the Lymphoprep (density = 1.077; with a 2.5:1 ratio) and centrifuged for 30 min at $400g$. The cells from the interface are washed twice and resuspended in TC199/FCS. After first centrifugation, LC enrichment reaches up to 7–15%. The cells are plated onto collagen type 1-coated dishes at a concentration of 5–6.10^6 cells ml^{-1} and incubated for 20–30 min at 37°C in order to deplete the basal keratinocytes. Non-adherent cells are collected and layered on the diluted Lymphoprep (a 3.4:1.6 ratio of Lymphoprep : distilled water) and centrifuged for 20 min at $400g$. The low-density fraction at the interface is collected and washed three times with 10% FCS-supplemented RPMI-1640 medium. This last enrichment procedure (depletion of basal keratinocytes and modified gradient sedimentation) leads to a suspension containing about 70–95% LC.

Further enrichment may be performed with a positive selection procedure using monoclonal antibodies (MoAb) against surface markers (e.g. anti-CD1a) and magnetic beads. Briefly, cells are suspended in phosphate buffered saline (PBS) containing 10 mM EDTA and 0.5% bovine serum albumin (PBS/EDTA/BSA) and incubated with anti-CD1a MoAb at a final concentration of 10^7 cells/ml for 15 min at 4°C. After washes, the cells are incubated with goat anti-mouse IgG coated Miltenyi microbeads according to manufacturer recommendations. After 15 min at 4°C, EC are washed once and isolation of CD1a-positive cells is performed using Minimacs separation columns. EC are suspended in PBS/EDTA/BSA at a concentration of 2.10^6 ml^{-1}, layered on the top of the column fixed under the magnet. After washing, the column is removed from separator and placed on a suitable tube, the retained cells are eluted with 2 ml of PBS/EDTA/BSA buffer using the plunger. The cells are centrifuged then resuspended in 10% FCS-supplemented RPMI-1640 medium. The isolated cells contain up to 95–99% LC as determined by staining with FITC-conjugated anti-CD1a MoAb and flow cytometry.

From human peripheral blood

Characteristic features

Two subsets of DCs can be identified in the blood, each representing a small fraction (~0.3%) of the mononuclear cells (O'Doherty et al., 1994) (Fig. 1). The CD11c$^+$ CD123$^-$ myeloid DC (mDC) subset can give rise to interstitial DC, LC and macrophages (Ho et al., 1999; Robinson et al., 1999) and promptly differentiate into mature DC in response to inflammatory stimuli (O'Doherty et al., 1994). The CD11c$^-$ CD123$^+$ plasmacytoid DCs (pDCs) subset was first observed by pathologists (Muller-Hermelink et al., 1973, 1983; Lennert et al., 1975; Vollenweider and Lennert, 1983) who saw, within the T-cell areas of lymphoid organs, cells that looked like plasma cells. Their isolation from tonsils revealed unique features (Grouard et al.,

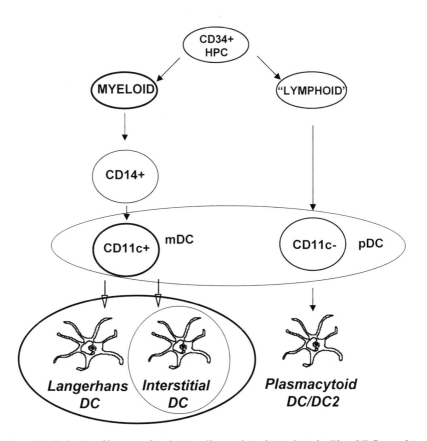

Figure 1. Subsets of human dendritic cells used in clinical trials. Blood DCs, mobilized by FLT3 ligand, contain both CD11c+ myeloid DC and CD11c− plasmacytoid DC. Most studies to date have been carried out with DC made by culturing monocytes with GM-CSF and IL-4 (intDCs). These DC are immature and require exogenous factors (CD40 ligand or macrophage cytokines) for maturation. DCs can also be generated by culturing CD34+ HPC with GM-CSF and TNF-α that permits to obtain two DC subsets: LCs and intDC (solid lines). Adding IL-4 to CD34 cultures with GM-CSF/TNF skews differentiation towards intDCs and inhibits generation of LCs.

1997): (a) an abundant rough endoplasmic reticulum and an eccentric nucleus, (b) expression of CD123 (IL-3 receptor) conferring responsiveness to IL-3 (Olweus *et al.*, 1997), and (c) lack of myeloid markers. pDCs can be further distinguished from CD11c+ DCs by differential expression of Immunoglobulin-Like Transcripts (ILTs), with pDCs being ILT1−/ILT3+ and CD11c+ DCs being ILT1+/ILT3+ (Cella *et al.*, 1999). These cells are also unique in their function, i.e. in their capacity to (a) rapidly make large amounts of IFN-α (Siegal *et al.*, 1998, 1999), (b) mature, when triggered by virus, into DCs able to induce T cells to produce IFN-γ and IL-10 (Cella *et al.*, 2000; Kadowaki *et al.*, 2000), and (c) mature, when triggered by IL-3 and CD40 ligand, into DCs that induce T cells to make type 2 cytokines, i.e. IL-4 and IL-5. Recent studies demonstrated that systemic administration of both Flt3-L and G-CSF increase the number of pDC in blood

(Arpinati *et al.*, 2000; Maraskousky *et al.*, 2000; Pulendran *et al.*, 2000), and established culture systems to generate these cells *in vitro* (Blom *et al.*, 2000; Spits *et al.*, 2000).

Most recently, three new molecules BDCA-2, BDCA-3, and BDCA-4, differentially expressed on human blood DCs were identified (Dzionek *et al.*, 2000). In blood, BDCA-2 and BDCA-4 are expressed on pDCs, whereas BDCA-3 is expressed on small population of CD11c$^+$CD123$^-$ DCs. BDCA-4 is also expressed on *in vitro* generated mDCs. BDCA molecules are involved in DCs functions as BDCA-2 is rapidly internalized at 37°C after mAb labelling and is critical for the release of IFN-α by pDCs (Dzionek *et al.*, 2001).

Technical aspect

Isolation of blood mononuclear cells

A sample of 450 ml blood is collected over CPD and diluted 50% with PBS. The suspension is layered on Ficoll-Hypaque and after 30 min centrifugation, the mononuclear cells are recovered and washed with PBS.

Depletion of T, B, NK cells and monocytes

Cells are resuspended at 2×10^7 cells ml^{-1} in PBS containing 2% human serum (HS) and 0.5 mM EDTA and stained with a cocktail of MoAb for depletion of cells expressing lineage markers CD3, CD14, CD16, CD19, CD56 and glycophorin A. After 20–30 min at 4°C under gentle shaking, cells are washed three times with PBS containing 2% FCS and 0.5 mM EDTA, and resuspended at 5×10^7 cells ml^{-1}. Goat anti-mouse Ig beads (Dynabeads or Miltenyi) are added to the cell suspension, and, after 20–30 min at 4°C under gentle shaking, cells are incubated over the magnet for 10 min. Supernatants containing unbound cells are harvested. Beads are washed twice under the magnet. Collected supernatants are mixed and after centrifugation cells may be submitted to another round of bead depletion. The resulting suspension which contains 20–60% of dendritic cells (mean ± SD = 35 ± 25) represents 2.5–7.5% of the original one (450 ml blood yields $4.5–10 \times 10^8$ MNC which yields $8–30 \times 10^6$ enriched cells).

FACS-sorting of peripheral blood DC subsets

The available methods are modifications of the original method described by O'Doherty *et al.* (1994). Cells are resuspended at 2×10^7 cells ml^{-1} in PBS/HS/EDTA and stained with biotinylated anti-HLA-DR, PE coupled anti-CD11c and a cocktail of FITC coupled MoAbs against lineage markers: CD3, CD14, CD15, CD16, CD20, CD57 and CD34. After 20–30 min at 4°C under gentle shaking, cells are washed three times with PBS/FCS/EDTA, resuspended at 2×10^7 cells ml^{-1}, and incubated with fluorochrome conjugated Streptavidin (FL3 fluorescence, e.g. Quantum red or Tricolor) for 20–30 min at 4°C under gentle shaking. After three washes cells are resuspended at $1–2 \times 10^6$ cells ml^{-1} and filtered. Cells are sorted according to lack of FITC staining and expression of high level of HLA-DR and CD11c into

FITC-HLA-DR$^+$CD11c$^+$ and FITC-HLA-DR$^+$CD11c$^-$ fractions. The CD11c$^+$ DC subpopulation represents 35–60% of the DR$^+$ pool. After sorting, 0.5 to 1.5×10^6 of each DC subset is recovered out of 450 ml blood. Alternatively, DC isolation kits from Miltenyi can be used.

Isolation of DC subsets and DC precursors from human tonsils

In human tonsils, part of mucosal-associated lymphoid tissues, different subsets of DCs and DC precursors have been identified by immunohistology, including CD1a$^+$ Langerhans cells in mucosal epidermis, CD40$^+$CD80$^+$ CD83$^+$CD86$^+$ interdigitating cells (IDC), CD4$^+$CD3$^-$CD11c$^-$ plasmacytoid DC precursors within the T-cell rich areas and CD4$^+$CD3$^-$CD11c$^+$ DCs in germinal centres (GCDC). In this section, the methods for isolation of IDC, GCDC and CD4$^+$CD3$^-$CD11c$^-$ plasmacytoid DC precursors are described.

Interdigitating cells of T-cell area

Characteristic features

IDCs represent mature DCs, characterized by long interdigitating dendrites, expression of high levels of MHC Class II, CD40, CD80, CD83 and CD86, moderate levels of CD4 and very low levels of myeloid antigen CD13 and CD33 (Bjorck *et al.*, 1997). They may be derived from epidermal Langerhans cells, that migrate into the T-cell areas of draining lymphoid tissues after capture and processing of antigens. These cells are thought to play a key role in priming antigen-specific naive T cells.

Technical aspect

Isolation of IDC

Tonsils are cut into small pieces and digested for 12 min at 37°C with collagenase IV (1 mg ml^{-1}) and deoxyribonuclease I (50 KU ml^{-1}) in RPMI 1640. The cells, pooled from two rounds of tissue digestion, are centrifuged over 50% Percoll (Pharmacia) for 20 min at 400g. The resulting low density cells are collected and subjected to magnetic bead depletion of T, B, NK and monocytes as described earlier. The resulting cells are labelled with a biotinylated antibody to CD40 (mAb89) and FITC-conjugated antibodies to CD19, CD3, CD20, CD14 and CD57. Cells are further incubated with tricolour-labelled streptavidin (Caltag, San Francisco, CA, USA). The CD40-only expressing cells are sorted as IDCs.

CD4$^+$CD3$^-$CD11c$^+$ germinal centre DCs (GCDC)

Characteristic features

A population of large dendritic CD4$^+$CD3$^-$ cells can be found within germinal centres (GC) of human lymph nodes, tonsils and spleens

(Grouard *et al.*, 1996). Double-colour immunohistochemistry reveals that these CD3⁻CD4⁺ cells are CD1a⁻, CD40^{low}B7^{low} and DRC-1⁻ (Liu *et al.*, 1997), indicating that these cells are not Langerhans cells (CD1a⁺), nor inter-digitating cells (CD40^{high}, B7.1/CD80^{high}, B7.2/CD86^{high}), nor FDCs (DRC-1⁺, KIM4⁺). These cells can be isolated from tonsillar mononuclear cells after magnetic bead depletion of T, B and NK, monocytes, followed by three-colour FACS sorting of CD4⁺CD11c⁺lin⁻ cells. The isolated cells are characterized by a strong MHC class II expression, DC morphology and potent stimulatory activity on CD4⁺ T cells. These DCs likely stimulate the germinal centre T cells that are required for the generation of memory B cells and may contribute to the unique feature of HIV pathology within GCs. The presence of similar CD4⁺CD11c⁺ DCs in blood suggests that GCDCs were derived from precursors in blood.

Technical aspect

Isolation of GCDC

Tonsillar cells are depleted of T, B, NK cells and monocytes and stained with mouse anti-CD4-PE-Cy5 (Immunotech), anti-CD11c-PE (Becton Dickinson) and a cocktail of FITC labelled mAbs, including anti-CD34, -CD3, -CD20, -CD57, -CD7, -CD14, -CD16 and -CD1a. Then CD4⁺CD11c⁻ lin⁻ cells are isolated by cell sorting.

CD4⁺CD3⁻CD11c⁻ plasmacytoid DC precursors

Isolation of CD4⁺CD3⁻CD11c⁻ plasmacytoid DC precursors

This population of cells can be isolated at the same time as the CD4⁺CD3⁻CD11c⁺ cells (see above).

◆◆◆◆◆◆ PROPAGATION OF DENDRITIC CELLS *IN VITRO*

Considerable progresses in the generation of DC have recently been accomplished. Consequently culture systems are now available for the *in vitro* generation of high numbers of DC.

From peripheral blood monocytes

Characteristic features

More than a decade ago, Knight *et al.* (1986) described monocytes that have acquired a veiled and dendritic appearance after separation. More recently, monocytes were induced to acquire features of DCs after treat-ment with GM-CSF and/or IL-4 (Porcelli *et al.*, 1992; Kasinrerk *et al.*, 1993; Peters *et al.*, 1993). It is now well established that monocytes can be

induced, without any proliferation, to differentiate into CD1a⁺ DC, upon
culture with GM-CSF and IL-4 (Romani *et al.*, 1994; Sallusto and
Lanzavecchia, 1994) or IL-13 (Piemonti *et al.*, 1995). The monocyte-derived
DC display a phenotype of immature DC including low CD80, CD86,
CD58 expression, expression of MHC class II within intracytoplasmic
compartments, and expression of monocyte markers (CD11b, CD36, CD68,
c-fms). The cells display an efficient antigen uptake by macropinocytosis
or by receptor-mediated endocytosis through the mannose-receptor and a
weak capacity to prime naive T cells. These DC can undergo maturation
when stimulated by signals such as LPS, TNF-α, IL-1 (signals also inducing
DC migration) or by T-cell signals such as CD40L. Then DC display a
mature phenotype including a typical morphology with extended
dendrites, a loss of monocyte markers, a loss of antigen uptake, an up-
regulation of accessory molecules (CD80, CD86, CD58), a translocation of
MHC class II onto the cell surface, and a strong capacity to prime naive T
cells (Fig. 2). Monocytes can also be induced to differentiate into DCs with
features of Langerhans cells when TGF-β is added (Geissman *et al.*, 1998) or
when IL-4 is replaced by IL-15 (Mohamadzadeh *et al.*, 2001).

Technical aspect

Purification of monocytes

Following a Ficoll-Hypaque (see earlier), monocytes are enriched
through a 50% Percoll. The interface population usually contains 50–80%

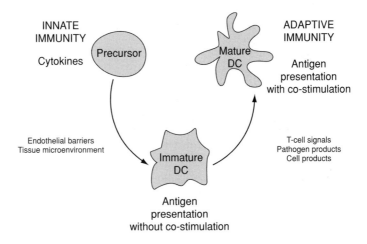

Figure 2. Dendritic cells have different functions at different maturation stages.
DCs originate from haematopoietic progenitors. Circulating precursors have at
least two functions: 1. Innate effector function, they make cytokines such as IFN-
α, and 2. they give rise to immature tissue-residing antigen-capturing DCs.
Immature DCs that have not been activated present tissular antigens to
induce/maintain peripheral tolerance. Mature antigen-presenting DC display
peptide–MHC complexes and co-stimulatory molecules, alowing selection,
expansion and differentiation of antigen-specific lymphocytes.

of CD14$^+$ monocytes. Monocytes are further purified by depletion of contaminating T, B, and NK cells using MoAbs anti-CD3, -CD16, -CD19, -CD56, -glycophorin A and goat anti-mouse Ig beads (see earlier). After two rounds of bead depletion the resulting suspension contains 95–99% monocytes as judged by anti-CD14 staining 450 ml blood yield 30–150 × 10^6 monocytes.

Generation of dendritic cells from monocytes

Cultures are established in endotoxin-free medium consisting of RPMI 1640 supplemented with 10% (v/v) heat-inactivated fetal bovine serum (FBS), 10 mM Hepes, 2 mM L-glutamine, 5×10^{-5} M β-2-mercaptoethanol, penicillin and streptomycin (referred to as complete medium). Purified monocytes are seeded at 2.5×10^5 cells ml^{-1} in RPMI complete medium (10% FCS) containing 100 ng ml^{-1} GM-CSF (specific activity: 2.10^6 U mg^{-1}, Schering-Plough Research Institute, Kenilworth, NJ) and 50 U ml^{-1} IL-4 (specific activity: 10^7 U mg^{-1}, Schering-Plough Research Institute, Kenilworth, NJ, USA). Cells are usually diluted by half at day 4–5 and recovered at day 8–10. At the end of the culture the number of DC represents 40–60% of the number of seeded monocytes (Sallusto and Lanzavecchia, 1994).

From human CD34$^+$ haematopoietic progenitor cells

Characteristic features

TNF-α, in association with GM-CSF or IL-3 induces development of dendritic cells from CD34$^+$ cells

Tumour necrosis factors strongly potentiate the proliferation induced by either IL-3 or GM-CSF of CD34$^+$ haematopoietic progenitor cells (HPC), isolated from cord blood or bone marrow mononuclear cells (Caux et al., 1990; Moore, 1991; Jacobsen et al., 1992; Reid et al., 1992). In these culture conditions the cooperation between TNF-α and GM-CSF/IL-3 is critical for the development of DC from CD34$^+$ HPC (Caux et al., 1992; Santiago-Schwarz et al., 1992; Szaboles et al., 1995; Strunk et al., 1996). Within 8 days in liquid cultures of CD34$^+$ HPC, addition of TNF-α to GM-CSF yields a six- to eight-fold increase in cell number. At day 12, 50–80% cells express CD1a thus yielding 10–30 × 10^6 CD1a$^+$ cells from 10^6 CD34$^+$ HPC. Moreover SCF or FLT3-L increase by three- to six-fold the yield of CD1a$^+$ cells (Siena et al., 1995; Young et al., 1995). These CD1a$^+$ cells are dendritic cells according to (a) a typical morphology; (b) a phenotype of DC (expression of high MHC class II, CD4, CD40, CD54, CD58, CD80, CD86, CD83, and lack of CD64/Fc_RI and CD35/CR1); (c) the presence of Birbeck granules (characteristic of LC) in 20% of cells; (d) a high capacity to stimulate proliferation of naive T cells and efficient presentation of soluble antigen to CD4$^+$ T cell clones (Caux et al., 1995, 1996). CD1a$^+$ cells are CD45RO$^+$ and express the myeloid markers CD13 and CD33. Recently, for cultures performed under serum free conditions, TGF-β was shown to be required for the development of DC with characteristics of LC (Birbeck granules and Lag molecule) (Strobl *et al.*, 1996).

Using semi-solid cultures, DC were shown to arise within single colonies together with monocytes/macrophages, suggesting the existence of a common precursor cell (Reid *et al.*, 1992; Santiago-Schwarz *et al.*, 1992).

Although GM-CSF or IL-3, in association with TNF, appear critical to support DC development from CD34$^+$ progenitors, other pathways of development have been reported. In this respect, CD40L induces a GM-CSF independent DC development from CD34$^+$ HPC (Flores-Romo *et al.*, 1997).

Identification of two pathways of dendritic cell development

While most cells are CD1a$^+$CD14$^-$ after 12 days of culture, at early time points (day 5–7) of the culture, two subsets of DC precursors, identified by the exclusive expression of CD1a and CD14, emerge independently (Caux *et al.*, 1996, 1997). Both precursor subsets mature at day 12–14 into DC with typical morphology, phenotype (CD80, CD83, CD86, CD58, high HLA class II) and function. CD1a$^+$ precursors give rise to cells with Langerhans cell characteristics (Birbeck granules, Lag antigen and E-cadherin). In contrast, the CD14$^+$ precursors mature into CD1a$^+$ DC lacking Birbeck granules, E-cadherin and Lag antigen but expressing CD2, CD9, CD68 and the coagulation factor XIIIa described in dermal dendritic cells. Interestingly, the CD14$^+$ precursors, but not the CD1a$^+$ precursors, represent bipotent cells that can be induced to differentiate, in response to M-CSF, into macrophage-like cells, lacking accessory function for T cells. Furthermore CD14- but not CD1a-derived DC express IL-10 mRNA and protein. These two pathways of development have been documented and further characterized by others. In particular, the commitment into either pathway has already occurred at the level of CD34$^+$ cells (Strunk *et al.*, 1997). Peripheral blood CD34$^+$ which express CLA (cutaneous lymphocyte-associated antigen) differentiate in response to GM-CSF+TNF-α into CD1a$^+$, Birbeck granule$^+$, Lag$^+$ Langerhans cells, while, CLA$^-$ progenitors differentiate into CD1a$^+$ Birbeck granule$^-$, Lag$^-$ interstitial DC.

While the two populations are equally potent in stimulating naive CD45RA cord blood T cells, each also displays specific activities (Caux *et al.*, 1996, 1997). In particular CD14-derived DC demonstrate a potent and long lasting (from day 8 to day 13) antigen uptake activity (FITC-dextran or peroxidase) that is about 10-fold higher than that of CD1a$^+$ cells. The antigen capture is exclusively mediated by receptors for mannose polymers. The high efficiency of antigen capture of CD14-derived cells is co-regulated with the expression of non-specific esterase activity, a tracer of the lysosomial compartment. In contrast, the CD1a$^+$ population never expresses non-specific esterase activity. A striking difference between the two populations is the unique capacity of CD14-derived DC to induce naive B cells to differentiate into IgM secreting cells, in response to CD40 triggering and IL-2. Thus, while the two populations can allow T-cell priming, initiation of humoral responses might be preferentially regulated by the CD14-derived DC.

Thus, different pathways of DC development might exist *in vivo*: (a) the Langerhans cell type which might be mainly involved in cellular immune responses, and (b) the CD14-derived DC related to dermal DC or circulating blood DC which may be more dedicated to the development of humoral immune responses.

Technical aspect

Collection and purification of cord blood CD34+ HPC

Umbilical cord blood samples are diluted by 2/3 in PBS and layered on Ficoll-Hypaque. The mononuclear cells are resuspended at 2×10^7 cells ml^{-1} in PBS/HS/EDTA and stained with anti-CD34. After 20–30 min at 4°C under gentle shaking, cells are washed three times with PBS/FCS/EDTA, and resuspended at 10^8 cells ml^{-1} in presence of goat anti-mouse IgG coated Miltenyi microbeads. After 20–30 min at 4°C under gentle shaking, cells are washed twice and isolation of CD34+ progenitors is achieved using Minimacs separation columns. The isolated cells are 80% to 99% CD34+ as judged by staining with anti-CD34 MoAb. After purification CD34+ cells are cryopreserved in 10%DMSO.

DC generation from CD34+ HPC

Cultures are established in RPMI 1640 complete medium (see earlier) supplemented with 100 ng/ml rhGM-CSF (specific activity: 2.106 U mg^{-1}, Schering-Plough Research Institute, Kenilworth, NJ, USA), 2.5 ng ml^{-1} rhTNF-α (specific activity: 2×10^7 U mg^{-1}, Genzyme, Boston, MA, USA) and 25 ng ml^{-1} rhSCF (specific activity 4×10^5 U mg^{-1}, R&D Abington, UK) (Caux *et al.*, 1990, 1993).

After thawing, CD34+ cells are seeded for expansion in 25 to 75 cm^2 culture vessels (Linbro, Flow Laboratories, Mc Lean, VA, USA) at 2×10^4 cells ml^{-1}. Optimal conditions are maintained by splitting these cultures at day 4 with medium containing fresh GM-CSF and TNF (cell concentration: $1–3 \times 10^5$ cells ml^{-1}). For most experiments, cells are routinely collected after 5–6 days of culture for FACS-sorting. Culture medium is supplemented with 2.5% AB+ pooled human serum at the initiation of the cultures, by day 5–6 human serum is washed away (Caux *et al.*, 1996).

Isolation of CD1a and CD14 DC precursors by FACS-sorting

After 5–6 days of culture in presence of SCF, GM-CSF and TNF, cells are collected and labelled with FITC-conjugated CD1a and PE-conjugated CD14 MoAbs. Cells are sorted into CD14+CD1a−, CD14−CD1a+ fractions. All the procedures of staining and sorting are performed in the presence of 0.5 mM EDTA in order to avoid cell aggregation. Reanalysis of the sorted populations shows a purity higher than 98%, the other cells being immature myeloid cells (T cells could never be detected even using PCR amplification of T-cell receptor components).

Sorted cells are seeded in presence of GM-CSF+TNF ($1–2 \times 10^5$ cells ml^{-1}) for 6–7 additional days, a last medium change being performed at day 10. Cells are routinely collected between day 11 and day 14. Adherent cells are eventually recovered using a 0.5 mM EDTA solution.

◆◆◆◆◆◆ *IN VITRO* GENERATION OF PLASMACYTOID DCS

Two recent papers described the *in vitro* generation of this intriguing subset of human DCs. The group of Yong Jun-Liu demonstrated the existence of CD34^{low} cells with the phenotype and function of pDCs, as well as a simple culture system in which early CD34$^+$ progenitors yield pDCs (Blom *et al.*, 2002). The group of Hergen Spits provided another culture system in which early CD34$^+$ progenitors yield pDCs, and demonstrated that their differentiation can be blocked by overexpression of Id2 and Id3 proteins (Spits *et al.*, 2000).

CD34$^+$ cells with features of pDCs

As described above, several features permit the identification of pDCs: the expression of CD4, CD45RA, CD123; the production of high levels of IFN-α upon viral triggering or CD40 ligation (Cella *et al.*, 1999; Siegal *et al.*, 1999; Kadowaki *et al.*, 2000), and the morphology of plasma cells. Recently, a minor subset of CD34^{low} CD45RA$^+$CD4$^+$CD123$^+$ cells was identified in cord blood, adult blood, fetal liver and most abundantly fetal bone marrow (Blom *et al.*, 2000). These cells produce large amounts of IFN-α in response to viruses and can differentiate into mature DCs in response to IL3 and CD40L.

In vitro generation of pDCs from early CD34$^+$ HPC

Culturing early HPCs (CD34$^+$ CD45RA$^-$) with Flt3-L gives rise to cells (a) able to secrete IFNα upon viral exposure and (b) displaying the typical DC2 phenotype (CD4$^+$CD123$^+$CD45RA$^+$) (Blom *et al.*, 2000). The pDCs are first observed after 10 days and reach a maximal frequency (~10%) after 3 weeks of culture. A different approach is based on culturing early uncommitted CD34$^+$CD38$^-$ fetal liver HPCs on the murine stromal cell line S17, which is known to support the development of human B cell and myeloid progenitors (Collins and Dorshkind, 1987; Rawlings *et al.*, 1997; Hao *et al.*, 1998). In these cultures, it takes only a few days for pDCs to emerge, but then the cells quickly disappear. The newly generated cells display phenotypic and functional features of pDCs. Cells with the same properties can be obtained from thymic CD34$^+$ progenitors, but from the CD34$^+$CD1a$^-$ subset only.

Generation of pDCs from CD34$^+$CD45RA$^-$ early haematopoietic stem cells

Human CD34$^+$CD45RA$^-$ early haematopoietic stem cells are isolated from cord blood, bone marrow, or fetal liver mononuclear cells by cell sorting.

Briefly, fetal tissue (16–22 weeks of gestation) and cord blood were obtained from ABR (Advanced Bioscience Resources Inc., ABR, Alameda, CA, USA). Mononuclear cells (MNC) were isolated from these samples by Ficoll density gradient centrifugation (Lymphoprep, 1.077 g ml^{-1}, Amersham Pharmacia Biotech Inc., Piscataway, NJ, USA). MNC were washed three times in phosphate buffered saline (PBS; BioWhittaker Inc., Walkersville, MD, USA), and resuspended in PBS containing 2% (vol./vol.) human serum (HS) (Gemini Bioproducts, Woodland, CA, USA) and 2 mM EDTA (PBS/HS/EDTA). Magnetic bead depletion was performed to remove lineage positive cells. Briefly, mononuclear cells were incubated with a mixture of antibodies against CD3 (OKT-3 ascites), CD8 (OKT-8 ascites), CD14 (RPA-M1 ascites), CD16 (3G8) (Immunotech, Miami, FL, USA), CD19 (4G7 ascites), CD56 (My31 ascites), CD66B (80H3) (Immunotech), and Glycophorin A (10F7MN ascites). After two washes the cells were incubated with goat-anti-mouse IgG coupled to magnetic beads (Dynabeads® M-450, goat-anti-mouse IgG; Dynal Inc., Lake Success, NY, USA) and isolated according to the manufacturer's instructions. The enriched lineage marker negative cells were stained with CD34-APC (HPCA-2) and CD45RA-PE (Leu-18) from Beckton Dickinson. The CD34+CD45RA- cells were isolated by a cell sorter. 25,000–50,000 stem cells were cultured in 200 μl Yssel's medium containing 2% human serum and 100 ng human FLT3-ligand (R&D system) in 96-well round-bottomed culture plates. Cell cultures were refreshed every 5 days by demi-depletion and splitted if necessary. At 25–30 days of culture, the presence of pDCs was examined. Cells were analysed by three-colour flow cyto-metric analysis after staining with IL-3Rα-PE, HLA-DR-FITC, together with CD11c- CD4-CyChrome, or CD45RA-CyChrome, or CD11c-CytoChrome. About 4–10% of total cultured cells express the pDCs phenotype: IL-3Rα++HLA-DR+CD4+CD11c-CD45RA+. These cells are isolated by cells sorting and stimulated with HSV-1 for 24 h. IFN-α (pg ml^{-1}) was measured in the supernatants by ELISA (Biosource International).

◆◆◆◆◆◆ GENERATION OF CLINICAL GRADE DENDRITIC CELL VACCINES

Inaba and colleagues demonstrated that the injection of DCs, charged with antigen *ex vivo*, could sensitize normal mice to protein antigens (Inaba *et al.*, 1990). This seminal work prompted studies aimed at using DCs as a vaccine. The immunogenicity of antigens delivered on DCs has now been demonstrated in human studies. Indeed, single s.c. immunization of healthy volunteers with 2–4 × 10^6 antigen-loaded mature monocyte-derived DCs rapidly expanded CD8+ and CD4+ T-cell immunity. A single boost several months later led to expansion of CTL with increased affinity against viral peptide, an observation never made with any other vacci-nation strategy so far (Dhodapkar *et al.*, 2000).

There is a large literature involving animal models of tumour immunity in which DCs loaded with tumour-associated antigens (TAA) are able to induce protective anti-tumour responses. When tested, DCs can be superior to other vaccination strategies (Gilboa, 1999). A number of trials have now utilized TAA-loaded DCs as vaccines in humans (Hsu *et al.*, 1996; Nestle *et al.*, 1998). More recent DC vaccination studies demonstrated that T-cell immunity to both control antigens (viral peptide and bacterial protein) and melanoma peptide can be induced, even in patients with advanced stage IV melanoma, by vaccination with antigen-pulsed, mature monocyte-derived DC (Thurne *et al.*, 1999). Furthermore, when these DCs were loaded with MHC class II binding melanoma peptides, strong tumour-specific Th1 responses were elicited. CD34+ HPC-derived DCs, pulsed with control antigens and multiple melanoma peptides, induce primary and recall immune responses detectable directly in the blood in patients with stage IV melanoma (Banchereau *et al.*, 2001a). The level of immune responses in the blood correlated with early outcome at the tumour sites, thus providing further stimulus for the idea that the measurement of immune responses in the blood helps evaluate vaccine efficacy (Banchereau *et al.*, 2001a). Although the results with antigen-bearing DCs are encouraging, DC vaccination is at an early stage, and several parameters need to be established (Banchereau *et al.*, 2001b).

Parameters of dendritic cell vaccines

The subsets of DC

The majority of clinical studies to date have been carried out with *ex vivo* generated monocyte-derived DC, which resemble a single subset, i.e. interstitial DCs. The efficacy of distinct human DC subsets described above will need to be compared in clinical studies.

The optimal DC maturation state and stimulus

Immature DCs are weak immunogens. Indeed, intranodal injection of immature DCs does not lead to significant immune responses, contrary to the intranodal injection of mature DCs in the same patient. Furthermore, there is increasing experimental evidence that immature DCs can be tolerogenic while mature DCs (triggered for instance by a mix of macrophage products such as IL-1β/IL-6/TNFα/PGE2) induce functionally superior CD8+ T cells and polarize CD4+ T cells towards IFN-γ production.

DCs dose, frequency and route of injections

The optimal dose and frequency of immunization with DC is yet to be established. In human trials published so far the DCs were usually given at 2–4-week intervals, and at doses between 4–40 million without striking differences in results.

Source, preparation and antigen loading strategy

Several systems have been employed to load DCs with TAA. Loading MHC class I molecules with peptides derived from defined antigens is most commonly used, and is also applied to recently identified MHC class II helper epitopes. Although important for 'proof of concept' studies, the use of peptides has limitations coming from (a) their restriction to a given HLA type, (b) the limited number of defined TAA, and (c) the induction of a restricted repertoire of T-cell clones less able to control tumour antigen variation. Alternative strategies that provide both MHC class I and class II epitopes and lead to a diverse immune response involving many clones of CD4$^+$ T cells and CTL include: recombinant proteins, exosomes (vesicles rich in MHC/peptide complexes and heat shock proteins), viral vectors, plasmid DNA, RNA transfection and dead tumour cells.

CD34$^+$ HPC-derived DC vaccine

Harvest of DC progenitors

Patients receive recombinant G-CSF (Neupogen) 10 µg kg day^{-1} s.c. for 5 days, for peripheral blood stem cell mobilization, and then undergo leukapheresis for two consecutive days to collect mobilized CD34$^+$HPC. The cells are processed using either the CEPRATE SC stem cell concentration system (CellPro Inc., Seattle) or Isolex 300i Stem Cell Concentration System (Nexell, Inc.) to obtain an enriched population of CD34$^+$ HPC (purity 62 ± 17%; recovery 158 ± 133 × 10^6) which are then cryopreserved (Banchereau et al., 2001a).

Preparation of DC vaccine

All procedures are performed according to Good Laboratory Practice standards and FDA regulations. CD34$^+$ HPC are cultured at a concentration of 0.5 × 10^6 ml^{-1} culture medium (X-VIVO-15, BioWhittaker) supplemented with autologous serum, 10^{-5}M 2-β-mercaptoethanol and 1% L-glutamine. The following human recombinant cytokines, approved for clinical use, are used: GM-CSF (50 ng ml^{-1}, Immunex Corp.), Flt3-L (100 ng ml^{-1}, Immunex Corp.) and TNF (10 ng ml^{-1}, CellPro, Inc.). Cultures are conducted in a humidified incubator at 37°C and 5% CO$_2$ with a separate incubator being assigned to each patient. On day 8 of culture, all cells are pulsed overnight with KLH (2 µg ml^{-1}, Intracell), 20% of cells are pulsed separately with HLA-A*0201 restricted flu-matrix peptide (Flu-MP) GILGFVFTL$_{58-66}$ (2.5 µg ml^{-1}) and 80% of cells are pulsed overnight with a mix of tumour antigens. In our study there were 4 HLA-A201 restricted peptides (2.5 µg ml^{-1}) derived from melanoma antigens (MelanA/MART-1$_{27-35}$: AAGIGILTV,

(contd.)

gp100$_{g209-2M}$: IMDQVPFSV, Tyrosinase$_{368-376}$: YMDGTMSQV, MAGE-3$_{271-279}$: FLWGPRALV). After overnight loading, all DCs are washed three times with sterile saline, counted and resuspended in 10 ml of sterile saline. Vaccine release criteria include: (a) negative bacterial culture 48 h prior to DC injection, (b) negative Gram-staining after antigen pulsing, (c) dendritic cell morphology on Giemsa stained cytospins performed 2 h before DC administration, (d) cell viability >80%, (e) a minimum of 20% DC (CD1a$^+$ and CD14$^+$) in cell preparation as determined by phenotypic analysis, and (f) negative staining for the presence of mycoplasma contamination. The post-release testing includes (a) testing of mycoplasma by PCR and by 14 days culture, (b) sterility testing (bacterial and fungal) in 14 days culture, (c) reactivity with a panel of monoclonal antibodies and (d) determination of DC stimulatory capacity in mixed lymphocyte reactions (Plate 9).

◆◆◆◆◆◆ FUNCTIONAL CHARACTERIZATION OF DENDRITIC CELLS

Antigen uptake

Characteristic features

Immature DC are very efficient in antigen capture (Banchereau *et al.*, 2000) and can utilize several pathways such as (a) macropinocytosis; (b) receptor-mediated endocytosis via C-type lectin receptors (mannose receptor, DEC-205) (Reis and Sousa *et al.*, 1993; Jiang *et al.*, 1995; Sallusto *et al.*, 1995; Engering *et al.*, 1997; Tan *et al.*, 1997; Mommaas *et al.*, 1999) or Fcγ receptors type I (CD64) and type II (CD32) (uptake of immune complexes or opsonized particles) (Fanger *et al.*, 1996); (c) phagocytosis of particles such as latex beads, apoptotic and necrotic cell fragments (involving CD36 and αvβ3 or αvβ5 integrins) (Rubartelli *et al.*, 1997; Albert *et al.*, 1998a,b), viruses, bacteria including mycobacteria (Inaba *et al.*, 1993; Rescigno *et al.*, 2000), as well as intracellular parasites, and (d) DC can internalize heat shock proteins gp96 and Hsp70 (Arnold-Schild *et al.*, 1999).

Whereas mannosylated Ag are rapidly internalized by intDC and selectively targeted to class II processing/presentation pathway, the capacity of LC to capture mannosylated Ag is somewhat controversial. Freshly isolated murine epidermal LC can uptake both mannosylated and non-mannosylated Ag (Reis and Sousa *et al.*, 1993). However, human LC lack classical mannose receptor, have poor endocytic capacity and low level of lysosome markers (Mommaas *et al.*, 1999). These differences between LC and intDC extend to differential expression of Fcε (Maurer *et al.*, 1998) and Fcγ receptors further strengthening the notion of DC functional heterogeneity and its effect on the type of immune response induced.

Technical aspect

Quantitation of endocytosis using FITC dextran capture and flow cytometry (Sallusto et al., 1995)

DC are resuspended at 5×10^5 ml^{-1} in 10% FBS medium buffered with 25 mM HEPES at 37°C in a water bath. FITC-dextran is added at the final concentration of 0.1 mg ml^{-1} and for 15 (5–60) min to 10^5 cells in 5-ml polypropylene tubes (Falcon, Becton Dickinson). The cells are washed four times with cold PBS containing 1% FBS and 0.01% NaN$_3$. To further characterize the endocytosing cell population, double staining may be done using anti-CD1a-PE (Coulter, Hialeah, FL) or anti-CD14-PE (Becton Dickinson) or any other PE coupled MoAb. To analyse the fate of endo-cytosed material, cells may be pulsed at 37°C, washed four times in cold medium, and recultured at 37°C for different times in marker-free medium. After staining, cells are analysed using a FACS. When the results are expressed as MFI, the background (cells pulsed with FITC-dextran at 4°C) is subtracted.

Quantitation of endocytosis by horseradish peroxidase capture (Sallusto et al., 1995)

DC are resuspended at 5×10^5 ml^{-1} in 10% FBS medium buffered with 25 mM HEPES at 37°C in a water bath. Horseradish peroxidase (HRP, Sigma) is added to 10^5 cells at the final concentration of 0.1 µg ml^{-1} to 0.1 mg ml^{-1} for 15 min. The cells are washed four times with cold PBS containing 1% FBS and 0.01% NaN$_3$, lysed with 0.05% Triton X-100 in 10 mM Tris buffer, pH 7.4, for 30 min. After centrifugation (10 min, 600g), the enzymatic activity of the lysate is measured using ABTS (2.2 azinobis-3-ethylhenthi-azoline-6-sulfonic acid) as substrate at 1 mg ml^{-1} in buffer (0.1 M citric acid, 0.2 M di-sodium hydrogenophosphate) supplemented with 0.1 µl ml^{-1} H$_2$O$_2$ (30%). OD is read at 420 nm with reference to a standard curve.

Interactions between dendritic cells and T cells

Characteristic features

The capacity of DC in priming T cells *in vivo* has been directly demon-strated in cell transfer experiments. Thus, upon reinjection into foot-pad or blood or upon intratracheal instillation, mouse DC pulsed *in vitro* with protein antigen-induced MHC restricted antigen-specific T-cell responses (Inaba *et al.*, 1990; Havenith *et al.*, 1993; Levin *et al.*, 1993; Liu *et al.*, 1993). Also, mouse DC pulsed with viral or tumour restricted peptides have been shown to induce CD8$^+$ cytotoxic responses *in vivo* and tumour clearance (Flamand *et al.*, 1994; Boczkowski *et al.*, 1996; Celluzzi *et al.*, 1996; Zitvogel *et al.*, 1996; Mayordomo *et al.*, 1997). The availability of mice expressing transgenic T-cell receptors has allowed the analysis of the capacity of DC to induce a primary antigen specific T-cell response to soluble antigens *in vitro*. In such systems DC appear to be 100- to 300-fold more efficient than any other APC (Croft *et al.*, 1992; Macatonia *et al.*, 1993).

The most commonly used functional assessment of DC is their remarkable ability to drive allogeneic mixed lymphocyte reaction (MLR). DC are about 100-fold more efficient than any other APC population including B cells and monocytes (Steinman, 1991). DC can also stimulate allogeneic CD8+ T cells though higher APC numbers are required (Steinman, 1991). The use of superantigens has further demonstrated the efficiency of DC in primary T-cell activation (Bhardwaj et al., 1992, 1993).

DC generated in vitro from CD34+ HPC induce a strong proliferation of allogeneic naive CD4+ T cells and of autologous naive CD4+ T cells in the presence of low concentration of superantigens (Caux et al., 1992, 1995). The proliferation of allogeneic CD8+ T cells is weaker than that of CD4+ T cells when cultured with DC alone, but reaches comparable levels in the presence of cytokines such as IL-2, IL-4 or IL-7. Accordingly, allospecific CD4+ or CD8+ T-cell lines can be generated by repeated culturing of T cells on DC, and the CD8+ T-cell lines display a high cytotoxic potential in a MHC restricted manner. DC generated from CD34 (Caux et al., 1992, 1995) HPC are also able to present soluble antigen to MHC matched tetanus toxoid specific T-cell clones (Caux et al., 1995).

Technical aspect

T-cell proliferation assay

After 12 days of culture, CD34+ HPC derived cells are used, after 30 Gy irradiation, as stimulator cells for (a) resting allogeneic adult or cord blood T cells (CD3, CD4, CD8, CD45RA, 2.5×10^4 per well), (b) resting syngeneic cord blood CD4+ T cells (2.5×10^4 per well) in presence or absence of superantigen, (c) syngeneic CD4+ T cells from specific T-cell clones (10^4 per well) (Caux et al., 1994a,b). Naive T cells are purified by negative selection using MoAbs and magnetic beads. From 6 to 3×10^4 stimulator cells are added to the T cells in 96-well round-bottomed microtest tissue-culture plates (Nunc, Roskilde, Denmark). Cultures last 5 days for resting CD4+ T cells and 3 days for T-cell clones. After incubation, cells are pulsed with 1 µCi of 3H-TdR (specific activity 25 Ci mmol^{-1}) per well, for the last 8 h, harvested and counted. 3H TdR uptake by stimulator cells alone is always below 100 cpm.

Role of CD40/CD40L in DC/T-cell interactions

Characteristic features

The CD40 antigen, that is of critical importance in T-cell dependent B-cell growth, differentiation and isotype switch, is also functional on other cell types (Banchereau et al., 1994; van Kooten and Banchereau et al., 2000). In particular CD40 is expressed on LC, blood DC, interdigitating cells as well as in vitro generated DC (Banchereau et al., 2000). Transferring in vitro generated DC into a medium devoid of GM-CSF results in their prompt death unless their CD40 antigen is engaged. In addition, CD40 triggering induces changes in morphology and phenotype. Thus, CD40 activated DC upregulate accessory molecules such as CD58, CD80 and CD86 and

609

secrete cytokines (TNF-α, IL-8 and MIP1α). CD40 engagement also turns on DC maturation as illustrated by upregulation of CD25 and down-regulation of CD1a that, respectively, appear and disappear on LC/veiled cells when they enter secondary lymphoid organs to become interdigitating dendritic cells. As T cells that are activated by DC upregulate CD40-L, it is likely that CD40 activation of DC represents an important physiological interaction between DC and T cells. The role of CD40 engagement on DC activation/maturation has also been clearly demonstrated using monocyte-derived DC. CD40 triggering on GM-CSF+IL-4-derived DC induces upregulation of co-stimulatory molecules, translocation of MHC class II from intracellular compartment to cell surface, and importantly secretion of IL-12 (Cella *et al.*, 1996). The production of IL-12 by DC following interaction with T cells results in the production of IFN-γ by the primed T cells (Banchereau *et al.*, 2000).

Technical aspect

CD40 activation

L cells transfected with human CD40 ligand (116) (CD40L-L cells), are used to induce CD40 triggering on D-Lc. CD32 transfected L cells (CD32 L cells) are used for control cultures. 10^4 L cells are seeded together with 4×10^5 cells (either CD34 derived GM-CSF+TNF DC, or monocytes derived DC) per well (24-well culture plate, Linbro) in presence or absence of GM-CSF (100 ng ml^{-1}) in 1 ml culture medium and cultured for 2 to 4 days. Cell survival is monitored by enumeration of cells excluding Trypan blue. For determination of cytokine production, supernatants are recovered after 48 h. For phenotypic and morphological analysis, cells are recovered after 4 days of culture.

Interactions between dendritic cells and B cells

Characteristic features

T-cell dependent primary B-cell activation is known to be dependent on DC (Banchereau *et al.*, 2000), and to occur in the extrafollicular area of secondary lymphoid organs. However, DC can also directly activate naive and memory B cells. Indeed, differentiation of CD40-activated memory B cells towards IgG secreting cells is enhanced by DC in a process mediated by soluble IL-6- gp80/IL-6 complexes (Dubois *et al.*, 1997). DC-derived IL-6 also mediates differentiation of activated-naive B cells into plasma cells in concordance with IL-12. In the presence of IL-10 and TGFβ, DC skew CD40-activated naive B cells towards the secretion of both IgA1 and IgA2 subclasses (Fayette *et al.*, 1997). These results suggest the DC-mediated direct activation of naive B cell during the initiation of the immune response and the involvement of DC in the development of mucosal/humoral immune responses.

DC are also relevant in germinal centre reaction where they select the rare antigen-specific B cells. As recently demonstrated *in vivo* in the rat,

DC can also capture and retain unprocessed antigen, then transfer it to naive B cells to initiate a specific Th2-associated antibody response (Wykes *et al.*, 1998). While several molecules may be involved in DC/B-cell interactions two recently described candidates appear important, OX40L and BAFF/Blys. CD40 activation upregulates OX40L expression on DC and B cells (Stuber *et al.*, 1995; Ohshima *et al.*, 1997) and early OX40 ligation promotes Th2 cytokines secretion (Flynn *et al.*, 1998) and causes CD4 T-cell migration within B-cell follicles (Brocker, 1999). BAFF/Blys (for B-cell activating factor belonging to the TNF family), found on DC and T cells, binds to a receptor restricted to B cells (Mackay *et al.*, 1999; Moore *et al.*, 1999; Schneider *et al.*, 1999; Gross *et al.*, 2000; Laabi *et al.*, 2000) and induces both proliferation and immunoglobulin secretion by different B-cell subsets. Thus, BAFF may represent an important co-stimulator through which DC regulate B-cell proliferation and function.

Technical aspect

Coculture of B cells and DC

10^4 sIgD$^+$ B lymphocytes, purified by preparative magnetic cell sorter (Miltenyi Biotec), are seeded together with 2.5×10^3 irradiated transfected L cells (75 Gy), with or without irradiated (30 Gy) DC (3×10^3 per well) in 96 well-culture plates (Nunc). For measurement of proliferation, cells are pulsed, after 5 days of incubation, with 1 µCi of 3H-TdR (specific activity 25 Ci mmol^{-1}) per well, for the last 8 h, harvested and counted. 3H TdR uptake by stimulator cells alone is always below 100 cpm. For determination of Ig production, supernatants are harvested after 15 days and used for indirect ELISA.

◆◆◆◆◆◆ *IN VIVO* DENDRITIC CELL MOBILIZATION

Vaccination with *ex vivo* generated DCs is not feasible for large-scale immunization either in cancer or infectious diseases, e.g. malaria. Thus, there is a need to develop strategies that can provide protective/therapeutic immune responses with minimal amounts of vaccine and limited boosting. Research in this area will benefit from the recently discovered ability of some cytokines to mobilize large numbers of DCs *in vivo* (DC-poietins). DC-poietins are the cytokines that mobilize DCs *in vivo*: either increase their numbers (G-CSF, GM-CSF and FLT3 ligand) or activate them (IFN-α). In particular, administration of either Flt3-L or G-CSF to healthy volunteers dramatically increases distinct DC subsets in the blood (Pulendran *et al.*, 2000). Flt3-L increases both the CD11c$^+$ DCs (48-fold) and CD11c$^-$CD123$^+$ DCs (13-fold). In contrast, G-CSF only increases CD11c$^-$CD123$^+$ DCs (> seven-fold). These two DC subsets elicit distinct cytokine profiles in CD4$^+$ T cells, with the CD11c$^-$CD123$^+$ DCs inducing higher levels of the Th2 cytokine IL-10. Such differential mobilization of distinct DC subsets is likely to determine the type of *in vivo* induced immune responses. Furthermore, FLT3L mobilized CD11c[+] blood DCs represent 'emergency cells' able to very quickly differentiate into potent APCs.

DCs may be targeted *in vivo* by 'Intelligent missile', a generic vaccine equipped with (a) the immunogens, the optimal antigenic preparations that can be targeted to desired MHC molecules; (b) DC activation molecules and specific ligands that would permit targeting of desired DC subset. One might take advantage of molecules that bind to pattern recognition receptors, like Toll, through which the immune system senses microbial products and/or tissue damage. For example, CpG oligo-nucleotides can activate DCs *in vivo* while heat shock proteins can both activate DCs and chaperone the peptide. Furthermore, as the Toll receptors are differentially expressed on DC subsets, their ligands could serve as targeting molecules; and (c) molecules determining the quality and the type of the immune response, for instances, chemokines attracting naive T cells, costimulatory molecules as well as type 1/type 2 skewing molecules.

◆◆◆◆◆◆ CONCLUSIONS

DCs are an attractive target for therapeutic manipulation of the immune system to enhance insufficient immune responses, in infectious diseases and cancer, or attenuate excessive immune responses, in allergy and autoimmunity. However, the complexity of the DC system brings about the necessity for their rational manipulation to achieve protective or therapeutic immunity. Immunization with *ex vivo* generated DC has proven feasible and permits the enhancement as well as the dampening of antigen-specific immune responses in man. These *ex vivo* strategies should help identify the parameters for DC targeting *in vivo*. Today's studies are also considering the relationship of DCs to the correction of pathologic or undesired immune responses, like allergic and autoimmune diseases, graft rejection and graft versus host disease.

References

Albert, M. L., Pearce, S. F., Francisco, L. M., Sauter, B., Roy, P., Silverstein, R. L. and Bhardwaj, N. (1998a). Immature dendritic cells phagocytose apoptotic cells via alphavbeta5 and CD36, and cross-present antigens to cytotoxic T lympho-cytes. *J. Exp. Med.* **188**, 1359.

Albert, M. L., Sauter, B. and Bhardwaj, N. (1998b). Dendritic cells acquire antigen from apoptotic cells and induce class I- restricted CTLs. *Nature* **392**, 86.

Arnold-Schild, D., Hanau, D., Spehner, D., Schmid, C., Rammensee, H. G., de la Salle, H. and Schild, H. (1999). Cutting edge: receptor-mediated endocytosis of heat shock proteins by professional antigen-presenting cells. *J. Immunol.* **162**, 3757.

Arpinati, M., Green, C. L., Heimfeld, S., Heuser, J. E. and Anasetti, C. (2000). Granulocyte-colony stimulating factor mobilizes T helper 2-inducing dendritic cells [see comments]. *Blood* **95**, 2484.

Banchereau, J., Bazan, F., Blanchard, D., Briere, F., Galizzi, J. P., van Kooten, C., Liu, Y. J., Rousset, F. and Saeland, S. (1994). The CD40 antigen and its ligand. *Ann. Rev. Immunol.* **12**, 881.

Banchereau, J., Briere, F., Caux, C., Davoust, J., Lebecque, S., Liu, Y., Pulendran, B. and Palucka, K. (2000). Immunobiology of dendritic cells. *Ann. Rev. Immunol.* **18**, 767.

Banchereau, J., Palucka, A. K., Dhodapkar, M., Burkeholder, S., Taquet, N., Rolland, A., Taquet, S., Coquery, S., Wittkowski, K. M., Bhardwaj, N., Pineiro, L., Steinman, R. and Fay, J. (2001a). Immune and clinical responses in patients with metastatic melanoma to CD34(+) progenitor-derived dendritic cell vaccine. *Cancer Res.* **61**, 6451.

Banchereau, J., Schuler-Thurner, B., Palucka, A. K. and Schuler, G. (2001b). Dendritic cells as vectors for therapy. *Cell* **106**, 271.

Bhardwaj, N., Friedman, S. M., Cole, B.C. and Nisianian, A. J. (1992). Dendritic cells are potent antigen-presenting cells for microbial superantigens. *J. Exp. Med.* **175**, 267.

Bhardwaj, N., Young, J. W., Nisianian, A. J., Baggers, J. and Steinman, R. M. (1993). Small amounts of superantigen, when presented on dendritic cells, are sufficient to initiate T cell responses. *J. Exp. Med.* **178**, 633.

Bjorck, P., Flores-Romo, L. and Liu, Y. J. (1997). Human interdigitating dendritic cells directly stimulate CD40-activated naive B cells. *Eur. J. Immunol.* **27**, 1266.

Blom, B., Ho, S., Antonenko, S. and Liu, Y. J. (2000). Generation of interferon alpha-producing predendritic cell (Pre-DC)2 from human CD34(+) hematopoietic stem cells. *J. Exp. Med.* **192**, 1785.

Boczkowski, D., Nair, S. K., Snyder, D. and Gilboa, E. (1996). Dendritic cells pulsed with RNA are potent antigen-presenting cells in vitro and in vivo. *J. Exp. Med.* **184**, 465.

Brocker, T. (1999). The role of dendritic cells in T cell selection and survival. *J. Leukoc. Biol.* **66**, 331.

Caux, C., Saeland, S., Favre, C., Duvert, V., Mannoni, P. and Banchereau, J. (1990). Tumor necrosis factor-alpha strongly potentiates interleukin-3 and granulocyte-macrophage colony-stimulating factor-induced proliferation of human CD34+ hematopoietic progenitor cells. *Blood* **75**, 2292.

Caux, C., Dezutter-Dambuyant, C., Schmitt, D. and Banchereau, J. (1992). GM-CSF and TNF-alpha cooperate in the generation of dendritic Langerhans cells. *Nature* **360**, 258.

Caux, C., Durand, I., Moreau, I., Duvert, V., Saeland, S. and Banchereau, J. (1993). Tumor necrosis factor alpha cooperates with interleukin 3 in the recruitment of a primitive subset of human CD34+ progenitors. *J. Exp. Med.* **177**, 1815.

Caux, C., Massacrier, C., Vanbervliet, B., Barthelemy, C., Liu, Y. J. and Banchereau, J. (1994a). Interleukin 10 inhibits T cell alloreaction induced by human dendritic cells. *Int. Immunol.* **6**, 1177.

Caux, C., Massacrier, C., Vanbervliet, B., Dubois, B., Van Kooten, C., Durand, I. and Banchereau, J. (1994b). Activation of human dendritic cells through CD40 cross-linking. *J. Exp. Med.* **180**, 1263.

Caux, C., Massacrier, C., Dezutter-Dambuyant, C., Vanbervliet, B., Jacquet, C., Schmitt, D. and Banchereau, J. (1995). Human dendritic Langerhans cells generated in vitro from CD34+ progenitors can prime naive CD4+ T cells and process soluble antigen. *J. Immunol.* **155**, 5427.

Caux, C., Vanbervliet, B., Massacrier, C., Dezutter-Dambuyant, C., de Saint-Vis, B., Jacquet, C., Yoneda, K., Imamura, S., Schmitt, D. and Banchereau, J. (1996). CD34+ hematopoietic progenitors from human cord blood differentiate along two independent dendritic cell pathways in response to GM-CSF+TNF alpha. *J. Exp. Med.* **184**, 695.

Caux, C., Massacrier, C., Vanbervliet, B., Dubois, B., Durand, I., Cella, M., Lanzavecchia, A. and Banchereau, J. (1997). CD34+ hematopoietic progenitors from human cord blood differentiate along two independent dendritic cell pathways in response to granulocyte-macrophage colony-stimulating factor plus tumor necrosis factor alpha: II. Functional analysis. *Blood* **90**, 1458.

Caux, C., Ait-Yahia, S., Chemin, K., de Bouteiller, O., Dieu-Nosjean, M. C., Homey, B., Massacrier, C., Vanbervliet, B., Zlotnik, A. and Vicari, A. (2000). Dendritic cell biology and regulation of dendritic cell trafficking by chemokines. *Springer Semin. Immunopathol.* **22**, 345.

Cella, M., Scheidegger, D., Palmer-Lehmann, K., Lane, P., Lanzavecchia, A. and Alber, G. (1996). Ligation of CD40 on dendritic cells triggers production of high levels of interleukin-12 and enhances T cell stimulatory capacity: T-T help via APC activation. *J. Exp. Med.* **184**, 747.

Cella, M., Jarrossay, D., Facchetti, F., Alebardi, O., Nakajima, H., Lanzavecchia, A. and Colonna, M. (1999). Plasmacytoid monocytes migrate to inflamed lymph nodes and produce large amounts of type I interferon [In Process Citation]. *Nat. Med.* **5**, 919.

Cella, M., Facchetti, F., Lanzavecchia, A. and Colonna, M. (2000). Plasmacytoid dendritic cells activated by influenza virus and CD40 ligand drive a potent Th1 polarization. *Nat Immunol* **1**, 305.

Celluzzi, C. M., Mayordomo, J. I., Storkus, W. J., Lotze, M. T. and Falo, L. D., Jr. (1996). Peptide-pulsed dendritic cells induce antigen-specific CTL-mediated protective tumor immunity [see comments]. *J. Exp. Med.* **183**, 283.

Collins, L. S. and Dorshkind, K. (1987). A stromal cell line from myeloid long-term bone marrow cultures can support myelopoiesis and B lymphopoiesis. *J. Immunol.* **138**, 1082.

Croft, M., Duncan, D. D. and Swain, S. L. (1992). Response of naive antigen-specific CD4+ T cells in vitro: characteristics and antigen-presenting cell requirements. *J. Exp. Med.* **176**, 1431.

Crotta, S., Stilla, A., Wack, A., D'Andrea, A., Nuti, S., D'Oro, U., Mosca, M., Filliponi, F., Brunetto, R. M., Bonino, F., Abrignani, S. and Valiante, N. M. (2002). Inhibition of natural killer cells through engagement of CD81 by the major hepatitis C virus envelope protein. *J. Exp. Med.* **195**, 35.

Dezutter-Dambuyant, C., Schmitt, D. A., Dusserre, N., Hanau, D., Kolbe, H. V., Kieny, M. P., Cazenave, J. P., Schmitt, D., Pasquali, J. L., Olivier, R. *et al.* (1991). Interaction of human epidermal Langerhans cells with HIV-1 viral envelope proteins (gp 120 and gp 160s) involves a receptor-mediated endocytosis independent of the CD4 T4A epitope. *J. Dermatol.* **18**, 377.

Dhodapkar, M. V., Krasovsky, J., Steinman, R. M. and Bhardwaj, N. (2000). Mature dendritic cells boost functionally superior CD8(+) T-cell in humans without foreign helper epitopes. *J. Clin. Invest.* **105**, R9.

Dubois, B., Vanbervliet, B., Fayette, J., Massacrier, C., Van Kooten, C., Briere, F., Banchereau, J. and Caux, C. (1997). Dendritic cells enhance growth and differentiation of CD40-activated B lymphocytes [see comments]. *J. Exp. Med.* **185**, 941.

Dzionek, A., Fuchs, A., Schmidt, P., Cremer, S., Zysk, M., Miltenyi, S., Buck, D. W. and Schmitz, J. (2000). BDCA-2, BDCA-3 and BDCA-4: three markers for distinct subsets of dendritic cells in human peripheral blood. *J. Immunol.* **165**, 6037.

Dzionek, A., Sohma, Y., Nagafune, J., Cella, M., Colonna, M., Faccheti, F., Gunther, G., Johnston, I., Lanzavecchia, A., Nagasaka, T., Okada, T., Vermi, W., Winkels, G., Yamamoto, T., Zysk, M., Yamaguchi, Y. and Schmitz, J. (2001). BDCA-2, a novel plasmacytoid dendritic cells-specific type II C-type lectin, mediates antigen capture and is a potent inhibitor of interferon alpha/beta induction. *J. Exp. Med.* **194**, 1823.

Engering, A. J., Cella, M., Fluitsma, D., Brockhaus, M., Hoefsmit, E. C., Lanzavecchia, A. and Pieters, J. (1997). The mannose receptor functions as a high capacity and broad specificity antigen receptor in human dendritic cells. *Eur. J. Immunol.* **27**, 2417.

Enk, A. H. and Katz, S. I. (1992). Identification and induction of keratinocyte-derived IL-10. *J. Immunol.* **149**, 92.

Fanger, N. A., Wardwell, K., Shen, L., Tedder, T. F. and Guyre, P. M. (1996). Type I (CD64) and type II (CD32) Fc gamma receptor-mediated phagocytosis by human blood dendritic cells. *J. Immunol.* **157**, 541.

Fayette, J., Dubois, B., Vandenabeele, S., Bridon, J. M., Vanbervliet, B., Durand, I., Banchereau, J., Caux, C. and Briere, F. (1997). Human dendritic cells skew isotype switching of CD40-activated naive B cells towards IgA1 and IgA2. *J. Exp. Med.* **185**, 1909.

Ferlazzo, G., Tsang, M. L., Moretta, L., Melioli, G., Steinman, R. M. and Munz, C. (2002). Human dendritic cells activate resting natural killer (NK) cells and are recognized via the NKp30 receptor by activated NK cells. *J. Exp. Med.* **195**, 343.

Finkelman, F. D., Lees, A., Birnbaum, R., Gause, W. C. and Morris, S. C. (1996). Dendritic cells can present antigen in vivo in a tolerogenic or immunogenic fashion. *J. Immunol.* **157**, 1406.

Flamand, V., Sornasse, T., Thielemans, K., Demanet, C., Bakkus, M., Bazin, H., Tielemans, F., Leo, O., Urbain, J. and Moser, M. (1994). Murine dendritic cells pulsed in vitro with tumor antigen induce tumor resistance in vivo. *Eur. J. Immunol.* **24**, 605.

Flores-Romo, L., Bjorck, P., Duvert, V., van Kooten, C., Saeland, S. and Banchereau, J. (1997). CD40 ligation on human cord blood CD34+ hematopoietic progenitors induces their proliferation and differentiation into functional dendritic cells. *J. Exp. Med.* **185**, 341.

Flynn, S., Toellner, K. M., Raykundalia, C., Goodall, M. and Lane, P. (1998). CD4 T cell cytokine differentiation: the B cell activation molecule, OX40 ligand, instructs CD4 T cells to express interleukin 4 and upregulates expression of the chemokine receptor, Blr-1. *J. Exp. Med.* **188**, 297.

Garrone, P., Neidhardt, E. M., Garcia, E., Galibert, L., van Kooten, C. and Banchereau, J. (1995). Fas ligation induces apoptosis of CD40-activated human B lymphocytes. *J. Exp. Med.* **182**, 1265.

Geissmann F., Prost, C., Monnet, J. P., Dy, M., Brousse, N. and Hermine, O. (1998). Transforming growth factor beta1, in the presence of granulocyte/macrophage colony-stimulating factor and interleukin 4, induces differentiation of human peripheral blood monocytes into dendritic Langerhans cells. *J. Exp. Med.* **187**, 961.

Gerosa, F., Baldani-Guerra, B., Nisii, C., Marchesini, V., Carra, G. and Trinchieri, G. (2002). Reciprocal activating interaction between natural killer cells and dendritic cells. *J. Exp. Med.* **195**, 327.

Gilboa, E. (1999). The makings of a tumor rejection antigen. *Immunity* **11**, 263.

Grabbe, S., Beissert, S., Schwarz, T. and Granstein, R. D. (1995). Dendritic cells as initiators of tumor immune responses: a possible strategy for tumor immunotherapy? *Immunol. Today* **16**, 117.

Gross, J. A., Johnston, J., Mudri, S., Enselman, R., Dillon, S. R., Madden, K., Xu, W., Parrish-Novak, J., Foster, D,. Lofton-Day, C., Moore, M., Littau, A., Grossman, A., Haugen, H., Foley, K., Blumberg, H., Harrison, K., Kindsvogel, W. and Clegg, C. H. (2000). TACI and BCMA are receptors for a TNF homologue implicated in B-cell autoimmune disease [see comments]. *Nature* **404**, 995.

Isolation and Propagation of Human Dendritic Cells

Grouard, G., Durand, I., Filgueira, L., Banchereau, J. and Liu, Y. J. (1996). Dendritic cells capable of stimulating T cells in germinal centres. *Nature* **384**, 364.

Grouard, G., Rissoan, M. C., Filgueira, L., Durand, I., Banchereau, J. and Liu, Y. J. (1997). The enigmatic plasmacytoid T cells develop into dendritic cells with interleukin (IL)-3 and CD40-ligand. *J. Exp. Med.* **185**, 1101.

Hao, Q. L., Smogorzewska, E. M., Barsky, L. W. and Crooks, G. M. (1998). In vitro identification of single CD34+CD38- cells with both lymphoid and myeloid potential. *Blood* **91**, 4145.

Hart, D. N. (1997). Dendritic cells: unique leukocyte populations which control the primary immune response. *Blood* **90**, 3245.

Havenith, C. E., Breedijk, A. J., Betjes, M. G., Calame, W., Beelen, R. H. and Hoefsmit, E. C. (1993). T cell priming in situ by intratracheally instilled antigen-pulsed dendritic cells. *Am. J. Respir. Cell. Mol. Biol.* **8**, 319.

Heath, W. R. and Carbone, F. R. (2001). Cross-presentation, dendritic cells, tolerance and immunity. *Ann. Rev. Immunol.* **19**, 47.

Hsu, F. J., Benike, C., Fagnoni, F., Liles, T. M., Czerwinski, D., Taidi, B., Engleman, E. G. and Levy, R. (1996). Vaccination of patients with B-cell lymphoma using autologous antigen- pulsed dendritic cells. *Nat. Med.* **2**, 52.

Huang, F. P., Platt, N., Wykes, M., Major, J. R., Powell, T. J., Jenkins, C. D. and MacPherson, G. G. (2000). A discrete subpopulation of dendritic cells transports apoptotic intestinal epithelial cells to T cell areas of mesenteric lymph nodes. *J. Exp. Med.* **191**, 435.

Inaba, K., Metlay, J. P., Crowley, M. T. and Steinman, R. M. (1990). Dendritic cells pulsed with protein antigens in vitro can prime antigen- specific, MHC-restricted T cells in situ [published erratum appears in *J. Exp. Med.* 1990 Oct 1;172(4):1275]. *J. Exp. Med.* **172**, 631.

Inaba, K., Inaba, M., Romani, N., Aya, H., Deguchi, M., Ikehara, S., Muramatsu, S. and Steinman, R. M. (1992). Generation of large numbers of dendritic cells from mouse bone marrow cultures supplemented with granulocyte/macrophage colony-stimulating factor. *J. Exp. Med.* **176**, 1693.

Inaba, K., Inaba, M., Naito, M. and Steinman, R. M. (1993). Dendritic cell progenitors phagocytose particulates, including bacillus Calmette-Guerin organisms, and sensitize mice to mycobacterial antigens in vivo. *J. Exp. Med.* **178**, 479.

Ito, T., Inaba, M., Inaba, K., Toki, J., Sogo, S., Iguchi, T., Adachi, Y., Yamaguchi, K., Amakawa, R., Valladeau, J., Saeland, S., Fukuhara, S. and Ikehara, S. (1999). A CD1a+/CD11c+ subset of human blood dendritic cells is a direct precursor of Langerhans cells. *J. Immunol.* **163**, 1409.

Jacobsen, S. E., Ruscetti, F. W., Dubois, C. M. and Keller, J. R. (1992). Tumor necrosis factor alpha directly and indirectly regulates hematopoietic progenitor cell proliferation: role of colony-stimulating factor receptor modulation. *J. Exp. Med.* **175**, 1759.

Jiang, W., Swiggard, W. J., Heufler, C., Peng, M., Mirza, A., Steinman, R. M. and Nussenzweig, M. C. (1995). The receptor DEC-205 expressed by dendritic cells and thymic epithelial cells is involved in antigen processing. *Nature* **375**, 151.

Kadowaki, N., Antonenko, S., Lau, J. Y. and Liu, Y. J. (2000). Natural interferon alpha/beta-producing cells link innate and adaptive immunity. *J. Exp. Med.* **192**, 219.

Kasinrerk, W., Baumruker, T., Majdic, O., Knapp, W. and Stockinger, H. (1993). CD1 molecule expression on human monocytes induced by granulocyte-macrophage colony-stimulating factor. *J. Immunol.* **150**, 579.

Knight, S. C., Farrant, J. and Bryan, A. (1986). Non-adherent, low density cells from human peripheral blood contain dendritic cells and monocytes, both with veiled morphology. *Immunology* **57**, 595.

Kripke, M. L., Munn, C. G., Jeevan, A., Tang, J. M. and Bucana, C. (1990). Evidence that cutaneous antigen-presenting cells migrate to regional lymph nodes during contact sensitization. *J. Immunol.* **145**, 2833.

Laabi, Y. and Strasser, A. (2000). Immunology. Lymphocyte survival—ignorance is BLys. *Science* **289**, 883.

Lennert, K., Kaiserling, E. and Muller-Hermelink, H. K. (1975). Letter: T-associated plasma-cells. *Lancet* **i**, 1031.

Levin, D., Constant, S., Pasqualini, T., Flavell, R. and Bottomly, K. (1993). Role of dendritic cells in the priming of CD4+ T lymphocytes to peptide antigen in vivo. *J. Immunol.* **151**, 6742.

Liu, L. M. and MacPherson, G. G. (1993). Antigen acquisition by dendritic cells: intestinal dendritic cells acquire antigen administered orally and can prime naive T cells in vivo. *J. Exp. Med.* **177**, 1299.

Liu, Y. J., Xu, J., de Bouteiller, O., Parham, C. L., Grouard, G., Djossou, O., de Saint-Vis, B., Lebecque, S., Banchereau, J. and Moore, K. W. (1997). Follicular dendritic cells specifically express the long CR2/CD21 isoform. *J. Exp. Med.* **185**, 165.

Macatonia, S. E., Hsieh, C. S., Murphy, K. M. and O'Garra, A. (1993). Dendritic cells and macrophages are required for Th1 development of CD4+ T cells from alpha beta TCR transgenic mice: IL-12 substitution for macrophages to stimulate IFN-gamma production is IFN-gamma-dependent. *Int. Immunol.* **5**, 1119.

Mackay, F., Woodcock, S. A., Lawton, P., Ambrose, C., Baetscher, M., Schneider, P., Tschopp, J. and Browning, J. L. (1999). Mice transgenic for BAFF develop lymphocytic disorders along with autoimmune manifestations. *J. Exp. Med.* **190**, 1697.

Maraskovsky, E., Brasel, K., Teepe, M., Roux, E. R., Lyman, S. D., Shortman, K. and McKenna, H. J. (1996). Dramatic increase in the numbers of functionally mature dendritic cells in Flt3 ligand-treated mice: multiple dendritic cell subpopulations identified. *J. Exp. Med.* **184**, 1953.

Maraskovsky, E., Daro, E., Roux, E., Teepe, M., Maliszewski, C. R., Hoek, J., Caron, D. Lebsack, M. E. and McKenna, H. J. (2000). In vivo generation of human dendritic cell subsets by Flt3 ligand. *Blood* **96**, 878.

Maurer, D., Fiebiger, E., Reininger, B., Ebner, C., Petzelbauer, P., Shi, G. P., Chapman, H. A. and Stingl, G. (1998). Fc epsilon receptor I on dendritic cells delivers IgE-bound multivalent antigens into a cathepsin S-dependent pathway of MHC class II presentation. *J. Immunol.* **161**, 2731.

Mayordomo, J. I., Zorina, T., Storkus, W. J., Zitvogel, L., Garcia-Prats, M. D., DeLeo, A. B. and Lotze, M. T. (1997). Bone marrow-derived dendritic cells serve as potent adjuvants for peptide-based antitumor vaccines. *Stem Cells* **15**, 94.

Mohamadzadeh, M., Berard, F., Essert, G., Chalouni, C., Pulendran, B., Davoust, J., Bridges, G., Palucka, A. K. and Banchereau, J. (2001). Interleukin 15 skews monocyte differentiation into dendritic cells with features of Langerhans cells. *J. Exp. Med.* **194**, 1013.

Mommaas, A. M., Mulder, A. A., Jordens, R., Out, C., Tan, M. C., Cresswell, P., Kluin, P. M. and Koning, F. (1999). Human epidermal Langerhans cells lack functional mannose receptors and a fully developed endosomal/lysosomal compartment for loading of HLA class II molecules. *Eur. J. Immunol.* **29**, 571.

Moore, M. A. S. (1991). Clinical implications of positive and negative hematopoietic stem cell regulators. *Blood* **78**, 1.

Moore, P. A., Belvedere, O., Orr, A., Pieri, K., LaFleur, D. W., Feng, P., Soppet, D., Charters, M., Gentz, R., Parmelee, D., Li, Y., Galperina, O., Giri, J., Roschke, V., Nardelli, B., Carrell, J., Sosnovtseva, S., Greenfield, W., Ruben, S. M., Olsen, H. S., Fikes, J. and Hilbert, D. M. (1999). BLyS: member of the tumor necrosis factor family and B lymphocyte stimulator. *Science* **285**, 260.

Muller-Hermelink, H. K., Kaiserling, E. and Lennert, K. (1973). [Pseudofollicular nests of plasmacells (of a special type?) in paracortical pulp of human lymph nodes (author's transl)]. *Virchows Arch. B Cell Pathol.* **14**, 47.

Muller-Hermelink, H., Stein, H., Steinman, G. and Lennert, K. (1983). Malignant lymphoma of plasmacytoid T cells. Morphologic and immunologic studies characterizing a special type of T cell. *Am. J. Surg. Pathol.* **8**, 849.

Nestle, F. O., Alijagic, S., Gilliet, M., Sun, Y., Grabbe, S., Dummer, R., Burg, G. and Schadendorf, D. (1998). Vaccination of melanoma patients with peptide- or tumor lysate-pulsed dendritic cells [see comments]. *Nat. Med.* **4**, 328.

O'Doherty, U., Peng, M., Gezelter, S., Swiggard, W. J., Betjes, M., Bhardwaj, N. and Steinman, R. M. (1994). Human blood contains two subsets of dendritic cells, one immunologically mature and the other immature. *Immunology* **82**, 487.

Ohshima, Y., Tanaka, Y., Tozawa, H., Takahashi, Y., Maliszewski, C. and Delespesse, G. (1997). Expression and function of OX40 ligand on human dendritic cells. *J. Immunol.* **159**, 3838.

Olweus, J., BitMansour, A., Warnke, R., Thompson, P. A., Carballido, J., Picker, L. J. and Lund-Johansen, F. (1997). Dendritic cell ontogeny: a human dendritic cell lineage of myeloid origin. *Proc. Natl. Acad. Sci. USA* **94**, 12551.

Peters, J. H., Xu, H., Ruppert, J., Ostermeier, D., Friedrichs, D. and Gieseler, R. K. (1993). Signals required for differentiating dendritic cells from human monocytes in vitro. *Adv. Exp. Med. Biol.* **329**, 275.

Piccioli, D., Sbrana, S., Melandri, E. and Valiante, N. M. (2002). Contact-dependent stimulation and inhibition of dendritic cells by natural killer cells. *J. Exp. Med.* **195**, 335.

Piemonti, L., Bernasconi, S., Luini, W., Trobonjaca, Z., Minty, A., Allavena, P. and Mantovani, A. (1995). IL-13 supports differentiation of dendritic cells from circulating precursors in concert with GM-CSF. *Eur. Cytokine Netw.* **6**, 245.

Porcelli, S., Morita, C. T. and Brenner, M. B. (1992). CD1b restricts the response of human CD4-8- T lymphocytes to a microbial antigen. *Nature* **360**, 593.

Pulendran, B., Lingappa, J., Kennedy, M. K., Smith, J., Teepe, M., Rudensky, A., Maliszewski, C. R. and Maraskovsky, E. (1997). Developmental pathways of dendritic cells in vivo: distinct function, phenotype, and localization of dendritic cell subsets in FLT3 ligand-treated mice. *J. Immunol.* **159**, 2222.

Pulendran, B., Banchereau, J., Burkeholder, S., Kraus, E., Guinet, E., Chalouni, C., Caron, D., Maliszewski, C., Davoust, J., Fay, J. and Palucka, K. (2000). Flt3-ligand and granulocyte colony-stimulating factor mobilize distinct human dendritic cell subsets in vivo. *J. Immunol.* **165**, 566.

Rawlings, D. J., Quan, S., Hao, Q. L., Thiemann, F. T., Smogorzewska, M., Witte, O. N. and Crooks, G. M. (1997). Differentiation of human CD34+CD38- cord blood stem cells into B cell progenitors in vitro. *Exp. Hematol* **25**, 66.

Ray, A., Schmitt, D., Dezutter-Dambuyant, C., Fargier, M. C. and Thivolet, J. (1989). Reappearance of CD1a antigenic sites after endocytosis on human Langerhans cells evidenced by immunogoldrelabeling. *J. Invest. Dermatol.* **92**, 217.

Reid, C. D., Stackpoole, A., Meager, A. and Tikerpae, J. (1992). Interactions of tumor necrosis factor with granulocyte-macrophage colony-stimulating factor and other cytokines in the regulation of dendritic cell growth in vitro from early bipotent CD34+ progenitors in human bone marrow. *J. Immunol.* **149**, 2681.

Reis e Sousa, C., Stahl, P. D. and Austyn, J. M. (1993). Phagocytosis of antigens by Langerhans cells in vitro. *J. Exp. Med.* **178**, 509.

Rescigno, M., Granucci, F. and Ricciardi-Castagnoli, P. (2000). Molecular events of bacterial-induced maturation of dendritic cells [In Process Citation]. *J. Clin. Immunol.* **20**, 161.

Robinson, S. P., Patterson, S., English, N., Davies, D., Knight, S. C. and Reid, C. D. (1999). Human peripheral blood contains two distinct lineages of dendritic cells. *Eur. J. Immunol.* **29**, 2769.

Romani, N., Schuler, G. and Fritsch. P. (1991). Identification and phenotype of epidermal Langerhans cells in vivo and in vitro. In *Epidermal Langerhans Cells* (G. Schuler, Ed.), p. 49. Boca Raton, FL, CRC Press.

Romani, N., Gruner, S., Brang, D., Kampgen, E., Lenz, A., Trockenbacher, B., Konwalinka, G., Fritsch, P. O., Steinman, R. M. and Schuler, G. (1994). Proliferating dendritic cell progenitors in human blood. *J. Exp. Med.* **180**, 83.

Rosenzwajg, M., Canque, B. and Gluckman, J. C. (1996). Human dendritic cell differentiation pathway from CD34+ hematopoietic precursor cells. *Blood* **87**, 535.

Rubartelli, A., Poggi, A. and Zocchi, M. R. (1997). The selective engulfment of apoptotic bodies by dendritic cells is mediated by the alpha(v)beta3 integrin and requires intracellular and extracellular calcium. *Eur. J. Immunol.* **27**, 1893.

Sallusto, F. and Lanzavecchia, A. (1994). Efficient presentation of soluble antigen by cultured human dendritic cells is maintained by granulocyte/macrophage colony-stimulating factor plus interleukin 4 and downregulated by tumor necrosis factor alpha. *J. Exp. Med.* **179**, 1109.

Sallusto, F., Cella, M., Danieli, C. and Lanzavecchia, A. (1995). Dendritic cells use macropinocytosis and the mannose receptor to concentrate macromolecules in the major histocompatibility complex class II compartment: downregulation by cytokines and bacterial products [see comments]. *J. Exp. Med.* **182**, 389.

Santiago-Schwarz, F., Belilos, E., Diamond, B. and Carsons, S. E. (1992). TNF in combination with GM-CSF enhances the differentiation of neonatal cord blood stem cells into dendritic cells and macrophages. *J. Leukoc. Biol.* **52**, 274.

Schneider, P., MacKay, F., Steiner, V., Hofmann, K., Bodmer, J. L., Holler, N., Ambrose, C., Lawton, P., Bixler, S., Acha-Orbea, H., Valmori, D., Romero, P., Werner-Favre, C., Zubler, R. H., Browning, J. L. and Tschopp, J. (1999). BAFF, a novel ligand of the tumor necrosis factor family, stimulates B cell growth. *J. Exp. Med.* **189**, 1747.

Siegal, F. P., Kadowaki, N., Shodell, M. and Liu, Y. J. (1998). Evidence for identity of the natural interferon producing cells (NIPC) and the CD11c- precursor of DC2. *J, Leukoc, Biol,* Supplement 2.

Siegal, F. P., Kadowaki, N., Shodell, M., Fitzgerald-Bocarsly, P. A., Shah, K., Ho, S., Antonenko, S. and Liu, Y. J. (1999). The nature of the principal type 1 interferon-producing cells in human blood [In Process Citation]. *Science* **284**, 1835.

Siena, S., Di Nicola, M., Bregni, M., Mortarini, R., Anichini, A., Lombardi, L., Ravagnani, F., Parmiani, G. and Gianni, A. M. (1995). Massive ex vivo generation of functional dendritic cells from mobilized CD34+ blood progenitors for anti-cancer therapy. *Exp. Hematol.* **23**, 1463.

Spits, H., Couwenberg, F., Bakker, A. Q., Weijer, K. and Uittenbogaart, C. H. (2000). Id2 and Id3 inhibit development of CD34+ stem cells into pre-DC2 but not into pre-DC1: Evidence for a lymphoid origin of pre-DC2. *J. Exp. Med.* **192**(12) 1775–1784.

Steinman, R. M. (1991). The dendritic cell system and its role in immunogenicity. *Ann. Rev. Immunol.* **9**, 271.

Steinman, R. M. and Cohn, Z. A. (1973). Identification of a novel cell type in peripheral lymphoid organs of mice. I. Morphology, quantitation, tissue distribution. *J. Exp. Med.* **137**, 1142.

Steinman, R. M., Pack, M. and Inaba, K. (1997). Dendritic cells in the T-cell areas of lymphoid organs. *Immunol. Rev.* **156**, 25.

Steinman, R. M., Turley, S., Mellman, I. and Inaba, K. (2000). The induction of tolerance by dendritic cells that have captured apoptotic cells. *J. Exp. Med.* **191**, 411.

Stingl, G. and Shevach, E. (1991). Langerhans cells as antigen-presenting cells. In *Epidermal Langerhans Cells* (G. Schuler, Ed.), p. 159. Boca Raton, FL, CRC Press.

Strobl, H., Riedl, E., Scheinecker, C., Bello-Fernandez, C., Pickl, W. F., Rappersberger, K., Majdic, O. and Knapp, W. (1996). TGF-beta 1 promotes in vitro development of dendritic cells from CD34+ hemopoietic progenitors. *J. Immunol.* **157**, 1499.

Strunk, D., Rappersberger, K., Egger, C., Strobl, H., Kromer, E., Elbe, A., Maurer, D. and Stingl, G. (1996). Generation of human dendritic cells/Langerhans cells from circulating CD34+ hematopoietic progenitor cells. *Blood* **87**, 1292.

Strunk, D., Egger, C., Leitner, G., Hanau, D. and Stingl, G. (1997). A skin homing molecule defines the langerhans cell progenitor in human peripheral blood. *J. Exp. Med.* **185**, 1131.

Stuber, E., Neurath, M., Calderhead, D., Fell, H.P. and Strober, W. (1995). Cross-linking of OX40 ligand, a member of the TNF/NGF cytokine family, induces proliferation and differentiation in murine splenic B cells. *Immunity* **2**, 507.

Szabolcs, P., Moore, M. A. and Young, J. W. (1995). Expansion of immunostimu-latory dendritic cells among the myeloid progeny of human CD34+ bone marrow precursors cultured with c-kit ligand, granulocyte-macrophage colony-stimulating factor, and TNF-alpha. *J. Immunol.* **154**, 5851.

Tan, M. C., Mommaas, A. M., Drijfhout, J. W., Jordens, R., Onderwater, J. J., Verwoerd, D., Mulder, A. A., van der Heiden, A. N., Scheidegger, D., Oomen, L. C., Ottenhoff, T. H., Tulp, A., Neefjes, J. J. and Koning, F. (1997). Mannose receptor-mediated uptake of antigens strongly enhances HLA class II-restricted antigen presentation by cultured dendritic cells. *Eur. J. Immunol.* **27**, 2426.

Thurner, B., Haendle, I., Roder, C., Dieckmann, D., Keikavoussi, P., Jonuleit, H., Bender, A., Maczek, C., Schreiner, D., von den Driesch, P., Brocker, E. B., Steinman, R. M., Enk, A., Kampgen, E. and Schuler, G. (1999). Vaccination with mage-3A1 peptide-pulsed mature, monocyte-derived dendritic cells expands specific cytotoxic T cells and induces regression of some metastases in advanced stage IV melanoma. *J. Exp. Med.* **190**, 1669.

Valladeau, J., Duvert-Frances, V., Dezutter-Dambuyant, C., Vincet, C., Pin, J., Massacrier, C., Vincent, J., Davoust, J. and Saeland, S. (1998). A monoclonal antibody against Langerin, a protein specific of Langerhans cells, is internalized in coated pits and Birbeck granule. *J. Leukocyt. Biol.* Suppl **2**, Abstr A35.

Valladeau, J., Ravel, O., Dezutter-Dambuyant, C., Moore, K., Kleijmeer, M., Liu, Y., Duvert-Frances, V., Vincent, C., Schmitt, D., Davoust, J., Caux, C., Lebecques, S. and Saeland, S. (2000). Langerin, a novel C-type lectin specific to Langerhans cells, is an endocytic receptor that induces the formation of Birbeck granules. *Immunity* **12**, 71.

van Kooten, C. and Banchereau, J. (2000). CD40-CD40 ligand. *J. Leukoc. Biol.* **67**, 2.

Vollenweider, R. and Lennert., K. (1983). Plasmacytoid T-cell clusters in non-specific lymphadenitis. *Virchows Arch. B Cell Pathol. Incl. Mol. Pathol.* **44**, 1.

Wykes, M., Pombo, A., Jenkins, C. and MacPherson, G. G. (1998). Dendritic cells interact directly with naive B lymphocytes to transfer antigen and initiate class switching in a primary T-dependent response. *J. Immunol.* **161**, 1313.

Young, J. W., Szabolcs, P. and Moore, M. A. (1995). Identification of dendritic cell colony-forming units among normal human CD34+ bone marrow progenitors that are expanded by c-kit-ligand and yield pure dendritic cell colonies in the presence of granulocyte/macrophage colony-stimulating factor and tumor necrosis factor alpha. *J. Exp. Med.* **182**, 1111.

Zitvogel, L., Mayordomo, J. I., Tjandrawan, T., DeLeo, A. B., Clarke, M. R., Lotze, M. T. and Storkus, W. J. (1996). Therapy of murine tumors with tumor peptide-pulsed dendritic cells: dependence on T cells, B7 costimulation, and T helper cell 1-associated cytokines [see comments]. *J. Exp. Med.* **183**, 87.

2 Isolation of T Cells and Establishment of T-cell Lines and Clones

Elisabeth Märker-Hermann
Innere Medizin IV mit Schwerpunkt Rheumatologie, Immunologie und Nephrologie, Dr.-Horst-Schmidt-Kliniken (HSK) Wiesbaden, Germany

Rainer Duchmann
Medizinische Klinik I mit Schwerpunkt Gastroenterologie, Infektiologie und Rheumatologie, Universitätsklinikum Benjamin Franklin, Freie Universität Berlin, Germany

◆◆

CONTENTS

Isolation of T Cells

◆◆◆◆◆◆ INTRODUCTION

The study of human T cells is best performed using purified cells, since the presence of other cell types may have indirect effects on T-cell function. However, for any kind of functional assays on T-cell specificity, antigen-presenting cells are necessary. In this chapter we will describe the basic procedures for isolating mononuclear cell fractions containing monocytes, T and B lymphocytes from different sources, i.e. human peripheral blood, body fluids or diseased tissues. Further experimental methods will describe the separation of T cells and T-cell subsets from mononuclear cell populations, and the generation of T-cell lines and T-cell clones with specificity for bacterial antigens.

METHODS IN MICROBIOLOGY, VOLUME 32
ISBN 0–12–521532–0

Copyright © Elsevier Science Ltd
All rights of reproduction in any form reserved

◆◆◆◆◆◆ ISOLATION OF MONONUCLEAR CELLS (MNC)

Mononuclear cell isolation from peripheral blood

The mononuclear cell fraction containing monocytes, T and B lymphocytes is separated from polymorphonuclear and red blood cells by density gradient centrifugation.

Equipment and reagents

- Suppl. RPMI-1640 medium = RPMI-1640 (e.g. Invitrogen, Life Technologies, Gaithersburg, MD) containing 2 mM L-glutamine (Biochrom AG, Berlin, Germany), 25 mM N-(2-hydroxyethyl) piperazine-N´-(2-ethanesulfonic acid) (HEPES) (Biochrom AG), 100 U ml^{-1} penicillin, 100 µg ml^{-1} streptomycin (Pen-Strep, Biochrom AG).
- Fetal calf serum (FCS) (e.g. Invitrogen, Life Technologies) which has been inactivated by heat (56°C, 30 min) before use.
- Heat-inactivated human serum, blood group AB (HUS, obtained from a local blood transfusion centre); Ficoll-Hypaque (Histopaque-1077^R, Sigma-Aldrich, St Louis, MO, USA).
- 50-ml and 15-ml conical centrifuge tubes (e.g. Greiner, Nürtingen, Germany), temperature-controlled centrifuge with GH-3.7 horizontal rotor (e.g. Heraeus or Beckman).
- Trypan blue, haemocytometer.

Procedure (Modified after Boyum, 1968)

Peripheral blood is collected in sterile heparinized tubes. Heparinized whole blood (10 ml) is mixed with 15 ml suppl. RPMI-1640. The mixture is carefully layered over 15 ml of Ficoll-Hypaque in a 50-ml conical centrifuge tube. Spin for 20 min at 2000 rpm (900g, 4°C). The layer between Ficoll and the upper layer (containing RPMI-1640 and serum) contains the mononuclear cell (MNC) fraction (Fig. 1). Using a pipette remove 80% of the upper layer and recover the interface (MNC) layer. Transfer the latter to a new 50-ml conical tube, and fill tube with suppl. RPMI-1640/5% fetal calf serum (FCS) and centrifuge 10 min at 1300 rpm (400g, 18°C). After the supernatant has been removed, the MNC pellet is resuspended in suppl. RPMI-1640/5% FCS, and the wash is repeated twice. For the last wash, 15-ml conical tubes can be used. Finally, the cells are resuspended in 1 ml suppl. RPMI-1640/10% heat-inactivated human AB serum (HUS).

Counting and markers of cell death

Cell suspension (20 µl) is diluted with 20 µl 0.5% aqueous Trypan blue. The stained (dead) and non-stained (viable) cells are counted in a haemocytometer.

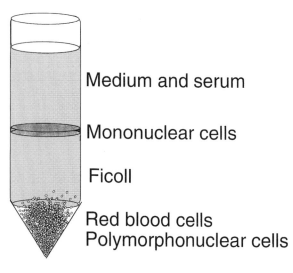

Figure 1. Separation of mononuclear cells (MNC) by Ficoll-Hypaque density gradient centrifugation.

Isolation of mononuclear cells from synovial fluid or synovial membrane

There is growing evidence that bacteria and other micro-organisms play an important role in the aetiopathogenesis of several inflammatory rheumatic diseases such as reactive arthritis, rheumatic fever, Lyme disease, and the so-called HLA-B27 associated seronegative spondyloarthropathies. Furthermore, T cells appear to play a major role in the development, maintenance and also resolution of these forms of bacteria-associated arthritides. Recent developments in understanding the processes involved in T-cell activation now allow us to examine the synovial fluid and synovial membrane T-cell responses to 'arthritogenic' micro-organisms in terms of antigen specificity, epitope identification, cytokine secretion patterns, cytotoxicity and HLA restriction (Hermann, 1993). The relative importance of the αβ-TCR+ CD4+ and CD8+ and the γδ-TCR+ T-cell populations in the pathogenesis of these inflammatory rheumatic diseases has yet to be determined.

Equipment and reagents

- Heparinized synovial fluid (obtained as part of the routine diagnostic and therapeutic management), synovial tissue from surgical synovectomies.
- Sterile PBS; suppl. RPMI-1640; heat-inactivated FCS (e.g. Invitrogen Life Technologies); heat-inactivated HUS; HEPES (Biochrom AG); Collagenase (Type CLS II) (Biochrom AG); Deoxyribonuclease (Dnase Type II) (Sigma-Aldrich); Trypsin-EDTA mixture (0.05%

(contd.)

Isolation of T Cells

w/v Trypsin, 0.02% w/v EDTA) (e.g. ICN Biomedicals, Costa Mesa, CA, USA); Ficoll-Hypaque (Histopaque-1077^R, Sigma-Aldrich).
- Sterile scissors; 50-ml glass spinner flask; sterile steel sieves (mesh 80); sterile petri dishes; 50-ml conical centrifuge tubes (e.g. Greiner); temperature controlled centrifuge; humidified 37°C, 5% CO_2 incubator (e.g. Heraeus).

Procedure (Abrahamsen et al., 1975; Burmester et al., 1981; Klareskog et al., 1982)

- *Synovial fluid derived MNCs.* These are isolated by Ficoll-Hypaque density-gradient sedimentation in exactly the same way as described for peripheral blood MNC (see below).
- *Isolation of synovial membrane MNCs.* Synovial tissues from routine synovectomies are collected in sterile PBS, stored at 4°C, and transported to the laboratory within 2 h. Fat and fibrous material is carefully removed from the synovial lining layer. The resulting synovium is minced into small pieces with sterile scissors, suspended in balanced salt solution containing 20 mM Hepes, pH 7.4, 0.2 mg ml^{-1} collagenase, and 0.15 mg ml^{-1} deoxyribonuclease, and stirred for 1 h at 37°C in a 50-ml glass spinner flask. The digested mixture is filtered through a sterile steel sieve (mesh 80) to remove large clumps of debris. The resulting single-cell suspension is further cleared from dead cells and subcellular material by standard Ficoll-Hypaque centrifugation as described above, washed and suspended in suppl. RPMI-1640/10% FCS.
- *Preparation of adherent synovial cell.* The purified cell suspension is subsequently seeded onto sterile petri dishes at an approximate cell concentration of 5×10^5 cells ml^{-1}. After 45 min of incubation at 37°C, 5% CO_2, the non-adherent cells are removed from the adherent cell layer by gently washing the petri dishes three times with warm RPMI-1640. Cells adhering to the petri dishes are further incubated in suppl. RPMI-1640/10% FCS overnight and then washed two times with PBS containing 0.02% EDTA and subjected to treatment with a trypsin-EDTA mixture (0.05% w/v trypsin, 0.02% w/v EDTA) at 37°C until virtually all cells are in suspension as visualized by phase-contrast microscopy (usually after 15–30 min). This cell suspension is washed twice in suppl. RPMI-1640/5% FCS and finally resuspended in suppl. RPMI-1640/10% HUS. It has been shown that this preparation of adherent synovial cells contains type A synoviocytes (macrophage-like synovial cells) and type B synoviocytes (fibroblast-like synovial cells). The cells can be used as stimulator cells (irradiated antigen-presenting cells with soluble bacterial antigen, or as stimulators in an autologous or allogeneic mixed leukocyte culture reactions). Furthermore, they can be further propagated and passaged *in vitro*, and characterized by immunofluorescence or immunoperoxidase staining, or by the secretion of cytokines and enzymes into the supernatant.

(contd.)

> • *Preparation of synovial membrane derived T lymphocytes.* The non-adherent cells obtained above are washed twice and further fractionated into T and non-T non-adherent cells by separating the cells that form spontaneous rosettes with neuraminidase treated sheep red blood cells (see page 630).

Mononuclear cell (MNC) isolation from gut mucosa

In humans, access to gut associated lymphoid tissue (GALT) is usually obtained through intestinal resections or procurement of biopsy specimens. The decision to use either one of these will be influenced by the experimental setting. Thus, the requirement to study early or focal intestinal lesions will make it necessary to use biopsy specimens whereas the requirement to obtain large numbers of cells will change the approach to using surgical resections.

Another important question that needs to be settled before starting an isolation procedure, is to define the cell population of interest. This decision will be aided by the knowledge that lymphoid cells within the GALT reside in several compartments, which are morphologically and functionally distinct (Fig. 2). These include an afferent or inductive limb constituted by cells localized in organized lymphoid tissues in the form of Peyer's patches (PP), mesenteric lymph nodes (MLN) and the appendix as well as single follicles beneath the intestinal epithelium. PP are a sampling site for intestinal antigens and play a major role in the induction of a mucosal immune response. Lymphocytes primed within the PP enter the circulation via the mesenteric lymph node and thoracic duct. From here, they recirculate to less organized lymphoid effector regions within the lamina propria and intestinal epithelium. T lymphocytes within the lamina propria (LP-T) secrete B-cell helper factors upon specific stimulation and perform important effector functions. They are mostly CD4+ (60–70%), αβ-TCR+ (95%) and display a memory phenotype. Intraepithelial lymphocytes (IELs) in contrast are located between the epithelial cells. They are mostly CD8+ T cells and display a variety of phenotypic differences compared with lamina propria or peripheral blood T cells+. IELs may be a first line of defence but their physiological function is still being investigated.

Since most groups seem to be interested in studying effector T cells, the present unit will describe methods to isolate LPMCs and IELs from intestinal resections. The described procedure is a modification of procedures previously described (Bull *et al.*, 1977; Fiocchi *et al.*, 1979). Further modifications for isolation of IELs have been reported (Cerf-Bensussan *et al.*, 1985; Lundquist *et al.*, 1992). The procedure to isolate mononuclear cells from biopsy specimen follows the same principle. If pure separations of LPMCs or IELs are important, cross-contamination may be a problem with biopsy specimens and can be controlled by using FACS-Analysis.

Although it may be required for some experiments, the present unit will not contain a detailed description of methods used for isolation of

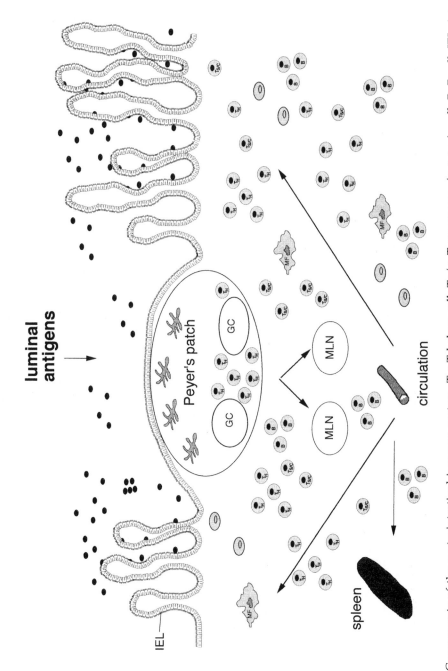

Figure 2. Components of the gastrointestinal immune system. T_H: T-helper cell; Ts/c: T suppressor/cytotoxic cell; B: B cell; IEL: intraepithelial lymphocytes; GC germinal centre; MLN: mesenteric lymph node.

luminal antigens

Peyer's patch

GC

GC

MLN

MLN

circulation

spleen

IEL

immune cells from the afferent limb of the intestinal immune system, i.e. cells localized in organized lymphoid tissue in the form of Peyer's patches (PP) and mesenteric lymph nodes (MLN). For those, however, who are interested in the isolation of these cells, it should be stated that lymphocytes from PP, which extend through the mucosa and into the submucosa, can be isolated from small bowel resections or ileal biopsy specimens (MacDonald *et al.*, 1987) using collagenase digestion. Lymphocytes from MLN, which can be obtained from the mesentery of surgically resected specimens, can be isolated by teasing through a steal sieve (mesh 80) after MLN have been cleaned from fat and vessels. In general, isolation of mononuclear cells from MLN is not different from isolation of MNC from other lymph nodes.

Lamina propria lymphocytes

Equipment and reagents for isolation of lamina propria lymphocytes from surgical resections

- Fresh mucosal tissue from surgical resection.
- PBS; Hanks w/o Ca/Mg; RPMI-1640 (Biochrom); HEPES (Biochrom); Penicillin/Streptomycin (Pen-Strep, Biochrom); Gentamycin (Invitrogen Life Technologies); Amphotericin B (Sigma-Aldrich); DTT (Sigma-Aldrich); EDTA (Sigma-Aldrich); 2-ME (Sigma-Aldrich); collagenase Typ IV (Sigma-Aldrich); Dnase Type II (Sigma-Aldrich); Ficoll-Hypaque (Histopaque-1077^R, Sigma-Aldrich); Percoll (Amersham Pharmacia Biotech, Piscataway, NY).
- Surgical scissors; nylon strainer (100 µm); sterile Erlenmeyer flasks; sterile steel sieves (mesh 80); magnetic stirrer (IKAMAG); 50-ml conical centrifuge tubes (e.g. Greiner); temperature controlled, refrigerated centrifuge (e.g. Heraeus).

Procedures

1. Fresh mucosal tissue from surgical resections is washed with sufficient amounts of PBS (500 ml) to remove gross contaminants and the mucosa is dissected from the underlying muscular layer with scissors.
2. To remove mucus, tissue is then incubated in 150 ml of Isolation Solution A (Hanks w/o Ca/Mg 500 ml containing 1× Pen-Strep, 500 ml Gentamycin, Amphotericin B 1.25 mg, 5 ml L-Glutamine, 12.5 ml HEPES, 500 mg DTT) in a sterile Erlenmeyer flask with a magnetic stirrer for 20 min at 200 rpm and 37°C. To increase yield, this step is repeated twice or until the tissue surface remains free of mucus. Before proceeding to step 3, supernatant is discarded and tissue is rinsed as in step 1.

(contd.)

3. To remove epithelial cells, tissue is incubated in 150 ml isolation solution B (Hanks w/o Ca/Mg 500 ml containing 1 × Pen-Strep, 500 ml Gentamycin, 5 ml L-Glutamine, Amphotericin B 1.25 mg, 12.5 ml HEPES, 25 µl 2-ME, 2.5 ml EDTA 0.2 M) in a sterile Erlenmeyer flask with magnetic stirrer for 30 min at 200 rpm and 37°C. Supernatant is saved for isolation of IEL and tissue is rinsed vigorously as in step 1. Step 3 is repeated three times or until clear supernatant indicates that all epithelial cells have been removed.

4. To liberate LPMCs, the tissue is now cut into 1-mm pieces using sharp surgical scissors, and the pieces are incubated in approximately 150 ml pre-warmed enzyme solution (500 ml RPMI-1640 containing 5 ml Pen-Strep, 500 µl Gentamycin, 5 ml L-Glutamine, 12.5 ml HEPES, 50 mg collagenase Typ IV, and 50 mg DNase for 60 min at 200 rpm and 37°C.

5. Supernatant containing the LPMC is poured off and RPMI-1640 is added 1:1 to dilute the enzymes. Tissue is resubmitted to step 4, while keeping isolated cells on ice.

6. To collect the cells, supernatants are transferred to 50-ml conical tubes, and centrifuged for 10 min at 1300 rpm (400g). Sedimented LPMCs are resuspended in 100 ml RPMI-1640 and passed through a nylon strainer to remove clumps and residual debris.

7. To isolate the mononuclear cells, 25-ml aliquots of cells are layered onto each of 4–8 Ficoll-Hypaque gradients for density sedimentation (1450 rpm, 15 min). If necessary, recovered LPMCs can be stored overnight on ice in 25–50 ml suppl. RPMI-1640.

An experienced worker can isolate LPMCs from intestinal resections within 8–10 h.

Pitfalls

Isolation of LPMCs from intestinal resections may be complicated due to low cell viability, low cell yield or cross-contamination with other cell types (epithelial cells, IELs). In addition, it is a common experience that it is difficult to standardize the procedure, and the length of the isolation procedure certainly increases the risk for technical errors. If low cell yield is a problem, different causes should be considered. First, resections that contain increased ratios of collagenous tissue (stenosed or bulky resections from Crohn's disease) may contain and thus yield lower numbers of LPMCs. Second, low cell yield may be due to incomplete dissection of the mucosa or incomplete removal of fat and mucus from the tissue, producing sticky clumps which entrap LPMCs. Furthermore, cell yield and cell viability are often inversely correlated. Thus, increasing digestion time and increasing amounts of enzymes may give better cell yields at the cost of decreased cell viability and vice versa. Other enzymes such as collagenase (e.g. DISPASE) have been tested, but collagenase seems to give the best yield of viable cells with unaltered expression of cell surface molecules. Fortunately, bacterial contamination is rarely a problem because of the many washing steps and Percoll fractionations involved in the protocol. When beginning to isolate LPMCs, significant contamination with other cell types, especially epithelial cells and IELs, should be excluded by FACS analysis.

Intraepithelial lymphocytes

Equipment and reagents

- Fresh mucosal tissue from surgical resection.
- PBS, Hanks w/o Ca/Mg; RPMI-1640 (Biochrom); HEPES (Biochrom); Pen-Strep (Biochrom), Gentamycin (Invitrogen Life Technologies); Amphotericin B (Sigma-Aldrich); DTT (Sigma-Aldrich); EDTA (Sigma-Aldrich); 2-ME (Sigma-Aldrich); Percoll (Amersham Pharmacia Biotech).
- Surgical scissors; sterile Erlenmeyer flasks; magnetic stirrer; 50-ml centrifuge tubes; temperature controlled, refrigerated centrifuge (Heraeus).

Procedures

The first three steps for isolating IELs are identical to those described above for isolation of LPMCs. Following incubation with isolation solution B, supernatants containing the IELs are diluted 1:1 in RPMI-1640. The cell suspension is transferred via a pipette to 50-ml centrifuge tubes. Epithelial cells are then collected from the solution by centrifugation for 10 min at 1300 rpm. After suspension in suppl. RPMI-1640, cells are layered onto Percoll gradients containing 10-ml layers of 70%, 40%, 20% and 10% Percoll. After centrifugation for 30 min at 1600 rpm and 4°C, IELs can be collected between the 70% and 40% Percoll steps, whereas epithelial cells will sediment at the interface between the 40% and 20% Percoll steps.

An experienced worker can isolate IEL from intestinal resections within 4 h.

Isolation of T Cells

Pitfalls

When beginning to isolate IELs, significant contamination with other cell types, especially LPMCs, should be excluded by FACS analysis demonstrating low numbers of CD20$^+$ and CDIgh$^+$ cells. If purity from epithelial cells is important, modifications of the procedure may be necessary (Lundquist *et al.*, 1992). When standardizing the isolation procedure, the source of IELs (i.e. small bowel vs. large bowel) should be considered, as different sites may contain different numbers of IELs and phenotypes.

◆◆◆◆◆◆ SEPARATION OF T AND NON-T CELLS FROM MNC

The E-rosetting technique

The E-rosetting technique describes a procedure for separating T cells and non-T cells from a population of MNCs (e.g. peripheral blood or synovial fluid MNCs). This method is based on the ability of human T cells to bind to sheep erythrocytes via their CD2 molecule. Neuraminidase treatment of sheep red blood cells (SRBCs) enhances the binding of SRBCs to T lymphocytes (Weiner *et al.*, 1973). In the first step, Neuraminidase-treated SRBCs are prepared. Secondly, SRBCs and MNCs are mixed to form rosettes (E+), which are then isolated from the non-rosetting population (E-, i.e. B cells and monocytes) by Ficoll-Hypaque gradient centrifugation. In the last step, bound SRBCs are separated from the rosetted T cells by hypotonic lysis (Gmelig-Meyeling and Ballieux, 1977).

Equipment and reagents for E-rosetting

- Sheep red blood cells (SRBCs), e.g. from Biologische Arbeitsgemeinschaft Hessen, Germany); sterile PBS, suppl. RPMI-1640; FCS, heat inactivated; Test-Neuraminidase (Aventis Behring, King of Prussia, PA), Ficoll density 1.09 (Biochrom AG)
- 15-ml conical centrifuge tubes (e.g. Greiner or Falcon), a temperature-controlled centrifuge (e.g. Beckman or Heraeus).

Preparation of neuraminidase-treated SRBC

A suspension of SRBCs (2 ml) and sterile PBS (10 ml) are placed in a 15-ml conical centrifuge tube and spun at 2000 rpm (900g) for 10 min, whereafter the PBS supernatant is removed and the cells are resuspended in PBS. This washing procedure is repeated twice. Before treatment with neuraminidase, washed SRBCs can be stored at 4°C for 3 days. Part of the dry SRBC pellet (300 μl) is incubated with 4.6 ml RPMI-1640 and 100 μl of neuraminidase in a water bath (37°C, 30 min), washed twice with RPMI-1640 (2000 rpm, 10 min.), and finally resuspended in RPMI-1640 to a total volume of 5 ml. The suspension is stored at 4°C until use.

Rosette formation and Ficoll density gradient centrifugation

1. MNCs are prepared by standard Ficoll-Hypaque centrifugation as described (see page 622), washed, counted, and suspended in suppl. RPMI-1640/10% HUS (10 × 10^6 cells ml^{-1}). The neuraminidase treated SRBCs are mixed with the MNCs (20–30 min, room temperature) to allow E-rosette formation, whereafter the

(contd.)

mixture is layered over a Ficoll solution (density 1.09) in a 15-ml conical centrifuge tube. The volumes of SRBCs, medium and Ficoll used in this protocol depend on the number of MNCs to be separated (Table 1). The tubes are centrifuged for 30 min at 2800 rpm.

2. Remove and decant about 80% of the upper layer (RPMI-1640/10% HUS) from the centrifuged suspension. The E-rosette-negative (monocyte/B cell enriched) layer (E⁻) is recovered from the interface layer with a pipette, transferred to a 15-ml conical tube, and washed with suppl. RPMI-1640/5% FCS.

3. The E-rosette-positive (T cell) pellet (E+) is suspended in 1 ml RPMI-1640/10% FCS in the 15-ml tube. Cold distilled water (2 ml) is added for hypotonic lysis of SRBCs and mixed gently. After a few seconds, add 8 ml of RPMI-1640/10% FCS. Transfer this suspension to a 50-ml tube containing 40 ml RPMI-1640/10% FCS and centrifuge for 10 min at 1300 rpm.

Table I. Volumes of neuraminidase-treated SRBCs, Ficoll (density 1.09), and RPMI-1640/10% HUS for density gradient centrifugation depending on the number of MNCs to be separated

	No. of MNCs		
MNCs in RPMI-1640/10% HUS	10×10^6 cells in 1.5 ml	$20–30 \times 10^6$ cells in 2 ml	$40–50 \times 10^6$ cells in 3 ml
Neuraminidase-treated SRBCs	1.5 ml	2 ml	2.5 ml
Ficoll 1.09	2 ml	3 ml	4 ml

Pitfalls

After T cells have been exposed to SRBCs some functional activities may be enhanced, as T lymphocytes can be activated via the sheep erythrocyte receptor protein (CD2) to proliferate in an accessory-cell-independent manner (Meuer *et al.*, 1984).

◆◆◆◆◆◆ SEPARATION OF T-CELL SUBSETS

Purification of T-cell populations by indirect antibody panning

T cells expressing particular cell surface markers such as the CD4, the CD8, the αβ-TCR or the γδ-TCR molecules, or TCR BV (Vβ) receptors belonging to a specific TCR BV family can be selected by their capacity to bind to antibody-coated plastic plates (Wysocki and Sato, 1978). For example, to purify CD8⁺ T cells, isolated T cells (E+ cells) are treated with a mouse anti-human monoclonal antibody against the CD4 molecule, and then incubated on plastic dishes that have been coated with an anti-mouse immunoglobulin G (IgG) antibody. The T-cell population which is not CD4 positive (i.e. the αβ–TCR⁺ CD8⁺ and the γδ–TCR⁺ subpopulations) and does not therefore bind the mouse anti-human CD4 antibody will not adhere to the coated plate. These CD4⁻ cells can be selected physically from the adherent CD4⁺ subpopulation. Another example for positive selection of particular T-cell subpopulations is the enrichment of T-cell bearing a TCR belonging to a specific TCR BV (Vβ) family (May *et al.*, 2000).

Equipment and reagents

- T cell population (E+ cells).
- Appropriate mouse monoclonal antibody (e.g. OKT4 or OKT8 hybridoma supernatant containing anti-CD4 or anti-CD8 antibodies; commercially available anti-CD4 or anti-CD8 antibodies; TCR VB-specific antibodies) (e.g. Coulter Immunotech; goat anti-mouse IgG antibody (e.g. DAKO Medac GmbH, Hamburg, Germany), suppl. RPMI-1640; FCS, heat inactivated; PBS, sterile).
- Plastic, six-well plates (Macroplate Standard, Greiner); Plastic 24-well plates (Greiner); 15-ml conical centrifuge tubes (e.g. Falcon); sterile rubber scraper; temperature-controlled centrifuge (e.g. Beckman or Heraeus).

Procedure for the separation of T cells into CD4⁺ and CD8⁺ T cells

1. *Preparation of the panning plate* Goat anti-mouse Ig is diluted to 10 μg ml⁻¹ in suppl. RPMI-1640 and added to the wells of a plastic six-well plate (1.5 ml per well). To separate 2×10^6 to 3×10^6 T-cells, one well of the panning plate is needed. Incubate over night at 4°C or 60 min at room temperature. Remove unbound Ig by using a sterile pipette and gently wash the plate by adding 3 ml PBS to each well (wash three times). Add 2 ml suppl. RPMI-1640/5% FCS and keep the plate at 4°C until the T cells are added to the plate (at least 30 min).
2. *Pretreatment of the T cells* Prepare the monoclonal antibody (e.g. sterile filtered OKT4 hybridoma supernatant containing these anti-bodies, or commercially available anti-CD4 antibody diluted in ster-ile PBS at a concentration appropriate for flow cytometry according to the manufacturer´s instructions). Count the T-cell population and place the cells in a 15-ml centrifuge tube in suppl. RPMI-1640/5% FCS, spin for 10 min at 1300 rpm (400*g*) and 4°C. Decant the super-natant and resuspend the cell pellet in 1–2 ml of the monoclonal antibody hybridoma supernatant or in 0.5 ml of the anti-CD4 mon-oclonal antibody diluted in PBS. Incubate the tube with the cells 30 min on ice (iced water) and then fill with suppl. RPMI-1640/10%
(contd.)

FCS. Centrifuge 10 min at 1300 rpm (400g) and 4°C. After the supernatant is removed, the cell pellet is resuspended in suppl. RPMI-1640/10% FCS, and the wash is repeated once. Finally, the cells are resuspended in suppl. RPMI-1640 (1.5 ml per 2×10^6 to 3×10^6 cells).

3. *Incubation of the coated plate with the pretreated T cells* Remove the RPMI-1640/5% FCS from the coated wells of the six-well panning plate with a sterile pipette and immediately add the pretreated T cells in RPMI-1640 (1.5 ml per well). Spin the plate 10 min at 300 rpm and 4°C. Carefully remove the plate from the centrifuge, and incubate for another 30 min at 4°C.

4. *Collection of the negatively selected cells* Gently swirl the plates for 1 min and collect the supernatant containing the non-adherent cells using a sterile pipette. The negatively selected, non-adherent (i.e. CD4-negative) T cells are washed twice with suppl. RPMI-1640/10% FCS in a 15-ml conical tube, counted, and resuspended in suppl. RPMI-1640/10% human serum or interleukin-2 supplemented medium depending on further culture proceedings. This non-adherent population of cells should be 90–95% pure.

5. *Collection of positively selected adherent CD4+ T cells* Wash the plates gently with 3 ml suppl. RPMI-1640/5% FCS per well (2–3 washes) until all non-adherent cells have been removed (check by using an inverted microscope). The adherent T cells are then gently scraped from the wells of the plate using a sterile rubber scraper. This cell population should be approximately 98% pure.

6. *Positive selection of other T-cell subpopulations* To enrich T cells carrying TCRs belonging to a distinct TCR VB family, monoclonal antibodies with specificity for TCR VB families can be used. In this case, smaller panning plates (24-well plates) are useful, since the numbers of T-cells to be separated are usually smaller than 1×10^6. The goat anti-mouse Ig is diluted to 10 µg ml^{-1} in suppl. RPMI-1640 at 1 ml per well. Accordingly, the pre-treatment of T-cells with monoclonal antibodies (step 2) and the resuspension of the pre-treated T cells in suppl. RPMI-1640 (step 2) before adding them to the coated wells of the 24-well-plate (step 3) is performed in a volume of 1 ml. The other steps of the protocol are identical to the steps described above.

Pitfalls

The purity of the adherent cell population is greater than that of the non-adherent population. However, it must be considered that the function of the adherent T-cell population may be altered by the binding of specific antibodies to surface molecules to be positively selected.

Immunomagnetic negative selection of CD4+ T cells

The protocol below describes another cell separation technique mediated by antibody–antigen reactions. T cells (E+ cells) are incubated with specific monoclonal antibodies to surface molecules (here: anti-CD8) to coat unwanted T cells. Magnetic beads coated with goat anti-mouse IgG are then applied to the cell suspension to bind the antibody-coated cells. After binding, the target cells can be recovered using a strong magnetic field. Negative isolation is a method by which the CD4+ subset is purified from the CD8+ subset binding to the coated magnetic beads. Furthermore, in a positive selection step, the beads can be removed from the CD8+ target cells by a process of detachment.

Equipment and reagents

- T cell population (E+ cells).
- Appropriate monoclonal antibody (e.g. anti-CD8 antibody by BD Biosciences Pharmingen); goat anti-mouse IgG-coated magnetic beads (Dynabeads M-450, Dynal Biotech, Oslo, Norway); sterile PBS; FCS, heat inactivated; coating medium (Hanks balanced salt solution (HBSS) without Ca^{2+}, Mg^{2+}, or Phenol Red, supplemented with 10% FCS, 20 mM HEPES); suppl. RPMI-1640; HUS, heat inactivated.
- Magnetic separation device (Dynal MPC-1); Mixing device (Dynal MX1, 2 or 3), 15-ml centrifugation tubes (e.g. Falcon); vortex mixer; temperature-controlled centrifuge (e.g. Beckman or Heraeus).

Procedure (Funderud et al., 1987; DYNAL Handbook, 1996)

All steps in the protocol are done at 4°C.
1. *Prewash of Dynabeads M-450.* Transfer the required amount of Dynabeads M-450 from the vial to a polypropylene washing tube containing PBS/2% FCS (washing buffer), and place on the Dynal MPC-1 for 2 min. Decant the supernatant, resuspend in excess washing buffer, and replace on the Dynal MPC-1. Finally, resuspend in a small volume of coating medium (e.g. the volume originally pipetted from the vial).
2. *Antibody coating of CD8+ T cells.* Resuspend washed T cells in 10 ml coating medium at 2×10^7 cells ml^{-1} in a 15-ml conical tube, and add 1 ml anti-CD8 monoclonal antibody at a 10× saturating concentration. Incubate for 30 min at 4°C with gentle tilting and rotation (e.g. in the mixing device).
3. Wash twice in coating buffer (centrifugation at 1000 rpm, 4°C) to remove unbound antibody.
4. Add the suspension of washed Dynabeads and incubate for 30 min at 4°C with gentle tilting and rotation (e.g. in the mixing device) to keep cells and beads in suspension.
5. Place the tube in a Dynal MPC and leave it to rest for 2 min to magnetically remove the CD8+ cells labelled by antibody and coated with beads. Transfer the unbound cells to a fresh tube, perform a second magnetic separation, count, and resuspend in suppl. RPMI-1640/10% HUS. Negatively selected cells obtained by this method are unstimulated, pure and of high yield.

(contd.)

6. For recovery of positively selected CD8$^+$ T cells, remove the tube from the Dynal MPC, and wash the rosetted cells by resuspending in RPMI-1640/10% HUS. Repeat step 5 two times. These positively selected cells can be removed from the beads by a process of detachment (see DETACHaBEAD system according to the manufacturer's (Dynal) instructions).

◆◆◆◆◆◆ ESTABLISHMENT OF T-CELL LINES

T-cell lines are cultures of T-lymphocytes grown by repeated cycles of stimulation, usually with antigen, antigen-presenting cells and growth factors such as interleukin-2 (IL-2). In the primary reaction, antigen is encountered and processed by the antigen-presenting cells (monocytes, macrophages, dendritic cells, B-cells). The processed antigen is presented to T-lymphocytes, which then secrete a variety of T-cell growth factors, primarily IL-2, which are required for the expansion of both CD4$^+$ and CD8$^+$ T cells. For the *in vitro* induction of bacteria-specific T-cell lines, it is important to note that different preparations of the bacterial antigens (e.g. whole viable or killed bacteria, soluble bacterial proteins, usage of bacteria infected antigen-presenting cells) will selectively or predominantly expand different subsets of T-lymphocytes. To generate distinct T-cell ligands, the host cell can process microbial pathogens from two major compartments within the cell. Bacteria that normally proliferate outside the cell may secrete bacterial toxins and soluble proteins, which can be internalized by antigen-presenting cells via endocytosis. These and other bacterial degradation products in the vesicular compartment (endosomes and lysosomes) will deliver peptides that are bound to major histocompatibility complex (MHC) class II molecules. The peptide–MHC class II complex is recognized by CD4$^+$ T cells. When using experimental strategies for identification of immunologically important proteins of a pathogen, it should be remembered that the native conformation of the bacterial proteins need not be preserved during their isolation, since the T-cells are stimulated with linear peptides of proteins complexed with MHC molecules (Shimonkevitz *et al.*, 1983). CD8$^+$ T cells appear to be a critical component of host resistance to intracellular bacteria. Some bacteria (e.g. *Listeria monocytogenes*) replicate in the cytosol or in the contiguous nuclear spaces. Bacterial peptides thus originating in the cytosol are loaded on MHC (HLA) class I molecules in the endoplasmic reticulum. A stable complex of MHC class I molecule, β2-microglobulin and peptide is expressed on the cell surface, where it is recognized by the antigen-specific receptor (TCR) of CD8$^+$ T cells. Interestingly, even *Salmonella typhimurium*, a facultative intracellular bacterium confined to vacuolar compartments with no active egress into the cytosol, can induce CD8$^+$ T-cell responses. This has been shown by Pfeifer *et al.* (1993) who demonstrated an alternative class I processing pathway of *Salmonella* antigens suggesting that post-Golgi class I MHC molecules might be used for passing antigenic peptides to the cell surface and presenting them to CD8$^+$ CTL. In human HLA class I-associated disease, we have characterized and propagated HLA-B27 restricted CD8$^+$ T-cell lines and clones with specificity for *Yersinia* and *Salmonella* spp. (Hermann *et al.*, 1993). In the latter experiments, bacteria infected cells can be used as reagents to propagate or screen lymphocytes for responsiveness.

After PBL stimulation with live mycobacteria (Kabelitz *et al.*, 1990; Havlir *et al.*, 1991) or with other live bacteria such as *Salmonella* spp. (Havlir *et al.*, 1991), γδ-TCR$^+$ cells are found to be the most prominent population in the resulting T-cell cultures. An enrichment of TCR-γδ$^+$ cells in T-cell lines upon stimulation with live bacteria is even more prominent in synovial lymphocyte cultures as compared with peripheral blood T-cell cultures (Hermann *et al.*, 1992).

It is the purpose of this chapter to describe strategies to stimulate human T cells within a mononuclear cell population, and to obtain T-cell lines with specificity for bacteria or their distinct products. Different protocols will lead to predominant expansion of either αβ-TCR$^+$ CD4$^+$, αβ-TCR$^+$ CD8$^+$, or γδ-TCR$^+$ lymphocytes. In all protocols described here, the culture medium is supplemented with human AB serum instead of fetal calf serum, since bovine serum proteins may stimulate human T lymphocytes and therefore induce FCS specific T-cell lines.

Preparation of bacteria for *in vitro* stimulation of MNC

For *in vitro* stimulation of MNCs it is possible either to use whole bacteria (viable or inactivated) or to use bacterial fractions. The approach used to generate bacterial preparations will have important implications on the experiment.

We focus here on methods for generation of bacterial preparations from *Yersinia enterocolitica* that are straightforward and may be applied to a wide range of experimental situations (Probst *et al.*, 1993a; Duchmann *et al.*, 1995, 1996). In the first two sections, the generation of preparations from whole *Y. enterocolitica* by heat inactivation, sonication or shearing force using beads is described. If available, a French press can be used to disrupt bacterial cells by applying a shearing force. We then give an example of how bacterial cell fractions can be obtained and further purified in order to characterize a clonal T-cell response in more detail, i.e. eventually to define specific antigen(s). When trying to establish your own approach, it may be necessary to consult detailed references (Apicella *et al.*, 1994; Ferro-Luzzi Ames, 1994; Fischetti, 1994; May and Chakrabartry, 1994; Nikaido, 1994; Rosenthal and Dziarski, 1994; Vann and Freese, 1994). Attention should also be paid to the fact that any procedure which includes the use of substances that are toxic to cellular assays needs to be excluded.

Heat-killed bacteria

Equipment and reagents

- Bacterium of interest.
- Appropriate reagents and equipment for culture, e.g. CIN agar (Invitrogen, Life Technologies); Brain heart infusion (BHI) broth (Merck, Industrial Chemicals, Hawthorne, NY, USA), Autoclave, Erlenmeyer flask; spectrophotometer.

Procedures

Y. enterocolitica 03 or *09* are first grown on *Yersinia*-selective agar plates (CIN), cultivated in BHI broth for 24 h at 27°C, and harvested at the late exponential phase. Suspend the bacteria in 10 ml PBS and wash twice. Resuspend the washed bacteria in 10 ml PBS and autoclave for 20 min at 121°C. After short vortexing of autoclaved bacterial extract, allow the crude components to sediment. Collect the supernatant and dilute with PBS to an OD $_{(600\,nm)}$ of 0.5.

Pitfalls

It has been shown that the cultivation temperature influences the protein expression of *Y. enterocolitica* (Brubaker, 1991) and that heat (1 h, 42°C) induces increased production of 'heat-shock' proteins (Gething and Sambrook, 1992). Heat-killed bacteria may activate rather γδ-TCR$^+$ than αβ-TCR$^+$ CD4$^+$ T cells in some blood donors (Munk *et al.*, 1990) particularly in synovial MNC cultures (Hermann *et al.*, 1992).

Bacterial sonicates

Equipment and reagents

- Bacterium of interest.
- Appropriate reagents and equipment for culture; BHI broth (Merck, Industrial Chemicals); buffer S (Hepes 0.02 M, DNase 0.05 mg ml^{-1} (Sigma-Aldrich), RNase (0.05 mg ml^{-1}) (Sigma), equilibrated at pH 7.5 using 5N NaOH); sonicator (e.g. Branson Sonifier 250/450 or Labsonic L/U, B. Braun Biotech International); ice, sterile tubes (e.g. Falcon), sterile filter (45 µm).

Procedures

1. For bacterial culture, incubate 4 ml bacteria in 50 ml BHI and 1.5% agar (BD Biosciences) sterilized at 121°C for 15 min, and grow
   (contd.)

overnight in a bacterial shaker at 37°C. The next day, centrifuge the bacterial cultures at 9000 rpm and wash twice in 0.02 M Hepes buffer pH 7.5.

2. To disrupt bacteria, dissolve the bacterial pellet in 5-ml ice-cold buffer S in a 15-ml tube. Surround the tube with ice and sonicate the bacterial suspension four times for 30 s.

3. Sediment non-disrupted cells and cell detritus by centrifugation at 9000 rpm, and filter the supernatant using a 45 μm sterile filter. Determine the protein concentration by the Lowry method (e.g. Bio-Rad protein assay, Bio-Rad Laboratories, Hercules, CA, USA). The use of 50 μg ml^{-1} in lymphocyte assays is a good starting point.

Pitfalls

Sonication is a quick and efficient method for disintegration of cells and avoids use of reagents that might be toxic in cellular assays. However, in some situations, one may find that the procedure is influenced by many variables and therefore difficult to standardize. Factors like the density of the bacterial suspension (which can be standardized using a spectro-photometer), the type of sonicator, the type of tip used to transduce the oscillations into the buffer, the shape of the tube containing the bacterial suspension, the temperature during sonication, may influence the completeness of the cellular disintegration process and thus the particle size and antigenicity. Incomplete disintegration of cells and contamination of lymphocyte cell cultures by sonicates that are not sterile can be a problem. Using 4×30 s or 6×20 s may be a good start, but titration of conditions for the individual bacterium and experiment may be necessary to achieve optimal results.

Bacterial disintegration using beads

Equipment and reagents

- Bacterium of interest.
- Appropriate reagents and equipment for culture; buffer S (Hepes 0.02 M, DNase 0.05 mg ml^{-1} (Sigma-Aldrich), RNase (0.05 mg ml^{-1}) (Sigma-Aldrich), equilibrated at pH 7.5 using 5N NaOH), Beads (e.g. 0.1 mm), bead-beater (Biospec Products); ice; appropriate tube or vial; sterile filter (45 μm).

Procedures

1. Using an apparatus for disintegration of small amounts of cells (Mini-Bead Beater Model 3110BX, Biospec Products), growth of

(*contd.*)

bacteria described in step 1 for preparation of bacterial sonicates (see above) will yield more bacteria than needed. Using an apparatus for larger amounts of cells (Biospec Products), the procedure should be upscaled.

2. Using the Mini-Bead beater, fill a 2-ml screw cap vial half to one-third full with beads (0.1 mm). Fill the remaining vial volume with buffer S and bacterial cells (up to 40% wet weight of the solution volume).

3. Using a speed of 5000 rpm the procedure is performed for 6×30 s with intervening cooling on ice.

4. Sediment non-disrupted cells, cell detritus and beads by centrifugation at 9000 rpm, and filter the supernatant using a 45 µm sterile filter. Determine the protein concentration using the Lowry method (Bio-Rad protein assay, Bio-Rad). Use of 50 µg ml^{-1} in lymphocyte assays may be a good starting point.

Pitfalls

Pitfalls encountered using beads are similar to those that arise using a sonicator. Variables like density of the bacterial cell suspension, the size of the beads, the speed used in centrifugation, the total time of the procedure and the choice of time intervals may be optimized for the individual experiment.

Bacterial cell fractions

This section gives an example of how bacterial cells that have been disintegrated using beads or sonication can be fractionated further. In this protocol, ultracentrifugation results in separation of cytoplasm/periplasm from membranes. The cytoplasm/periplasm fraction is then purified further by precipitation with ammonium sulfate, gel chromatography and reverse-phase HPLC (Probst *et al.*, 1993b). Membranes are purified into outer and inner membranes by saccharose density gradient centrifugation (Hancock and Nikaido, 1978).

Equipment and reagents

- Equipment and reagents for disintegration of cells, as described above.
- HEPES (Biochrom AG), lysozyme (Sigma-Aldrich), saccharose (Sigma-Aldrich), ammonium sulfate ($(NH_4)_2SO_4$).
- Centrifuge (SW41 and 60 Ti rotor, Beckmann); chromatography unit.

Procedures

1. Disintegrate the bacteria (e.g. *Y. enterocolitica*) using beads or sonication, as described above.
2. Ultracentrifuge the crude bacterial extract (200 000*g*, 60 min, SW41 rotor) and fractionate into cytoplasm/periplasm (supernatant) and membranes (pellet).
3. Fractionate membranes (pellet) into outer and inner membranes by saccharose density-gradient centrifugation. Incubate the pellet containing the membranes in 0.02 M HEPES with 0.05 mg ml^{-1} lysozyme for 30 min, and equilibrate in 11% (v/v) saccharose. Centrifuge a first gradient consisting of 5 ml suspension containing the membranes/11% (v/v) saccharose, 6 ml 15% (v/v) saccharose and 1 ml 70% (v/v) saccharose (39 000 rpm, 1 h, 40°C, SW41 rotor). Overlay the lower 2 ml of the gradient (white-yellow ring) on a second gradient consisting of 1.2 ml 70%, 3 ml 64%, 3 ml 58% and 3 ml 52% saccharose (v/v). Centrifuge this gradient (39 000 rpm, 16 h, 40°C, SW41 rotor) and separate it into four fractions. Wash the total membranes and fractions obtained from the second gradient twice in 0.02 M Hepes pH 7.5 (41 000 rpm, 1 h, 40°C, 60 Ti rotor). Resuspend the pellets in 1–2 ml 0.02 M Hepes pH 7.5, sterile filter and store frozen.
4. Further fractionate cytoplasm/periplasm (supernatant) by sequential ammonium sulfate precipitation using saturations of 0–26%, 26–38%, 38–51%, 51–54% and 64–78% ammonium sulfate. The different concentrations of ammonium sulfate are prepared using saturated $(NH_4)_2SO_4$ solution in 0.05 M Tris/HCl pH 7.5. Precipitate the proteins for 15 min at 40°C while stirring. After centrifugation (10 000 *g*, 20 min, 40°C), resuspend the pellet in 0.05 M Tris/HCL pH 7.5 and desalte using Sephadex G25 columns (Amersham Pharmacia Biotech).
5. For further biochemical fractionation, desalted proteins can be subjected to gel-chromatography using, for example, Sephacryl S300 Superfine columns (Amersham Pharmacia Biotech), anion exchange chromatography (e.g. using Mono Q columns, Amersham Pharmacia Biotech) or subjected to reverse-phase HPLC.

In vitro generation and restimulation of antigen-specific T-cell lines

Protocol for predominant expansion of CD4$^+$ (and γδ-TCR$^+$) bacteria specific T-cell lines

Equipment and reagents

- MNC population isolated by Ficoll-Hypaque density centrifugation from heparinized peripheral blood, synovial fluid, synovial membrane, gut mucosa, or from other source.

(contd.)

- Suppl. RPMI-1640; human AB serum (HUS, heat inactivated); recombinant Interleukin-2 (rIL2, e.g. Invitrogen Life Technologies); bacterial antigen preparation of interest (heat-killed bacteria, bacterial sonicates, or bacterial cell fractions, see pages 636–640).
- Sterile 24-well flat-bottomed microtitre plates (e.g. Costar, Cambridge, USA); 15-ml conical centrifuge tubes (e.g. Greiner or Falcon); humidified 37°C, 5% CO_2 incubator (e.g. Heraeus). ^{137}Cs source for irradiation.

Note: When using whole heat-killed bacteria or viable bacteria (which will be killed by antibiotics after short incubation with the MNC), γδ-TCR$^+$ rather then αβ-TCR$^+$ CD4$^+$ cells may be stimulated and expanded in some donors.

Procedure (Probst *et al.*, 1993a)

1. *Antigen specific stimulation – I.* Resuspend unfractionated, washed MNCs in suppl. RPMI-1640/10% HUS (1.5×10^6 cells ml^{-1} of culture medium). Place 3×10^6 MNCs in the wells of a 24-well microtitre plate at a volume of 2 ml suppl. RPMI-1640/10% HUS and add the soluble bacterial antigen at an optimal concentration. (This optimal concentration (usually between 5 and 50 μg of protein ml^{-1}) should have been determined before starting the generation of T-cell lines, using a standard lymphocyte proliferation assay.) Incubate plates for 8 days in a humidified 37°C, 5% CO_2 incubator. After 3 days of culture, gently pipette off the medium fluid from the wells. Replace the medium by rIL-2 (20 U ml^{-1}) supplemented RPMI/10% HUS (without antigen). Inspect wells for growing cells microscopically and for the colour of the medium every 2 days. If necessary (colour changing to yellow), split the cultures using fresh rIL-2 (20 U ml^{-1}) supplemented RPMI/10% HUS culture medium.
2. *Antigen-specific restimulation – II.* Restimulate the resulting T-cell line with antigen and with autologous irradiated peripheral blood MNCs as feeder cells. (To obtain the feeder cells, prepare blood MNCs from the same T-cell donor by Ficoll-Hypaque density centrifugation as described above.) Count the washed MNCs, irradiate (4000 rad from a ^{137}Cs source), suspend in RPMI-1640/10% HUS/20 U ml^{-1} rIL-2 ($2–3 \times 10^6$ cells ml^{-1}), and keep on ice until needed. (For ease of manipulation without having to draw blood several times from the same donor, larger numbers of blood MNCs can be irradiated, aliquoted, cryopreserved and recovered as needed.) Pool cells from all wells of the 24-well microtitre plate in 15-ml conical centrifuge tubes. Fill tubes with suppl. RPMI-1640/5% HUS and centrifuge 10 min at 1300 rpm (400g) and 18°C. After removing the supernatant, resuspend the cell pellet in suppl. RPMI-1640/10% HUS/20 U ml^{-1} r-IL2 at 1×10^6 cells ml^{-1}.

(contd.)

Isolation of T Cells

> Pipette 1 ml into each well of a 24-well microtitre plate and co-incubate these responder T cells with $2–3 \times 10^6$ irradiated autologous feeder cells at a total volume of 2 ml. Add the optimal concentration of the bacterial antigen preparation and incubate the plates in a humidified 37°C, 5% CO_2 incubator. Feed the cultures with suppl. RPMI-1640/10% HUS/20 U ml^{-1} r-IL2 medium, or split the cultures if necessary.
>
> 3. After 8 days, wash the T-cell line and restimulate it as described in step 2. By checking the wells microscopically, one may detect many dead cells or detritus. The detritus can then be removed by purifying viable T cells by Ficoll-Hypaque density gradient centrifugation. Co-incubate the T cells with autologous feeder cells and antigen for an additional 8 days.
>
> 4. At this point (day 24 of T-cell culture), analysis of phenotypes (e.g. FACS analysis) and antigen specificity (e.g. lymphocyte proliferation assay) of the T-cell line can be done. Furthermore, T-cell clones can be prepared from the T cell line (see page 646).

Protocol for bacteria-specific MHC class I restricted CD8+ specific T-cells

Certain human rheumatic diseases, namely the so-called spondyloarthropathies, are strongly associated with the MHC class I molecule HLA-B27 and with antecedent infection with Gram-negative bacteria. The latter is proven in *Yersinia*, *Shigella*, *Salmonella* and *Chlamydia*-induced reactive arthritis, but is also strongly implicated in ankylosing spondylitis. The function of HLA-B27 in presentation of bacterial and self-peptides is considered to be crucial to the pathogenesis of the spondyloarthropathies. This theory has been supported by the identification of synovial fluid cytotoxic T-lymphocyte (CTL) clones that specifically killed *Yersinia* or *Salmonella* infected HLA-B27+ target cells (Hermann *et al.*, 1993). It is known, however, that the stimulation of synovial fluid MNCs with soluble bacterial antigens produces CD4+ T-cell lines, and that stimulation with live bacteria leads to an enrichment of TCR-$\gamma\delta^+$ cells (up to 90% of all CD3+ T cells), whereas bacteria-specific MHC class I restricted cytotoxicity is not detectable in these T-cell cultures (Hermann *et al.*, 1992). Therefore, to generate MHC class I-restricted *Yersinia* or *Salmonella* responsive CTL lines, one has to apply specific culture conditions that are specific to the facultatively intracellular nature of the bacteria, and the antigen should be presented optimally for class I, that is endogenously, for example by infection of human MHC class I transfected cell lines with live bacteria. We describe here the generation of HLA-B27 restricted CTL lines with specificity for *Y. enterocolitica* as an example for the propagation of MHC class I-restricted bacteria specific CTL.

Equipment and reagents

- Synovial fluid MNCs derived from an inflamed joint of a patient (HLA-B27[+]) with *Yersinia*-induced reactive arthritis (for the isolation of synovial fluid MNC by Ficoll-Hypaque density gradient centrifugation, see page 624).
- HLA-B*2705- and human β2-microglobulin-transfected mouse L-cells ([B27-L-cells] described by Kapasi and Inman, 1992).
- A patient isolate of *Y. enterocolitica O3* or *O9*, grown for 24 h at 27°C in BHI broth (Merck, Industrial Chemicals).
- PBS; DMEM (Invitrogen, Life Technologies) supplemented with 100 U ml^{-1} penicillin, 100 μg ml^{-1} streptomycin (Pen-Strep, Biochrom), and HAT-supplement (Invitrogen Life Technologies). RPMI-1640 medium without antibiotics; gentamycin (Merck, Industrial Chemicals), suppl. RPMI-1640 (see page 622); FCS, heat inactivated; human AB serum (HUS, heat inactivated); rIL2, (e.g. Invitrogen Life Technologies); Ficoll-Hypaque (Histopaque-1077^R, Sigma-Aldrich).
- 15-ml conical centrifuge tubes (e.g. Greiner); sterile 12-well flat-bottomed plates (e.g. Costar), sterile 96-well round-bottomed microtitre plates (e.g. Greiner); sterile rubber scraper; temperature-controlled centrifuge with GH-3.7 horizontal rotor (e.g. Heraeus or Beckman); humidified 37°C, 5% CO_2 incubator (e.g. Heraeus).

Procedures

1. *Growth and infection of stimulator cell line* (Hermann *et al.*, 1993; Huppertz and Heesemann, 1996). Grow HLA-B*2705- and human β2-microglobulin-transfected mouse L-cells (B27-L-cells) in 12-well plates in DMEM containing antibiotics (penicillin/streptomycin), 10% FCS and HAT supplement. Cultivate *Y. enterocolitica* BHI broth for 24 h at 27°C, and harvest at the late exponential phase. Wash a bacterial suspension and resuspend in PBS, diluted to an OD (600 nm) of 0.1. Thereafter dilute 1:20 in RPMI-1640 without antibiotics (at a multiplicity of infection of 50 yersiniae per L-cell, or, 5×10^6 yersiniae per 10^5 L-cells). Wash subconfluent adherent monolayers of B27-L-cells with RPMI-1640 (without antibiotics). For B27-L-cell infection, add 2 ml of the suspension of live *Y. enterocolitica* in RPMI-1640 (without antibiotics) to each well of the 12-well plate, and incubate at 37°C, 5% CO_2 in a humidified atmosphere. After 2 h of incubation, add gentamycin at 100 μg ml^{-1} to the cultures to kill extracellular bacteria. After an additional 1 h of culture, carefully remove the supernatant from the adherent cells by using a sterile pipette. Gently wash the wells three times by adding and removing 3 ml RPMI-1640.

(*contd.*)

Isolation of T Cells

2. *Stimulation of synovial fluid MNCs* (Hermann *et al.*, 1993): The *Yersinia*-infected B27-L cells are used as stimulators for synovial fluid MNCs derived from a HLA-B27$^+$ patient. Add 2×10^6 MNCs in 2 ml suppl. RPMI-1640/10%HUS to each of the wells containing *Yersinia*-infected B27-L cells, and incubate in a humidified 37°C, 5% CO_2 incubator. After 24 h of co-culture, L-cell detritus must be removed from the human responder cells. The supernatant containing the non-adherent human cells, killed bacteria and cell detritus is pooled from all wells of the 12-well plate in 15-ml conical centrifuge tubes. In addition, gently scrape the remaining adherent cells (i.e. L-cells with attached human cells) from the wells using a sterile rubber scraper. Pool the cells and centrifuge the suspensions through a Ficoll-Hypaque density gradient to recover the viable human cells at interface, and wash twice in suppl. RPMI-1640/5% FCS. Resuspend the cell pellet in suppl. RPMI-1640/10%HUS/20 U ml^{-1} rIL2 at 1×10^6 cells ml^{-1}. Pipette 100 µl cell suspension into the wells of a 96-well round-bottomed microtitre plate and incubate for 8 days. If necessary, feed the cultures with suppl. RPMI-1640/10% HUS/20 U ml^{-1} rIL2 medium or split the cultures.
3. *Restimulation of the T-cell line.* Prepare *Yersinia*-infected B27-L cells as described in step 1. Pool the cells from all wells of the 96-well plate in 15-ml conical centrifuge tubes, and fill tubes with suppl. RPMI-1640/5% HUS. Centrifuge for 10 min at 1300 rpm (400*g*) and 18°C. After removing the supernatant, resuspend the cell pellet in suppl. RPMI-1640/10% HUS/20 U ml^{-1} rIL2 at 1×10^6 cells ml^{-1}. Add 2 ml of this cell suspension to each of the wells containing *Yersinia*-infected B27-L cells and incubate in a humidified 37°C, 5% CO_2 incubator. After 24 h of co-culture, purify the T-cell lines by Ficoll-Hypaque density gradient centrifugation, and wash and seed them into the wells of a 96-well round-bottomed plate as described in step 2.
4. After another 8 days of culture, CD8$^+$ T cells are negatively selected from the T-cell line by indirect antibody panning (see page 632). These CD8$^+$ T cells can now be cloned using a non-specific cloning protocol (see page 645); an antigen-specific protocol is not useful in this instance.

◆◆◆◆◆◆ ESTABLISHMENT OF T-CELL CLONES

A T-cell clone is a continuously growing line of T cells derived from a single progenitor cell. The generation of T-cell clones helps in the identification of even minor fractions of T-cell populations and the investigation of effector functions that are stable within each line. Similar to T-cell lines, T-cell clone cells must be stimulated periodically with antigen and antigen presenting cells to maintain growth. However, if the antigen is not yet known or not available in sufficient quantities, non-specific methods of restimulation of T-cell clones can be used (see pages 648–650).

Representative cloning protocol

This protocol provides direct cloning of T cells (from unseparated MNCs) without prior antigen-specific *in vitro* stimulation or selection for certain T-cell phenotypes (Fleischer and Bogdahn, 1983; Hermann *et al.*, 1989).

Equipment and reagents

- MNCs to be cloned (from peripheral blood or other sources), isolated by Ficoll-Hypaque density gradient centrifugation (as described on page 622), or purified T-cell population (E+ cells, see page 630).
- Allogeneic blood MNCs (feeder cells), isolated by Ficoll-Hypaque density gradient centrifugation from a buffy coat.
- Suppl. RPMI-1640; human AB serum (HUS), heat-inactivated; rIL-2 (e.g. Invitrogen Life Technologies), Phytohaemagglutinin (PHA-P) (e.g. Biochrom AG).
- A 450-nm filter; plastic flasks; 15-ml conical centrifuge tubes (e.g. Greiner or Falcon); Terasaki microtest plates (Nunc, Roskilde, Denmark); 96-well roundbottom microtitre plates (Nunc), 24-well flatbottom plates (e.g. Costar, Cambridge, USA); humidified 37°C, 5% CO_2 incubator (e.g. Heraeus); ^{137}Cs source for irradiation.

Procedure

1. *Preparation of conditioned medium (T-cell growth factor).* T-cell growth factor (TCGF) is generated from PHA-P activated allogeneic blood MNCs. Suspend MNCs at 1×10^6 ml^{-1} in suppl. RPMI-1640/2% HUS containing 1 µg ml^{-1} PHA-P and incubate the cells in plastic flasks in a humidified 37°C, 5% CO_2 incubator, each flask containing 200 ml cell suspension. After 24 h, collect the supernatant and pass it through a 0.45 µm filter.
2. *Preparation of feeder cells.* Irradiate allogeneic peripheral blood MNCs (feeder cells) (4000 rad from a ^{137}Cs source) and resuspend in TCGF at a cell concentration of 1×10^6 cells ml^{-1}. To this cell suspension add 1 µg ml^{-1} PHA-P. Plate 10 µl of the cell suspension into each well of 60-well Terasaki plates. (We recommend that at least 25 plates (i.e. 1500 wells) are prepared in order to achieve a sufficient number of growing T-cell clones.)
3. *Preparation and seeding of cells to be cloned.* Count the MNCs (T cells) to be cloned several times, resuspend in suppl. RPMI-1640/10% HUS/20 U ml^{-1}, rIL-2 at 5×10^2 cells ml^{-1} (= 5 cells per 10 µl) and mix vigorously before plating at limiting dilution in the prepared Terasaki plates. (The limiting dilution can be determined by seeding the T cells at 5, 1, and 0.5 cells per well in 10 µl medium.)
4. *Incubation.* Incubate the plates 8–12 days in a humidified 37°C, 5% CO_2 incubator. Inspect the wells for growing clonal T-cell colonies microscopically.

(contd.)

5. Transfer growing T-lymphocyte clones (TLCs) to larger plates (96-well microtitre plates) and restimulate. Prepare irradiated allogeneic feeder cells at 1×10^6 cells ml^{-1} of TCGF/1 µg ml^{-1} PHA-P and pipette 100 µl into the wells of 96-well microtitre plates (1×10^5 cells per well). Each growing TLC is transferred to one of the prepared wells. During the following days of incubation, it is important to inspect the wells microscopically every 2 days in order to feed or split the colonies with culture medium (suppl. RPMI-1640/10% HUS/20 U ml^{-1} rIL-2) if necessary, or to transfer them to 24-well plates with additional culture medium.
6. After 10 days of culture, test clones for antigen specificity or phenotype (see page 647).
7. TLCs of interest are propagated further by restimulation with rIL-2 and oxidized stimulator cells (see page 651).

Establishment of T-cell clones from bacteria-specific T-cell lines

Equipment and reagents

- Antigen-specific T-cell line to be cloned (generated as described on page 640). Autologous blood MNCs (feeder cells), isolated by Ficoll-Hypaque density gradient centrifugation from heparinized peripheral blood.
- Suppl. RPMI-1640; bacterial antigen at an optimal concentration; HUS, heat inactivated; rIL-2 (Invitrogen Life Technologies).
- 15-ml conical centrifuge tubes (e.g. Greiner or Falcon); 96-well round-bottomed microtitre plates (e.g. Nunc), 24-well flat-bottomed plates (e.g. Costar, Cambridge, USA); humidified 37°C, 5% CO_2 incubator (e.g. Heraeus); ^{137}Cs source for irradiation.

Procedure

1. *Preparation of feeder cells.* Irradiate (4000 rad from a ^{137}Cs source) autologous peripheral blood MNCs (feeder cells) and resuspend them in suppl. RPMI-1640/10% HUS/20 U ml^{-1} rIL-2 at a cell concentration of 1×10^6 cells ml^{-1}. Add the bacterial antigen preparation at optimal concentration and pipette 100 µl cell suspension into the wells of 96-well microtitre plates (1×10^5 cells per well).
2. *Preparation and seeding of cells to be cloned.* Count the T-cells to be cloned several times, resuspend them in suppl. RPMI-1640/10% HUS/20 U ml^{-1}, rIL-2 at 1×10^2 cells ml^{-1} (i.e. 5 cells per 50 µl), add the bacterial antigen preparation at optimal concentration, and mix vigorously before plating them at limiting dilution in the prepared 96-well plates. (The limiting dilution can be determined by seeding the T-cells at 5, 1, and 0.5 cells per well in 50 µl culture medium.)

(*contd.*)

3. *Incubation.* Incubate plates 8 days in a humidified 37°C, 5% CO_2 incubator. After that time, inspect wells for growing clonal T-cell colonies microscopically. Between day 8 and day 12 of culture, growing TLCs can be fed with medium (suppl. RPMI-1640/10% HUS/20 U ml^{-1} rIL-2) or split in a new 96-well microtitre plate.
4. Growing TLCs are then transferred to larger (24-well) plates and specifically restimulated with irradiated autologous MNCs and antigen (see page 641). Split and transfer to 12-well plates if necessary.
5. After another 10 days of culture, test clones for antigen specificity and/or phenotype (see page 647).
6. TLCs of interest are propagated further by non-specific restimulation with rIL-2 and oxidized stimulator cells (see page 650).

Specific and non-specific restimulation of T-lymphocyte clones

Long-term growth of human TLCs with proliferative activity is not possible in IL-2 containing medium alone, since with increasing time after stimulation of the TLCs the expression of activation antigens such as IL-2 receptor and HLA-DR antigens decreases, and the cells lose their blastoid appearance and return to a resting state (Fleischer, 1988). Further periodical stimulation with specific antigen and autologous antigen-presenting cells is then required in order to obtain IL-2 receptor expression and a response to IL-2, and thus maintain long-term growth *in vitro*. Alternatively, methods for a non-antigen-specific stimulation of antigen-dependent proliferative or cytotoxic human TLCs can be applied.

Specific restimulation of TLC

In cases in which the specificity of the TLCs to be propagated is known and the appropriate antigen is available (e.g. crude preparations of heat-killed or sonicated bacteria, or defined bacterial antigens), periodic stimulation of the TLCs is achieved by recognition of antigen and MHC-encoded molecules. Autologous irradiated PBLs are used as antigen-presenting cells. If HLA restriction molecules of antigen recognition are known, HLA-matched allogeneic PBLs can be used as well. TLCs are restimulated in intervals of 7–9 days.

> **Equipment and reagents: preparation of autologous antigen presenting cells and for antigen-specific stimulation of TLCs**
>
> - Heparinized autologous peripheral blood; cloned T-cells to be restimulated.
> - Ficoll-Hypaque (Histopaque-1077^R, Sigma-Aldrich); suppl. RPMI-1640; FCS, heat-inactivated; HUS, heat-inactivated; rIL-2 (e.g. Invitrogen, Life Technologies).
>
> *(contd.)*

- 50-ml conical centrifuge tubes (e.g. Greiner); 15-ml conical centrifuge tubes (e.g. Falcon); 24-well and 12-well flat-bottomed plates (Costar); ^{137}Cs source for irradiation; temperature-controlled centrifuge; humidified 37°C, 5% CO_2 incubator (e.g. Heraeus).

Procedure

MNCs are obtained from heparinized autologous whole blood by Ficoll-Hypaque density gradient centrifugation and washed as described on page 622. The cells are counted, resuspended in suppl. RPMI-1640/10% HUS/20 U ml^{-1} rIL-2 (1×10^7 cells ml^{-1} in a 15-ml conical centrifuge tube) and kept on ice. Before use as stimulator cells, the PBLs are irradiated (4000 rad from a ^{137}Cs source). The cloned T cells are washed with suppl. RPMI-1640/5% FCS, counted and resuspended in IL-2-medium (1×10^6 cells ml^{-1}). These TLCs (1×10^6) are incubated with 4×10^6 irradiated autologous or HLA-matched antigen-presenting cells in 3 ml of rIL-2-medium in 12-well plates. For smaller numbers of cloned T cells, 0.5×10^6 T-cells are incubated with 2×10^6 to 3×10^6 irradiated antigen-presenting cells in 2 ml suppl. RPMI-1640/10% HUS/20 U ml^{-1} rIL-2 medium in 24-well plates. The appropriate bacterial antigen is added in an optimal concentration. (The optimal concentration is determined in a standard lymphocyte proliferation assay using antigen concentrations of 1–100 µg ml^{-1} in suppl. RPMI-1640/10% HUS.) The plates are cultured in an incubator at 37°C, 5% CO_2 in a humidified atmosphere. IL-2 medium is added to the cultures every 2–3 days and growing cultures are splitted with IL-2 medium if necessary.

Pitfalls

It has been reported that CD4$^+$ TLCs may lose their proliferative function when cultivated for longer than 30–35 population doublings (Pawelec *et al.*, 1986). The growth rate usually slows down after 8–10 weeks if stimulated with antigen and autologous PBLs alone. In many cases, such clones can be readily expanded using the non-specific stimulation protocol with oxidized stimulator cells (see below, page 650).

Non-specific propagation of T-lymphocyte clones

In many cases, such as during the investigation of functional properties of T cells cloned from biological fluids, the specificities of the T lymphocytes to be propagated are not known. Another major limitation of successful long-term culture of human antigen-specific, HLA-restricted TLCs can be the supply of sufficient autologous or HLA-typed cells for use as antigen-presenting cells. Therefore, alternative methods of TLC restimulation are required. In this section we describe the method of non-specific propagation of $\alpha\beta$-TCR$^+$CD4$^+$, $\alpha\beta$-TCR$^+$CD8$^+$, or $\gamma\delta$-TCR$^+$ TLC by mitogenic lectins (phytohaemagglutinin or concanavalin A), by anti-CD3 mAb, or by oxidized stimulator cells. For these methods, irradiated allogeneic peripheral blood MNCs are required as stimulator cells.

Equipment and reagents: preparation of allogeneic stimulator cells

- One buffy coat; Ficoll-Hypaque (Histopaque-1077^R, Sigma-Aldrich), RPMI-1640; FCS, heat-inactivated.
- 50-ml conical centrifuge tubes (e.g. Greiner); 15-ml conical centrifuge tubes (e.g. Falcon); a ^{137}Cs source for irradiation; temperature-controlled centrifuge; humidified 37°C, 5% CO_2 incubator (e.g. Heraeus).

Procedure

MNCs are obtained from buffy coats by Ficoll-Hypaque density gradient centrifugation and washed as described on page 622. Washed MNCs are counted, irradiated (4000 rad from a ^{137}Cs source), suspended in RPMI-1640/10% HUS/20 U ml^{-1} rIL-2 (2×10^6 to 3×10^6 cells ml^{-1}), and kept on ice until needed.

Non-specific stimulation of T-lymphocyte clones by mitogenic lectins or anti-CD3 moAb

Mitogenic lectins such as phytohaemagglutinin (PHA) or concanavalin A (ConA) as well as the mitogenic monoclonal antibody anti-CD3 can be used to stimulate TLCs in the presence of accessory cells.

Equipment and reagents

- Irradiated allogeneic MNCs (prepared as described above).
- Suppl. RPMI-1640; PHA-P (e.g. Biochrom AG); ConA (e.g. Biochrom AG); anti-CD3 monoclonal antibody (e.g. BD Biosciences Pharmingen); HUS, heat-inactivated; FCS, heat-inactivated.
- 12-well flat-bottomed plates (e.g. Costar); 15-ml conical centrifuge tubes (e.g. Falcon); humidified 37°C, 5% CO_2 incubator (e.g. Heraeus).

Procedure

Irradiated stimulator cells (3×10^6) are incubated with 5×10^5 cloned T-cells in suppl. RPMI-1640 in the presence of either PHA-P (1 µg ml^{-1}) or ConA (1 µg ml^{-1}) or anti-CD3 (0.1 µg ml^{-1}). Use HUS as medium supplement with lectins, and FCS with anti-CD3.

Pitfalls

The use of lectins or anti-CD3 to obtain long-term growth of TLC may have several disadvantages (Fleischer, 1988): (a) lectin accumulation on the surface of the responder T-cells may have toxic effects; (b) lectin or anti-CD3 antibody can lead to mutual killing if the clones possess cytotoxic activity; (c) lectin or anti-CD3 antibody bound on the cell surface can be carried into experimental assay systems and have disturbing effects in functional tests. In some cases, loss of function of restimulated TLC can be traced back to mycoplasma contamination of the T-cell clones. The risk of mycoplasma infection is especially high if tumour cell lines grown in FCS are incorporated into the stimulation cultures of TLCs or if FCS has to be used for stimulation, e.g. with the anti-CD3 monoclonal antibody (Padula et al., 1985).

Non-specific propagation of T-lymphocyte clones by oxidized stimulator cells

This method (Fleischer, 1988) is based on the selective modification of the stimulator cells by oxidation of surface galactose residues to reactive aldehydes. The stimulation of responder T cells in this system is due to a covalent cross-linking of stimulator and responder cells via Schiff bases. The oxidizing enzyme galactose oxidase is removed before adding the responding T cells and, therefore, has no disturbing effects in later assay systems.

Equipment and reagents

- Irradiated allogeneic MNCs (prepared as described on p. 649).
- RPMI-1640, Test-neuraminidase (Aventis Behring, King of Prussia, PA); galactose oxidase (Sigma-Aldrich); galactose (Sigma-Aldrich).
- 15-ml conical centrifuge tubes (e.g. Falcon); 12-well flat-bottomed plates (e.g. Costar); a temperature controlled centrifuge; humidified 37°C, 5% CO_2 incubator (e.g. Heraeus).

Procedure

Irradiated (4000 rad) allogeneic MNCs (4×10^7 ml^{-1}) are incubated with 0.02 U ml^{-1} neuraminidase and 0.05 U ml^{-1} galactose oxidase in serum-free RPMI-1640 for 90 min at 37°C (15-ml conical tubes). The cells are washed three times in suppl. RPMI-1640 containing 0.01M galactose to remove residual galactose oxidase bound to galactose residues on the cells. After the last wash, resuspend the cell pellet in suppl. RPMI-1640/10% HUS/rIL-2 20 U ml^{-1}. Oxidized stimulator cells (3×10^6) are then incubated with 5×10^5 cloned T-cells in 12-well plates at a total volume of 2 ml. Incubate the plates for 8 days, feed or split the cultures with suppl. RPMI-1640/10% HUS/rIL-2 20 U ml^{-1} medium, if necessary.

Phenotypic analysis of high numbers of T-cell clones

FACS analysis according to standard procedures with mouse anti-human monoclonal antibodies (e.g. anti-CD3, anti-$\alpha\beta$-TCR, anti-$\gamma\delta$-TCR, anti-CD4, anti-CD8) and FITC labelled second antibodies (goat anti-mouse immunoglobulin) is often used to phenotypically characterize T-cell clones and identify certain T-cell subsets of interest. When high numbers (>100) of T-cell clones are produced by the method described above, and when only a limited number of cloned T cells is available for phenotypic analysis, we recommend an enzyme-linked immunoassay (ELISA) for the detection of surface antigens on individual cells (Holzmann and Johnson, 1983), which has a sensitivity comparable to immunofluorescence and can be readily used for the investigation of T-cell populations.

Equipment and reagents

- Cloned T-cells suspended in PBS (approximately 1.5×10^4 cells in 100 μl).
- Poly-L-lysine (Sigma-Aldrich); glutaraldehyde (Merck, Industrial Chemicals); monoclonal antibodies (anti-CD3, anti-$\alpha\beta$-TCR, anti-$\gamma\delta$-TCR, anti-CD4, anti-CD8) either as undiluted antibody containing supernatant from hybridomas or as antibody dilutions from purchased reagents prepared in PBS/10% FCS; peroxidase coupled rabbit anti-mouse Ig (e.g. BD Biosciences); 3-amino-9-ethylcarbazol (Sigma-Aldrich); 0.002% H_2O_2; 0.05 M Tris-HCl pH 7.6; dimethylsulfoxide (DMSO) (Merck, Industrial Chemicals); storage buffer (PBS containing 60% glycerine) (Sigma-Aldrich) and 0.1% NaN_3 (Sigma-Aldrich).
- Terasaki microtest plates (e.g. Nunc).

Procedure

Terasaki microtest plates (one plate per monoclonal antibody tested) are treated with poly-L-lysine by flooding the plate with 10 ml PBS containing 25 μg ml^{-1} of poly-L-lysine and then washed with PBS (Holzmann and Johnson, 1983). All washing steps are performed by dipping and flicking, whereafter the plates are carefully blotted dry with a paper towel to avoid subsequent dilution of reagents. From each T-cell clone to be investigated, 10 μl of cells (approximately 1500 cells per well) are added to the wells of the Terasaki plates. The plates are centrifuged for 5 min at 90g (4°C) to produce an even monolayer of attaching cells, and fixed for 5 min at room temperature by flooding the plate with 10 ml of 0.20% glutaraldehyde in PBS. After three washes, 10 ml suppl. RPMI-1640/10% FCS are added for 1 h (37°C, 5% CO_2) to block any remaining protein binding sites. After another six washes, 10 μl undiluted monoclonal antibody-containing hybridoma supernatant or monoclonal antibody dilutions from purchased reagents prepared in PBS/10% FCS are added to the wells for 1 h at 4°C, after which the plates are washed 10 times.

(contd.)

A 1 : 50 dilution (total volume 10 µl) of peroxidase coupled goat anti-mouse Ig (in PBS/10% FCS) is added to each well for 15 min at room temperature. The plates are washed 10 times and then flooded with 10 ml of substrate solution consisting of 0.07 mg ml^{-1} 3-amino-9-ethylcarbazol, 0.002% H_2O_2 in 0.05 M Tris-HCl, pH 7.6. (The carbazol must initially be dissolved in 1/30 volume of DMSO.) It is important to note that the substrate solution should be freshly prepared (e.g. during the incubation time of the peroxidase coupled goat anti-mouse Ig). The cells are allowed to react for 30–45 min and the plates are then washed three times in PBS and flooded with storage buffer (PBS containing 60% glycerine, 0.1% NaN$_3$). Positive-staining cells can be detected by examining the wells with an inverted microscope.

References

Abrahamsen, T. G., Frøland, S. S., Natvig, J. B. and Pahle, J. (1975). Elution and characterization of lymphocytes from rheumatoid inflammatory tissue. *Scand. J. Immunol.* **4**, 823–830.

Apicella, M. A., Griffiss, J. M. and Schneider, H. (1994). Bacterial cell fractionation: isolation and characterization of lipolysaccharides, lipooligosaccharides, and lipid A. *Methods in Enzymology* **235**, 242–252.

Boyum, A. (1968). Isolation of mononuclear cells and granulocytes from human blood. *Scand. J. Lab. Invest.* **21**, 77–80.

Brubaker, R. R. (1991). Factors promoting acute and chronic diseases caused by Yersiniae. *Clin. Microbiol. Rev.* **4**, 309-324.

Bull, D. M. and Bookman, M. A. (1977). Isolation and functional characterization of human intestinal mucosal mononuclear cells. *J. Clin. Invest.* **59**, 966–974.

Burmester, G. R., Yu, D. T. Y., Irani, A.-M., Kunkel, H. G. and Winchester, R. J. (1981). Ia+ T cells in synovial fluid and tissues of patients with rheumatoid arthritis. *Arthritis Rheum.* **24**, 1370–1376.

Cerf-Bensussan, N., Guy-Grand, D. and Griscelli, C. (1985). Intraepithelial lymphocytes of human gut: isolation, characterisation and study of natural killer activity. *Gut* **26**, 81.

Duchmann, R., Kaiser, I., Hermann, E., Mayet, W., Ewe, K. and Meyer zum Büschenfelde, K.-H. (1995). Tolerance exists towards resident intestinal flora but is broken in active inflammatory bowel disease (IBD). *Clin. Exp. Immunol.* **102**, 448–455.

Duchmann, R., Märker-Hermann, E. and Meyer zum Büschenfelde, K.-H. (1996). Bacteria-specific T-cell clones are selective in their reactivity towards different enterobacteria or *H. pylori* and increased in inflammatory bowel disease. *Scand. J. Immunol.* **44**, 71–79.

Ferro-Luzzi Ames, G. (1994). Bacterial cell fractionation: Isolation and purification of periplasmic binding proteins. *Methods in Enzymology* **235**, 234–241.

Fiocchi, C., Battisto, J. R. and Farmer, R. G. (1979). Gut mucosal lymphocytes in inflammatory bowel disease. Isolation and preliminary functional characterization. *Dig. Dis. Sci.* **24**, 705–717.

Fischetti, V. A. (1994). Bacterial cell fractionation: purification of Streptococcal M protein. *Methods in Enzymology* **235**, 286–294.

Fleischer, B. (1983). Activation of human T lymphocytes. I. Requirements of mitogen-induced proliferation of antigen-specific T lymphocyte clones. *Eur. J. Immunol.* **13**, 970–976.

Fleischer, B. (1988). Non-specific propagation of human antigen-dependent T lymphocyte clones. *J. Immunol. Methods* **109**, 215–219.

Fleischer, B. and Bogdahn, U. (1983). Growth of antigen specific, HLA restricted T lymphocyte clones from cerebrospinal fluid. *Clin. Exp. Immunol.* **52**, 38–44.

Funderud, S., Nustad, K., Lea, T., Vartdal, F., Guardernack, G., Stensted, P. and Ugelstad, J. (1987). Fractionation of lymphocytes by immunomagnetic beads. In *Lymphocytes: A Practical Approach* (G. G. B. Klaus, Ed.) pp. 55–61. Oxford University Press, New York.

Gething, M. J. and Sambrook, J. (1992). Protein folding in the cell. *Nature* **355**, 33–45.

Gmelig-Meyling, F. and Ballieux, R. E. (1977). Simplified procedure for the separation of human T and non-T cells. *Vox Sang.* **33**, 5–10.

Hancock, R. E. W. and Nikaido, H. (1978). Outer membranes of gram-negative bacteria. XIX. Isolation from *Pseudomonas aeruginosa* PAO1 and use in reconstitution and definition of the permeability barrier. *J. Bacteriol.* **136**, 381–390.

Havlir, D. V., Ellner, J. J., Chervenak, K. A. and Boom, W. H. (1991). Selective expansion of human γδ T cells by monocytes infected with live *Mycobacterium tuberculosis*. *J. Clin. Invest.* **87**, 729–733.

Hermann, E. (1993). T cells in reactive arthritis. *APMIS* **101**, 177–186.

Hermann, E., Fleischer, B., Mayet, W.-J., Poralla, T. and Meyer zum Büschenfelde, K.-H. (1989). Response of synovial fluid T-cell clones to Yersinia enterocolitica antigens in patients with reactive Yersinia arthritis. *Clin. Exp. Immunol.* **75**, 365–372.

Hermann, E., Lohse, A. W., Mayet, W.-J., van der Zee, R., van Embden, J. D. A., Poralla, T., Meyer zum Büschenfelde, K.-H. and Fleischer, B. (1992) Stimulation of synovial fluid mononuclear cells with the human 65kD heat shock protein or with live enterobacteria leads to preferential expansion of TCR-γδ+ lymphocytes. *Clin. Exp. Immunol.* **89**, 427–433.

Hermann, E., Yu, D. T. Y., Meyer zum Büschenfelde, K.-H. and Fleischer, B. (1993). HLA-B27-restricted CD8 T cells derived from synovial fluids of patients with reactive arthritis and ankylosing spondylitis. *Lancet* **342**, 646–650.

Holzmann, B. and Johnson, J. J. (1983). A beta-galactosidase linked immunoassay for the analysis of antigens on individual cells. *J. Immunol. Methods* **60**, 359–367.

Huppertz, H.-I. and Heesemann, J. (1996). Experimental *Yersinia* infection of human synovial cells: Persistence of live bacteria and generation of bacterial antigen deposits including 'ghosts', nucleic acid-free bacterial rods. *Infect. Immun.* **64**, 1484–1487.

Kabelitz, D., Bender, A., Schondelmaier, S., Schoel, B. and Kaufmann, S. H. E. (1990). A large fraction of human peripheral blood γ/δ+ T cells is activated by *Mycobacterium tuberculosis* but not by its 65-kD heat shock protein. *J. Exp. Med.* **171**, 667–679.

Kaposi, K. and Inman, R. D. (1992). HLA-B27 expression modulates gram-negative bacterial invasion into transfected L cells. *J. Immunol.* **148**, 3554–3560.

Klareskog, L., Forsum, U., Kabelitz, D., Plöen, L., Sundström, C., Nilsson, K., Wigren, A. and Wigzell, H. (1982). Immune functions of human synovial cells. Phenotypic and T cell regulatory properties of macrophage-like cells that express HLA-DR. *Arthritis Rheum.* **25**, 488–501.

Lundquist, C., Hammarström, M. L., Athlin, L. and Hammarström, S. (1992). Isolation of functionally active intraepithelial lymphocytes and enterocytes from human small and large intestine. *J. Immunol. Methods* **152**, 253–263.

MacDonald, T. T., Spencer, J., Viney, J. L., Williams, C. B. and Walker, S. J. (1987). Selective biopsy of human Peyer's patches during ileal endoscopy. *Gastroenterology* **93**, 1356–1362.

May, T. B. and Chakrabartry, A. M. (1994). Isolation and assay of *Pseudomonas aeruginosa* alginate. *Methods in Enzymology* **235**, 295–303.

Isolation of T Cells

May, E., Märker-Hermann, E., Wittig, B., Zeitz, M., Meyer zum Büschenfelde, K. H. and Duchmann, R. (2000). Identical T-cell expansions in the colon mucosa and the synovium of a patient with enterogenic spondyloarthropathy. *Gastroenterology* **119**, 1724–1755.

Meuer, S. C., Hussey, R. E., Fabbi, M., Fox, D., Acuto, O., Fitzgerald, K. A., Protentis, J. C., Schlossmann, S. F. and Reinherz, E. L. (1984). An alternative pathway of T cell activation: a functional role for the 50kD T11 sheep erythrocyte receptor protein. *Cell* **36**, 897–906.

Munk, M. E., Gatrill, A. J. and Kaufmann, S. H. E. (1990). Target cell lysis and IL-2 secretion by γ/δ T lymphocytes after activation with bacteria. *J. Immunol.* **145**, 2434–2439.

Murphy, T. F. (1996). *Branhamella catarrhalis*: Epidemiology, surface antigenic structure, and immune response. *Microbiol. Rev.* **60**, 267–279.

Nikaido, H. (1994). Bacterial cell fractionation: Isolation of outer membranes. *Methods in Enzymology* **235**, 225–233.

Padula, S. J., Pollard, K. M., Lingenheld, E. G. and Clark, R. B. (1985). Maintenance of antigen-specificity by human interleukin-2 dependent T cell lines. *J. Clin. Invest.* **75**, 788–797.

Pawelec, G., Schneider, E. M. and Wernet, G. (1986). Acquisition of suppressive activity and natural-killer-like cytotoxicity by human alloproliferative 'helper' T cell clones. *J. Immunol.* **136**, 402–411.

Pfeifer, J. D., Wick, M. J., Roberts, R. L., Findlay, K., Normark, S. J. and Harding, C. V. (1993). Phagocytic processing of bacterial antigens for class I MHC presentation to T cells. *Nature* **361**, 359–362.

Probst, P., Hermann, E., Meyer zum Büschenfelde, K.-H. and Fleischer, B. (1993a). Multiclonal synovial T cell response to *Yersinia enterocolitica* in reactive arthritis. The Yersinia 61kD heat shock protein is not the major target antigen. *J. Infect. Dis.* **167**, 385–391.

Probst, P., Hermann, E., Meyer zum Büschenfelde, K.-H. and Fleischer, B. (1993b). Identification of the *Yersinia enterocolitica* urease β-subunit as a major target antigen for human synovial T cells in reactive arthritis. *Infect. Immun.* **61**, 4507–4509.

Rosenthal, R. and Dziarski, R. (1994). Bacterial cell fractionation: Isolation of peptidoglycan and soluble peptidoglycan fragments. *Methods in Enzymology* **235**, 253–285.

Shimonkevitz, R., Kappler, J., Marrack, P. and Grey, H. (1983). Antigen recognition by H-2 restricted T cells. I. Cell free antigen processing. *J. Exp. Med.* **158**, 303–316.

Vann, W. F., Freese, S. J. (1994). Purification of *Escherichia coli* K antigens. *Methods in Enzymology* **235**, 304–314.

Weiner, M. S., Bianco, C. and Nussenzweig, V. (1973). Enhanced binding of neuraminidase treated sheep erythrocytes to human T lymphocytes. *Blood* **42**, 939–946.

Wysocki, L. J. and Sato, V. L. (1978). 'Panning' for lymphocytes: A method for cell separation. *Proc. Natl. Acad. Sci. USA* **75**, 2840–2848.

Appendix: List of commercial suppliers

Amersham Pharmacia Biotech Inc.   Percoll
800 Centennial Avenue   Sepadex, Sephacryl and Mono Q columns
P.O. Box 1327
Piscataway, NJ 08855-1327, USA

Tel.: + Fax: +1 800 526 3593
www.apbiotech.com

Aventis Behring
1020 First Avenue
King of Prussia PA 19406-1310, USA

Tel.: +1 610 878 4000
Fax: +1 610 878 4009
www.AventisBehring.com

Test-neuraminidase

BD Biosciences
10975 Torreyana Road
San Diego, CA 92121, USA

Tel.: +1 858 812 8800
Fax: +1 858 812 8888
www.bdbiosciences.com

Monoclonal antibodies, agar

BD Immunocytometry Systems
2350 Qume Drive
San Jose, CA 95131-1807, USA

Tel.: +1 800 448 2347
Fax: +1 800 408 954 2347
www.Bdfacs.com

Monoclonal antibodies

Biochrom AG
Leonorenstrasse 2-6
D - 12247 Berlin, Germany

Tel.: +49 - 30 7799060
Fax: +49 - 30 77990666
www.biochrom.de

Ficoll density 1.09, media supplement

Bio-Rad Laboratories
2000 Alfred Nobel Drive
Hercules CA 94547, USA

Tel.: +1 510 741 1000
Fax: +1 510 741 5800
www.bio-rad.com

Protein assay

Biospec Products
P.O. Box 788
Bartlesville, OK 74005-0788, USA

Tel.: +1 918 336 3363
Fax: +1 918 336 6060
www.biospec.com

Bead-beater

Isolation of T Cells

DAKO Corp.
6392 Via Real
Carpinteria, CA 93013, USA

Tel.: +1 805 566 6655
Fax: +1 805 566 6688
www.dakousa.com

Antibodies

DYNAL Biotech
PO Box 114, Smestad
N-0309 Oslo, Norway

Tel.: +47 2206 1000
Fax: +47 2250 7015
www.dynal.no

Immunomagnetic separation (beads, magnetic
separation and mixing device). Useful technical
handbook provided

ICN Biomedicals
3300 Hyland Avenue
Costa Mesa CA 92626, USA

Tel.: +1 714 545 0100
Fax: +1 714 557-4872
www.icnbiomed.com

Biochemical products, EDTA, Trypsin

Invitrogen Corporation
Life Technologies
P.O. Box 6482
Carlsbad
CA, 92008, USA

Tel.: +1 760 603-7200
Fax: +1 760 602-6500
www.invitrogen.com

Cell culture products, media

Merck, Industrial Chemicals
EU Industries Inc.
5 Skyline Drive
Hawthorne, NY 10532, USA

Tel.: +1 592 4660
Fax: +1 592 9469
www.merck.com

Brain heart infusion (BHI) broth medium,
biochemical products

Sigma-Aldrich Corp.

St Louis, MO, USA
Tel.: 314-771-5765
Fax: 314-771-5757
www.sigma-aldrich.com

Biochemical products

3 Growth Transformation of Human T Cells

Helmut Fickenscher
Abteilung Virologie, Hygiene Institut, Ruprecht-Karls-Universität Heidelberg, Germany

Bernhard Fleckenstein
Institut für Klinische und Molekulare Virologie, Friedrich-Alexander-Universität Erlangen-Nürnberg, Erlangen, Germany

◆◆◆

CONTENTS

Introduction
Procedures for viral lymphocyte transformation
Phenotype of transformed T cells
Vector applications of herpesvirus saimiri

◆◆◆◆◆◆ INTRODUCTION

Molecular and biochemical studies of human T lymphocytes are limited for mainly two reasons. First, T-lymphoblastic tumour-cell lines such as Jurkat (Schneider *et al.*, 1977) have a strongly altered phenotype in comparison to primary cells with respect to signal transduction (Bröker *et al.*, 1993) and gene regulation. Secondly, primary T-cell cultures are limited in their lifespan. It is laborious and frequently impossible to grow primary T lymphocytes to large cell numbers, and it requires considerable effort to amplify the T cells in periodic response to a specific antigen and accessory cells expressing the appropriate major histocompatibility complex (MHC) restriction elements. Moreover, impurities due to the addition of irradiated feeder cells may cause difficulties in interpretation of results.

The immortalization of human T cells should be the ideal way to solve such problems. Various approaches have been applied for this purpose. The technique of T-cell fusion hybridomas was successful in the murine system. However, this method proved to be much more difficult with human T cells and to be hampered by genomic instability of the clones. Virus-mediated transformation is an established routine method for human B cells, which are efficiently immortalized by Epstein-Barr virus

Copyright © Elsevier Science Ltd
All rights of reproduction in any form reserved

(EBV; Nilsson and Klein, 1982; Tosato, 1997). EBV-transformed lymphoblastoid B cells retain their antigen-presenting capability and are widely used in studying the antigen specificity of T-cell clones. Under special experimental conditions, EBV is even able to immortalize human T cells *in vitro* to a rather autonomous phenotype (Groux *et al.*, 1997). Human T-cell leukaemia virus type 1 (HTLV-1) offers a solution to the problem for human T lymphocytes: HTLV-1 transforms human T cells to stable growth in culture (Miyoshi *et al.*, 1981; Yamamoto *et al.*, 1982; Yoshida *et al.*, 1982; Popovic *et al.*, 1983; Faller *et al.*, 1988). For many applications, this approach proved useful, although the transformation is largely confined to CD4[+] T cells. Retrovirus-transformed T lymphocytes produce HTLV-1 virions regularly. Unfortunately, the cells tend to lose their T-cell receptor (TcR) complex, their cytotoxic activity, and their dependence on interleukin-2 (IL-2) after prolonged cultivation (Inatsuki *et al.*, 1989; Yssel *et al.*, 1989). Recent observations suggest that retroviral expression vectors for the cellular telomerase reverse transcriptase can be used as an alternative method to support the survival of CD8[+] cytotoxic T-cell clones (Hooijberg *et al.*, 2000). In this chapter, we summarize the various procedures for the viral transformation of human lymphocytes with the main emphasis on using cell-free herpesvirus saimiri strain C488 for the targeted transformation of human T lymphocytes to antigen-independent proliferation and to an activated phenotype (Biesinger *et al.*, 1992).

Herpesvirus saimiri (HVS; Melendez *et al.*, 1968) is the prototype of the genus rhadinoviruses γ_2-herpesviruses. The official taxonomic designation 'saimirine herpesvirus type 2' (SHV-2; Roizman *et al.*, 1992) occurs only rarely in the literature. The closest relative of HVS in humans is the Kaposi's sarcoma-associated herpesvirus (KSHV) or human herpesvirus type 8 (HHV-8) which is regularly present in Kaposi's sarcoma (Chang *et al.*, 1994). HVS is not pathogenic in its natural host, the squirrel monkey (*Saimiri sciureus*), and can easily be isolated from the peripheral blood of most specimens (Falk *et al.*, 1972; Wright *et al.*, 1976; Greve *et al.*, 2001). In other New World monkey species, such as common marmosets (*Callithrix jacchus*), cotton top tamarins (*Saguinus oedipus*), and owl monkeys (*Aotus trivirgatus*), as well as in certain rabbits (Ablashi *et al.*, 1985; Medveczky *et al.*, 1989), HVS causes fulminant polyclonal T-cell lymphomas and acute lymphatic leukaemias (reviewed in Fleckenstein and Desrosiers, 1982; Fickenscher and Fleckenstein, 2001). Furthermore, marmoset T lymphocytes can be transformed *in vitro* by HVS (Schirm *et al.*, 1984; Desrosiers *et al.*, 1986; Kiyotaki *et al.*, 1986; Szomolanyi *et al.*, 1987; Chou *et al.*, 1995).

The nucleotide sequence of HVS strain A11 was entirely determined (Albrecht *et al.*, 1992). The viral genome consists of 112 930 bp of L-DNA with 75 tightly packed major open reading frames, flanked by approximately 35 non-coding, GC-rich H-DNA repeat units of 1444 bp in tandem orientation. The genome organization is co-linear to that of the human lymphocryptovirus (γ_1-herpesvirus) EBV. However, the conserved gene blocks are rearranged and the well-studied transformation- and persistence-associated genes of EBV (EBNA, LMP) are lacking. Several open reading frames of HVS display strong sequence homologies to known cellular genes, among them thymidylate synthase, dihydrofolate reductase, complement control proteins, apoptosis inhibitors, cyclins, superantigens, IL-17, and G-protein coupled IL-8-receptors (IL-8R). Most of these viral homologues to cellular genes were shown to encode functional molecules. However, most of them are not expressed in transformed human T cells. In most of the cell-homologous genes, a functional contribution to T-lymphocyte transformation has been excluded by the generation of deletion-mutant viruses (Knappe et al., 1997, 1998a,b; Ensser et al., 1999, 2001; Glykofrydes et al., 2000). The cell-homologous genes are typically expressed during lytic replication and may play a role in virus replication and during apathogenic persistence in the natural host (reviewed in Fickenscher and Fleckenstein, 2001).

The genomic region which is essential for transformation of monkey T lymphocytes and for oncogenicity was mapped to a terminal genomic region of the L-DNA (Desrosiers et al., 1984, 1985a, 1986; Koomey et al., 1984; Murthy et al., 1989). HVS strains of different subgroups (A, B, C) vary in oncogenicity and terminal genomic sequence (Desrosiers and Falk, 1982; Medveczky et al., 1984, 1989; Biesinger et al., 1990; Choi et al., 2000; Hör et al., 2001). This genomic region of virus strains from the distinct HVS subgroups encodes different herpesvirus saimiri trans-formation-associated proteins (Stp). The respective proteins of subgroup A and C (StpA and StpC) were able to transform rodent fibroblasts which then caused tumours in nude mice (Jung et al., 1991). Moreover, *stpC*-transgenic mice developed epithelial tumours of the salivary gland, bile ducts and thymus (Murphy et al., 1994). *StpA*-transgenic animals, in contrast, suffered from peripheral T-cell lymphoma (Kretschmer et al., 1996). StpA and StpB interact as substrates with the tyrosine kinase Src (Lee et al., 1997; Choi et al., 2000; Hör et al., 2001). StpC488 contains a stretch of 18 collagen triplet repeats and a specific charged N-terminal region of 17 amino acids (Biesinger et al., 1990). StpC was shown in the perinuclear compartment of transformed rodent fibroblasts and to be phosphorylated in the transformed fibroblasts on a serine residue close to the N-terminus (Jung and Desrosiers, 1991, 1992). The N-terminal part of StpC is relevant for its transforming function (Jung and Desrosiers, 1994). The association of StpC with cellular Ras and with tumour necrosis factor receptor-associated factors (TRAF) is suggestive for its mechanism of transformation (Jung and Desrosiers, 1995; Lee et al., 1999).

The neighbouring open reading frame in subgroup C viruses codes for a protein, that specifically interacts as a substrate with the T-cell specific tyrosine kinase p56^{lck}. The tyrosine kinase-interacting protein (Tip) inter-feres with the Lck-dependent signal transduction and blocks or stimulates its enzymatic activity depending on the assay system (Biesinger et al., 1995; Jung et al., 1995a,b; Lund et al., 1995, 1996, 1997b; Wiese et al., 1996; Fickenscher et al., 1997). In addition, Tip–Lck complexes were found to interact with signal transducer and activator of transcription (STAT) factors leading to the activation of STAT-dependent transcription (Lund et al., 1997a; Hartley and Cooper, 2000). Conditional *tip*-transgenic mice developed peripheral T-cell lymphoma (Wehner et al., 2001). The properties of the transformation-associated proteins have been recently reviewed in detail (Fickenscher and Fleckenstein, 2001).

Primary human T lymphocytes are transformed by wild-type subgroup C strains of HVS to stably proliferating T-cell lines with the phenotype of mature CD4+ or CD8+ cells. The growth-transformed human T lymphocytes usually remain IL-2 dependent, but do not need a periodic restimulation with antigen or mitogen. They carry multiple non-integrated viral episomes and do not produce virion particles (Biesinger *et al.*, 1992). In transformed human T cells, viral gene expression was absent in most genomic regions. StpC and Tip are the only virus proteins which have been detectable in HVS-transformed human T cells (Biesinger *et al.*, 1995; Fickenscher *et al.*, 1996a).

Both, *stpC* and *tip*, are essential for pathogenicity *in vivo* and T-cell transformation *in vitro* (Knappe *et al.*, 1997; Duboise *et al.*, 1998a). StpC and Tip are expected to act in synergy in T-cell activation and trans-formation (Merlo and Tsygankov, 2000). Transformed human T lympho-cytes expressing *stpC/tip* did not induce tumourigenesis in nude or SCID mice in xenogeneic transplantation experiments (Huppes *et al.*, 1994). Transfused autologous HVS-transformed T cells did not cause tumours or disease in macaques, but persisted over months (Knappe *et al.*, 2000a). When antigen-specific T cells were transformed by HVS, the MHC-restricted antigen-specific reaction was retained (Behrend *et al.*, 1993; Bröker *et al.*, 1993; De Carli *et al.*, 1993; Mittrücker *et al.*, 1993; Weber *et al.*, 1993). The transformed T lymphocytes maintain early signal transduction patterns of primary cells, express activation surface markers, show inducible cytotoxicity, and secrete Th1-type cytokines (Biesinger *et al.*, 1992; Bröker *et al.*, 1993; De Carli *et al.*, 1993; Mittrücker *et al.*, 1993; Weber *et al.*, 1993; Klein *et al.*, 1996). Thus, HVS provides novel tools to study T-cell biology (reviewed in Fickenscher and Fleckenstein, 2001; Meinl *et al.*, 1995a).

◆◆◆◆◆◆ PROCEDURES FOR VIRAL LYMPHOCYTE TRANSFORMATION

Propagation and manipulation of herpesvirus saimiri

Owl monkey kidney cells

Epithelial owl monkey kidney (OMK) cells from healthy *Aotus trivirgatus* (Owl monkeys) can be kept in primary culture over long time periods (Daniel *et al.*, 1976b) if they are not contaminated with lytically replicating viruses, such as herpesvirus aotus type 1 or 3. The American Type Culture Collection (ATCC; Manassas, VA, USA) offers the cell line OMK-637-69 (CRL-1556). This cell line has been intensively used in several laboratories over decades without any hints of viral contaminations or latent agents. The monolayer cells are maintained in Earle's Minimal Essential Medium (MEM) or Dulbecco's Modified Eagle Medium (DMEM) supplemented with 10% heat-inactivated fetal bovine serum (FCS), glutamine (350 µg ml⁻¹) and antibiotics (100 µg ml⁻¹ gentamycin or penicillin/streptomycin at 120 µg ml⁻¹ each). The cells are detached with trypsin and split only once a week onto a doubled area of tissue culture plastic ware. This low ratio maintains the original status of the primary OMK cells and does not enforce the selection for fast-growing subtypes, which are no longer fully permissive. The medium is changed on the fourth day after splitting. The cells should not be used for more than 50 passages. OMK cells are the typical propagation system for herpesviruses of New World primates, such as HVS, herpesvirus aotus, and herpesvirus ateles (for review: Fleckenstein and Desrosiers, 1982). It should be kept in mind, that *Aotus trivirgatus* monkeys are a highly endangered species. If OMK cells have to pass the customs during shipment, appropriate official documentation according to the convention on international trade in endangered species (CITES) is needed.

Virus cultures

The tissue culture medium of freshly confluent OMK monolayer cultures is removed on day one or two after splitting, and infectious virion suspension is added in a minimal volume (such as 2 ml for a 25 cm^2 flask, 5 ml for an 80 cm^2 flask). Adsorption is allowed to take place at 37°C for 2 h, during which the monolayer must not suffer from drying out. Afterwards, medium is added, and the incubation is continued. Alternatively, we did not observe relevant differences in the time course of infection if the virus suspension was simply added to the culture medium without the adsorption step at low volumes. After one to 14 days, initial cytopathic effects (CPE) are found. Typically, focal rounding of cells is observed. Later, plaques appear, bordered by rounded cells. Several days later, the whole cell layer will be lysed by the virus. When high titred supernatants are to be prepared, it is essential to use a low multiplicity of infection (0.1–0.5). Ideally, it should last 1–2 weeks, until a complete CPE is developed. Fast CPE at high multiplicity often results in low titres of infectious particles. In such cases, a high proportion of viruses carries repetitive H-DNA only, without coding L-DNA. After completion of the CPE, infected OMK cultures including the OMK cell debris are used as HVS stocks, without any further treatment. The virus stocks are stable and can be stored at +4°C for several months without loss of infectivity. Small volumes of supernatant should be frozen in liquid nitrogen or at −80°C; the titre may decrease by approximately one order of magnitude upon freeze-thawing.

The lytic OMK cell system is used to estimate concentration of infectious virion particles from the supernatant of lytic cultures (see p. 664 virus titration) and also to monitor HVS-transformed T lymphocytes for their (non)-producer status. Typically, approximately 10^6 transformed lymphocytes are added to a 25 cm^2 flask (5 ml MEM or DMEM with 10% FCS, but without IL-2) of fresh confluent OMK cells. Co-cultivation allows the infection of OMK cells with small amounts of cell-associated virions from the lymphocytes, as close cell contact is achieved. In many cases, however, the transformed human T lymphocytes are activated by the OMK cell contact, presumably by the CD2/CD58 interaction (Mittrücker *et al.*, 1992). As a side-effect of the activated status, especially CD8⁺ transformed lympho-cytes may exhibit a strong non-specific cytotoxic effect on OMK cells. Thus, the OMK cells are sometimes killed by non-specific cytotoxicity in co-cul-tivation tests. The cytotoxic activity will decrease during further passage of co-cultivation on fresh OMK cells. If cytotoxicity problems occur, cell-free culture supernatants can also be checked for virus particles. For this purpose, large volumes of supernatant are collected (up to 50 ml). Residual T cells are removed by low-speed centrifugation (1000 rpm, 10 min). Potentially present virus particles are sedimented from the supernatant by high-speed centrifugation in a Sorvall centrifuge at 20 000 rpm and 4°C for at least 4 h in a SS34 rotor using screw-capped tubes (35 ml; Nalgene). The sediment is resuspended in medium and transferred onto fresh OMK cul-tures. Traditionally, co-cultivations have been followed up during three passages, which is approximately 6 weeks. We have good experience with virus isolation attempts if the cultures receive fresh medium only on every fourth day during 6 weeks without further passaging. Under these condi-tions, low amounts of infectious particles are easily detected. If question-able, a weak CPE can be confirmed by transferring an aliquot of sterile filtered supernatant to a fresh OMK culture, where a typical CPE should be visible after a few days. If there was virus growth only after several pas-sages, very low titres of infectivity have been amplified. Several weeks after infection, virions can no longer be demonstrated in human T-lymphocyte cultures (Biesinger et al., 1992; Fickenscher et al., 1996a). The sensitivity of the co-cultivation method was estimated to be one virus-producing cell per 10^6 cells (Wright et al., 1976).

Virus strain C488 (Desrosiers and Falk, 1982; Biesinger et al., 1990) was submitted to ATCC (Manassas, VA, USA; VR-1396 and VR-1414) after only a few passages on OMK cells. Moreover, ATCC offers virus strain S295C, which, however, does not belong to subgroup C. This subgroup B virus is unable to transform human T cells, but is oncogenic with low efficiency in marmoset monkeys (Melendez et al., 1968; Fleckenstein et al., 1977).

Virion purification

Both, the supernatant and the cell remnants from lytic OMK cultures, contain high amounts of virion particles (approximately 10^9 virion particles per ml in the supernatant). Cultures with complete CPE (e.g. one or two litres) are bottled at 200 ml each into GSA centrifuge flasks (Sorvall) with canted neck to avoid any spilling. A first centrifugation is carried out for 15 min at 4°C and 3000 rpm to collect the cell-associated and the cell-free fractions of the virus. The cell sediment is resuspended in 3 ml hypotonic 'virus standard buffer' (VSB; pH 7.8, 50 mM Tris, 12 mM KCl, 5 mM EDTA). The supernatant is transferred to clean GSA bottles and spun for a second time at high speed (12 000 rpm) and 4°C for 4 h. Usually, clearly visible virus sediments are found. Virions are re-suspended in 3 ml of VSB. Sucrose gradients are prepared in ultra-clear Beckman-Coulter ultracentrifuge tubes (SW27) using a gradient mixer with 15 ml of 30% (w/w) sucrose in VSB solution (stirring) close to the outlet and 16 ml of 15% (w/w) solution in the distal container. The rotor and the buckets are cooled to 4°C prior to centrifugation. Both virus and cell-debris suspensions are dounced in a glass homogenizer by 20 strokes. Approximately 3 ml of the homogenized material can be loaded on one gradient. The buckets are weight adjusted with VSB and centrifuged in a SW27 rotor for 30 min at 4°C with 20 000 rpm. In case the virus band is not clearly visible, a strong focused apical light source should be used. The band material is collected into a Beckman-Coulter SW27 polyallomer tube, diluted with VSB, and spun for 90 min at 4°C with 17 000 rpm. Alternatively, 35 ml Nalgene tubes with screw cap can be used in a Sorvall SS34 rotor at 20 000 rpm and 4°C for 4 h. The virions are then resuspended in a small volume of VSB, e.g. 100 to 500 µl. This method is based on the procedures described by Fleckenstein and Wolf (1974) and Fleckenstein *et al.* (1975). The small final volume should contain nearly all infectious particles of the starting material and is concentrated by several orders of magnitude. Therefore special care should be taken for the biosafety of all laboratory staff. It may be wise to perform large-scale virion purifications of transforming HVS strains in a P3 facility, although transforming HVS strains are generally classified to P2/L2 containment level.

Purification of virion DNA

Purified virion DNA of transforming HVS strains is infectious. When virion DNA had been injected intramuscularly into marmoset monkeys, the animals died from lymphoma after a few weeks (Fleckenstein *et al.*, 1978). Large-scale purification of virion DNA starts with concentrated virion suspensions as described above. Sodium lauryl sarcosinate (200 µl of 20%) is added to 1.8 ml purified virions in VSB. The lysis of virions is performed for 1 h at 60°C. The lysate (2 ml) is diluted with 6 ml VSB, and 10 g CsCl are added. The solution should have a refraction index of 1.412 at 25°C. The DNA/CsCl solution is loaded without ethidium bromide in 50TI Beckman-Coulter quickseal tubes (polyallomer) which are filled to completion with paraffin. The gradients are spun at 35 000 rpm and 20°C for 60 h. Two syringe needles are placed into the gradient tube: the first one on top into the paraffin to allow pressure compensation, the second one at the bottom. The flow rate can be reduced, when a syringe filled with paraffin is plugged to the upper needle. Approximately 20 fractions of 10 drops each are collected in Eppendorf caps. One µl of each fraction is diluted to 10 or 15 µl with water and coloured loading buffer and run slowly on a conventional agarose test gel. One single clean high-molecular weight band at limiting mobility should be visible. The pooled DNA-containing fractions are dialysed against 20 mM Tris pH 8.5, extracted with phenol/chloroform/isoamylalcohol (25/24/1) and chloroform/isoamylalcohol (24/1), ethanol precipitated, and resuspended in water. The concentration is determined by spectrophotometry. The yields are approximately 40 µg from 200 ml of supernatant from infected cultures. Virion DNA isolated from the cleared supernatant will be rather pure and show less than 5% contamination by cellular DNA. The cell-bound virus may lead to varying quality and amounts of DNA yields. However, large amounts of pure virion DNA are often obtained also from the cell fraction. These procedures are modified from Fleckenstein and Wolf (1974), Fleckenstein et al. (1975) and Bornkamm et al. (1976).

A short protocol is useful for preparation of virus DNA at small scale. When small amounts of virion DNA are sufficient, 1 ml of cell-free supernatant is centrifuged at 13000 rpm in a cooled benchtop microcentrifuge for several hours, using screw cap Eppendorf caps and a rotor with appropriate tight lid to exclude any aerosol formation. To the virus sediment, 100 µl SDS buffer is added (1% SDS, 50 mM TrisHCl pH 8.0, 100 mM EDTA, 100 mM NaCl, 1 mg ml^{-1} proteinase K). The lysates are incubated at 56°C for 2 h, diluted with water to 300 µl, phenol extracted, chloroform extracted and ethanol precipitated. The yield from 1 ml is approximately 50 ng and sufficient for several Southern blot analyses (modified after Grassmann et al., 1989).

Virus titration

The limiting dilution method is the easiest way to estimate virus titres. OMK cells are detached with trypsin, split by 1:5 per area, e.g. into 24-well plates, and incubated under CO_2. The cell-free virus suspension is diluted in DMEM at 10^{-4} to 10^{-9}. To each 24-plate well, 1 ml of virus dilution is added one day after splitting. It is important to perform the assay at least in triplicate and to run control cultures. The plates are observed for at least 14 days. The progress of CPE should be monitored daily.

The more laborious plaque assay is also based on serial dilution. Methyl cellulose (4.4 g; e.g. Fluka 64630 Methocel MC 4000) is suspended in 100 ml H_2O in a 500 ml bottle and autoclaved (45 min, 121°C). One hundred ml of two-fold concentrated DMEM, 20 ml FCS (heat-inactivated), antibiotics, and glutamine (350 µg ml^{-1}) are added. The mixture is stirred at 4°C overnight. Subconfluent OMK cultures in 6 cm dishes are infected with each 1 ml of cell-free virus dilutions (10^{-3} to 10^{-8} in complete DMEM) to three dishes each, one day after splitting the cells. After 1 h at 37°C, the supernatant medium is aspirated, and approximately 5 ml of 2% methyl cellulose in complete DMEM are added. If this approach is used for virus plaque purifications, single plaques are picked in approximately 10 µl with a micropipette tip and transferred to a fresh OMK culture. To exclude clumps of virus particles, the supernatant should be sterile filtered before starting the plaque cloning cultures. This will allow virus cloning with low contamination rate by other virus clones. For titre estimations, the medium is poured off carefully after plaque formation has been observed. The cells are washed mildly with phosphate-buffered saline (PBS) and stained with 1% crystal violet in 20% ethanol/80% water (v/v). Subsequently, the surplus dye is washed out carefully with water without destroying the cell layer. The cells are air dried, and the plaques are counted. Routine OMK cell cultures should reach titres between 10^5 to 10^7 plaque forming units (p.f.u) per ml (Daniel *et al.*, 1972).

Demonstration of episomes and virion DNA

In order to demonstrate and distinguish persistent non-integrated episomal (slowly migrating) or replicative linear (higher mobility) viral DNA, *in situ* lysis gel electrophoresis can be applied. The method has first been described by Gardella *et al.* (1984). A simplified version of that protocol is presented (modified after Ablashi *et al.*, 1985). Typical examples for 'Gardella gels' are shown in various reports (Gardella *et al.*, 1984; Ablashi *et al.*, 1985; Biesinger *et al.*, 1992; Bröker *et al.*, 1994; Meinl *et al.*, 1997). The gel system is a vertical 1% agarose gel in $1 \times$ Tris borate EDTA buffer (TBE) with the gel size 20×20 cm, 0.5 cm thick, at 4°C. The wells have 0.5×0.5 cm area and are 1.0 cm deep; they are separated by 0.3–0.5 cm agarose teeth. The thick gel tends to slip out from the gel plates during assembly of the gel apparatus. A large 3% agarose block is put into the lower buffer chamber, so that the gel and the gel plates can rest on it. The gel system is placed into a refrigerator or cold room; the upper chamber is still left without buffer. TBE ($1\times$) for the chambers is precooled. Frozen aliquots of buffers A and B (composition listed below) without enzymes have been prepared earlier, because ficoll is very sticky. The enzymes RNase A and Protease K are added just before use. Approximately 2×10^6 lymphocytes (e.g. transformed with HVS C488) are washed in PBS, resuspended in 50 µl blue sample buffer A (200 mg ml^{-1} Ficoll 400, 1× TBE, 0.25 mg ml^{-1} bromo phenol blue, 50 µg ml^{-1} RNase A) and loaded into the very bottom of the slots. As controls, conventional λ-*Hind*III marker and a small amount of virions are used, the latter obtained by centrifuging 1.0 ml virus suspension at 13 000 rpm in a cooled benchtop centrifuge, as described above for the miniprep protocol for virion DNA. Such a small virus particle preparation should be sufficient for several gels. The controls are also loaded in buffer A. Green buffer B (120 µl per sample; 50 mg ml^{-1} Ficoll 400, 1× TBE, 1% SDS, 0.25 mg ml^{-1} xylene cyanole, 1 mg ml^{-1} protease K) is laid over the sample, nearly filling the gel slots. Some 1× TBE buffer is added carefully to fill the slots completely, and more buffer to fill the upper chamber. Electrophoresis is slowly performed at 4°C. Proteins and RNA are allowed to be degraded enzymatically during a first period of at least 3 h at 10 V. Subsequently, electrophoresis is continued for at least 12 h at 100 V, until the green colour reaches the bottom of the gel. The agarose gel is stained in TBE with ethidium bromide and photographed. However, this staining is usually not sufficient for demonstrating episomal bands. The DNA is transferred onto Nylon membranes using the alkali DNA transfer protocol. Viral DNA bands will be visualized after stringent hybridization and subsequent autoradiography. There are several alternatives to choosing the hybridization probe. If maximal sensitivity is necessary, a cloned H-DNA fragment can be used (Bankier *et al.*, 1985). H-DNA is highly repetitive in the non-integrated episomes and, thus, amplifies the signals. If higher specificity is needed, e.g. to prove that the correct virus strain was used, a fragment from the variable genomic end should be applied (Biesinger *et al.*, 1990). If band intensities have to be compared from cultures infected with different virus strains, it might be advantageous to take a probe from a less variable genomic region. Insert DNA as well as flanking sequences should be used to probe for recombinant viral vectors.

Insertion of foreign genes into the viral genome

Foreign genes can efficiently be inserted behind open reading frame 75 into the terminal genomic junction region of viral L- and H-DNA, where no hints of viral transcription had been found (Bankier *et al.*, 1985; Stamminger *et al.*, 1987). The neomycin resistance marker gene was initially inserted into that position by Grassmann and Fleckenstein (1989). Cloning into HVS is achieved via homologous recombination. As herpesviruses are assumed to replicate according to the rolling circle model, a single cross-over event suffices for insertion of the foreign gene into the terminal L-DNA region (Grassmann *et al.*, 1989). Therefore, the respective recombination plasmids provide in series the terminal genomic sequences, the gene of interest to be inserted, and a selection marker. Two kb of homologous terminal sequences were sufficient for homologous recombination. Alt *et al.* (1991) constructed the plasmid pRUPHy containing 2 kb of homologous sequence, a multilinker stretch for the insertion of the gene of interest, and the hygromycin B resistance gene under the transcription control of the human cytomegalovirus major immediate early enhancer-promoter. An altenative recombination plasmid pRecosac combines 5 kb of homologous DNA with unique restriction sites and the neomycin resistance gene under SV40 enhancer control (Hiller *et al.*, 2000). The targeting plasmid (4 µg) is linearized upstream of the viral segment, mixed at approximately 200-fold molar excess with purified infectious virion DNA (200 ng), and co-transfected into permissive OMK cells by the calcium phosphate method. Virion DNA is prepared at small scale from the supernatant at complete CPE (after 2–3 weeks), and tested by Southern hybridization. Clean recombinant virus stocks are obtained using selection media (200 µg ml^{-1} hygromycin B or G418) and by plaque purification.

In an alternative protocol, the linearized recombination plasmid is transfected into OMK cells using lipofectamine (Invitrogen). One day after transfection, the cells are infected with wild-type virus. After completion of CPE in such cultures, recombinant viruses are enriched in serial passage under drug selection. It is important to use a low multiplicity of infection such as 0.1 to 0.5 in order to avoid wild-type virus overgrowing the culture before selection can take place. The CPE should need at least 10 days until completion, in order to achieve efficient selection. When the recombination has been confirmed by Southern hybridization from such supernatants, plaque purification of sterile-filtered material is performed. Sterile filtration removes virus clumps from the suspension, enhances cloning efficiency and reduces the probability of contamination with wild-type virus. The plaque purification procedure is described above in the section on virus titration. Under selection pressure, HVS recombinants with G418 resistance were shown to infect a broad spectrum of cell types (Simmer *et al.*, 1991). Cloned virus mutants can also be used to introduce foreign genes into T cells, e.g. the transforming genes of HTLV-1 (Grassmann *et al.*, 1989, 1992).

The transformation-associated region of HVS at the other genomic end was used for vector purpose as well. As the transformation-associated genes are not essential for virus replication, they can be replaced by foreign genes. Such an approach was first used for the expression of bovine growth hormone (Desrosiers *et al.*, 1985b). Moreover, virus mutants were constructed carrying the genes for secreted alkaline phosphatase or green fluorescent protein instead of the transforming region. This vector system uses unique *Asc*I and *Not*I restriction enzyme sites introduced into the viral genome. A major advantage of this system is the easy manipulation by restriction enzyme digestion and conventional ligation into purified virion DNA (Duboise *et al.*, 1996). The use of green fluorescent protein and secreted alkaline phosphatase provide two alternative methods for screening for virus recombinants. In the first case, the enzyme is detected from the supernatant by a simple ELISA assay. In the second case, fluorescent plaques are detected by conventional fluorescence microscopy (Duboise *et al.*, 1996).

A third method has been developed using overlapping cosmids of C488 virus DNA (Ensser *et al.*, 1999). Mutations or insertions can be performed easily by conventional cloning procedures in *E. coli*. Appropriate sets of cosmids are then co-transfected into OMK cells using lipofectamine. By homologous recombination, overlapping cosmids reconstitute replication-competent virus DNA. The resulting populations are free of contaminating wild-type virus (Ensser *et al.*, 1999).

Growth-transformation of human T cells by herpesvirus saimiri

The infection of primary human T lymphocytes with HVS subgroup C strains, but not with strains of subgroups A or B, yields stably proliferating mature T-cell lines that exhibit the CD4$^+$/CD8$^-$ or the CD4$^-$/CD8$^+$ phenotype (Biesinger *et al.*, 1992; reviewed in Fickenscher and Fleckenstein, 2001; Meinl *et al.*, 1995a). Stably growing T-cell lines have been obtained from primary cells of various sources. The degree of purity did not relevantly influence the outcome. Polyclonal preparations of mononuclear cells from adult peripheral blood or from cord blood, from thymus or bone marrow, as well as characterized T-cell clones, or flow-cytometry sorted T cells are suitable for the procedure. The CD4/CD8 and Th1/Th2 phenotypes do not influence the susceptibility to growth transformation. Transformed cell lines with αβ-TcR represent a broad variety of Vβ-chains (Fickenscher *et al.*, 1996b). The transformation of bulk cultures is polyclonal (Behrend *et al.*, 1993; Fickenscher *et al.*, 1996b; Knappe *et al.*, 1997; Saadawi *et al.*, 1997). Although γδ-cell lines were never obtained when large amounts of polyclonal cells were infected, they were generated either after lysis of αβ-T cells, or from small cultures at microwell scale, or even upon differentiation of virus-infected CD34+ thymocytes (Yasukawa et al., 1995; Klein et al., 1996; Pacheco-Castro et al., 1996; Fickenscher et al., 1997). If not, fresh T cells are to be used, proliferating cells are needed with good viability and with the morphology of activated T lymphocytes. The primary cells may be purified on Ficoll density gradients (1.077 g ml^{-1}; Biochrom, Berlin) or by the dextran sedimentation protocol (selective sedimentation of erythrocytes at 37°C and 1.2% dextran 250.000 [w/v]/30 mM NaCl for 45 min). Prestimulation of fresh primary cells for two days with phytohaemagglutinin (PHA, 4 μg ml^{-1}, Murex), concanavalin A (ConA, 4 μg ml^{-1}, Sigma), or the monoclonal anti-CD3 antibody OKT-3 (10 to 200 ng ml^{-1}, Ortho or Janssen-Cilag) does not give significant advantage over untreated samples. It depends mainly on the number of available cells whether mitogenic stimulation should be done. Antigen-dependent T-cell clones should be restimulated with antigen and feeder cells 3 to 5 days prior to infection. If the antigen specificity of an established clone is not known, a mitogen stimulation in presence of irradiated feeder cells should be performed 3 to 5 days before infection. The T-cell cultures are kept at a density of approximately 1 (0.5 to 1.5) × 10^6 cells ml^{-1}. Different types of cell culture plastic ware have been tested. The cells retain best viability in 25 cm^2 flasks (3–10 ml with 3–10 × 10^6 cells), 24-well plates (1–2 ml with 1–2 × 10^6 cells), or 96 round-bottom microwell plates (100 to 200 μl with 10^4 to 10^5 cells). The culture flasks should be incubated with slightly elevated top (at an angle of approximately 15°) to allow close cell contacts in the lower edge of the culture flask.

Various media formulations have been used. The cell lines CB-15, PB-W, Lucas (Biesinger et al., 1992), and CB-23 (Nick et al., 1993) were even isolated without prestimulation, without IL-2 supplementation, and with a standard medium formulation (RPMI-1640 80%, FCS 20%, 50 μM 2-mercaptoethanol, glutamine, gentamycine). During further experiments, the addition of supplements enhanced the transformation efficiency considerably. The quality of FCS batches even from the same supplier varies remarkably. It is very important to compare different serum batches and to take a batch with a low endotoxin level. FCS is inactivated at 56°C for 30 min to avoid mainly complement-depending problems with both virus and lymphocytes. The addition of IL-2 (at 20–50 U ml^{-1} in the medium stock) activates cell proliferation and enhances the transformation frequency. Since usually approximately one-third of the culture medium is added or replaced twice a week, a final concentration in the culture of less than 20 U ml^{-1} IL-2 is presumed. Comparison of many different IL-2 batches from different suppliers revealed consistently in our experiments, that pretested batches with low endotoxin levels of the human recombinant IL-2, isolated from E. coli and offered by Roche Diagnostics (Cat. No. 1011456 and 1147528), are very active and provide optimal conditions for a high transformation efficiency. For financial reasons it may be applied mainly for the initial transformation procedure itself and for maintaining small culture volumes. When stable growth is established and when large cell-culture volumes of transformed lines are required, the less expensive and less active Proleukin (Chiron) is recommended at 50 to 100 U ml^{-1} in the medium stock. It is a human recombinant IL-2 for clinical use, which carries 1-des-Ala and 125-Ser amino acid substitutions. Efforts to enhance the transformation efficiency led to the observation that adding 45% of Panserin 401 medium was beneficial (PAN Systems GmbH, Aidenbach, Germany). Panserin 401 medium is a synthetic medium for the serum-free culture. Several types of synthetic media were tested as supplement, but only Panserin 401 and AIM-V (Invitrogen) improved transformation efficiency significantly in our experiments. Panserin 401 allows to judge the metabolic activity by colour, whereas there is little colour difference using AIM-V medium. Using pretested IL-2 and medium supplement, approximately 90% of primary T-cell cultures were transformed to antigen- and mitogen-independent growth. The failure rate of approximately 10% is mostly explained by trivial reasons such as low viability of the cells prior to infection. According to our experience, the optimal medium formulation is: RPMI 1640 45%, Panserin 401 45%, FCS 10%, glutamine, gentamycine, and IL-2 (Roche Diagnostics, 20–40 U ml^{-1}). In order to block autoxidation of thiols and the resulting toxicity, the addition of 1 mM sodium pyruvate (Invitrogen, 043-01360, 100× stock solution at 100 mM), 50 μM α-thioglycerol (Sigma M-6145, 1000× stock solution at 50 mM), and 20 nM bathocuproine disulfonic acid (Sigma B-1125; 1000× stock solution at 20 μM) may be advantageous, however, is not essential (M. Falk and G. Bornkamm, personal communication). α-Thioglycerol is a more stable alternative to β-mercaptoethanol. Copper ions which are relevant for autoxidation are chelated by bathocuproine disulfonic acid. Addition of sodium pyruvate reduces H_2O_2 concentrations (O'Donnell-Tormey et al., 1987).

Most experience in T-cell transformation is based on using the HVS strain C488 (Biesinger et al., 1992). However, also different strains called C484 (Desrosiers and Falk, 1982; Medveczky et al., 1984, 1989, 1993a; Biesinger et al., 1992; Fickenscher et al., 1997) and virus strain C139 are able to transform human T cells, at least to a certain degree and phenotype. Strain C139 is an isolate from our laboratory (Klein et al., 1996; Fickenscher et al., 1997). C488 seems to be the most reliable HVS strain for transforming human T cells (Fickenscher et al., 1997). Typically, 10% (v/v) of infectious OMK supernatant are added to a vital lymphocyte culture, for example 500 μl of 10^6 p.f.u. ml^{-1} (5×10^5 p.f.u.) to a 5 ml culture volume with 5×10^6 cells ml^{-1}. During the following weeks and sometimes months, the cells need patient care; medium should be partially exchanged at regular intervals, such as twice a week. Good negative controls are non-infected cells and cultures infected with a virus strain of the same titre, such as strain 11 (subgroup A) or SMHI (subgroup B), which are not able to transform human T cells. During the first weeks of culture, the normal growth of T lymphocytes is observed. This normal mitotic phase should not be misinterpreted as transformation events. It may last up to several months until the viral infected T cells will start to proliferate. At least 10^5 viable cells are necessary for obtaining a transformed culture (Fickenscher et al., 1997). The efficiency seems to vary considerably between different donors. When using a few million vital cells for infection, the efficiency rate reaches 90%. Several criteria were found to indicate a transformed phenotype: (a) doubling of the cell number constantly once to four times a week over several months, independently of antigen and mitogen; (b) morphology of T lymphoblasts, enlargement of cells, good contrast upon microscopy, irregular shape; (c) death of control cultures which have been treated identically to the transformed ones; (d) persistence of viral non-integrated episomes without virion production (PCR, Gardella gel, and co-cultivation test on OMK); (e) CD2 hyperreactivity against membrane-bound CD58, or cross-linked CD2 antibodies, or foreign cells (Mittrücker *et al.*, 1992), and non-specific cytotoxicity. Cultures which seem to be growth-transformed can often be subjected to a gradual withdrawal of the medium supplement. Addition of 10 mM Hepes pH 7.4 (Invitrogen) may help to avoid cell degeneration caused by low pH. In CD4$^+$ cells, exogenous IL-2 supplementation could also be reduced gradually and even be terminated in several cases. However, withdrawing IL-2 from CD8$^+$ transformed cell lines for longer periods was not successful in our experiments.

Support of T-cell survival with telomerase vectors

Recently, the exogenous expression of human telomerase reverse transcriptase via retroviral vectors was reported to immortalize cytotoxic T-cell clones, however, without alteration of the phenotype and without loss of antigen-specificity or functionality (Hooijberg *et al.*, 2000). The senecence of T cells had been correlated with the erosion and shortening of telomere ends and the counter-acting telomerase reverse transcriptase activity is thought to regulate the life span of primary T cells. The term immortalization is difficult in this context: although the stability of CD8$^+$ cytotoxic T-cell clones was enhanced, the proliferation remained dependent on cytokines and periodic restimulations (Hooijberg *et al.*, 2000; Migliaccio *et al.*, 2000; Rufer *et al.*, 2001). Since these observations of other research groups are recent, we refer to the original literature for experimental details.

Transformation by human T-cell leukaemia virus type I

Human T-cell leukaemia virus type 1 (HTLV-1) has been used extensively for immortalizing human T cells. Here, the method is described based on a report by Nutman (1997). HTLV-transformed T cells do release virus particles. Moreover, HTLV is a human leukaemia virus with containment level P3. Mononuclear cells from cord blood or adult peripheral blood or T-cell clones are applied at 10^6 cells ml^{-1} in complete RPMI-1640 medium with 10 to 100 U ml^{-1} recombinant IL-2. The cells are activated with mitogen (PHA or ConA) or with the specific antigen and incubated in 24-well plates for 2 to 3 days. Activated cells are seeded at 10^6 cells ml^{-1} in

24-well plates. The HTLV-1 producer cells (10^6 cells ml^{-1}; e.g. MT-2; or HUT-102, available from ATCC as TIB-162) are irradiated at 6 krad in a γ-irradiator (^{137}Cs source, e.g. Gammacell 1000). Alternatively, the cells are inactivated with 50 μg ml^{-1} mitomycin C for 1 h, followed by extensive washing with buffered saline. One million inactivated HTLV-producer cells are added to one million stimulated non-transformed cells. The co-cultures are incubated for 2 to 6 weeks with feeding IL-2 containing medium twice weekly. After 6 weeks, rapidly growing cells should be observed, which are already IL-2 independent in many cases. The co-cultivation is more efficient and reliable in comparison to infection with cell-free supernatants. HTLV-transformed T cells express CD25/IL-2Rα and MHC class II molecules abundantly on their surface. After a few months of culture, HTLV-transformed human T cells tend to lose the IL-2 dependence, cytotoxicity, and TcR expression (Inatsuki *et al.*, 1989; Yssel *et al.*, 1989).

Immortalization of human B lymphocytes

In contrast to viral transformation of human T lymphocytes, the B-cell transformation by EBV is widely applied and a reliable, highly repro-ducible and easy procedure, which does not require much care about the cultures. This protocol is based on our own experience and on the recom-mendations by Tosato (1997). Similarly to the viruses used in the other protocols, EBV is considered a human pathogen with containment level P2. EBV-seronegative persons should not have access to EBV-containing cultures. One should keep in mind that EBV is easily reactivated from most EBV-transformed lymphoblastoid cell lines.

In order to generate EBV stocks, exponentially growing cells of the EBV-transformed marmoset producer-cell line B95-8 (ATCC CRL-1612) are seeded at 1×10^6 cells ml^{-1} in RPMI 1640 with 10% FCS, glutamine and antibiotics. After three days of culture, the cells are sedimented at low speed (1000 rpm, 10 min). The supernatant is sterile filtered and stored in portions at −80°C. Such supernatants contain up to 1000 transforming units per ml.

Fresh gradient-purified peripheral blood cells (10^7) in 2.5 ml complete RPMI medium are infected by addition of 2.5 ml of thawed supernatant from B95-8 cells (as described above). After 2 hours at 37°C, 5 ml of complete medium are added. EBV-specific cytotoxic T lymphocytes may overgrow such cultures. For this reason, cyclosporin A can be added to such cultures at a final concentration of 0.5 μg ml^{-1}. During the following 3 weeks, the cultures do not need much care. Once weekly, some medium may be exchanged (1/5 volume) without removing cells. After 3 weeks and depending on the metabolic activity, the cells can be expanded in complete medium. In most cases, rapid proliferation is established soon, which allows splitting the cells twice a week at ratio 1:2 per volume. The yielding lymphoblastoid B-cell lines (Nilsson and Klein, 1982) are latently infected by EBV and can be induced to production of virus antigens and particles in many cases.

◆◆◆◆◆◆ PHENOTYPE OF TRANSFORMED T CELLS

Persistence of herpesvirus saimiri

HVS, a monkey virus, can infect human cells. An early analysis of the HVS strain SMHI revealed a weak productive replication on primary human fetal cells (Daniel *et al.*, 1976a). Selectable HVS recombinants, derived from strain A11, were later used to study the spectrum of cells which can be infected by the virus. A broad range of epithelial, mesenchymal and haematopoetic cells became infected and carried non-integrated episomal DNA of the recombinant viruses. The pancreatic carcinoma line Panc-1 and human foreskin fibroblasts produced infectious virus under selection conditions (Simmer *et al.*, 1991). These findings suggest that the receptor used by HVS is widely distributed among various tissues. The receptor seems to be well conserved, as also rabbit T cells can be infected and transformed by HVS strains (Ablashi *et al.*, 1985; Medveczky *et al.*, 1989). Cell lines which had been infected with the recombinants under selection pressure retained the viral non-integrated episomes after withdrawal of the selecting drug for long time periods. The lack of counterselection against cells with persisting viral episomes may suggest that the virus persists with mostly suppressed viral gene expression (Simmer *et al.*, 1991). This model is also supported by the observations that the persisting non-integrated viral episomes are heavily methylated (Desrosiers *et al.*, 1979) and may carry extensive genomic deletions (Schirm *et al.*, 1984). These observations were made using mainly HVS strain A11. It is likely that they are also valid for subgroup C strains and for transformed human T cells. Although monkey T lymphocytes produce HVS particles in most cases, it was not possible to isolate virus from transformed human T-cell cultures which carry non-integrated viral episomes in high copy number (Biesinger *et al.*, 1992). Even after treatment with phorbolesters, nucleoside analogues and other drugs known to cause reactivation of other viruses like EBV, or after specific or non-specific stimulation of the T cells, virion production could not be demonstrated (Fickenscher *et al.*, 1996a). Nevertheless, it will be difficult to provide a formal proof that the virus can never be reactivated from transformed human T lymphocytes.

Viral gene expression in rhadinovirus-transformed human T cells

In contrast to transformed New World monkey T cells which regularly express all the virus genes and produce virus particles, viral transcription in HVS-transformed human T cells was mainly demonstrated from the terminal genomic L-DNA region where the viral oncogene *stpC/tip* is situated (Fickenscher *et al.*, 1996a). In transformed human T cells, the mRNA of the viral oncogene *stpC/tip* is subjected to cellular regulation. Two proteins are translated from the bicistronic transcript. The abundant cytoplasmic phosphoprotein StpC has a predicted size of approximately 10 kDa and migrates at 21 kDa on SDS protein gels (Jung and

Desrosiers, 1991; Fickenscher *et al.*, 1996a). p40Tip is expressed at extremely low levels and is only detectable after Lck immunoprecipitation with subsequent Lck phosphotransferase assay (Biesinger *et al.*, 1995; Jung *et al.*, 1995a,b; Lund *et al.*, 1995, 1996, 1997b; Wiese *et al.*, 1996; Fickenscher *et al.*, 1997). Similarly to the reported observations for virus strain C488, the homologous *stpC484* transcript is expressed in transformed cells (Geck *et al.*, 1990, 1991; Medveczky *et al.*, 1993b). In strain C484-infected tumour cells, a tricistronic transcript was described encoding dihydrofolate reductase and the transformation-associated genes (Whitaker *et al.*, 1995). Most likely, this transcript is correlated with lytic conditions (Fickenscher *et al.*, 1996a). Both, *stpC* and *tip* are essential for pathogenicity and T-cell transformation (Knappe *et al.*, 1997; Duboise *et al.*, 1998a).

Moreover, the adjacent non-coding viral U-RNA genes (HSUR, herpesvirus saimiri U-RNA) are abundantly expressed, similarly to the EBER-RNAs of EBV (Biesinger *et al.*, 1990; Albrecht and Fleckenstein, 1992; Albrecht *et al.*, 1992; Myer *et al.*, 1992). However, the HSUR genes are dispensable for virus-induced T-cell transformation (Ensser *et al.*, 1999). By subtractive hybridization, another virus gene was found to be inducibly and heavily transcribed in HVS-transformed human T cells (Knappe *et al.*, 1997). The immediate-early gene *ie14/vsag* shows homology to murine superantigens and codes for a secreted factor which binds to MHC class II molecules and is able to support T-cell proliferation (Nicholas *et al.*, 1990; Yao *et al.*, 1996). In common marmoset monkeys, the deletion of *ie14/vsag* was reported to affect virus persistence, T-cell transformation, and pathogenicity (Duboise *et al.*, 1998b). However, in human T-cell cultures, a selection for certain Vβ chains could be observed, and *ie14/vsag* deletion mutants did neither influence human T-cell transformation in culture, nor pathogenicity in cotton top tamarins (Knappe *et al.*, 1998b). In transformed human T cells, only a few other HVS genes have been reported to be transcribed at very low amounts. These comprise the immediate-early gene *ie57* which is a post-transcriptional regulator and a homologue to ICP27 of herpes simplex virus, the early transcriptional regulator *orf50*, and the bicistronic transcript for the viral cyclin and the apoptosis inhibitor vFLIP. By deletion mutagenesis, *ie14/vsag*, *cyclin*, and *vFLIP* have been excluded to contribute to pathogenesis or T-cell transformation (Knappe *et al.*, 1997, 1998b; Glykofrydes *et al.*, 2000; Thurau *et al.*, 2000; Ensser *et al.*, 2001).

The replication origins of HVS have not yet been clearly defined. Preliminary work by Schofield (1994) assigned the origin for lytic replication to the region upstream of the thymidylate synthase gene. Kung and Medveczky (1996) identified a viral DNA fragment that permits stable episomal replication in transformed T cells. This observation, however, is not sufficient to define the plasmid replication origin: There is pronounced sequence variation between virus strains in this region, and viral deletion mutants of this region have been described, which are still capable of episomal persistence. It is noteworthy that strain C139 persists at low copy number in comparison to the high copy number of all other previous studies on HVS (Fickenscher *et al.*, 1997).

The phenotype of human growth-transformed T lymphocytes was surveyed by Fickenscher and Fleckenstein (2001) and Meinl et al. (1995a). Human T lymphocytes which are transformed to stable growth by HVS C strains are mostly dependent on exogenous IL-2 and do not need restimulation with antigen or mitogen. The cell numbers increase by factors of two to four per week. The cells show the morphology of T blasts with irregular shape and express surface molecules that are typical for activated mature T lymphocytes. The cell lines exhibit the CD4$^+$/CD8$^-$ or the CD4$^-$/CD8$^+$ phenotype. Mixed populations may occur when polyclonal populations are infected. The surface antigens CD2, CD3, CD5, CD7, CD25 (IL-2Rα), CD30, CD69, MHC class II and CD56 (an NK-cell marker) are expressed, while the NK-cell markers CD16 and CD57 are lacking (Biesinger et al., 1992). The RO and RB isoforms of the membrane-bound phosphatase CD45, which are typically found on mature memory T cells, were both present on CD4$^+$ transformed T lymphocytes. CD8$^+$ transformed cell lines, however, expressed the CD45 isoform RA additionally, which is typical for naive T cells and their precursors.

Cytokines are produced by the transformed human T blasts after activation. IL-2 (Mittrücker et al., 1992) and IL-3 (De Carli et al., 1993) are secreted by the cells in response to mitogenic or antigenic stimuli. Antibodies against the IL-2Rα chain (Mittrücker et al., 1992) and against IL-2 and IL-3 (De Carli et al., 1993) suppressed the growth rate; both cytokines seem to support autocrine growth. IL-4 and IL-5 are secreted only at low rates by transformed Th2 cells (De Carli et al., 1993). Transformed CD4$^+$ T cells secrete interferon-γ (IFN-γ), tumour necrosis factor-α (TNF-α), TNF-β and granulocyte-macrophage colony-stimulating factor (GM-CSF) after specific or non-specific stimulation (Bröker et al., 1993; De Carli et al., 1993; Weber et al., 1993; Meinl et al., 1995b). Both, Th1 and Th2 clones were transformed by HVS. The cytokine pattern of the Th1 clones was enhanced, while the profile of Th2 clones switched to a mixed phenotype after transformation (De Carli et al., 1993). The viral transformed human T cells show an inducible non-specific cytotoxic activity. When tested on K562 cells, CD8$^+$ lines and to a lesser extent CD4$^+$ cells showed NK-like cytolytic activity (Biesinger et al., 1992). The lectin-dependent cytolytic activity of Th1 clones against P815 target cells was enhanced after transformation, while Th2 clones showed this activity only in the transformed state (De Carli et al., 1993). The cytotoxic activity of a transformed $\gamma\delta$-T-cell clone on K562 was inducible by stimulation with IL-12 (Klein et al., 1996). HVS-transformed T cells are capable of delivering non-specific B-cell help via membrane-bound TNF-α or via CD40 ligand (Del Prete et al., 1994; Hess et al., 1995; Saha et al., 1996b).

The karyotypes of a series of cell lines have been analysed in detail and found to be normal (Troidl et al., 1994). When early signal transduction properties of the transformed cells were compared with those of the non-infected parental cells, no significant differences were encountered. After stimulation with IL-2, anti-CD3 and/or anti-CD4, similar patterns of tyrosine phosphorylation and calcium mobilization were observed in primary clones or in transformed lines (Bröker *et al.*, 1993; Mittrücker *et al.*, 1993). In contrast, Jurkat cells (Schneider *et al.*, 1977) behaved differently (Bröker *et al.*, 1993). HVS-transformed cell lines were strongly stimulated by cell-bound CD58 which is expressed on cells of various origin. This effect was mediated by CD2/CD58 interaction and led to IL-2 production and enhanced proliferation (Mittrücker *et al.*, 1992). The functionality of CD3, CD4 and IL-2R was shown after stimulation by signal transduction parameters, by proliferation, and by IFN-γ production (Bröker *et al.*, 1993; Weber *et al.*, 1993). The IL-2 dependent proliferation of transformed lymphocytes was strongly inhibited by soluble CD4 antibodies. This effect could be overcome by high doses of IL-2. In parallel, the activity and abundance of the CD4 bound fraction of the tyrosine kinase p56^{lck} was diminished by anti-CD4 treatment (Bröker *et al.*, 1994). A comparison of non-infected clones and their HVS-transformed derivatives indicated that the growth-transformation by HVS is not based on a resistance to apoptosis (Kraft *et al.*, 1998). Neither the viral antiapoptotic gene *vFLIP*, nor the viral Bcl2-homologue appear to contribute to T-cell transformation (Thome *et al.*, 1997; Derfuss *et al.*, 1998; Glykofrydes *et al.*, 2000). HVS transformed T cells can be driven into activation-dependent cell death after anti-CD3- or TPA-stimulation. This form of cell death is independent of the CD95/CD95-ligand interaction (Bröker *et al.*, 1997).

Characterized T helper-cell clones reacting specifically to myelin basic protein (Weber *et al.*, 1993; Meinl *et al.*, 1995b), tetanus toxoid (Bröker *et al.*, 1993), bovine 70 kDa heat shock protein (HSP70), *Lolium perenne* group I antigen, *Toxocara canis* excretory antigen, and to purified protein derivative from *Mycobacterium tuberculosis* (De Carli *et al.*, 1993) were successfully transformed and retained their MHC-restricted antigen specificity. The high basal proliferation activity which is probably due to contact with CD58 bearing cells (Mittrücker *et al.*, 1992) may interfere with the demonstration of antigen specificity, since the antigen-presenting cells alone cause a stimulation of the transformed T cells. This antigen-independent proliferation of the transformed antigen-specific clones could be reduced by using monoclonal antibodies against CD58 and CD2 together with MHC-transfected mouse L-cells as antigen-presenting cells (Weber *et al.*, 1993), or by starving the cells prior to antigen presentation (Bröker *et al.*, 1993). In all three cases, clear responses to antigen contact were noticed, measured by proliferation and cytokine production. IFN-γ production is the best readout parameter for activation (Bröker *et al.*, 1993; Weber *et al.*, 1993). Behrend *et al.* (1993) reported that EBV-specific cytotoxic T lymphocytes (CTL) retained their antigen-specific reactivity after viral transformation. In another study, a transformed tumour-specific CTL line recognized the autologous target only if the tumour cells were pretreated with IFN-γ (Okada *et al.*, 1997).

Only a few specific cellular and biochemical features are changed in HVS transformed T cells: the spontaneous growth is associated with CD2 hyperreactivity. Whereas primary T cells need the engagement of two epitopes on the CD2 molecule for activation, ligation of a single CD2 monoclonal antibody leads to the stimulation of herpesvirus-transformed

675

T lymphocytes (Mittrücker *et al.*, 1992). IFN-γ secretion can be stimulated to very high levels (Bröker *et al.*, 1993; De Carli *et al.*, 1993; Weber *et al.*, 1993). The B-cell specific tyrosine kinase Lyn is expressed in T cells transformed by either HVS or HTLV-1 (Yamanashi *et al.*, 1989; Wiese *et al.*, 1996; Fickenscher *et al.*, 1997). By subtractive hybridization, AK155 or IL-26, a novel cellular sequence homologue to IL-10, was found over-expressed in HVS-transformed T cells (Knappe *et al.*, 2000b). AK155/IL-26 is a candidate for an autocrine growth factor, but may alternatively be functionally unrelated to T-cell transformation (Fickenscher *et al.*, 2002).

When polyclonal T-cell populations are transformed by HVS, T-cell cultures with αβ-TcR are routinely obtained. These cultures are usually polyclonal or oligoclonal. There is no preference for expression of particular TcR-Vβ genes. Several months after infection, a predominance of a reduced number of TcR Vβ gene families was found, which varied between parallel cultures. The most likely explanation is that the originally polyclonal T-cell cultures are overgrown by those clones which proliferate most rapidly. This is a general phenomenon in long-term cell culture which occurs in the absence of any specific stimulation (Fickenscher *et al.*, 1996b; Knappe *et al.*, 1997).

Various subgroup C virus strains are able to transform human T cells, however to a varying extent. Virus strain C484 was reported to transform human T cells to a short term, IL-2 independent growth (Medveczky *et al.*, 1993a). Different subgroup C strains (C488, C484, C139) were compared in growth transformation of human cord blood T cells. The resulting clonal T-cell lines were either CD4+ or CD8+, and expressed either αβ- or γδ-TcRs. If transformed by the same virus strain, the αβ- and γδ-clones were similar with respect to viral persistence, virus gene expression, proliferation and Th1 type cytokine production. However, major differences were observed in T cells transformed by different subgroup C strains. Strain C139 persisted at low copy number as compared with the high copy number of prototype C488. The transformation-associated genes *stpC* and *tip* of strain C488 were strongly induced after T-cell stimulation. The homologous genes of strain C139 were only weakly expressed and not induced after activation. After CD2 ligation of a single CD2 epitope, C488 transformed T cells produced IL-2, whereas C139 transformed cells did not. Correspondingly, C139-transformed T cells were less sensitive to cyclosporin A. IFN-γ production was induced to a similar extent in both C139- and C488-transformed T cells by the CD2 stimulus. Sequence comparison from different subgroup C strains revealed a variability of the *stpC/tip* promoter region and of the Lck-binding viral protein Tip. Thus, closely related subgroup C strains of HVS cause major differences in the functional phenotype of growth-transformed human T cells (Fickenscher *et al.*, 1997).

So far, all transformed T cells tested from normal donors could be stimulated via TcR or CD3. Surprisingly, cells with defects in TcR-dependent signalling can also be transformed by HVS. The defect of the CD3γ chain did not prevent transformation (Rodriguez-Gallego *et al.*, 1996; Pacheco-Castro *et al.*, 1998; Zapata *et al.*, 1999). Transformed T cells from a patient with atypical X-linked severe combined immunodeficiency showed a spontaneous partial reversion of the genetic defect affecting the IL-2Rα chain (Stephan *et al.*, 1996). Fas-deficient T cells of a human patient were efficiently transformed (Bröker *et al.*, 1997). Moreover, HVS-transformed T cells have been studied functionally from patients with defects in CD18, MHC class II, or ZAP-70, or with Wiskott Aldrich syndrome, ataxia telangiectasia, or X-linked lymphoproliferative syndrome (Gallego *et al.*, 1997; Alvarez-Zapata *et al.*, 1998; Allende *et al.*, 2000; Rivero-Carmena *et al.*, 2000; Meinl *et al.*, 2001; Nakamura *et al.*, 2001). Transformation with HVS seems promising for studying T cells from patients with severe combined immunodeficiencies.

HVS-transformed human CD4+ T cells provide a productive lytic system for T-lymphotropic viruses such as human herpesvirus type 6 (Neipel and Fleckenstein, unpublished observations) and human immunodeficiency virus (HIV) (Nick *et al.*, 1993; Saha *et al.*, 1996a; Vella *et al.*, 1997, 1999a,b). The prototype viruses HIV-1$_{IIIB}$ and HIV-2$_{ROD}$ replicated rapidly and caused cell death within 14 d. Also a poorly replicating HIV-2 strain and primary clinical isolates grew to high titres. HVS-transformed human CD4+ T cells can be used for poorly replicating HIV strains with narrowly restricted host cell range (Nick *et al.*, 1993). Moreover, HVS-transformed T cells can be persistently and productively infected with HIV. In comparison to conventional T-lymphoma cell lines, the down-regulation of surface CD4 molecules is delayed (Vella *et al.*, 1997). Similarly to cultures of primary peripheral blood cells, HVS-transformed T cells allow the propagation of macrophage-tropic HIV isolates without selecting for subtypes with changed phenotype or cell tropism (Vella *et al.*, 1999a,b). HVS-transformed CD8+ T cells were shown to secrete soluble HIV-inhibiting factors distinct from any known inhibitory cytokine (Copeland *et al.*, 1995, 1996; Lacey *et al.*, 1998; Leith *et al.*, 1999; Saha *et al.*, 1999). Transformed CD4+ and CD8+ T cells produced varying amounts of the cytokines IL-8, IL-10, TNF-α, TNF-β, RANTES, MIP-1α, and MIP-1β (Mackewicz *et al.*, 1997; Saha *et al.*, 1998). Surprisingly, transformed CD4+ T-cell clones from AIDS patients produced no RANTES and little or no MIP-1α or MIP-1β and were more readily infectable with HIV in comparison to T cells from non-progressors which produced high amounts of chemokines and were less infectable (Saha *et al.*, 1998). HVS-transformed human CD4+ T cells expressing CCR5 and CXCR4 were fully functional as antigen-presenting target cells for HIV-specific, MHC class I-restricted cytotoxic T-cell activity (Bauer *et al.*, 1998). Similarly, HVS-transformed CD4+ T cells were used to present plasmodium antigens and to generate antigen-reactive T-cells (Daubenberger *et al.*, 2001).

The use of HVS focused on T cells from humans during the last years. Its behaviour in various monkey systems is of importance, as on the one hand, HVS is a tumour virus of New World monkeys, and on the other hand, Old World monkeys like macaques are the most used animal model for the close-to-human situation. In New World primate T cells, HVS establishes a semi-permissive infection: the cells do produce virus for long time periods, and are transformed as well (summarized in Fickenscher *et al.*, 1996a). Transcription of IL-2 and activity of IL-4 has been shown from

677

such cultures (Chou *et al.*, 1995). Similarly to human T cells, the T lympho-cytes from macaque monkeys can be growth-transformed by HVS (Akari *et al.*, 1996; Feldmann *et al.*, 1997; Meinl *et al.*, 1997; Knappe *et al.*, 2000a). Some researchers observe IL-2 dependence, others IL-2 independence. With many respects, the transformed macaque T cells resemble their human counterparts. Antigen-specific T-cell lines against myelin basic protein or streptolysin O retained their reactivity after transformation, but in contrast to their non-infected progenitor cells, the antigen specificity of the transformants was obscured by increasing concentrations of irradiated autologous blood cells as antigen-presenting cells. The MHC class II expressing transformed cells were able to present the antigen to each other in the absence of autologous presenter cells (Meinl *et al.*, 1997). One major difference is the pronounced frequency of double positive CD4+/CD8+ T cells, which are uncommon in humans. T-cell immunology and T-cell transformation from macaques is greatly hampered by reactivation of foamy virus, of which most rhesus monkeys in primate centres are infected (Feldmann *et al.*, 1997; Knappe *et al.*, 2000a). In macaque monkeys which were transfused with autologous transformed T cells, leukaemias or lymphomas did not develop, and the animals were protected from disease after intravenous challenge infection with high-titred wild-type virus (Knappe *et al.*, 2000a). In naive macaques, HVS C488 caused an acute disease which is lethal within 2 weeks. Most probably, it is a peracute form of polyclonal T-cell leukaemia which has not been described for humans (Alexander *et al.*, 1997; Knappe *et al.*, 2000a).

Retrovirus-transformed human T cells

HTLV-1 provided the first method to immortalize human T lymphocytes. Similarly to the EBV method for B cells, HTLV-1 has been used success-fully for T-cell transformation (Miyoshi *et al.*, 1981; Yamamoto *et al.*, 1982; Yoshida *et al.*, 1982; Popovic *et al.*, 1983). The resulting cell lines were valuable in numerous scientific questions, which cannot be summarized in this article. However, the phenotype of HTLV-1 transformed human T cells needs to be discussed. In most cases the transformed T cells are CD4+/CD8−. However, Faller *et al.* (1988) reported that CD4−/CD8+ human T cells can also be immortalized. The transformed cells do release virus particles in most cases, and they express high amounts of CD25 (IL-2Rα chain) and HLA-DR molecules on their surface. Signal transduction studies revealed that the B-cell tyrosine kinase Lyn is aberrantly expressed, which does not normally occur in T cells (Yamanashi *et al.*, 1989). This observation is similarly made in HVS-transformed T cells (Wiese *et al.*, 1996; Fickenscher *et al.*, 1997). After a few months of culture, HTLV-1 transformed T cells tend to lose their antigen-specific reactivity, their TcR complex on the surface, their cytotoxic activity, as well as the dependence on exogenous IL-2 (Inatsuki *et al.*, 1989; Yssel *et al.*, 1989). This tendency reflects the development of autonomous growth in leukaemia development.

◆◆◆◆◆◆ VECTOR APPLICATIONS OF HERPESVIRUS SAIMIRI

Transformation assay vectors

Non-oncogenic HVS strains lacking the oncogene region have been used as eukaryotic expression vectors in order to investigate heterologous oncogenes. HVS deletion mutants without the terminal transformation-associated L-DNA region neither cause malignant disease in animals, nor transform simian lymphocytes in culture (Desrosiers *et al.*, 1984, 1986). Recombinant viruses can infect and persist in human T cells and in a broad range of haematopoietic, mesenchymal and epithelial cells under selection conditions (Simmer *et al.*, 1991).

The X region of the human T-cell leukaemia virus type 1 (HTLV-1) was inserted by homologous recombination into the HVS A11 derivate S4 (Grassmann *et al.*, 1989). The HTLV-1 X region in the recombinant herpesvirus vector was able to transform primary human T lymphocytes to IL-2 dependent growth, in a similar way to HTLV-1 transformed cells (Grassmann *et al.*, 1989). Similarly to HTLV-immortalized cells, the *tax* recombinant-transformed cells expressed CD4, MHC class II and CD25 (IL-2Rα) in large amounts. The transformed cells contained episomes of the recombinant rhadinoviruses in high copy number, but did not shed infectious virus. Deletion variants of the HTLV-1 X region were introduced into the viral vector. The Rex protein of HTLV-1 did not show transforming properties. The broad transcriptional activator p40Tax was found to be necessary and sufficient for lymphocyte immortalization in the context of the herpesviral vector, thus assigning the T-cell transforming function of the HTLV-1 X region to the Tax protein (Grassmann *et al.*, 1989, 1992; Rosin *et al.*, 1998; Schmitt *et al.*, 1998; reviewed in Grassmann *et al.*, 1994).

An improved HVS recombination vector system was used for studying *c-fos* function. Overexpression of the proto-oncogene *c-fos* is known to induce transformation in a broad range of primary cells of various mammalian species. *c-fos* recombinant HVS vectors expressed large amounts of the oncoprotein during persistent infection of human neonatal fibroblasts. However, these primary mesenchymal cells did not show any sign of transformation (Alt *et al.*, 1991; Alt and Grassmann, 1993). The T-cell transforming function of HVS deletion vectors was successfully reconstituted by the cellular *ras* gene, by the gene *K1* of KSHV/HHV-8, and by the gene *R1* of the rhesus rhadinovirus (Guo *et al.*, 1998; Lee *et al.*, 1998; Damania *et al.*, 1999). This system of complementing a transformation-deficient HVS strain to a transforming phenotype by heterologous oncogenes is also applicable as an oncogene trap in order to identify novel cellular or viral transforming genes.

Rhadinovirus vectors for the expression of heterologous genes

HVS-derived vectors were further used for growth hormone, for secreted alkaline phosphatase, for thymidine kinase, and for green fluorescent protein (Desrosiers *et al.*, 1985b; Duboise *et al.*, 1996; Stevenson *et al.*, 1999; Hiller *et al.*, 2000). An early preclinical gene therapy trial was performed with a non-transforming replication-competent HVS vector expressing the bovine growth hormone in genomic intron-containing configuration. Persistently infected simian T cells produced high amounts of growth hormone. Experimentally infected New World primates produced circulating bovine growth hormone and later developed a humoral immune response (Desrosiers *et al.*, 1985b). These observations suggested that persisting HVS vectors could be used to replace missing or defective genes in hereditary genetic disorders.

The original method of generating HVS expression vectors used homologous recombination via a single stretch of viral terminal L-DNA in the recombination construct (Desrosiers *et al.*, 1985b; Grassmann and Fleckenstein, 1989). A more elaborated procedure for the isolation of HVS mutants was developed by the insertion of an autofluorescent reporter gene flanked by single-cutter restriction endonuclease recognition sites into the viral genome instead of the transforming *stpC* gene. Thus, the reporter gene cassette could easily be replaced by other transgenes after simple restriction enzyme digestion and ligation (Duboise *et al.*, 1996). This vector has mainly been used for the expression of heterologous oncogenes in order to assay transforming activity in monkey T cells, as mentioned above. An alternative approach has been developed by cloning HVS C488 into cosmid vectors. The co-transfection of overlapping cosmids into permissive epithelial OMK cells led to the reconstitution of recombinant replication-competent virus (Ensser *et al.*, 1999). This approach is valuable for generating expression vectors for foreign genes because a contamination with wild-type virus is excluded.

Non-selectable recombinant viruses expressing an autofluorescent protein were able to transduce human haematopoietic progenitor cells in culture, but at low efficiency and with a tendency towards partially differentiated cells (Stevenson *et al.*, 1999). Moreover, active HVS replication was observed in certain human cell types (Stevenson *et al.*, 1999, 2000b,c), confirming similar results published previously (Daniel *et al.*, 1976a; Simmer *et al.*, 1991). HVS efficiently infected totipotent mouse embryonic stem (ES) cells under drug selection pressure. The infected ES cells stably maintained the viral episomal genome and could be terminally differentiated into mature haematopoietic cells, while the heterologous gene was rather stably expressed (Stevenson *et al.*, 2000a). This system may be of particular interest for studying transgene effects during cell differentiation *in vitro* and *in vivo*.

Episomal herpesvirus saimiri vectors for adoptive immunotherapy

Gene transfer into primary human T cells by transfection or retroviral transduction methods remains difficult and unreliable with respect to long-term transgene expression. The maintained functional phenotype of HVS-transformed T cells suggested the use of HVS as a vector for human T lymphocytes at least for cell culture experiments. The reactivation of recombinant or wild-type virus from transformed human T cells has not been observed, but cannot be formally excluded. The techniques of homologous recombination and cosmid complementation can be applied for constructing replication-defective, but transformation-competent deletion variants which preclude reactivation. Furthermore, additional genes can be introduced into subgroup C virus strains by these methods in order to express and to study those gene products in human T lymphocytes.

HVS might be useful as a gene vector for targeted amplification of functional human T cells even for therapeutic applications if a series of biosafety aspects are clarified. By reinfusion of autologous transformed T cells into the donor macaques, an intrinsic oncogenic phenotype could be excluded, because the animals did not develop disease while the infused T cells persisted for extended periods (Knappe et al., 2000a). As an example for a possible application, a HVS-transformed human CD4+ T-cell clone with specific cytolytic activity against mucin-expressing tumour cells inhibited the growth of a tumour-cell line in NOD/SCID mice (Pecher et al., 2001). In order to improve the biological safety of such vectors, the prodrug activating gene thymidine kinase of herpes simplex virus was inserted into the genome of HVS. Thymidine kinase-expressing transformed T cells were efficiently eliminated in the presence of low concentrations of ganciclovir over an observation period of 1 year. At any time during the course of a therapeutic application, thymidine kinase-expressing transformed human T cells might be eliminated after administration of ganciclovir. In principle, this function could be useful for the T-cell dependent immunotherapy of resistant blood cancer while avoiding the risk of uncontrolled graft-versus-host disease (Hiller et al., 2000). Replication-deficient vector variants are another step towards application. The use of HVS vectors for redirecting the antigen specificity of primary human T cells may provide an important tool for experimental cancer therapy.

Acknowledgements

Original work underlying this review was supported by the Sonderforschungsbereich 466 (Deutsche Forschungsgemeinschaft, Bonn), the Bayerische Forschungsstiftung (Munich), and the Wilhelm Sander-Stiftung (Neustadt/Donau).

References

Ablashi, D. V., Schirm, S., Fleckenstein, B., Faggioni, A., Dahlberg, J., Rabin, H., Loeb, W., Armstrong, G., Peng, J. W., Aulahk, G. and Torrisi, W. (1985). Herpesvirus saimiri-induced lymphoblastoid rabbit cell line: growth characteristics, virus persistence, and oncogenic properties. J. Virol. 55, 623–633.

Akari, H., Mori, K., Terao, K., Otani, I., Fukasawa, M., Mukai, R. and Yoshikawa, Y. (1996). In vitro immortalization of old world monkey T lymphocytes with herpesvirus saimiri: Its susceptibility to infection with simian immunodeficiency viruses. Virology 218, 382–388.

Growth Transformation
of Human T Cells

Albrecht, J. C. and Fleckenstein, B. (1992). Nucleotide sequence of HSUR 6 and HSUR 7, two small RNAs of herpesvirus saimiri. *Nucl. Acids Res.* **20**, 1810.

Albrecht, J. C., Nicholas, J., Biller, D., Cameron, K. R., Biesinger, B., Newman, C., Wittmann, S., Craxton, M. A., Coleman, H., Fleckenstein, B. and Honess, R. W. (1992). Primary structure of the herpesvirus saimiri genome. *J. Virol.* **66**, 5047–5058.

Alexander, L., Du, Z., Rosenzweig, M., Jung, J. U. and Desrosiers, R. C. (1997). A role for natural simian immunodeficiency virus and human immunodeficiency virus type 1 nef alleles in lymphocyte activation. *J. Virol.* **71**, 6094–6099.

Allende, L. M., Hernandez, M., Corell, A., Garcia-Perez, M. A., Varela, P., Moreno, A., Caragol, I., Garcia-Martin, F., Guillen-Perales, J., Olive, T., Espanol, T. and Arnaiz-Villena, A. (2000). A novel CD18 genomic deletion in a patient with severe leucocyte adhesion deficiency: a possible CD2/lymphocyte function-associated antigen-1 functional association in humans. *Immunology* **99**, 440–450.

Alt, M. and Grassmann, R. (1993). Resistance of human fibroblasts to c-fos mediated transformation. *Oncogene* **8**, 1421–1427.

Alt, M., Fleckenstein, B. and Grassmann, R. (1991). A pair of selectable herpesvirus vectors for simultaneous gene expression in human lymphoid cells. *Gene* **102**, 265–269.

Alvarez-Zapata, D., de Miguel Olalla, S., Fontan, G., Ferreira, A., Garcia-Rodriguez, M. C., Madero, L., van den Elsen, P. and Regueiro, J. R. (1998). Phenotypical and functional characterization of herpesvirus saimiri-immortalized human major histocompatibility complex class II-deficient T lymphocytes. *Tissue Antigens* **51**, 250–257.

Bankier, A. T., Dietrich, W., Baer, R., Barrell, B. G., Colbere-Garapin, F., Fleckenstein, B. and Bodemer, W. (1985). Terminal repetitive sequences in herpesvirus saimiri virion DNA. *J. Virol.* **55**, 133–139.

Bauer, M., Lucchiari-Hartz, M., Fickenscher, H., Eichmann, K., McKeating, J. and Meyerhans, A. (1998). Herpesvirus saimiri-transformed human CD4+ T-cell lines: an efficient target cell system for the analysis of human immunodeficiency virus-specific cytotoxic CD8+ T-lymphocyte activity. *J. Virol.* **72**, 1627–1631.

Behrend, K. R., Jung, J. U., Boyle, T. J., DiMaio, J. M., Mungal, S. A., Desrosiers, R. C. and Lyerly, H. K. (1993). Phenotypic and functional consequences of herpesvirus saimiri infection of human CD8+ cytotoxic T lymphocytes. *J. Virol.* **67**, 6317–6321.

Biesinger, B., Trimble J. J., Desrosiers, R. C. and Fleckenstein, B. (1990). The divergence between two oncogenic herpesvirus saimiri strains in a genomic region related to the transforming phenotype. *Virology* **176**, 505–514.

Biesinger, B., Müller-Fleckenstein, I., Simmer, B., Lang, G., Wittmann, S., Platzer, E., Desrosiers, R. C. and Fleckenstein, B. (1992). Stable growth transformation of human T lymphocytes by herpesvirus saimiri. *Proc. Natl. Acad. Sci. USA* **89**, 3116–3119.

Biesinger, B., Tsygankov, A. Y., Fickenscher, H., Emmrich, F., Fleckenstein, B., Bolen J. R. and Bröker, B. M. (1995). The product of the herpesvirus saimiri open reading frame 1 (tip) interacts with T cell-specific kinase p56lck in transformed cells. *J. Biol. Chem.* **270**, 4729–4734.

Bornkamm, G. W., Delius, H., Fleckenstein, B., Werner, F. J. and Mulder, C. (1976). Structure of herpesvirus saimiri genomes: Arrangement of heavy and light sequences in the M genome. *J. Virol.* **19**, 154–161.

Bröker, B. M., Tsygankov, A. Y., Müller-Fleckenstein, I., Guse, A. H., Chitaev, N. A., Biesinger, B., Fleckenstein, B. and Emmrich, F. (1993). Immortalization of human T cell clones by herpesvirus saimiri. Signal transduction analysis reveals functional CD3, CD4 and IL-2 receptors. *J. Immunol.* **51**, 1184–1192.

Bröker, B. M., Tsygankov, A. Y., Fickenscher, H., Chitaev, N. A., Schulze-Koops, H., Müller-Fleckenstein, I., Fleckenstein, B., Bolen, J. B. and Emmrich, F. (1994). Engagement of the CD4 receptor inhibits the interleukin-2-dependent proliferation of human T cells transformed by herpesvirus saimiri. *Eur. J. Immunol.* **24**, 843–850.

Bröker, B. M., Kraft, M., Klauenberg, U., Le Deist, F., de Villartay, J. P., Fleckenstein, B., Fleischer, B. and Meinl, E. (1997). Activation induces apoptosis in herpesvirus saimiri transformed T cells independent of CD95 (Fas, APO-1). *Eur. J. Immunol.* **27**, 2774–2780.

Chang, Y., Cesarman, E., Pessin, M. S., Lee, F., Culpepper, J., Knowles, D. M. and Moore, P. S. (1994). Identification of herpesvirus-like DNA sequences in AIDS-associated Kaposi's sarcoma. *Science* **266**, 1865–1869.

Choi, J. K., Ishido, S. and Jung, J. U. (2000). The collagen repeat sequence is a determinant of the degree of herpesvirus saimiri STP transforming activity. *J. Virol.* **74**, 8102–8110.

Chou, C.-S., Medveczky, M. M., Geck, P., Vercelli, D. and Medveczky, P. G. (1995). Expression of IL-2 and IL-4 in T lymphocytes transformed by herpesvirus saimiri. *Virology* **208**, 418–426.

Copeland, K. F., McKay, P. J. and Rosenthal, K. L. (1995). Suppression of activation of the human immunodeficiency virus long terminal repeat by CD8+ T cells is not lentivirus specific. *AIDS Res. Hum. Retroviruses* **11**, 1321–1326.

Copeland, K. F., McKay, P. J. and Rosenthal, K. L. (1996). Suppression of the human immunodeficiency virus long terminal repeat by CD8+ T cells is dependent on the NFAT-1 element. *AIDS Res. Hum. Retroviruses* **12**, 143–148.

Damania, B., Li, M., Choi, J. K., Alexander, L., Jung, J. U. and Desrosiers, R. C. (1999). Identification of the R1 oncogene and its protein product from the rhadinovirus of rhesus monkeys. *J. Virol.* **73**, 5123–5131.

Daniel, M. D., Melendez, L. V. and Barahona, H. H. (1972). Plaque characterization of viruses from south american nonhuman primates. *J. Natl. Cancer Inst.* **49**, 239–249.

Daniel, M. D., Silva, D., Jackmann, D., Sehgal, P., Baggs, R. B., Hunt, R. D., King, N. W. and Melendez, L. V. (1976a). Reactivation of squirrel monkey heart isolate (herpesvirus saimiri strain) from latently infected human cell cultures and induction of malignant lymphoma in marmoset monkeys. *Bibliotheca Haematologica* **43**, 392–395.

Daniel, M., Silva, D. and Ma, N. (1976b). Establishment of owl monkey kidney 210 cell line for virological studies. *In vitro* **12**, 290.

Daubenberger, C. A., Nickel, B., Hubner, B., Siegler, U., Meinl, E. and Pluschke, G. (2001). Herpesvirus saimiri transformed T cells and peripheral blood mononuclear cells restimulate identical antigen-specific human T cell clones. *J. Immunol. Methods* **254**, 99–108.

De Carli, M., Berthold, S., Fickenscher, H., Müller-Fleckenstein, I., D'Elios, M. M., Gao, Q., Biagiotti, R., Giudizi, M. G., Kalden, J. R., Fleckenstein, B., Romagnani, S. and Del Prete, G. (1993). Immortalization with herpesvirus saimiri modulates the cytokine secretion profile of established Th1 and Th2 human T cell clones. *J. Immunol.* **151**, 5022–5030.

Del Prete, G., De Carli, M., D'Elios, M. M., Müller-Fleckenstein, I., Fickenscher, H., Fleckenstein, B., Almerigogna, F. and Romagnani, S. (1994). Polyclonal B cell activation induced by herpesvirus saimiri-transformed human CD4+ T cell clones: Role for membrane TNF-alpha/TNF-alpha receptors and CD2/CD58 interactions. *J. Immunol.* **152**, 4872–4879.

Derfuss, T., Fickenscher, H., Kraft, M. S., Henning, G., Lengenfelder, D., Fleckenstein, B. and Meinl, E. (1998). Antiapoptotic activity of the herpesvirus saimiri-encoded Bcl-2 homolog: stabilization of mitochondria and inhibition of caspase-3-like activity. *J. Virol.* **72**, 5897–5904.

Desrosiers, R. C. and Falk L. A. (1982). Herpesvirus saimiri strain variability. *J. Virol.* **43**, 352–356.

Desrosiers, R. C., Mulder, C. and Fleckenstein, B. (1979). Methylation of herpesvirus saimiri DNA in lymphoid tumor cell lines. *Proc. Natl. Acad. Sci. USA* **76**, 3839–3843.

Desrosiers, R. C., Burghoff, R. L., Bakker, A. and Kamine, J. (1984). Construction of replication-competent herpesvirus saimiri deletion mutants. *J. Virol.* **49**, 343–348.

Desrosiers, R. C., Bakker, A., Kamine, J., Falk, L. A., Hunt, R. D. and King, N. W. (1985a). A region of the herpesvirus saimiri genome required for oncogenicity. *Science* **228**, 184–187.

Desrosiers, R. C., Kamine, J., Bakker, A., Silva, D., Woychik, R. P., Sakai, D. D. and Rottmann, F. M. (1985b). Synthesis of bovine growth hormone in primates by using a herpesvirus vector. *Mol. Cell. Biol.* **5**, 2796–2803.

Desrosiers, R. C., Silva, D. P., Waldron, L. M. and Letvin N. L. (1986). Non-oncogenic deletion mutants of herpesvirus saimiri are defective for in vitro immortalization. *J. Virol.* **57**, 701–705.

Duboise, S. M., Guo, J., Desrosiers, R. C. and Jung, J. U. (1996). Use of virion DNA as a cloning vector for the construction of mutant and recombinant herpesviruses. *Proc. Natl. Acad. Sci. USA* **93**, 11389–11394.

Duboise, S. M., Guo, J., Czajak, S., Desrosiers, R. C. and Jung, J.U. (1998a). STP and Tip are essential for herpesvirus saimiri oncogenicity. *J. Virol.* **72**, 1308–1313.

Duboise, M., Guo, J., Czajak, S., Lee, H., Veazey, R., Desrosiers, R. C. and Jung, J. U. (1998b). A role for herpesvirus saimiri orf14 in transformation and persistent infection. *J. Virol.* **72**, 6770–6776.

Ensser, A., Pfinder, A., Müller-Fleckenstein, I. and Fleckenstein, B. (1999). The URNA genes of herpesvirus saimiri (strain C488) are dispensable for transformation of human T cells in vitro. *J. Virol.* **73**, 10551–10555.

Ensser, A., Glykofrydes, D., Niphuis, H., Kuhn, E. M., Rosenwirth, B., Heeney, J. L., Niedobitek, G., Müller-Fleckenstein, I. and Fleckenstein, B. (2001). Independence of herpesvirus-induced T cell lymphoma from viral cyclin D homologue. *J. Exp. Med.* **193**, 637–642.

Faller, D. V., Crimmins, M. A. V. and Mentzer, S. J. (1988). Human T-cell leukemia virus type 1 infection of CD4+ or CD8+ cytotoxic T-cell clones results in immortalization with retention of antigen specificity. *J. Virol.* **62**, 2942–2950.

Falk, L. A., Wolfe, L. G. and Deinhardt, F. (1972). Isolation of herpesvirus aimiri from blood of squirrel monkeys (Saimiri sciureus). *J. Natl. Cancer Inst.* **48**, 1499–1505.

Feldmann, G., Fickenscher, H., Bodemer, W., Spring, M., Nisslein, T., Hunsmann, G. and Dittmer, U. (1997). Generation of herpesvirus saimiri-transformed T-cell lines from macaques is restricted by reactivation of simian spuma viruses. *Virology* **229**, 106–112.

Fickenscher, H. and Fleckenstein, B. (2001). Herpesvirus saimiri. *Philos. Trans. R. Soc. Lond. B Biol. Sci.* **356**, 545–567.

Fickenscher, H., Biesinger, B., Knappe, A., Wittmann, S. and Fleckenstein, B. (1996a). Regulation of the herpesvirus saimiri oncogene stpC, similar to that of T-cell activation genes, in growth-transformed human T lymphocytes. *J. Virol.* **70**, 6012–6019.

Fickenscher, H., Meinl, E., Knappe, A., Wittmann, S. and Fleckenstein, B. (1996b). TcR expression of herpesvirus saimiri immortalized human T cells. *Immunologist* **4**, 41–43.

Fickenscher, H., Bökel, C., Knappe, A., Biesinger, B., Meinl, E., Fleischer, B., Fleckenstein, B. and Bröker, B. M. (1997). Functional phenotype of transformed human alpha-beta and gamma-delta T cells determined by different subgroup C strains of herpesvirus saimiri. *J. Virol.* **71**, 2252–2263.

Fickenscher, H., Hör, S., Küpers, H., Knappe, A., Wittmann, S. and Sticht, H. (2002). The interleukin-10 family of cytokines. *Trends Immunol.*, in press.

Fleckenstein, B. and Desrosiers, R. C. (1982). Herpesvirus saimiri and herpesvirus ateles. In *The Herpesviruses. Volume 1* (B. Roizman Ed.), pp. 253–331. Plenum Press, New York.

Fleckenstein, B. and Wolf, H. (1974). Purification and properties of herpesvirus saimiri DNA. *Virology* **58**, 55–64.

Fleckenstein, B., Bornkamm, G. W. and Ludwig, H. (1975). Repetitive sequences and defective genomes of herpesvirus saimiri. *J. Virol.* **15**, 398–406.

Fleckenstein, B., Müller, I. and Werner, J. (1977). The presence of herpesvirus saimiri genomes in virus-transformed cells. *Int. J. Cancer* **19**, 546–554.

Fleckenstein, B., Daniel, M. D., Hunt, R., Werner, J., Falk, L. A. and Mulder, C. (1978). Tumour induction with DNA of oncogenic primate herpesviruses. *Nature* **274**, 57–59.

Gallego, M. D., Santamaria, M., Pena, J. and Molina, I. J. (1997). Defective actin reorganization and polymerization of Wiskott-Aldrich T cells in response to CD3-mediated stimulation. Blood **90**, 3089–3097.

Gardella, T., Medveczky, P., Sairenji, T. and Mulder, C. (1984). Detection of circular and linear herpesvirus DNA molecules in mammalian cells by gel electrophoresis. *J. Virol.* **50**, 248–254.

Geck, P., Whitaker, S. A., Medveczky, M. M. and Medveczky, P. G. (1990). Expression of collagenlike sequences by a tumor virus, herpesvirus saimiri. *J. Virol.* **64**, 3509–3515.

Geck, P., Whitaker, S. A., Medveczky, M. M. and Medveczky, P. G. (1991). Expression of collagenlike sequences by a tumor virus, herpesvirus saimiri. Erratum. *J. Virol.* **65**, 7084.

Glykofrydes, D., Niphuis, H., Kuhn, E. M., Rosenwirth, B., Heeney, J. L., Bruder, J., Niedobitek, G., Muller-Fleckenstein, I., Fleckenstein, B. and Ensser, A. (2000). Herpesvirus saimiri vFLIP provides an antiapoptotic function but is not essential for viral replication, transformation, or pathogenicity. *J. Virol.* **74**, 11919–11927.

Grassmann, R. and Fleckenstein, B. (1989). Selectable recombinant herpesvirus saimiri is capable of persisting in a human T-cell line. *J. Virol.* **63**, 1818–1821.

Grassmann, R., Dengler, C., Müller-Fleckenstein, I., Fleckenstein, B., McGuire, K., Dokhelar, M.-C., Sodroski, J. G. and Haseltine, W. A. (1989). Transformation to continuous growth of primary human T lymphocytes by human T-cell leukemia virus type I X-region genes transduced by a herpesvirus saimiri vector. *Proc. Natl. Acad. Sci. USA* **86**, 3351–3355.

Grassmann, R., Berchtold, S., Randant, I., Alt, M., Fleckenstein, B., Sodroski, J. G., Haseltine, W. A. and Ramstedt, U. (1992). Role of human T-cell leukemia virus type 1 X region proteins in immortalization of primary human lymphocytes in culture. *J. Virol.* **66**, 4570–4575.

Grassmann, R., Fleckenstein, B. and Desrosiers, R. C. (1994). Viral transformation of human T lymphocytes. *Adv. Cancer. Res.* **63**, 211–244.

Greve, T., Tamgüney, G., Fleischer, B., Fickenscher, H. and Bröker, B. M. (2001). Downregulation of p56 lck tyrosine kinase activity in T cells of squirrel monkeys (*Saimiri sciureus*) correlates with the nontransforming and apathogenic properties of herpesvirus saimiri in its natural host. *J. Virol.* **75**, 9252–9261.

Groux, H., Cottrez, F., Montpellier, C., Quatannens, B., Coll, J., Stehelin, D. and Auriault, C. (1997). Isolation and characterization of transformed human T-cell lines infected by Epstein-Barr virus. *Blood* **89**, 4521–4530.

Guo, J., Williams, K., Duboise, S. M., Alexander, L., Veazey, R. and Jung, J. U. (1998). Substitution of ras for the herpesvirus saimiri STP oncogene in lymphocyte transformation. *J. Virol.* **72**, 3698–3704.

Hartley, D. A. and Cooper, G. M. (2000). Direct binding and activation of STAT transcription factors by the herpesvirus saimiri protein tip. *J. Biol. Chem.* **275**, 16925–16932.

Hess, S., Kurrle, R., Lauffer, L., Riethmüller, G. and Engelmann, H. (1995). A cytotoxic CD40/p55 tumor necrosis factor receptor hybrid detects CD40 ligand on herpesvirus saimiri-transformed T cells. *Eur. J. Immunol.* **25**, 80–86.

Hiller, C., Wittmann, S., Slavin, S. and Fickenscher, H. (2000). Functional long-term thymidine kinase suicide gene expression in human T cells using a herpesvirus saimiri vector. *Gene Ther.* **7**, 664–674.

Hooijberg, E., Ruizendaal, J. J., Snijders, P. J. F., Kueter, E. W. M., Walboomers, J. M. M. and Spits, H. (2000). Immortalization of human CD8+ T cell clones by extopic expression of telomerase reverse transcriptase. *J. Immunol.* **165**, 4239–4245.

Hör, S., Ensser, A., Reiss, C., Ballmer-Hofer, K. and Biesinger, B. (2001). Herpesvirus saimiri protein StpB associates with cellular Src. *J. Gen. Virol.* **82**, 339–344.

Huppes, W., Fickenscher, H., 't Hart, B. A. and Fleckenstein, B. (1994). Cytokine dependence of human to mouse graft-versus-host disease. *Scand J. Immunol.* **40**, 26–36.

Inatsuki, A., Yasukawa, M. and Kobayashi, Y. (1989). Functional alterations of herpes simplex virus-specific CD4+ multifunctional T cell clones following infection with human T lymphotropic virus type I. *J. Immunol.* **143**, 1327–1333.

Jung, J. U. and Desrosiers, R. C. (1991). Identification and characterization of the herpesvirus saimiri oncoprotein STP-C488. *J. Virol.* **65**, 6953–6960.

Jung, J. U. and Desrosiers, R. C. (1992). Herpesvirus saimiri oncogene STP-C488 encodes a phosphoprotein. *J. Virol.* **66**, 1777–1780.

Jung, J. U. and Desrosiers, R. C. (1994). Distinct functional domains of STP-C488 of herpesvirus saimiri. *Virology* **204**, 751–758.

Jung, J. U. and Desrosiers, R. C. (1995). Association of the viral oncoprotein STP-C488 with cellular ras. *Mol. Cell. Biol.* **15**, 6508–6512.

Jung, J. U., Trimble, J. J., King, N. W., Biesinger, B., Fleckenstein, B. W. and Desrosiers, R. C. (1991). Identification of transforming genes of subgroup A and C strains of herpesvirus saimiri. *Proc. Natl. Acad. Sci. USA* **88**, 7051–7055.

Jung, J. U., Lang, S. M., Friedrich, U., Jun, T., Roberts T. M., Desrosiers, R. C. and Biesinger, B. (1995a). Identification of Lck-binding elements in Tip of herpesvirus saimiri. *J. Biol. Chem.* **270**, 20660–20667.

Jung, J. U., Lang, S. M., Jun, T., Roberts, T., Veillette, A. and Desrosiers, R. C. (1995b). Downregulation of Lck-mediated signal transduction by tip of herpesvirus saimiri. *J. Virol.* **69**, 7814–7822.

Kiyotaki, M., Desrosiers, R. C. and Letvin, N. L. (1986). Herpesvirus saimiri strain 11 immortalizes a restricted marmoset T8 lymphocyte subpopulation in vitro. *J. Exp. Med.* **164**, 926–931.

Klein, J. L., Fickenscher, H., Holliday, J. E., Biesinger, B. and Fleckenstein, B. (1996). Herpesvirus saimiri immortalized gamma-delta T cell line activated by IL-12. *J. Immunol.* **156**, 2754–2760.

Knappe, A., Hiller, C., Thurau, M., Wittmann, S., Hofmann, H., Fleckenstein, B. and Fickenscher, H. (1997). The superantigen-homologous viral immediate-early gene ie14/vsag in herpesvirus saimiri-transformed human T cells. *J. Virol.* **71**, 9124–9133.

Knappe, A., Hiller, C., Niphuis, H., Fossiez, F., Thurau, M., Wittmann, S., Kuhn, E. M., Lebecque, S., Banchereau, J., Rosenwirth, B., Fleckenstein, B., Heeney, J. and Fickenscher, H. (1998a). The interleukin-17 gene of herpesvirus saimiri. *J. Virol.* **72**, 5797–5801.

Knappe, A., Thurau, M., Niphuis, H., Hiller, C., Wittmann, S., Kuhn, E. M., Rosenwirth, B., Fleckenstein, B., Heeney, J. and Fickenscher, H. (1998b). T-cell lymphoma caused by herpesvirus saimiri C488 independently of ie14/vsag, a viral gene with superantigen homology. *J. Virol.* **72**, 3469–3471.

Knappe, A., Feldmann, G., Dittmer, U., Meinl, E., Nisslein, T., Wittmann, S., Mätz-Rensing, K., Kirchner, T., Bodemer, W. and Fickenscher, H. (2000a). Herpesvirus saimiri-transformed macaque T cells are tolerated and do not cause lymphoma after autologous reinfusion. *Blood* **95**, 3256–3261.

Knappe, A., Hör, S., Wittmann, S. and Fickenscher, H. (2000b). Induction of a novel cellular homolog of interleukin-10, AK155, by transformation of T lymphocytes with herpesvirus saimiri. *J. Virol.* **74**, 3881–3887.

Koomey, J. M., Mulder, C., Burghoff, R. L., Fleckenstein, B. and Desrosiers, R. C. (1984). Deletion of DNA sequences in a nononcogenic variant of herpesvirus saimiri. *J. Virol.* **50**, 662–665.

Kraft, M. S., Henning, G., Fickenscher, H., Lengenfelder, D., Tschopp, J., Fleckenstein, B. and Meinl, E. (1998). Herpesvirus saimiri transforms human T cell clones to stable growth without inducing resistance to apoptosis. *J. Virol.* **72**, 3138–3145.

Kung, S.-H. and Medveczky, P. G. (1996). Identification of a herpesvirus saimiri cis-acting DNA fragment that permits stable replication of episomes in transformed T cells. *J. Virol.* **70**, 1738–1744.

Kretschmer, C., Murphy, C., Biesinger, B., Beckers, J., Fickenscher, H., Kirchner, T., Fleckenstein, B. and Rüther, U. (1996). A herpes saimiri oncogene causing peripheral T-cell lymphoma in transgenic mice. *Oncogene* **12**, 1609–1616.

Lacey, S. F., Weinhold, K. J., Chen, C. H., McDanal, C., Oei, C. and Greenberg, M. L. (1998). Herpesvirus saimiri transformation of HIV type 1 suppressive CD8+ lymphocytes from an HIV type 1-infected asymptomatic individual. *AIDS Res. Hum. Retroviruses* **14**, 521–531.

Lee, H., Trimble, J. J., Yoon, D. W., Regier, D., Desrosiers, R. C. and Jung, J. U. (1997). Genetic variation of herpesvirus saimiri subgroup A transforming protein and its association with cellular src. *J. Virol.* **71**, 3817–3825.

Lee, H., Veazey, R., Williams, K., Li, M., Guo, J., Neipel, F., Fleckenstein, B., Lackner, A., Desrosiers, R. C. and Jung, J. U. (1998). Deregulation of cell growth by the K1 gene of Kaposi's sarcoma-associated herpesvirus. *Nat. Med.* **4**, 435–440.

Lee, H., Choi, J. K., Li, M., Kaye, K., Kieff, E. and Jung, J. U. (1999). Role of cellular tumor necrosis factor receptor-associated factors in NF-kappaB activation and lymphocyte transformation by herpesvirus saimiri STP. *J. Virol.* **73**, 3913–3919.

Leith, J. G., Copeland, K. F., McKay, P. J., Bienzle, D., Richards, C. D. and Rosenthal, K. L. (1999). T cell-derived suppressive activity: evidence of autocrine noncytolytic control of HIV type 1 transcription and replication. *AIDS Res. Hum. Retroviruses* **15**, 1553–1561.

Lund, T. Medveczky, M. M., Geck, P. and Medveczky, P. G. (1995). A herpesvirus saimiri protein requrired for interleukin-2 independence is associated with membranes of transformed T cells. *J. Virol.* **69**, 4495–4499.

Lund, T., Medveczky, M. M., Neame, P. J. and Medveczky, P. G. (1996). A herpesvirus saimiri membrane protein required for interleukin-2 independence forms a stable complex with p56lck. *J. Virol.* **70**, 600–606.

Lund, T. C., Garcia, R., Medveczky, M. M., Jove, R. and Medveczky, P. G. (1997a). Activation of STAT transcription factors by herpesvirus saimiri Tip-484 requires p56lck. *J. Virol.* **71**, 6677–6682.

Lund, T., Medveczky, M. M. and Medveczky, P. G. (1997b). Herpesvirus saimiri Tip-484 membrane protein markedly increases p56lck activity in T cells. *J. Virol.* **71**, 378–382.

Mackewicz, C. E., Orque, R., Jung, J. and Levy, J. A. (1997). Derivation of herpesvirus saimiri-transformed CD8+ T cell lines with noncytotoxic anti-HIV activity. *Clin. Immunol. Immunopathol.* **82**, 274–281.

Medveczky, P., Szomolanyi, E., Desrosiers, R. C. and Mulder, C. (1984). Classification of herpesvirus saimiri into three groups based on extreme variation in a DNA region required for oncogenicity. *J. Virol.* **52**, 938–944.

Medveczky, M. M., Szomolanyi, E., Hesselton, R., DeGrand, D., Geck, P. and Medveczky, P. G. (1989). Herpesvirus saimiri strains from three DNA subgroups have different oncogenic potentials in New Zealand white rabbits. *J. Virol.* **63**, 3601–3611.

Medveczky, M. M., Geck, P., Sullivan, J. L., Serbousek, D., Djeu, J. Y. and Medveczky, P. G. (1993a). IL-2 independent growth and cytotoxicity of herpesvirus saimiri-infected human CD8 cells and involvement of two open reading frame sequences of the virus. *Virology* **196**, 402–412.

Medveczky, M. M., Geck, P., Vassallo, R. and Medveczky, P. G. (1993b). Expression of the collagen-like putative oncoprotein of herpesvirus saimiri in transformed T cells. *Virus Genes* **7**, 349–365.

Meinl, E., Hohlfeld, R., Wekerle, H. and Fleckenstein, B. (1995a). Immortalization of human T cells by herpesvirus saimiri. *Immunol. Today* **16**, 55–58.

Meinl, E., 't Hart, B. A., Bontrop, R. E., Hoch, R. M., Iglesias, A., de Waal Malefyt, R., Fickenscher, H., Müller-Fleckenstein, I., Fleckenstein, B., Wekerle, H., Hohlfeld, R. and Jonker, M. (1995b). Activation of a myelin basic protein-specific human T cell clone by antigen-presenting cells from rehesus monkeys. *Int. Immunol.* **7**, 1489–1495.

Meinl, E., Fickenscher, H., Hoch, R. M., De Waal Malefyt, R., 't Hart, B. A., Wekerle, H., Hohlfeld, R. and Fleckenstein, B. (1997). Growth transformation of antigen-specific T cell lines from rhesus monkeys by herpesvirus saimiri. *Virology* **229**, 175–182.

Meinl, E., Derfuss, T., Pirzer, R., Blank, N., Lengenfelder, D., Blancher, A., Le Deist, F., Fleckenstein, B. and Hivroz, C. (2001). Herpesvirus saimiri replaces ZAP-70 for CD3- and CD2-mediated T cell activation. *J. Biol. Chem.* **276**, 36902–36908.

Melendez, L. V., Daniel, M. D., Hunt, R. D. and Garcia, F. G. (1968). An apparently new herpesvirus from primary kidney cultures of the squirrel monkey (*Saimiri sciureus*). *Lab. Animal Care* **18**, 374–381.

Merlo, J. J. and Tsygankov, A. Y. (2001). Herpesvirus saimiri oncoproteins Tip and StpC synergistically stimulate NF-kappaB activity and interleukin-2 gene expression. *Virology* **279**, 325–338.

Migliaccio, M., Amacker, M., Just, T., Reichenbach, P., Valmori, D., Cerottini, J. C., Romero, P. and Nabholz, M. (2000). Ectopic human telomerase catalytic subunit expression maintains telomere length but is not sufficient for CD8+ T lymphocyte immortalization. *J. Immunol.* **165**, 4978–4984.

Mittrücker, H.-W., Müller-Fleckenstein, I., Fleckenstein, B. and Fleischer, B. (1992). CD2-mediated autocrine growth of herpesvirus saimiri-transformed human T lymphocytes. *J. Exp. Med.* **176**, 909–913.

Mittrücker, H.-W., Müller-Fleckenstein, I., Fleckenstein, B. and Fleischer, B. (1993). Herpes virus saimiri-transformed human T lymphocytes: normal functional phenotype and preserved T cell receptor signalling. *Int. Immunol.* **5**, 985–990.

Miyoshi, I., Kubonishi, I., Yoshimoto, S., Akagi, T., Ohtsuki, Y., Shiraishi, Y., Nagata, K. and Hinuma, Y. (1981). Type C virus particles in a cord T-cell line derived by co-cultivating normal human cord leukocytes and human leukaemic T cells. *Nature* **294**, 770–771.

Murphy, C., Kretschmer, C., Biesinger, B., Beckers, J., Jung, J., Desrosiers, R., Müller-Hermelink H. K., Fleckenstein, B. W. and Rüther, U. (1994). Epithelial tumors induced by a herpesvirus oncogene in transgenic mice. *Oncogene* **9**, 221–226.

Murthy, S. C. S., Trimble, J. J. and Desrosiers, R. C. (1989). Deletion mutants of herpesvirus saimiri define an open reading frame necessary for transformation. *J. Virol.* **63**, 3307–3314.

Myer, V. E., Lee, S. I. and Steitz, J. A. (1992). Viral small nuclear ribonucleoproteins bind a protein implicated in messenger RNA destabilization. *Proc. Natl. Acad. Sci. USA* **89**, 1296–1300.

Nakamura, H., Zarycki, J., Sullivan, J. L. and Jung, J. U. (2001). Abnormal T cell receptor signal transduction of CD4 TH cells in X-linked lymphoproliferative syndrome. *J. Immunol.* **167**, 2657–2665.

Nicholas, J., Smith, E. P., Coles, L. and Honess, R. (1990). Gene expression in cells infected with gammaherpesvirus saimiri: properties of transcripts from two immediate-early genes. *Virology* **179**, 189–200.

Nick, S., Fickenscher, H., Biesinger, B., Born, G., Jahn, G. and Fleckenstein, B. (1993). Herpesvirus saimiri transformed human T cell lines: A permissive system for human immunodeficiency viruses. *Virology* **194**, 875–877.

Nilsson, K. and Klein, G. (1982). Phenotypic and cytogenetic characteristics of human B lymphoid cell lines and their relevance for the etiology of Burkitt's lymphoma. *Adv. Cancer Res.* **37**, 319–380.

Nutman, T. B. (1997). Generation of HTLV-I-transformed T cell lines. In *Current protocols in immunology* (J. E. Coligan, A. M. Kruisbeek, D. H. Margulies, E. M. Shevach, W. Strober, Eds), pp. 7.20.1–7.20.3. Wiley, New York.

O'Donnell-Tormey, J., Nathan, C. F., Lanks, K., DeBoer, C. J. and de la Harpe, J. (1987). Secretion of pyruvate. An antioxidant defense of mammalian cells. *J. Exp. Med.* **165**, 500–514.

Okada, K., Yasumara, S., Müller-Fleckenstein, I., Fleckenstein, B., Talib, S., Koldovsky, U. and Whiteside, T. L. (1997). Interactions between autologous CD4+ and CD8+ T lymphocytes and human squamous cell carcinoma of the head and neck. *Cell. Immunol.* **177**, 35–48.

Pacheco-Castro, A., Marquez, C., Toribio, M. L., Ramiro, A. R., Trigueros, C. and Regueiro, J. R. (1996). Herpesvirus saimiri immortalization of alpha-beta and gamma-delta human T-lineage cells derived from CD34+ intrathymic precursors in vitro. *Int. Immunol.* **8**, 1797–1805.

Pacheco-Castro, A., Alvarez-Zapata, D., Serrano-Torres, P. and Regueiro, J. R. (1998) Signaling through a CD3 gamma-deficient TCR/CD3 complex in immortalized mature CD4+ and CD8+ T lymphocytes. *J. Immunol.* **161**, 3152–3160.

Pecher, G., Harnack, U., Gunther, M., Hummel, M., Fichtner, I. and Schenk, J. A. (2001). Generation of an immortalized human CD4+ T cell clone inhibiting tumor growth in mice. *Biochem. Biophys. Res. Commun.* **283**, 738–742.

Popovic, M., Lange-Wantzin, G., Sarin, P. S., Mann, D. and Gallo, R. C. (1983). Transformation of human umbilical cord blood T cells by human T-cell leukemia/lymphoma virus. *Proc. Natl. Acad. Sci. USA* **80**, 5402–5406.

Rivero-Carmena, M., Porras, O., Pelaez, B., Pacheco-Castro, A., Gatti, R. A. and Regueiro, J. R. (2000). Membrane and transmembrane signaling in herpesvirus saimiri-transformed human CD4(+) and CD8(+) T lymphocytes is ATM-independent. *Int. Immunol.* **212**, 927–935.

Rodriguez-Gallego, C., Corell, A., Pacheco, A., Timon, M., Regueiro, J. R., Allende, L. M., Madrono, A. and Arnaiz-Villena, A. (1996). Herpesvirus saimiri transformation of T cells in CD3 gamma immunodeficiency: phenotypic and functional characterization. *J. Immunol. Meth.* **198**, 177–186.

Roizman, B., Desrosiers, R. C., Fleckenstein, B., Lopez, C., Minson, A. C. and Studdert, M. J. (1992). The herpesvirus study group of the international committee on taxonomy of viruses. The family herpesviridae: an update. *Arch. Virol.* **123**, 425–449.

Rosin, O., Koch, C., Schmitt, I., Semmes, O. J., Jeang, K. T. and Grassmann, R. (1998). A human T-cell leukemia virus Tax variant incapable of activating NF-kappaB retains its immortalizing potential for primary T-lymphocytes. *J. Biol. Chem.* **273**, 6698–6703.

Rufer, N., Migliaccio, M., Antonchuk, J., Humphries, R. K., Roosnek, E. and Landsdorp, P. M. (2001). Transfer of the human telomerase reverse transcriptase (TERT) gene into T lymphocytes results in extension of replicative potential. *Blood* **98**, 597–603.

Saadawi, A. M., L'Faqihi, F., Diab, B. Y., Sol, M. A., Enault, G., Coppin, H., Cantagrel, A., Biesinger, B., Fleckenstein, B. and Thomsen, M. (1997). Dominant clones in immortalized T-cell lines from rheumatoid arthritis synovial membranes. *Tissue Antigens* **49**, 431–437.

Saha, K., Sova, P., Chao, W., Chess, L. and Volsky, D. J. (1996a). Generation of CD4+ and CD8+ T-cell clones from PBLs of HIV-1 infected subjects using herpesvirus saimiri. *Nature Med.* **2**, 1272–1275.

Saha, K., Ware, R., Yellin, M. J., Chess, L. and Lowy, I. (1996b). Herpesvirus saimiri-transformed human CD4+ T cells can provide polyclonal B cell help via CD40 ligand as well as the TNF-alpha pathway and through release of lympho-kines. *J. Immunol.* **157**, 3876–3885.

Saha, K., Bentsman, G., Chess, L. and Volsky, D. J. (1998). Endogenous production of beta-chemokines by CD4+, but not CD8+, T-cell clones correlates with the clinical state of human immunodeficiency virus type 1 (HIV-1)-infected individuals and may be responsible for blocking infection with non-syncytium-inducing HIV-1 in vitro. *J. Virol.* **72**, 876–881.

Saha, K., Volsky, D. J. and Matczak, E. (1999). Resistance against syncytium-inducing human immunodeficiency virus type 1 (HIV-1) in selected CD4(+) T cells from an HIV-1-infected nonprogressor: evidence of a novel pathway of resistance mediated by a soluble factor(s) that acts after virus entry. *J. Virol.* **73**, 7891–7898.

Schirm, S., Müller, I., Desrosiers, R. C. and Fleckenstein, B. (1984). Herpesvirus saimiri DNA in a lymphoid cell line established by in vitro transformation. *J. Virol.* **49**, 938–946.

Schmitt, I., Rosin, O., Rohwer, P., Gossen, M. and Grassmann, R. (1998). Stimulation of cyclin-dependent kinase activity and G1- to S-phase transition in human lymphocytes by the human T-cell leukemia/lymphotropic virus type 1 Tax protein. *J. Virol.* **72**, 633–640.

Schneider, U., Schwenk, H.-U. and Bornkamm, G. (1977). Characterization of EBV-genome negative 'null' and T cell lines derived from children with acute lymphoblastic leukemia and leukemic transformed non-Hodgkin lymphoma. *Int. J. Cancer* **19**, 621–626.

Schofield, A. (1994). Investigations of the origins of replication of herpesvirus saimiri. Ph. D. Thesis, Open University, London.

Simmer, B., Alt, M., Buckreus, I., Berthold, S., Fleckenstein, B., Platzer, E. and Grassmann, R. (1991). Persistence of selectable herpesvirus saimiri in various human haematopoietic and epithelial cell lines. *J. Gen. Virol.* **72**, 1953–1958.

Stamminger, T., Honess, R. W., Young, D. F., Bodemer, W., Blair, E. D. and Fleckenstein, B. (1987). Organization of terminal reiterations in the virion DNA of herpesvirus saimiri. *J. Gen. Virol.* **68**, 1049–1066.

Stephan, V., Wahn, V., Le Deist, F., Dirksen, U., Bröker, B., Müller-Fleckenstein, I., Horneff, G., Schroten, H., Fischer, A. and De Saint Basile, G. (1996). Atypical X-linked severe combined immunodeficiency due to possible spontaneous reversion of the genetic defect in T cells. *New Engl. J. Med.* **335**, 1563–1567.

Stevenson, A. J., Cooper, M., Griffiths, J. C., Gibson, P. C., Whitehouse, A., Jones, E. F., Markham, A. F., Kinsey, S. E. and Meredith, D. M. (1999). Assessment of herpesvirus saimiri as a potential human gene therapy vector. *J. Med. Virol.* **57**, 269–277.

Stevenson, A. J., Clarke, D., Meredith, D. M., Kinsey, S. E., Whitehouse, A. and Bonifer, C. (2000a). Herpesvirus saimiri-based gene delivery vectors maintain heterologous expression throughout mouse embryonic stem cell differentiation in vitro. *Gene Ther.* **7**, 464–471.

Stevenson, A. J., Frolova-Jones, E., Hall, K. T., Kinsey, S. E., Markham, A. F., Whitehouse, A. and Meredith, D. M. (2000b). A herpesvirus saimiri-based gene therapy vector with potential for use in cancer immunotherapy. *Cancer Gene Ther.* **7**, 1077–1085.

Stevenson, A. J., Giles, M. S., Hall, K. T., Goodwin, D. J., Calderwood, M. A., Markham, A. F. and Whitehouse, A. (2000c). Specific oncolytic activity of herpesvirus saimiri in pancreatic cancer cells. *Br. J. Cancer* **83**, 329–332.

Szomolanyi, E., Medveczky, P. and Mulder, C. (1987). In vitro immortalization of marmoset cells with three subgroups of herpesvirus saimiri. *J. Virol.* **61**, 3485–3490.

Thome, M., Schneider, P., Hofmann, K., Fickenscher, H., Meinl, E., Neipel, F., Mattmann, C., Burns, K., Bodmer, J. L., Schröter, M., Scaffidi, C., Krammer, P. H., Peter, M. E. and Tschopp, J. (1997). Viral FLICE-inhibitory proteins (FLIPs) prevent apoptosis induced by death receptors. *Nature* **386**, 517–521.

Thurau, M., Whitehouse, A., Wittmann, S., Meredith, D. and Fickenscher, H. (2000). Distinct transcriptional and functional properties of the R transactivator gene orf50 of the transforming herpesvirus saimiri strain C488. *Virology* **268**, 167–177.

Tosato, G. (1997). Generation of Epstein–Barr Virus (EBV)-immortalized B cell lines. In *Current protocols in immunology* (J. E. Coligan, A. M. Kruisbeek, D. H. Margulies, E. M. Shevach, W. Strober, Eds), pp. 7.22.1–7.22.3. Wiley, New York.

Troidl, B., Simmer, B., Fickenscher, H., Müller-Fleckenstein, I., Emmrich, F., Fleckenstein, B. and Gebhart, E. (1994). Karyotypic characterization of human T-cell lines immortalized by herpesvirus saimiri. *Int. J. Cancer* **56**, 433–438.

Vella, C., Fickenscher, H., Atkins, C., Penny, M. and Daniels, R. (1997). Herpesvirus saimiri immortalized human T-cells support long-term, high titred replication of human immunodeficiency virus types 1 and 2. *J. Gen.Virol.* **78**, 1405–1409.

Vella, C., King, D., Zheng, N. N., Fickenscher, H., Breuer, J. and Daniels, R. S. (1999a). Alterations in the V1/V2 domain of HIV-2CBL24 glycoprotein 105 correlate with an extended cell tropism. *AIDS Res. Hum. Retroviruses* **15**, 1399–1402.

Vella, C., Zheng, N. N., Vella, G., Atkins, C., Bristow, R. G., Fickenscher, H. and Daniels, R. S. (1999b). Enhanced replication of M-tropic HIV-1 strains in herpesvirus saimiri immortalised T-cells which express CCR5. *J. Virol. Methods* **79**, 51–63.

Weber, F., Meinl, E., Drexler, K., Czlonkowska, A., Huber, S., Fickenscher, H., Müller-Fleckenstein, I., Fleckenstein, B., Wekerle, H. and Hohlfeld, R. (1993).

Transformation of human T-cell clones by herpesvirus saimiri: Intact antigen recognition by autonomously growing myelin basic protein-specific T cells. *Proc. Natl. Acad. Sci. USA* **90**, 11049–11054.

Wehner, L. E., Schröder, N., Kamino, K., Friedrich, U., Biesinger, B. and Rüther, U. (2001). Herpesvirus saimiri Tip gene causes T-cell lymphomas in transgenic mice. *DNA Cell Biol.* **20**, 81–88.

Whitaker, S., Geck, P., Medveczky, M. M., Cus, J., Kung, S.-H., Lund, T. and Medveczky, P. G. (1995). A polycistronic transcript in transformed cells encodes the dihydrofolate reductase of herpesvirus saimiri. *Virus Genes* **10**, 163–172.

Wiese, N., Tsygankov, A. Y., Klauenberg, U., Bolen, J. B., Fleischer, B. and Bröker, B. M. (1996). Selective activation of T cell kinase p56lck by herpesvirus saimiri protein Tip. *J. Biol. Chem.* **271**, 847–852.

Wright, J., Falk, L. A., Collins, D. and Deinhardt, F. (1976). Mononuclear cell fraction carrying herpesvirus saimiri in persistently infected squirrel monkeys. *J. Natl. Cancer Inst.* **57**, 959–962.

Yamamoto, N., Okada, M., Koyanagi, Y., Kannagi, M. and Hinuma, Y. (1982). Transformation of human leukocytes by cocultivation with an adult T cell leukemia virus producer cell line. *Science* **217**, 737–739.

Yamanashi, Y., Mori, S., Yoshida, M., Kishimoto, T., Inoue, K., Yamamoto, T. and Toyoshima, K. (1989). Selective expression of a protein-tyrosine kinase, p56lyn, in hematopoietic cells and association with production of human T-cell lymphotropic virus type I. *Proc. Natl. Acad. Sci. USA* **86**, 6538–6542.

Yao, Z., Maraskovsky, E., Spriggs, M. K., Cohen, J. I., Armitage, R. J. and Alderson, M. R. (1996). Herpesvirus saimiri open reading frame 14, a protein encoded by a T lymphotropic herpesvirus, binds to MHC class II molecules and stimulates T cell proliferation. *J. Immunol.* **156**, 3260–3266.

Yasukawa, M., Inoue, Y., Kimura, N. and Fujita, S. (1995). Immortalization of human T cells expressing T-cell receptor gamma-delta by herpesvirus saimiri. *J. Virol.* **69**, 8114–8117.

Yoshida, M., Miyoshi, I. and Hinuma, Y. (1982). Isolation and characterization of retrovirus from cell lines of human adult T-cell leukemia and its implication in the disease. *Proc. Natl. Acad. Sci. USA* **79**, 2031–2035.

Yssel, H., De Waal Malefyt, R., Duc Dodon, M., Blanchard, D., Gazzolo, L., De Vries, J. E. and Spits, H. (1989). Human T cell leukemia/ lymphoma virus type I infection of a CD4+ proliferative/cytotoxic T cell clone progresses in at least two distinct phases based on changes in function and phenotype of the infected cells. *J. Immunol.* **142**, 2279–2289.

Zapata, D. A., Pacheco-Castro, A., Torres, P. S., Ramiro, A. R., Jose, E. S., Alarcon, B., Alibaud, L., Rubin, B., Toribio, M. L. and Regueiro, J. R. (1999). Conformational and biochemical differences in the TCR.CD3 complex of CD8(+) versus CD4(+) mature lymphocytes revealed in the absence of CD3gamma. *J. Biol. Chem.* **274**, 35119–35128.

4 Generation and Characterization of Human Killer Cells

Graham Pawelec
Center for Medical Research, ZMF, University of Tübingen Medical School, Germany

Erminia Mariani
Istituto di Ricerca Codivilla Putti, Laboratorio di Immunologia e Genetica, Bologna, Italy

Rienk Offringa and Cornelius JM Melief
Department of Immunohematology and Blood Transfusion, Leiden University Medical Center, Leiden, The Netherlands

Raphael Solana
Servicio de Inmunología, Hospital Universitario Reina Sofía, Córdoba, Spain

◆◆

CONTENTS

◆◆◆◆◆◆ PREFACE

Many immunocytes have the capacity to recognize and kill other cells. In this chapter we will concentrate on killer cells of the lymphocyte type. These fall into two main categories distinguished by their MHC restriction or lack thereof, and their ability (and requirement) to be primed to the sensitizing antigen and generate immunological memory therefore MHC-restricted effectors are exclusively T lymphocytes and are commonly known as cytotoxic T lymphocytes (CTL). They usually recognize antigen in the context of self-MHC class I molecules and constitute the major category of anti-viral effector cells. They are almost always CD8+ due to the nature of the MHC molecule recognized. In contrast, CD4+ T lymphocytes recognize antigen in the context of MHC class II molecules and may sometimes be cytotoxic to the cells which they recognize. 'Natural killing' is the term given to the phenomenon of

Human Killer Cells

METHODS IN MICROBIOLOGY, VOLUME 32
ISBN 0–12–521532–0

Copyright © Elsevier Science Ltd
All rights of reproduction in any form reserved

MHC-unrestricted cytotoxicity mediated by a variety of lymphoid cell types without prior sensitization and without the generation of immunological memory. NK cells are also involved as a first line of defence in protection against viral infection. The majority of NK cells can be distinguished from T cells by their surface markers, but TCR1 cells form an important group also capable of mediating NK-like activity. In this chapter we will limit ourselves to a discussion of the generation and characterization of CD8+ 'classical' CTL and the CD16+ CD56+ 'classical' NK cells in human.

◆◆◆◆◆◆ NATURAL KILLER CELLS

Phenotyping and isolation of NK cells

Phenotyping of NK cells

Surface molecules and antibodies

Several surface markers are useful for distinguishing NK cells from other cell types, although many are shared with T cells (e.g. CD2, CD7, CD8, CD43 and CD122), and NK cells also express members of the α/β-integrin family shared with myelomonocytic cells (e.g. CD11b/CD18). However, other differentiation molecules show a restricted expression on NK cells when compared with other lymphocytes. Thus, NK cells express CD56 (N-CAM) and CD16 (the Fc RIII).

Human NK cells are CD3epsilon-negative, but the zeta chain, which in T cells is associated with the TCR, is coupled to the CD16 receptor in NK cells. Natural killer cells express inhibitory receptors specific for HLA class I proteins and stimulatory receptors with different specificities. In humans, NK cells express at least one inhibitory receptor and each inhibitory receptor is clonally distributed on only a subset of NK cells. These receptors discriminate between shared A, B or C allelic determinants of the classic class I molecules, as well as E and G alleles of non-classic class I. In general, class I HLA antigens binding to corresponding receptors on NK cells deliver inhibitory signals and block NK cell cytotoxicity. All NK cell inhibitory receptors are characterized by the presence of an immunoreceptor tyrosine-based inhibition motif (ITIM) in their cytoplasmic tail. Two structural types of inhibitory receptors have been found on human NK cells: immunoglobulin-like and lectin-like receptors. The killer cell Ig-like receptors (KIR; previously termed 'killer cell inhibitory' receptors), are encoded as a family of receptors on human chromosome 19, belong to the immunoglobulin superfamily (Ig-SF) and are characterized by two (p58) or three (p70/p140) extracellular Ig-like domains.

The anti-p58 (CD158a and 158b) mAbs GL183, EB6 and HP-3E4 interact with HLA-Cw4 and Cw3 respectively and related alleles, whereas p70 (NKB1) appears to be the NK receptor for the HLA-Bw4 public determinant present on the HLA-B*5101, -B*5801, and -B*2705 antigens and p140 the receptor for HLA-A alleles. The use of NK clones has defined three NK specificities according to their capacity to recognize Class I polymorphic motifs. Namely, NK1, defined by NK clones which are inhibited by HLA-C alleles with Lys80; NK2, by those HLA-C alleles with Ser77; and NK3, defined by NK clones inhibited by HLA-Bw4 alleles with Ile80.

Because of their KIR, NK cells have to be used as effector cells against HLA allele-negative cell lines. Non-transfected and HLA allele trans-fected C1R and 721.221 cell lines can be used to analyse HLA-induced inhibition of NK cytotoxicity. When cytotoxicity against an HLA class I transfectant is diminished by 50% or more compared with the un-transfected parental cell line, it can be concluded that the particular NK cell clone is able to recognize the HLA class I molecules expressed by the transfectant. Alternatively, NK cells can be directly typed for expression of KIR using monoclonal antibodies (mAb) as described in the following protocols. A list of KIR and their ligands can be found at the National Centre of Biotechnology Information Protein Reviews on the Website (www.ncbi.nlm.nih.gov/prow/guide).

The lectin-like inhibitory receptors for class I HLA antigens inhibit NK cells in a similar manner, even if they differ from KIR in both structure and specificity. They are encoded by a family of receptors (the NK complex) on human chromosome 12. In humans, only a limited number of HLA class I-specific lectin-like receptors have been identified. One of them is formed by the association of CD94 with a member of the NKG2 family, of which there are four genes (A, B, C and D). One of them, NKG2A, mediates an inhibitory signal to NK cells when CD94 recognizes an invariant receptor, namely, HLA-E, whereas the other members of the family appear to activate cells following ligand binding. CD94/NKG2A inhibitory receptors are expressed on KIR-negative NK cells and are responsible for HLA class I recognition in this subset. Close to KIR is another related family, the ILT (Ig-like transcripts) or LIR (leukocyte Ig-like receptors) family, that comprises eight genes and their products which are also found on T and B cells, macrophages and dendritic cells.

A number of surface molecules expressed by NK cells (CD2, CD16, CD69 and DNAM-1) trigger NK cell-mediated cytotoxicity in redirected killing assay. Moreover, the activating counterpart of the HLA-specific receptor p50 and CD94/NKG2C may play a role in the NK-mediated cytolytic activity against HLA class I-positive target cells.

Recently three new receptors responsible for NK-cell triggering in the process of non-HLA-restricted natural cytotoxicity have been identified: NKp46, NKp30 and NKp44. These molecules represent the first members of an emerging group of 'natural cytotoxicity receptors' (NCRs), belonging to the immunoglobulin superfamily. They are expressed exclusively by all NK cells, both resting and activated, apart from NKp44 that is absent in freshly isolated but is progressively induced only on IL-2 activated NK cells *in vitro*. There is a direct correlation between the surface density of NCR and the ability of NK cells to kill various tumours. NCR are coupled to different signal-transducing adaptor proteins, including CD3zeta, FcepsilonRIgamma and KARAP/DAP-12. Another triggering NK receptor is NKG2D. It appears to play either a complementary or a synergistic role with NCRs. Other triggering surface molecules including 2B4 and the novel NKp80 appear to function as co-receptors

Human Killer Cells

rather than as true receptors. Indeed, they can induce natural cytotoxicity only when co-engaged with a triggering receptor.

Other surface molecules also expressed in human NK cells or NK sub-populations and useful for their definition in health and disease are: CD26, CD27, CD29, CD45, CD57, CD69, CD81, p38 (C1.7.1), PEN5 and p75/AIRM-1. There are several standard methods for surface marker phenotyping.

Surface marker phenotyping

mAb are commonly used to identify surface antigens on viable cell suspensions by either: (a) direct immunofluorescence, this employs specific antibodies conjugated with fluorochromes; or (b) indirect immuno-fluorescence, which uses a two-step procedure, where the specific antibody is unlabelled and is tagged by a conjugated anti-Ig antibody.

(a) 2×10^5 PBMC are incubated with each labelled mAb in V-bottom plates for 30 min at 4°C, washed twice with PBS plus 2% FCS and 0.1% sodium azide (PBS-FCS-Az) and resuspended in 1% paraformaldehyde for the analysis. Mixtures of two or more differently labelled mAb can be incubated simultaneously with the same cells for multi-colour fluorescence. Negative control cells should be incubated with Ig isotype control mAb.

(b) 2×10^5 PBMC are incubated as for the direct method but using unlabelled antibodies. After washing and resuspension, cells are incubated again with labelled goat anti-mouse Ig for 30 min at 4°C. After final washing, the cells are resuspended in 1% paraformaldehyde for the analysis. Negative control cells are incubated only with the secondary labelled antibody.

Intracellular staining

This method is useful to evaluate the presence of intracellular granule content, in particular perforins or synthesized cytokines or chemokines in NK cells. 10^6 PBMC are resuspended in 1 ml of cold 2% paraformaldehyde solution in PBS, incubated for 1 h at 4°C in the dark, then centrifuged for 10 min at 250g at 4°C. The fixed cells are resuspended in 1 ml of 0.2% Tween 20 in PBS at room temperature and incubated for 15 min at 37°C. Three millilitres of PBS-FCS-Az are added and the suspension is centrifuged for 10 min at 250g at room temperature. The cell pellet is then incubated with an optimal dilution of an anti-human perforin mAb for 30 min at 4°C; positive cells are developed using a 1:20 diluted goat anti-mouse Ig conjugated with FITC for 30 min at 4°C. The cell pellet is washed with PBS-FCS-Az and analysed.

Flow cytometry

Optical/electronic instruments that measure cell size and phenotype by the presence of bound fluorochrome-labelled antibodies are widely used nowadays instead of a manual fluorescence microscope for analysis of cells labelled as described above. Several models are commercially available in two forms: only analytical or analytical and sorting machines ('cell sorter').

A single-cell suspension labelled with FITC, PE or tandem PE-Cy3 conjugated antibodies is forced through the nozzle of the machine under pressure. The cells confined to the axis of the fluid stream by a concentric sheath of cell-free fluid pass through a laser beam focused onto the stream. The light scattered and reflected by the cells and emitted by excited fluorochromes bound to the cell membrane is collected by a suitable equipment of lenses, optical filters and photo-electric devices. The electrical signals are analysed, processed and stored by a computer. When the machine is used to separate subpopulations of cells, these electrical signals are also used to activate the cell-sorting process. The present protocol was developed on the FACStar Plus machine (Becton Dickinson), equipped with a 4W argon ion laser (Spectra Physics), operating at a wavelength of 488 nm in order to excite both FITC and PE fluorochromes. Forward light scatter or 'forward scatter' (FLS) (2–15°) was collected with a 0.5% neutral density filter in front of the photodiode. This parameter is a measure of cell size and is useful in discriminating between viable and dead cells and between nucleated cells and erythrocytes. Perpendicular light scatter or 'side scatter' (PLS) (75–105°) was measured collecting 5% of the light through a 488-nm band pass filter with a photomultiplier tube (PMT). It is an indicator of the heterogeneity of cell structure ('granularity'). The simultaneous measurement of FLS and PLS allows the identification of different subpopulations of white blood cells. FITC green fluorescence was selected by a 515 nm LP filter and a BP 530 F1 10 nm filter. PE red fluorescence was selected by a BP 575 F1 15 nm filter. The PMT for the detection of green and red fluorescences was operating at 500 V. 10^5 labelled cells should be analysed for each sample. In order to apply optimal gating for the identification and exclusion of monocytes from lymphocyte populations, mononuclear cells are analysed on the scattergram using rectangular or polygonal computer-generated windows. The number of lymphocytes contained either in the regular or in the polygonal windows should be the same, but the number of contaminating monocytes can be decreased by two-fold in the polygonal one. Additionally, care must be exercised not to use a too restricted lymphocyte gate in an attempt to achieve lower monocyte contamination. This is because this can lead to a loss of lymphoid cells which does not appear to be random, but seems to affect mainly the CD16 population with natural killer activity. This point should be borne in mind when setting gates for analysing NK cells in whole PBMC. These cells produce forward and perpendicular scatter signals higher than other lymphocyte subsets, and, therefore, are mainly located in the area of the scattergram which divides lymphocytes from monocytes. These data are in accordance with the large granular lymphocyte morphology of natural killer cells. Therefore, the use of the polygonal windows seems to be beneficial to reduce monocyte contamination without selective loss of natural killer lymphocytes, and may be particularly helpful in the analysis of pathological samples.

Human Killer Cells

Isolation of NK cells

The relatively small amount of NK cells within the blood has made it difficult to obtain these cells in pure form and in a number large enough to perform functional tests. Cytotoxic NK cells can be purified by cell sorting, immunomagnetic purification or complement-dependent depletion of other cell types. In the past they have been isolated based on their physical properties ('large granular lymphocytes'), but this method has now been superseded. Because NK cells can be kept functionally active in culture for several days with IL-2, a new simple procedure to obtain NK cells is to isolate mononuclear cells from blood, remove the TCR$^+$ T cells with anti-CD3 antibody and culture the cells for several days with IL-2 and EBV transformed feeder cell lines. These cells are all CD3-negative but positive for NK-markers. NK cells can also be propagated with IL-2 for prolonged periods of time as lines or clones using appropriate stimulator or feeder cells.

Cell sorting

PBMC are incubated with different fluorochrome-labelled mAb specific for the cell to be purified. If the sorted cells are to be used in functional studies, mAb which do not activate the cells must be utilized. To sort the cells according to their detected characteristics, the fluid stream through the flow cytometer is induced to break up into separate droplets by a microscopically small vertical vibration of the nozzle assembly. At the end of the unbroken stream, the droplets are electrically charged by applying a charging pulse to the whole stream. The polarity of this pulse is predetermined by the experimenter. The charged drops pass through a transverse electric field formed between two metal plates differing in potential. Positively and negatively charged drops are deflected towards the corresponding opposite plates. The resultant separate droplets are collected in different left and right tubes, the central component of the stream is discarded.

Preparation of human NK cells by immunomagnetic separation

This technique also allows a positive and a negative system of selection. Both positive and negative selection with magnetic beads may be accomplished by indirect or direct antibody-labelling methods. With the indirect method, the cells are incubated with mouse mAb that subsequently adhere to anti-mouse IgG-coated beads. With the direct method the cells are directly incubated with beads coated with mAb. In the following protocol, human NK cells are purified by negative selection, using an indirect method.

Take PBMC at 5×10^6 PBMC ml^{-1}, in complete medium (CM, RPMI 1640 with Hepes buffer supplemented with 10% FCS, antibiotics and 2 mM L-Glutamine) and incubate in a petri dish for 60–90 min at 37°C in 5% CO$_2$, for monocyte depletion. Non-adherent cells are recovered and washed twice with PBS and diluted in 2 ml of medium. Incubate the cells in prewashed nylon wool for 45–60 min at 37°C in 5% CO_2, for B cell depletion. The non-adherent (T/NK cells) are recovered by washing the column with 20 ml of CM, resuspended in 500 µl and incubated with anti-CD3 mAb (10–25 µg) (anti-Leu 4, Becton-Dickinson) for 30 min at 4°C. $0.7–0.9 \times 10^8$ of goat anti-mouse (GAM)-coupled magnetic beads (Dynal, Oslo, Norway) are now added for 30 min at 4°C under gentle shaking for T-cell depletion. Add 9 ml of PBS and use a magnetic particle concentrator (MPC1, Dynal) to remove the cells rosetting with the beads. Collect the supernatant containing CD3-negative cells. Alternatively, whole PBMC (5×10^6 ml^{-1}) can be directly incubated with CD3, CD4, CD19 and CD14 mAb at 10 µg ml^{-1} for 30 min at 4°C, and then with GAM-beads for 45 min at 4°C. Cells and beads should be gently mixed throughout. After the incubation with beads, a magnet is used to separate beads with attached CD3/CD4 T cells, CD14 monocytes and CD19 B cells from NK cells. Free cells are recovered, centrifuged and washed with complete medium. The cell population obtained should be 75–95% CD56$^+$ and CD16$^+$, <10% CD3$^+$ and <5% CD14$^+$ and CD19$^+$ as routinely established by flow cytometry.

Alternatively, MACS cell separation reagents and equipment can be used as follows.

Resuspend 10^7 PBMC in 80 µl of buffer (PBS with 2 mM EDTA and 0.5% bovine serum albumin), add 20 µl of reagent A (cocktail of hapten-conjugated monoclonal CD3, CD4, CD19, CD33 antibodies) and incubate 15 min at 6–12°C. Wash the cell suspension carefully, resuspend again with 80 µl of buffer, add 20 µl of reagent B (colloidal super-paramagnetic MACS micro-beads conjugated to an anti-hapten antibody) and incubate as before. Wash cell suspension and resuspend the pellet in 500 µl of buffer. Apply cell suspension to a prefilled depletion column placed in the magnetic field of the VarioMACS magnetic cell separator (Miltenyi Biotec), and collect effluent cells representing the purified NK cell fraction.

Separation of NK cells by complement-mediated lysis

This is a classical, but frequently employed technique to obtain sub-populations of lymphocytes by negative selection, but is limited to complement-fixing mAb. The cells (5×10^6–1×10^7 ml^{-1}) are resuspended in medium containing the optimal dilution of antibody. Cells are maintained at 4°C to prevent capping. Sodium azide (10 mM) may also be used for the same purpose, but because of its interference with metabolism, it must be avoided when the cells are purified for functional studies. After incubation for 30 min at 4°C with periodical shaking, the cells are pelleted, washed with cold medium and resuspended in the original volume of warm medium containing complement at the optimal dilution. The suspension is incubated for 45 min at 37°C, with regular shaking. At the end of incubation time, the cells are centrifuged, washed twice with complete medium containing 5% serum or 0.2% BSA and tested for viability, purity and recovery.

Human Killer Cells

Activation of NK cells with IL-2

The cells obtained as indicated above can be cultured in 5% CO_2 at 37°C for 7–10 days in the presence of 500 U ml^{-1} IL-2, without further activation requirements. However for a better efficiency in NK expansion the addition of irradiated feeder cells is recommended. EBV transformed cell lines, in particular the RPMI 8226 cell line, are commonly used as feeder cells for NK cell activation.

Generation of NK clones

Isolated NK cells obtained as indicated above are incubated with autologous cells (10^5 NK cells) with 5×10^4 irradiated (40 Gy) autologous PBMC and irradiated (100 Gy) allogeneic EBV-transformed cells (1:1 mixture) in a total volume of 2 ml CM containing 250 U ml^{-1} IL-2 in a 24-well cluster plate. Incubate the plate 4 days in a humidified 37°C, 5% CO_2 incubator. Remove 1 ml medium and replace with 1 ml containing 500 U rIL-2 (final concentration 250 U ml^{-1}). Incubate plate for 3 days. Wash cells by centrifuging 5–10 min at 1500 rpm and count them. Place cells (at limiting dilution) into each well of a 96-well round-bottom microtitre plate containing 10^5 irradiated autologous mononuclear cells and 5×10^4 irradiated allogeneic EBV-transformed cells and 250 U ml^{-1} rIL-2 in a final volume of 200 μl. Incubate plates at 37°C, 5% CO_2. Remove 100 μl medium and replace with 100 μl complete medium with 500 U ml^{-1} rIL-2. Incubate for 3 additional days. Remove 100 μl medium. Add 5×10^4 irradiated allogeneic EBV-transformed cells in CM and 500 U ml^{-1} rIL-2. Incubate another 4 days. Remove 100 μl medium and again replace with 100 μl CM containing 500 U ml^{-1} rIL-2 and incubate another 3 days. Identify growing wells by visual microscopy. Wash cells in serum-free RPMI, resuspend at 10^6 cells ml^{-1} and place 2 ml of this suspension in each well of a 24-well microtitre plate. (At this point some cells can be cryo-preserved and the remainder cultured under different conditions to establish optimal parameters for each particular clone.) Remove 1 ml and add fresh medium containing rIL-2 only. Incubate 4–7 days. Control growth and feed the cells with fresh medium containing rIL-2 weekly depending of the growth rate.

◆◆◆◆◆◆ CYTOTOXIC T CELLS

Generation of CTL

One problem hindering the successful and reproducible generation of CD8$^+$ CTL is the common finding that the *in vitro* sensitization and culture techniques routinely employed in cellular immunology strongly favour the outgrowth of CD4$^+$ helper cells (although many of these, particularly 'Th1-like' cells also possess some cytotoxic potential). To obtain high-yield antigen-specific MHC class I-restricted CD8$^+$ cells it is necessary to enrich for CD8 cells or deplete CD4 cells at the beginning.

This can be effected using antibody-based techniques involving separation with magnetic beads, FACS or solid-phase absorption. For generating antigen-specific cells it is also of critical importance to use high-efficiency antigen-presenting cells (APC). For most purposes, dendritic cells (DC) are currently favoured. Space available here does not allow protocols for all these techniques to be given; instead we will present a hybrid technique for the generation of CTL to antigenic peptides *in vitro* which has been refined by the collaborative group EUCAPS (European Union Concerted Action on Peptide Sensitization, BMH4-CT98-3058; see www.medizin.uni-tuebingen.de/eucaps/).

Preparation of APC

The first step in the procedure is to prepare the APC and the responders, starting from normal human PBMC. These are plated in 6-well plates at 20×10^6 per well in 3 ml culture medium (we favour the use of BioWhittaker X-Vivo 15 serum-free medium without serum supplementation), incubated for 2 h at 37°C, gently swirled, washed, and then the medium containing non-adherent cells collected. The latter are cryopreserved for later use as responders. To prepare the DC that will serve as APC, add 2.5 ml medium containing 800 U ml^{-1} of GM-CSF (8×10^8 U mg^{-1}) and 500 U ml^{-1} of IL 4 (10^7 U mg^{-1}) to each well with adherent cells and culture for 2 days at 37°C. Then add 2.5 ml of fresh medium containing 1600 U ml^{-1} GM-CSF and 1000 U ml^{-1} IL 4, and 2 days later, remove 2.5 ml of culture medium from each well and replace with 2.5 ml of fresh medium containing 1600 U ml^{-1} GM-CSF and 1000 U ml^{-1} IL 4. Next day, remove 1 ml of culture medium from each well and replace with 1 ml of fresh medium containing 10 ng ml^{-1} of TNF-α. For a final 16-h culture period (overnight), add poly-IC at a final concentration of 12.5 µg ml^{-1}. Then harvest the DC and wash twice with medium, using tubes 'pre-coated' with the medium to prevent adherence. Record the phenotype of DC present using MHC class II-antibody (e.g. L243) and at least antibodies for CD1a, CD14, CD83 and CD86, for reference regarding the level of maturity of the DC. Resuspend in 1 ml medium containing 50 µg ml^{-1} of the desired peptide antigen and 3 µg ml^{-1} β$_2$m. Incubate at 37°C for 4 h; gently resuspend every hour. Finally, irradiate at 25 Gy and wash twice with medium, resuspend at 0.3×10^6 ml^{-1}.

Responder cells

Thaw non-adherent fraction (those previously frozen); keep 25% of these cells; deplete the remaining 75% of CD4$^+$ cells as follows: wash cells three times with cold medium, add three- to four-fold excess of CD4-Dynal-beads and incubate at 4°C for 30–45 min, wash cells three times with cold medium while using magnet; collect supernatant. Count resulting CD4-depleted population. Then reconstitute the responders with a small proportion of CD4 cells. To do this, consider the non-depleted cells as 40% CD4$^+$. Add non-depleted cells to the CD4-depleted population to give a final dose of ca. 10% CD4$^+$ cells.

Initiation of CTL-stimulation cultures

Use at least 15×10^6 responders prepared as above (and therefore 1.5×10^6 APC to give R:S ratio of 10:1) per induction culture. Resuspend these responders at 3×10^6 ml^{-1} in medium containing 20 ng ml^{-1} of IL-7 (2×10^8 U mg^{-1}) and 100 pg ml^{-1} of IL 12 (2×10^5 U mg^{-1}) and plate in 24-well plates (1 ml per well). Add 10% APC (i.e. 0.3×10^6 in 1 ml per well). Seven days later, remove 1 ml of medium and replace with 1 ml of fresh medium containing 20 ng ml^{-1} of IL-7. After a further 5 days, harvest the sensitized cells, separate over Ficoll/Hypaque, wash once and count viable cells. Resuspend at 1.5×10^6 ml^{-1} and keep tube at 37°C. Then thaw 4×10^6 autologous PBMC per 1.5×10^6 responder cells. Use serum-free RPMI for thawing, wash once in serum-free RPMI and irradiate at 60 Gy. Wash again in serum-free RPMI and resuspend at 4×10^6 ml^{-1} in X-Vivo 15. Plate in 24-well plates (1 ml per well) and incubate for 2 h at 37°C. Then swirl plates gently and collect medium containing non-adherent cells. Gently wash cells with 2 ml medium to remove remaining non-adherent cells and add 0.5 ml of medium containing 20 µg ml^{-1} of peptide and 3 µg ml^{-1} of β_2m to the adherent cells and incubate for 2 h at 37°C. After this, remove medium, gently wash once and add 1 ml of responder cell suspension (i.e. 1.5×10^6 ml^{-1}). After a total of 2 weeks, add 1 ml of medium containing 20 IU ml^{-1} of IL-2 (1.8×10^7 U mg^{-1}) to each well. After a further 5 days, restimulate as last time. Whether isolation of live cells over Ficoll/Hypaque is necessary will depend on the quality of the cultures and total number of cells. After a total of 3 weeks, add fresh medium + IL-2 as after 2 weeks and 5 days later restimulate again. Check the CD4/CD8 ratio of responder cells at this time. If CD4$^+$ cells are found to predominate in the cultures, CD4-depletion must be performed again, as at the beginning of the procedure. After 4 weeks feed with fresh medium + IL-2 and harvest responder cells 5 days later. Whether isolation of live cells over Ficoll/Hypaque is necessary will again depend on the quality of the cultures and total number of cells. The effector cells are now ready for testing in a cytotoxicity assay. At this point, the effectors may also be subjected to cloning.

Cloning CTL

Once populations of sensitized CTL have been obtained, they can be successfully cloned by limiting dilution under standard conditions.

---

1. Dilute the cells for cloning in the culture medium in which they had been sensitized, but to which 20 U ml^{-1} of IL-2 is added. Set the concentration so that 10 µl contain 45, 4.5 or 0.45 cells. Then plate 10 µl of the 0.45 suspension to 60×1 mm-diameter wells of culture trays ('Terasaki plates') and leave in a vibration-free area for an hour. Check the distribution of cells in the wells visually using an inverted microscope and being careful to look around the edges of

(contd.)

---

the wells. According to the Poisson distribution, only a maximum of 37% of the wells should contain cells. Readjust dilutions if necessary, and recheck.

2. Plate at least five trays with the 0.45 cells per 10 µl suspension, one with 4.5 and one with 45, and add a constant number of feeder cells to each well. Irradiated PBMC are commonly used as feeder cells at 1×10^4 per well. Use autologous PBMC, a mixture of autologous PBMC and autologous B-lymphoblastoid line cells, or other appropriate APC, in the presence of specific antigen. Alternatively, use an antigen non-specific stimulus such as 50 ng ml^{-1} of the anti-CD3 monoclonal antibody OKT3 or 2 µg ml^{-1} of the mitogen PHA, together with the same number of allogeneic or autologous PBMC.

3. Stack plates and wrap in aluminium foil for ease of handling and as a precaution against contamination. Incubate for about a week and then examine the plates using an inverted microscope. Transfer contents of positive wells (> one-third full) to 7 mm-diameter flat-bottom microtitre plate wells with fresh medium and 1×10^5 of the same feeder cells as before. Check Terasaki plates again at intervals of a few days up to 2–3 weeks of age to identify late developers and transfer these also. Check microtitre plates every few days, and identify wells becoming crowded within a week post-transfer. These must be split 1:1 into new culture wells and re-fed with medium (but not feeder cells). After 1 week in microtitre plates, contents of wells with growing cells are transferred to 16 mm-diameter cluster plate wells with 2–5 × 10^5 of the same feeder cells, and fresh medium. Observe after 3–4 days and establish which wells are already full or nearly full. The former should be divided into four, the latter into two, with fresh media, but no more feeders. After a total of 1 week in cluster plates, count the number of cells in each clone and split to 2×10^5 per well, again with 2–5 × 10^5 feeders per well and fresh medium. Feed after 3–4 days with fresh medium, and split again if necessary. Clones successfully propagated in cluster plate wells for this second week are taken to be established. At this point, some (or all) can be cryopreserved and the remainder cultured under different conditions to establish optimal parameters for each particular clone. Having a frozen stock enables one to test different culture conditions in order to optimize growth, without the fear of losing the whole clone.

4. Test whether established clones can be propagated with the most convenient feeder cells (80 Gy-irradiated B-lymphoblastoid cell lines) instead of PBMC feeders. Most TCC flourish on B-LCL alone, but some appear for unknown reasons to benefit from the presence of PBMC as well (this is especially true during cloning). Propagation of the TCC on PBMC feeders can of course also be continued, but many laboratories may find it easier to grow large amounts of B-LCL than to isolate the PBMC. Furthermore, PBMC from the autologous

(contd.)

Human Killer Cells

donor may not be freely available in sufficient amounts for large-scale propagation of numerous clones. The international availability of well characterized MHC homozygous B-LCL makes it possible to match the feeder cell to the specificity of the TCC being propagated and enhance the antigen-presentation function of the feeders.

5. As a matter of convenience, it is easier to grow TCC in scaled-up culture vessels than in cluster plates, but not all clones can be adapted to growth in flasks. This must also be tested for each clone, using 1×10^5 and 5×10^5 ml^{-1} TCC with an equal number of feeders in tissue culture flasks. Those clones not growing under these conditions can be rarely adapted to flask growth by altering the amounts or concentrations of TCC or feeders seeded or by increasing or decreasing the frequency of stimulation and/or feeding. It remains unknown why some TCC fail to flourish in flasks.

6. Establish restimulation parameters for each clone. T cells require periodic reactivation through the T-cell antigen receptor in order to retain responsiveness to growth factors. This can be accomplished specifically or non-specifically. All clones can be propagated with weekly restimulation; some but not all can be propagated with re-stimulation only every 2 weeks. Human T-cell clones can by readily cryopreserved using the same protocols as are suitable for freezing resting T cells. Clones developing with different kinetics can thus be collected and conveniently tested for cytotoxicity in the same experiment.

◆◆◆◆◆◆ CYTOTOXIC ASSAYS FOR NK CELLS AND CTL

Effector cells

Effector cells generated as described in the preceding sections (CTL, NK cells, clones or lines) are suspended in RPMI 1640 medium containing 10% FCS. The cell concentration required will depend on the nature of the effector population and its level of cytotoxic activity. A starting effector : target ratio of 50 or 100:1 will be required for assaying freshly isolated NK cells, whereas for highly active cloned effectors this may be at most 10:1. In addition to NK and CTL assays, antibody-dependent cellular cytotoxicity (ADCC) measures redirected lysis of either type of effector provided that an antibody specific for a target antigen can bind the target cell and also bind and activate the effector cell.

Target cells

Commonly used target cells for measuring resting and activated NK lysis are K562 and DAUDI lines, respectively. Other cell lines used as targets are C1R or 721.221 EBV transformed cell lines which do not express HLA class I antigens on the surface. A variety of NK resistant cell lines can be used to analyse ADCC by using NK cells and IgG antibodies specific for the target cells. P815 a mastocytoma murine cell line which is also resistant to lysis by resting NK cells is frequently used to analyse redirected lysis, e.g. lysis of the target in the presence of antibodies against NK triggering structures. B-lymphoblastoid cell lines are easy to culture, express MHC molecules at high density on the cell surface and are able to take up and process antigens. They can therefore act as appropriate targets for measuring allospecific or antigen-specific lysis. T cells simulated by mitogens such as PHA or Con A can also be used as targets as follows: for the generation of such lymphoblasts, PBMC at 2×10^6 cells ml^{-1} are cultured for 2–3 days in RPMI 1640 with 10% FCS, 2 mM L-glutamine, 1 mM Na-pyruvate and 1 mg ml^{-1} of PHA or 5 mg ml^{-1} of Con A. Before use, dead cells are removed by Lymphoprep separation. Most other types of large cells are also suitable for use in cytotoxicity assays, e.g. tumour cells or fibroblasts. However, for the chromium- or calcein-release assays described in the following, small resting cells are unsuitable because they do not take up sufficient label into their cytoplasm.

Chromium-release assay

To label the selected target cells, ca. 1×10^6 cells are incubated together with 3.7×10^6 Bq of sodium ^{51}-chromate in ca. 0.25 ml complete medium (behind lead shielding). Specific activity of the chromium should be around 2.22×10^{10} Bq (600 mCi) per mg chromium (used no later than 15 days after the reference day). After 90 min at 37°C, during which time the cells are shaken every 15 min, they are washed, counted and adjusted to an appropriate concentration for distribution into the wells of U-form or V-form microtitre plates. It is generally sufficient to use 5×10^3 labelled target cells per well, which can be added in 100 µl of complete medium. During all procedures after incubation with chrome, the target cells should be kept on ice to reduce spontaneous isotope release before the test. Having plated the appropriate number of wells with target cells, 100 µl aliquots of titrated numbers of effector cells are added in doubling or tripling dilutions in triplicate to give at least four different effector:target ratios. Target cells must also be incubated in medium alone, to measure spontaneous release of isotope, and lysis buffer (e.g. 1% Triton X-100 detergent) to measure the maximum release possible. For most assays, incubating the plates at 37°C for 4 hours is sufficient for measurement of specific lysis. At the beginning and at the end of the incubation period, the plates are centrifuged at 4°C for 7 min. After this, or for some assays, an extended time, 100 µl of cell-free supernatant are carefully removed with a pipette and transferred to a tube or other receptacle for measuring released radioactivity in a gamma counter. Alternatively, the radioactivity in approx. 85% of the supernatant can be collected in filters with the Titertek microplate semi-automatic harvesting system, which uses cellulose acetate absorption cartridges (Skatron, Lyerbyen, Norway). The percent specific chromium release (% lysis) is then calculated according to the simple formula:

$$\frac{\text{Experimental} - \text{spontaneous CPM}}{\text{Maximum} - \text{spontaneous CPM}} \times 100$$

We consider sufficient labelling to be at least 1 cpm per cell, measurable in the maximum release assay. An optimal spontaneous release should be less than 10% of the maximum, but up to 33% is commonly accepted, especially for 'difficult' targets (e.g. fresh leukaemia cells).

Calcein-release assay (CARE-LASS)

Because the chromium release assay involves unavoidable exposure of the experimenter to gamma-irradiation, attempts have been made repeatedly to establish non-radioactive cytotoxicity assays. Moreover, the chromium release assay suffers from the disadvantage that the spontaneous release may be very high, particularly with extended assays, and with difficult target cells (e.g. chronic myelogenous leukaemia blasts). Here we describe only one of these alternative assays, with which we have had the most favourable experience. In this case, the target cells are not loaded with radioactive chromium but with a fluorescent dye, which is released into the supernatant on lysis. Otherwise, the entire procedure is analogous to the chromium release assay, but requires less time to perform. Cells to be labelled are washed, resuspended and incubated for 30 min. At 37°C in a final concentration of 25 µM Calcein-AM (provided by Molecular Probes Inc. at a 2.5 mM concentration in DMSO). The cells are then washed twice, counted and resuspended at an appropriate concentration. Effector and target cells are then plated as in the chromium-release assay, and also including target cells in medium alone (spontaneous release) and in lysis buffer, which can be Triton X-100 for this technique as well (maximum release). Incubate for 2–4 h and then pipette cell-free supernatant into the corresponding wells of a new microtitre plate. The calcein fluorescence of each well can now be measured in an automated plate-reading fluorometer with an excitation filter setting of 485/20 and an emission filter setting of 530/25 (e.g. the Fluoroskan Ascent from Labsystems Oy, Helsinki, Finland).

5 Measuring Human Cytokine Responses

Hans Yssel
INSERM U454, Montpellier, France

Emmanuel Claret
Diaclone Research, Besançon, France

Rene de Waal Malefyt
Department of Immunobiology, DNAX Research Institute, Palo Alto, CA

Françoise Cottrez
INSERM U343, Nice, France

◆◆

CONTENTS

◆◆◆◆◆◆ **INTRODUCTION**

The term cytokine has been used to describe a diverse group of low molecular weight (generally <20 kDa) protein mediators, that have a broad spectrum of immunoregulatory effects and which are produced by a variety of cell types. Although many cytokines were originally defined and named after the particular biological functions that they display, the development of recombinant DNA technology has permitted the classification of a plethora of factors and activities into a growing list of well-defined proteins. From a functional point of view, cytokines have several features in common.

- Most, if not all, cytokines have redundant activities, as reflected by their ability to perform similar functions. Among the many examples are interleukin (IL)-2, IL-4, IL-7, IL-10, IL-12, IL-15, IL-21 and IL-23 which all have T-cell growth

METHODS IN MICROBIOLOGY, VOLUME 32
ISBN 0–12–521532–0

Copyright © Elsevier Science Ltd
All rights of reproduction in any form reserved

Measuring Human
Cytokine Responses

promoting activities, IL-4 and IL-13 which share most of their functional activities, such as induction of B-cell proliferation and differentiation, induction of IgG1 and IgE isotype switching, induction of mast cell differentiation and down-regulation of secretion of pro-inflammatory cytokines by macrophages, or IL-1, IL-6 and TNF-α that activate hepatocytes to synthesize acute phase proteins, induce bone marrow epithelium to release neutrophils and that act as endogenous pyrogens on the hypothalamus thereby raising the body temperature which is believed to help eliminate infections.

- Their effects are pleiotropic, affecting many different target cells, which is reflected by the functional expression of receptors for certain cytokines, such as IL-1, IL-2, IL-4, IL-6 or IL-10 on most cells of the immune system.
- Many cytokines are able to induce or inhibit each other's synthesis, including their own, resulting in the creation of regulatory networks. An example of the mutual inducing activity of cytokines resulting in a positive feedback is the effect of IL-12 and IFN-γ on each other's production. Secretion of IL-12 by activated monocytes strongly induces the production of IFN-γ by T cells and natural killer (NK) cells, which in turn enhances the production of IL-12. Moreover, at least in the mouse, IFN-γ induces a functional IL-12R at the cell surface of T cells, rendering them susceptible to the IFN-γ-inducing effect of IL-12. The inhibitory action of cytokines is exemplified by IL-4, IL-10 and IL-13 which all efficiently block the secretion of pro-inflammatory cytokines, such as IL-1, IL-6 and TNF-α, by activated macrophages and furthermore enhance the production of IL-1RA by these cells, thus amplifying a negative feedback mechanism. Another example is IL-10 a cytokine with relatively late production kinetics that is able to inhibit its own synthesis via an autocrine feedback mechanism.
- Mixtures of cytokines often have synergistic effects, as compared with the effect of each of them separately, resulting in an amplification of responses. For example, the pro-inflammatory cytokines IL-1, IL-6 and TNF-α which are involved in the so-called acute phase response synergize to mediate inflammation, shock and even death in response to infectious agents.
- Their secretion is highly regulated, because of the potential for tissue destruction and other adverse effects, and they often are secreted during bursts of immune responses.
- Cytokines interact with specific high-affinity cell surface receptors which is followed by a cascade of signal transduction events, resulting in mRNA synthesis and, eventually, protein secretion. For a large number of interleukin receptors these signal transduction events have been shown to be associated with the phosphorylation of specific tyrosine kinases, members of the Jak-Stat signal transduction system (reviewed in Ihle, 2001). In addition, most cytokines are glycosylated, giving rise to considerable molecular heterogeneity, with the different sugar molecules being important for receptor binding and modulation of receptor-mediated signal transduction.

Cytokines may resemble hormones at first sight, since both are soluble mediators which serve as means of intracellular communication. However, there are important differences between the effects of these two families of mediator molecules. Whereas hormones can be easily detected in the circulation, having endocrine (systemic) effects, most cytokines are released locally. Moreover, hormones are released as a result of internal physiologic variation and therefore are important in maintaining a situation of homeostasis. In contrast, cytokines are produced in short bursts, following external insults, as well as during developmental and effector phases of immune responses, thereby modulating the function of adjacent cells or of the cytokine-producing cells themselves. Due to a short half-life time in the circulation, cytokines are generally difficult to detect in serum or plasma, although there are a number of exceptions. For example, during acute inflammation or septic shock, IL-1, IL-6 and TNF-α, orchestrate a series of events that induce the acute phase response and they can readily be detected in human serum or plasma. The systemic effects of the latter cytokines, however, may differ significantly from the effects in the local sphere of influence. IL-1, when released by tissue macrophages, is a co-stimulatory factor for T cell activation, whereas high systemic levels of IL-1 result in fever, leukopaenia and, as mentioned above, even shock.

Cytokines can be classified into a limited number of categories (listed in Table 1), broadly based on their functional properties, i.e. cytokines:

- that are involved in innate immunity
- that are involved in the regulation of lymphocyte function
- that are involved in the regulation of haematopoiesis
- that have anti-inflammatory modes of action
- with pro-inflammatory modes of action
- with chemoattractant properties.

As is clear from this list, which is not exhaustive, many cytokines feature in more than one category, which is in line which the notion that redundancy serves the purpose of the immune system to mount an effective and rapid response following inflammation or antigenic challenge. The concomitant release of several cytokines will lead to a rapid mobilization of effector cells at the place of injury or insult to produce the desired biological effects. Cytokines with a broad spectrum of action, such as IL-1 and IL-6, are not only involved in innate immune responses, which form the first line of defence of the host against antigenic challenge, but also in adaptive immune responses by exerting costimulatory effects on T cells and furthermore induce differentiation and growth of B cells. Conversely, many cytokines have downregulatory effects on their own secretion, as well as on the production of other cytokines, in order to prevent exacerbated immune responses, which may result in tissue damage or tissue destruction. Examples of such regulatory cytokines are IL-10 and TGF-β which affect a wide range of target cells. Taken together, several cytokines are usually involved in, and required for, the generation of an appropriate immune response and it seems difficult to bestow a more or less prominent role on each of them in view of their interregulatory effects, although there may be a certain hierarchy in the action of cytokines especially with respect to kinetics of production. Therefore, the concomitant inducing and inhibitory effects of many cytokines constitute the basis for the creation of overlapping cytokine networks which are able to tightly regulate an appropriate immune response.

Finally, whereas cytokines play a pivotal role in the generation of immune responses, dysregulation of cytokine production has been shown to result in acute or chronic immunopathology. Thus, the possibility to accurately measure the production of cytokines is not only of great importance for scientific purposes, but also for diagnostic purposes and furthermore is useful for the monitoring of clinical therapies.

Table I. Classification of cytokines based on functional properties

Cytokine	Biological activities	Produced by
1. Cytokines involved in innate immunity		
IL-1α/β	Mediates host response to infectious agents	Mø, DC, Fibroblasts, Astrocytes
IL-1RA	Natural antagonist of IL-1, blocks IL-1-mediated signal	Mø
TNF-α	Mediates host response to infectious agents	Mø, T cells
IL-6	Mediates and regulates inflammatory responses	Mø, T cells, Fibroblasts
Chemokines	Mediate leukocyte chemotaxis and activation	
2. Cytokines involved in regulation of lymphocyte function		
IL-1	Mediates co-stimulation of T cells	T cells
IL-2	T cell growth factor, proliferation of B cells	T cells, NK cells
IL-4	T cell proliferation, Differentiation of Th2 cells, B cell differentiation, IgG4/IgE switching	T cell, mast cells
IL-5	B cell growth and activation	T cells
IL-6	B cell proliferation, enhances Ig secretion by B cells	
IL-7	Pre-T and pre-B cell proliferation, differentiation LAK and cytotoxic T cells	Bone marrow stromal cells
IL-9	Co-stimulator for mast cell and fetal thymocyte growth	T cells
IL-10-like cytokines: IL-10, IL-19, IL-20, IL-22, IL-24		
IL-10	Co-stimulator for T cell, B cell and mast cell growth, downregulation MHC class II on Mono	Mono, T cells, keratinocytes
IL-19	Unknown	Mono
IL-20	Keratinocyte proliferation and differentiation	Keratinocytes
IL-22	Stimulates Mono to make TNF-α	T cells, mast cells
IL-12	Differentiation of Th1 cells, induction of IFN-γ by T cells and NK cells	Mono, B cells
IL-13	B cell differentiation, IgG4-IgE switching	T cells, B cells, mast cells
IL-15	T cell, mast cell growth factor	Mono
IL-17	Induction of IL-6 and IL-8 production by monocytes	T cells
	Induction of CD54 expression by fibroblasts	
IL-18	Induction of IFN-γ production by T/NK cells	Mono
IL-21	Maturation and proliferation of NK cells from BM	T cells
	Proliferation of activated B cells	

Table I. Continued

Cytokine	Biological activities	Produced by
IL-23	Induction of memory T-cell proliferation Stimulation of IFN-γ production	Dendritic cells
TNF-β	Stimulates T-cell growth	T cell, B cells
IFN-γ	Mø/NK cell activation, upregulation MHC class I/II on Mono	T cells, NK cells

3. Cytokines involved in the regulation of haemopoiesis

IL-3	Synergistic action in haemopoiesis	T cells, thymic epithelial cells
IL-5	Growth/differentiation of eosinophils	T cells, mast cells
IL-6	Growth/diff. megakaryocytes	
G-CSF	Growth/diff. granulocytic lineage	Mø, T cells
GM-CSF	Growth/diff. myelomonocytic lineage	Mø, T cells
SCF (ckit-L)	Growth/diff. haemopoietic precursor cells	Bone marrow stromal cells
Erythropoietin	Growth/diff. erythroid progenitor cells	Kidney cells

4. Cytokines having anti-inflammatory effects

IL-4	Inhibition of pro-inflammatory cytokines by Mono	
IL-13	Inhibition of pro-inflammatory cytokines by Mono	
IL-10	Inhibition of pro-inflammatory cytokines by Mono Inhibition of MHC class II expression Mono	
TGF-β	Inhibition of pro-inflammatory cytokines by Mono Inhibition of T-cell growth	Chrondrocytes, Mono, T cells

5. Cytokines having pro-inflammatory effects

IL-1α	Participates in acute phase response	
IL-6	Induction acute phase proteins in the liver	
TNF-α	Participates in acute phase response	
IL-16	Migratory response in CD4+ T cells/ Mono/EO	CD8+ T cells

6. Chemokines
(see Zlotnik and Yoshie (2000) for classification of chemokines)

DE: dendritic cells, EO: eosinophils, Mono: monocytes, Mø: macrophages, BM: bone marrow

◆◆◆◆◆◆ **ASSAYS FOR MEASURING CYTOKINE PRODUCTION**

Several assays have been developed to detect and quantify cytokines in serum, body fluids and culture supernatants, including bioassays, immunoassays, such as the enzyme-linked immunosorbent assay

(ELISA), intracellular cytokine staining assay and ELISPOT assay, as well as the quantitative reverse transcriptase-polymerase chain reaction (RT-PCR) and real-time PCR assay, which are both used quantitatively to detect transcripts in cytokine-producing cells. An overview of the advantages and disadvantages of each of these assays, described in this chapter, is shown in Table 2.

Table 2. Advantages and disadvantages of assays for the measurement of cytokines

Bioassay
Very sensitive
Active form of cytokine is detected
Can be performed in the absence of standard: activity expressed in arbitrary biological Units
Not specific and requires use of neutralizing anti-cytokine antibodies to demonstrate specificity
Large series of dilutions required to fit results on dose-dependent S-shaped curve
Requires labour-intensive tissue culture for relevant responder cells
Presence of stimulating agent in culture supernatant may act on responder cells

ELISA
Specific
Results are easy to fit onto dose-dependent S-shaped curve
Different modes of activation, including antigen-specific stimulation, can be used
Less sensitive than bioassay: sensitivity is dependent on the capture antibody
Non-active form of cytokine is also detected
Results can be affected by sequestration of cytokines from biological samples by (soluble) cytokine receptors, by inducing/inhibitory effects of simultaneously produced cytokines or by false-positive values with heterophilic antibodies

ELISPOT assay
More sensitive than ELISA, because of local release and capture of cytokines
Determination of frequencies of cytokine producing cells although quantification is tedious
Simultaneous (maximum two) detection of cytokine production by the same cell
Possibility to analyse cytokine production by freshly isolated clinical material without prior *in vitro* stimulation
No possibility to combine flow cytometry and cytokine production
Same technical advantages and constraints as ELISA

For Bioassay, ELISA and ELISPOT
Accumulation of secreted cytokines is measured: no constraints regarding kinetics of production

Table 2. Continued

Intracellular staining
Specific, but not very sensitive for certain cytokines: sensitivity is
    dependent on antibody
Requires selection of suitable antibodies that detect chemically fixed
    cytokines
Use in flow cytometry permits analysis of frequency and phenotype of
    cytokine-producing cells, as well as kinetics of cytokine production
Simultaneous detection of several cytokines produced by same cell
Results are not affected by inducing/inhibitory effects of simultaneously
    produced cytokines
Although theoretically possible, difficult to use under conditions of
    antigen-specific stimulation
Requires viable cells at the end of the stimulation
Requires exogenous (recombinant) cytokine to confirm specificity and
    meaningful interpretation of the results might require kinetics
    measurements

Competitive RT-PCR
Specific and extremely sensitive
Technically demanding and calibration difficult

Real-time RT-PCR
Quantitation of cytokine transcripts
Suitable for high through-put
Expensive, technically demanding and calibration difficult

For RT-PCR in general
Detection of cytokine transcripts only and not (secreted) protein

Activation procedure

Since the production of cytokines is tightly regulated, at the mRNA, as
well as the protein level, cell populations need to be appropriately
activated for the required amount of time to allow detection of cytokine
transcripts and/or production of protein. Activation procedures for each
of the cytokine detection assays, described in this chapter, are nearly
identical and will be described first.

> **Protocol**
>
> 1. Stimulate 10^6–2.10^6 ml^{-1} peripheral blood mononuclear cells (PBMC)
>   or 10^6–4.10^6 ml^{-1} T cells ml^{-1} with either of the following agents (see
>   notes 1 and 2):
>
> - Soluble (for PBMC) or plate-bound (for purified T cells or T cell
>   lines) anti-CD3 monoclonal (m) antibody (Ab) (SPV-T3b, UCHT-L1,
>   OKT3, B-B11 or equivalent) and soluble anti-CD28 mAb (L293, 9.3,
>   *(contd.)*

BT-3; 1 µg ml^{-1}). Coat plates with 10 µg ml^{-1} anti-CD3 mAb diluted in PBS and incubate for 18 h at 4°C or for 4 h at 37°C, remove mAb solution, wash twice with PBS and once with culture medium and use in experiment (see note 3).

- A combination of mitogenic anti-CD2 mAb (mAb D66 (Rosenthal-Allieri et al., 1995) combined with any sheep red blood cell binding anti-CD2 mAb).
- The combination of PMA (1 ng ml^{-1}) and the calcium ionophore A23187 (500 ng ml^{-1}) (Calbiochem: 524400 and 100105, respectively) or Ionomycin (Calbiochem: 524400), respectively.

Use 24-well (final volume 1 ml), 48-well (500 µl) or 96-well (200 µl) plates.

2. When using plate-bound anti-CD3 mAb, spin the culture plates at 100g for 2 min (see note 3).
3. Incubate the cells at 37°C, 5% CO_2.
4. For *ELISPOT*: gently harvest the cells after 6 h of incubation, transfer to ELISPOT plates and incubate for an additional 20 or 40 h at 37°C, 5% CO_2. Analyse filters for presence of cytokines.
5. For *ELISA* (see note 4): harvest the supernatants after 24–48 h of incubation, spin at 200g for 10 min to remove residual cells and analyse for cytokine production.
6. For *intracellular staining* (see notes 4 and 5): after 4 h of incubation, add 10 µl of Brefeldin A (A stock solution of 10 mg ml^{-1} Brefeldin A (Epicentre Technologies: B905MG, or Sigma: B7651) is made in DMSO, diluted 1:10 in PBS and Brefeldin A is added at a final concentration of 10 µg ml^{-1} to the cell suspension) and incubate the cells for an additional 2 h at 37°C, 5% CO_2. Harvest cells and put on ice, prior to analysis for cytokine production.
7. For *RT-PCR*: harvest the cells after 4–8 h of incubation, spin at 200g for 10 min at 4°C, carefully remove supernatant, and process cells for RNA preparation, as described on pages 735–736.

Notes and recommendations

1. The nature of the cytokine production profile of T cells depends on the mode of activation and it is therefore recommended to compare different stimulation protocols. Most importantly, these should include activation of the cells with anti-CD3 and anti-CD28 mAbs which resembles antigen-specific stimulation conditions. Another possibility is the use of superantigens, such as Staphylococcus enterotoxin B (SEB), which bind to certain TCR Vβ gene products. When using PBMC and specific, soluble antigen, use longer incubation periods (up to 4 days) to obtain optimal cytokine production levels. Modes of stimulation involving polyclonal activators, such as the combination of phorbol ester and calcium ionophore or ionomycine, *(contd.)*

respectively, should be used only as control activations to ensure intrinsic capacity of the cells to produce cytokines or as a positive control in analysis of cytokine production in the intracellular staining assay. Lipopolysaccharide (LPS, 1 µg ml^{-1} Sigma ref L-2880) is the stimulation of choice for monocytes. To measure the production of IL-12 by these cells, pre-incubate freshly isolated cells with 10 ng ml^{-1} IFN-γ for 18h and stimulate with LPS at a concentration of 100 ng ml^{-1}.

2. Use Iscove's Modified Dulbecco Medium supplemented with 10% human or fetal calf serum or Yssel's medium (Yssel *et al.*, 1984), supplemented with 1% human serum. (Detailed procedure for preparation sent upon request.)

3. Spinning cells onto anti-CD3-coated plates will enhance magnitude of stimulation (H. Yssel, unpublished results). This is especially important for short-time kinetics in RT-PCR assays or for the first incubation of cells for the ELISPOT assay prior to transfer to the filters.

4. Since cytokines are secreted with different kinetics, T cells should be stimulated for at least 24 h (48 h is recommended) before harvesting the supernatants for measurement by ELISA. When not immediately analysed, culture supernatants should be aliquotted and stored at −80°C. Similarly, cytokine standards should be aliquotted and kept frozen. Repetitive freeze/thaw cycles should be avoided, since most cytokines, in particular IFN-γ, will degrade rapidly. For intracellular staining and RT-PCR analysis, 6–8 h of stimulation is optimal for most cytokines (see also page 725).

5. Agents which block intracellular protein transport, such as Brefeldin A and Monensin have dose- and time-dependent cytotoxic effects and it is not recommended to have both agents included in the cultures for >6 h. Since Monensin has been found to induce the intracellular production of IL-1 and TNF-α in monocytes within 30 min of activation, the use of Brefeldin A, instead of Monensin, is preferred. A stock solution of 10 mg ml^{-1} Brefeldin A (Epicentre: B905MG, or Sigma: B7651) is made in DMSO, diluted 1:10 in PBS and Brefeldin A is added at a final concentration of 10 µg ml^{-1} to the cell suspension. Brefeldin A is toxic: avoid contact with skin, eyes and mucous membranes.

Measuring Human Cytokine Responses

◆◆◆◆◆◆ ELISA

In this section, the antibody sandwich ELISA system is described which is used for the detection of cytokines present in the culture supernatants of *in vitro* activated cells of the immune system or in clinical samples, such as serum, plasma or bronchoalveolar fluid. Although generally not as sensitive as a bioassay, the major advantage of the

ELISA is its specificity, enabling the detection of soluble proteins in complex mixtures, which often have overlapping, or, in contrast, counter-regulatory activities and which are therefore extremely difficult to detect, based on their biological activities alone. To measure a cytokine in solution, plates are coated with an anti-cytokine mAb which functions as a catcher to bind the cytokine. After removal of unbound cytokine, a second anti-cytokine mAb is added (detection mAb), which is usually biotinylated and which recognizes a different epitope on the cytokine, followed by a washing step and the addition of an enzyme-streptavidin conjugate. The use of such a biotin-streptavidin conjugate in the assay enables amplification of the signal. After removal of unbound conjugate, substrate is added, the hydrolysis oxidation of which by the enzyme-conjugate is proportional to the amount of cytokine present in the solution.

Reagents and equipment

- Cytokine-containing supernatants, cytokine-specific capture mAb; biotin-labelled, cytokine-specific tracer mAbs; PBS; dH$_2$O.
- Culture medium to dilute cytokine-containing samples and cytokine standards:

Coating buffer: Carbonate-bicarbonate buffer pH 7.2 to 7.4.

Washing buffer: PBS, supplemented with Tween-20 (0.05%).

Tween buffer: PBS, supplemented with BSA (0.1%) and Tween-20 (0.05%).

Blocking buffer: PBS, supplemented with BSA (2%).

Conjugate: Strepavidine-alkaline phosphatase (AP) conjugate (Southern Biotechnology Associates: 7100-04) or streptavidin-HRP (Prozyme: CJ30H).

ELISA substrate: Sigma 104 phosphatase substrate (Sigma: 104-0) or ready-to-use 3,3',5,5'-tetramethylbenzidine (TMB) (Moss: TMBUS).

Substrate buffer: Diethanol amine: 97 ml in 800 ml d H$_2$O.
    0.2 g NaN$_3$.
    0.1 g MgCl$_2$.6H$_2$O.
    Adjust to pH 9.8 with HCl and adjust volume to 1 l with dH$_2$O.

- H$_2$O$_2$; PVC U bottom 96-well plates; Immunolon I U bottom 96-well plates; Immunolon II U bottom 96-well plates (Dynatech: 011-010-3450); Microtitre plate reader-spectrophotometer with 405 nm filter or spectrofluorometer (Dynatech: 011-970-1900) with 365 nm excitation filter and 450 nm emission filter.

All reagents should be at room temperature before use in the ELISA. The optimal working concentrations of all Abs should be determined for each ELISA. Usually catcher Ab concentrations range from 1 to 10 µg ml^{-1}. A list of mAbs and polyclonal Abs available for the detection of cytokines by ELISA is shown in Table 3.

Table 3. Cytokine-specific mAbs for ELISA assay of human cytokines

Cytokine	Coating mAb Designation	Detection mAb Designation	Source * coat/det.
IL-1α	B-Z3	B-Z5	1/1
IL-1β	B-A15	B-Z8	1/1
IL-2	MQ1-17H12	BG-5	2/1
IL-2	Goat polyclonal Ab	BG-5	1/1
soluble (s) CD25	BG-3	B-F2	1/1
IL-3	BVD8-3G11	BVD3-IF9	2/2
IL-4	B-R14	B-G28	1/1
sIL-4R	S456C9	BB4N1	1/1
IL-5	JES-39D10	JES1-5A10	2/2
IL-5	Goat polyclonal Ab	B-Z25	1/1
IL-6	MQ2-39C3	MQ2-13A5	2/2
IL-6	B-E8	B-E4	1/1
sIL-6R	B-N12	B-R6	1/1
IL-7	BVD10-40F6	BVD10-11C10	2/2
IL-7	B-N18	B-S16	1/1
IL-8	B-K8	Rabbit polyclonal Ab	1/1
IL-10	JES8-9D7	JES3-12G8	2/2
IL-10	B-N10	B-T10	1/1
IL-12p70	B-T21	B-P24	1/1
IL-12p40	B-P40	B-P24	1/1
IL-13	JES10-35G12	JES10-2E10	2/2
IL-13	B-B13	B-P6	1/1
IL-18	Goat polyclonal Ab	Goat polyclonal Ab	1/1
G-CSF	BVD13-3A5	BVD11-37G10	2/2
GM-CSF	BVD2-23B6	BVD11-21C11	2/2
TNF-α	MP9-20A4	GMO1-1782	2/2
IFN-γ	A35	B27	1/1
IFN-γ	B-B1	B-G1	1/1

* This list is a selection of (m)Ab for ELISA purposes, but is not exhaustive. (1) Diaclone, Besançon, France. (2) Generated at DNAX Research Institute, Palo Alto, CA, USA; Relevant hybridomas can be obtained via American Tissue Culture Collection; permission must be obtained from DNAX for acquisition of cell lines from this source.

> **Protocol**
>
> The procedure for activating cells is given on pages 713–714.
> 1. Coat 96-well assay plate with 100 µl catcher Ab, diluted in carbonate buffer.
> 2. Seal plate with lid or parafilm and incubate for 2 h at 37°C or overnight at 4°C. Plates can be stored in a sealed pouch with a desiccant bag at 4°C, Before storage, a blocking solution should be added. Avoid storing plates for periods longer than several weeks.
> *(contd.)*

3. Wash wells three times with washing buffer and flick plates dry on absorbent tissue.
4. Add 300 µl of bocking buffer, incubate for 30 min and flick plates dry on absorbent tissue. Do not wash.
5. Dilute the samples and cytokine standard (aliquots stored at −80°C) in Tween buffer, mix well and add 100 µl per well in duplicate at appropriate dilutions.
- Standard curve: dilute in separate plate in 1:2 dilutions to cover a 1000–20 pg ml⁻¹ range and add 100 µl per well in duplicate.
- Unknown samples: dilute as needed (see notes) and add 50 µl per well.
6. Incubate plates for overnight at room temperature. When incubation time for the standard or sample is unknown, it is recommended to incubate the plate overnight at 4°C. Alternatively the biotinylated tracer Ab together with the standard should be incubated overnight at 4°C.
7. Wash three times with washing buffer and flick plates dry on absorbent tissue.
8. Dilute the biotin-conjugated anti-cytokine detection mAb in Tween buffer and add 100 µl per well.
9. Incubate for 1–2 h at room temperature.
10. Wash three times with PBS and flick plates dry on absorbent tissue.
11. Prepare strepavidin-alkaline phosphatase or streptavidin-HRP conjugates in Tween buffer and add 100 µl per well.
12. Incubate for 1 h at room temperature.
13. Wash three times with washing buffer and flick plates dry on absorbent tissue.
14. Prepare the ELISA substrate (Dilute 1 mg ml⁻¹ ELISA substrate in 100 ml of substrate buffer. Prewarm at 37°C and add 100 µl per well. TMB substrate is ready to use, should be at room temperature before addition.
15. Incubate for 30 min at 37°C to let colour develop and read OD at 405–430 nm using a spectrophotometer or spectrofluorometer (Dynatech).
    For TMB block the reaction with 1N sulfuric acid and read absorbence at 450 nm with a reference filter set at 630 nm.
16. Use the Softmax program (Molecular Devices Corporation) or comparable program to analyse the data.

Notes and recommendations

It is recommended to make several dilutions of cytokine-containing supernatants to be able to measure at the linear part of the standard curve.

It is important to note that the presence or absence of detectable levels of cytokines in clinical samples does not always correlate with their functional activity. Cytokines are often bound to non-specific serum proteins or to soluble cytokine receptors, which will sequester free cytokine from the peripheral circulation and might result in a decrease in detectable cytokine levels, as measured with ELISA. The biological significance of free versus complexed cytokines however is not yet clear. In addition, circulating anti-cytokine Abs, notably in the serum of those who have received cytokine or anti-cytokine therapy may decrease levels of free cytokine, although they may still be detectable by cytokine-specific mAbs. Conversely, cytokine receptor antagonists may specifically compete with the cytokine for binding to its receptor. For example, despite high levels of IL-1 in certain serum samples, possible receptor-occupancy by IL-1RA will prevent IL-1-mediated responses and therefore will prevent a meaningful analysis about its functional activity (Ahrendt, 1993). This situation is similar for IL-18 and the IL-18BP (Novick *et al.*, 1999).

The half-life time of many cytokines in serum is very short, generally in the range of minutes, which is one of the reasons that cytokine levels in the peripheral circulation are usually very low in healthy individuals, as measured by ELISA. However, certain cytokines, notably IL-1, IL-6, IL-10 and TNF-α may be detectable in the serum following sepsis, trauma or an acute or chronic inflammatory state. It should however be kept in mind that, due to existing cytokine networks, even in above-mentioned pathological situations, the presence in excess of certain cytokines may result in the suppressed production of another and a failure to detect the latter cytokine.

Due to the possible presence of proteases or other (unknown) factors, present in clinical samples, as well as culture supernatants, cytokines may be unstable and subject to degradation. Therefore, it is recommended to aliquot and store samples for cytokine measurement at −80°C prior to analysis, especially when prolonged periods of storage are required. Serum may contain certain factors such as heterophilic Abs that have the ability to cross-bridge the capture Ab attached to the solid phase with the biotinylated tracer Ab which may result in false-positive results. The reported incidence of human anti-mouse Abs in the serum of normal individuals is around 3%. Human anti-mouse Abs develop with high incidence in subjects who are in contact with animals or animal products or in patients who have been treated with murine mAbs. Rheumatoid factors (RF) are polyreactive IgM Abs produced by a subset of B lymphocytes and bind to the Fc portion of the IgG molecule. RF may also interfere in the ELISA assay by bridging the tracer to the capture Ab in the absence of analyte.

Another important issue in immunoassays in general is the calibration of the assay. Since the same amount of cytokine will not give the same signal in different assays, it is very important to calibrate the standard to the available international standard in order to be able to compare results.

One last point that should be emphasized is that different monoclonal or polyclonal Ab pairs sometimes give different results which is likely to depend on the different epitopes that are recognized by these Abs. For example, some epitopes might be hidden or denatured by certain cytokine-binding proteins or proteases. Furthermore the variable

level of glycosylation of some cytokines, such as IL-6, may also interfere with the capacity of Abs to recognize fully the natural cytokine.
Finally, the capacity of various Abs to detect monomers rather than
naturally occurring polymers (TNFα, IFN-γ, IL-10, IL-12) should also be
taken into account when considering the variability between
immunoassays.

◆◆◆◆◆◆ ELISPOT ASSAY

The ELISPOT assay is based on immuno-enzyme technology originally
developed for the enumeration of Ab-secreting cells (Czerkinsky *et al.*,
1983, 1988; Sedwick and Holt, 1983). The assay was subsequently adapted
to measure cytokine production at the single-cell level. The ELISPOT
assay is easy to perform and requires compared with other approaches
such as limiting dilution or intracellular staining, a minimum *in vitro* cell
manipulation, allowing analysis of cells close as in the *in vivo* situation.
It can be used successfully to measure ongoing cytokine production by
freshly isolated mononuclear cells. The technique is designed to determine the frequency of cytokine-producing cells under a given stimulation, and it enables the follow-up of these frequencies during a treatment
and/or different stages of disease.

Cells can either be stimulated directly in wells containing Ab-
coated membrane (direct method) or first stimulated for 6 h in 24-well
plates, harvested, and transferred to ELISPOT plates (indirect method).
The choice between the two methods depends on the type of cell to be
assayed, the expected frequency of cytokine-producing cells and the
mode of stimulation. Stimulation of cells with immobilized anti-CD3
mAb can only be carried out using the indirect stimulation procedure.

The technique includes the following steps: Single cell suspensions are
distributed in membrane-bottomed wells previously coated with an anti-
cytokine Ab. During cell stimulation, locally produced cytokine molecules will be caught by surrounding capture Ab and accumulate in the
close environment of the cell. After cell lysis, trapped cytokine molecules
are revealed by a secondary biotinylated detection Ab, which is in turn
recognized by streptavidin conjugated to alkaline phosphatase. The enzymatic reaction in the presence of the substrate BCIP/NBT produces a precipitate. Cytokine secreted by individual cells is visualized by sharp
blue/purple spots that can be counted on an inverted microscope or a
reading system, allowing the numeration of cytokine-producing cells
following activation.

Recently developed ELISPOT assays allow to monitor the simultaneous production of two cytokines by the same cell population (Fig. 1). In
this method, PVDF filters in 96-well plates are coated with two different
Abs each one specific for one of two cytokines to be analysed. After
blocking of non-specific binding sites on the membrane, cells are incubated for the appropriate length of time, in the presence of a stimuli, i.e.
protein, peptide or tumour cells. After lysing of the cells, cytokine

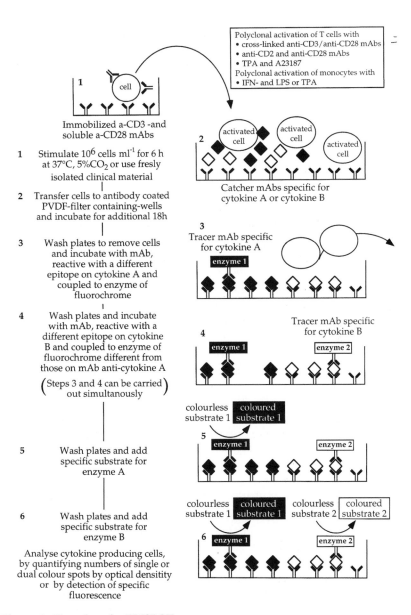

Measuring Human Cytokine Responses

Polyclonal activation of T cells with
• cross-linked anti-CD3/anti-CD28 mAbs
• anti-CD2 and anti-CD28 mAbs
• TPA and A23187
Polyclonal activation of monocytes with
• IFN- and LPS or TPA

1   cell

Immobilized a-CD3 -and
soluble a-CD28 mAbs

1   Stimulate 10^6 cells ml^{-1} for 6 h
at 37°C, 5%CO_2 or use fresly
isolated clinical material

2   Transfer cells to antibody coated
PVDF-filter containing-wells
and incubate for additional 18h

3   Wash plates to remove cells
and incubate with mAb,
reactive with a different
epitope on cytokine A and
coupled to enzyme of
fluorochrome

4   Wash plates and incubate
with mAb, reactive with a
different epitope on cytokine
B and coupled to enzyme of
fluorochrome different from
those on mAb anti-cytokine A

(Steps 3 and 4 can be carried
out simultanously)

5   Wash plates and add
specific substrate for
enzyme A

6   Wash plates and add
specific substrate for
enzyme B

Analyse cytokine producing cells,
by quantifying numbers of single or
dual colour spots by optical densitity
or by detection of specific
fluorescence

2   activated cell   activated cell   activated cell

Catcher mAbs specific for
cytokine A or cytokine B

3   Tracer mAb specific
for cytokine A

enzyme 1

Tracer mAb specific
for cytokine B

4   enzyme 1   enzyme 2

colourless substrate 1   coloured substrate 1

5   enzyme 1   enzyme 2

colourless substrate 1   coloured substrate 1   colourless substrate 2   coloured substrate 2

6   enzyme 1   enzyme 2

Figure 1. Flowchart for ELISPOT assay.

molecules trapped by the catcher Ab are recognized by two differently
tagged secondary Abs. For one cytokine, Ab are labelled with biotin, for
the other, Abs are conjugated with FITC molecules. These Ab are in turn
detected by streptavidin-alkaline phosphatase conjugates, or FITC peroxidase conjugates for the biotinylated and by FITC conjugated Ab,
respectively. Finally, addition of BCIP/NBT and AEC, the respective
substrates of alkaline phosphatase and peroxidase, will result in
blue/purple and red/brownish spots in the presence of the enzyme,
allowing the visualization of cytokine molecules produced by the cells.
In a recent development, the enzymatic revelation system has been

replaced by fluorescent dye conjugates such as streptavidin-phycoery-thrin and anti-FITC green fluorescent dye. Fluorescent spots are then visualized under a UV light beam on a UV detection system. Examples of ELISPOT assays for the detection of either IFN-γ or IL-4 and concomitant detection of IFN-γ and IL-10 are shown in Plate 10 and Plate 11, respectively.

Reagents and reagent preparation

- Capture anti-cytokine mAbs.
- Biotin-conjugated anti-cytokine detection mAbs.
- FITC-conjugated anti-cytokine detection mAbs (for dual cytokine ELISPOT).
- Streptavidin alkaline phosphatase conjugate (Prozyme). Dilute 1/1000 in PBS-1% Bovine serum albumin (BSA).
- PBS BSA (Sigma: A2153), skimmed dry milk (2% in PBS), ethanol (70% in H_2O), Tween 20 (0.1% in PBS).
- Anti-FITC HRP-conjugated mAb (for dual cytokine ELISPOT). Dilute 1/500 in PBS–1% BSA.
- AEC substrate buffer (for dual ELISPOT). For one plate mix 1 ml of AEC buffer A with 9 µl of distilled H_2O. Then add 200 ml of AEC buffer B.
- BCIP/NBT substrate buffer.
- 96-well PVDF filter-containing ELISPOT plates (Millipore Multi-Screen; MAIPN4510) are recommended.
- ELISPOT reading system (Carl Zeiss).

Reagent storage

- If not used within a short period of time, reconstituted detection Ab should be aliquoted and stored at −20°C. In these conditions the reagent is stable for at least one year.
- Substrate buffers to be stored at 4°C.
- Streptavidin-alkaline phosphatase as well as anti-FITC HRP conjugates should be stored at 4°C.

Protocol

The procedure for activating cells is given on pages 713–714 (see notes 2 and 3).

1. Incubate PVDF filter-containing wells of an ELISPOT plate with 100 µl of 70% ethanol for 10 min at room temperature.
2. Empty wells (flick the plate and gently tap it on absorbent paper). Wash three times with 100 µl PBS.

(contd.)

3. Pipette 100 µl of each of the capture mAbs in 10 ml of PBS. Mix and dispense 100 µl into each well, cover the plate and incubate overnight at 4°C.
4. Empty wells and wash plate once with 100 µl of PBS.
5. Dispense 100 µl of 2% skimmed dry milk in PBS into wells, cover and incubate for 2 h at room temperature.
6. Empty wells and wash plate once with 100 µl of PBS.
7. Add to a well of an ELISPOT plate between 1.10^5 to 2.5×10^4 of pre-activated cells (indirect method) or cells to be activated with the appropriate stimulator (direct method) in a final volume of 100 µl. Cover the plate with a plastic lid and incubate for 10–15 h at 37°C, 5% CO_2. *During incubation do not disturb the plate.*
8. Empty wells, add 100 µl of PBS-0.1% Tween 20 and incubate for 10 min at 4°C.
9. Empty wells and wash plate three times with 100 µl of PBS-0.1% Tween 20.
10. For 1 plate dilute 100 µl of cytokine 1 detection Ab and 100 µl of cytokine 2 detection Ab into 10 ml of PBS containing 1% BSA. Dispense 100 µl into wells, cover the plate and incubate 1 h 30 min at 37°C.
11. Empty wells and wash three times with 100 µl of PBS-0.1% Tween 20.
12. For 1 plate dilute 20 µl of anti-FITC HRP and 10 µl of streptavidin-Alkaline phosphatase conjugates into 10 ml of PBS-1% BSA. Dispense 100 µl of the dilution into wells. Seal the plate and incubate for 1 h at 37°C.
13. Empty wells and wash three times with 100 µl of PBS-0.1% Tween 20.
14. Prepare AEC buffer (see reagents preparation) and add 100 µl of solution to each well.
15. Let colour reaction proceed for 5–15 min at room temperature. When the spots have developed, remove the buffer and dispose of appropriately.
16. Wash both sides of the membrane thoroughly with distilled H_2O and carefully remove residual H_2O by tapping the plate on absorbent paper.
17. Dispense 100 µl of ready-to-use BCIP/NBT buffer into wells.
18. Let the colour reaction proceed for about 5–15 min at room temperature. When spots have developed, empty the buffer into an appropriate tray.
19. Wash thoroughly both sides of the membrane with distilled water.
20. Dry the membrane by repeatedly tapping the plate on absorbent paper. Store the plate upside down to avoid remaining liquid to flow back onto the membrane. Read spots once the membrane has dried. Note that spots may become more distinct following incubation of the filters at 4°C for 24 h. Store the plate at room temperature away from direct light.

Notes and recommendations

1. AEC and BCIP/NBT buffers are potentially carcinogenic and should be disposed of appropriately. Caution should be taken while handling those reagents. Always wear gloves.
2. While using MAIPN4510 plates during incubation steps, reagents are leaking through the membrane by capillary action. As this liquid is generally not properly removed during the various washing steps, it will increase the background signal. To avoid this, it is recommended that the plate bottom is peeled off at the end of step 12 of the procedure and that the reverse of the membrane is washed with running distilled water, as well as the regular washes with PBS-0.1% Tween 20. For the following incubation in step 14, as the plate bottom has been removed, it is suggested to put the MAIP4510 plate on top of an empty 96-well plate for use as a tray.
3. Automatization of the ELISPOT technique makes it suitable for use in clinical and hospital settings (Herr *et al.*, 1997). The detection limit of the technique is between one and five cytokine-producing cells for a total of 10^5 cells which makes it one of the most sensitive techniques for the detection of antigen-specific T cells. However, in contrast to the tetramer-MHC-class II technique, cytokine-producing T cells are only detected following antigen-specific activation of the cells. Unless the cells are sorted before the assay, the ELISPOT method does not allow to phenotype neither to sort cytokine-producing cells. A list of monoclonal and polyclonal Abs available for the detection of cytokines by ELISPOT is shown in Table 4.

Table 4. Human cytokine-specific (m)Abs for ELISPOT

Cytokine	Tracer (phase) mAb	Catcher (m)Ab
IL-1β	B-A15	B-Z8
IL-2	B-G5	Rabbit polyclonal Ab
IL-4	B-R14	B-G28
IL-5	B-Z25	Goat polyclonal Ab
IL-6	B-E8	B-E4
IL-10	B-N10	B-T10
IL-12p70	B-T21	B-P24
IFN-γ	B-B1	B-G1
TNF-α	B-F7	B-C7

At present, the most complete, well-characterized set of (m)Ab for use in the ELISPOT assay is available from Diaclone, Besançon, France.

◆◆◆◆◆◆ INTRACELLULAR CYTOPLASMIC STAINING

The method of staining with cytokine-specific mAbs for the analysis of intracellular cytokines in suspension by immunofluorescence using a UV microscope has originally been described by Anderson *et al.* (1990). Subsequently, it was shown that the presence of intracellular cytokines could be detected using a FACS flow cytometer (Jung *et al.*, 1993; Assenmacher *et al.*, 1994) and recent improvements in the fixation procedure (Openshaw *et al.*, 1995) and the use of Brefeldin as a protein transport inhibitor (Picker *et al.*, 1995) has made the method even more suitable for multiparameter flow cytometry analysis. The advantage of this method over the measurement of cytokines in culture supernatants of activated cells is that it enables the determination of the frequency of cytokine-producing cells, as well as kinetics of cytokine production. Moreover, in combination with cell surface staining, the method can be used to identify the phenotype of cytokine-producing cells in a population of non-separated cells, whereas the development of fluorochromes with different emission wavelength and flow cytometers, equipped with dual lasers, has permitted the simultaneous detection of several cytokines, provided the mAbs are directly conjugated with the appropriate fluorescent dyes.

Critical parameters for successful intracellular cytokine staining include the choice of a proper fixation and permeabilization protocol, the inclusion of a protein transport inhibitor, cell type and activation protocol and kinetics of activation (Fig. 2).

The principle and sensitivity of the method depends on the availability of mAb which recognize natural cytokines, in spite of a fixation and permeabilization procedure. Fixation of the cells is required for subsequent treatment with detergent and ideal fixatives preserve the morphology of the cells and antigenicity of the cytokines with minimal cell loss. Subsequent permeabilization of the cell membranes with the detergent saponin allows fluorochrome-labelled cytokine-specific mAb to penetrate through the cell membrane, cytosol and the membranes of the Golgi apparatus and endoplasmic reticulum. It should be kept in mind, however, that the method provides a 'snapshot' of cytokine production and the observed frequency of cytokine-producing cells is affected by multiple variables, such as activation conditions and state of cell-cycle of the cells.

Reagents and equipment

- *Washing buffer*. PBS without Ca^{2+} and Mg^{2+}, supplemented with 2% heat-inactivated FCS and 0.1% (w/v) sodium azide (NaN_3) (see note 1). Adjust buffer to pH 7.4 and store at 4°C.
- *Permeabilization buffer*. Washing buffer, supplemented with 0.5% saponin (Sigma; S7900). Make stock solution of 10% saponin in PBS (pH 7.4), dissolve at 37°C, filter solution through a 0.2 μm filter and store at 4°C (see note 1).

(contd.)

Measuring Human Cytokine Responses

725

Stimulate 10^6 cells ml^{-1} for 4 h
at 37°C, 5%CO$_2$

Control cells are cultured in
medium for the same period of
time and treated similarly during
the protocol

> Polyclonal activation of T cells with
> • cross-linked anti-CD3/anti-CD28 mAbs
> • anti-CD2 and anti-CD28 mAbs
> • TPA and A23187
> Polyclonal activation of monocytes with
> • IFN-γ and LPS or TPA

Add Brefeldin A (10
mg ml^{-1}) and incubate for
an additional 2 h

> Agents which block intracellular protein transport have dose-
> and time-dependent cytotoxic effects. It is not recommended to
> include Brefeldin in the cultures >6 h
> Monensin will induce both intracellular TNF and IL-1
> production within 30 min of stimulation

Harvest the cells, wash with
PBS/2% FCS and stain for
surface molecule expression

> Activation with TPA causes a down-regulation of CD4 expression
> Some mAb to cell surface Ag may not bind to fixed, denatured
> proteins, and cell surface staining should be performed prior to
> fixation
> Protect cells from light during staining procedures
> Non-specific FcR-binding can be inhibited by preincubation of the
> cells with 2% Ig from species in which mAb was generated

Wash the cells with PBS

> Cells should be thoroughly washed to remove any protein, prior
> to fixation

Fix the cells for 20 min
with 2% Formaldehyde
Wash twice with PBS/2% FCS

> Vortex gently to avoid aggregation of the cells
> Fixed cells can be stored at 4°C for at least a week and for
> several weeks following storage at -80°C
> When methods involving simultaneous fixation/
> permeabilization are used, stored cells should be re-
> permeabilized prior to staining

Permeabilize the cells with
saponin (0.5% in PBS/2% FCS)
for 10 min in the presence or
absence of relevant animal
serum

> Because its effects are reversible, saponin should be
> maintained in the mAb solution and staining buffer to prevent
> closure of the membranes.
> Cells, spun in the presence of saponin will nor show clear
> pellet. However, if properly spun, no cells loss will occur

Stain with cytokine-
specific mAb diluted in
PBS/saponin/FCS

> Non-stimulated, identically treated, cells serve as negative staining
> control
> Each mAb should be titrated for optimal staining on stimulated
> cells and absence of staining on control cells
> Specificity of staining can be controlled by
> • Preincubation of the anti-cytokine mAb with recombinant cytokine
> • Preincubation of the cells with unlabelled mAb, followed by
>   staining with the same, fluorochrome-labelled, mAb

Wash with PBS/2%FCS

> Saponin should be removed prior to flow cytometric analysis

Analyse on the FACS

> If two- or more colour analysis is performed, PMT voltage and
> compensation should be set on cell surface controls
> Always include a positive control (T-cell clone) which should be
> prepared, aliquotted and stored at -80°C
> Quadrant markers should be set on the negative control (non-
> stimulated cells)

> Perform control analysis using UV microscope

Figure 2. Flowchart for Intracellular cytokine staining.

- *Formalin* (=37% formaldehyde solution, Sigma F1635). Dissolve 10.8
  ml of Formalin in 89.2 ml PBS, filter to remove particles and store at
  4°C (see note 1).
- *Cytokine-specific mAbs* (preferentially conjugated directly with a fluorochrome (FITC, Phycoerythrin (PE), or Cy5). A list of mAbs suitable
  for intracellular staining is shown in Table 5.
- *Second step fluorochrome-labelled Abs.* Horse-anti-mouse IgG-FITC/PE
  (Vector; FI 2000/EL 2000). Rabbit-anti-rat IgG-FITC (Vector; FI
  4000).

(contd.)

- *For analysis by UV microscope.* Bio-Rad adhesion slides (Bio-Rad Laboratories; 180-7001); plastic box with humid paper tissue, dH_2O; slideholder to wash slides; Vector stain enhancer; cover slides; Nail polish.
- *For analysis by FACScan flow cytometer.* Vortex; Centrifuge with rotor to spin microtitre plates; 96-well V-bottom plates (Falcon, Becton Dickinson); FACS tubes (Becton Dickinson).

Table 5. Human cytokine-specific mAbs for intracellular staining

Cytokine	mAb	Isotope*	Source†
IL-1β	B-A15	mIgG1	1
IL-2	MQ1-17H12	rIgG2a	2/3
	BG-5	mIgG1	1
IL-3	BVD8-3G11	rIgG2a	2/3
	BVD3-IF9	rIgG2a	2/3
IL-4	8D4-8	mIgG1	2/3
	MP4-25D2	rIgG1	2/3
	B-G28	mIgG1	1
	3010.2	mIgG1	3
IL-5	JES-39D10	rIgG2a	2/3
IL-6	MQ2-6A3	rIgG2a	2/3
	B-E8	mIgG1	1
IL-8	B-K8	mIgG1	1
IL-10	JES8-12G8	rIgG2a	2/3
	B-N10	mIgG1	1
	JES3-19F1	rIgG2a	2/3
IL-12p70	B-T21	mIgG1	1
IL-12p40	B-P24	mIgG1	1
JIL-13	JES8-30F11	rIgG2a	2
	JES8-5A2	mIgG1	2/3
	B-B13	mIgG1	1
IL-15	B-T15	mIgG1	1
GM-CSF	BVD2-21C11	rIgG2a	2/3
G-CSF	BVD13-3A5	rIgG1	2/3
TNF-α	BVD11-37G1	mIgG1	2/3
	B-D9	mIgG1	1
	MP9-20A4	rIgG1	3
IFN-γ	B27	rIgG1	2
	4S.B3	mIgG1	3
	25723.11	mIgG1	3
	B-B1	mIgG1	1

* m: mouse mAb, r: rat mAb.

† mAb sources are: (1) Diaclone, Besançon, France, (2) Generated at DNAX Research Institute, Palo Alto, CA, USA; Relevant hybridomas can be obtained via American Tissue Culture Collection; permission must be obtained from DNAX for acquisition of cell lines from this source, (3) Becton Dickinson/PharMingen, San Diego, CA, USA.

Measuring Human Cytokine Responses

Protocol: Staining for flow cytometry

The procedure for activating cells is given on pages 713–714 (see notes 2 and 3).

If cell-surface staining is performed, start with step 1; for intracellular cytokine staining without cell-surface staining, start with step 7 (see note 4).

1. Transfer the stimulated cells to 15-ml centrifuge tube and wash twice with 1 ml of ice-cold washing buffer. Washing with larger volumes will result in cell loss. Spin the cells at 200g for 5 min.
2. To block non-specific binding of FcR, add 20 µl of 2% serum (from same animal species as the mAbs used for the immunofluorescence staining) to the cell pellet and incubate cells for 15 min.
3. Wash cells twice with 1 ml of cold washing buffer.
4. Add 20 µl of directly conjugated anti-cell surface mAb to the cell pellet.
5. Incubate for 15 min on ice.
6. Wash cells twice with 1 ml ice-cold PBS and proceed to step 8.
7. Transfer the stimulated cells to 15-ml centrifuge tube and wash twice with ice-cold PBS.
8. Resuspend the cells at 2×10^6 cells ml^{-1} in cold PBS and add an equal volume of 4% formaldehyde (see note 5). Mix well to prevent aggregates.
9. Incubate for at least 20 min at room temperature.
10. Spin the cells at 200g for 5 min and wash once more with 1 ml of cold PBS. At this point the cells can be resuspended in PBS and stored at 4°C in the dark for one week or aliquoted and stored at −80°C for at least one month prior to analysis (see note 6).
11. Resuspend the cells in washing buffer and transfer ± 10^5 cells per well of a 96-well V-bottom microtitre plate and spin at 200g for 2 min.
12. Remove supernatant by flicking the plate, blot the plate dry on paper tissue and resuspend the cells by gently vortexing.
13. Add 150 µl of permeabilization buffer to the cells and incubate for 10 min at room temperature. A final concentration of 2% serum (as described under 2 above) can be added to block non-specific FcR binding of mAbs.
14. Spin the cells at 200g for 2 min and remove supernatant.
15. Add 20 µl of cytokine-specific mAb solution to the cell pellet (see note 7).
16. Incubate cells for 20 min at room temperature.
17. When using non-conjugated mAbs, proceed to step 18;
    for FITC, PE or Cy5-conjugated mAbs, proceed to step 22.
18. Add 150 µl of permeabilization buffer, centrifuge the plate at 200g for 2 min, remove supernatant and resuspend the cells.
19. Repeat washing step.

(contd.)

20. Add 20 µl per well of an appropriate dilution of the second conjugated mAb.
21. Incubate cells for 30 min at room temperature.
22. Add 150 µl of permeabilization buffer, centrifuge the plate at 200*g* for 2 min, remove supernatant and resuspend the cells.
23. Repeat washing step.
24. Wash once with *washing* buffer, resuspend in 200–300 µl PBS and analyse on the FACS flow cytometer as soon as possible.
25. For instrument control and compensation setting, refer to the manufacturer's instructions (see Lanier and Recktenwald (1991) and notes 6–9).

Protocol: Staining for microscopic analysis, using Bio-Rad slides

1. Wash off the protective green layer from the Bio-Rad adhesion slide with dH$_2$O and incubate the slide for 5' in PBS.
2. Add between 2×10^4 and 10^5 cells in a volume of 15 µl to each slot and incubate for 10 min at room temperature. Check cell density using inversion microscope.
3. Gently wash the slide with washing buffer to remove non-adherent cells.
4. Add 20 µl of fixation buffer to each slot and incubate for 15 min.
5. Gently wash the slide with washing buffer.
6. Staining procedure used for Bio-Rad slides is identical to that for flow cytometric analysis (see note 9).

Notes and recommendations

1. NaN$_3$ is known to be toxic, saponin is an irritant and formaldehyde is a suspected carcinogen. Avoid contact with skin, eyes and mucous membranes. Dispose of formaldehyde in container for treatment, do not put formaldehyde-containing waste in sink. Always wear gloves.
2. Since cytokines are produced with different kinetics, optimal time points for the analysis of activated cytokine-producing cells may vary, depending not only on the type of cytokine to be analysed, but also on the mode of stimulation. The peak production of most cytokines, including IL-2, IL-4 and TNF-α, by T-cell clones, following stimulation with TPA and A23187, is within the first 6–7 h, whereas others, such as IFN-γ and IL-10 have more long-lasting kinetics. Interleukin-13 is produced early (2 h) following activation, but this cytokine also continues to be produced up to 72 h. Analysis of the cells 4–6 h after activation, results in a characteristic staining of the Golgi apparatus/endoplasmic reticulum, which is bright for cytokines such as IL-2, TNF-α and IFN-γ.

(contd.)

Measuring Human
Cytokine Responses

3. The addition of agents which inhibit cellular transport systems, such as Brefeldin A or Monensin, results in an accumulation of newly synthesized protein in the Golgi complex. Although the cellular integrity changes following treatment with intracellular protein transport inhibitors, resulting in a disappearance of Golgi staining and a homogeneously fluorescing cells, their use generally gives a brighter staining pattern and enables the detection of intracellular cytokines that stain only weakly with anti-cytokine mAbs, such as IL-4 and IL-5. However, since these compounds have dose- and time-dependent cytotoxic effects, their use should be limited. Brefeldin A is less toxic than Monensin and should be added to the cultures between 2 and 6 h before harvesting the cells. Extensive analysis of peripheral blood mononuclear T cells and cloned T-cell lines, in order to determine optimal kinetics, has shown that activation for 6 h, including 2 h of Brefeldin A treatment, results in a higher frequency of IL-2, IL-4 and IFN-γ-producing cells, than activation for 24 h including 2, 6 or 10 h Brefeldin A treatment (H. Yssel, unpublished results).

4. The staining procedure for flow cytometric analysis is performed in 96-well V-bottom microtitre plates. However, since it is preferable to fix the cells in 15-ml centrifuge tubes, it is recommended to perform staining of cell-surface molecules in tubes, followed by fixation and transfer of the cells to microtitre plates. All steps during cell-surface staining are carried out at 4°C, in the absence of saponin. Since some mAbs may not bind to fixed, denatured, protein cell-surface staining should be carried out before fixation of the cells. Staining for intracellular cytokines is performed at room temperature in the presence of saponin. All incubation periods during the staining procedures are carried out in the dark.

- Make the Ab dilutions in PBS/2%FCS/NaN$_3$/0.5%saponin analysis. For each mAb, the optimal titre has to be determined. Generally, final concentrations of mAbs are 1–5 μg per 10^6 cells. PE-conjugated mAbs are not recommended for analysis on a UV microscope, because of the quick fading of the dye: use Rhodamin-conjugated Abs instead.
- Centrifugation of the Abs at high speed in a microfuge to remove aggregates will improve the quality of the staining.
- If non-conjugated anti-cytokine mAbs are used, obtained from different species, two-colour analysis can be performed. Although cytokine-specific staining signals generally tend to be higher, often increased fluorescence backgrounds are observed, requiring rigid blocking of non-specific mAb binding.

5. Optimal results have been obtained with formaldehyde (Openshaw et al., 1995) and although this fixative is less efficient to cross-link and change the tertiary structure of protein as compared with glutaraldehyde, its effects are more gentle and fixation with formaldehyde generally preserves a high degree of antigenicity.

(contd.)

6. *Positive control.* Samples of a resting and activated T clone with a Th0 cytokine production profile can be fixed, aliquotted in PBS, stored at −80°C for several weeks and used in the staining procedure as negative control (to adjust flow cytometer instrument settings) and positive controls, respectively.

7. *Negative control.* Non-stimulated cells, which have been identically treated during the staining procedure serve as negative control. Prior to its use in the staining procedure, each anti-cytokine mAb should be titrated at a concentration which gives optimal staining on activated cells and absence of staining on non-stimulated control cells. Furthermore, it is recommended to use one of the following controls to confirm specificity of the staining:

- *Ligand-blocking control.* Pre-incubate mAb with excess of recombinant cytokine in permabilization buffer at 4°C for 30 min. It is recommended to use at least a 50-fold Molar excess of cytokine over mAb. Proceed to step 15 of staining protocol.

- *Non-conjugated mAb control.* Pre-incubate cells after step 14 of staining protocol with non-conjugated mAb, diluted in permeabilization buffer, at 4°C for 30 min, wash cells twice and proceed to step 15 of staining protocol.

- *Isotype control.* Use an irrelevant mAb of the same isotype as the anti-cytokine mAbs in the staining procedure. Isotype controls can be used to adjust instrument settings, including quadrant markers and compensation of flow cytometer. In cases where isotype controls give brighter staining than anti-cytokine mAb, rely on non-activated cells as negative control.

8. It should be stressed that when the intracellular staining method, using flow cytometry, is introduced in the laboratory, the results should be confirmed by immunofluorescence using a UV microscope. Intracellular cytokine staining of cells that have been activated less than 6 h in the absence of Brefeldin A will result in distinct staining of the Golgi apparatus and endoplasmatic reticulum, whereas after longer activation periods also cell-surface cytokine staining may be detected. Microscopic analysis of cells, activated in the presence of Brefeldin A show homogeneous staining throughout the cytoplasm, as a result of disintegration of Golgi apparatus and endoplasmatic reticulum.

9. All incubations using Bio-Rad slides are carried out at room temperature.

When using Bio-Rad adhesion slides, keep the slides in a wet chamber (filter paper soaked with dH$_2$O) during incubations to prevent drying out of the slots. After each washing step, remove remaining medium between the slots with a piece of absorbent paper to prevent spill over between adjacent slots. Add 15 µl of Ab solution per slot. After staining procedure, add 10 µl Vector stain enhancer, cover the slots with a cover slide, seal off with nail polish and analyse the cells using a UV microscope. (Slides can be stored for months in the dark, if properly sealed.) Slides that have been fixed can be stored at −80°C for several weeks.

◆◆◆◆◆◆ COMPETITIVE RT-PCR

The reverse transcription polymerase chain reaction (RT-PCR) is a powerful technique to analyse gene regulation at the cellular level (Ferre, 1992). However, the advantage of this sensitivity is also a limitation due to the two enzymatic reaction steps. The reverse transcriptase first synthesizes a complementary strand from the RNA with specific, random or oligo dT primers, and this newly synthesized cDNA is subsequently amplified by the Taq polymerase with two specific primers for a sufficient number of times to be analysed. Moreover, due to the exponential nature of the PCR step, the effects of minor variations in reaction conditions can modify greatly the results from sample to sample (Hengen, 1995). Finally, the quantification of a specific cDNA present in the sample is only possible in the exponential phase of the enzymatic reaction, where the amount of amplified targets is a linear function of the starting template (Gilliland *et al.*, 1990).

In order to monitor all these potential variations, many controls have been proposed, from the amplification of an internal housekeeping gene, along with the unknown target (Chelly *et al.*, 1988), to serial dilutions of the samples aside with a standard curve of a known template (Murphy *et al.*, 1993), but the most precise and accurate method remains the competitive PCR (Zimmermann and Mannhalter, 1996). In this latter technique, an external template is mixed at different dilutions with a constant amount of cDNA previously equilibrated on the relative amount of a constitutively expressed gene. This competitor shares the same primer recognition sites with the cDNA of interest, but has a different amplification size in order to easily separate the two amplified products after electrophoresis (Becker-André and Hahlbrock, 1989; Wang *et al.*, 1989). To be really accurate, the competitor must be amplified with the same efficiency as the cDNA of interest (Cottrez *et al.*, 1994). This setting has to be carefully carried out before the adoption of the routine procedure of this method. After the PCR step, the samples are analysed on an electrophoresis gel, the amount of each amplified PCR product is measured, and the respective ratios plotted on a graph against the molar concentrations of the competitor. The amount of the mRNA present in the sample corresponds to the concentration of the competitor when the ratio of the two amplified molecules is equal to one.

A known amount of competitor can also be introduced at the reverse transcription step in a single reaction tube along with the other samples in order to control this other enzymatic reaction step (Wang *et al.*, 1989).

Methods

Designing specific primers

The generation of specific primers is a crucial factor to succeed in a very accurate competitive PCR (Rappolee, 1990; He *et al.*, 1994). The primers for all cytokines should ideally be designed to amplify PCR products which are about the same size. Indeed, since the competitor is amplified by the different primer pairs to give products of same size, the size difference of the PCR products, amplified from the wild-type template and the competitor, should not exceed 150 bp in order to minimize differences in PCR efficiencies between the competitor and the wild-type template, (Zimmermann and Mannhalter, 1996).

The primers need to have a high annealing temperature (at least 60°C) and should not give rise to non-specific products which will compete for the specific templates during the PCR. In the same manner, they should distinguish the DNA from the RNA, this should be ensured by designing at least one or both primers across a splicing site.

Validation of primers

The first step is to monitor the expected size of the amplified product. Then, ideally a specific hybridization with an internal sequence has to be performed.

Construction of an internal competitor for quantitative PCR

Rationale

There are multiple ways to construct PCR competitors (Reiner *et al.*, 1993; Forster, 1994), but the most convenient one when the number of templates is relatively high, as is often the case for cytokines, is to create a multi-competitor standard by linking together all 5' primer sequences spaced by a non-specific sequence from all linked 3' primers (Wang *et al.*, 1989; Siebert *et al.*, 1992), as shown in Fig. 3A. Although there are different ways to link nucleotide sequences, it is recommended to use the overlapping primers-PCR strategy (Fig. 3B).

The first step is to draw the map of the final construct. Ideally, the number of cytokines should not exceed five or six per competitor (Fig. 3A), it is also better to design some competitors for a specific purpose, like pro-inflammatory cytokines or Th1/Th2 cytokines for example. The second step is to design the primers for the construct itself. These primers have 15 nucleotides at the 3' end which hybridize specifically on the DNA to be extended and 25 nucleotides which consist of the primer for the cytokine of interest which needs to be incorporated in the construct. At the final step, it is possible to add a poly A sequence to mix the competitor with the mRNA. Another possibility is to add restriction sites, included in the last primers, in order to make a directional cloning. Finally, it is easier to use the TA cloning vector system commercialized by Invitrogen directly to clone PCR products. The competitor is then excised from the plasmid as a double-strand DNA and can be used at the PCR level, alternatively or it could be synthesized as a cRNA from the RNA polymerase recognition site in the plasmid and be used at the RNA or cDNA steps.

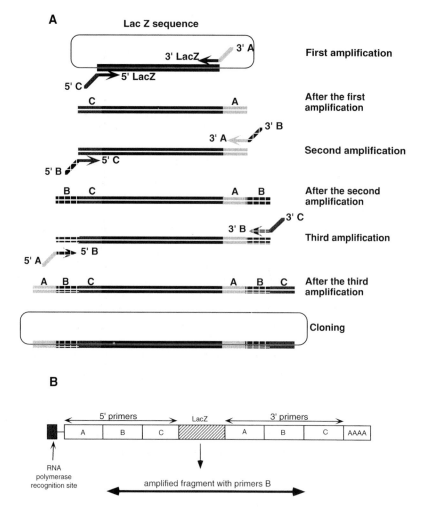

Figure 3. Construction of an internal competitor for competitive RT-PCR.

Protocol

Plasmid (1 ng), containing the Lac-Z sequence, is added to a total reaction volume of 25 µl of the PCR mix. After an initial denaturation step at 94°C for 5 min, two PCR cycles (30 s at 94°C, 30 s at 55°C and 30 s at 72°C), followed by 35 cycles (30 s at 94°C, 30 s at 60°C and 30 s at 72°C), and by a final elongation step at 72°C for 7 min. The amplified product is loaded on a 2% agarose gel in TAE buffer and stained with ethidium bromide. The specific product is recovered by freezing the agarose band at −20°C for 30 min in a 0.5-ml tube which is then introduced into another clean 1.5-ml Eppendorf tube (a little hole has been previously made on the bottom of the 0.5-ml tube, containing the agarose band with a small needle) and then centrifuged at 13 000 rpm for 5 min. One µl out of the 500 µl adjusted volume of the recovered DNA is then used for the second step of amplification. This process is repeated enough times to incorporate all primer pairs in the construct. The final fragment is cloned using the plasmid of choice. After bacterial transformation and amplification, the specific band is cut out of the plasmid followed by a purification step on an agarose gel using a commercial kit for a clean recovery (Wizard PCR prep, Promega). The excised band is then quantified by optical density reading and diluted to be used as an internal competitor.

It must be emphasized that all these amplifications and purification steps are to be carried out preferentially in a room different from the one that will be used for RT-PCR analysis to avoid plasmid contamination and PCR product carry-over.

Analysis of PCR efficiency for both templates

As mentioned above, the wild-type template and the competitor have to be amplified by PCR with the same efficiency. To monitor the different efficiencies, several methods using labelled primers have been described previously (Celi *et al.*, 1993; Cottrez *et al.*, 1994). Briefly, the wild-type and the competitor DNAs are amplified with labelled primers, the respective PCR products are purified, quantified, mixed at different ratios and diluted. The different mixtures are re-amplified and the ratios obtained after amplification are measured and compared with the initial ones. An equivalent efficiency should give rise to the same ratios before and after amplification, even after the plateau phase of the PCR.

RNA preparation

There are two preferential ways to purify the RNA, depending mostly on the amount of available material.

More than 10⁶ cells

Total RNA is isolated and purified using RNAzol (Quantum-Appligene) according to the manufacturer's instructions and quantified by optical density readings. Pelleted cells are lysed with 1 ml RNAzol and then 0.2 ml chloroform is added. After centrifugation, the aqueous phase is recovered and precipitated with isopropanol (v/v). The pellet is re-suspended in 10 µl of RNAse free water. It is sometimes necessary to add glycogen or t-RNA as a carrier for better RNA recovery.

Less than 10⁶ cells

Total RNA is prepared using an affinity column technique (Glassmax from Qiagen, Invitrogen, Gibco). The Glassmax technique from Invitrogen is used according to the manufacturer's instructions. Briefly, the cells are pelleted and homogenized into 400 µl of the lysis buffer, then stored at −20°C until use. The homogenate is thawed on ice and pelleted 5 min at

13 000 rpm with 280 µl of cold ethanol. The pellet is resuspended in 450 µl of the binding buffer and 40 µl of ammonium acetate and applied on a column, spun for 20 s at 13 000 rpm, and washed successively in the same manner with the wash buffer followed by 80% ethanol in water. The RNA is then eluted with 30 to 50 µl of RNAse-free water, previously heated at 70°C.

Reverse transcription

One microgram of total RNA (a lower concentration can be used, although for an easier comparison of the different samples, a constant number of cells is recommended) is resuspended in 20 µl of H_2O and 1 µl of oligo-dT at 1 mg ml^{-1} is added as long as 0.1 µl of 40 U ml^{-1} RNAsin (Promega). The samples are heated at 70°C for 10 min and cooled down at room temperature; the samples are briefly spun down in a microfuge after which 15 µl of the enzyme mixture (7 µl buffer (5X Superscript buffer) (Invitrogen), 5 µl of 10 mM dNTP, 1.5 µl of 1 M DTT (Invitrogen), 0.1 µl of 40 U ml^{-1} RNAsin (Promega) and 1.5 µl Superscript II (Invitrogen)) is added. The samples are then incubated at 42°C for 1 h, heat denatured at 95°C for 3 min and after rapid cooling on ice, the volume is adjusted to 200 µl with H_2O. The samples are then stored frozen at −20°C to −80C° until use for RT-PCR amplification.

PCR amplification

For PCR amplification, 18 µl of the PCR mix is distributed in each PCR reaction tube (the total reaction volume is 25 µl: 1X PCR buffer, 200 µM of each dNTP, 0.5 µM of each primer, 2 mM $MgCl_2$, and 2.5 U/100 µl Taq polymerase). Then, 2 µl of cDNA (0.2 to 1 µl of the reverse transcriptase reaction which corresponds to about the equivalent of 2000 cells) is added on the top of each PCR tube. Then, 5 µl of each competitor dilution is added to the corresponding reactions tubes, preferentially in the reversed caps, which are then carefully returned, closed and the tubes spun for 20 s at 13 000 rpm, to allow all the reagents to mix. All PCR samples should be maintained at 4°C. It is also essential to work in separate rooms for the competitor and the first step of the PCR. In order to avoid primer–dimer formation which competes for the specific PCR products, it is recommended to work with the Anti-Taq Ab (Clontech), or to use another strategy to ensure a good hot start PCR.

PCR conditions

An initial denaturation step at 94°C for 5 min (necessary for Anti Taq Ab use), followed by 35 cycles of PCR cycles 30 s at 94°C, 30 s at 60°C, 30 s at 72°C, and by a final elongation step at 72°C for 7 min.

Adjustments of cDNA concentrations

The cDNA is adjusted according to a housekeeping gene by performing competitive PCR for β-actin or HPRT between the cDNA of interest and the internal competitor DNA (HPRT is expressed at approximately the same levels as IL-2 or IFN-γ, and thus is preferentially used for cytokine gene expression studies). First, each cDNA is amplified in the presence of 10-fold serial dilutions of the internal competitor, in order to evaluate the amount required to achieve equal band intensities for both PCR fragments. This step is followed by two-fold serial dilutions of the internal competitor to determine precisely the amount of cDNA which gives the equivalence for both fragments. Volumes of cDNA are adjusted to obtain equal HPRT (or β-actin) gene amplification in each sample and used to quantify genes of interest, using serial dilutions of competitors in each reaction, in the presence of the specific primers for cytokines.

Analysis of PCR products

The first step of the competitive PCR which consists of the 10-fold dilutions of the competitor is analysed on a regular agarose gel for a visual evaluation of the corresponding 1/2 dilution working zone. For a fine evaluation, the PCR products can be analysed by two different ways.

The easiest and quickest way is to analyse PCR products on an agarose gel, stained with Sybr-green (Molecular probes), a DNA intercalate dye which is excited at 497 nm and can be analysed either on a scanner Storm™ (Molecular Dynamics), or with a 300 nm UV transilluminator linked to an image quantification software. The Sybr-green is 10-fold more sensitive than the ethidium bromide and the volumes of each PCR loading have to be carefully determined for each cytokine. It is also recommended to stain the gel after electrophoresis (as specified in the manufacturer's instructions). The scan or the image are then analysed with a specific software like Image Quant (Molecular Dynamics) which gives an accurate evaluation of the bands intensities by creating a graph with the height and areas of the respective peaks analysed. The ratios are then plotted on a graph against the concentrations of the competitor, and the concentration of each unknown sample is determined for a ratio equal to 1 (Fig. 4). If the molarity of the competitor has not been determined, the results can be expressed as fold increase over the lowest sample that has been arbitrarily given a value of one. The results are expressed on a histogram.

Another attractive and elegant method because of its great accuracy and sensitivity is the use of a DNA sequencer for the analysis of the PCR products labelled with different fluorochromes on their forward primer (Porcher et al., 1992; Cottrez et al., 1994). This method of choice allows the multiplex of the different dyes in a single lane of the gel, which enhances the number of loadings per gel. The analysis is carried out with specific software, like Genescan which integrates the fluorescent peaks, as described above.

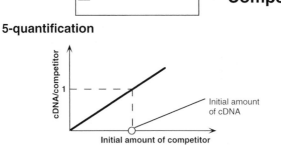

Figure 4. Competitive RT-PCR analysis.

◆◆◆◆◆◆ REAL-TIME QUANTITATIVE PCR 'TAQMAN'

Real-time quantitative PCR is another, more recently developed, PCR technology (Higuchi *et al.*, 1992, 1993) that has been applied to the detection of cytokine gene expression profiling. This technique allows for the detection in real time of a PCR product as it accumulates through the successive cycles of a PCR reaction. As a consequence, it is possible to measure product accumulation at the logarithmic amplification phase of a PCR reaction, independently of the amount of target sequences present at the onset of the reaction. This has the absolute advantage that quantitation of the PCR product is not performed as an endpoint measurement where availability of reaction components or activity of the Taq polymerase become rate limiting. Moreover, this technology allows quantitation with a dynamic range of seven logs and a sensitivity of less than 10 target sequences per sample. Three different types of chemistry are now applied to detect the accumulation of PCR products: fluorogenic probes, intercalating dyes, such as Sybr Green, and MGB (minor groove binding) probes. All three chemistries are based on the detection of changes in fluorescence levels which are proportional to accumulation of the PCR product.

The first type of chemistry makes use of the 5′ fluorogenic assay (Holland *et al.*, 1991) and measures the amount of PCR product formed as a function of the cleavage of a fluorogenic probe containing a reporter dye and a quencher dye. These probes are called Taqman probes, or FRET (Föster Resonance Energy Transfer) probes. While the probe is intact, the proximity of the quencher dye reduces the fluorescence of the reporter. During the extension phase of PCR, the probe anneals to its target sequence, if present, and is cleaved by the 5′ nuclease activity of Taq DNA polymerase. The resulting separation of the reporter dye from the quencher generates increased fluorescence from the reporter. Since probe cleavage is dependent on polymerization, the increase in fluorescence is proportional to the amount of PCR product formed.

A variant of this chemistry, which has been applied more recently, makes use of modified internal probes that contain a reporter dye at the 5′ end of the sequence, and at the 3′ end a non-fluorescent quencher and a minor groove binder (MGB). There are various names for these probes including MGB probes and Blackhole quenchers. The minor groove binder increases the melting temperature (Tm) and allows for the use of shorter probes. The use of these probes is also dependent on the 5′ nuclease activity of Taq polymerase to generate a signal from the reporter dye that is proportional to the accumulation of the PCR product in the 5′ fluorogenic reaction. The advantage of this approach is the use of shorter probes and the absence of fluorescence from the quencher. The calculations of the increase of fluorescence derived from the reporter dye are more accurate since there is no contribution of fluorescence from the quencher, leading to an increased specificity. This also simplifies the detection of more than one target in a single tube (multiplex reactions) and up to four individual reactions can be monitored simultaneously with four different reporters.

The second type of chemistry is based on the changes in fluorescence of a intercalating dye when it interacts with double-stranded DNA. Again, the accumulation of a PCR product is measured directly by the increase in fluorescence caused by the binding of the Sybr Green dye to double-stranded DNA. This type of reaction is independent of the sequence of the amplified product. It could thus detect amplification of non-specific or contaminant reactions. However, a dissociation or melting curve can be generated as a post-amplification step which determines the Tm of the PCR product and provides information on the amplification and specificity. The advantage of this approach is the elimination of (costly) probes for each different target.

The third type of chemistry is the use of molecular beacons (Tyagi *et al.*, 1996, 1998). These are also internal probes that contain the target sequence in a hairpin loop between with reporter and quencher dyes on short flanking sequences. These probes will hybridize to the target, if present, which will disrupt the hairpin structure and allow for separation of the reporter and quencher dyes, resulting in a fluorescent signal. The fluorescent signal in this application is recorded during the annealing phase of the PCR reaction. Alternative ways of applying and quantitating gene expression based on these three types of chemistries have also been designed and include incorporation of fluorochromes during amplification and addition of quenched probes as hairpin structures at the 5′ or 3′ end of primers (Didenko, 2001). In addition to the wide dynamic range and sensitivity of this type of quantitative PCR a major advantage of the technology is that post-amplification steps are no longer required, making it very suitable for high throughput analyses. Indeed, 96 and 384 plate formats are currently in use. We will focus in this part on the application of Taqman and Sybr Green assays.

Methods

Designing specific primers

The design of specific primer and probe pairs is dependent on the chemistry that is used in the application. In general, the melting temperature of the primers in all three chemistry reactions must be between 58 and 60°C and the maximal difference between the Tm of forward and reverse primer is 2° C. The GC can be variable, but must be between 20% and 80%. More important is that there are no runs of four or more identical bases, in particular no Gs, and that the last five bases of the primers contain no more than 2 G/C residues. Primer length can be between 9 and 40 residues with an optimal length of 20. Complementary stretches at the 3′ ends should be avoided to prevent primer–dimer formation, as well as hairpin structures in the primers, which decrease the efficiency of primer annealing. The Taqman probe should have a Tm of 68–70°C, which guarantees annealing of the probe prior to that of primers in the annealing phase of the PCR reaction. Again, runs of four or more identical bases should be avoided, as well as a G residue at the 5′ end. Any G residue adjacent to the reporter dye will quench reporter fluorescence, even after cleavage. Preferably, the strand should be chosen where the number of C residues exceeds the number of G residues. MGB probes should have a Tm of 65–67°C, be as short as possible and contain less than 20 nucleotides.

Since the PCR products do not have to be visualized on gel post amplification, the size of the amplicon can be short, in many cases just encompassing primer and probe sequences. This results in amplicon sizes that are generally less than 150 bp with a variable Tm. All these criteria are difficult to take into account without the use of computation. Primer Express (Perkin Elmer) and other software packages are available to design primers and probes that fit these criteria.

Once primer and probe sequences are designed, an additional check for specificity can be done by blast search against the non-redundant GenBank database (http://www.ncbi.nlm.nih.gov/). Primer design is an empirical process, as even after using the computational design algorithms the primer and probe combination may not function efficiently.

Thus, it may be worthwhile to design and test several primer/probe pairs for a particular target.

Genes encoding cytokines or cytokine receptors, as well as many other genes in the genome, are usually composed of exons, containing coding sequences which are separated by non-coding intron sequences. To prevent amplification of genomic DNA it is recommended to design one of the primers or the probe over an exon/intron boundary. Primers and probes detecting human cytokine and cytokine receptor expression can be obtained as pre-developed assay reagents (PDARs) (Perkin Elmer).

Primer and probe purification

Due to the high constraints on the efficiency of the PCR reaction it is worthwhile to purify primers and probes from contaminants. For primer purification a simple desalting step by spin chromatography is usually sufficient. Probes need to be purified by an additional HPLC (High Pressure Liquid Chromatography) step and preferably a PAGE (Poly-Acrylamide Gel Electrophoresis) step to eliminate any fluorochrome residues that are not attached and have to be checked for proper dual labelling. These procedures are usually performed by the manufacturer. Primers and probe concentrations are optimized for use in real-time PCR following design and purification.

Protocol

Primers from DNA synthesis facility (20–30 OD) are dissolved in H_2O in 500 µl. Spin Sephadex G25 column (Microspin G-25 Columns: Amersham/Pharmacia 27-5325) for 1 min at $1700g$. Change collecting tube, add 200 µl primer solution and spin Sephadex G25 column for 2 min at $1700g$. Remove tube and read the OD of an aliquot which was diluted 100-fold. Calculate the molarity (M) according to:

$$\text{Absorbance (260 nm) (OD)} = \begin{array}{c}\text{sum extinction coefficient contributions of}\\ \text{each of the individual bases} \times \text{cuvette}\\ \text{pathlength} \times \text{concentration/dilution}\end{array}$$

The extinction coefficients of the individual bases are

A 15.200 / **C** 7.050 / **G** 12.010 / **T** 8.400

The concentration in M is thus equal to $100 \times OD$/sum extinction coefficient contributions of each of the individual bases when using a pathlength of 1 cm. Dilute primers to 45 µM.

The concentrations of primers and probes are important parameters for the efficiency of the PCR reaction. Apart from the design itself these are the only parameters that are not standardized in a Taqman PCR reaction. Buffer components are optimized for use with a wide variety of primer/probe combinations with respect to Mg^{2+} concentration in the Taqman and Sybr Green Mastermix™ reaction buffers. Primer concentrations for Taqman applications are optimized in a draughts board configuration at 50, 300 and

900 nM *in duplo*. Prepare in a tube 300 μl mastermix, 5 μl positive control target DNA (stock 10 μg ml⁻¹), 12 μl probe (Stock 10 μM, final 200 nM) and 83 μl H_2O. In separate tubes make serial dilutions of forward and reverse primers from the 45 μM Stock solution to yield final concentrations of 50, 300, and 900 nM. (Add 6 μl of stock primer to 44 μl H_2O, from this add 16.3 μl to 32.6 μl H_2O and from this add 8.3 μl to 42.6 μl H_2O.) Assemble 40 μl target/mastermix/probe and 10 μl of each dilution of forward and reverse primers in nine separate tubes, mix and transfer 25 μl to the reaction plate in duplicate wells. Run reactions and select the primer combination that results in the lowest Ct value and the highest ΔRn (see below). Repeat this process with optimized primer concentrations, now varying the probe concentration from 100, 150, 200, 250 to 300 nM. Generally primer concentrations of 900 nM and probe concentration of 250 nM give optimal results.

RNA isolation and cDNA reactions

RNA isolation and cDNA reaction protocols are in principle identical to those, described on pages 735-736 with solely the addition of 50 ng p(dN6) (Amersham/Pharmacia) to the cDNA reaction to enhance reverse transcription of large mRNA species.

PCR amplification

Ten to 50 ng of cDNA per reaction are used as templates for quantitative PCR and analysis for the expression of human cytokine and cytokine receptor or transcription factor genes by the fluorogenic 5'-nuclease PCR assay using the GeneAmp 5700, ABI Prism 7700 or 7900 Sequence Detection Systems (Perkin-Elmer), Light Cycler (Roche), iCycler iQ (Biorad) or the Stratacycler (Stratagene). PCR reactions are assembled using 2× concentrated buffer solutions Universal PCR Master Mix (Taqman)- or SYBR Green PCR master mix (ABI)-containing enzymes, dNTPs and an inert fluorochrome (ROX) to yield final concentrations of 1× PCR buffer, 200 mM dATP, dCTP, dGTP and 400 mM dUTP, 5.5 mM $MgCl_2$, 1.25 U *AmpliTaq* Gold DNA polymerase, Passive Reference I, and 0.5 U Amp-Erase Uracil-N-glycocylase (UNG). Forward and reverse primers are added at 900 nM each and FAM labelled probes at 250 nM in Taqman assays and forward and reverse primers are used at 200 nM in SYBR Green assays.

In a multiplex Taqman assay it is possible to measure additional targets in the same tube (actin, GAPDH, Ubiquitin, 18S rRNA) which can serve as internal controls for amount of input cDNA. For multiplex assays containing 18S rRNA detection reagents, forward primer, reverse primer and VIC labelled probe are all added at primer limiting concentrations of 50 nM in the PCR reaction. The thermal cycling conditions include at step 1 an incubation of 2 min at 50°C to eliminate any PCR-derived DNA contaminants by UNG, at step 2 an incubation of 10 min at 95°C to activate the 'hotstart' Taq Polymerase, and at step 3, 40 cycles of amplification alternating between denaturing conditions at 95°C for 15 s, and annealing-extension conditions at 55°C for 1 min.

Protocol

This protocol is for a multiplex Taqman reaction with a final reaction volume of 25 µl. To ensure rigorous mixing and equal loading, 30 µl reactions are prepared of which 25 µl is transferred to the final 96-well reaction plate. For 100 reactions assemble in a 5 ml polypropylene tube 1500 µl 2× buffer (Mastermix), 60 µl forward primer target (45 µM Stock), 60 µl reverse primer target (45 µM Stock), 75 µl FAM labelled probe target (10 µM Stock), 15 µl forward primer control (10 µM Stock), 15 µl reverse primer control (10 µM Stock) and 15 µl VIC labelled probe control (10 µM Stock) and 60 µl H_2O. Mix well and distribute 18 µl per well with an electronic multipipetter (Biohit, Eppendorf or Gelman). Add 12 µl of sample DNA solution (containing 12–60 ng cDNA) or standard DNA solutions to the wells, mix well, spin plate 2 min at 770g and transfer 25 µl to the final reaction plate using a 12 channel electronic pipetter. Try to avoid as much as possible the formation of air bubbles during transfer. Close the plate with optical caps or adhesive optical cover and run.

To make DNA standard solutions start with a stock solution of plasmid or other DNA containing the appropriate target sequence at 10 µg ml^{-1}. In Eppendorf tubes make at least seven 10-fold dilutions of the 10 µg ml^{-1} stock in H_2O. From the 10 µg ml^{-1} dilution (third dilution) take 12 µl as the first point of the standard curve, which equals 100 pg in the final reaction. Do not use the first two dilutions as these are too concentrated for the detection system. You can make stocks of these plasmid dilutions.

This protocol can be easily adapted for Sybr Green reactions by replacing the Taqman mastermix by Sybr Green Mastermix and addition of only 13.3 µl target forward primer and 13.3 µl target reverse primer with 274 µl H_2O.

Analyses of results and calculations

Following the amplification, the measurements of fluorescence intensity taken at each PCR cycle are collected and normalized by the software. The normalized reporter value Rn is obtained by dividing the emission intensity of the reporter dye (FAM, VIC, Sybr) by the emission intensity of the passive reference (ROX). This step allows for the correction of small changes in volume present in each individual well. If a positive reaction occurs, then the Rn will increase with an increase in cycle number. Rn+ is the Rn value of a reaction that contains all the components whereas Rn- is the Rn value of an unreacted sample that is obtained from the early cycles in a PCR run (baseline) or from a reaction that did not contain template. The difference in Rn over time in the reaction is given by ΔRn = (Rn+) − (Rn−). ΔRn is thus the increase in fluorescent signal, specific for accumulation of the PCR product. Results of a quantitative PCR reaction are expressed as the threshold cycle or C_T values. The C_T is the first cycle number at which the reporter fluorescence generated by cleavage of the probe passes a fixed threshold above baseline. The C_T value is most accurately obtained when from ΔRn values following log transformation. The software allows for easy visualization of Rn, ΔRn and log-transformed ΔRn

Measuring Human
Cytokine Responses

values which is needed to set the baseline and threshold. The software will then give the CT values for the PCR reactions. It is important to realize that C_T number is inversely correlated with the amount of target DNA in the sample, i.e. the C_T equals 40 if no target is present.

Using the CT values, it is possible to perform calculations on absolute and relative quantitation. When a standard curve of target DNA was run, then the software calculates the absolute amount of target DNA present in the samples based on the C_T values of standard curve and samples. These values can even be corrected based on the amount of internal control present by multiplication of normalized log-transformed C_T values of the internal control reactions. Finally, fold expression over a calibrator (for example a timepoint = 0, or an untreated sample) can be calculated without the need for absolute quantitation by the formula: fold = $2^{(-\Delta\Delta C_T)}$. The $\Delta\Delta C_T$ is equal to (sample target C_T – sample internal control C_T) – (calibrator sample C_T – calibrator internal control C_T). A full derivation and theory of this latter can be found at http://www2.perkin-elmer.com/ab/about/pcr. These calculations can also be done using CT values obtained using Sybr Green if the internal control is run on a separate plate. One additional step that is recommended using Sybr Green assays is to analyse the dissociation curve of the PCR products and discount those that do not have the expected Tm.

Acknowledgements

The authors would like to thank Drs Hervé Groux (INSERM U343, Nice, France), Ulf Andersson (Karolinska Hospital, Stockholm, Sweden), Eric Tartour (Hôpital Européen Georges Pompidou, Paris, France and Jérôme Pène (INSERM U454, Montpellier, France) for sharing information. The DNAX Research Institute is supported by Schering-Plough Corporation.

References

Andersson, U., Andersson, J. Lindfors, A., Wagner, K., Möller, G. and Heusser, C. H. (1990). Simultaneous production of interleukin 2, interleukin 4 and interferon-γ by activated human blood lymphocytes. *Eur. J. Immunol.* **20**, 1591–1596.

Arend, W. P. (1993). IL-1 antagonism in inflammatory arthritis. *Lancet* **341**, 155–156.

Assenmacher, M., Schmitz, J. and Radbruch, A. (1994). Flow cytometric determination of cytokines in activated murine T helper lymphocytes: expression of interleukin-10 in interferon-γ and in interleukin-4-expressing cells. *Eur. J. Immunol.* **24**, 1097–1101.

Becker-André, M. and Hahlbrock, K. (1989). Absolute mRNA quantification using the polymerase chain reaction (PCR). A novel approach by a PCR aided transcript titration assay (PATTY). *Nucleic Acids Research* **17**, 9437–9446.

Celi, F. S., Zenilman, M. E. and Shuldiner, A. R. (1993). A rapid and versatile method to synthesize internal standard for competitive PCR. *Nucleic Acids Res.* **21**, 1047.

Chelly, J., Kaplan, J. C., Maire, P., Gautron, S. and Khan, A. (1988). Transcription of the dystrophin gene in human muscle and non-muscle tissues. *Nature* **333**, 858–860.

Cottrez, F., Auriault, C., Capron, A. and Groux, H. (1994). Quantitative PCR: validation of the use of a multispecific internal control. *Nucleic Acids Res.* **22**, 2712–2713.

Czerkinsky, C., Nilsson, L.-A., Nygren, H., Ouchterlony, Ö. and Tarkowski, A., (1983). A solid-phase enzyme-linked immunospot (ELISPOT) assay for enumeration of specific Ab-secreting cells. *J. Immunol. Methods* **65**, 109.

Czerkinsky, C., Andersson, G., Ekre, H. P., Nilsson, L.-A., Klareskog, L. and Ouchterlony, Ö. (1988). Reverse ELISPOT assay for clonal analysis of cytokine production. I. Enumeration of γ-interferon-secreting cells. *J. Immunol. Methods.* **110**, 29.

Didenko, V. V. (2001) DNA probes using fluorescence resonance transfer (FRET): Designs and Applications. *Biotechniques* **31**, 1106–1121.

Ferre, F. (1992). Quantitative or semi-quantitative PCR: reality versus myth. *PCR Methods Appl.* **2**, 1–9.

Forster, E. (1994). Rapid generation of internal standards for competitive PCR by low stringency primer annealing. *BioTechniques* **16**, 1006–1008.

Gilliland, G., Perrin, S., Blanchard, K. and Bunn, H. F. (1990). Analysis of cytokine mRNA and DNA: detection and quantitation by competitive polymerase chain reaction. *Proc. Natl. Acad. Sci. USA* **87**, 2725–2729.

He, Q., Marjamarki, M., Soini, H., Mertsola, J. and Viljanen, M. K. (1994). Primers are decisive for sensitivity of PCR. *BioTechniques* **17**, 82–87.

Hengen, P. N. (1995). Quantitative PCR: an accurate measure of mRNA? *TIBS* **20**, 476–477.

Herr, W., Linn, B., Leister, N., Wandel, E., Meyer zum Buschenfelde, K. H. and Wolfel, T. (1997). The use of computer-assisted video image analysis for the quantification of CD8+ T lymphocytes producing tumor necrosis factor a spots in response to peptide antigens. *J. Immunol. Methods* **203**, 141–152.

Higuchi, R., Dollinger, G. Walsh, P. S. and Griffith, R. (1992). Simultaneous amplification and detection of specific DNA sequences. *Biotechnology* **10**, 413–417.

Higuchi, R., Fockler, C., Dollinger, G. and Watson, R. (1993). Kinetic PCR: Real-time monitoring of DNA amplification reactions. *Biotechnology* **11**, 1026–1030.

Holland, P. M., Abramson, R. D., Watson, R. and Gelfand, D. H. (1991). Detection of specific polymerase chain reaction product by utilizing the 5′ to 3′ exonuclease activity of *Thermus aquaticus* DNA polymerase. *Proc. Natl. Acad. Sci. USA* **88**, 7276–7280.

Ihle, J. N. (2001). The stat family in cytokine signaling. *Curr. Opin. Cell Biol.* **13**, 211–217.

Jung, T., Schauer, U., Heusser, C., Neumann, C. and Rieger, C. (1993). Detection of intracellular cytokines by flow cytometry. *J. Immunol. Methods* **159**, 197–207.

Lanier, L. L. and Recktenwald, D. J. (1991). Multicolor immunofluorescence and flow cytometry. *Methods: A Companion to Methods in Enzymology* **2**, 192–204.

Murphy, E., Hieny, S., Sher, A. and O'Garra, A. (1993). Detection of in vivo expression of interleukin-10 using a semi-quantitative polymerase chain reaction method in *Schistosoma mansoni* infected mice. *J. Immunol. Methods* **162**, 211–223.

Novick, D., Kim, S. H., Fantuzzi, G., Reznikov, L. L., Dinarello, C. A. and Rubinstein, M. (1999). Interleukin-18 binding protein: a novel modulator of the Th1 cytokine response. *Immunity* **10**, 127–136.

Openshaw, P., Murphy, E. E., Hosken, N. A., Maino, V., Davis, K., Murphy, K. and O'Garra, A. (1995). Heterogeneity of intracellular cytokine synthesis at the single-cell level in polarized T helper 1 and T helper 2 populations. *J. Exp. Med.* **182**, 1–11.

Picker, L., Singh, M. K., Zdraveski, Z., Treer, J. R., Waldrop, S. L., Bergstresser, P. R. and Maino. V. C. (1995). Direct demonstration of cytokine synthesis heterogeneity among human memory/effector T cells by flow cytometry. *Blood* **86**, 1408–1419.

Porcher, C., Malinge, M. C., Picat, C. and Grandchamps, B. (1992). A simplified method for determination of specific DNA or RNA copy number using quantitative PCR and an automatic DNA sequencer. *BioTechniques* **13**, 106–113.

Rappolee, D. A. (1990). Optimizing the sensitivity of RT-PCR. *Amplifications* **4**, 5–7.

Reiner, S. L., Zheng, S., Corry, D. B. and Locksley R. M. (1993). Constructing poly-competitor cDNAs for quantitative PCR. *J. Immunol. Methods* **165**, 37–56.

Rosenthal-Allieri, M. A., Ticchioni, M., Deckert, M., Breittmayer, J. P., Rochet, N., Rouleaux, M., Senik, A. and Bernard, A. (1995). Monocyte-independent T cell activation by simultaneous binding of three CD2 monoclonal antibodies (D66 + T11.1 + GT2). *Cell. Immunol.* **163**, 88–95.

Sedwick, J. D. and Holt, P. G. (1983). A solid-phase immunoenzymatic technique for the enumeration of specific Ab-secreting cells. *J. Immunol. Methods* **57**, 301.

Siebert, P. D. and Larrick, J. W. (1992). Competitive PCR. *Nature* **359**, 557–558.

Tyagi, S. and Kramer, F. R. (1996). Molecular beacons: probes that fluoresce upon hybridization. *Nature Biotechnology* **14**, 303–308.

Tyagi, S., Bratu, D. and Kramer, F. R. (1998) Multicolor molecular beacons for allele discrimination. *Nature Biotechnology* **16**, 49–53.

Wang, A., Doyle, M. V. and Mark, D. F. (1989). Quantification of mRNA by the polymerase chain reaction. *Proc. Natl. Acad. Sci.* **86**, 9717–9721.

Yssel, H., de Vries, J. E., Koken, M., Van Blitterswijk, W. and Spits, H. (1984). Serum-free medium for the generation and propagation of functional human cytotoxic and helper T cell clones. *J. Immunol. Methods* **72**, 219–227.

Zimmermann, K. and Mannhalter, J. W. (1996). Technical aspects of quantitative competitive PCR. *BioTechniques* **21**, 268–279.

Zlotnik, A. and Yoshie, O. (2000). Chemokines: a new classification system and their role in immunity. *Immunity* **12**, 121–127.

Manufacturers

American Type Culture Collection
http://www.atcc.org
Hybridomas and antibodies for ELISA, ELISPOT and intracellular staining

Amersham/Pharmacia
http://www.apbiotech.com
Columns for primer and probe purification for use in real-time PCR

Applied Biosystems
http://home.appliedbiosystems.com
Reagents for quantitative RT-PCR

BD PharMingen
http://www.bdbiosciences.com/pharmingen
(m) Abs for cellular activation, intracellular staining, flowcytometer equipment

BioRad Laboratories
http://www.bio-rad.com
Adhesion slides for microscopy, PCR machines, reagents for quantitative RT-PCR

Biosource International
http://www.biosource.com
Reagents for quantitative RT-PCR

Calbiochem-Novabiochem
http://www.calbiochem.com
Reagents for T-cell activation

Caltag
http://www.caltag.com
Fluorochrome-conjugated antibodies

Carl Zeiss
http://www.zeiss.de
ELISPOT detection system

Clontec
http://www.bdbiosciences.com/pharmingen
Reagents for RT-PCR

DAKO
http://www.dako.com
Antibodies for ELISPOT assay

Diaclone Research
http://www.diaclone.com
(m)Abs for cellular activation, ELISA, ELISPOT and intracellular staining

Dynatech Corporation
http://www.dynatech.com/html
ELISA plates and spectrophotometer equipment

Epicentre Technologies
http://www.epicentre.com
Brefeldin A

Eppendorf
http://www.eppendorf.com
PCR machines, multipippeter equipment

Intergen Company
http://www.intergen.com
Reagents for quantitative RT-PCR

In Vitrogen (Gibco)
http://www.invitrogen.com
Reagents for RT-PCR

Millipore
http://www.millipore.com
ELISPOT plates

Molecular Devices
http://www.moleculardevices.com
Equipment for ELISA, software

Molecular Dynamics
http://www.mdyn.com
Scanning equipment

Molecular Probes
http://www.probes.com
Reagents for RT-PCR

Moss Substrates Inc.
http://www.MossSubstrates.com
Substrates for ELISA

Perkin-Elmer
http://www.perkinelmer.com
PCR machines, reagents for quantitative RT-PCR, software for primer sequence determination

PharMingen
http://www.bdbiosciences.com/pharmingen
(m)Abs for cellular activation, ELISA, ELISPOT and intracellular staining

Promega
http://www.promega.com
Reagents for RT-PCR

Prozyme
http://www.prozyme.com
Enzymes and substrates for ELISA

Qiagen
http://www.qiagen.com
Reagents for RT-PCR, Reagents for quantitative RT-PCR

Quantum-Appligene
http://www.quantum-appligene.com
Products for RNA isolation

R&D Systems
http://www.rndsystems.com
mAbs for cellular activation, ELISA and intracellular staining

Roche
http://www.roche.com
PCR machines, reagents for quantitative RT-PCR

Sigma-Genosys
http://www.genosys.com
Reagents for quantitative RT-PCR

Southern Biotechnology Associates
http://southernbiotech.com
Enzymes and substrates for ELISA

Stratagene
http://www.stratagene.com
PCR machines, Reagents for quantitative RT-PCR, Molecular Beacons

Vector Laboratories
http://www.vectorlabs.com
Fluorochrome-conjugated antibodies

6 Measuring Immune Responses *In Situ*: Immunofluorescent and Immunoenzymatic Techniques

Ulrike Seitzer, Elmer Endl and Johannes Gerdes
Department of Immunology and Cell Biology, Research Center Borstel, Borstel, Germany

Christine Hollmann
AdnaGen AG, Hannover, Germany

◆◆

CONTENTS

◆◆◆◆◆◆ **INTRODUCTION**

The indirect immunofluorescence technique developed by Coons and Kaplan in 1950 was the first to describe the localization of antigen in tissues using fluorescently labelled antibody. Since then, considerable progress in immunohistochemical procedures has been made, most notably on the application of enzyme-labelled antibodies, signal enhancement and development of new fluorochromes. Enzyme coupled detection methods were developed primarily because of the limitations of fluorescence techniques, such as naturally occurring autofluorescence, the lack of permanence of the preparations which tend to fade, the difficult correlation to morphological details and the need for expensive equipment. Of a variety of enzymes assayed, the enzymes of choice for direct and indirect procedures became horseradish peroxidase (PO) and calf intestinal alkaline phosphatase (AP) (Nakane and Pierce, 1967). The sensitivity of these immunoenzymatic detection methods was greatly enhanced by the development of the peroxidase anti-peroxidase (PAP) (Sternberger *et al.*, 1970) and the alkaline phosphatase anti-alkaline phosphatase (APAAP) methods (Cordell *et al.*, 1984), which are based

Measuring Immune Responses In Situ

Copyright © Elsevier Science Ltd
All rights of reproduction in any form reserved

on the utilization of non-covalently bound enzyme to enzyme-specific antibodies.

With the advent of monoclonal antibody production techniques developed by Köhler and Milstein in 1975, problems associated with polyclonal antisera could be overcome in immunochemical techniques. Serum derived antibodies may react with many irrelevant antigens, sometimes making interpretation of results difficult. Monoclonal antibodies help to overcome these difficulties since they recognize only one specific epitope. Due to the homogeneity of monoclonal antibodies, a standardization from laboratory to laboratory has become possible. In addition, hybridoma cell lines provide an unlimited supply of antibodies in contrast to the limited supply of a polyclonal antiserum.

Great effort was also put into making paraffin embedded material more amenable to immunohistological techniques. Although formalin remains the most popular fixative used, it is not always the best of choice for preserving antigenicity of tissues, due to the intermolecular cross-linking formed between formalin and proteins. Approaches to unmask antigenic sites hidden by cross-linked proteins have been to develop antibodies that can recognize formalin-resistant epitopes, to choose the correct fixative and optimize the duration of fixation, and the application of protease digestion. A major advance in antigen retrieval was achieved by Shi et al. in 1991 based on microwave heating of tissue sections attached to microscope slides to temperatures up to 100°C in the presence of metal ion solutions. This resulted in the possibility of visualizing antigens that were otherwise undetectable in formalin-fixed, paraffin-embedded tissues, and also greatly simplified the method for antigen retrieval. A study undertaken by Cattoretti et al. (1993) optimized antigen retrieval techniques for a large number of antibodies to be used on formalin-fixed, paraffin-embedded tissue sections.

The developments from then on were focused on methods for signal amplification, an important factor when one is confronted with the detection of low antigen levels. Methods which have proven to be of value are the tyramine amplification technique, which may be applied to achieve higher sensitivity and/or to reduce costs for primary antibody (Wasielewski et al., 1997), or the EnVision (DAKO) polymeric conjugate system (Kammerer et al., 2001). Rapid progress has also been made in the development of fluorescence techniques with the advancement of confocal laser scanning microscopy (CLSM) and the production of new and more stable fluorochromes (Kumar et al., 1999), which may allow the simultaneous description of seven parameters (Tsurui et al., 2000). The challenge in this field is to minimize fading of the utilized fluorochromes by the high energy laser beams (Ono et al., 2001).

Key aspects to consider in establishing immunohistology methods are, however, unchanged since the first description of this technology: fixation procedures with respect to epitope and morphology conservation, specificity of the antibody, antigen retrieval, specificity of the secondary reagents and the properties of the staining reaction. Methods should be chosen and applied which lead to reproducible, reliable and specific results (Burry, 2000). The methods described below are based on the reports found in the literature. They have been developed to yield optimal signal to background ratios and are used routinely in our laboratory with consistently good results.

◆◆◆◆◆◆ SPECIMEN PREPARATION

For immunoenzymatic techniques, cryostat preparations generally have a superior preservation of antigens than paraffin sections. Morphological details, however, are more readily destroyed in cryostat sections and the great majority of archival material accessible for immunostaining is paraffin-embedded. In the following, the emphasis is on the treatment of cryostat and paraffin sections prior to staining. For other tissue pre-treatments and more details on paraffin embedding and sectioning further reading is recommended (Watkins, 1989; Zeller, 1989).

Paraffin sections

For antigen retrieval it may be necessary to try several approaches with respect to the method used or the buffers necessary in preparing paraffin sections for immunoenzymatic staining (Cattoretti *et al.*, 1993; Pileri *et al.*, 1997; Shi *et al.*, 2000).

For no treatment for antigen retrieval, routinely processed paraffin sections are dewaxed by submerging for 10 min in xylene followed by 10 min in acetone and 10 min in a $1:1$ mixture of acetone and tris buffered saline (TBS). The sections are kept in TBS until staining.

For microwave oven pretreatment, paraffin sections are deparaffinated for 10 min in xylene and rehydrated in a series of ethanols or acetones at 10-min intervals (100%, 70%, 40%, 0%). The slides are transferred to a plastic staining jar filled with 10 mM sodium-citrate buffer, pH 6.0. The plastic staining jars are put in the microwave oven and heated for 5 min at 720 W. After 5 min it is essential that the staining jar is refilled with H_2O to prevent drying of the specimens, before heating for another 5 min. The frequency of the heating steps depends on the fixation and embedding procedures performed previously and should be optimized. After the microwave treatment the slides are cooled in the staining jar for approximately 20 min at room temperature and washed briefly in TBS before proceeding with immunostaining.

For pressure cooker pretreatment, sections are dewaxed and hydrated as for microwave treatment. Using a normal household pressure cooker (with a 15 psi valve), boil 10 mM citric acid, pH 6.0 before adding the sections. Make up enough buffer to totally submerge the sections (approx. 2 l). Rinse the sections briefly in distilled water and place them into the boiling buffer. Close the lid and heat until top pressure is reached and boil the sections at this pressure. The boiling time must be optimized (approx. 1–10 min). After cooking, immediately cool the pressure cooker under running cold water taking extreme care at this step and absolutely following the instructions of the manufacturer before opening the cooker.

The sections are transferred immediately into cold running tap water before proceeding with the immunohistological staining.

Frozen sections

Several steps in the preparation of frozen material are crucial for obtaining samples with well-preserved morphological structure and antigenicity. The following procedure may be followed.

Fresh surgical specimens are submerged in flat-bottomed polyethylene tubes (Cat. no. 619-x, Brand Laboratory Equipment Manufacturers, Wertheim, Germany) filled with sterile physiological NaCl or PBS, snap frozen in liquid nitrogen and stored at −70°C. If tissues have not been appropriately snap-frozen, ice crystals will be present in the tissue causing artifactual staining along the fracture lines caused by these crystals. Tissues should always be maintained in a frozen state, since thawing and refreezing will result in extensive damage to the tissue and loss of antigenicity.

Cryostat sections of 4–5 μm are air dried for 4 to 24 h and then fixed for 15–30 min in acetone followed by 15–30 min in chloroform. If the cryostat sections are not meant to be stained the following day, also air dry 4 to 24 h, fix for 10 min in acetone and store at −70°C. When needed, thaw the sections covered with a paper towel to prevent water condensation on the slides. Fix in acetone and chloroform as mentioned above.

For cytospin specimens, 1 to 5×10^5 cells are applied per slide and centrifuged for 5 min at 220g. The cytospins are air dried as above for 4 to 24 h and stored at −70°C. Before staining, samples are thawed and fixed solely in acetone for 10 min.

For some staining procedures concerning primarily cytokines, the fixing of specimens with paraformaldehyde/saponin may be preferred (Sander et al., 1991). Air-dried slides are fixed for 15 min in 4% paraformaldehyde in phosphate buffered saline without NaCl, pH 7.4–7.6, followed by a 15-min treatment in 0.1% saponin/PBS to elute cholesterol from the membranes. Wash thoroughly with PBS before immunostaining.

New and alternative fixation procedures exist in the literature and may be applied if the above procedures do not lead to satisfactory results, such as for example the recently described alternative to acetone fixation using pararosaniline (Schrijver et al., 2000).

◆◆◆◆◆◆ IMMUNOFLUORESCENCE DETECTION

Fluorescent probes allow the detection of particular components of cells and subcellular structures with a high sensitivity and selectivity. The obtained information can be analysed statistically, quantitatively, temporally and with a high spatial resolution. The use of multicolour probes allows the simultaneous monitoring of different antigens at the same time and can provide information about co-localization of antigens or co-expression of molecules on a single-cell level. The innovative potential of immunofluorescence is reflected by the rising number of applications in fluorescence microscopy, laser scanning microscopy, laser scanning cytometry (Clatch *et al.*, 1998) and flow cytometry (Baumgarth and Roederer, 2000).

A limitation for the combination of fluorochromes is the fact that the emission spectra of the fluorochromes are usually broad and therefore overlap. This drawback can be overcome in part by choosing fluorochromes that have a well-separated emission spectrum and using individual combinations of excitation and emission filters. FITC, TRITC and rhodamine derivatives are the most common dyes used for commercially available directly conjugated antibodies. New product lines like the ALEXA™ (Panchuk-Voloshina *et al.*, 1999) or CY™ dyes are similar in their spectral properties but are improved in brightness and photostability. These dyes can also be conjugated to the antibody of choice using the procedures described by the manufacturer. Counterstaining of DNA is usually performed with UV excitable dyes like DAPI, Hoechst 33258 and Hoechst 33342, with Hoechst 33342 being one of the exceptional dyes that can stain DNA in living cells. On laser-based microscopes that lack an expensive UV light source, these DNA dyes can be replaced by dyes that are excited with a red laser line such as TOTO-3 and TOPRO-3 (Suzuki *et al.*, 1997).

A problem one is confronted with in using fluorochromes in tissue is the presence of autofluorescence. Methods to overcome this problem have been described (Schnell *et al.*, 1999) and additionally, excitation and emission in the red spectral region also has the advantage of lowering the background of autofluorescence. Therefore fluorochromes that have red-shifted spectral properties, like Cy5, and can be coupled to antibodies are also under current development.

Fluorochrome-labelled antibodies allow a high resolution to be obtained, since subcellular structures can be studied at magnifications beyond the limit of resolution of the light microscope. Because fluorochromes can be chosen that do not have overlapping emission spectra, double immunofluorescence permits the study of two different antigens in the same specimen even if they have identical subcellular distributions, thus allowing co-localization studies.

Even though conventional fluorescence microscopy is being increasingly substituted by CLSM, it is nonetheless possibile to analyse multiple fluorochromes and to use far-red fluorochromes by appropriate selection of filters (Ferri *et al.*, 1997a,b). A possibility to attain CLSM quality in conventional microscopy was described using ultrathin cryosections of skin, which provided convincing comparable images to CLSM (Ishiko *et al.*, 1998).

Direct and indirect staining

Fluorescence staining can be performed using a primary antibody which is conjugated with a fluorochrome (direct staining) or with an unlabelled primary and a fluorochrome-conjugated secondary antibody directed against the primary immunoglobulin. In both cases, the staining procedure can be performed in the following fashion (Fig. 1a, b).

The fixed specimen is incubated for 10 min in TBS/10% bovine serum albumin to block unspecific binding. Wash twice with TBS before applying the directly labelled or unlabelled primary antibody. Incubate for 30 min and wash twice with TBS. When using a directly labelled antibody, proceed with the optional DNA-staining and mounting, before viewing under a fluorescence microscope. For the indirect detection procedure, continue with the incubation of the flourochrome conjugated secondary antibody for 30 min. For optional DNA-staining of nuclei, wash the specimen twice and incubate for 10 min with the bisbenzimide dye Hoechst 33342 (6 μg ml^{-1}) or 1 μg ml^{-1} DAPI (4′,6-diamidino-2-phenylindole). At this stage it has proven to be useful to fix the stained specimen in 4% paraformaldehyde for 10 min before washing twice in TBS and mounting in 10 μl DABCO (1,4-Diazabicyclo [2,2,2] octane) anti-fading solution (Johnson *et al.*, 1982) and viewing under a fluorescence microscope. The specimens can be stored for further viewing at 4°C for several months.

Multicolour immunofluorescence staining

Two and more fluorochromes can be combined for staining cells and tissue sections, whereby the repertoire used is not restricted to the most commonly used fluorochromes and depends on the microscope and filtersets available. To combine fluorochromes in order to detect multiple antigens several possible strategies present themselves. First, the use of direct staining with a fluorochrome-conjugated primary antibody combined with indirect staining procedures. The advantages of this approach are that it is easy to perform, however a main drawback is the availability of suitably fluorochrome-conjugated antibodies and the fact that the signal intensity may be too weak. A second approach is to use primary antibodies of different isotypes combined with isotype specific fluorochrome-conjugated antibodies, bringing the advantage of signal amplification and the choice of up to five different isotypes to be distinguished simultaneously. The inherent restrictions of this method are the availability of primary antibodies of different isotype and of suitable fluorochrome-conjugated isotype specific secondary antibodies. A third approach is to use indirect staining with biotin-conjugated antibodies and fluorochrome-conjugated streptavidin.

Critical to any one chosen approach is the staining sequence which is required to combine the chosen primary and secondary antibodies in order to avoid unwanted false-positive results. In addition, it is advised to test the isotype specificity of secondary reagents.

The following staining procedure is an example for the distinction of different cell subsets (T cells, macrophages, dendritic cells and follicular dendritic cells) in cryosections of tonsilar tissue by multicolour immunofluorescence staining (Plate 12). After blocking, sections were incubated with an antibody against the CD4 antigen (BD Biosciences, clone Leu3a, IgG$_1$) followed by detection with an Alexa-568-conjugated goat-anti-

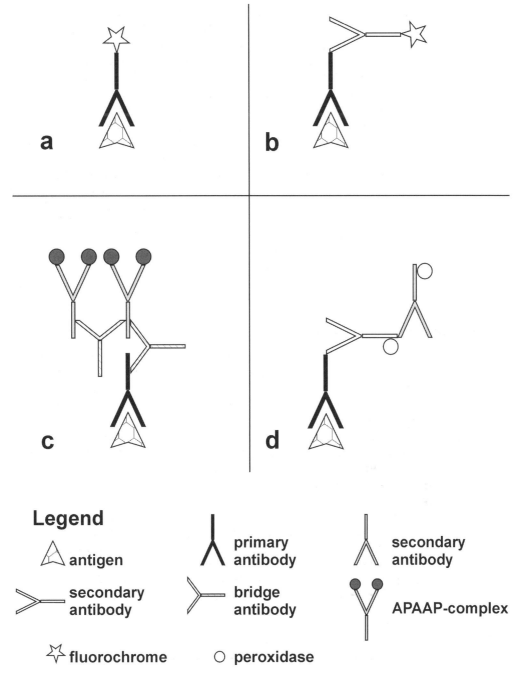

Figure 1. Scheme of the principle of the methods described for immunofluorescence and immunoenzymatic detection of antigens: (a) direct fluorescence, using a fluorochrome coupled primary antibody; (b) indirect fluorescence with a fluorochrome-coupled secondary antibody; (c) alkaline phosphatase anti-alkaline phosphatase (APAAP)-procedure, performed twice with the bridging antibody and the APAAP-complex; and (d) indirect immuno-peroxidase-method using two different PO-conjugated secondary antibodies.

Legend content within figure:
- antigen
- primary antibody
- secondary antibody
- secondary antibody
- bridge antibody
- APAAP-complex
- fluorochrome
- peroxidase

Measuring Immune
Responses *In Situ*

mouse-IgG antibody (Molecular Probes). To detect follicular dendritic cells, the DRC-1 antibody (Dako, IgM) was applied (Naiem *et al.*, 1983) followed by incubation with a biotinylated goat-anti-mouse-IgM and streptavidin-Cy5 (Dianova). Lastly, a directly labelled FITC-CD11c (Dako, clone KB90, IgG_1) was applied before fixing the specimens in 4% paraformaldehyde and mounting in DABCO anti-fading medium. The results of the staining procedure were analysed with a confocal microscope (Leica TCS SP, Bensheim, Germany) equipped with a Kr/Ar ion laser. Using this protocol, CD4 is detected on T cells, macrophages and dendritic cells by red fluorescence, DRC-1 reactivity on follicular dendritic cells was shown by blue and CD11c expression by green fluorescence on macrophages and dendritic cells. Only the merged fluorescences allow the clear depiction of red-fluorescing CD4-positive T cells, since macrophages and dendritic cells then appear yellow due to the co-expression of CD11c (Plate 12).

◆◆◆◆◆◆ IMMUNOENZYMATIC DETECTION

The choice of the system to use (alkaline-phosphatase, (AP), or peroxidase (PO)) depends on several circumstances, such as the species in which the antigen is to be detected and the availability of appropriate primary and secondary antibodies, as well as the presence of non-inhibitable endogenous enzyme activity (e.g. intestinal AP is not inhibited by levamisol). For the APAAP procedure with human tissues, the primary antibody is usually a mouse monoclonal. When using a rabbit antiserum an additional incubation step has to be included to make the antiserum detectable with the system.

In the following procedures, incubation steps are generally followed by washing twice in TBS. Slides are incubated with 100 µl of antibody dilution for 30 min at room temperature in a level humid chamber. Excessive humidity should be avoided since water condensation on the slides will interfere with the staining reaction. Insufficient humidity on the other hand will dry the antibody on the sections, resulting in false-positive staining. Drying is usually most apparent at the edge of sections (rim effect).

Alkaline phosphatase anti-alkaline phosphatase (APAAP) method

This procedure is described according to the method of Cordell *et al.* (1984) and is delineated exemplarily for human tissue and with reagents used routinely in our laboratory (Fig. 1c). It must be emphasized that the dilutions given may not apply in all cases and should be optimized before routine use.

Incubation for 30 min with the appropriately diluted primary antibody in 10% fetal calf serum (FCS)/TBS is applied to prepared cryosections and cytopreparations directly after the last fixation step and to paraffin sections after washing in TBS. For washing after the first incubation, separate containers are recommended for different antibodies and negative controls.

If the primary antibody is a rabbit antiserum, the following incubation must be performed: 30 min with a monoclonal mouse anti-rabbit immunoglobulin (Cat. no. M 0737, DAKO Diagnostika GmbH, Hamburg, Germany).

The secondary (bridging) antibody (rabbit anti-mouse IgG H+L antiserum (Cat. no. Z 259, DAKO Diagnostika GmbH, Hamburg, Germany)) is diluted 1:20 in TBS, 1:8 inactivated human serum and incubated for 30 min.

The APAAP-complex (Cat. no. M 800, dianova, Hamburg, Germany) is diluted 1:40 in TBS/FCS and also incubated for 30 min.

The incubations with the secondary (bridging) antibody and APAAP-complex are repeated once for 15 min each and may be repeated *ad libitum* to increase the detection limit.

For the visualization of the staining, the following developing solution must be prepared freshly each time. Amounts are for one staining jar and the order of addition is essential for results.

- Solution A: 35 ml APAAP-buffer, 12.5 ml AP buffer, 20 mg levamisole
- Solution B: dimethylformamide with 8.3% (w/v) Naphthol-As biphosphate
- Solution C: 250 µl $NaNO_3$ solution with 100 µl Newfuchsin solution

After 1 min of reaction time 125 µl of solution C are added to 47.5 ml solution A, then 300 µl of solution B are added and the pH is adjusted to pH 8.8. The solution is filtered before adding the slides and incubating for 20 min on a shaker. Counterstaining is performed in haematoxylin for 90 s, the specimens are washed and left in tap water for 5 min before being mounted with pre-warmed (56°C) Kaiser's glycerol gelatine.

Indirect immuno-peroxidase (PO) method

The advantage of peroxidase staining is a greater variability and flexibility in the availability and combinations of antibody as well as a less time-consuming procedure. Again, the procedure delineated is exemplarily for human tissue and a mouse or rabbit primary antibody (Fig. 1d).

Before proceeding with primary antibody incubations, it may be advisable to block endogenous peroxidase activity in the tissue by pre-incubating the slides for 20 min in a light protected staining jar with 1% H_2O_2 in TBS.

The mouse (rabbit) primary antibody is applied in the appropriate dilution in TBS/10% FCS, and incubated for 30 min. The first secondary PO-conjugated goat anti-mouse (rabbit) antibody is applied diluted in TBS/10% FCS and inactivated human serum. The second secondary PO-conjugated rabbit anti-goat is also incubated for 30 min. In contrast to the APAAP-method, this method is limited to two signal enhancing steps due to an increase of background staining with further incubation steps.

For development, prepare the developing buffer as described and incubate the slides with 100 µl each. Incubate in the dark in the humid chamber for 3 to 15 min. The degree of development can be checked microscopically and the incubation stopped at the desired point. Slides are counterstained in haematoxylin and mounted as described for the APAAP procedure.

In analogy to the PO-method a combination of two AP-coupled secondary antibodies may also be used applying the detection procedure for AP-activity.

Double-staining procedure

If suitable primary antibodies are available for the antigens of interest (e.g. mouse and rabbit) double immunoenzymatic staining may be performed. Ideally, the rabbit primary antibody is detected using PO-conjugated secondary antibodies while the detection of the mouse monoclonal antibody is subsequently performed using the APAAP procedure. Alternatively, the streptavidin-biotin-complex/HRP detection method offered by DAKO may be used for the primary rabbit antibody followed by APAAP detection of the primary mouse antibody. Plate 12 depicts an example of double immunoenzymatic staining in tonsil tissue performed by a different approach. The first antibody used was a mouse monoclonal antibody against the Ki-67 antigen (MIB-1, Dianova) detected with PO-labelled secondary antibodies (PO-goat-anti-mouse, PO-rabbit-anti-goat). The chromogenic substrate was only incubated for 10 min. The subsequent staining procedure for the rabbit-anti-CD3 antiserum (DAKO) was performed using only one secondary AP-conjugated goat-anti-rabbit antibody. The incubation with the chromogenic substrate must then be carefully monitored by microscopic control since the PO-reaction tends to continue, causing signal overlap. This is often a drawback in evaluating double-stained specimens, however digital image analysis may offer the possibility of circumventing this difficulty (Lehr *et al.*, 1999).

◆◆◆◆◆◆ APPENDIX

Antibodies

In general, care should be taken in choosing the correct combination of antibodies for the choice of detection system, i.e. compatibility of antibodies (isotype, species) and prevention of cross-reactivity (species from which the antibody originates, species of secondary antibodies, species for which the antigen is to be detected), and choice of serum used to block nonspecific reactions. Each first antibody should be titred against a known tissue such as normal lymph node or spleen, before its use in the laboratory. Most antibodies are used in a concentration of 20–40 μg ml^{-1}, and most commercially available antibodies are used at titres from 1:20 to 1:200. In our experience, a first evaluation of the correct titre is obtained by using dilution steps of 1:10, 1:30, 1:100, 1:300, 1:1000 and 1:3000. For storage, sterility of antibodies should be maintained, and freezing and thawing should be avoided. Most antibodies are stable for months at 4°C. An important source of information for cluster of differentiation (CD) antibodies is the Proceedings (Leukocyte Typing I–VII) of the International Workshop and Conference on White Cell Differentiation Antigens (Oxford University Press). This information is also partly accessible by internet (Protein Reviews On The Web http://www.ncbi.nlm.nih.gov/prow/).

Chemicals and solutions

Chemicals used for fixation procedures should be of p.a. grade.

Tris buffered saline (TBS): 50 mM Tris (tris(hydroxymethyl)-aminomethane) 150 mM NaCl, pH 7.5.

Phosphate buffered saline (PBS): 150 mM NaCl, 10 mM $NaH_2PO_4xH_2O$, pH 7.2.

Haematoxylin stock solution: 1 g haematoxylin, 0.2 g $NaJO_3$, 50 g aluminium potassium sulfate dodecahydrate ($KAl(SO_4)_2·12H_2O$), add 1000 ml H_2O. Finally, add 50 g chloralhydrate and 1 g of citric acid.

APAAP-Developing buffer: 1.21 g Tris, 5.85 g NaCl, 1000 ml H_2O.

AP-buffer: 12.1 g Tris, 5.85 g NaCl, 1000 ml H_2O.

Newfuchsin solution: 5 g Newfuchsin in 100 ml 2N HCl, store in a dark glass vessel at 4°C.

$NaNO_3$-Solution: 6% (w/v) in H_2O.

PO-Developing buffer: 6 mg 3,3'-diaminobezidinetetrahydrochloride are dissolved in 10 ml TBS and are mixed with 100 µl H_2O_2 directly before use.

DABCO anti-fading solution: 2.5% DABCO (1,4-Diazabicyclo [2,2,2] octane) is dissolved in 90% glycerol overnight. Adjust the pH with 2N HCl to pH 8.6. Store at room temperature.

Controls

Negative controls should include antibody of the same immunoglobulin class at an equivalent concentration (isotype control). The negativity of secondary reagents is confirmed by incubating a sample without the primary antibody (TBS control).

Positive controls should be included where possible to exclude negative results due to incorrect staining procedures and to control specificity of the staining.

Specificity of the primary antibody may be analysed by neutralization experiments, to check for blocking of antibody binding resulting in negative staining. This is done prior to the application in immunoenzymatic staining, by preincubating the antibody with recombinant or purified antigen for 30 min at 37°C. It should be kept in mind, however, that the absorption control does not prove the specificity of the antibody for the protein in the tissue (Burry, 2000).

Measuring Immune Responses *In Situ*

Fluorochromes

http://facs.scripps.edu/spectra/
Java script generator of spectra of commonly used dyes in microscopy and flow cytometry.

http://www.fluorescence.bio-rad.com/
Interactive database of commonly used fluorochromes.

Immunohistological detection of cytokines

Cytokines as central mediators of the immune response are at the focus of interest with regard to detection, however the investigator will be confronted with problems in obtaining satisfactory results, mainly attributable to the low concentration of cytokines in tissue preparations, the fact that cytokines are usually secreted and that available antibodies are often unsuitable for immunohistology. Nonetheless, appropriate staining procedures with suitable antibodies may lead to good results, as shown for instance for interferon-gamma (Scheel-Toellner *et al.*, 1995; van der Loos *et al.*, 2001), tumour necrosis factor alpha (Kretschmer *et al.*, 1990), interleukin-1 beta (Seitzer *et al.*, 1997) and interleukin-15 (Maeurer *et al.*, 1999).

References

Baumgarth, N. and Roederer, M. (2000). A practical approach to multicolor flow cytometry for immunophenotyping. *J. Immunol. Methods.* **243**, 77–97.

Burry, R. W. (2000). Specificity controls for immunocytochemical methods. *J. Histochem. Cytochem.* **48**, 163–165.

Cattoretti, G., Pileri, S., Parravicini, C., Becker, M. H. G., Poggi, S., Bifulco, C., Key, G., D'Amato, L., Sabattini, E., Feudale, E., Reynolds, F., Gerdes, J. and Rilke, F. (1993). Antigen unmasking on formalin-fixed, paraffin-embedded tissue sections. *J. Path.* **171**, 83–98.

Clatch, R. J., Foreman, J. R. and Walloch, J. L. (1998). Simplified immunophenotypic analysis by laser scanning cytometry. *Cytometry* **34**, 3–16.

Coons, A. H. and Kaplan, M. H. (1950). Localization of antigen in tissue cells. II. Improvements in a method for the detection of antigen by means of fluorescent antibody. *J. Exp. Med.* **91**, 1–13.

Cordell, J. L., Falini, B., Erber, W. N., Ghosh, A. K., Abdulaziz, Z., MacDonald, S., Fulford, K. A. F., Stein, H. and Mason, D. Y. (1984). Immunoenzymatic labeling of monoclonal antibodies using immune complexes of alkaline phosphatase and monoclonal anti-alkaline phosphatase (APAAP complexes). *J. Histochem. Cytochem.* **32**, 219–229.

Ferri, G.-L., Gaudio, R. M., Castello, I. F., Berger, P. and Giro, G. (1997a). Quadruple immunofluorescence: a direct visualization method. *J. Histochem. Cytochem.* **45**, 155–158.

Ferri, G.-L., Isola, J., Berger, P. and Giro. G. (1997b). Direct eye visualization of Cy5 fluorescence for immunocytochemistry and in situ hybridization. *J. Histochem. Cytochem.* **48**, 437–444.

Ishiko, A., Shimizu, H., Masunaga, T., Kurihara, Y. and Nishikawa, T. (1998). Detection of antigens by immunofluorescence on ultrathin cryosections of skin. *J. Histochem. Cytochem.* **46**, 1455–1460.

Johnson, G. D., Davidson, R. S., McNamee, K. C., Russel, G., Goodwin, D. and Holborow, E. J. (1982). Fading of immunofluorescence during microscopy: a study of the phenomenon and its remedy. *J. Immunol. Methods* **55**, 231–242.

Kammerer, U., Kapp, M., Gassel, A. M., Richter, T., Tank, C., Dietl, J. and Ruck, P. (2001). A new rapid immunohistochemical staining technique using the EnVision antibody complex. *J. Histochem. Cytochem.* **49**, 623–630.

Köhler, G. and Milstein, C. (1975). Continuous cultures of fused cells secreting antibody of predefined specificity. *Nature* **256**, 495–497.

Kretschmer, C., Jones, D. B., Morrison, K., Schlüter, C., Feist, W., Ulmer, A. J., Arnoldi, J., Matthes, J., Diamantstein, T., Flad, H.-D. and Gerdes J. (1990). Tumor necrosis factor α and lymphotoxin production in Hodgkin's disease. *Am. J. Path.* **137**, 341–351.

Lehr, H.-A., Loos, C. van der Teeling, P. and Gown, A. M. (1999). Complete chromogen separation and analysis in double immunohistochemical stains using photoshop-based image analysis. *J. Histochem. Cytochem.* **47**, 119–125.

Loos, C. M. van der, Houtkamp, M. A., Boer, O. J. de, Teeling, P., Wal, A. C. van der and Becker, A. E. (2001) Immunohistochemical detection of interferon-gamma: fake or fact? *J. Histochem. Cytochem.* **49**, 699–710.

Nakane, P. K. and Pierce, G. B. Jr (1967). Enzyme-labeled antibodies for the light and electron microscopic localization of tissue antigens. *J. Cell Biol.* **33**, 307–318.

Ono, M., Murakami, T., Kudo, A., Isshiki, M., Sawada, H. and Segawa, A. (2001). Quantitative comparison of anti-fading mounting media for confocal laser scanning microscopy. *J. Histochem. Cytochem.* **49**, 305–311.

Panchuk-Voloshina, N., Haugland, R. P., Bishop-Stewart, J., Bhalgat, M. K., Millard, P. J., Mao, F., Leung, W.-Y. and Haugland, R. P. (1999). Alexa dyes, a series of new fluorescent dyes that yield exceptionally bright, photostable conjugates. *J. Histochem. Cytochem.* **47**, 1179–1188.

Pileri, S. A., Roncador, G., Ceccarelli, C., Piccioli, M., Briskomatis, A., Sabattini, E., Ascani, S., Santini, D., Piccaluga, P. P., Leone, O., Damiani, S., Ercolessi, C., Sandri, F., Pieri, F., Leoncini, L. and Falini, B. (1997). Antigen retrieval techniques in immunohistochemistry: comparison of different methods. *J. Pathol.* **183**, 116–123.

Kretschmer, C., Jones, D. B., Morrison, K., Schlüter, C., Feist, W., Ulmer, A. J., Arnoldi, J., Matthes, J., Diamantstein, T., Flad, H.-D. and Gerdes, J. (1990). Tumor necrosis factor α and lymphotoxin production in Hodgkin's disease. *Am. J. Path.* **137**, 341–351.

Kumar, R. K., Chapple, C. C. and Hunter N. (1999). Improved double immunofluorescence for confocal laser scanning microscopy. *J. Histochem. Cytochem.* **47**, 1213–1217.

Maeurer, M., Seliger, B., Trinder, P., Gerdes, J. and Seitzer, U. (1999). Interleukin-15 in mycobacterial infection of antigen-presenting cells. *Scand. J. Immunol.* **50**, 280–288.

Naiem, M., Gerdes, J., Abdulazizz, Z., Stein, H. and Mason, D. Y. (1983). Production of a monoclonal antibody reactive with human dendritic reticulum cells and its use in the immunohistological analysis of human lymphoid tissue. *J. Clin. Path.* **36**, 167–175.

Sander, B., Andersson, J. and Andersson, U. (1991). Assessment of cytokines by immunofluorescence and the paraformaldehyde-saponin procedure. *Immunol. Rev.* **119**, 65–93.

Scheel-Toellner, D., Richter, E., Toellner, K. M., Reiling, N., Wacker, H. H., Flad, H. D. and Gerdes J. (1995). CD26 expression in leprosy and other granulomatous diseases correlates with the production of interferon-gamma. *Lab. Invest.* **73**, 685–690.

Schnell, S. A., Staines, W. A. and Wessendorf, M. W. (1999). Reduction of lipofuscin-like autofluorescence in fluorescently labeled tissue. *J. Histochem. Cytochem.* **47**, 719–730.

Schrijver, I. A., Melief, M.-J., Meurs, M. van, and Companjen, A. R. (2000). Pararosaniline fixation for detection of co-stimulatory molecules, cytokines, and specific antibody. *J. Histochem. Cytochem.* **48**, 95–103.

Seitzer, U., Scheel-Toellner, D., Toellner, K. M., Reiling, N., Haas, H., Galle, J., Flad, H. D. and Gerdes, J. (1997). Properties of multinucleated giant cells in a new in vitro model for human granuloma formation. *J. Pathol.* **182**, 99–105.

Shi, S.-R., Key, M. E. and Karla, K. L. (1991). Antigen retrieval in formalin-fixed, paraffin-embedded tissues: an enhancement method for immunohistochemical staining based on microwave oven heating of tissue sections. *J. Histochem. Cytochem.* **39**, 741–774.

Shi, S.-R., Gu, J. and Taylor, C. R. (Eds) (2000). *Antigen Retrieval Techniques: Immunohistochemistry and Molecular Morphology*. BioTechniques Press, Natick MA, Eaton Pub.

Sternberger, L. A., Hardy, P. H., Cuculis, J. J. and Meyer, H. G. (1970). The unlabeled antibody-enzyme method of immunohistochemistry. Preparation and properties of soluble antigen-antibody complex (horseradish peroxidase-anti-horseradish peroxidase) and its use in identification of spirochetes. *J. Histochem. Cytochem.* **18**, 315–333.

Suzuki, T., Fujikura, K., Higashiyama, T. and Takata, K. (1997). DNA staining for fluorescence and laser confocal microscopy. *J. Histochem. Cytochem.* **45**, 49–53.

Tsurui, H., Nishimura, H., Hattori, S., Hirose, S., Okumura, K. and Shirai, T. (2000). Seven-color fluorescence imaging of tissue samples based on Fourier spectroscopy and singular value decomposition. *J. Histochem. Cytochem.* **48**, 653–662.

Wasielewski, R. von, Mengel, M., Gignac, S., Wilkens, L., Werner M. and Georgii, A. (1997). Tyramine amplification technique in routine immunohistochemistry. *J. Histochem. Cytochem.* **45**, 1455–1459.

Watkins, S. (1989). Cryosectioning. In *Current Protocols in Molecular Biology* (F. M. Ausubel, R. Brent, R. A. Kingston, D. D. Moore, J. G. Seidman, J. A. Smith and K. Struhl, Eds), pp. 14.2.1–14.2.8. Greene Publishing and Wiley-Interscience, New York.

Zeller, R. (1989). Fixation, embedding, and sectioning of tissues, embryos and single cells. In *Current Protocols in Molecular Biology* (F.M. Ausubel, R. Brent, R.A. Kingston, D.D. Moore, J.G. Seidman, J.A. Smith, K. Struhl, Eds), pp. 14.1.1–14.1.8. Greene Publishing and Wiley-Interscience, New York.

List of suppliers

Antibody resource page:
http://www.antibodyresource.com/

Linscott's Directory:
http://www.linscottsdirectory.com/

MSRS Catalog of Primary Antibodies:
http://www.antibodies-probes.com/

BD Biosciences
2350 Qume Drive
San Jose, CA 95131-1807, USA

Tel: (800) 223-8226
Fax: (408) 954-2347
http://www.bdbiosciences.com/

DAKO Corporation
6392 Via Real
Carpinteria, CA 93013, USA

Tel.: +1 805 566 6655
Fax: +1 805 566 6688
http://www.dako.com

DIANOVA GmbH
Mittelweg 176
20148 Hamburg, Germany

Tel: +49 (40) 45 06 70
Fax : +49 (40) 45 06 73 90
http://www.dianova.de

Molecular Probes, Inc.
PO Box 22010
Eugene, OR 97402-0469, USA
4849 Pitchford Ave.
Eugene, OR 97402-9165, USA

Tel: (541) 465-8300
Fax: (541) 344-6504

Molecular Probes Europe BV
PoortGebouw,
Rijnsburgerweg 10,
2333 AA Leiden,
The Netherlands

Tel: +31-71-5233378
Fax: +31-71-5233419
http://www.probes.com

7 Isolation, Characterization and Cultivation of Human Monocytes and Macrophages

Stefan W Krause, Michael Rehli and Reinhard Andreesen
Department of Hematology and Oncology, University of Regensburg, Germany

◆◆

CONTENTS

◆◆◆◆◆◆ INTRODUCTION

Monocytes (MO) and the different types of tissue macrophages (MAC) are grouped together as the 'mononuclear phagocyte system' (MPS). As described for other species, it is assumed that mature human tissue MAC arise from circulating blood MO upon leaving the vasculature (van Furth, 1989). The circulating pool of MO is supplied by proliferating progenitor cells in the bone marrow. The proliferative capacity of MO and tissue MAC is quite low, although some proliferation may contribute to cell homeostasis in different tissues. The cells of the MPS have a prominent role in host defence against microbial pathogens and malignant cells due to their capacity of phagocytosis, secretion of cytokines, enzymes, oxygen radicals and other soluble products, and direct cellular cytotoxicity. Furthermore, they have an important position at the bridge between innate and specific immunity in presenting antigen to T cells in the context of MHC class II antigens.

In experiments with mice, peritoneal or spleen MAC are usually used as prototype members of the MPS. In the human system, blood MO are the starting cell population which is most easily obtained. These cells can be cultured *in vitro* in order to obtain prototypic 'tissue' or 'exudate-type' MAC (Musson *et al.*, 1980; Andreesen *et al.*, 1983a) (for further details see below). Researchers in a hospital department sometimes may have access to clinical specimens of bronchial lavages that contain alveolar MAC or of peritoneal dialysis fluids that contain peritoneal MAC.

METHODS IN MICROBIOLOGY, VOLUME 32
ISBN 0–12–521532–0

Copyright © Elsevier Science Ltd
All rights of reproduction in any form reserved

If large numbers of cells are required, permanent cell lines may also be considered. Of course, these permanently dividing cells are of malignant origin and in many aspects quite different from non-proliferating normal human MO or MAC. However, they may be acceptable depending on the experiments that are intended. We have personal experience with two cell lines that share quite a number of features with MO or MAC: MonoMac 6 (Ziegler Heitbrock *et al.*, 1988) is CD14 positive and morphologically monocytic. It is used by many researchers in the field to study the effects of microbial stimuli on the activation of monocytic cells. The second cell line, THP-1 (obtainable from ATCC), is carboxypeptidase M positive and cells transform to morphologically MAC-like cells when stimulated by phorbol esters. These cells are our preferred targets for the analysis of gene transcription regulation (see later).

◆◆◆◆◆◆ ISOLATION OF PERIPHERAL BLOOD MONOCYTES

Human blood MO are usually purified from mononuclear cells (MNC) prepared by standard density centrifugation over Ficoll/Hypaque. MNC can be purified from blood, from buffy coats or from white cell products obtained by leukapheresis. A leukapheresis procedure is the optimal method to obtain large numbers of cells. If you have access to this method in a blood banking or haematology department, attention has to be paid to the fact, that standard settings for harvesting of lymphocytes or peripheral blood progenitor cells will only lead to low MO yields. High numbers of MO will be obtained, if cells are collected deep in the interface (containing rather high numbers of red blood cells). With appropriate settings, a leukapheresis product containing $5–8 \times 10^9$ MNC can be expected from a healthy donor, with 10^9 of these cells being MO. If in doubt, contact the supplier of your aphaeresis system. In many labs, buffy coats from local blood banking services are used as starting material. These should be freshly prepared, a requirement quite frequently leading to logistical problems. If preparation of MO from buffy coats gives low yields or other technical problems occur, freshly drawn blood anticoagulated by preservative-free heparin should be tried as starting material in comparison.

MO can be obtained from MNC preparations by several different methods. Sorting by FACS or magnetic beads is described in Chapter I.1 (pp. 25–58) of this book. Density gradient centrifugation (e.g. with percoll) is another possibility (Davies and Lloyd, 1989), yet MO will be contaminated by platelets that have to be removed by an additional adherence step. Two methods will be described here in detail: (a) separation by cell adherence, which is simple and does not require expensive equipment, (b) counter-current elutriation, which is the technique of choice for laboratories, where an elutriation centrifuge is accessible and large numbers of cells have to be processed or preactivation of MO by adherence (Haskill *et al.*, 1988; Krause *et al.*, 1996) has to be avoided.

MO purification by cell adherence

MO and MAC have an intrinsic avidity to adhere to cell culture plastic or glass surfaces. They will attach to 'bacterial grade' or 'tissue culture grade' polystyrene, somewhat weaker than Teflon, but not or only very weakly compared with polypropylene or polyethylene. This avidity for cell adhesion can be used to separate them from other cells. Adhesion of MO/MAC is different from the adhesion of fibroblasts or cancer cell lines. It occurs within a few minutes at 37°C but not in the cold and it cannot be reversed by trypsination.

Standard protocol for MO purification by adherence

- *Complete medium*: RPMI 1640 with 2% human serum, $NaHCO_3$, 2 mM glutamine, 50 μM 2-mercaptoethanol, 100 U ml^{-1} penicillin, 100 μg ml^{-1} streptomycin, MEM non-essential amino acids, MEM vitamins, MEM pyruvate. (MEM supplements are diluted 1/100 from 100x stock solutions, Gibco-BRL, Paisley, UK)

1. MNC are prepared by Ficoll/Hypaque density centrifugation followed by three washing steps in PBS and resuspended in cell culture medium containing 2% human serum at a density of 5×10^6 MNC ml^{-1}. Keep cells in the cold.
2. About 3 ml of cell suspension (15×10^6 MNC) containing an estimated number of 3×10^6 MO are transferred into a 60-mm petri dish with a hydrophilic Teflon surface (e.g. Petriperm, Vivascience, Hannover, Germany) and placed in an incubator at 37°C, 7% CO_2. In parallel, prepare about 20 ml of complete medium prewarmed to 37°C.
3. After 30 to 45 min, attachment of MO to the surface of the culture vessel can be examined using an inverted microscope (Fig. 1A). If the number of adherent MO is insufficient, the incubation time can be extended up to 90 minutes although usually this is not necessary.
4. After incubation, swirl the plates very gently to suspend the nonadherent cells, and remove the cell suspension containing mainly lymphocytes and platelets. Rinse the plates gently with 3 ml of prewarmed medium without loosening the attached MO. Swirl gently and remove the cell suspension. Repeat this twice. It is essential to keep dishes and culture medium warm during the washing steps.
5. After three washing steps, cells are examined under the microscope (Fig. 1B). If many nonadherent cells are still present, the procedure can be repeated once or twice.
6. When a sufficient purity is achieved, add 3 ml of medium or PBS and place the dish on ice for one hour to detach cells.
7. Subsequently, MO can be harvested from the dishes by vigorous pipetting, counted in trypan blue and used for further experiments. Alternatively, MO-containing dishes can be incubated over night in complete medium and harvested on the next day taking advantage of the fact, that MO adhesion usually slightly decreases after one day of culture.

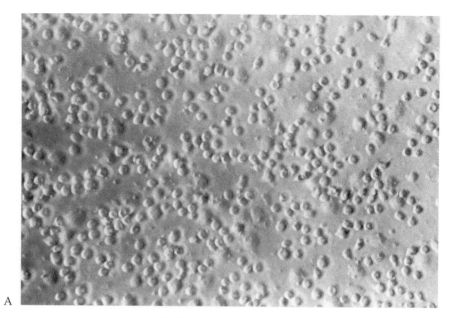

A

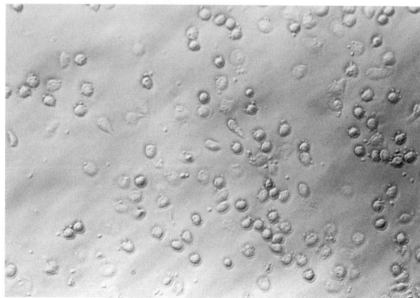

B

Modifications of the protocol and comments

The protocol described above has the advantage that MO can reliably be detached from the Teflon dishes for further experiments. As a disadvantage, compared with polystyrene dishes, more MO will be lost with the non-adherent cell fraction due to the less avid cell adhesion on Teflon surfaces. Furthermore, Teflon dishes are quite expensive. If polystyrene plates are used instead of Teflon, MO will attach more tightly, but getting them off for further experiments may be difficult. If cells are to be

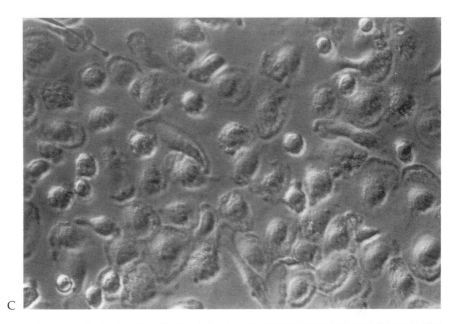

C

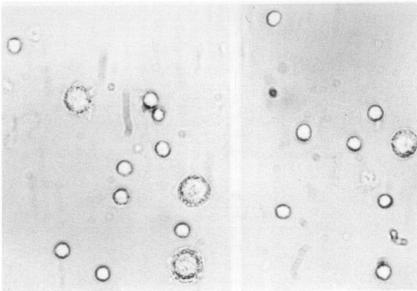

D

Figure 1. Phase contrast microscopy of MO and MAC. (A): MNC including adherent MO in cell culture dishes. (B): Same dishes after washing away most of the non-adherent cells. (C): Same dishes after one week of culture: MAC differentiation has occurred leading to much larger cells (MAC with a similar morphology will develop in Teflon bags) (D): A mixed population of MAC and lymphocytes harvested from a Teflon bag is placed in a chamber for cell counting. 200× magnification for all figures.

detached from normal plastic, we suggest an overnight incubation of the washed MO at 37°C in complete medium before placing them in a refrigerator or on ice and trying to detach them by vigorous pipetting. Some authors recommend the use of EDTA for detaching adherent MO, but in our hands this is not satisfactory.

If long-term cultures to induce MAC maturation are intended (see below), MO can also be cultured as MNC suspension without initial separation. Washing away the lymphocytes and platelets is then performed after MAC maturation has occurred.

Counter-current elutriation

Counter-current elutriation is the method of choice for separating large numbers of unstimulated monocytes. Starting material for counter-current elutriation are peripheral blood mononuclear cells (MNC) that can be purified from white blood cell concentrates or whole blood (see above). The separation is performed with a centrifuge equipped with a special elutriator rotor system and a separation chamber. We use a Beckman J6M-E centrifuge with a JE-5 rotor system (Beckman, Munich, Germany), that can be equipped with a small separation chamber for processing 10^8–10^9 MNC or a large chamber for processing 10^9–10^{10} MNC, in combination with a peristaltic pump (e.g. masterflex, Barnant, Barrington, Illinois, USA).

The principle of the separation is the balance between an outward directed centrifugal force and inward directed hydrodynamic force created by pumping fluid continuously through the rotor in a centripetal direction (Fig. 2). While the rotor is spinning in the centrifuge, a suspension of cells is pumped at a preset flow rate from outside the centrifuge into the separation chamber. Cells migrate to positions in the separation

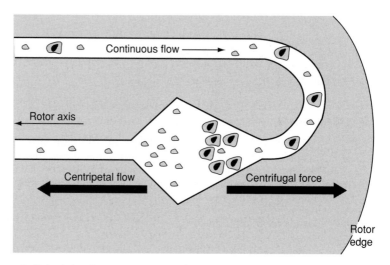

Figure 2. Principle of counter-current elutriation of cells of different size. Smallest cells are already washed out with the fluid flow, while large cells are still held back by the centrifugal force.

chamber where both forces are balanced, e.g. they equilibrate at different radii due to their size and density (Sanderson *et al.*, 1977). Different fractions of cells can be sequentially eluted from the centrifuge by increasing the fluid flow rate. Smaller cells (i.e. platelets, then lymphocytes) leave the system first, monocytes which are the largest cells are washed out last (Table 1). Platelets and MO can be obtained in a rather high purity, whereas the cell distribution in the different lymphocyte fractions is widely overlapping. The purity of the monocyte fraction is about 90%. In general, about 90% of all cells loaded into the system are recovered and the viability of the cells in all fractions is about 99% as determined by trypan blue exclusion.

Table 1. Separation of MNC into Different Cell Fractions by Counter Current Elutriation

Fraction	Flow Rate (ml/min) (small chamber)	(large chamber)	Prominent Cell Type
I	52	12	Platelets, few small lymphocytes
II	57	13	Lymphocytes (many B cells)
III	64	14.5	Lymphocytes (mainly T cells)
IV	74	16.5	Lymphocytes (T cells, NK cells)
V	82	18	T cells, NK cells, MO
VI	111	25	MO

Settings are given for a Beckman J6M-E centrifuge with a JE-5 rotor system, using either a large or small separation chamber and a constant centrifuge speed of 2500 rpm.

Counter-current elutriation protocol

Solutions needed:

- Hank's balanced salt solution without Ca^{2+} and Mg^{2+}
- 6% H_2O_2 in sterile pyrogen-free water
- human serum albumin (clinical grade) or autologous human plasma
- 70% ethanol in pyrogen-free water

1. Before each run the separation chamber is assembled and sterilized by running about 500 ml of a 6% hydrogen peroxide solution through the system in a closed circuit for about 20 min. Avoid strong light exposure.
2. During this procedure air bubbles may become trapped in the system. These air bubbles have to be removed by starting and slowing down the centrifuge for several times (see instruction manual of the elutriation system for details), then the centrifuge is set at the intended speed (a speed of 2500 rpm is used in our laboratory, all flow rates are given for this setting) and kept at this speed throughout the whole procedure.
3. Following the sterilization the system is flushed with 1000 ml of sterile Hank's balanced salt solution or PBS to remove the hydrogen peroxide.

4. Now the peristaltic pump system has to be calibrated. For this purpose the amount of fluid pumped through the system per minute at different settings of the peristaltic pump is measured and a calibration curve (pump setting versus fluid flow in ml min^{-1}) is drawn.

5. The system is then flushed with Hank's solution supplemented with either 2% albumin or 6% human plasma and is now ready for loading the MNC suspension. If the MNC preparation is not yet ready to be loaded, a closed circuit with centrifuge and peristaltic pump continuously running can be installed.

6. MNC are loaded into the system at a flow rate of 52 ml min^{-1} for the large chamber. At the same speed of 52 ml min^{-1} small lymphocytes and platelets are removed by collecting 1000 ml of fluid (see Table 1).

7. By increasing the flow rate, cells of increasing size are collected in fractions of 1000 ml. At a flow rate of 82 ml min^{-1} a mixed fraction (fraction V) of lymphocytes and monocytes is washed out. A higher flow rate for this fraction will give higher purity of MO in fraction VI, but at the expense of a considerable loss of MO, because more MO will be present in fraction V. Finally, at a flow of 111 ml min^{-1} the monocyte fraction (fraction VI) is obtained.

8. When all cells have been collected, stop the rotor first and then the pump. The tubing should be flushed with sterile water followed by 70% isopropanol to remove all buffer, debris and remaining cells. The chamber must be disassembled, carefully cleaned and dried.

◆◆◆◆◆◆ CULTURE OF HUMAN MONOCYTES

Short-term culture

The majority of MO will survive for a time period of one day on a wide variety of culture substrates in different types of culture media. However, if no dead cells are seen microscopically in MO cultures, this does not mean that no cells are dying at all. Dead MO are phagocytosed by their neighbours, therefore, down to a survival of about 30%, the culture will be looking 'healthy'. Overall, the researcher is quite free in his choice of a culture system for short time analysis of MO. Our standard cultures are in complete medium containing 2% human serum on Teflon or tissue culture plastic at 37°C and 7% CO_2. Serum-free conditions in standard medium are also possible for short-term cultures.

As described above, MO will adhere to almost every common tissue culture lab ware or to 'bacterial grade' polystyrene. In fact it is almost impossible to cultivate MO as non-adherent cells. They will attach on Teflon, albeit reversibly, and thus cannot be called 'non-adherent' in this case. If cultured on polypropylene surfaces, they will form big clumps of cells sticking tightly together. However, it is possible to keep MO alive and non-adherent in polypropylene tubes on a roller bottle device inside the incubator.

Generation of differentiated macrophages in long-term culture

Mature human tissue MAC are not easily available for functional analysis. For many studies, *in vitro* MO-derived MAC can be used instead. If blood MO are cultured *in vitro* in the presence of human serum, cells will increase in size and acquire some of the features of 'prototype' tissue or exsudate MAC: the expression of many cell surface antigens will change (Andreesen *et al.*, 1990), the ability for phagocytosis, tumour cytotoxicity (Andreesen *et al.*, 1983b) and procoagulant activity (Scheibenbogen *et al.*, 1992) are increased, whereas the capacity to stimulate unprimed T-lymphocytes decreases (Schlesier *et al.*, 1994). Furthermore the pattern of secretion products changes, for example, MO produce high amounts of IL-1 upon stimulation, whereas mature MAC secrete much more TNF but no IL-1 (Scheibenbogen and Andreesen, 1991).

Standard protocol for MAC differentiation in Teflon bags

1. Rectangular bags are prepared in advance from Teflon foil (Biofolie 25, Vivascience, Hannover, Germany) by cutting pieces in an appropriate size, e.g. 10 × 20 cm, folding them once with the hydrophobic side inside and sealing them on two sides with a heat sealing device (e.g. Fermant 400, Joisten & Kettenbaum, Bergisch Gladbach, Germany), resulting in bags with an inside size of about 3.5 × 18 cm. Use unpowdered gloves while preparing the bags. Appropriate settings of the sealing device have to be determined empirically. Test the quality of the sealing by filling a test bag with water before using the bags for cell culture. Sterilize in ethylene oxide.

2. Under a sterile hood, prepare a clamp with self-adhesive tape attached to a small stand to hold the Teflon bag with the short unsealed side upwards (Fig. 3A). Prepare a suspension of 10^6 MO ml^{-1} in complete medium and fill it into the bags. Between 10 and 20 ml are appropriate for a small bag as described above. Seal the remaining side of the bag (Fig. 3B). Incubate at 37°C, 7% CO_2 for about one week. Refeeding is not necessary during this time period.

3. After an appropriate culture period, check under an inverted microscope, if MAC maturation has occurred (Fig. 1C). Teflon bags have to be placed on a thin transparent support tray, e.g. lid of a multiwell plate, for microscopy.

4. Place the bag at 4°C for 60 min and afterwards gently move back and forth the culture medium inside the bag with your fingertips for detaching cells. Detachment of MAC can be observed macroscopically but again should be checked under an inverted microscope.

5. One corner of the bag is then treated with 80% ethanol and cut off with scissors that have been disinfected over a gas flame. Transfer the contents of the bag into a polypropylene tube and wash once by centrifugation and resuspension in new medium or PBS.

6. Count the cells by trypan blue exclusion. Differentiated MAC will be considerably larger (about twice the diameter) compared to contaminating lymphocytes that are eventually present (Fig. 1D).

Isolation of Monocytes and Macrophages

(A)   (B)

Figure 3. Teflon bags for long-term culture of MO. (A): A bag is attached to a stand with self-adhesive tape and filled with the cell suspension. (B): The bag is closed with a sealing device before incubation.

Comments and modifications of the protocol

The protocol described above has several advantages: differentiated MAC can reliably be harvested, counted and transferred in defined cell numbers to subsequent experiments. Furthermore, the size of the Teflon bags can easily be scaled up for the processing of large cell numbers. For researchers, who do not want to prepare their own bags, ready to use bags can be obtained from commercial suppliers (e.g. from Vivascience, Göttingen, Germany). These bags do not need to be sealed, but can be closed with a Luerlock cap. Alternatively, 'Petriperm' dishes (Vivascience) can be used but in our hands are not as good as the bags (cells tend to stick firmly to the hydrophilic type of dishes and maturation does not take place as well in the hydrophobic type).

MO to MAC differentiation will also occur on glass surfaces or on polystyrene. Any brand of lab ware has to be individually tested for being suitable to support this process. In most instances, dishes especially prepared for tissue culture are inferior to standard 'bacterial grade' polystyrene. We have found that bacterial dishes or ELISA plates from Greiner, Frickenhausen, Germany, work quite well. If differentiating MO/MAC are examined microscopically from day to day, freshly plated MO will tightly adhere, after 1 to 2 days the adherence may decrease and if the tissue culture substrate is not suitable, most cells will detach during the following days and subsequently die. Attachment and differentiation work somewhat more reliably, if MO are seeded in serum-free medium and serum is only added after an overnight incubation period. On a 'good' substrate, MO/MAC will attach firmly, spread out and increase in size day by day. Differentiated MAC after 1 week will then be more adherent than freshly isolated MO (Fig. 1C).

Complete RPMI 1640 medium containing 2% human serum is our standard condition for differentiation of MO from MAC *in vitro*. AB group serum is usually used, but this is not necessary, nor is heat-inactivation of the serum. If significantly lower concentrations of serum are used, the full phenotypic change will not take place, if higher concentrations of serum are used, MAC will acquire a somewhat rounder and more granular appearance, but the functional differences to MAC from 2% serum cultures are not very considerable.

No specific growth factors have to be added, but we know that an autocrine secretion of M-CSF takes place and is important for MAC survival and differentiation (Brugger *et al.*, 1991a). Other factors can of course modulate the differentiation process.

Note on contamination by LPS or other agents

Human MO and MAC are extremely sensitive to stimulation with even minute amounts of endotoxin from Gram-negative bacteria, i.e. lipopolysaccharide (LPS). A few picograms of LPS per ml may be sufficient to obtain a significant biological response, as for example characterized by the production of IL-1, IL-6 or TNF. One ng ml^{-1} may already lead to a close to maximum cytokine production. It has even been suggested that this response is used as a sensitive assay system to detect LPS contamination (Northoff *et al.*, 1987). If LPS is present in the cultures, this will obscure the response of cells to intentional stimulation by LPS or other stimuli. Moreover, normal MO to MAC differentiation will be disturbed after LPS stimulation (Brugger *et al.*, 1991b) and MO/MAC will not respond to a second stimulation for several days (a phenomenon called endotoxin tolerance) (Ziegler Heitbrock *et al.*, 1995). Endotoxin is the most frequent unwanted stimulus in MO cultures. However, other contaminations can be a problem. Heat-killed micro-organisms, including Gram-positive bacteria, may still activate the cells. Furthermore, as MO or MAC are potent phagocytosing cells, they may clear an inapparent infection by living bacteria in the cell culture and at the same time become stimulated.

Cell culture media, media supplements and serum should be ordered endotoxin free. Furthermore, regular measurements of spontaneous cytokine secretion should be performed. Production of IL-6 or TNF in unstimulated MO cultures should be below 2–5% of cultures stimulated with 100 ng ml^{-1} of LPS. Cultures with higher rates of 'spontaneous' secretion should be regarded as pre-activated.

◆◆◆◆◆◆ CHARACTERIZATION OF MONOCYTES AND MONOCYTE-DERIVED MACROPHAGES

Surface antigens

The expression of many cell surface antigens changes during differentiation from MO to MAC and is modulated by the special environment *in vivo* or by culture conditions *in vitro*. Typical surface markers found on most MO or MAC populations are, for example, CD11c, CD14, CD33 and HLA-DR among many others. Examples of surface antigens present on mature MAC are carboxypeptidase M (detected by MAX.1 and MAX.11 antibodies (Rehli *et al.*, 1995)), CD84/MAX.3 (Krause *et al.*, 2000), CD16, CD52, CD71. No single marker can 'define' a cell as being a MO or MAC, instead several features have to be considered together (Andreesen *et al.*, 1990).

MO or MO-derived MAC harvested from Teflon bags can be analysed by flow cytometry or by fluorescence microscopy without special difficulties (see Chapter II.1). A higher background fluorescence is observed in MO as well as in MAC compared with lymphocyte populations. MAC are quite large cells: in flow cytometry the forward and sideward scatter sensitivity has to be set lower as compared with freshly prepared MO.

MO or MAC as adherent cells are especially suitable for analysis by cell ELISA. The density of a given antigen on adherent cells in a microculture well is determined by an enzyme-coupled antibody sandwich with a soluble substrate. Of course, average antigen density of the whole-cell population is measured, but the method has the advantage of allowing a high throughput of different samples in a reasonable period of time. The detailed protocol for such a procedure is described by Andreesen *et al.* (1988).

Phagocytosis

The ability of phagocytosis is a prominent feature of mononuclear phagocytes. Whereas many cells will endocytose fluids or take up very small particles, MO and MAC are also able to ingest particles up to a size of several μm, including inert material, micro-organisms, cell debris and dead cells. Many protocols exist to test phagocytosis *in vitro*. A very simple method is to add latex particles (e.g. Sigma LB-11, St. Louis, MO, USA) to cell cultures and examine phagocytosis microscopically. A flow cytometric method is given below.

Flow cytometry of fluorescent microspheres

Reagents

- Latex beads: fluoresbrite carboxylated YG beads, diameter 1.7 μm, Polysciences, 400 Valley Road, Warrington, PA 18976, USA
- Phagocytosis medium: RPMI 1640 containing 1% w/v BSA, 2% human serum, 20 mM Hepes, pH 7.2
- PBS/EDTA: phosphate buffered saline containing 10 mM EDTA and 0.1% Na-azide
- Bovine serum albumin 2% w/v in PBS/EDTA

(contd.)

- Formaldehyde solution: Dissolve paraformaldehyde to a final concentration of 1% (w/v) in PBS at about 50°C. Adjust pH to 7.2 with 1 M NaOH (Paraformaldehyde will only dissolve while adding NaOH). Store at 4°C for a maximum of 2 weeks.

1. Prepare phagocytosis medium. Add 10 µl of beads (about $1–2 \times 10^8$ particles) to 1 ml of phagocytosis medium and incubate for 10 min at room temperature. Do not store these opsonized beads.
2. Spin down 10^6 MO or MAC harvested from Teflon cultures in 2 ml microcentrifuge tubes. Prepare a similar sample as control without beads.
3. Resuspend the cell pellet in 300 µl of phagocytosis medium containing beads. Incubate for 60 min at 37°C in an incubator. Agitate the tube every 10 min to keep cells in suspension or place it on a slowly rotating device. After incubation, add 1.5 ml of cold phosphate buffered saline/EDTA (PBS/EDTA). Perform all following steps on ice.
4. Carefully layer the sample on top of a 3-ml cushion of 2% bovine serum albumin in PBS/EDTA (w/v) in a 10-ml centrifuge tube. Centrifuge for seven minutes at 150g, 4°C. A large part of the non-phagocytosed beads will remain at the interphase.
5. Carefully remove the fluid from the top and resuspend the cell pellet in PBS/EDTA. Wash twice in PBS/EDTA by centrifugation (counter-staining with a red-fluorescent antibody can be done at this time point before fixation, see Chapter II.1 and comments below).
6. Resuspend in 500 µl formaldehyde solution and store at 4°C until flow cytometric evaluation.
7. For flow cytometry, use a linear scale for forward scatter (FSC) and sideward scatter (SSC) and log scale for green fluorescence. First adjust FSC and SSC sensitivity using control cells without beads. Control cells should appear in the lower quarter of SSC range, because SSC will increase after phagocytosis (Fig. 4A,B). Save the appropriate settings.
8. Measure a sample of beads only. With the settings adjusted of MO or MAC, no fluorescence signal should be detected, because beads alone are below the FSC detection limit. If you increase FSC sensitivity and decrease the detection threshold, a sharp peak of the fluorescent beads will appear (Fig. 4C). Adjust green fluorescence sensitivity in order to place the peak between 10 and 50 on a 10^4 scale.
9. Store the measurement of beads without cells in a list file in order to document the fluorescence intensity of single beads.
10. Keep the fluorescence sensitivity unchanged and return FSC sensitivity and threshold to the settings determined before with cells only. Now only cells should be detected as events, because beads without cells are below the FSC threshold.
11. Measure phagocytosis samples and control sample.

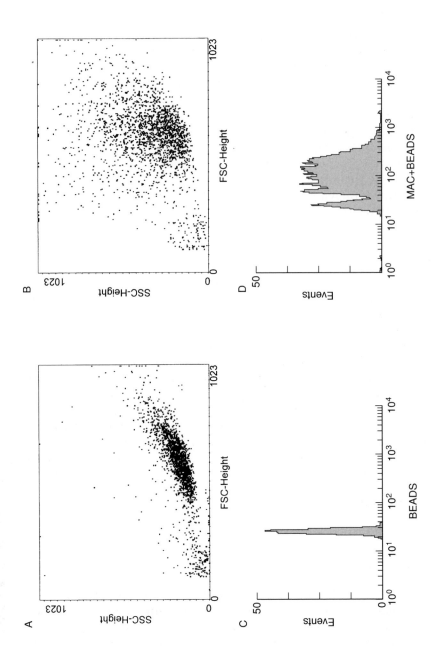

Figure 4. Flow cytometric evaluation of phagocytosis. (A) Scatter diagram of MAC harvested from Teflon bags. (B) Scatter diagram of MAC after phagocytosis. (C) Histogram of beads without cells. (D) Histogram of MAC with ingested beads.

Krause et al Fig 4a–d Methods in Microbiology

The protocol described below is a modified version of the technique first published by Steinkamp *et al.* in 1982. Fluorescence intensity of the cells will correlate with the number of beads ingested. Cells having ingested no beads at all or one, two or three beads can clearly be separated from each other (Fig. 4D). In higher fluorescence ranges, the number of ingested beads can be estimated by comparing the fluorescence level of a single bead with the fluorescence of the bead-containing cells, assuming an exact logarithmic scale of the flow cytometer.

The results obtained by flow cytometry should always match the microscopical examination. A fluorescence microscope is not necessary, beads can also easily be seen using phase contrast. If high amounts of protein are still present after washing cells, remaining beads may start to aggregate into small clumps. These clumps may then be above the FSC threshold and falsely be detected as phagocytosing cells by flow cytometry. This phenomenon can be avoided, if cell preparations are washed sufficiently and measured without long delay on the day of preparation.

Protocol modification

Instead of short-time incubation of harvested cells, MO or MAC can be incubated with the beads in cell culture in Teflon bags or dishes for 24 h. In this case the bead to cell ratio should not be above 50 : 1. If the bead to cell ratio is below 20 : 1, centrifugation over 2% BSA can be omitted and cells directly washed in PBS/EDTA after harvesting.

If counterstaining of surface antigens with a PE-labelled marker is intended, proper compensation of two-colour fluorescence (Chapter II.1.1) may become a problem, since fluorescence intensity of the beads is very high. Green fluorescence of beads will strongly appear in the PE channel and high compensation settings to attenuate for this have to be used. Counterstaining works best, if cells do not ingest too many particles (below 10 to 15 – if necessary cut down the bead number in the reaction) and particle size is rather low. Furthermore, choose an antigen to be analysed by counterstaining that is expressed intensely. Alternatively, try PerCP or PE-Cy5 for counterstaining.

Production of cytokines and other soluble products

MO/MAC can destroy invading microorganisms by reactive oxygen metabolites and nitric oxide. Methods to measure these effector molecules are given in Chapter II 2.5. However, whereas oxygen is quite similarly produced by human and mouse MAC, human MAC only weakly produce nitric oxide. Furthermore MO and MAC produce a vast pattern of pro- and anti-inflammatory cytokines depending on their stage of differentiation and stimulation. See above for some comments on LPS and Chapter II 2.4 on protocols for cytokine measurements.

Cytotoxicity against tumour cells

MAC can kill tumour cells by direct cellular cytotoxicity or by the production of cytotoxic effector molecules. Additionally, they can act as effector cells in antibody-dependent cytotoxicity (Andreesen *et al.*, 1983b; Munn und Cheung, 1990). On the other hand, MAC may help tumour cell growth by supporting angiogenesis or by secreting growth factors (Mantovani *et al.*, 1992). The myeloid cell line U937 and the bladder carcinoma line RT4 (both can be obtained from ATCC) are examples for targets sensitive for MAC cytotoxicity *in vitro*. Cellular cytotoxicity can be tested by a multitude of different methods. Chromium release that is used for T-cell or NK-cell mediated cytotoxicity (described in Chapter II 4), is not very suitable for MAC, probably, because MAC cytotoxicity is slower and because MAC ingest whole damaged cells.

Tumour cytotoxicity measured by post-labelling with ³H-Thymidine

A system that can easily be adapted for evaluation of effects exerted by MO or MAC is a post-labelling assay measuring ^{3}H-Thymidine incorporation into the tumour target. This test cannot distinguish between real target cell death and growth arrest, but it is a very simple screening assay to detect a net impact of effector cells upon tumour cell growth. Tumour cell proliferation is measured with and without co-culture with MAC. Human MO and MAC do not proliferate *in vitro* and thereby do not contribute background incorporation.

Technical details of measuring thymidine incorporation are described in Chapter I 2, therefore only the principle of the assay is outlined: 10^4 tumour cell targets/well/100 μl of a tumour cell line under evaluation (e.g. U937) are placed in a microtitre plate. MO or MAC as effector cells are added in triplicate in serial dilutions from 10^3 to 10^5 cells/well/100 μl. Control well triplets are left without effector cells and without target cells, respectively. Effector cells can additionally be stimulated with interferon-gamma (200 U ml^{-1}) or other MAC-activating agents (in that case control targets should be treated with the same stimuli). After a co-culture period of 48 h, cells are pulsed with 0.5 μCi of ^{3}H-Thymidine/well and harvested 4–16 h later. Growth inhibition is calculated as

$$1 - (cpm_{\text{co-culture}} - cpm_{\text{MAC only}})/cpm_{\text{tumour only}}$$

Transient DNA transfections of THP-1 cells for reporter assays

Efficient transfer of foreign DNA molecules (plasmids) into the nucleus of a primary MO or MAC is an extremely difficult task. In order to study gene expression regulation by reporter gene-based assay systems we therefore use THP-1 cells (see Introduction) as a model system. THP-1 cells can be transfected using DEAE-dextran which gives optimal reproducibility in our hands. You may have to optimize DEAE-dextran concentration and time of incubation with your line of THP-1. To standardize for transfection efficiency of individual transfections, firefly luciferase reporters are co-transfected with a renilla luciferase construct. Luciferase activity of the latter construct can be assayed independently from the same cell lysate. Note that the transfection procedure itself will stimulate the cells to some degree.

Transient transfection protocol

Reagents

- Culture medium: complete RPMI 1640 as described above, but with 10% fetal calf serum instead of 2% human serum
- STBS buffer: 25 mM Tris/HCl (pH 7.4), 137 mM NaCl, 5 mM KCl, 0.6 mM Na_2HPO_4, 0.7 mM $CaCl_2$, 0.5 mM $MgCl_2$
- DEAE-Dextran (Amersham Biosciences): Stock solution of 10 mg ml^{-1} in STBS, sterile-filtered. 800 µg ml^{-1} in STBS for transfection

1. The day before transfection, THP-1 cells are seeded into tissue culture flasks at a density of 3.5×10^5 cells ml^{-1} culture medium. On the next day, for every single transfection, 6 ml cell suspension are washed twice with STBS (5 ml) and pelleted.
2. 200 ng firefly luciferase reporter plasmid and 20 ng renilla control vector (Promega) in 70 µl STBS buffer are mixed with 70 µl of DEAE-Dextran (800 µg ml^{-1} in STBS) and immediately added to the pelleted THP-1 cells. We perform duplicate transfections for each plasmid.
3. The cells are incubated at 37°C for 20 min, washed twice with STBS, resuspended in 6 ml of culture medium and placed into 60 mm petri dishes. (If required, differentiation may be induced by the addition of PMA (10^{-8} M) or Vitamin D3 (10^{-7} M).)
4. The transfected cell lines are cultivated for 48 h, harvested and cell lysates assayed for firefly and renilla luciferase activity using the Dual-Luciferase Reporter Assay System (Promega) on a standard luminometer. Firefly luciferase activity of individual transfections can be normalized against renilla luciferase activity.

References

Andreesen, R., Picht, J. and Lohr, G. W. (1983a). Primary cultures of human blood-borne macrophages grown on hydrophobic teflon membranes. *J. Immunol. Methods* **56**, 295–304.

Andreesen, R., Osterholz, J., Bross, K. J., Schulz, A., Luckenbach, G. A. and Lohr, G. W. (1983b). Cytotoxic effector cell function at different stages of human monocyte-macrophage maturation. *Cancer Res.* **43**, 5931–5936.

Andreesen, R., Mackensen, A., Osterholz, J., Brugger, W. and Lohr, G. W. (1988). Microculture assay for human macrophage maturation in vitro. Cell-ELISA analysis of differentiation antigen expression. *Int. Arch. Allergy Appl. Immunol.* **86**, 281–287.

Andreesen, R., Brugger, W., Scheibenbogen, C., Kreutz, M., Leser, H. G., Rehm, A. and Lohr, G. W. (1990). Surface phenotype analysis of human monocyte to macrophage maturation. *J. Leukoc. Biol.* **47**, 490–497.

Brugger, W., Kreutz, M. and Andreesen, R. (1991a). Macrophage colony-stimulating factor is required for human monocyte survival and acts as a cofactor for their terminal differentiation to macrophages in vitro. *J. Leukoc. Biol.* **49**, 483–488.

Isolation of Monocytes and Macrophages

Brugger, W., Reinhardt, D., Galanos, C. and Andreesen, R. (1991b). Inhibition of in vitro differentiation of human monocytes to macrophages by lipopolysaccharides (LPS): phenotypic and functional analysis. *Int. Immunol.* **3**, 221–227.

Davies, D. E. and Lloyd, J. B. (1989). Monocyte-to-macrophage transition in vitro. A systematic study using human cells isolated by fractionation on percoll. *J. Immunol. Methods* **118**, 9–16.

Haskill, S., Johnson, C., Eierman, D., Becker, S. and Warren, K. (1988). Adherence induces selective mRNA expression of monocyte mediators and proto-oncogenes. *J. Immunol.* **140**, 1690–1694.

Krause, S. W., Kreutz, M. and Andreesen, R. (1996). Differential effects of cell adherence on LPS-stimulated cytokine production by human monocytes and macrophages. *Immunobiology* **196**, 522–534.

Krause, S. W., Rehli, M., Heinz, S., Ebner, R. and Andreesen, R. (2000). Characterization of MAX.3 antigen, a glycoprotein expressed on mature macrophages, dendritic cells and blood platelets: identity with CD84. *Biochem. J.* **346**, 729–736.

Mantovani, A., Bottazzi, B., Colotta, F., Sozzani, S. and Ruco, L. (1992). The origin and function of tumor-associated macrophages. *Immunol. Today* **13**, 265–270.

Munn, D. H. and Cheung, N. K. (1990). Phagocytosis of tumor cells by human monocytes cultured in recombinant macrophage colony-stimulating factor. *J. Exp. Med.* **172**, 231–237.

Musson, R. A., Shafran, H. and Henson, P. M. (1980). Intracellular levels and stimulated release of lysosomal enzymes from human peripheral blood monocytes and monocyte-derived macrophages. *J. Reticuloendothel. Soc.* **28**, 249–264.

Northoff, H., Gluck, D., Wolpl, A., Kubanek, B. and Galanos C. (1987). Lipopolysaccharide-induced elaboration of interleukin 1 by human monocytes: use for detection of lipopolysaccharide in serum and the influence of serum–lipopolysaccharide interactions. *Rev. Infect. Dis.* **9** (Suppl 5), S599–601.

Rehli, M., Krause, S. W., Kreutz, M. and Andreesen, R. (1995). Carboxypeptidase M is identical to the MAX.1 antigen and its expression is associated with monocyte to macrophage differentiation. *J. Biol. Chem.* **270**, 15644–15649.

Sanderson, R. F., Shepperdson, F. T., Vatter, A. E. and Talmage, D. W. (1977). Isolation and enumeration of peripheral blood monocytes. *J. Immunol.* **118**, 1409–1414.

Scheibenbogen, C. and Andreesen, R. (1991). Developmental regulation of the cytokine repertoire in human macrophages: IL-1, IL-6, TNF-alpha, and M-CSF. *J. Leukoc. Biol.* **50**, 35–42.

Scheibenbogen, C., Moser, H., Krause, S. and Andreesen, R. (1992). Interferon-gamma-induced expression of tissue factor activity during human monocyte to macrophage maturation. *Haemostasis* **22**, 173–178.

Schlesier, M., Krause, S., Drager, R., Wolff Vorbeck, G., Kreutz, M., Andreesen, R. and Peter, H. H. (1994). Monocyte differentiation and accessory function: different effects on the proliferative responses of an autoreactive T cell clone as compared to alloreactive or antigen-specific T cell lines and primary mixed lymphocyte cultures. *Immunobiology* **190**, 164–174.

Steinkamp, J. A., Wilson, J. S., Saunders, G. C. and Stewart, C. C. (1982). Phagocytosis: flow cytometric quantitation with fluorescent microspheres. *Science* **215**, 64–66.

van Furth, R. (1989). Origin and turnover of monocytes and macrophages. *Curr. Top. Pathol.* **79**, 125–150.

Ziegler Heitbrock, H. W., Thiel, E., Futterer, A., Herzog, V., Wirtz, A. and Riethmuller, G. (1988). Establishment of a human cell line (Mono Mac 6) with characteristics of mature monocytes. *Int. J. Cancer* **41**, 456–461.

Ziegler Heitbrock, H. W., Frankenberger, M. and Wedel, A. (1995). Tolerance to lipopolysaccharide in human blood monocytes. *Immunobiology* **193**, 217–223.

List of suppliers

Dako
Postbox 1359
DK 2600 Glostrup, Denmark

Tel: +45-44 92 0044
Fax: +45-42 841822
Antibodies, animal sera

Polysciences
400 Valley Road
Warrington, PA 18976, USA

Tel: +215-343 6484
Fax: +215-343 0214
Beads for phagocytosis experiments

Vivascience AG
Feodor-Lynen-Straße 21
30625 Hannover, Germany

Tel: +49 511 / 524 875-0
Fax: + 49 511 / 524 875-19
E-mail: info@vivascience.com
Teflon foil, cell culture ware, rotating device for cell cultures

Index

Epitope-driven vaccine
  design 108–111
  optimizing 111–114
Epitopes
  conserved 114
  enhancement 113–114
  promiscuous 112–113
  spacers between 111
Epstein-Barr virus (EBV) 142, 657–658
E-rosetting technique 630–631
Escherichia coli 374
  Bir A in 128
Euthanizing mice 447
Experimental infection 364–382, 403
  animal management 218, 218–21, 368–372
  following course of infection 382–390
  see also Leishmaniasis model; Tuberculosis model
Experiments, as stress factors 185
Expression profile libraries 172
Extended MHC binding motifs 102
External standard DNA,Th1/Th2 cytokines 316–317
Extraction efficiency of epitopes 291

Face shields 445
Faeces, collection for IgA detection 391
Fas/FasL pathway 138–139
Fas-ligand (*Fas-L*) 10, 142
Federation of European Laboratory Animal Science
    Associations (FELASA) 188, 190, 192
Feeding tubes, lung intubation 375–376
Fetal thymocyte fusions 267
Fibroblasts, Stpc488 transformed 659
Ficoll-Hypaque 237, 239
  density gradient centrifugation 622–623, 630–631,
    668
Filter
  cabinets 190
  of air 203
Fixation of cells 752, 754
  affinity labelling 37
  intracellular cytokine staining 725
Flow cytometers
  compensation circuit, spectral overlap 37–38
  fluorescent light detection 24–26
  light scatter detection 24
  setting up 45–47
Flow cytometry 407–411
  analysis of cellular response 456–458
  cell numbers and cell divisions 86–89
  cytokine secretion assay 70
  elicited peritoneal macrophages
  intracellular staining, cytokines 728–729, 730, 731
  of lymphocyte division, with CFSE labelling
    275–276
  macrophages and monocytes 777–778
    fluorescent microspheres 778–781
  phenotyping NK cells 696–697
  quantitation of endocytosis 608
  of secretion 50
FITC 32, 755
  coupling to proteins 345–346
  dextran capture, quantitation of endocytosis 608
  versus PE display, quadrant statistic analysis 92
Fluorescein (FL) 24, 32
  spectral overlap with phycoerythrin 38

see also FITC
Fluorescence
  analysis 38
  detection 24–26
  intensity, statistical analysis 42
Fluorescence-activated cell sorting (FACS) 5, 9, 24,
    50, 51, 141, 725
  CD1a and CD14 DC precursors 601, 602–603
  PB-DC,IDC, GCDC and CD11c- precursors
    596–597, 598
  phenotypic characterization of T-cell clones
    651–652
  standard cell dilution assay (SCDA) 89, 90, 91, 92,
    93
Fluorescent labelling, affinity-based 24–50
  basic considerations 24–29
    quantification 28–29
    sensitivity 28
  data acquisition and analysis 38–44
    plotting and presentation of data 40
  of proteins or particles 345–346
  staining parameters 33–37
    controls 37
  staining reagents 29–33
    direct staining 30, 46
    indirect staining 30–31, 46
  standard protocols, staining and instrumental set-
    up 44–47
Fluorescent microspheres, flow cytometry 778–781
  see also Latex beads
Fluorochrome-labelled antibodies 755
Fluorochromes 762
  conjugation to proteins 32, 33
  limitations of 755
Foot-pad thickness, dial-gauge caliper 394, 453, 476
Foreign genes, insertion into viral genome 666–667
Formaldehyde 726, 730, 779
Formalin 726, 752
Fortified diet 205
Forward scatter (FSC), flow cytometry 24
  evaluation of phagocytosis 779
Francisella tularensis 365
Free cytokines 717
FRET (Föster Resonance Energy Transfer) probes
    739
Freund's adjuvant 556, 574
Frozen
  bacteria 366
  tissues, preparation 404, 754
  see also Cryopreservation
Fungi
Fusion lines, hybridomas 265–269
  cells 267–268
  cloning 268

Galactose oxidase 650
'Gardella gels' 665
Gastrointestinal immune system, components 626
Gate setting, SCDA data analysis 92
Gene
  gun 528, 529, 534
  gene immunization 527–532
  insertion into viral genome 666–667
  knockout mice (GKO) 371
  mutations affecting immune system 199–200

Index

Index

Index

procedures, personnel infection prevention 218–219
Saimiri herpesvirus type 2 *see* Herpesvirus saimiri
Salivary Gland Virus (SGV) 495–496
Salmonella 199, 373, 642, 642
toxin, induced Thl and Th2 responses 309
Salmonella typhimurium 158, 374, 635
Sample
animal monitoring 188–189
buffer, SCDA 91, 92, 94
Saponins 561–563, 730
dosing 562–563
incorporation into iscoms 568–569
type of immune modulation 562
SAS PROC IML program 473
Scatter plots 39, 40
Scavenger receptor (SR-A) 344
SCID
human transformed Fas-deficient T cells 677
mice 365, 370
Secreted molecules, staining 27–28
Secretion Assay *see* Cytokine Secretion Assay
Seed cultures, mycobacteria 437–439
Self Organizing Maps 165–166, 167
Sendai virus
first isolations 186
histopathological changes, regenerative phase 186
Sentinels/'control' animals 189–190
Separation
of cells 237–242
leukocyte populations 23–55
columns, MACS sorting 53
Serine esterase release 296
Serologic tests, mice and rats 193
Serotonin, human/mouse differences 363
Serum
collection of 388–389
composition of murine 389
cytokine levels 395
macrophage cultures 341
Set-up, colonies 211
Severe combined immunodeficiency *see* SCID
Shigella flexneri 374
Shipping containers, mycobacteria cultures 434–435
SHPM *see* Single-hit Poisson model (SHPM)
Sideward scatter (S5C), flow cytometry 24
evaluation of phagocytosis 779
Simian immunodeficiency virus (SIV) 146, 147
non-ionic block polymers 559
Simple tandem repeats (STRs) 211
Single cell suspensions, obtaining 45
Single-hit Poisson model (SHPM) 473, 474
Skin graft rejection 593
isolation of dendritic cells from 592–594
Soluble antigen, preparation 394–395
Sonication of bacteria 637–638
ultrasonication, heat-killed bacteria 394
Specific immune response *see* Adaptive immune system
Specified pathogen-free (SPF) status 188, 370
Specimen
collection 387–390
preparation 753–754

Spectral overlap, flow cytometry dyes 38–39
'Speed congenic' production 208
Sperm freezing 210
SPF-Unit 204–5
Spironucleus muris 189
Spleen
antigen-induced cytokine production 395
cellular components of murine 398
human/mouse differences 362
isolation of macrophages 338
isolation of T cells from 251
as source of lymphocytes 234
Spondylarthropathies, HLA-B27 623, 623, 642
Spontaneous infections, animals 186
Spot-forming cells (SFCs), cytokine 308
Staining
antigen, by single-colour IHC 505–507
Bio-Rad slides, protocol for 729
cell viability
differential staining 244–245
propidium iodide 25, 36, 47, 243
vital dye 25, 47, 243–244
DNA, by two-colour ISH 507–509
double, procedure 760
immunofluorescence detection 755–758
conventional surface staining 25, 46–47
counterstaining surface antigens 781
intracellular staining 27, 37, 47–48, 712, 713, 725–731
multicolour immunofluorescence staining 756–758
staining parameters 32, 33–37
staining reagents 29–33
standard cell dilution assay (SCDA) 91, 92–93
intracellular cytokine 9, 312–313, 457–458
Ki67 137
MHC class I tetramer 129–134, 141
peroxidase method, immunochemistry 405–407
SYBR-Green, analysis PCR products 737
T cells, by two-colour IHC 510–512
tissue 386
volume (cells) 34–35
see also Dyes
Staining reagents 29–33
Stainless steel screens, narrow mesh 236
Standard cell dilution assay (SCDA) 78, 89–95
applications 94–95
assay 91–92
data analysis 92
equipment and reagents 90–91
principle 89
troubleshooting 92–94
Standardization, microbiological 184–199
Standardized diet, sterilization 205
Staphylococcal enterotoxins, T-cell subset stimulation 94
Staphylococcus aureus 191, 193
Static micro-isolators 201–202
Statistical analysis
cytometric populations 41–44
of microarrays (SAM) 171
see also Single-hit Poisson model (SHPM)
Status of laboratory animals 197–199
Sterilization, standardized diet 205
Stock suspensions, preservation 365

Index

805

Vaccines (*cont.*)
  epitope-driven design 108–111
    advantages over whole-protein vaccines
        110–111
    animal models 114–115
    language analogy 108–109
    proof of principle 109–110
    tetramers 115
  epitope-driven, optimizing 111–114
    co-stimulatory molecules 111–112
    conserved epitopes 114
    epitope enhancement 113–114
    promiscuous epitopes 112–113
    spacers between 111
    targeting peptides to MHC I and MHC II 112
  pDNA
    allergic diseases 540–542
    applications 538–542
    immune response to 528–532
    infections diseases 538–540
    neoplastic disorders 541
    T-cell, *in vivo* assessment 114–115
    tuberculosis 13, 106
Vagina, bacterial application 378
Vectors
  herpesvirus saimiri
    transformation-associated region 667
    vector applications 679–681
Vegetable oils as adjuvants 557
Ventilated isolators 202
Vermin, as source of infection 197
Viability
  of cells, assessing 243–244
  of seeded cultures, random sampling 439
Viral
  genome, insertion of foreign genes 666–667
  proteins, degradation 9
  supernatants, determining viral titre 272–273
Virion
  DNA 663–664
    demonstration of 665–666
    purification 662–663
Virion protein target cells (VP) 516
Virulence
  bacterial
    bacterial strains 365
    clinical isolates 366
    culture conditions 365–366

frozen bacteria 366
    LD_{50} and time of death 382
    Leishmania parasites 468
Viruses 1, 184–185
  commercial breeding colonies 197
  contamination of biological materials 195–196
  dendritic cells as target 592
  in immunodeficient animals 195
  latency 493
  new rodent 192
  organotropism 185
  particles, checking culture supernatants for 662
  plaques 499–500
  titration 664–665
Virus-mediated transformation, human T cells
      657–671
Vital dye staining 25, 47, 136, 243–244
Vitamin A deficient diet 370

Washing
  of cells, affinity labelling 36
  procedures, effect on bacterial virulence 365–366
Waste management
  infected animals 219–220, 221
  mycobacteria waste materials 438
Water-in-oil emulsions 556–557
Water-in-squalene emulsions 559
Water/oil/water emulsions 558
Wire matrices 53
Work surfaces, decontamination 221

X-linked severe combined immunodeficiency,
      transformed T cells 676–677
XTT assay 244
Xylacine 375

Yersinia 373, 636, 642
Yersinia enterocolitica 374
  generation of HLA-B27 restricted CTL lines
      642–644
  preparation for in vitro stimulation of
      mononuclear cells 636–638
Yssel's medium 715

Zero differential (valley) method 43, 44
Z-LLF, host cell proteolysis inhibition 295
Zoonotic risk 218
Zymosan 348–349